Sodium-cooled Fast Reactors

AF739542

JSME Series in Thermal and Nuclear Power Generation

Sodium-cooled Fast Reactors

Volume 3

Edited by

Masaki Morishita
Hiroyuki Ohshima

ELSEVIER

ACADEMIC PRESS
An imprint of Elsevier

Academic Press is an imprint of Elsevier
125 London Wall, London EC2Y 5AS, United Kingdom
525 B Street, Suite 1650, San Diego, CA 92101, United States
50 Hampshire Street, 5th Floor, Cambridge, MA 02139, United States
The Boulevard, Langford Lane, Kidlington, Oxford OX5 1GB, United Kingdom

Copyright © 2022 Elsevier Inc. All rights reserved.

No part of this publication may be reproduced or transmitted in any form or by any means, electronic or mechanical, including photocopying, recording, or any information storage and retrieval system, without permission in writing from the publisher. Details on how to seek permission, further information about the Publisher's permissions policies and our arrangements with organizations such as the Copyright Clearance Center and the Copyright Licensing Agency, can be found at our website: www.elsevier.com/permissions.

This book and the individual contributions contained in it are protected under copyright by the Publisher (other than as may be noted herein).

Notices
Knowledge and best practice in this field are constantly changing. As new research and experience broaden our understanding, changes in research methods, professional practices, or medical treatment may become necessary.

Practitioners and researchers must always rely on their own experience and knowledge in evaluating and using any information, methods, compounds, or experiments described herein. In using such information or methods they should be mindful of their own safety and the safety of others, including parties for whom they have a professional responsibility.

To the fullest extent of the law, neither the Publisher nor the authors, contributors, or editors, assume any liability for any injury and/or damage to persons or property as a matter of products liability, negligence or otherwise, or from any use or operation of any methods, products, instructions, or ideas contained in the material herein.

ISBN 978-0-12-824076-2

For information on all Academic Press publications
visit our website at https://www.elsevier.com/books-and-journals

Publisher: Charlotte Cockle
Editorial Project Manager: Sara Valentino
Production Project Manager: Anitha Sivaraj
Cover Designer: Miles Hitchen

Typeset by STRAIVE, India

ELSEVIER Book Aid International
Working together to grow libraries in developing countries
www.elsevier.com • www.bookaid.org

Contents

Contributors

Kosuke Aizawa Sector of Fast Reactor and Advanced Reactor Research and Development, Japan Atomic Energy Agency, Ibaraki, Japan

Masanori Ando Sector of Fast Reactor and Advanced Reactor Research and Development, Japan Atomic Energy Agency, Ibaraki, Japan

Takashi Ashida Sector of Fast Reactor and Advanced Reactor Research and Development, Japan Atomic Energy Agency, Ibaraki, Japan

Yoshitaka Chikazawa Sector of Fast Reactor and Advanced Reactor Research and Development, Japan Atomic Energy Agency, Ibaraki, Japan

Norihiro Doda Sector of Fast Reactor and Advanced Reactor Research and Development, Japan Atomic Energy Agency, Ibaraki, Japan

Yasuhiro Enuma Sector of Fast Reactor and Advanced Reactor Research and Development, Japan Atomic Energy Agency, Ibaraki, Japan

Toshiki Ezure Sector of Fast Reactor and Advanced Reactor Research and Development, Japan Atomic Energy Agency, Ibaraki, Japan

Yoshitaka Fukano Sector of Fast Reactor and Advanced Reactor Research and Development, Japan Atomic Energy Agency, Ibaraki, Japan

Yoji Gondai Sector of Fast Reactor and Advanced Reactor Research and Development, Japan Atomic Energy Agency, Ibaraki, Japan

Erina Hamase Sector of Fast Reactor and Advanced Reactor Research and Development, Japan Atomic Energy Agency, Ibaraki, Japan

Ryuta Hashidate Sector of Fast Reactor and Advanced Reactor Research and Development, Japan Atomic Energy Agency, Ibaraki, Japan

Taira Hazama Sector of Fast Reactor and Advanced Reactor Research and Development, Japan Atomic Energy Agency, Ibaraki, Japan

Shun Hirooka Sector of Fast Reactor and Advanced Reactor Research and Development, Japan Atomic Energy Agency, Ibaraki, Japan

Tomoyuki Hiyama Sector of Fast Reactor and Advanced Reactor Research and Development, Japan Atomic Energy Agency, Ibaraki, Japan

Masaki Inoue Sector of Fast Reactor and Advanced Reactor Research and Development, Japan Atomic Energy Agency, Ibaraki, Japan

Katsunori Ishii Sector of Fast Reactor and Advanced Reactor Research and Development, Japan Atomic Energy Agency, Ibaraki, Japan

Hiroki Ishikawa Sector of Fast Reactor and Advanced Reactor Research and Development, Japan Atomic Energy Agency, Ibaraki, Japan

Masaru Ishimaru Sector of Fast Reactor and Advanced Reactor Research and Development, Japan Atomic Energy Agency, Ibaraki, Japan

Kazunori Isozaki Sector of Fast Reactor and Advanced Reactor Research and Development, Japan Atomic Energy Agency, Ibaraki, Japan

Wataru Itagaki Sector of Fast Reactor and Advanced Reactor Research and Development, Japan Atomic Energy Agency, Ibaraki, Japan

Chikara Ito Sector of Fast Reactor and Advanced Reactor Research and Development, Japan Atomic Energy Agency, Ibaraki, Japan

Hiroyuki Kageyam Sector of Fast Reactor and Advanced Reactor Research and Development, Japan Atomic Energy Agency, Ibaraki, Japan

Takeji Kaito Sector of Fast Reactor and Advanced Reactor Research and Development, Japan Atomic Energy Agency, Ibaraki, Japan

Masatoshi Kameyama Sector of Fast Reactor and Advanced Reactor Research and Development, Japan Atomic Energy Agency, Ibaraki; Projects Development Department, Japan Atomic Power Company, Tokyo, Japan

Hideki Kamide Sector of Fast Reactor and Advanced Reactor Research and Development, Japan Atomic Energy Agency, Ibaraki, Japan

Kenji Kamiyama Sector of Fast Reactor and Advanced Reactor Research and Development, Japan Atomic Energy Agency, Ibaraki, Japan

Atsushi Kato Sector of Fast Reactor and Advanced Reactor Research and Development, Japan Atomic Energy Agency, Ibaraki, Japan

Hirotaka Kawahara Sector of Fast Reactor and Advanced Reactor Research and Development, Japan Atomic Energy Agency, Ibaraki, Japan

Shodai Kawakami Sector of Fast Reactor and Advanced Reactor Research and Development, Japan Atomic Energy Agency, Ibaraki, Japan

Norihiro Kikuchi Sector of Fast Reactor and Advanced Reactor Research and Development, Japan Atomic Energy Agency, Ibaraki, Japan

Jun Kobayashi Sector of Fast Reactor and Advanced Reactor Research and Development, Japan Atomic Energy Agency, Ibaraki, Japan

Shoji Kotake Sector of Fast Reactor and Advanced Reactor Research and Development, Japan Atomic Energy Agency, Ibaraki; Japan Atomic Power Company, Tokyo, Japan

Shigenobu Kubo Sector of Fast Reactor and Advanced Reactor Research and Development, Japan Atomic Energy Agency, Ibaraki, Japan

Akikazu Kurihara Sector of Fast Reactor and Advanced Reactor Research and Development, Japan Atomic Energy Agency, Ibaraki, Japan

Shigetaka Maeda Sector of Fast Reactor and Advanced Reactor Research and Development, Japan Atomic Energy Agency, Ibaraki, Japan

Shuhei Maruyama Sector of Fast Reactor and Advanced Reactor Research and Development, Japan Atomic Energy Agency, Ibaraki, Japan

Kentaro Matsushita Sector of Fast Reactor and Advanced Reactor Research and Development, Japan Atomic Energy Agency, Ibaraki, Japan

Masanobu Michino Sector of Fast Reactor and Advanced Reactor Research and Development, Japan Atomic Energy Agency, Ibaraki, Japan

Takero Mori Sector of Fast Reactor and Advanced Reactor Research and Development, Japan Atomic Energy Agency, Ibaraki, Japan

Kazuki Morishita Sector of Fast Reactor and Advanced Reactor Research and Development, Japan Atomic Energy Agency, Ibaraki, Japan

Koji Morita Sector of Fast Reactor and Advanced Reactor Research and Development, Japan Atomic Energy Agency, Ibaraki; Department of Applied Quantum Physics and Nuclear Engineering, Kyushu University, Fukuoka, Japan

Hiroyuki Naito Sector of Fast Reactor and Advanced Reactor Research and Development, Japan Atomic Energy Agency, Ibaraki, Japan

Hiroaki Ohira Sector of Fast Reactor and Advanced Reactor Research and Development, Japan Atomic Energy Agency, Ibaraki, Japan

Shigeo Ohki Sector of Fast Reactor and Advanced Reactor Research and Development, Japan Atomic Energy Agency, Ibaraki, Japan

Shuji Ohno Sector of Fast Reactor and Advanced Reactor Research and Development, Japan Atomic Energy Agency, Ibaraki, Japan

Masaki Okagaki Sector of Fast Reactor and Advanced Reactor Research and Development, Japan Atomic Energy Agency, Ibaraki, Japan

Satoshi Okajima Sector of Fast Reactor and Advanced Reactor Research and Development, Japan Atomic Energy Agency, Ibaraki, Japan

Yasushi Okano Sector of Fast Reactor and Advanced Reactor Research and Development, Japan Atomic Energy Agency, Ibaraki, Japan

Takashi Onishi Sector of Fast Reactor and Advanced Reactor Research and Development, Japan Atomic Energy Agency, Ibaraki, Japan

Takashi Onizawa Sector of Fast Reactor and Advanced Reactor Research and Development, Japan Atomic Energy Agency, Ibaraki, Japan

Ayako Ono Sector of Fast Reactor and Advanced Reactor Research and Development, Japan Atomic Energy Agency, Ibaraki, Japan

Shoji Ono Sector of Fast Reactor and Advanced Reactor Research and Development, Japan Atomic Energy Agency, Ibaraki, Japan

Takayuki Ozawa Sector of Fast Reactor and Advanced Reactor Research and Development, Japan Atomic Energy Agency, Ibaraki, Japan

Hiroto Saito Sector of Fast Reactor and Advanced Reactor Research and Development, Japan Atomic Energy Agency, Ibaraki, Japan

Aaru Sano Sector of Fast Reactor and Advanced Reactor Research and Development, Japan Atomic Energy Agency, Ibaraki, Japan

Yuichi Sano Sector of Fast Reactor and Advanced Reactor Research and Development, Japan Atomic Energy Agency, Ibaraki, Japan

Shinji Sasaki Sector of Fast Reactor and Advanced Reactor Research and Development, Japan Atomic Energy Agency, Ibaraki, Japan

Hiroshi Seino Sector of Fast Reactor and Advanced Reactor Research and Development, Japan Atomic Energy Agency, Ibaraki, Japan

Akiyuki Seki Sector of Fast Reactor and Advanced Reactor Research and Development, Japan Atomic Energy Agency, Ibaraki, Japan

Yuta Shizukawa Sector of Fast Reactor and Advanced Reactor Research and Development, Japan Atomic Energy Agency, Ibaraki, Japan

Misao Takamatsu Sector of Fast Reactor and Advanced Reactor Research and Development, Japan Atomic Energy Agency, Ibaraki, Japan

Kazuya Takano Sector of Fast Reactor and Advanced Reactor Research and Development, Japan Atomic Energy Agency, Ibaraki, Japan

Takashi Takata Sector of Fast Reactor and Advanced Reactor Research and Development, Japan Atomic Energy Agency, Ibaraki; Department of Nuclear Engineering and Management, School of Engineering, The University of Tokyo, Tokyo, Japan

Shigeru Takaya Sector of Fast Reactor and Advanced Reactor Research and Development, Japan Atomic Energy Agency, Ibaraki, Japan

Masayuki Takeuchi Sector of Fast Reactor and Advanced Reactor Research and Development, Japan Atomic Energy Agency, Ibaraki, Japan

Masaaki Tanaka Sector of Fast Reactor and Advanced Reactor Research and Development, Japan Atomic Energy Agency, Ibaraki, Japan

Hiroshi Taninaka Sector of Fast Reactor and Advanced Reactor Research and Development, Japan Atomic Energy Agency, Ibaraki, Japan

Takashi Tanno Sector of Fast Reactor and Advanced Reactor Research and Development, Japan Atomic Energy Agency, Ibaraki, Japan

Kodai Toyota Sector of Fast Reactor and Advanced Reactor Research and Development, Japan Atomic Energy Agency, Ibaraki, Japan

Akihiro Uchibori Sector of Fast Reactor and Advanced Reactor Research and Development, Japan Atomic Energy Agency, Ibaraki, Japan

Tomoyuki Uwaba Sector of Fast Reactor and Advanced Reactor Research and Development, Japan Atomic Energy Agency, Ibaraki, Japan

Takashi Wakai Sector of Fast Reactor and Advanced Reactor Research and Development, Japan Atomic Energy Agency, Ibaraki, Japan

Tomoyoshi Watakabe Sector of Fast Reactor and Advanced Reactor Research and Development, Japan Atomic Energy Agency, Ibaraki, Japan

Hiroki Yada Sector of Fast Reactor and Advanced Reactor Research and Development, Japan Atomic Energy Agency, Ibaraki, Japan

Tomohiko Yamamoto Sector of Fast Reactor and Advanced Reactor Research and Development, Japan Atomic Energy Agency, Ibaraki, Japan

Hidemasa Yamano Sector of Fast Reactor and Advanced Reactor Research and Development, Japan Atomic Energy Agency, Ibaraki, Japan

Keisuke Yokoyama Sector of Fast Reactor and Advanced Reactor Research and Development, Japan Atomic Energy Agency, Ibaraki, Japan

Kenji Yokoyama Sector of Fast Reactor and Advanced Reactor Research and Development, Japan Atomic Energy Agency, Ibaraki, Japan

Akihiro Yoshida Sector of Fast Reactor and Advanced Reactor Research and Development, Japan Atomic Energy Agency, Ibaraki, Japan

Ryuji Yoshikawa Sector of Fast Reactor and Advanced Reactor Research and Development, Japan Atomic Energy Agency, Ibaraki, Japan

Kazuo Yoshimura Sector of Fast Reactor and Advanced Reactor Research and Development, Japan Atomic Energy Agency, Ibaraki, Japan

About the authors

Kosuke Aizawa earned a master's degree from the Tokyo Institute of Technology in 2005 after receiving a bachelor's degree from Tokyo University of Science in 2003. He then joined the Japan Atomic Energy Agency (JAEA). His main research areas are thermal hydraulics and ultrasonic technology in sodium-cooled fast reactors (SFRs).

Masanori Ando has worked for more than 20 years in the field of elevated temperature design for fast reactors (FRs) after receiving bachelor's and master's degrees from Waseda University. He has gained valuable experience in the new structural material development for FRs, experimental FR plant operation, applying for permission of restarting and decommissioning in FRs on-site, and contributing to the Japan Society of Mechanical Engineers (JSME) to improve the elevated temperature design code. He has served as Chair of the working group for elevated temperature design code in the JSME, and was a member of the American Society of Mechanical Engineers (ASME) code committee, Section III, the working group on Creep-Fatigue and Negligible Creep. He earned a Doctor of Engineering from the University of Tokyo in 2016.

Takashi Ashida is a Manager in the Fast Reactor Irradiation Section, Experimental Fast Reactor Department, Fast Reactor Cycle System Research and Development Center, Oarai Research and Development Institute (ORDI), Sector of Fast Reactor and Advanced Reactor Research and Development in the Japan Atomic Energy Agency (JAEA). He started his career at the Power Reactor and Nuclear Fuel Development Corporation (PNC, now the JAEA) in 1986. He has worked on the design, installation, replacement, testing, and in-service inspection for Joyo, and the development of plant operation and maintenance technologies. He led the MK-III project to modify the reactor cooling system and successfully completed this project in 2003. He also joined in the MARICO-2 Subassembly retrieving project.

Yoshitaka Chikazawa started his research career at the Japan Nuclear Cycle Development Institute (JNC, now the Japan Atomic Energy Agency, or JAEA) in 2001 after earning a Doctor of Engineering from the University of Tokyo in the same year, having already received bachelor's and master's degrees from the same university. He is now working as a Manager of the Fast Reactor Design Group in the Sector of Fast Reactor and Advanced Reactor Research and Development of the Japan Atomic Energy Agency (JAEA).

Norihiro Doda has been working for more than 20 years in the field of thermal hydraulics for sodium-cooled fast reactors (SFRs) after receiving a Bachelor of Science and Technology from the Tokyo University of Science, and a Master and Doctor of Engineering from the Tokyo Institute of Technology. He contributed to the development of a statistical safety evaluation method for the core hot spot during natural circulation decay heat removal operations. He is a member of the subgroup on the statistical safety evaluation method in the Atomic Energy Society of Japan (AESJ). At present, he is carrying out research on a coupled analysis method between neutronics, thermal hydraulics, and structural mechanics for evaluation of core deformation reactivity, and uncertainty quantification of a simulation model for code validation.

Yasuhiro Enuma earned a bachelor of science degree from the Department of Mathematics, Faculty of Science and Technology, Tokyo University of Science in 1992, and has been working for the Power Reactor and Nuclear Fuel Development Corporation (PNC, now the Japan Atomic Energy Agency, or JAEA) since then. He is currently a group leader of the Fast Reactor Life-Cycle Evaluation and Knowledge Integration Group at the Japan Atomic Energy Agency (JAEA). He is in charge of systematizing and preserving knowledge on fast reactor technology, promoting the use of systematized knowledge, research, and development of structural materials and standards for high-temperature materials and maintenance.

Yoshitaka Fukano started his research career at the Power Reactor and Nuclear Fuel Development Corporation (PNC, now the Japan Atomic Energy Agency, or JAEA) in 1993 after receiving bachelor's degree from the University of Tokyo and earned a Doctor of Engineering from the University of Tokyo in 2010. He also stayed and continued his research at the Institut de Radioprotection et de Sûreté Nucléaire (IRSN) in France from 1999 to 2000, and at Argonnne National Laboratory (ANL) in the United States from 2004 to 2005. His main research areas are analyses and evaluations of severe accidents in fast reactors. At present, he is managing research on severe accidents in fast reactors.

Yoji Gondai is a Chief Engineer in the Experimental Fast Reactor Department, Fast Reactor Cycle System Research and Development Center, Oarai Research and Development Institute (ORDI), Sector of Fast Reactor and Advanced Reactor Research and Development in the Japan Atomic Energy Agency (JAEA). He received his Master of Science from Kyushu University in 2012. He started his career at the JAEA in 2012 for the operation of Joyo. His current task is safety design for the conformity review by the Nuclear Regulation Authority (NRA) on the new regulatory requirements.

Erina Hamase is a Researcher for the Fast Reactor Cycle System Research and Development Center at the Japan Atomic Energy Agency (JAEA). She received her Doctor of Engineering from the Tokyo Institute of Technology in 2012. Her research interests include the development of an analysis method for core thermal hydraulics, numerical modeling of reactor systems, and verification, validation, and uncertainty quantification (VVUQ) of numerical models.

Ryuta Hashidate is an Engineer in the Fast Reactor Cycle System R&D Center of the Japan Atomic Energy Agency (JAEA). He received a master's degree from the University of Tsukuba in 2012 after receiving a Bachelor of Science from the University of Tokyo. He worked on the maintenance of the Japanese prototype fast breeder reactor (FBR) Monju from 2012 to 2018. His main areas of research are maintenance engineering for sodium-cooled fast reactors (SFRs). He also participates in activities for leak-before-break (LBB) assessment guidelines for SFRs in the Japan Society of Mechanical Engineers (JSME).

Taira Hazama is a Senior Researcher at the Japan Atomic Energy Agency (JAEA). His main research areas are the development of calculation systems and experimental data validation analyses on fast reactor physics. At present, he is engaged in knowledge preservation activities on Monju.

Shun Hirooka started his research career at the Nuclear Fuel Cycle Engineering Laboratory, Japan Atomic Energy Agency (JAEA) in 2010, and worked on mixed oxide (MOX) fuel properties. At present, he is working on expanding the database of minor actinide (MA)-bearing MOX properties by an experimental approach, which contributes to the standardization of the fuel properties in the working group of the Organisation for Economic Co-operation and Development/Nuclear Energy Agency (OECD/NEA) and fuel performance code for fast reactors under international collaboration with Idaho National Laboratory (INL) and Los Alamos National Laboratory (LANL). His work on the study of the fuel properties is also dedicated to the manufacturing technology of fuel pellets.

Tomoyuki Hiyama earned a master's degree from Osaka University in 2009 after receiving a bachelor's degree from Osaka University in 2007. He subsequently joined the Japan Atomic Energy Agency (JAEA). His main research areas are thermal hydraulics in sodium-cooled fast reactors (SFRs).

Masaki Inoue is an Assistant Principal Engineer in the Experimental Fast Reactor Department, Fast Reactor Cycle System Research and Development Center, Oarai Research and Development Institute (ORDI), Sector of Fast Reactor and Advanced Reactor Research and Development in the Japan Atomic Energy Agency (JAEA). He started his research career at the JAEA in 1989, focusing on nuclear fuel and cladding materials. He received his PhD degree from Kyushu University in 2005. He obtained his certification for Chief Engineer of Nuclear Fuel from the Nuclear Regulation Authority (NRA) in 2015, and his national qualification for Professional Engineer (Nuclear and Radiation) from the Ministry of Education, Culture, Sports, Science and Technology in 2016. He supervises operational safety concerning the handling of nuclear fuel material at the usage facilities in the southern part of the ORDI.

Katsunori Ishii started his research career at the Tokai Works of the Power Reactor and Nuclear Fuel Development Corporation (PNC, now the Japan Atomic Energy Agency, or JAEA) in 1992 and earned a Doctor of Engineering from Ibaraki University in 2005 after receiving bachelor's and master's degrees from Kyoto University. His main research area is mixed oxide (MOX) fuel fabrication technologies for fast reactors. At present, he is carrying out research on advanced fuel fabrication methods.

Hiroki Ishikawa is an Engineer at the Experimental Fast Reactor Department, Fast Reactor Cycle System Research and Development Center, Oarai Research and Development Institute (ORDI), Sector of Fast Reactor and Advanced Reactor Research and

Development in the Japan Atomic Energy Agency (JAEA). He received his bachelor's degree from the Fukui University of Technology in 2016. He started his career as an Engineer at the JAEA in 2016 to carry out operation of Joyo. His major technical specialty is operation engineering of sodium-cooled fast reactors (SFRs). His current task is operation management of Joyo.

Masaru Ishimaru is an Engineer in the Experimental Fast Reactor Department, Fast Reactor Cycle System Research and Development Center, Oarai Research and Development Institute (ORDI), Sector of Fast Reactor and Advanced Reactor Research and Development in the Japan Atomic Energy Agency (JAEA). He received his bachelor's degree from Tokyo City University in 2017. He started his engineering career at the JAEA in 2017 to engage in compliance with the new regulatory requirements at Joyo. His current tasks are the maintenance of the plant data acquisition system and core management operations, in addition to engaging in compliance with the new regulatory requirements at Joyo.

Kazunori Isozaki is a Contract Principal Engineer in the Maintenance Engineering Section, Experimental Fast Reactor Department, Fast Reactor Cycle System Research and Development Center, Oarai Research and Development Institute (ORDI), Sector of Fast Reactor and Advanced Reactor Research and Development in the Japan Atomic Energy Agency (JAEA). He started his career at the Power Reactor and Nuclear Fuel Development Corporation (PNC, now the JAEA) in 1977. He has worked on the design, installation, replacement, testing, and in-service inspection for Joyo, and the development of plant operation and maintenance technologies. He led the MK-III project to modify the reactor cooling system and successfully completed this project in 2003. He also conducted the aging review and established a long maintenance program of Joyo. His current job is the seismic back-check and evaluation of Joyo for the conformity review by the Nuclear Regulation Authority (NRA) on the new regulatory requirements.

Wataru Itagaki is an Assistant Principal Engineer in the Experimental Fast Reactor Department, Fast Reactor Cycle System Research and Development Center, Oarai Research and Development Institute (ORDI), Sector of Fast Reactor and Advanced Reactor Research and Development in the Japan Atomic Energy Agency (JAEA). He received his Master of Science from Hokkaido University in 2005, and started his research career at the JAEA in the same year for the irradiation test planning at Joyo. His main research areas are the design of irradiation test rigs and evaluation of irradiation conditions. His current research focuses on developing irradiation techniques for future irradiation tests in Joyo.

Chikara Ito is a Principal Research Engineer in the Fuels and Materials Department, Fast Reactor Cycle System Research and Development Center, Oarai Research and Development Institute (ORDI) of the Japan Atomic Energy Agency (JAEA). He received his PhD degree in quantum engineering from Nagoya University in 2009. He started his research career at the Power Reactor and Nuclear Fuel Development Corporation (PNC, now the JAEA) in 1993. He worked on the development of irradiation examination technology and fast breeder reactor (FBR) plant instrumentation technology in the experimental fast reactor Joyo. From 2005 to 2006, he stayed in the Marcoule Center, Commissariat à l'énergie atomique et aux énergies alternatives (CEA), France to research FBR operational experience in the Phénix Nuclear Power Plant. He also worked on the development of sensing technology to survey and retrieve the fuel debris inside the reactor pressure vessel (RPV) and the primary containment vessel in the Fukushima Daiichi Nuclear Power Plant after 2011. Currently, he manages the postirradiation examination to develop high-performance materials for FBRs.

Hiroyuki Kageyam is a Chief in the Maintenance Engineering Section, Experimental Fast Reactor Department, Fast Reactor Cycle System Research and Development Center, Oarai Research and Development Institute (ORDI), Sector of Fast Reactor and Advanced Reactor Research and Development in the Japan Atomic Energy Agency (JAEA). He received his AS degree from the National Institute of Technology, Ibaraki College in 2008. He worked on the operation and maintenance of Monju from 2008 to 2018. He has been in charge of maintaining the sodium cooling system and equipment of Joyo since 2018. His current interest is the aging evaluation and life management of the Joyo sodium cooling system.

Takeji Kaito started his research career at Oarai Engineering Center, Power Reactor and Nuclear Fuel Development Corporation (PNC, now the Japan Atomic Energy Agency, or JAEA) in 1990, and worked on the postirradiation examination for core structural materials irradiated in the fast reactor at the Materials Monitoring Facility (MMF). Since then, he has mainly worked on the evaluation of the irradiation behavior of core materials and the development of long-life core materials such as oxide dispersion strengthened (ODS) steel cladding. At present, he is responsible to promote the entire development of mixed oxide (MOX) fuel and core materials.

Masatoshi Kameyama has been working on sodium-cooled fast reactor (SFR) development since 2012. He participated in the performance testing preparation of the Monju for 3 years. He was also involved in the study of the concept of commercial fast breeder reactors (FBRs) as a member of the Japan Atomic Power Company for 6 years.

Hideki Kamide has worked for 30 years in the field of thermal hydraulics of sodium-cooled fast reactor (SFRs), especially experimental studies on natural circulation, high cycle thermal fatigue, and safety related issues. In 1985, he joined the Power Reactor and Nuclear Fuel Development Corporation (PNC, now the Japan Atomic Energy Agency, or JAEA). From April 2015 until March 2016, he was the Chair of the Thermal Hydraulics Division of Atomic Energy Society of Japan (AESJ). Since 2018, he has been the Deputy Director General of Sector of Fast Reactor and Advanced Reactor Research and Development, JAEA, and in March 2020, he became a Fellow of the AESJ. His current concerns are advanced reactor development and also international collaboration on this issue. He has also served for 3 years as the Chair of the Generation IV International Forum (GIF) Policy Group since 2019.

Kenji Kamiyama started his research career at the Power Reactor and Nuclear Fuel Development Corporation (PNC, now the Japan Atomic Energy Agency, or JAEA), in 1994, and earned his Doctor of Engineering from the Tokyo Institute of Technology in 2013. His main research area is multicomponent multiphase thermal-hydraulic phenomena with phase changes in severe accidents of nuclear reactors. At present, he is carrying out experimental research on the relocation of molten-core materials and their cooling in the sodium-cooled condition using the facilities including the Impulse Graphite Reactor (IGR) in the National Nuclear Center of the Republic of Kazakhstan and the MELT facility in the Oarai Research and Development Institute (ORDI) of JAEA.

Atsushi Kato started his research career at the Japan Nuclear Cycle Development Institute (JNC, now the Japan Atomic Energy Agency, or JAEA) in 2002 after earning a master's degree from the Graduate School of Science and Faculty of Science, Tohoku University. His main research areas are sodium-cooled fast reactor (SFR) plant design and cost evaluation of a system. At present, he is working on a conceptual design and construction evaluation of a pool-type SFR in Japan. He is also in charge of the development of a hydrogen detector for a steam generator.

Hirotaka Kawahara is an Assistant Principal Engineer in the Maintenance Engineering Section, Experimental Fast Reactor Department, Fast Reactor Cycle System Research and Development Center, Oarai Research and Development Institute (ORDI), Sector of Fast Reactor and Advanced Reactor Research and Development in the Japan Atomic Energy Agency (JAEA). He received his AS degree from Ibaraki University in 1990. He has participated in the maintenance of Joyo. He joined the MK-III project to modify the reactor cooling system, including the main intermediate heat exchangers and air-cooled dump heat exchangers. He also worked at the Phénix Nuclear Power Plant of the Commissariat à l'énergie atomique et aux énergies alternatives (CEA), in Marcoule, France from 2006 to 2007. Since 2011, he has worked on the seismic evaluation of Joyo.

Shodai Kawakami is a Chief in the Maintenance Engineering Section, Experimental Fast Reactor Department, Fast Reactor Cycle System Research and Development Center, Oarai Research and Development Institute (ORDI), Sector of Fast Reactor and Advanced Reactor Research and Development in the Japan Atomic Energy Agency (JAEA). He received his Master of Science from Kanagawa University in 2006, and worked on plant management and maintenance of Monju from 2006 to 2017. His specialized field is the maintenance of fuel handling systems for fast breeder reactors (FBRs). His current job is maintenance and operation of the fuel handling system of Joyo. He is also interested in advanced fuel handling and spent fuel storage systems of FBRs.

Norihiro Kikuchi started his research career at the Japan Atomic Energy Agency (JAEA) in 2014 after receiving bachelor's and master's degrees from the University of Fukui. His main research areas are thermal hydraulics in sodium-cooled fast reactors (SFRs) and computational fluid dynamics (CFD) analysis. At present, he is carrying out research on thermal hydraulics in fuel subassembly.

Jun Kobayashi joined the Power Reactor and Nuclear Fuel Development Corporation (PNC, now the Japan Atomic Energy Agency, or JAEA) in 1991 after receiving a bachelor's degree from Gunma University in 1991. His main research area is thermal hydraulics. After he joined the PNC, he was involved in research on decay heat removal by natural circulation, local blockage issue in a fuel assembly, and measurement of flow velocity distribution in elbows. He is currently researching the thermal striping phenomenon in reactor vessels.

Shoji Kotake has been involved in fast reactor development projects, first for the Japan Atomic Power Company as a manager of safety design and evaluation group since 1989, then at the Japan Nuclear Cycle Development Institute (JNC, now the Japan Atomic Energy Agency, or JAEA) as the manager of the section responsible to various fast reactor designs and the related technology developments section since 1999. Since 2016, he has worked as a subleader of the task force for fast reactor development in the JAEA as a collaborative member from an electric power company.

Shigenobu Kubo has worked on sodium-cooled fast reactor (SFR) development since 1989. He has been working for the Japan Atomic Energy Agency (JAEA) in the field of SFR systems design, safety design, and related R&Ds. He was involved in the development of safety design criteria and safety design guidelines for SFRs in the Generation IV International Forum (GIF). He participated in the Feasibility Study on Commercialized Fast Reactor Cycle Systems (1999–2006) and the Fast Reactor Cycle Technology Development project (2006–2011). He also participated in the France-Japan Advanced Sodium Technological Reactor for Industrial Demonstration (ASTRID) collaboration as a Design Task Leader and Severe Accident Task Leader. One of the most impressive projects that he has worked on is the Experimental Acquisition of Generalized Logic to Eliminate Re-Criticalities (EAGLE) project (SFR severe accident experiments using the Impulse Graphite Reactor (IGR) and out-of-pile experimental facility in Kazakhstan). In 1989, he completed a Master of Nuclear Engineering from Nagoya University.

Akikazu Kurihara started his research career at the Power Reactor and Nuclear Fuel Development Corporation (PNC, now the Japan Atomic Energy Agency, or JAEA) in 1994. His specialty is the safety of steam generators in sodium-cooled fast reactors (SFRs) (sodium-water reaction, or SWR). In particular, he was involved in the design and construction of many test facilities related to the SWR and experimental research. In the evaluation of the overheating tube rupture, he formulated the criteria for damage to the heat transfer tubes under internal pressure load at high temperature, evaluated the thermal influences of the reaction jet, and received his doctoral degree from Saga University regarding the unsteady heat transfer characteristics inside the heat transfer tube during rapid heating. In recent years, he has been working on the chemical reaction process of the SWR and the derivation of the new wastage correlations based on the phenomenological theory.

Shigetaka Maeda is a Principal Research Engineer in the Experimental Fast Reactor Department, Fast Reactor Cycle System Research and Development Center, Oarai Research and Development Institute (ORDI) of the Japan Atomic Energy Agency (JAEA). He has focused on nuclear engineering over the past 21 years, especially fast reactor physics for the experimental fast reactor Joyo, characterization of radiation environment, core physics tests, core modification design, and nuclear measurements. He is currently working on the licensing activities for Joyo's restart, as well as versatile uses of Joyo such as radioisotope production. He holds a PhD degree in nuclear engineering from Nagoya University.

Shuhei Maruyama earned a Master of Science from Nagoya University in 2006 and started his current research career at the Japan Atomic Energy Agency (JAEA). His main research areas are reactor physics, and methodologies of core neutronics analysis and design in fast reactors. At present, he is expanding his research field to data assimilation methods using integral experimental data.

Kentaro Matsushita received bachelor's and master's degrees from Kyoto University, and started his research career at the Japan Atomic Energy Agency (JAEA) in 2017. His main research area is a numerical analysis of gas-liquid two-phase flow. At present, he is carrying out research on the development of advanced analysis methods on an unsteady vortex, the evaluation of gas entrainment phenomenon caused by a vortex at the free surface, and the analysis of gas behavior in the primary coolant systems of sodium-cooled fast reactors (SFRs).

Masanobu Michino is a Senior Engineer in the Joyo Operation Engineering Section in the Experimental Fast Reactor Department, Fast Reactor Cycle System Research and Development Center in Oarai Research and Development Institute (ORDI), Sector of Fast Reactor and Advanced Reactor Research and Development in the Japan Atomic Energy Agency (JAEA). He graduated from Kagoshima Prefectural Ei High School in 1977. He started his career as an engineer at the Power Reactor and Nuclear Fuel Development Corporation (PNC, now the JAEA) in 1977 to carry out development work of sodium-steam generator for fast breeder reactors (FBRs). He subsequently worked on the operation of Joyo and served as the Shift Supervisor. His major technical specialty is operation engineering of sodium-cooled fast reactors (SFRs). His current tasks are operation management of Joyo and training instruction for operators of Joyo.

Misao Takamatsu is a General Manager of the Fast Reactor Technology Section, Experimental Fast Reactor Department, Fast Reactor Cycle System Research and Development Center, Oarai Research and Development Institute (ORDI), Sector of Fast Reactor and Advanced Reactor Research and Development in the Japan Atomic Energy Agency (JAEA). He received his Master of Science from Waseda University in 1996. He has worked on the operation and maintenance of the experimental fast reactor Joyo. He is responsible for the conformity review by the Nuclear Regulation Authority (NRA) on the new regulatory requirements, based on the lessons learned from the Fukushima Daiichi Nuclear Power Plant accident.

Takero Mori joined the Japan Atomic Energy Agency (JAEA) in 2002 after receiving a master's degree from the Graduate School of Science and Engineering, Ibaraki University, and started a research career focused on fast reactors. He evaluated the design margin of Monju using a one-dimensional plant dynamics code named Super-COPD, which is a one-dimensional plant dynamics code, and also evaluated the coolability of the reactor core in Monju considering earthquakes and tsunamis after the accident at the Fukushima Daiichi Nuclear Power Plant. At present, he is developing analytical models of plant dynamics code to evaluate plant behavior under natural circulation conditions more accurately, and is also developing an optimization system for thermal transient design to develop the next fast reactor.

Kazuki Morishita started his engineering career at Oarai Research and Development Institute (ORDI), Japan Atomic Energy Agency (JAEA) in 2017, and worked on chemical analysis of spent nuclear fuels of sodium-cooled fast reactor (SFRs). His main research areas are radiochemistry, chemical separation, and mass spectrometry of actinide elements.

Koji Morita started his research career at the Power Reactor and Nuclear Fuel Development Corporation (PNC, now the Japan Atomic Energy Agency, or JAEA), in 1989 and earned his Doctor in Engineering from Kyushu University in 1998. He then moved to the Department of Applied Quantum Physics and Nuclear Engineering at Kyushu University, where he became a Professor in 2012. His main research interests are thermal-hydraulic phenomena in severe accidents of nuclear reactors and next-generation nuclear reactor systems with enhanced safety features. He develops numerical simulation techniques for multicomponent multiphase flow phenomena and passive safety systems to prevent severe accidents.

Hiroyuki Naito is an Assistant Principal Engineer in the Experimental Fast Reactor Department, Fast Reactor Cycle System Research and Development Center, Oarai Research and Development Institute (ORDI), Sector of Fast Reactor and Advanced Reactor Research and Development in the Japan Atomic Energy Agency (JAEA). He received his Master of Science from Kyoto University in 2008, and started his research career at the JAEA in the same year for the research and development of measurement, observation, and irradiation techniques. His main research areas are the design of irradiation test rigs and the evaluation of irradiation conditions. His current tasks are the safety design of irradiation rigs and the seismic analysis of the Joyo core for the conformity review by the Nuclear Regulation Authority (NRA) on the new regulatory requirements.

Shigeo Ohki started his research career at the Power Reactor and Nuclear Fuel Development Corporation (PNC, now the Japan Atomic Energy Agency, or JAEA) in 1993 after receiving a bachelor's degree from Ibaraki University and a master's degree from Hiroshima University. He earned a Doctor of Engineering from Tohoku University in 2012. His research areas are fast reactor core design, reactor core physics, and nuclear transmutation.

Shuji Ohno has been working on liquid metal sodium-related safety studies for more than 30 years since he graduated from the University of Tokyo and joined the Power Reactor and Nuclear Fuel Development Corporation (PNC, now the Japan Atomic Energy Agency, or JAEA), in 1987. He earned a Doctor of Engineering from the Department of Nuclear Engineering and Management at the University of Tokyo in 2008. He is now serving as the Director of Reactor Systems Design Department in the Sector of Fast Reactor and Advanced Reactor Research and Development of the JAEA.

Masaki Okagaki is an Engineer in the Experimental Fast Reactor Department, Fast Reactor Cycle System Research and Development Center, Oarai Research and Development Institute (ORDI), Sector of Fast Reactor and Advanced Reactor Research and

Development in the Japan Atomic Energy Agency (JAEA). He earned his Bachelor and Master of Engineering from the University of Fukui in 2019 and 2021, respectively. He carried out his research on nuclear thermal propulsion rockets at the university. He started his career in the Fast Reactor Technology Section of the JAEA in 2021, and still works there. He engages in adaptation tasks for the new regulatory requirement of Joyo and research on a novel On-Line Gamma Monitor using TlBr detectors.

Satoshi Okajima has been working for more than 10 years in the field of elevated temperature design for fast reactors (FRs) after receiving a Bachelor, Master, and Doctor of Engineering from thc University of Tokyo. He also worked for the Organisation for Economic Co-operation and Development/Nuclear Energy Agency (OECD/NEA) for 2 years as a technical secretariat for a committee for sodium-cooled fast reactors (SFRs) on the Generation IV International Forum (GIF). He has contributed to developing codes and standards for elevated temperature design and reliability of mechanical products.

Yasushi Okano is a group leader of the Reactor Safety Analysis & Evaluation Group, Sector of Fast Reactor & Advanced Reactor R&D of the Japan Atomic Energy Agency (JAEA). He leads the R&D activities on the sodium-cooled fast reactor (SFR), especially on safety evaluations, and he also actively deals with international corporations on advanced reactors. As for his education, he graduated from the University of Tokyo in 1992 and received a Ph.D. from the University of Tokyo in 1997, both in nuclear engineering. His research backgrounds are thermal-hydraulics, reactor core design, external hazard PSA, and safety requirement harmonization on advanced reactors. He began his professional career as an assistant professor at the University of Tokyo, and in 1997 he started work for the JAEA, mainly on R&D related to the SFR. He used to be a visiting engineer at the Massachusetts Institute of Technology during 2001–2002, an international technical secretary of OECD/NEA in France during 2005–2009, and a deputy director of the "Nuclear Energy Policy Planning Division" of the "Ministry of Economy, Trade and Industry of Japan (METI)" during 2019–2021. He also served for a decade as a member of the ANS 54.1 SFR Standard Committee, a vice-chair of the PRA Standard Subcommittee of the Atomic Energy Society of Japan, and a co-chair of the Risk and Safety Working Group of the Generation IV International Forum.

Takashi Onishi started his research career at Oarai Research and Development Institute (ORDI), Japan Atomic Energy Agency (JAEA) in 2008, and worked on the development of advanced nuclear fuel cycles and disposal of radioactive wastes. His main research areas are nuclide analysis of fission products and actinides. At present, he is working on the development of nuclide analysis methods for nuclear fuel debris and precise quantitative analysis of minor actinides. His work is also dedicated to the evaluation of leaching behaviors of spent nuclear fuels in deep underground environments.

Takashi Onizawa has been working for more than 20 years in the field of high-temperature structural materials for fast reactors (FRs) after receiving a Bachelor, Master, and Doctor of Engineering from Ibaraki University. He has gained valuable experience in the new structural material development for FRs, experimental FR plant operation, and contributing to the code committee in the Japan Society of Mechanical Engineers (JSME) to improve the material strength standard of structural material for FRs. He has served as Chair of the working group for elevated temperature material standards in the JSME, and as a member of the subgroup on elevated temperature codes in the JSME.

Ayako Ono is an Assistant Principal Researcher at the Japan Atomic Energy Agency (JAEA). Her main research activities are in the field of thermal-hydraulic phenomena in nuclear power plants. In 2008, she received her PhD degree from Hokkaido University for research on the study of the critical heat flux (CHF) mechanism for saturated and subcooled pool boiling. Since she joined the JAEA in 2008, she has devoted her research efforts to the development project of the sodium fast reactor such as the evaluation of flow-induced vibration for the pipings and the development of the heat removal system driven by natural circulation. Recently, her research interests are the establishment of an evaluation method for CHF in light water reactors based on the CHF mechanism. Her approach is to combine the model to predict CHF and the detailed two-phase flow numerical analysis model.

Shoji Ono is an Engineer at the Joyo Operation Engineering Section in the Experimental Fast Reactor Department, Fast Reactor Cycle System Research and Development Center at Oarai Research and Development Institute (ORDI), Sector of Fast Reactor and Advanced Reactor Research and Development in the Japan Atomic Energy Agency (JAEA). He graduated from Ibaraki Prefectural Hitachi Technical High School in 2005. In the same year, he started his career as an engineer at the Japan Nuclear Cycle Development Institute (JNC, now the Japan Atomic Energy Agency, or JAEA), and carried out operation of Monju. He subsequently worked on the operation of Joyo. His major technical specialty is operation engineering of sodium-cooled fast reactors (SFRs). His current task is the development of operation engineering of Joyo.

Takayuki Ozawa started his research career at Oarai Engineering Center, Power Reactor and Nuclear Fuel Development Corporation (PNC, now the Japan Atomic Energy Agency, or JAEA) in 1988, and worked on the postirradiation examination for mixed oxide (MOX) fuels irradiated in the fast reactor at the Alpha-Gamma Facility (AGF). He moved to Tokai Works in 1991, and took charge of Monju fuel design and development of fuel performance code to be used for fast reactor fuel design. At present, he is working on the development of a fuel design method for the future fast reactor and fuel performance code verification and validation (V&V) under the international collaboration with the Commissariat à l'énergie atomique et aux énergies alternatives (CEA), and developing a multidimensional fuel performance code for MOX, with Idaho National Laboratory (INL) in the framework of the Civil Nuclear Energy Working Group (CNWG) between Japan and the United States, by applying the JAEA's knowledge regarding MOX fuel behavior. He is also participating in the expert groups in the Organisation for Economic Cooperation and Development/Nuclear Energy Agency (OECD/NEA) and the International Atomic Energy Agency (IAEA) for benchmark exercises among fuel performance codes in the world.

Hiroto Saito is a Chief Engineer in the Experimental Fast Reactor Department, Fast Reactor Cycle System Research and Development Center, Oarai Research and Development Institute (ORDI), Sector of Fast Reactor and Advanced Reactor Research and Development in the Japan Atomic Energy Agency (JAEA). He received his Master of Science from Tohoku University in 2011. In the same year, he started his career at the JAEA for the core management of Joyo. His current tasks are core management and design of Joyo, and safety design for the conformity review by the Nuclear Regulation Authority (NRA) on the new regulatory requirements.

Aaru Sano is an Engineer in the Fast Reactor Irradiation Section, Fast Reactor Cycle System Research and Development Center, Oarai Research and Development Institute (ORDI), Sector of Fast Reactor and Advanced Reactor Research and Development in the Japan Atomic Energy Agency (JAEA). He received a bachelor's degree from Tokai University in 2020, and started his research career at the JAEA in the same year. His interest is the flow of coolant, which affects the irradiation test. His current task is to assess the intensity of the irradiation assembly related to the review.

Yuichi Sano started his research career at the Tokai Works of the Power Reactor and Nuclear Fuel Development Corporation (PNC, now the Japan Atomic Energy Agency, or JAEA) in 1994 and earned a Doctor of Engineering from Kyoto University in 2004 after receiving bachelor's and master's degrees from Osaka University. His main research area is separation chemistry in the aqueous reprocessing process for spent mixed oxide (MOX) fuel. At present, he is carrying out research on the recovery of minor actinides by solvent extraction and extraction chromatography techniques.

Shinji Sasaki started his research career at Oarai Research and Development Institute (ORDI), Japan Atomic Energy Agency (JAEA) in 2001, and worked on the development of core fuels and materials of sodium-cooled fast reactors (SFRs). His main research areas are microstructure change of mixed oxide (MOX) fuel and corrosion of stainless-steel cladding tubes. At present, he is working on developing postirradiation examination techniques.

Hiroshi Seino joined the Power Reactor and Nuclear Fuel Development Corporation (PNC, now the Japan Atomic Energy Agency, or JAEA) in 1988 and started his research career. His specialty is mainly safety studies on ex-vessel events that could occur during severe accidents in sodium-cooled fast reactors (SFRs). In particular, he has conducted experiments such as hydrogen combustion and sodium-concrete reaction, and has worked on the development and verification of related analysis codes. In recent years, he has been conducting ex-vessel event analyses in response to new regulatory standards for the restart of the experimental fast reactor Joyo, and has also been working on source term research for the fast reactors.

Akiyuki Seki studied nuclear physics by theoretical and numerical methods at Ibaraki University and received his bachelor's and master's degrees. He then worked on introduction projects of supercomputer systems in the Japan Atomic Energy Research Institute (JAERI), the Institute of Physical and Chemical Research (RIKEN), and the Japan Atomic Energy Agency (JAEA). After the Fukushima Daiichi nuclear accident in 2011, he became a member of the JAEA's environment monitoring team and developed a database system on the distribution of radioactive substances. At present, he belongs to the group developing the knowledge base of the "Advanced Reactor Knowledge and AI-aided Design Integration Approach" (ARKADIA). His research interests are principal component analysis for the results of research and development of sodium-cooled fast reactors (SFRs) and reinforcement learning for plant safety processes.

Yuta Shizukawa started his research career at Oarai Research and Development Institute (ORDI), Japan Atomic Energy Agency (JAEA) in 2016, and worked in research on decommissioning of the Fukushima Daiichi Nuclear Power Plant and development of ceramic neutron absorber materials for fast reactors. His main research areas are corrosion of steels under conditions containing seawater and microstructure of boron carbide. At present, he is working on developing miniature mechanical test techniques.

Kazuya Takano started his research career at the Japan Nuclear Cycle Development Institute (JNC, now the Japan Atomic Energy Agency, or JAEA) in 2004 after receiving a master's degree from the Tokyo Institute of Technology. His main research areas are fast reactor core design and experimental core analysis.

Takashi Takata achieved a Master of Engineering from the Tokyo Institute of Technology in 1994 and completed a PhD degree at Osaka University in 2007. He joined Kawasaki Heavy Industries, Ltd. in 1994. In 2005, he moved to Osaka University and then worked for the Japan Atomic Energy Agency (JAEA) from 2015 to 2021. In July 2021, he moved to the University of Tokyo. He is now a Professor at the Department of Nuclear Engineering and Management, School of Engineering. Research areas include nuclear safety, risk analysis, thermal hydraulics, and computational fluid dynamics (CFD), especially on sodium-cooled fast reactors (SFRs).

Shigeru Takaya is a Principal Researcher of Fast Reactor Cycle System R&D Center of the Japan Atomic Energy Agency (JAEA). He received a PhD degree in engineering from the University of Tokyo in 2003. His main areas of research interest are structural integrity evaluation and in-service inspection for sodium-cooled fast reactors (SFRs). He also participates in activities for the development of nuclear codes and standards in the Japan Society of Mechanical Engineers (JSME) and the American Society of Mechanical Engineers (ASME).

Masayuki Takeuchi started his research career at the Tokai Works of the Power Reactor and Nuclear Fuel Development Corporation (PNC, now the Japan Atomic Energy Agency, or JAEA) in 1992 and earned a Doctor of Engineering from Osaka University in 2004 after receiving bachelor's and master's degrees from Aoyama Gakuin University. His main research areas are equipment development and material corrosion in the aqueous reprocessing process for spent mixed oxide (MOX) fuel. At present, he is involved in research and development on processes and equipment for spent MOX fuel reprocessing.

Masaaki Tanaka completed a Master of Engineering at the University of Tsukuba in March 1996. In April 1996, he joined the Power Reactor and Nuclear Fuel Development Corporation (PNC, now the Japan Atomic Energy Agency) and participated in an experimental investigation of thermal-hydraulic phenomena in a sodium-cooled fast reactor (SFR). Since August 2003, he has been working on the development of thermal-hydraulic numerical simulation code and verification, validation, and uncertainty quantification (VVUQ) methodology for SFRs. In March 2010, he took a doctor's degree of Engineering at the University of Tsukuba.

Hiroshi Taninaka started his research career at the Japan Atomic Energy Agency (JAEA) in 2010 after earning a Doctor of Engineering from Kinki University for a study of subcriticality measurement with reactor noise and the inverse kinetics method. He participated in the prototype fast breeder reactor (FBR) Monju system start-up "zero-power reactor physics test" in 2010. Currently, he is building his career in reactor physics analysis and evaluation of actual fast reactors—such as Monju, China Experimental Fast Reactor (CEFR), Joyo, and EBR-II—to develop a verification and validation (V&V) method for next-generation fast reactor core design.

Takashi Tanno started his research career at Oarai Research and Development Institute (ORDI), Japan Atomic Energy Agency (JAEA) in 2011, and worked on the development of core materials of sodium-cooled fast reactors (SFRs). His main research areas are mechanical property, microstructure, corrosion, and manufacturing process of oxide dispersion strengthened (ODS) and ferritic steels. At present, he is working on expanding the database of properties of aged and irradiated core materials, and on optimizing the manufacturing process of ODS cladding tubes. His work is also dedicated to developing miniature mechanical test techniques.

Toshiki Ezure received his bachelor's and master's degrees from Kyoto University, and started his research career at the Japan Nuclear Cycle Development Institute (JNC, now the Japan Atomic Energy Agency, or JAEA), in 2004. In 2016, he received his PhD degree from Kyoto University. His main research areas are flow optimization issues on thermal hydraulics in sodium-cooled fast reactors (SFRs), especially gas entrainment due to the free surface vortex, vortex cavitation, and thermal stratification. At present, he is carrying out research on decay heat removal experiments using the plant dynamics test loop (PLANDTL-2) which is a large-scale sodium test facility.

Kodai Toyota started his career in the field of elevated temperature design of fast reactors (FRs) after receiving bachelor's and master's degrees from Hokkaido University. During his 4-year career to date, he has been conducting research on high cycle fatigue of structural materials for FRs and maintenance of FR plants.

Akihiro Uchibori received his Doctor of Engineering from Kyushu University in 2003 after completing bachelor's and master's degrees there. He has been the Research Engineer at the Japan Nuclear Cycle Development Institute (JNC, now the Japan Atomic Energy Agency, or JAEA) since 2003. He is currently the Principal Researcher at the JAEA. His research interests include numerical simulation of thermal-hydraulic problems in sodium-cooled fast reactors (SFRs). He has studied melting/solidification appearing in the nuclear fuel cycle, gas entrainment appearing in the reactor vessel, and multiphase flow with sodium-water chemical reaction appearing in a steam generator. At present, he is developing an integrated severe accident analysis code including thermal hydraulics, core disruption, sodium combustion, and sodium-debris-concrete interaction.

Tomoyuki Uwaba started his research career at the Power Reactor and Nuclear Fuel Development Corporation (PNC, now the Japan Atomic Energy Agency, or JAEA) in 1992 and earned a Doctor of Engineering from Hokkaido University in 2016 after receiving bachelor's and master's degrees from Waseda University. His main research areas are modeling and simulations of irradiation behaviors of fast reactor fuels. At present, he is carrying out research on simulations of irradiation behaviors of fuel subassembly by coupling computer codes of a fuel pin, a fuel pin bundle, and thermal hydraulics.

Takashi Wakai has been working for more than 30 years in the field of fracture mechanics at elevated temperature for structural integrity assessment of fast reactor (FR) components after receiving his Bachelor and Master of Engineering from Ritsumeikan University. He was awarded a Doctor of Engineering for his investigations on fracture mechanics as well as material mechanics in 2013. He has gained valuable experience in establishing a leak-before-break (LBB) assessment procedure for FR components and contributing to publication of the guidelines of LBB assessment for FR components as a secretary of the Japan Society of Mechanical Engineers (JSME) working group.

Tomoyoshi Watakabe started his research career at the Japan Atomic Energy Agency (JAEA) in 2009 after receiving a Doctor of Engineering from Tokyo Denki University in the same year. He then worked at the Agency for Natural Resources and Energy, Ministry of Economy, Trade and Industry of Japan, from 2019 to 2021, and came back to the JAEA in 2021. His main research areas are the seismic design of mechanical components of sodium-cooled fast reactors (SFRs). At present, he is conducting research on seismic evaluation of piping, three-dimensional isolation systems, and related seismic design code for SFRs.

Hiroki Yada is an Assistant Principal Researcher of Fast Reactor Cycle System R&D Center of the Japan Atomic Energy Agency (JAEA). He received a PhD degree in engineering from the University of Fukui in 2013. His main areas of research are structural integrity evaluation and in-service inspection for sodium-cooled fast reactors (SFRs). He also participates in activities for the development of fitness-for-service codes for SFRs and leak-before-break (LBB) assessment guidelines for SFRs in the Japan Society of Mechanical Engineers (JSME).

Tomohiko Yamamoto started his research career at the Japan Atomic Energy Agency (JAEA) in 2010 after receiving a Doctor of Engineering from the University of Tokyo. His main research areas are sodium-cooled fast reactor (SFR) plant design and seismic evaluation of reactor vessels and buildings. At present, he is carrying out research on seismic isolation systems and seismic evaluation of reactor vessels for SFRs.

Hidemasa Yamano completed a Master of Mechanical Engineering from Hiroshima University in 1996, and earned his PhD degree in nuclear engineering from Kyushu University in 2009. He joined the Japan Atomic Energy Agency (JAEA) in 1996 for a severe accident study in sodium-cooled fast reactors (SFRs). Currently, he is involved mainly in the development of safety design criteria and guideline, multiphysics simulation including core deformation, severe accident experimental and simulation studies, fission product transport study probabilistic risk assessment study of SFRs, and a renewable and nuclear harmonization strategy.

Keisuke Yokoyama started his research career at Oarai Research and Development Institute (ORDI), Japan Atomic Energy Agency (JAEA) in 2017, and worked on the development of nuclear fuels materials of sodium-cooled fast reactors (SFRs). His main research areas are mechanical properties, thermal properties, and irradiation behavior of nuclear fuel materials.

Kenji Yokoyama started his research career at the Power Reactor and Nuclear Fuel Development Corporation (PNC, now the Japan Atomic Energy Agency, or JAEA) in 1997 after receiving bachelor's and master's degrees of engineering from Osaka University. He earned a Doctor of Engineering from Osaka University in 2019. His main research areas are neutronics experimental data analysis for fast reactors and its application to reactor core design based on nuclear data adjustment techniques.

Akihiro Yoshida is a Director in the Experimental Fast Reactor Department, Fast Reactor Cycle System Research and Development Center, Oarai Research and Development Institute (ORDI), Sector of Fast Reactor and Advanced Reactor Research and Development in the Japan Atomic Energy Agency (JAEA). He received his Master of Science from Hokkaido University in 1988. He has worked on the core management, design, and review works of the Joyo core modification, and maintenance of the Joyo plant since 1988. His current work, as a representative of Joyo, involves responding to the conformity review by the Nuclear Regulation Authority (NRA) on the new regulatory requirements.

Ryuji Yoshikawa started his research career at the Institute of Nuclear Energy Technology of Tsinghua University in the People's Republic of China in 1996 after receiving a PhD degree from this university. He then moved to the Nuclear Engineering Research Laboratory at the University of Tokyo in 1999, and took up his present position at the Japan Atomic Energy Agency (JAEA) in 2007. His main research areas are thermal hydraulics of nuclear reactors, computational fluid dynamics (CFD), and multiphase flow. At present, he is performing R&D on numerical simulations of thermal hydraulics in fuel subassembly and heat exchangers of sodium-cooled fast reactors (SFRs).

Kazuo Yoshimura has worked mainly in the field of thermal-hydraulic analyses of nuclear power plants since receiving his bachelor's and master's degrees from Kyoto University. He has gained valuable experience in safety analyses in design extended condition (DEC) and applying for permission of restarting and decommissioning in the fast reactors. At present, he is contributing to developing the coupling method of thermal-hydraulic analysis codes.

Preface of JSME Series in Thermal and Nuclear Power Generation

Electric power supply is a fundamental and principal infrastructure for modern society. Modern society uses power generation through heat. This series of books consists of eight volumes describing thermal and nuclear power generation, taking Japan as the example, and referring to other countries. Volume 1 discusses how power supply has been attained historically, focusing on thermal and nuclear power generation along with the minimum scientific and technological fundamentals required to understand this series of books. The present state of the thermal and nuclear power generation technique is set out in detail in Volumes 2–8.

The rehabilitation and reconstruction of Japan after World War II were initiated through the utilization of a large amount of coal for boilers of thermal power plants. Meanwhile, environmental pollution caused by coal combustion became serious, and then oil was introduced to the boilers. Due to two worldwide oil crises and because of carbon dioxide issues, natural gas has also begun to be used for boilers. Current thermal power generation in Japan is based on coal and gas utilization. As a result of having a sufficient power supply, Japan has become one of the leading countries, economically and technologically, in the world. Thermal power technology, which started with introducing technology from abroad, has transformed Japan into one of the most advanced countries in the world through the research and development of its industry, government, and academia during this process.

Global warming related to excess carbon dioxide emissions has become a worldwide issue in recent years. Reducing carbon dioxide emissions in thermal power generation is an important part of dealing with this issue. One option is to change the fuel of a boiler from coal to gas, which exhausts less carbon dioxide. Another important approach is to enhance the thermal efficiency of coal thermal power plants as well as oil and gas. Many developing countries will need more thermal power plants in future. Although oil and/or gas thermal power plants may be introduced in these countries, it is believed that coal thermal power plants will continue to be used, for economic reasons. Bearing all this in mind, the publication of this series of books that sets out and explains the history and current status of the most advanced thermal power plants in Japan and other advanced countries is a timely aid to planning for engineers and researchers in developed countries to pursue further advancements, and for engineers and researchers in developing countries to acquire this knowledge.

Nuclear power generation technology in Japan started after being introduced from abroad approximately 60 years ago. Then, it reached the matured nuclear power technology through tireless efforts in research and development. However, nuclear power plants at the Fukushima Daiichi nuclear power station were heavily damaged by huge tsunamis caused by the Great East Japan Earthquake in 2011, which resulted in contamination of a large area around the power station. Measures to make nuclear power reactors more robust are currently being implemented by analyzing the factors that caused this serious situation. Technical vulnerability can be solved by technology. Nuclear power generation technology is undoubtedly a promising technology that should be used in the future. Nuclear power generation, as well as thermal power generation, is still expected as one of the main ways to supply the electricity in the framework of the basic energy plan of Japan. It implies that new power reactors will need to be constructed to replace the nuclear reactors that have reached the end of their useful lifetimes. Elsewhere, many developing countries are introducing nuclear power generation technology as a safe and economically excellent way to obtain electricity. The transfer of nuclear power generation technology developed and matured in Japan to these countries is considered by some to be the obligation of Japan. In these situations, the necessity of human resource development in the field of nuclear power generation technology in developing countries, as well as in Japan, is undisputed. Thus, it is an urgent task to summarize the nuclear power generation technology acquired by Japan and to provide it for new entrants to the nuclear technology field.

The Power and Energy Systems Division (PESD) of the Japan Society of Mechanical Engineers is celebrating its 30th anniversary from establishment in 1990. This department is entrusted with handling power supply technology in mechanical engineering. Responding to the above-mentioned is truly requested. This task can only be carried out by the PESD, as it is composed of leading engineers and researchers in this field in Japan. Therefore, summarizing Japan's and other countries' power generation technology and disseminating it not only in Japan but also overseas are important tasks. Thus, this book series was developed, to be published as one of the 30th anniversary events of the PESD.

The authors of this book series have engaged in the most advanced research and development for thermal power and nuclear power generation in Japan and Canada. Their experience and knowledge are reflected in their writing. This book is not an introductory book about what others did, but it describes living knowledge based on their own experiences and thoughts. We believe that this series will provide learning material that is not otherwise in existence in this field, and hope that readers will acquire a way of thinking as well as thorough and detailed knowledge with this book series as a guide.

This series is the joint effort of many authors who have generously shared their expertise, and their efforts are deeply appreciated. We are very grateful for the unbiased and thoughtful comments provided by many reviewers to make this series superb. Special thanks go to Maria Convey and Sara Valentino of the editorial staff at Elsevier.

Yasuo Koizumi, Editor in Chief
Graduate School of Information and Engineering, The University of Electro-Communications, Chofu, Tokyo, Japan

Mamoru Ozawa, Editor in Chief
Kansai University, Takatsuki, Osaka, Japan

Preface to Volume 3: Sodium-cooled fast reactors

A fast reactor, also known as a fast neutron reactor, is a nuclear reactor whose energy source is a fission reaction by fast neutrons. To avoid the deceleration of neutrons by scattering, liquid metal such as sodium is used as a coolant (sodium fast reactor, SFR). One of the most important features of fast reactors is that they can produce more nuclear fuel than they consume, which is achieved by converting nonfissile U_{238} to fissile material such as Pu_{239}. Thus, a quantum leap is expected in the efficient use of this valuable nuclear resource by fast reactors. Due to the above feature, they are also called fast breeder reactors (FBRs). Another important feature is that fast reactors can use long-lived and harmful minor actinides contained in high-level radioactive wastes fuel and can transmute them to short half-life radionuclides or stable nuclides, which lessens the burden on the global environment.

The history of fast reactor development dates back to the 1940s. Electricity was first generated by a small sodium-cooled fast reactor EBR-I in the United States in 1951. In Japan, the development of fast reactors started in the early 1960s. The experimental reactor Joyo achieved its first criticality in 1977, and the prototype reactor Monju achieved its first criticality in 1994. Currently, design studies and related research and development activities are being conducted for future commercial SFRs.

This book is a collection of the past experience of design, construction, and operation of two reactors, the latest knowledge and technology for SFR designs, and the future prospects of SFR development in Japan. It is intended to provide the perspective and the relevant knowledge to enable readers to become more familiar with SFR technology.

The anticipated readers of this book are postgraduate university students, laboratory researchers, and industry engineers, whose main disciplines include mechanical engineering and nuclear engineering. Of course, this book is not intended for a Japanese readership alone; the book is wide open to the world's students, researchers, and engineers.

Masaki Morishita, Editor
Hiroyuki Ohshima, Editor
Japan Atomic Energy Agency, Ibaraki, Japan

Introduction

1

Hideki Kamide
Sector of Fast Reactor and Advanced Reactor Research and Development, Japan Atomic Energy Agency, Ibaraki, Japan

Chapter outline

1.1 Sodium-cooled fast reactor

A fast neutron reactor (FNR) is a sustainable energy source that uses natural uranium resources very efficiently as a fission energy source. In addition to this, major long-lived radioactive fission products, that is, minor actinides (MAs), in spent fuels of nuclear power plants can be used as fissile material in FNR, yet are reduced through spent fuel recycling. A sodium-cooled fast reactor (SFR) is one of the most promising FNR systems. SFRs have been developed in many countries since the beginning of fission reactor development. Its sustainability significantly serves Japan's energy industries relying on imported natural resources for energy production. Peaceful uses of nuclear and reduction of the volume and toxicity of radioactive waste are also key issues for Japan, the only nation that has experienced a nuclear bombing.

This volume describes Japan's SFR cycle development, including

- Objectives and background of SFR cycle development as peaceful uses of nuclear
- How SFR is developed in Japan
- Overview of developed SFR technologies
- Future plans

The following briefly discusses SFR functions, development needs, policies, and histories.

1.2 Functions of SFR cycle and technology

The physics of FNRs is described in several books (e.g., Walter et al. [1]), so a detailed explanation is unnecessary here. However, the benefit of an SFR is notable and thus a

Sodium Fast Reactors. https://doi.org/10.1016/B978-0-12-824076-2.00001-8
Copyright © 2022 Elsevier Inc. All rights reserved.

demand for it in Japan. An SFR uses liquid sodium as a coolant of the primary cooling system within a temperature range of 350–550°C and has a high-energy-rich neutron spectrum resulting from sodium used as a coolant instead of water. As you know, neutron speed can be reduced by collision with nuclides in the reactor core, and its efficiency depends on the mass number (weight) of nuclides. The hydrogen atom of water has a mass number of unity and nearly the same weight as the neutron and reduces the neutron speed significantly. The sodium atom has a mass number of 23 as a major isotope and keeps the neuron at higher energy. In this sense, the lead atom has a mass number of around 207. Thus, lead-cooled fast reactor (LFR) is also a good candidate for FNR.

(1) Efficient use of uranium

An SFR core with high-energy neutrons produces more neutrons in one fission reaction than a light water reactor (LWR) does. Some excess neutrons from a fission reaction can be used to convert uranium-238 (U-238) into plutonium-239 (Pu-239) through double beta decay. U-238, the most common isotope of uranium that makes up 99.3% of naturally occurring uranium, is not fissile but can be well converted and used as fissile material in an SFR. High-energy neutrons induce more fission reactions of Pu-240 and Pu-241 than thermal neutrons that are dominant in an LWR. Spent fuel from LWRs and SFRs will be reprocessed to extract Pu including these high number Pu isotopes. Using them, SFRs can efficiently use limited natural uranium resources than LWRs.

(2) Fission reactions of minor actinides

Management of radioactive waste from spent fuels is a key issue to be addressed for nuclear energy use. MAs such as americium (Am) and neptunium (Np) in spent fuels have a long half-life of radioactivity, and Am is especially a major heat source concerned for the long-term storage of high-level radioactive waste (HLW) through vitrification. Thus, if we can treat MAs separately and reduce them, the management of HLW could be easier. High-energy neutrons in an SFR core can induce fission reactions of Am and Np, whereas thermal neutrons induce little reactions because of the neutron energy thresholds in fission reactions of Am and Np. In other words, with the separation of MAs through spent fuel reprocessing and fission reactions of MAs, an SFR can reduce the volume and toxicity of radioactive waste.

(3) High-temperature source of energy

In recent years, heat production using nuclear energy has received attention as a measure against global warming and for carbon neutrality. Electricity generation accounts for about 30% of carbon dioxide emissions in Japan [2], and burning fossil fuels for industrial heat applications is a major cause of emissions in not only Japan but also many parts of the world. Nuclear power, a carbon-free energy source, can generate both heat and electricity and contribute to achieving carbon neutrality by 2050. An SFR uses sodium at around 500°C, a temperature range higher than conventional LWRs. This high temperature efficiently spins steam turbines, generates more electricity, and brings about more heat use in chemical plants and other applications such as cogeneration and seawater desalination.

(4) Inherent safety features and design measures

Sodium is a common metal used in a wide range of industries because of its excellent characteristics such as high heat conductivity (higher than that of water by a factor of 100), a high boiling point (881°C) at atmospheric pressure, and lower dynamic viscosity (lower than that of water by a factor of 3). These characteristics contribute to the inherent safety features of an SFR.

An SFR is operated at low pressure (near ambient) thanks to the high boiling point of sodium. An average core outlet temperature of a typical SFR design is from 500°C to 550°C, which offers a significant temperature margin (about 300°C) to the boiling point. These low-pressure conditions contribute to keeping the coolant within a reactor vessel even if a coolant leakage occurs, maintaining the core cooling function.

Decay heat removal is crucial to a nuclear reactor. Recently, natural circulation is incorporated in the design of decay heat removal systems of various reactors, including small modular reactors (SMRs). As long as the coolant level is maintained in a reactor vessel, naturally circulated coolant removes heat from the core without a pump, eliminating failure of the pump itself and its power lines and control systems. This enhances the passive and reliable safety feature of the decay heat removal system. The driving force of the natural circulation is the fluid density difference between the hot upward flow and cold downward flow as well as the height difference between the heat source and the heat sink ($\Delta\rho \times \Delta h$). Thus, the natural circulation requires a heat sink in an elevated position to increase the flow rate.

The good heat conductivity and high operation temperature (~550°C) of sodium provide high heat exchange between sodium and the air at an air cooler. Unlike a seawater cooling system, an SFR uses the atmosphere as an infinite heat sink; the air cooler, therefore, can be installed at a high position in its reactor building. The high boiling point also allows the low-pressure cooling systems to maintain the liquid level and single-phase flow of the coolant as well as simple dynamic behavior. Thus, an SFR design is easy to take into account the natural circulation and its passive safety feature for the decay heat removal.

However, liquid sodium should be handled with great care since it has high chemical reactivity. It reacts with oxygen in the air and burns. It reacts with water, producing hydrogen and heat. An SFR, therefore, must be designed to prevent sodium leakage and tube break especially in a sodium-water heat exchanger (steam generator), which generates high-temperature steam to rotate the turbine. Also, an SFR should be designed to have mitigation measures against chemical reactions. For example, all rooms of the primary cooling system should have a nitrogen atmosphere, at least during power operation.

1.3 Needs for SFRs

Japan imports most natural resources of energy, such as coal, natural gas, oil, and also uranium. This makes our energy mix difficult in terms of energy security. An SFR fuel cycle will enable us to use the uranium already stored in spent fuel and depleted uranium as domestic resources, contributing to energy security.

When nuclear energy makes up a larger portion of the energy mix of Japan, we should consider its sustainability. In general, the impacts of power plants over their life cycles on the environment are also issues to be considered, such as waste of power generation, carbon dioxide emissions, water pollution, and land use. Generation-IV International Form (GIF) issued Nuclear Energy—An ESG Investable Asset Class [3] in 2021 showing the evaluation of nuclear energy from the environment, social, and governance (ESG) points of view. It clearly describes the impacts of energy industries' activities on the environment, such as land use. The volume of waste from nuclear power plants is much smaller (nearly zero carbon dioxide emission) than thermal power plants burning coal, natural gases, and oil. However, we must properly manage radioactive waste, especially spent fuel. Japan has built Rokkasho reprocessing plant to achieve the nuclear fuel cycle. Spent fuel is a source of energy production when recycled. Uranium and plutonium in spent fuel are recovered, and other highly radioactive fission products are turned into a solid, stable glass form through vitrification for safe storage. As discussed earlier, Am and Np are of major concern because of their long half-lives and heat-emitting nature in working toward the arrangement on a final disposal site. An SFR can effectively make fission reactions of Am and Np, reducing the volume and toxicity of radioactive waste. Also, land use by nuclear power plants is much smaller than that for photovoltaic panels for solar power generation.

When nuclear is considered as an energy source, its peaceful use is a prerequisite. The nuclear fuel cycle using U-238 and Pu can provide energy sources for a long time. However, Pu should be appropriately managed for peaceful use, and the Pu isotopes discussed earlier have long-term radioactivity and toxicity. SFR systems use Pu efficiently and play an important role in nuclear proliferation resistance.

Many international organizations including IAEA, IEA, OECD, and Clean Energy Ministerial (CEM) recognize nuclear energy as an important carbon-free energy source for carbon neutrality to be achieved by 2050. Flexible nuclear provides stable electricity output, compensating variable renewables such as solar and wind that possibly cause a power shortage in distribution grids at night and when the wind is low; eventually the long period of weeks. Nuclear energy can also serve industries and local communities by providing process heat and hydrogen production. SFR operating temperature is about 200°C higher than that of LWRs, which enables the system to have sufficient capability for applications of nuclear heat and heat storage. This increases the flexibility of nuclear power.

1.4 FBR development policy

Since 2003, the government has issued the Strategic Energy Plan [4] every 3–4 years, and the 6th edition was issued in October 2021. These show the national policy on nuclear energy over the past 17 years.

The first edition issued in 2003 sees nuclear power as a stable source of energy and a countermeasure against global warming, encouraging nuclear power use. It highlights the importance of nuclear fuel cycle development for the next-generation systems that offer higher economy, safety, and proliferation resistance.

The second edition (2007) states that nuclear power is a stable carbon-free energy source and a key solution to energy security and global warming. Nuclear power including the nuclear fuel cycle should be promoted while ensuring safety. It also shows the nation's expectation about early commercialization of sodium-cooled fast breeder reactor (FBR) cycle since FBR systems contribute to supplying stable energy by efficient uranium use and to reducing the volume of high-level radioactive waste by the use of fission reactions of MAs.

Although all Strategic Energy Plans acknowledge the nuclear fuel cycle as an important strategy for the country, the priority of required functions changed after the TEPCO's Fukushima Daiichi nuclear power plants (1F) accident in 2011. Nevertheless, the fourth edition issued in 2014 shows that the nuclear fuel cycle is an important process for spent fuel management, and the technology should be established for the volume and radiotoxicity reduction. Moreover, it declares that the developments of fast reactors are advanced through cooperation with international partners such as the United States and France.

The sixth plan issued in October 2021 links nuclear energy to carbon neutrality as shown in the Green Growth Strategy [5], a national policy on the economy and the environment, issued in June 2021. The Green Growth Strategy identifies hydrogen, ammonia fuel, carbon recycle, and nuclear power as solutions to carbon neutrality. To achieve it, Japan will continue to use nuclear power on an appropriate scale under a consensus on enhanced reliability and safety. The Green Growth Strategy clarifies that nuclear energy technology is advanced especially toward the reduction of the volume and toxicity of radioactive waste, resource recycling by the efficient use of limited natural resources, harmonization with renewables, carbon-free hydrogen, and nuclear heat for various applications. It encourages nuclear innovation to improve safety, reliability, and efficiency for the future. The development of fast reactors, SMRs, and high-temperature gas-cooled reactors (HTGRs) should be accelerated through international cooperation and innovations provided by private sectors by 2030. Remarkable descriptions about fast reactors development are as follows. "The nuclear fuel cycle should be promoted to reduce the volume and toxicity of high-level radioactive waste. Fast reactors development is promoted through international cooperation with fast reactors developing countries such as the United States and France, based on the Strategic Roadmap of Fast Reactor (2018) [6] and the Policy of Fast Reactor Development (2016) issued by the Inter-Ministerial Council for Nuclear Power Introduction."

The cabinet established the 2016 policy, which decided to build the Strategic Roadmap that specifies development activities to be conducted over the next decade. Then, the Strategic Roadmap was defined to set the fundamental concepts of the roadmap of fast reactors development at the Inter-Ministerial Council for Nuclear Power on December 21, 2018. It shows the significance of the fast reactors development and the roles of the policy and stakeholders as follows:

- The significance of fast reactors development in Japan
  - Effective use of uranium resources, reduction of high-level radioactive waste, and potential toxicity.
  - A fast reactor on a practical scale will start operation at the appropriate time in the mid-21st century.
- The major roles of the policy on nuclear technology are
  - To primarily ensure the highest level of safety and
  - To develop approaches by which uncertainties in the future can be flexibly dealt with.
- The roles of each party involved over the next decade
  - The government shows the basic direction of the development of the fast reactor and sets the objective of the development. It also provides appropriate financial support for technology development for the private sector.
  - Stakeholders including electric utilities conclusively decide technologies to be applied to fast reactors. Before this technology decision, they should participate in the development of promising technologies from the early stage.

- JAEA maintains and advances the development while addressing private sectors' needs. It also establishes global standards of design techniques and safety criteria.
- Manufacturers further innovate the technology by using their creativity and originality.

1.5 History of FBR development in Japan

Japan's FBR development started in the 1960s and has been progressed since then. A schematic of the development history is shown in Fig. 1.1. Now Japan has two FBRs.

In 1966, the Japanese government issued the Basic Policy on Power Reactor Development describing two policies on its FBR development. One is "independent development" and the other is "promotion of the development of experimental and prototype reactors." Based on these, the government established a four-step development plan: an experimental reactor, a prototype reactor, a demonstration reactor, and a commercial reactor, in which the experimental and prototype reactors were to be developed and operated by the government and demonstration reactor by electric utilities.

For the experimental and prototype reactor development, the government established Power Reactor and Nuclear Fuel Development Corporation (PNC), a predecessor of Japan Atomic Energy Agency (JAEA), and PNC started the development of an FBR and a nuclear fuel cycle. The government also established Oarai Engineering Center (OEC) and started the development of MOX fuel, fuel materials, FBR components, and safety researches, for example, sodium fire, sodium-water reaction, severe accidents.

PNC built the first SFR, experimental fast reactor Joyo in OEC, which started operation in 1977. For Joyo development, PNC and Japanese industries researched all technologies related to sodium components, MOX fuel, materials, and safety. Joyo has had four-core configurations. Mark-I core was for the demonstration of breeding ability with a reactor output of 75 MWth, and Mark-II core (100 MWth) was for the irradiation experiments on core materials. Then, the cooling systems of Joyo were

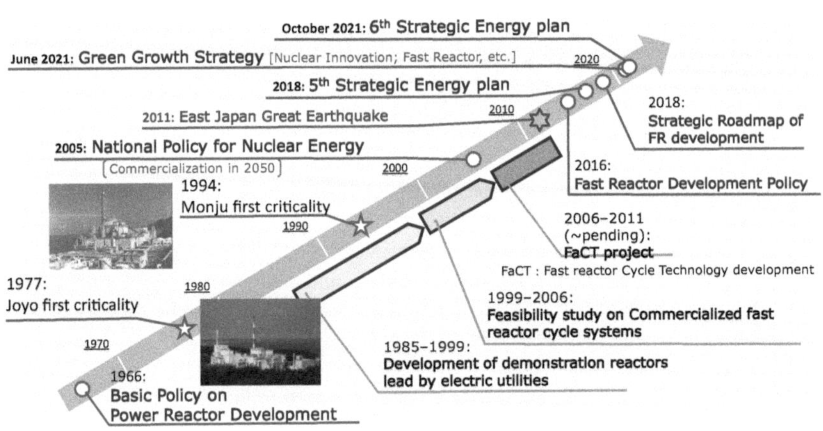

Fig. 1.1 History of fast reactor cycle development in Japan.

modified to ramp up its power output, increasing its irradiation capability. Mark-III core has 140 MWth for higher neutron fluences and efficient irradiation experiments. Now, certification processes for Mark-IV core are ongoing.

After Joyo started operation, PNC developed the prototype FBR Monju, which has 280 MW electric power. OEC developed the core and components larger than Joyo and also steam generators. Monju had the first criticality in 1994. Its system start-up tests and operation provided us with a wealth of data and experiences. In 1997, a sodium leak accident occurred and caused Monju to stop operation. Restarted in 2010, Monju had 250 days of operation in 22 years, but finally, the government decided Monju decommissioning in 2016.

Meanwhile, Japan Atomic Power Company (JAPC) started demonstration FBR (DFBR) development in 1987 with JAEA's technical support based on the development, construction, and operation of Monju. However, after the sodium leak accident in Monju, the DFBR development project was discontinued.

After the DFBR project, JAEA and electric utilities started an FBR feasibility study for demonstration in 1998. In this study, they designed and evaluated many types of FBR systems and selected a loop-type SFR as a future main energy source in Japan. After this study, JAEA and JAPC started the Fast reactor Cycle Technology development (FaCT) project led by the government. Main studies conducted in the FaCT project include the design concept of MOX-fueled Japan sodium-cooled loop-type fast reactor (JSFR), advanced aqueous reprocessing, and simplified pelletizing fuel fabrication systems. JAEA aimed to propose a design concept by 2015 that allows us to achieve FBR demonstration and that fulfills requirements for improved safety, economy, environmental impacts, resource use, and proliferation resistance. For safety and economy, 13 innovative technologies were proposed and developed in the design study. Finally, these innovative technologies were adopted to JSFR design, and the design and the technologies were evaluated by Japan Atomic Energy Commission in cooperation with related organizations. However, the Great East-Japan Earthquake and 1F accident in 2011 led to the termination of the FaCT project.

To obtain operational experience and technological maturity of a fast reactor, the Strategic Roadmap issued in 2018 (see the previous subsection) expects a fast reactor to be commercialized during the latter half of the 21st century and a fast reactor on a practical scale in the middle of this century. It also clarifies the importance of private sectors' development activities supported by JAEA and through international cooperation. Private sectors having creative ideas for fast reactor commercialization and development are financially supported by the government and technically supported by JAEA.

1.6 Overview of this volume

The contents and chapter structure of this volume are as follows.

Chapter 2: Experimental reactor Joyo

Chapter 2 describes the purposes, the plant configuration, and the construction and operation histories of the experimental fast reactor Joyo. It also shows the experiences of operation, maintenance, and notable irradiation tests as well as its future plans.

Chapter 3: Prototype reactor Monju
Chapter 3 discusses Monju's design features, research and development activities, histories of operation, outputs, and maintenance. Research results and achievements of Monju are also explained.

Chapter 4: Demonstration and commercial plant design study
DFBR project is explained as a reactor development program led by Japanese electric utilities. Chapter 4 also shows the detailed descriptions of the FaCT project, which aimed to build the demonstration reactor, JSFR, and of design features and improvements for enhancing the safety of JSFR. Lastly, the design and evaluation made for a pool-type SFR are also described.

Chapter 5: Key technologies for future sodium-cooled fast reactors
Based on the extensive experience of development activities, JAEA has studied a wide variety of innovative technologies to deploy commercial SFRs to achieve carbon neutrality by 2050. Chapter 5 shows JAEA's activities in the following research fields: (1) safety, (2) sodium component development, (3) reactor core physics, (4) fuel and materials, (5) thermal hydraulics, (6) high-temperature-resistant materials and structural design, (7) seismic design, (8) operation and maintenance, (9) fuel cycle technology, and (10) design assistance tool, ARKADIA (advanced reactor knowledge- and AI-aided design integration approach through the whole plant life cycle).

References

[1] A.E. Walter, D.R. Todd, P.V. Tsverkov, Fast Spectrum Reactor, Springer US, 2012.
[2] Energy White Paper 2021, Agency for Natural Resources and Energy (in Japanese). https://www.enecho.meti.go.jp/about/whitepaper/2021/pdf/.
[3] Nuclear Energy—An ESG Investable Asset Class, GIF website. https://www.gen-4.org/gif/jcms/c_179264/nuclear-energy-an-esg-investable-asset-class.
[4] Strategic Energy Plan, Agency for Natural Resources and Energy. https://www.enecho.meti.go.jp/en/category/others/basic_plan/.
[5] Green Growth Strategy Through Achieving Carbon Neutrality in 2050, Ministry of Economy, Trade, and Industry (in Japanese). https://www.meti.go.jp/english/policy/energy_environment/global_warming/ggs2050/index.html.
[6] Strategic Roadmap of Fast Reactor, Agency for Natural Resources and Energy. https://www.enecho.meti.go.jp/about/special/johoteikyo/kosokuro_roadmap.html https://www.cas.go.jp/jp/ seisaku/genshiryoku_kakuryo_kaigi/pdf/h301220_siryou.pdf.

Experimental reactor Joyo

2

Takashi Ashida, Yoji Gondai, Masaki Inoue, Hiroki Ishikawa, Masaru Ishimaru, Kazunori Isozaki, Wataru Itagaki, Chikara Ito, Hiroyuki Kageyam, Hirotaka Kawahara, Shodai Kawakami, Shigetaka Maeda, Masanobu Michino, Kazuki Morishita, Hiroyuki Naito, Masaki Okagaki, Takashi Onishi, Shoji Ono, Hiroto Saito, Aaru Sano, Shinji Sasaki, Yuta Shizukawa, Misao Takamatsu, Takashi Tanno, Keisuke Yokoyama, and Akihiro Yoshida
Sector of Fast Reactor and Advanced Reactor Research and Development, Japan Atomic Energy Agency, Ibaraki, Japan

Chapter outline

The experimental fast reactor Joyo is the first liquid sodium-cooled fast reactor in Japan. The purpose of constructing Joyo was to obtain technical information about sodium-cooled fast reactors (SFRs) through experience with its design, construction, and operation, and to use the reactor as a fast neutron irradiation facility for the development of fuels, materials, and other components required for the SFR program. Through design, construction, testing, operation, and maintenance experience, Joyo has contributed much to the SFR development program. In addition to providing operating experience, many kinds of irradiation tests have been conducted for the development of fuels and materials under the conditions of higher fast neutron flux and temperature than those in light water reactors. Through its successful operation over a quarter-century, Joyo has demonstrated the safety and reliability of the sodium-cooled fast reactor.

Sodium Fast Reactors. https://doi.org/10.1016/B978-0-12-824076-2.00002-X
Copyright © 2022 Elsevier Inc. All rights reserved.

1. **Introduction:** As an introduction, this section describes the history of the construction and operation of Joyo, as well as the general configuration of the plant.
2. **Operation and maintenance experience:** In Joyo, various SFR technologies were demonstrated in the operation experience. This section describes fast reactor technology such as nuclear calculation technology and handling of sodium components cultivated in Joyo.
3. **Irradiation test experience and future plan:** Joyo and the surrounding laboratories have various functions for irradiation tests. This section describes the irradiation test capability of Joyo, which has contributed to the development of fuels and materials for SFR.

2.1 Introduction

2.1.1 Construction and operation history of Joyo

To develop an "experimental fast reactor" of Japan, a design study was initiated in 1962 by the Japan Atomic Energy Research Institute (JAERI, presently, JAEA), and in 1967, "Long-Term Program for Research, Development, and Utilization of Nuclear Energy" was authorized by the Japan Atomic Energy Commission to construct an experimental fast reactor to obtain engineering knowledge on SFRs and, after construction, to use the reactor for irradiation required for fuel and material R&D. The program stated the adoption of a sodium-cooled-type reactor that was evaluated as the most promising based on the global R&D status, development of plutonium-uranium mixed oxide (MOX) fuel with high priority for the time being, and thermal power of approximately 100 MW, etc.

JAERI conducted the preliminary design (1964), the first conceptual design (1965–66), and the second conceptual design (1968), and the achievements were taken over by the Power Reactor and Nuclear Fuel Development Corporation (PNC, presently, JAEA) that was established in October 1967. It was decided to construct the experimental fast reactor as Japan's first national project centered on the PNC, and with the participation of domestic manufacturers, the third conceptual design (1967–68) and the adjustment design (1968) were performed. Subsequently, after completion of a design check, so-called French check, that was conducted by the French Commissariat a l'Energie Atomique (CEA), the application for reactor establishment license was submitted in June 1969, and the license was issued in February 1970. Japan's experimental fast reactor was named Joyo after a traditional name of the area where the construction site Oarai is located. An aerial photograph of Joyo taken recently is shown in Fig. 2.1.

In 1970, the first excavation started in January, and excavation of the areas for the reactor building and reactor auxiliary building finished in August. In 1971, construction of the reactor containment vessel (C/V, Fig. 2.2) started, and an early-phase pressure leakage test was performed in August. The C/V is a cylindrical airtight vessel with a diameter of approximately 30 m, a height of 55 m, and a top dome, that is similar to the C/V for PWRs. In December of the same year, two dump tanks for the primary sodium manufactured at Rinkai Factory of Hitachi Ltd. were landed from the seashore below the present Minami Junior High School in Oarai Town (Fig. 2.3).

Fig. 2.4 is a photograph showing the construction status of the buildings in March 1973. In the photograph, the operations management building, the C/V dome

Fig. 2.1 Aerial photograph of Joyo.

Fig. 2.2 Construction of reactor containment vessel.

surrounded by the reactor auxiliary building, and the main cooler building are seen from left to right. Construction work went well with the reactor vessel (R/V) installed in September, and the reactor building and reactor auxiliary building completed in November 1973.

Components near the core, including the internal core structure and the control rod drive mechanism, were installed in succession, and Emperor Showa and Empress Kojun visited Joyo to observe the inside of the C/V in October 1974. Fig. 2.5 shows the inside of Joyo's R/V as of 1975, before charging sodium. The upper core structure (UCS) is seen obliquely in the center upper area, and the core under configuration is seen below.

Then, approximately 136 and 73 t of sodium were charged into the primary and secondary cooling systems, stainless steel dummy subassemblies in the core region

Fig. 2.3 Landing of dump tank for primary sodium.

Fig. 2.4 Construction status of the buildings.

were replaced step by step with the core fuel subassemblies, and at 11:07 on April 24, 1977, Joyo achieved the first criticality of MK-I core (breeding core). Fig. 2.6 is a photograph taken at that time in the main control room.

MK-I core consists of the core fuel subassemblies having the blanket regions above and below the fuel part and the radial blanket fuel subassemblies in their surroundings.

Fig. 2.5 The inside of Joyo's reactor vessel.

Fig. 2.6 Photograph taken in the main control room at the moment of first criticality.

The breeding performance of Joyo was demonstrated through the operation of the MK-I core. During the operation period until 1981, the rated power was increased from 50 to 75 MW, and two periodic inspections and a test to demonstrate the natural circulation heat removal function were performed.

Subsequently, according to the original program, the modification was done to configure the MK-II core (irradiation core) by replacing the blanket region with the stainless steel reflector region, and the first criticality of the MK-II core was achieved in November 1982. For the MK-II core fuel subassemblies, PNC316 steel, which was then planned to be applied to the Monju fuel, was adopted as fuel cladding material. Since then, operation at a thermal power of 100 MW had been successfully performed for 18 years, and, in parallel, technical development for fast reactors focused on fuel and material irradiation tests was conducted.

From 2000 to 2003, the core and cooling systems were modified under the MK-III project (Fig. 2.7). In the MK-III core, the number of the core fuel subassemblies was increased to radially expand the fuel region and thereby moderate the negative effect of additional irradiation test subassemblies with less fuel inventory on reactivity. In addition, positions, where irradiation test subassemblies can be installed without being adjacent to any control rod position, were assured by relocating the 2 of 6 control rods to the periphery of the fuel region, while maintaining the reactor shutdown margin required by design. As the control rod upper guide tubes of the relocated control rods were removed from the UCS, it became possible to perform a test using an instrumented test assembly, which had been feasible only in the core periphery before, at a position surrounded by the core fuel subassemblies, and hence at a higher neutron flux level. Moreover, the adoption of a two-region core allowed to flatten the power distribution, increase the thermal power from 100 to 140 MWt, and specifically increase the fast neutron flux in the outer core region. PNC1520 steel that was then developed for the demonstration SFR was adopted as fuel cladding. Due to the increase in thermal power, the main intermediate heat exchangers (IHXs) and coolers that had been used before were all updated. After modification, the rated power of the MK-III core, 140 MW, was achieved on October 28, 2003, through steps of the integrated functional test, performance test, and power-up test.

After starting the MK-III core operation, in 2007, during the preparation for refueling under reactor shutdown, an integrated-type experimental device penetrating

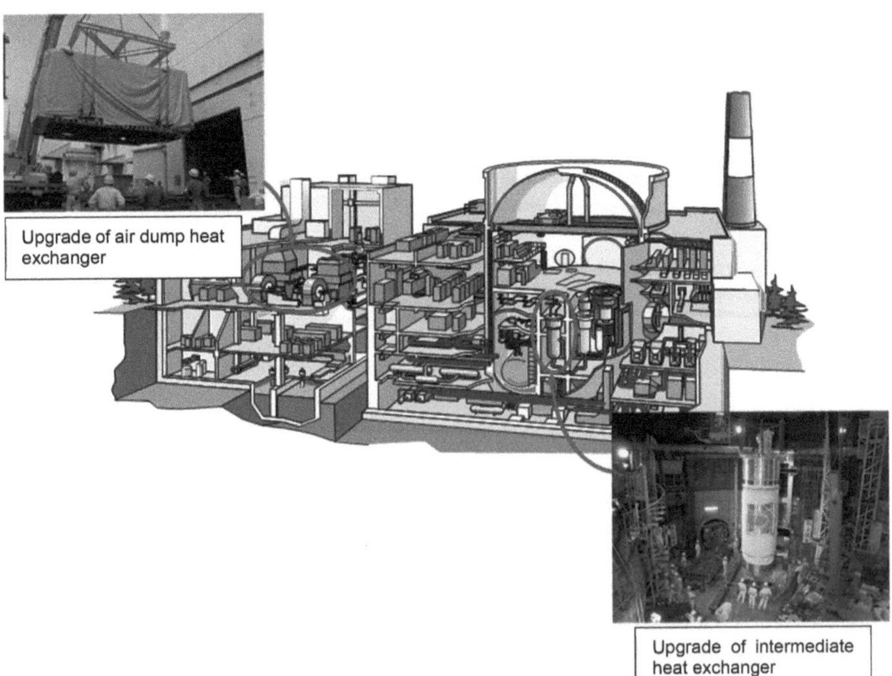

Fig. 2.7 Plant modification for MK-III project.

through the core and UCS was damaged. Internal inspection of the R/V using a radiation-resistant fiber scope revealed that the damaged portion of the experimental device damaged the bottom surface of the UCS. Consequently, a new UCS and equipment for retrieving the experimental device were manufactured, the UCS was replaced with the new one, and the experimental device was retrieved in 2014.

Based on the lessons learnt from the Tokyo Electric Power Company (TEPCO) Fukushima Daiichi Nuclear Power Plant's accident that occurred following the Great East Japan Earthquake in 2011, measures were taken to enhance the safety standards for nuclear facilities (hereinafter, "new regulatory standards"). For Joyo, an application to comply with the new regulatory standards was submitted to the Nuclear Regulation Authority (NRA) on March 30, 2017, and the review has been in progress. This application includes the modification to the MK-IV core with thermal power of 100 MWt. After completion of the safety review, it is planned to conduct plant upgrades, including the seismic work, and then resume operation.

The main achievements and operational data of Joyo are listed in Table 2.1, and the operational performance of Joyo is shown in Fig. 2.8.

2.1.2 Plant description of Joyo

Joyo is a loop-type sodium-cooled experimental fast reactor using plutonium-uranium mixed oxide (MOX) fuel. Two main cooling systems and an auxiliary cooling system constitute the heat transfer system. These systems contain over 200 t of sodium in total. The auxiliary cooling system is activated to remove decay heat during reactor shutdown and when both main cooling systems are tripped. Each of the cooling systems consists of the primary sodium loop, IHX, secondary sodium loop, and air dump heat exchangers (DHXs). Fig. 2.9 shows an outline of the cooling systems in Joyo. With the MK-IV core, sodium enters the core at 350°C at a flow rate of 1350 t/h per loop and exits the reactor vessel at 456°C through 500-mm-dia. piping. The IHX separates radioactive sodium in the primary system from nonradioactive sodium in the secondary system. The secondary sodium transports the reactor heat via the IHXs to the DHXs. All the heat generated is removed by these DHXs as Joyo has no steam generator.

Since its first criticality, Joyo has operated with four types of core designs: MK-I, II, III, and IV. Fig. 2.10 and Table 2.2 show the core maps and properties of the MK-I through the MK-IV. With the MK-I, the initial core, after completion of the low-power and power ascension tests, Joyo operated for two cycles with a nominal thermal power of 50 MWt and six cycles with 75 MWt. In this period, the number of reactor start-ups reached about 260 to perform many kinds of tests related to reactor physics, reactor dynamics, power ascension, and transients. These tests confirmed that the plant system satisfied the design objectives including the following: (i) core physics characteristics (control rod worth, excess reactivity, etc.) agreed with design values within 10%; (ii) core thermohydraulic characteristics (fuel outlet temperatures, flow rate, etc.) agreed with design values within 5%; and (iii) plant parameters (inlet and outlet temperatures of the cooling system, heat balance, etc.) agreed with design values

Table 2.1 Milestones of Joyo.

MK-I (Breeder core) (50 / 75 MWt)	
April 1977 ~December 1981	First criticality achieved Accumulate technical experience through planning, construction and operation • Breeding ratio 1.03±0.03 (design value 1.01)
MK-II (Irradiation core) (100 MWt)	
July 1982 ~September 1997	First Criticality of MK-II Core FBR fuels and materials irradiation tests • Power-to-melt test (PTM) • Fuel failure simulation test • High burn-up test (peak burn-up of 144 GWd/t, collaboration with CEA France) • Carbide and nitride fuel irradiation test
MK-III (Upgraded irradiation core) (140 MWt)	
July 2003 ~May 2007	First criticality of MK-III core • Minor actinide bearing fuel irradiation test • Demonstration of Self Actuated Shutdown System (SASS) • In-pile creep test using instrumented test subassembly
MK-IV (Improving Safety and Reliability) (100 MWt)	
December 2021	Undergoing safety review by NRA Japan

Operation Time: 70,796 h	**Heat Generation: 6,224 GWh**
Number of driver fuel subassemblies: • MK-I fuel : 116 • MK-II fuel : 342 • MK-III fuel : 130	Number of irradiation test subassemblies: • Fuel irradiation subassembly : 27 • Material irradiation subassembly : 68 • Instrumented test subassembly : 5 • SASS test equipment : 1

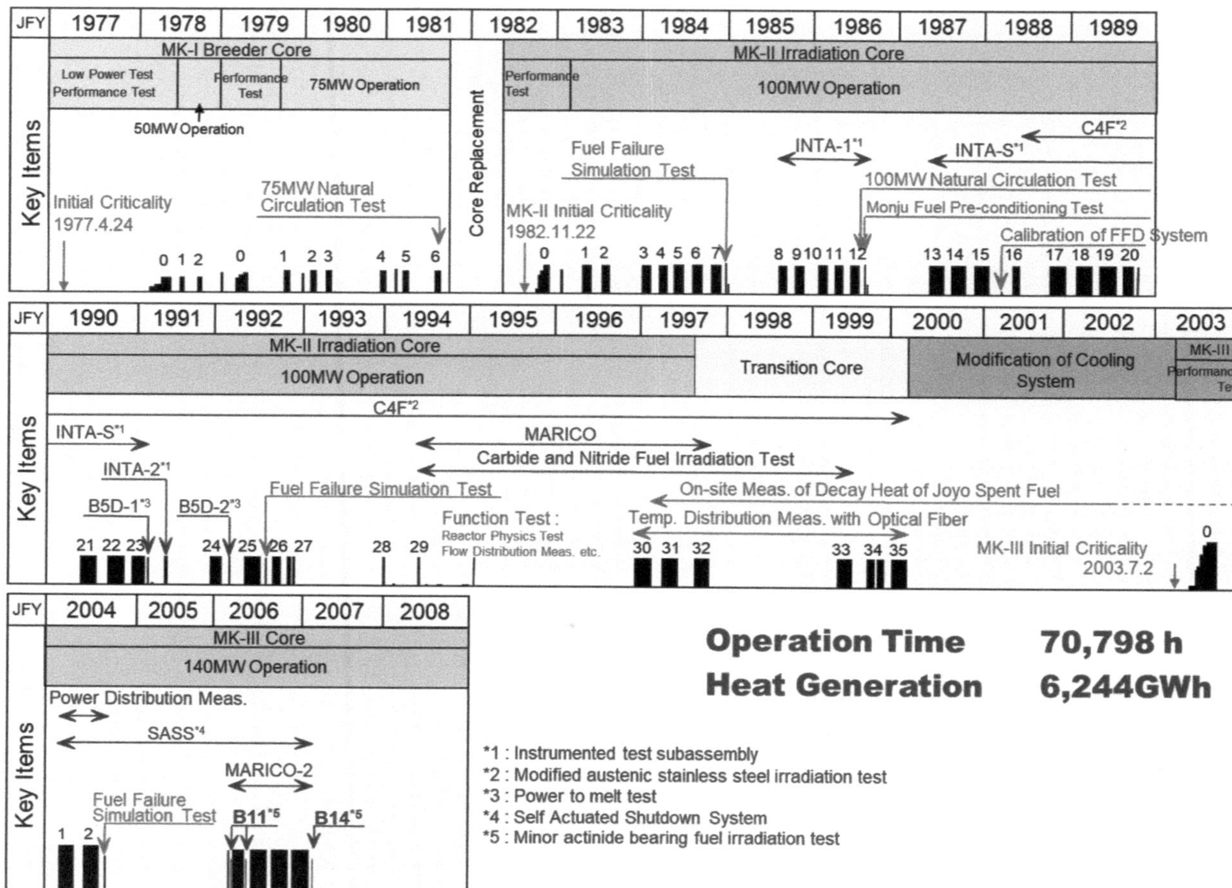

Fig. 2.8 Operational performance of Joyo.

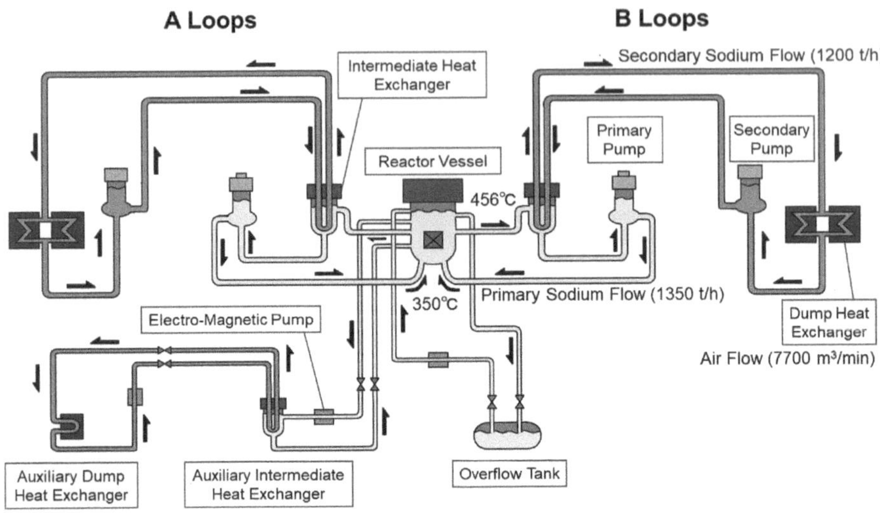

Fig. 2.9 Heat transfer system of Joyo with MK-IV core.

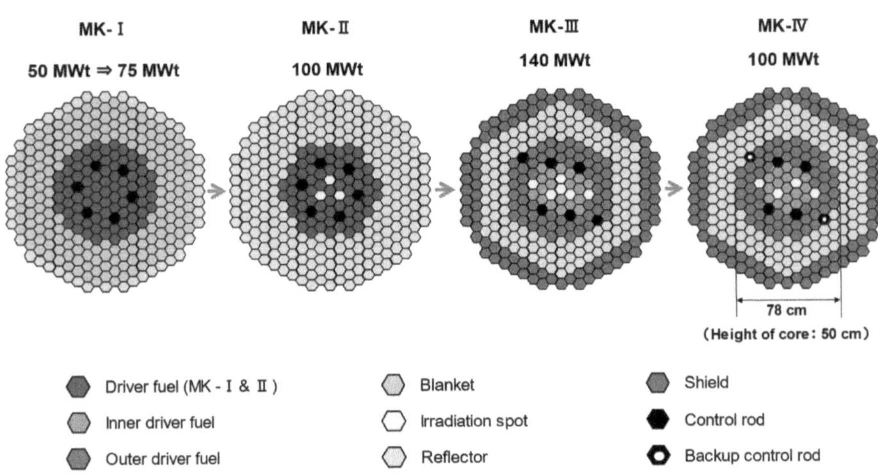

Fig. 2.10 Core maps of MK-I through MK-IV.

within 5%. In addition, the breeding ratio was verified by postirradiation examination (PIE) of the core and blanket fuels.

Later, core conversion work was carried out to replace the MK-I breeder core with the MK-II irradiation-test-bed core. The main features of the MK-II driver fuel are as follows: (a) increased Pu enrichment, (b) shortened core zone height, and (c) replacement of the axial and radial blanket by stainless steel reflectors. Improvement of various irradiation characteristics, including (a) increased neutron flux, (b) increased power density, and (c) enlarged fast neutron zone, was achieved by the above modifications.

Table 2.2 Properties of each core design.

Items	MK-I	MK-II	MK-III	MK-IV
Reactor thermal power (MWt)	50 → 75	100	140	100
Max. number of driver fuel	82	67	85	79
Max. number of irradiation test fuel	0	9	21	4
Blanket	Depleted UO_2	–	–	–
Reflector	Stainless steel	Stainless steel	Stainless steel	Stainless steel
Core diameter (cm)	80	73	80	78
Core height (cm)	60	55	50	50
^{235}U enrichment (wt%)	~23	18	18	18
Pu content (wt%) (inner/outer)				
Total	~18	28	23/29	22/29
Fissile	~14	20	16/21	16/21
Max. liner heat rate (W/cm)	320	400	420	330
Max. neutron flux (n/(cm^2 s))				
Total	3.2×10^{15}	4.9×10^{15}	5.7×10^{15}	4.2×10^{15}
Fast (>0.1 MeV)	2.2×10^{15}	3.2×10^{15}	4.0×10^{15}	2.9×10^{15}
Max. pin average burnup (GWd/t)	42	75	90	90
Primary coolant flow rate (t/h)	2200	2200	2700	2700
Reactor vessel inlet temp. (°C)	370	370	350	350
Reactor vessel outlet temp. (°C)	430 → 450	500	500	456

Joyo is expected to play a greater role in providing irradiation test fields to develop SFRs and for various materials used in other areas. Subsequently, an upgrading project named MK-III was conducted. The main objectives of this project are the increase in neutron flux, the modification of the cooling systems related to increased reactor power, the increase of irradiation periods, and the upgrading of irradiation technology. The MK-III standard core design and basic core specifications were determined such that the fast neutron flux would be 1.3 times higher than that in the MK-II core. The active core is cylindrical with a diameter of ~80 cm and a height of 50 cm. It is a conventional two-region core where the contents of fissile material in the inner and outer cores differ. The reflector region of stainless steel with a thickness of 25–30 cm surrounds the core. The shielding subassemblies with B_4C are loaded in the outermost two rows of the reactor grid. During conversion to the MK-III core, the MK-III fuels were gradually loaded into the MK-II core without causing any delay in the fuels and materials irradiation test programs.

In the 2010s, Japanese nuclear regulations have dramatically changed. To adopt these new regulations, several upgrades are undergoing at Joyo. Its representative upgrade is the replacement of the MK-III core by a novel MK-IV core. MK-IV decreases thermal power to 100 MWt to maintain the excess reactivity at a low level. The basic core design of the MK-IV core is the same as the MK-III core, but there are two remarkable changes: the addition of the backup control rod system and a decrease in the number of driver fuel subassemblies. Six control rods were originally driven by a single system at Joyo, whereas the MK-IV has an additional backup system that drives two of six control rods. This system improves the reliability of the reactor shutdown function. Regarding another change, the maximum number of driver fuel subassemblies is decreased from 85 to 79 because of the decreased thermal power.

2.2 Operation and maintenance experience

2.2.1 Operation and maintenance

2.2.1.1 Core management

A breeding ratio of 1.03 was confirmed during the operation of the MK-I breeder core. On the other hand, the MK-II irradiation core was operated with a Pu conversion ratio of 0.3. These results demonstrate that both breeding and consumption of Pu are feasible in the same SFR only with the replacement of the core.

Based on the irradiation results, the pellet peak burnup of driver fuel was increased step by step from 25 to 90 GWd/t by improving the fuel design by optimizing the safety design margins and using new cladding materials that are antiswelling and have higher strength at high temperatures. The linear heat rate (LHR) of the driver fuel was also increased by rationalizing the safety margin of design with revision of the maximum temperature of the fuels, based on the performance test evaluation. By the end of MK-II operation, 58,000 fuel pins were irradiated without any fuel pin failure to a maximum burnup of 87 GWd/t. This result indicates suitable management of the core and fuels, as well as high quality of fuel fabrication.

The core management code system named MAGI [1] was developed to predict the reactor parameters required for core and fuel management of the Joyo MK-II core. Fig. 2.11 outlines the MAGI system. MAGI is a coupled neutronics and thermo-hydraulics code system that calculates the excess reactivity, power distribution, fuel burnup, coolant flow rate, and temperature condition of each subassembly. MAGI uses the diffusion theory with seven neutron energy groups for nuclear calculation. The neutron cross-sections were collapsed from the 70-grouped Japan Atomic Energy Research Institute Fast Set version 3 (JFS-3-J2) cross-section set processed from the Japanese Evaluated Nuclear Data library version 2 (JENDL-2). It was updated to the JFS-3-J3.2 cross-section set based on the JENDL-3.2 in 2001. And then, it was revised to the JFS-3-J3.2R cross-section set in 2003. It was confirmed that the influence of this revision was minor [2]. The gamma-ray cross-sections, which include delayed fission

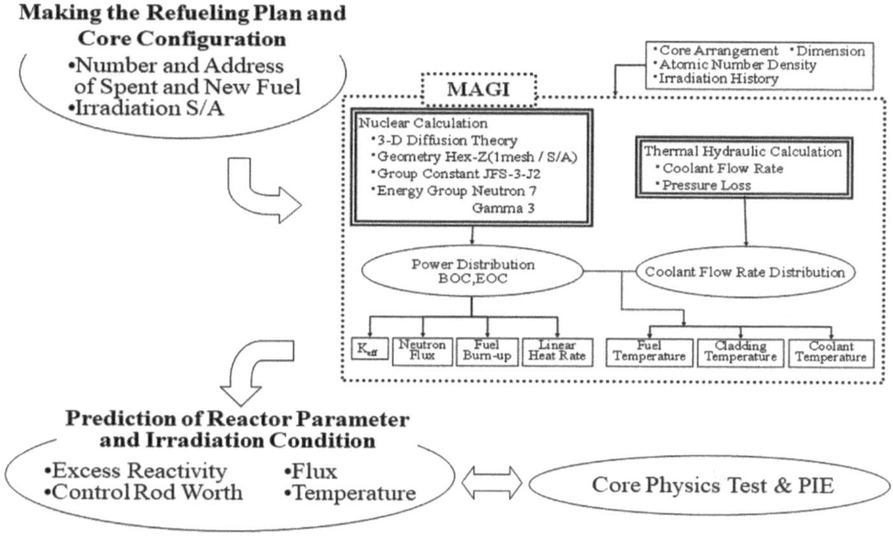

Fig. 2.11 MAGI system outline.

gamma rays, were processed from the JENDL-2 to improve the calculations of gamma heating in stainless steel. The core configuration was modeled in three-dimensional Hex-Z geometry for each operation cycle. The actual reactor power history was used in the burnup calculation.

MAGI supported the refueling plan for each operation cycle. The number of fresh driver fuel subassemblies and their positions were determined within design limitations of neutronic and thermal parameters with consideration for the following items: (a) burnup of fuel subassemblies, (b) operation period of the cycle, and (c) neutron fluence and temperature conditions of the irradiation test subassemblies. The number of spent-fuel subassemblies discharged at the end of a cycle (EOC) ranged from 10 to 12, about one-sixth of the total driver fuel subassemblies.

Since the frequent refueling significantly affected the core characteristics and irradiation conditions, it is necessary to verify the MAGI calculation results at each operation cycle by comparison with measured data. The calculated core characteristics, such as excess reactivity, control rod worth, reactivity coefficients, and fuel burnup, were verified by core physics tests in each cycle operation and PIE results. A comparison of core calculations with measurements in MK-II is shown in Table 2.3. The calculation was performed with the JFS-3-J3.2 cross-section set.

For the MK-III core, the core management code system named HESTIA was developed [3]. HESTIA is also a coupled neutronics and thermo-hydraulics code system that calculates the excess reactivity, power distribution, fuel burnup, coolant flow rate, and temperature condition of each subassembly. Fig. 2.12 outlines the HESTIA system. The calculation geometry was changed from Hex-Z to Tri-Z geometry. The number of calculation meshes per subassembly was increased to 24 in the radial direction. The number of neutron energy groups was increased from 7 to 18, and that

Table 2.3 Comparison of core characteristic calculations with measurements in MK-II.[a]

Core characteristics	Calculation over experiment
Excess reactivity (%$\Delta k/kk'$)	0.99 ± 0.03 (±0.1%$\Delta k/kk'$)
Control rod worth	
Symmetrical (%$\Delta k/kk'$ per total)	0.99 ± 0.02
Asymmetrical (%$\Delta k/kk'$ per total)	0.98 ± 0.01
Burnup reactivity coefficient (%$\Delta k/kk'$ per MWd/t)	0.96 ± 0.05

[a] Cycles 16 through 35.

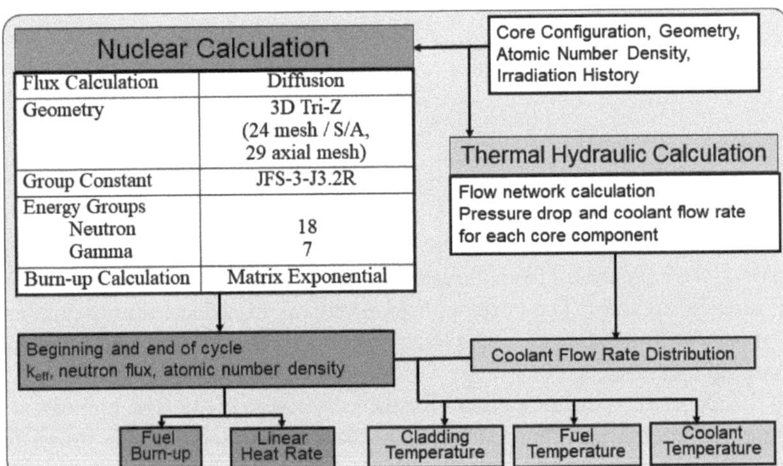

Fig. 2.12 HESTIA system outline.

of gamma energy groups was increased from 3 to 7. The neutron cross-sections were collapsed from the JFS-3-J3.2R cross-section set.

In addition, the JUPITER standard analysis method was also used for certain analyses [4].

Specifically, the method was used for the control rod worth calculation using its following advantages: the capability of addressing the heterogeneity of fuel and control rods, and the transport calculation [3,5]. A comparison of core calculations with measurements in MK-III is shown in Table 2.4.

2.2.1.2 Demonstration of Pu fuel recycle

Plutonium fuel recycling was demonstrated in 1984. Two MK-I driver subassemblies unloaded in 1981 were transferred to the Fuel Monitoring Facility (FMF) adjacent to Joyo. The FMF has large stainless-steel-lined, nitrogen-gas-tight hot cells with remote maintenance systems for PIE. The subassemblies were dismantled in FMF to remove the fuel pins, and 10 fuel pins were transported to the Chemical Processing Facility

Table 2.4 Comparison of core characteristic calculations with measurements in MK-III[a] [3].

Core characteristics	Calculation (HESTIA)	Calculation (JUPITER)	Experiment
Excess reactivity at 250°C (%$\Delta k/kk'$)	3.13	3.36	2.99 ± 0.10
Burnup reactivity coefficient (%$\Delta k/kk'$ per MWd)	2.12×10^{-4}	–	$(2.12 \pm 0.09) \times 10^{-4}$
Control rod worth (%$\Delta k/kk'$)			
Rod No. 1	–	2.01	2.09 ± 0.07
Rod No. 2	–	0.79	0.80 ± 0.03
Rod No. 3	–	1.97	2.03 ± 0.07
Rod No. 4	–	2.01	2.08 ± 0.07
Rod No. 5	–	0.79	0.78 ± 0.03
Rod No. 6	–	1.97	2.06 ± 0.07

[a] Cycle 0.

(CPF) of Tokai Works for the SFR fuel reprocessing tests in 1982. About 56 g of Pu was extracted in the CPF, and 25 g of Pu was used for processing MOX fuel pellets at the Plutonium Fuel Development Facility (PFDF) of Tokai Works. PDF fabricates MOX fuel for reactor physics tests and irradiation tests at various reactors both inside and outside Japan. A fuel pin was fabricated at PFDF using these new fuel pellets. This pin was transported to the Irradiation Rig Assembling Facility (IRAF), adjacent to Joyo, where an irradiation test subassembly was assembled using this pin. Fig. 2.13 depicts the demonstration of the Pu fuel recycling.

2.2.1.3 Chemical analysis of sodium and cover gas

When using sodium as the reactor coolant, corrosion of all components in contact with sodium, such as vessels, piping, and fuel cladding, poses a problem. Factors significantly influencing material corrosion include the solubility of nonmetal impurities (oxygen, hydrogen, etc.) in sodium, temperature gradient, and mass transfer. To minimize material corrosion for safe reactor operation, it is essential to always maintain high purity of the coolant sodium, so-called purity control. Also, for argon cover gas (CG) that covers the free surface of liquid sodium, to prevent the oxidation of sodium, it is necessary to set the concentration standard of impurities (oxygen, hydrogen, etc.) contained in CG and control the purity of CG at a high level. It is assumed that the main causes of impurities mixed in the system are elution of impurities on the surface of the material and in-leakage of air and moisture from the boundary seal with the outside of the system when replacing the core components of the reactor. Furthermore, in Joyo, the mixture of alcohol used for cleaning sodium and oil components of mechanical pumps, etc. is also assumed. In addition, there are impurities caused by fuel damage and corrosion, and impurities such as tritium generated from fuel and

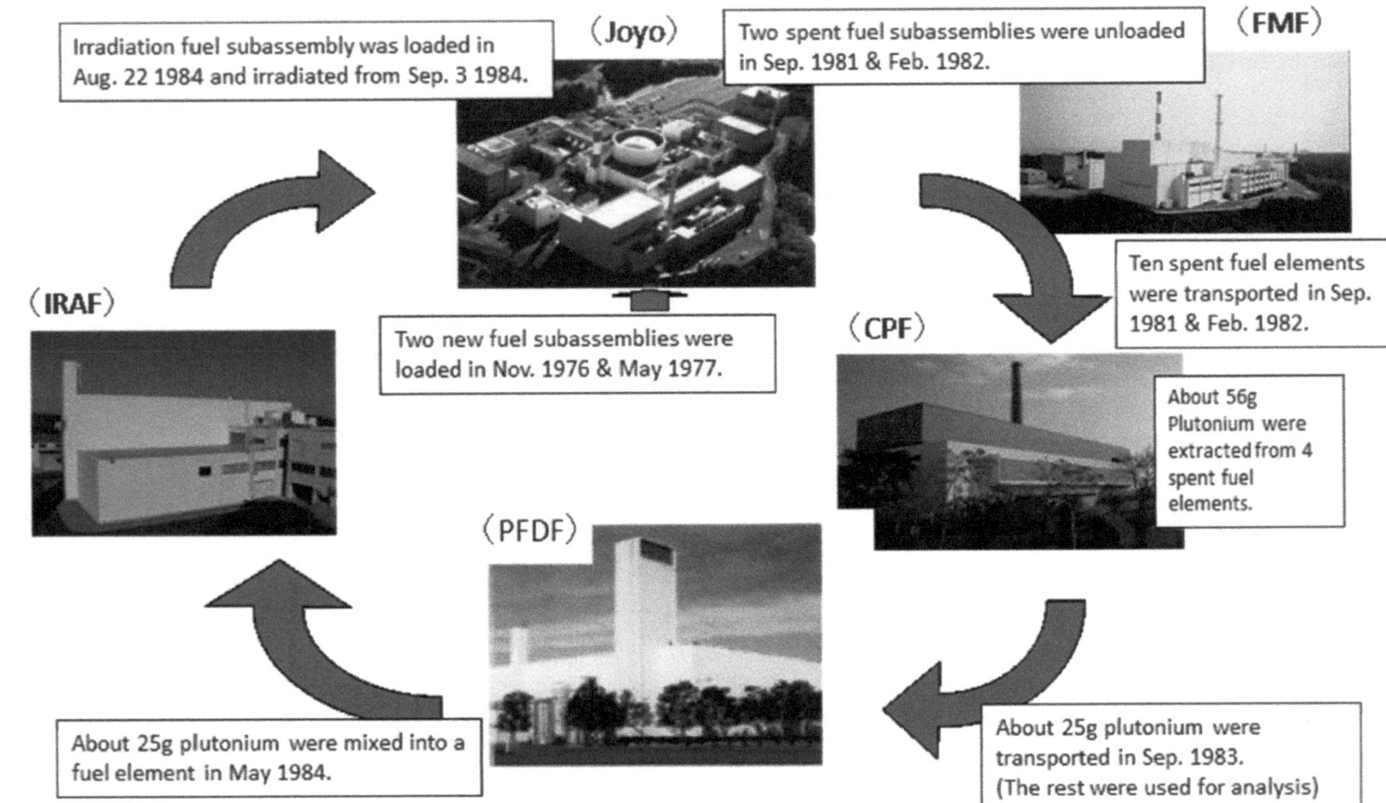

Fig. 2.13 Demonstration of the Pu fuel recycle around Joyo.

control rods. Therefore, impurities in sodium that must be monitored include oxygen, hydrogen, nitrogen, chlorine, tritium, metallic elements (Fe, Ni, Cr), and radioactive material such as ^{110m}Ag, ^{22}Na, and ^{137}Cs. Impurities in the CG that must be monitored include O_2, N_2, CO, CO_2, H_2, CH_4, He, ^3H, and the radioactive isotopes of Xe and Kr.

In Joyo, for the purity control of the coolant sodium, the cold traps (CTs) that can remove impurities are installed (see Fig. 2.14). The CT is an equipment that removes the impurities in sodium by reducing the sodium temperature lower than the saturated temperature of the target impurities to precipitate and trap them with stainless mesh, using the principle that the solubility of impurities in sodium varies with temperature.

To monitor the impurity concentration in sodium, the plugging indicator allowing for real-time measurement is used. The plugging meter measures the temperature below which impurities in sodium start precipitating (saturation temperature) and estimate the impurity concentrations in sodium through comparison with the saturation solubility curve of each impurity. The control standards for the plugging temperature of the primary and secondary coolant sodium are no greater than 200°C and 225°C (corresponding to an oxygen concentration of approximately 10 ppm and 20 ppm), respectively, and the control target value is conservatively set at 150°C. The CTs are normally operated at 130°C in consideration of the balance between further inhibition of material corrosion and stable measurement by plugging meter. In addition, sampling analysis measurement of the impurity concentrations is performed for more precise control because the purity control by plugging meter can measure only limited components although it allows for continuous monitoring. For such analysis measurement in Joyo, chemical analysis, gas chromatography, beta-ray scintillation, and gamma-ray spectrometry are used. As an example, the transition of analyzed oxygen and hydrogen (concentrations) in the primary coolant sodium during the MK-II and MK-III operation period including the modification work period in between is shown in Fig. 2.15. The oxygen concentration in the primary coolant sodium has been maintained at 1–7 ppm, sufficiently lower than the control standard value, 10 ppm, over the whole period. It should be noted that the analyzed data showed relatively high values in the early phase of MK-II operation because it was required to increase the CT operation temperature due to the increase in the CT pressure loss during the initial phase. Since the replacement of the CT in 1987, both oxygen and hydrogen concentrations in the primary coolant sodium have been stably maintained at sufficiently low levels.

To monitor the impurity concentrations in CG, gas chromatography is used. In addition, sampling analysis measurement of the impurity concentrations in CG is periodically performed, as is done for sodium. Analysis of impurities in the CG showed an occasional increase in hydrogen concentration. This is likely to be caused by the dissociation in the reactor of a small amount of alcohol that was brought in with the fuel handling machine, etc., that was cleaned by alcohol to remove sodium.

Due to these control measures, the purity control of sodium has been appropriately performed and the impurity concentrations have been stably maintained at low levels in Joyo. Consequently, corrosion of structural and fuel cladding materials, etc., has been prevented, and no sodium leak has occurred due to fuel failure or corrosion or blockage of piping.

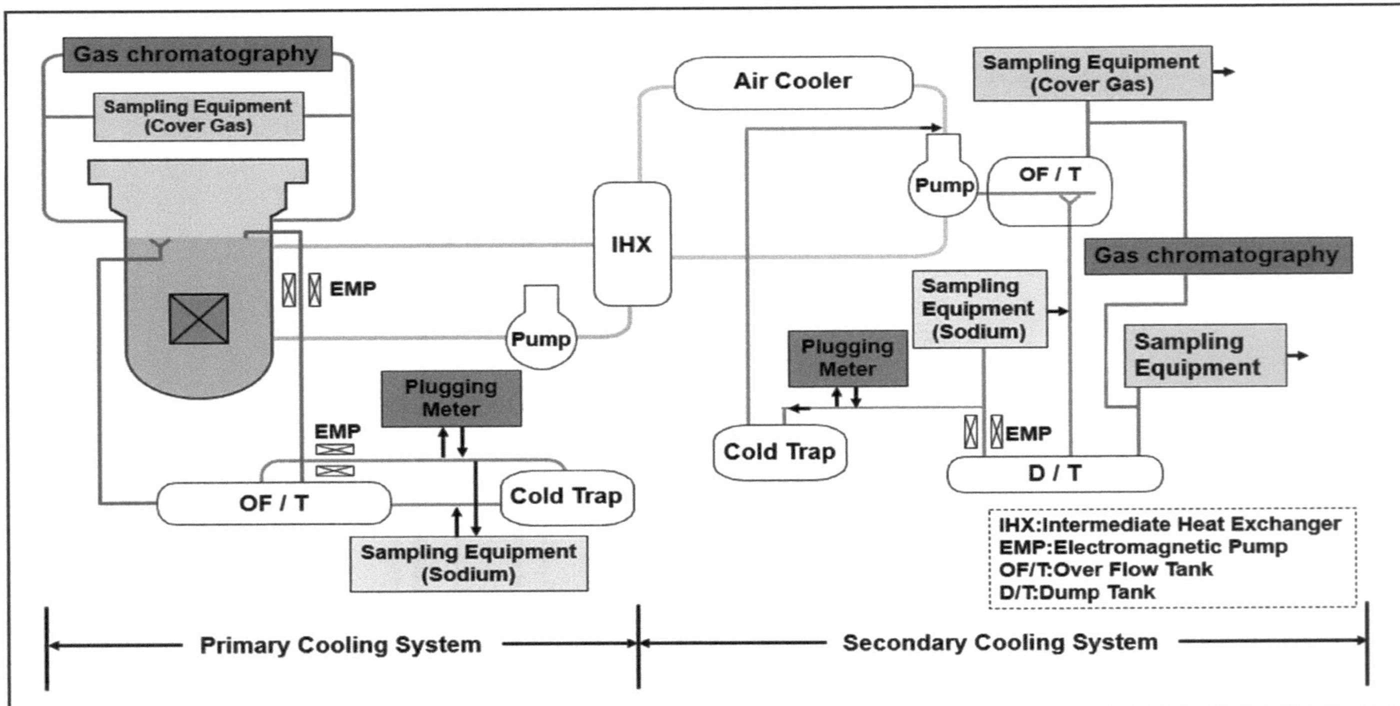

Fig. 2.14 Sodium and cover gas purity control systems in Joyo.

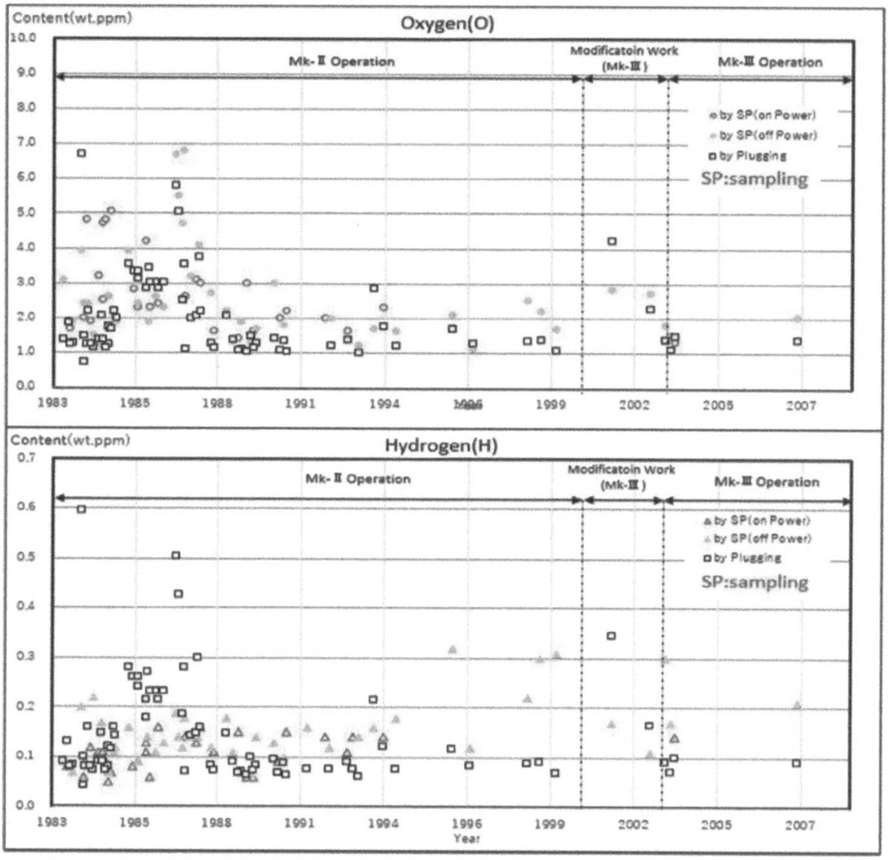

Fig. 2.15 Transition of oxygen and hydrogen concentrations as impurities in the primary coolant sodium in Joyo.

2.2.1.4 Reliability of sodium components

(1) Primary and secondary main cooling pumps

In Joyo, mechanical and electromagnetic pumps are used to circulate the liquid metal sodium coolant. Both the primary and secondary main circulation pumps are vertical free-surface centrifugal pumps, and each of the four pumps is placed in the cold leg of each main cooling system (of two loops for the primary and secondary systems, respectively). In the free-surface centrifugal pump used for Joyo, sodium flows into the suction nozzle located at the lower end of the pump and is pressurized by the impeller and discharged from the discharge nozzle through the diffuser. The free surface of sodium is covered with the argon CG, which is sealed from the outside air by mechanical seals. The primary main circulation pump is equipped with a direct current (DC) motor (pony motor) that is operated by an uninterruptible power supply to ensure a sufficient flow rate to remove the core decay heat even in the case where the main motor is unavailable due to the loss of offsite power, etc. The total operating times of the primary main circulation pumps A and B are 218,460 and 218,552 h. The total operating times of the primary and secondary main circulation pumps are listed in Table 2.5.

Table 2.5 Total operating time of the primary and secondary main circulation pumps (as of August 2021).

		Operation (main motor) (h)	Operation (pony motor) (h)	Operation (total) (h)
Primary system main circulation pump	A loop	212,156	6304	218,460
	B loop	212,156	6396	218,552
Secondary system main circulation pump	A loop	233,364		233,364
	B loop	233,301		233,301

Concerning the free-surface centrifugal pump used for fast reactors, it is known that natural convection of the CG occurs in the gap between the inner and outer casings, causing the circumferential temperature difference around the casing. This circumferential temperature difference causes thermal deformation that may lead to contact between the pump shaft and bearing. The occurrence of natural convection was also confirmed in the primary main circulation pump of Joyo. The increase in bearing load due to thermal deformation was insignificant, causing no problem with the pump performance; however, the argon gas anticonvection baffle was installed as a mechanism to prevent the natural convection from the perspective of preventive maintenance during the second and third periodic maintenance periods. As a result, the circumferential temperature difference was reduced to 0–4°C, and the pumps have been successfully operated. The structure of the primary main circulation pump and a sketch of the argon gas anticonvection baffle are shown in Fig. 2.16.

For the MK-III project, the coolant heat removal capacity was improved with increased core power by increasing the primary coolant flow rate from 1085 to 1335 t/h, while the main motor power of the primary main circulation pump was not changed from the original power, 330 kW, because the increase in flow rate was achievable within the existing capacity. In contrast to this, the main motor power of the secondary main circulation pump was changed from 180 to 220 kW due to the increase in pressure loss on the sodium side resulting from the increase in the secondary coolant flow rate from 1085 to 1186 t/h and the adoption of sigma-type heat transfer tubes for the DHXs. Furthermore, in the MK-III project, since the difference in sodium temperature between the reactor inlet and outlet increases, the run back control, where the primary main circulation pump reduces the rotation rate in a condition similar to the free coast down without stopping the main motor and subsequently runs at a low flow rate, was incorporated into to moderate thermal transient in sodium components after reactor scram.

(2) Electro-magnetic pump

Five electromagnetic pumps (EMPs) are installed in Joyo: three units in the auxiliary, overflow, and sodium purification systems of the primary cooling system, and two units in the auxiliary and purification systems of the secondary cooling system. Of these, the EMP of the secondary purification system is an AC Faraday pump, and the other EMPs are the flat linear induction pumps (FLIPs). The operating principle of both types of EMPs follows Fleming's left-hand rule. Schematic diagrams of FLIP and AC Faraday EMPs are shown in Fig. 2.17.

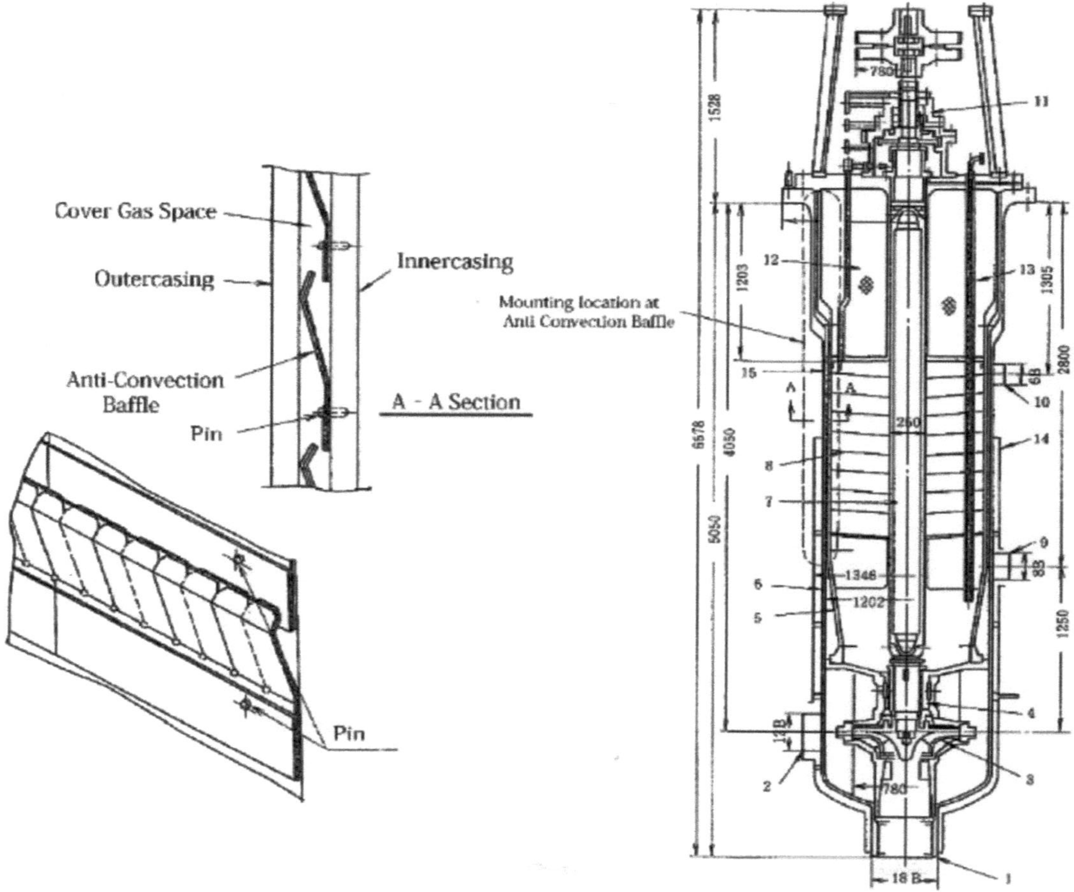

15	Anti-Convection Baffle
14	Leak Jacket
13	Sodium Level Detector
12	γ -Ray Shielding
11	Roller Bearing
10	Argon Gas Nozzle
9	Over Flow Nozzel
8	Thermal Shelding
7	Shaft
6	Outercasing
5	Innercasing
4	Hydrostatic Bearing
3	Impeller
2	Outlet Nozzel
1	Inlet Nozzel

Fig. 2.16 Structure of primary main circulation pump and a sketch of analysis model for the argon gas anticonvection baffle.

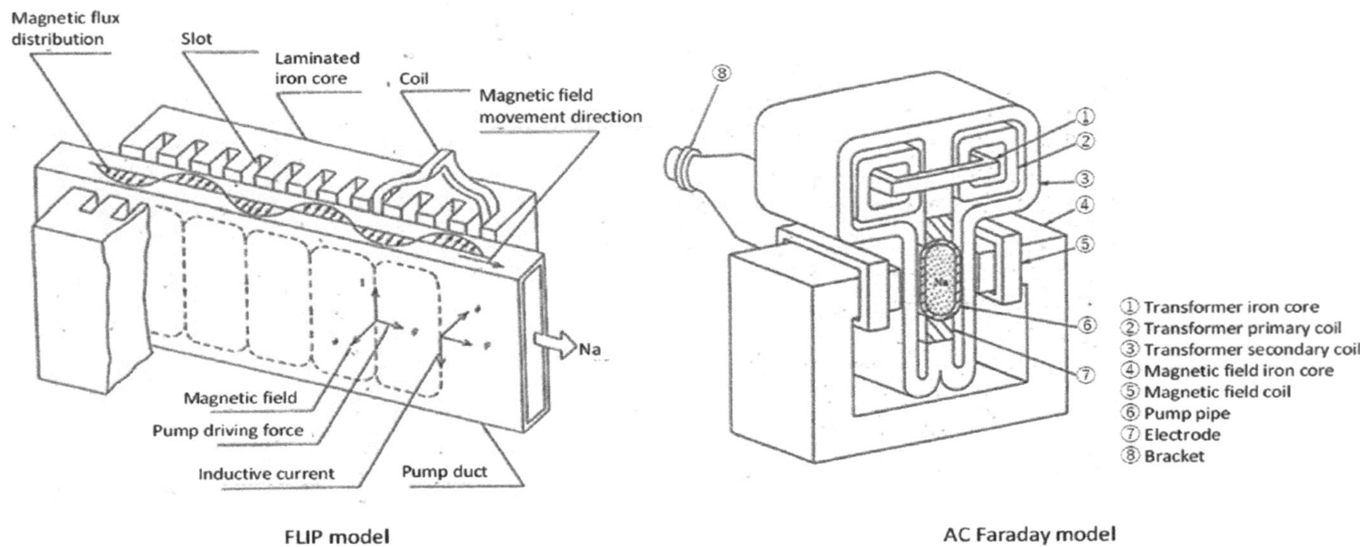

Fig. 2.17 Schematic diagrams of FLIP and AC Faraday EMPs.

Since EMPs have no moving parts, they have the advantage that failures caused by moving parts and deterioration due to wear occur less frequently and that there is no part sealed by packing, etc., allowing for a completely sealed structure; however, the pump efficiency of EMPs, a dozen or so percent, is lower than that of mechanical pumps. Therefore, EMPs were used for systems with a low flow rate at Joyo.

The structural feature of FLIP EMPs includes the copper end bar connected with the upper and lower pump ducts for improving the power factor, as shown in Fig. 2.18. An induced current forms a loop and flows in both upper and lower pump ducts. Due to this, the directions of driving force are different between the upper and lower pump ducts. Accordingly, when an induced current flows in sodium, sodium flow is disturbed. As a countermeasure to this, the copper end bar is connected with the upper and lower pump ducts, and the sodium flow disturbance is eliminated by applying an induced current in the copper end bar. EMPs installed in the secondary cooling system have a horizontal structure; however, the above-mentioned feature is quite similar to that of the vertical structure except for the copper end bar, which is connected with the left and right pump ducts. In addition, since the AC Faraday EMP is smaller than the FLIP EMP and has a different structure, a copper end bar is not required.

For each EMP, the visual appearance, operating conditions, insulation resistance, and the control unit have been inspected during every facility's periodic inspection. In addition, an overhaul inspection of the EMP units has been performed based on the operational performance, long-term maintenance plan, etc. The operating time of each EMP is listed in Table 2.6.

Insulation deterioration of the coil, the main deterioration factor of EMP, is caused by a combination of various factors, including electrical, thermal, and environmental (radiation, dust, etc.) factors and mechanical force. Through experience, it was known that insulation

Fig. 2.18 Copper end bar connected with the upper and lower pump ducts in FLIP EMP.

Table 2.6 Operating time of each EMP (as of August 2021).

Electromagnetic pump	Area	Model	Operation (h)
Primary auxiliary electromagnetic pump	Controlled area	FLIP	38,495
Over flow system electromagnetic pump	Controlled area	FLIP	221,823
Primary purification system electromagnetic pump	Controlled area	FLIP	225,943
Secondary auxiliary electromagnetic pump	Noncontrolled area	FLIP	233,302
Secondary purification system electromagnetic pump	Noncontrolled area	AC Faraday	244,124

deterioration due to radiation is an especially important factor, and therefore, an inspection focused on radiation has been performed. The EMPs of the primary cooling system are installed in areas under the floor of the reactor containment vessel, and the gamma dose rate in the auxiliary system during reactor power operation is approximately 1.5 Gy/h, while the gamma dose rate in the overflow and purification systems is one order of magnitude higher. For this reason, a state of deterioration is checked by hammering test of the coil during overhaul inspection of the EMPs of the overflow and purification systems. Since an overhaul inspection revealed slight insulation deterioration due to dust attachment although no anomaly in insulating material was observed, coil cleaning and insulation reinforcement were performed.

Due to the high operating temperature of the EMP's copper end bar, high-temperature oxidation becomes a problem as the operating time increases. Corrosion and scale deposition due to high-temperature oxidation of the copper end bar would reduce the sodium transport efficiency; however, since the EMPs in Joyo have a 20% energization margin concerning the sodium transport efficiency, operation at the normal flow rate can be done if the sodium transport efficiency reduces from the initial level to 80% or so. In the overhaul inspection of EMP performed during the 13th periodic inspection period in 2000, the size measurement of the copper end bar was performed. From the ratio between the effective and initial cross-sections, the reduction in the sodium transport efficiency was estimated to be approximately 5% compared with that in the initial stage of use of EMP in 1976. Since the energization margin is 20%, the reduction in efficiency due to high-temperature oxidation of the copper end bar has not interfered with the operation of EMP. Furthermore, while the operating voltage of EMP tends to increase with the decrease in sodium transport efficiency, this tendency has not yet been observed.

(3) Cold trap (CT)

The CT uses the principle that the solubility of impurities in sodium depends on temperature. The CT is designed to remove the impurities by reducing the sodium temperature lower than the saturated temperature of the target impurities to trap them with stainless mesh or through precipitation.

For the CTs installed in both primary and secondary systems of Joyo, a target operating temperature of 150°C was recommended to prevent the corrosion of structural material, and different types of CTs were adopted, respectively, for the primary and secondary systems, and developed.

The CT for purifying the primary sodium is called the "Fermi-type," in which the sodium temperature becomes the lowest at the inlet of mesh to precipitate impurities and trap them with mesh. As for the primary system, a reduced-temperature operation was performed to decrease the radioactive corrosion product (CP) in sodium by lowering the target operating temperature from 150°C to 120°C. Due to this, the tendency of blockage at the mesh inlet on the lower part of the CT became more apparent with the continued operation, resulting in a gradual increase in the plugging temperature. Therefore, a new-type CT capable of reduced-temperature operation was developed, and the existing CT was replaced with the new one during the sixth periodic inspection.

Conceptual diagrams of the old/new CT are shown in Fig. 2.19. The new CT has the following features compared with the existing one:

(i) An inflow model through the whole surface was adopted to prevent the blockage at the mesh inlet.

(ii) The sodium flow path in the annulus between the outer shell and the adiabatic gas layer was widened by reasonably minimizing the number of heat transfer tubes for cooling.

(iii) The taper angle of the cone in the lower part of the adiabatic gas layer was increased to form a structure where impurities are unlikely to accumulate.

(iv) To prevent compressive deformation caused by mesh clogging, the mesh wire diameter was increased. Additionally, the filling factor was decreased by changing the mesh type from stockinette stitch to plain weave to increase the capacity of trapped impurities.

(v) To prevent heat input from sodium in the annulus, a double-wall structure was adopted by providing the adiabatic gas layer over the whole mesh area and enlarging the bottom part to maintain the adiabatic effect.

After installation of the new CT, it was confirmed that stable low-temperature operation at the target operating temperature of 120°C could be performed as designed without freezing at the mesh inlet or the increase in discharge pressure of the purification system EMP due to local blockage.

The CT for purifying the secondary sodium is called the HNPF model, in which sodium is cooled in the mesh area to precipitate impurities and trap them at the mesh outlet controlled at the lowest temperature. The secondary CT has been operated at low temperatures by lowering the target operating temperature from 150°C to 130°C to confirm the balance of sodium purity and the difference in purification function between the primary and secondary systems.

OLD C.T

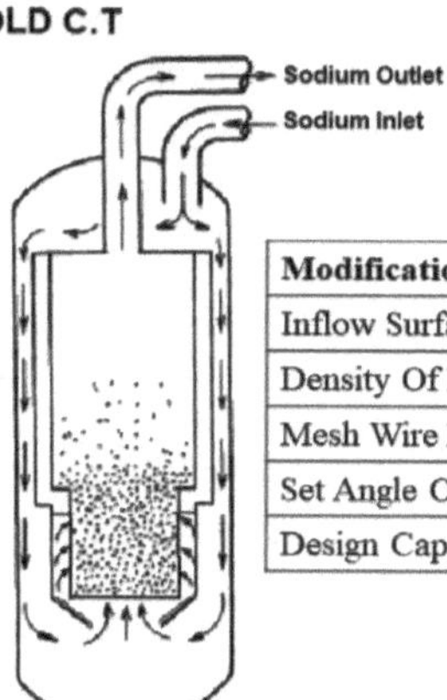

Modification Points	OLD C.T	NEW C.T
Inflow Surface Area	0.4m^2	5.4m^2
Density Of Mesh	200kg / m^3	150kg / m^3
Mesh Wire Diameter	0.127mmφ	0.5mmφ
Set Angle Of Cone	45°	52°
Design Capacity	11.5kgO_2	141kgO_2

NEW C.T

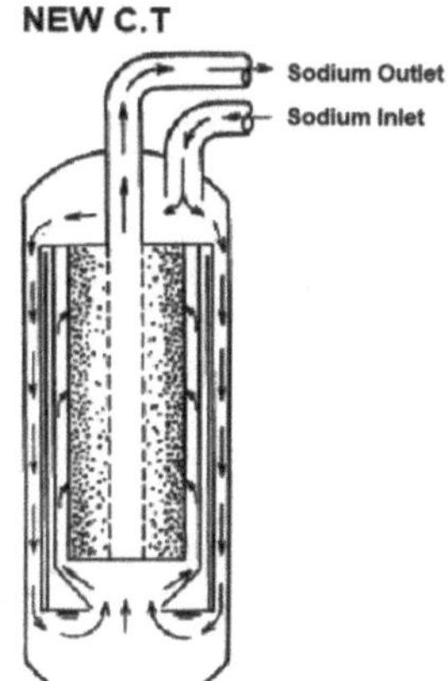

Fig. 2.19 Schematic diagrams of old and new primary cold traps.

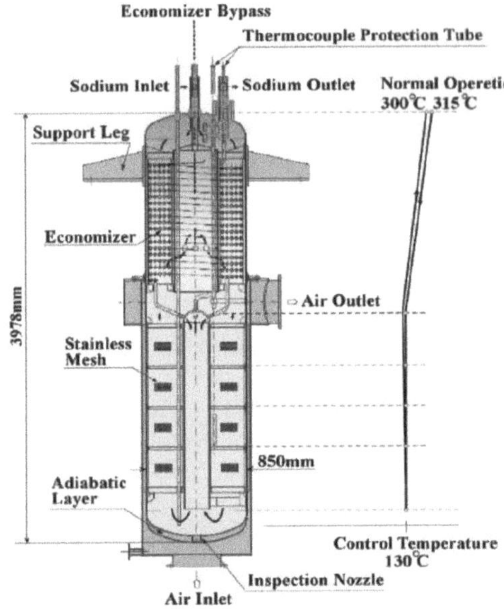

Main Principal of Secondary Cold Trap	
Type	Forced Air Cooling Type Including Built-in Economizer
Whole Body	2¼Cr−1Mo Steel(SCMV4-2)
Mesh Type	Austenitic Stainless Steel(SUS304)
Maximum Operation Pressure	0.79MPa
Maximum Operation Temperature	400℃
Flow	5.2ton/h
Purification Target Value	Below Level of 150℃
Purification Foundation Value	Below Level of 225℃
Oxigen Capture Capacity	18kgO_2

Fig. 2.20 Schematic diagram of secondary cold traps.

A conceptual diagram of the secondary CT is shown in Fig. 2.20. Concerning the secondary CT, the tendency of blockage was not found with the continued operation, different from the primary CT. However, since the amount of impurities that would be trapped during modification work for the MK-III project was expected to exceed the design amount of impurities that could be trapped, the secondary CT was replaced with that having the same structure as the existing one.

2.2.1.5 In-service inspection of the coolant boundary

(1) Surveillance test of reactor structural materials

Equipment comprising the reactor coolant boundary is designed to have sufficient fracture toughness to prevent instantaneous destruction during normal operation, abnormal transient during operation, and design basis accidents. For equipment including the R/V and the main primary cooling system, the effect of neutron irradiation on the material is monitored using surveillance materials.

Structural material of the R/V (SUS304) showed the following tendencies. Those tendencies agree with the general properties of neutron-irradiated austenitic stainless steel (0.2% offset yield strength: increase, tensile strength: increase slightly, breaking elongation: decrease). The mechanical strength properties of the material of the R/V, etc., satisfy the allowable values under neutron fluence ($E > 0.1$ MeV) beyond the value (3.48×10^{19} n/cm^2) estimated from the design life of the R/V (design operation period at rated power: 131,500 h).

- The 0.2% offset yield strength increases at neutron fluence ($E > 0.1$ MeV) higher than approximately 1×10^{20} n/cm^2, and the tendency becomes more apparent beyond approximately 5×10^{20} n/cm^2 (Fig. 2.21).

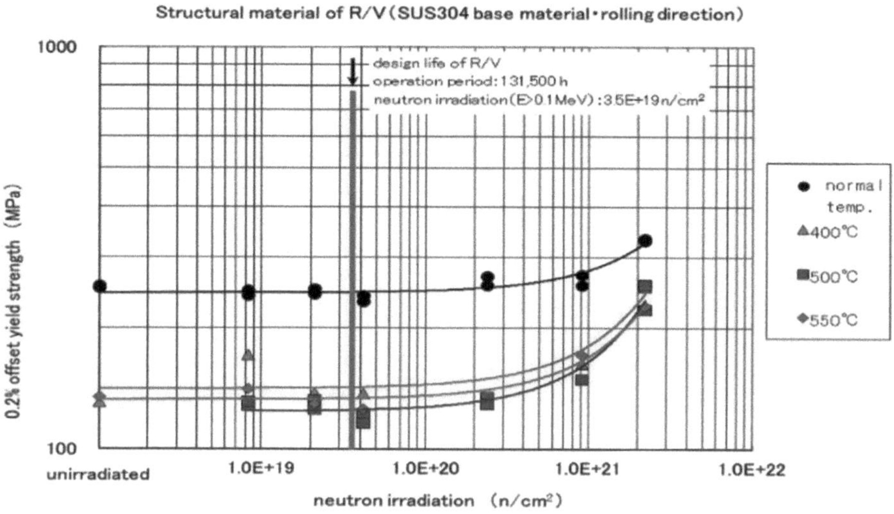

Fig. 2.21 0.2% offset yield strength of structural material for R/V.

- The tensile strength also increases at neutron fluence ($E > 0.1$ MeV) higher than approximately 1×10^{21} n/cm^2.
- The uniform elongation and breaking elongation decrease at neutron fluence ($E > 0.1$ MeV) higher than approximately 1×10^{20} n/cm^2, and the tendency becomes more apparent beyond 5×10^{20} n/cm^2. However, as described below, these two values are both greater than 10% under neutron fluence ($E > 0.1$ MeV) beyond the value (3.48×10^{19} n/cm^2) estimated from the design life of the R/V, suggesting sufficient fracture toughness (Fig. 2.22).

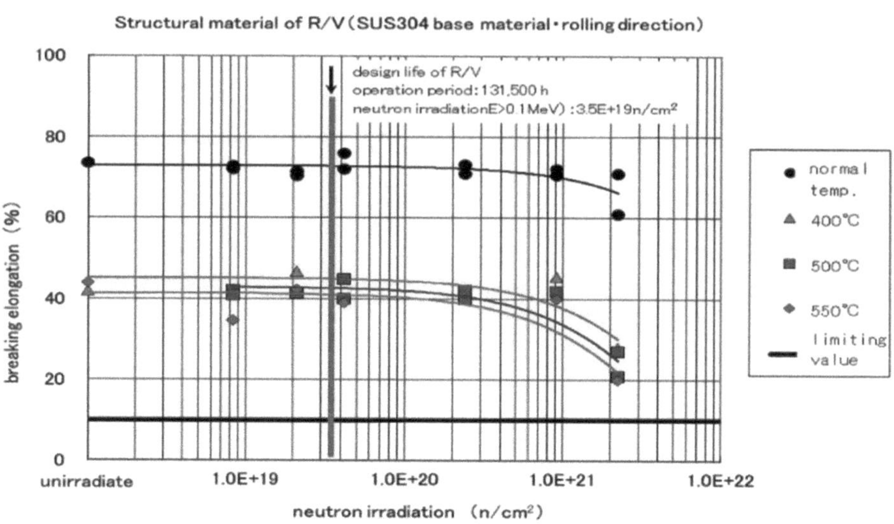

Fig. 2.22 Breaking elongation of structural material for R/V.

(2) In-service inspection of primary cooling system piping

In Joyo, visual appearance inspection has been made at four locations on the inner piping of the primary main cooling system piping (hereinafter, "primary main piping"), comprising the reactor coolant boundary.

As the primary main piping has a double-walled structure, there is a problem with the accessibility of the inner piping (the reactor coolant boundary) when conducting the ISI of the primary main piping, and the number of areas subject to inspection and applicable inspection methods are only limited. For this reason, as a result of pipe stress evaluation, elbow portions that are subject to the maximum stress and have higher stress concentration factors were selected for the ISI, and only four inspection holes were provided in the outer piping at the reactor and main pump outlets of A and B loops in consideration of pipe diameter and workability. The structure of the pipe inspection hole is shown in Fig. 2.23. During periodic inspections, using the pipe inspection hole, it is visually confirmed that there are no cracks, corrosion, deformation, or sodium leak on the pipe surface. An external view of the pipe inspection hole and a photograph of the pipe surface are shown in Fig. 2.24 and Fig. 2.25, respectively.

(3) In-service inspection of secondary cooling system piping

Visual inspection of the secondary cooling system can be done on the heat transfer tubes (see Fig. 2.26) of the air cooler as the representative points. For the air cooler heat transfer tubes, the wall thickness is measured during periodic inspections taking "the corrosion under air environment" into consideration. In addition, the main air coolers were replaced with new ones in the MK-III cooling system modification. Measurement of the wall thickness of the previous main cooler heat transfer tubes revealed that corrosion had progressed under an air environment while the corrosion depth was smaller than the assumed corrosion allowance and the required wall thickness was maintained (see Fig. 2.27). In MK-III cooling system modification work, wall thickness measurement was performed at 24 locations (4 points per location, 96 points in total) along piping (the secondary cooling system pipe sections where the outer surface is covered with a metallic interior plate and further covered with heat insulator: see Fig. 2.28). The measured values exceeded the reference values (minimum wall thickness which allows for the pipe manufacturing tolerance specified in JIS (plus/minus 10% of the nominal value)), as shown below.

- 12B pipe: 10.4–11.1 mm Reference value: 9.27 mm
- 10B pipe: 8.8–9.5 mm Reference value: 8.37 mm

Wall thickness measurement was also performed at 4 elbow portions (20 points per portion, 80 points in total), where the flow velocity is increased due to drift current, of the main IHX inlet/outlet pipes. The measured values exceeded the reference values (minimum wall thickness which allows for the elbow manufacturing tolerance specified in JIS (plus/minus 12.5% of the nominal value)), as shown below.

- 12B elbow: 10.9–14.9 mm Reference value: 9.01 mm
- 10B elbow: 9.5–12.8 mm Reference value: 8.14 mm

Concerning the pipe sections where heat insulator is installed, no significant thinning was observed for approximately 61,000 h of reactor operation under a high-temperature flowing sodium environment, and therefore, operational management by measuring the wall thickness of the main air cooler heat transfer tubes, that are subject to the most severe thinning condition during the design life of Joyo, was determined to be appropriate.

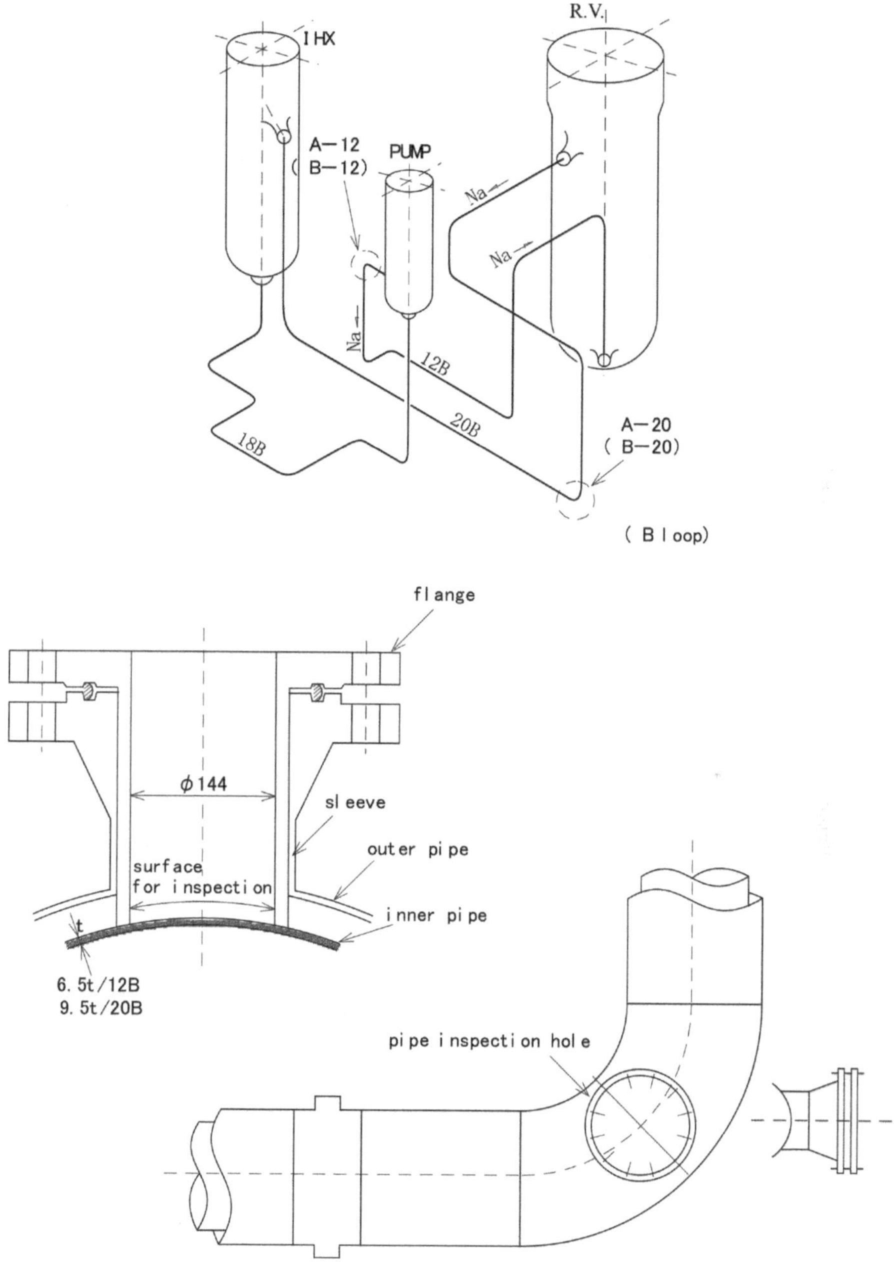

Fig. 2.23 Structure of pipe inspection hole in primary cooling system piping.

(before inspection) (during periodic inspection)

Fig. 2.24 External view of pipe inspection hole in primary cooling system piping.

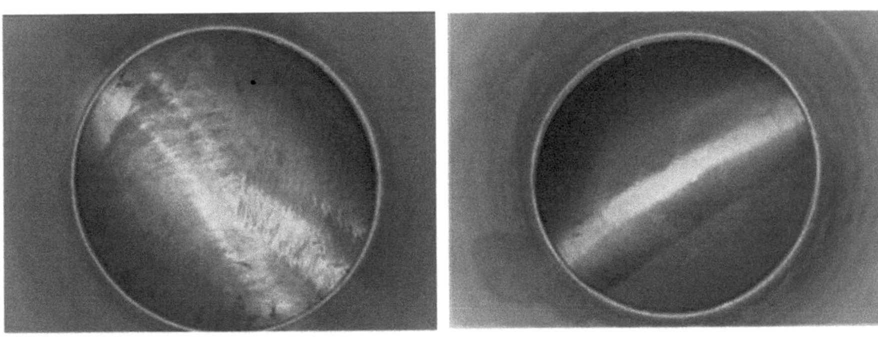

Dec.12, 1980 (second periodic inspection of Joyo)

Jul.21,2005 (14th periodic inspection of Joyo)

Fig. 2.25 Photographs of pipe surface in primary cooling system piping.

(thickness measurement points

of heat transfer tubes)

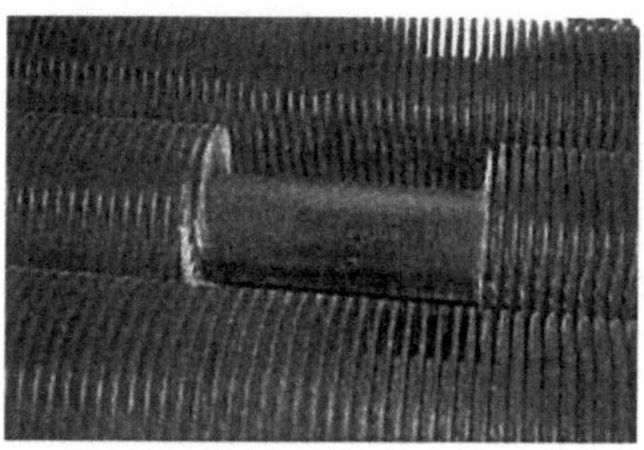

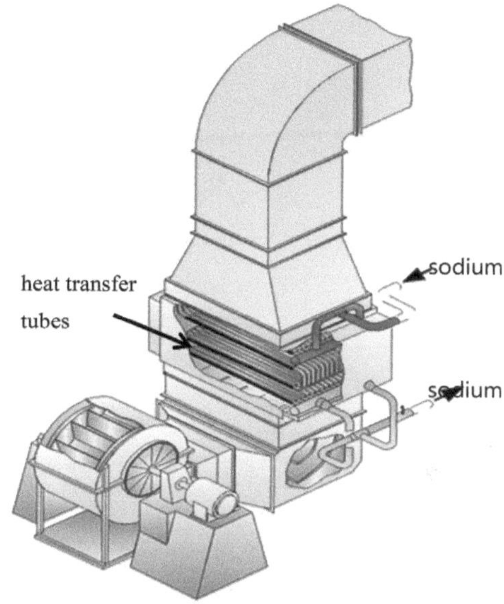

(thickness measurement position)

Fig. 2.26 Outline of heat transfer tubes of air cooler (DHXs) in secondary cooling system.

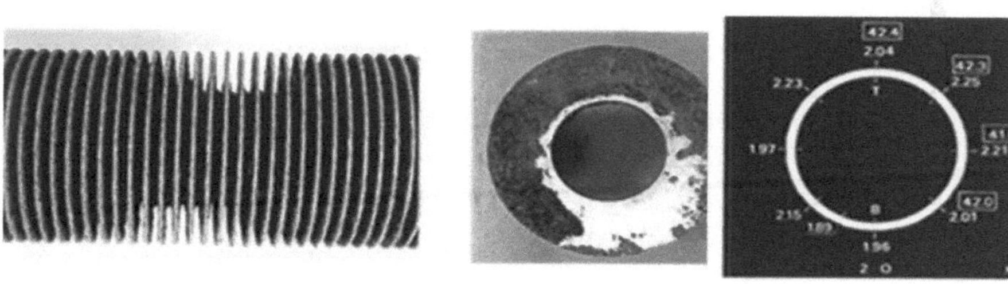

Fig. 2.27 Measurement of wall thickness of previous main cooler heat transfer tubes.

2.2.2 *Demonstration of SFR technologies*

2.2.2.1 *Natural circulation test*

The inherent safety of a fast reactor can be demonstrated by proving the natural circulation capability and establishing analytical techniques based on experiments. The natural circulation test from 100 MWt was carried out at the end of the 12th cycle of the MK-II core in 1986.

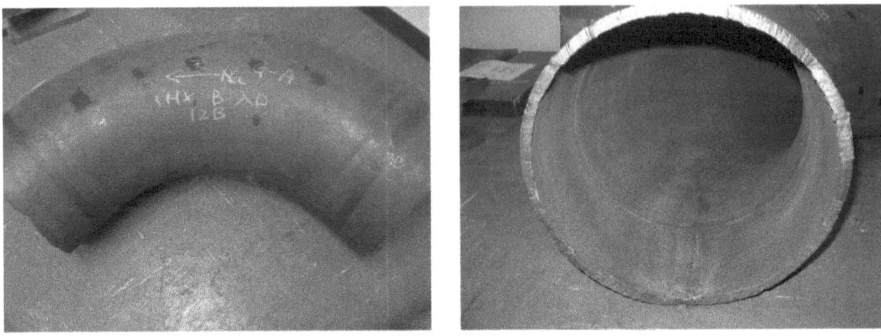

Elbows of secondary cooling outlet piping (12B) of IHX

Elbows of secondary cooling inlet piping (12B) of IHX

Fig. 2.28 Outer surface in secondary cooling system piping removed in MK-III cooling system modification work.

The test was initiated by manually tripping the primary and secondary sodium pumps without pony motor operation. This caused a reactor scram, and the blowers of the DHXs were stopped immediately upon receipt of the reactor scram signal. The central driver fuel subassemblies underwent the most severe temperature transition during the tests. Fig. 2.29 shows the outlet temperature of the central driver fuel subassembly, together with the transient of the primary coolant flow rate. The peak temperature reached 519°C due to reduced coolant flow rate; however, it is significantly below the initial temperature, 548°C. Thus, it was shown that the temperature increase will not cause any safety-related problems, such as fuel cladding failure. The posttest analysis results obtained by the plant-wide dynamics code MIMIR-N2 are in excellent agreement with the experimental data. MIMIR-N2 was developed to analyze Joyo plant kinetics and includes all plant system models (core, and primary and secondary cooling systems). In MIMIR-N2, the dynamic characteristics in both steady and unsteady states were obtained from one-point approximate reactor dynamics calculations and one-dimensional thermohydraulic calculations based on the finite difference method and treating the coolant sodium as a one-dimensional incompressible fluid.

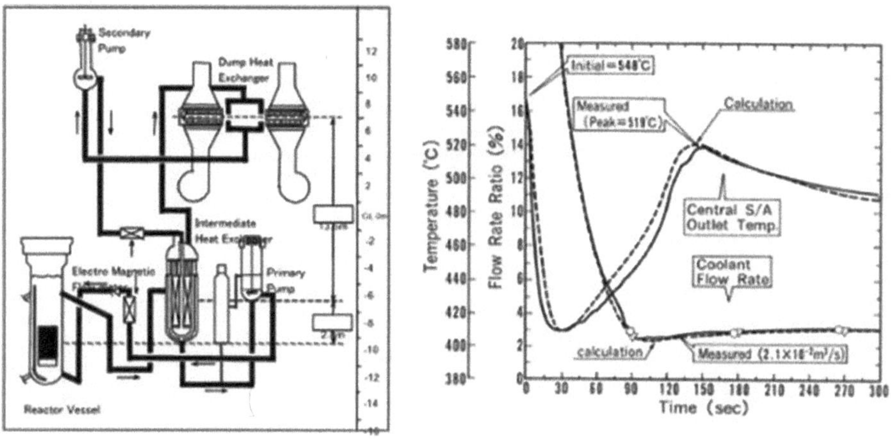

Components layout Test result

Fig. 2.29 Results of natural circulation test compared with Mimir-N2 calculation.

Various key parameters were clarified to improve the calculation accuracy through the study. Regarding the short-term analysis, the evaluation of thermohydraulic behavior in the core is largely affected by the interassembly heat transfer effect, the pump flow coast characteristics, and the coolant flow distribution. Regarding the long-term analysis, it is important to assess precisely how the buoyant head effect in the IHX affects the flow distribution between the heat transfer part and the bypass flow part of the IHX under natural circulation. This flow distribution is normally constant under forced circulation.

The test has also been used for validation of the plant dynamics analysis code "Super-COPD." Super-COPD, which can simulate the plant dynamics of the sodium-cooled fast reactor (SFR), has been developed for the prediction of plant behavior during plant tests, the safety analysis of actual plants, evaluation of the design margin, and design study of the advanced SFR named JSFR. Its basic analytical model and discretization method are almost the same as those of MIMIR-N2, and Super-COPD can apply to the modeling of several reactor types because the analytical models are modularized by calculation function. Moreover, the models simulating phenomena such as thermal stratification and flow redistribution among subassemblies that occur during decay heat removal by natural circulation were developed to immediately analyze the plant behavior. These models were validated by comparing them with the test result. An example of the validation is indicated below. The calculated subassembly outlet coolant temperature and core outlet flow rate are in good agreement with those of the test result as shown in Fig. 2.30. As a result, it was concluded that the primary circuit model including the developed model could simulate heat transfer characteristics, pressure loss, and natural circulation head. In the end, knowledge from the natural circulation test of Joyo is utilized for the development of analytical codes.

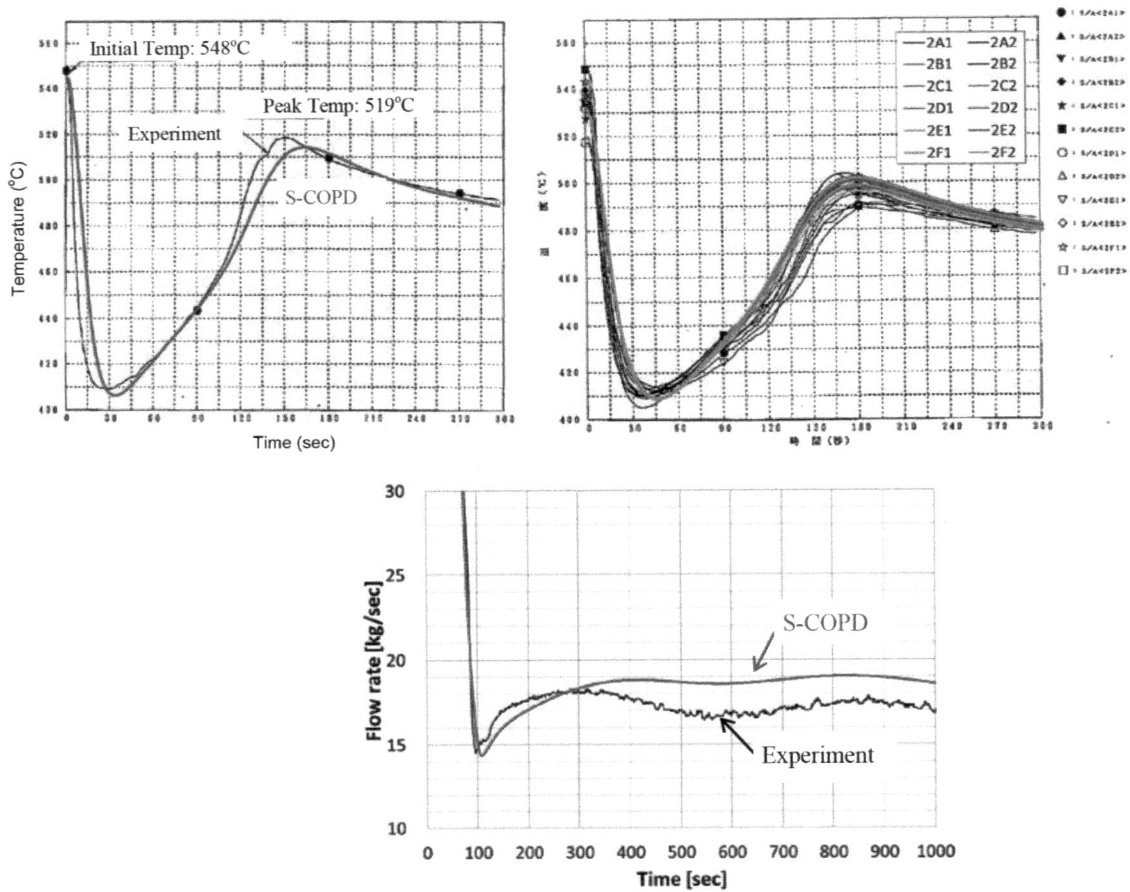

Fig. 2.30 Calculation results using Super-COPD for natural circulation test.

2.2.2.2 Fuel failure simulation tests

For SFR plants, to achieve high availability and operational reliability, failed fuel detection (FFD) and location (FFDL) are important. To conduct safe run-beyond-cladding-breach tests, the fission product (FP) traps are important for safety reasons.

The Joyo FFD system consists of both delayed neutron (DN) monitoring systems and a CG precipitator system. Two DN monitoring systems are located adjacent to the primary cooling loops to detect the DNs emitted from precursors released into the coolant sodium. The CG precipitator system detects an FP ^{88}Rb, which is produced by beta decay of ^{88}Kr released into the CG argon.

An in-pile simulation test was carried out at the end of the seventh operation cycle using two artificially defected (slit) fuel pins to verify the performance of the Joyo FFDL system. This system uses the sodium sipping method. The slit was a 1.0-mm-long, 0.1-mm-wide perforation located in the fuel cladding at the gas plenum level. In the FFDL operation, the signal level from the test subassembly was several hundred times higher than the background level from other core subassemblies, as shown in Fig. 2.31. This test confirmed that the FFDL system is capable of identifying failed fuel with a defect at a gas plenum level.

An FP source using U-Ni alloy tubes was irradiated at the end of the 15th operation cycle. The test provided two important results: Both the CG and DN monitoring systems were successfully calibrated, and the CG monitor was confirmed to have adequate sensitivity to the FP gas released from failed fuel pins. The major constants for the CG and DN behavior models were determined. The disengagement rate constant of the FP gas from the sodium coolant to the CG region varied depending on the flow rate of the primary cooling system.

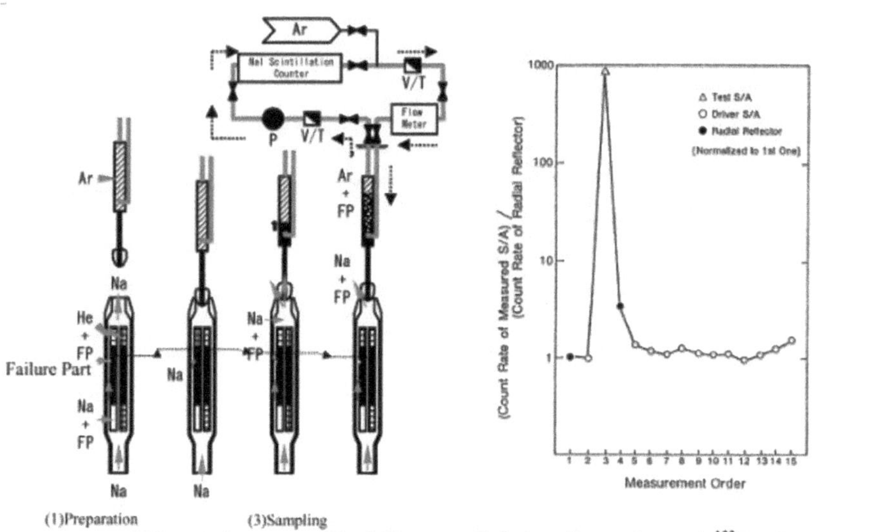

Sipping and Sniffing Mechanism of FFDL

Relative Count Rate of ^{133}Xe Gamma-ray (Photo Peak at 81keV) from Examined Subassemblies

Fig. 2.31 Failed fuel simulation test for FFDL system (S/A: subassembly).

2.2.2.3 *Demonstration test of self-actuated shutdown system (SASS) with a Curie point electromagnet*

The SASS with a Curie point electromagnet (CPEM) has been developed for use in a large-scale SFR to establish the passive shutdown capability against anticipated transient without scram (ATWS) events. The basic characteristics of SASS have already been investigated by various out-of-pile tests for material elements. As the final stage of the development, the stability of SASS needs to be confirmed under the actual reactor operating environment at high temperatures, neutron flux, and sodium flow rate to ensure high plant availability.

For this purpose, a demonstration test of holding stability using reduced-scale experimental equipment of SASS was conducted in the first and second operation cycles of the MK-III. Fig. 2.32 shows the SASS experimental equipment. This structure is the same as the SASS for a large-scale SFR. The SASS experimental equipment

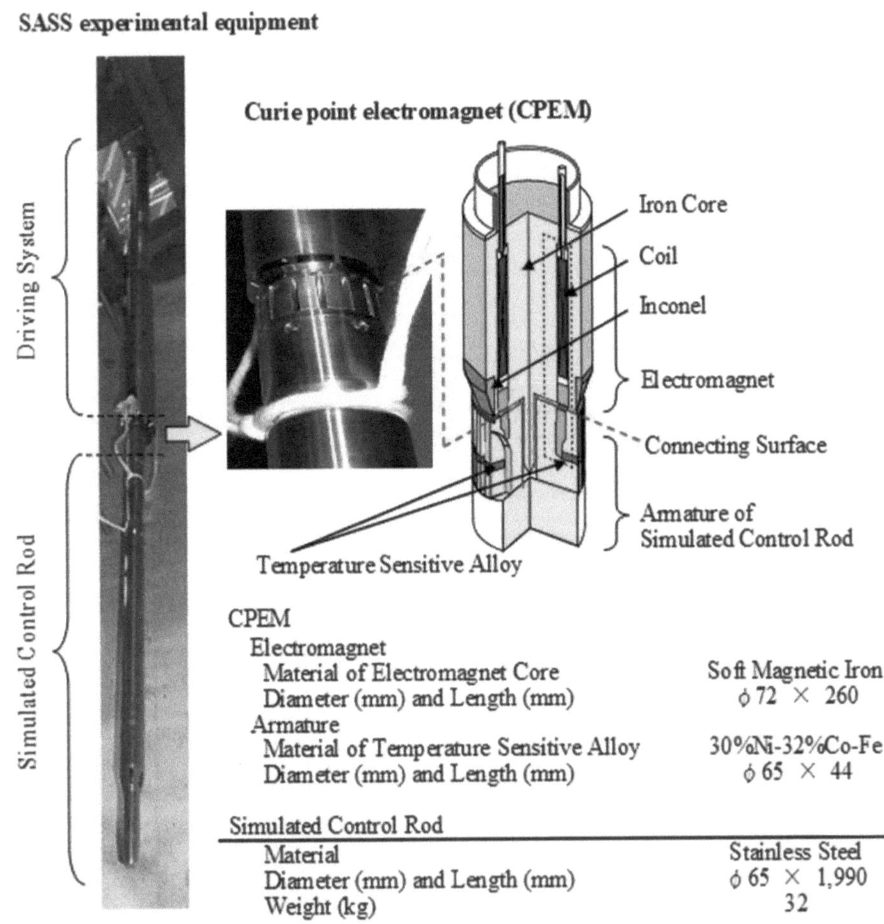

Fig. 2.32 SASS experimental equipment.

is composed of the CPEM, the driving system, and a simulated control rod. And the CPEM is composed of an electromagnet that is located at the bottom of the driving system and an armature that is located on the top of the control rod. The control rod is held by the electromagnet. The CPEM is made of soft magnetic iron. The temperature-sensitive alloy made of 30%Ni-32%Co-Fe is included in the armature. The R&D results described below (refer to Fig. 2.33) were incorporated into the design of the SASS experimental equipment.

(1) Test to improve the control rod holding stability during normal operation
   (i) The gap between the soft iron connecting surface of the electromagnet and the armature of the simulated control rod was kept at 0.1 mm as an Inconel barrier to prevent the connecting surface from self-welding and the free intrusion of impurities.
   (ii) A guide pin is fixed to the CPEM for adjustment with the connecting surface of the electromagnet and the armature of the simulated control rod.
(2) Test to improve the control rod drop performance in accidents
   (i) To improve the response of temperature-sensitive alloy to the coolant temperature ascent, the contact area between the temperature-sensitive alloy and coolant is expanded by slits installed in the temperature-sensitive alloy of the control rod armature.
   (ii) The segment-type simulated control rod is used to improve the control rod insertion performance against core bowing and earthquake.

The rod-holding stability and the rod-recovering functions of the driving system to reconnect and pull out the separated control rod were fully confirmed. The results also indicate there is no essential problem for the practical use of SASS, including the operational trouble involving the unexpected drop during reactor operation.

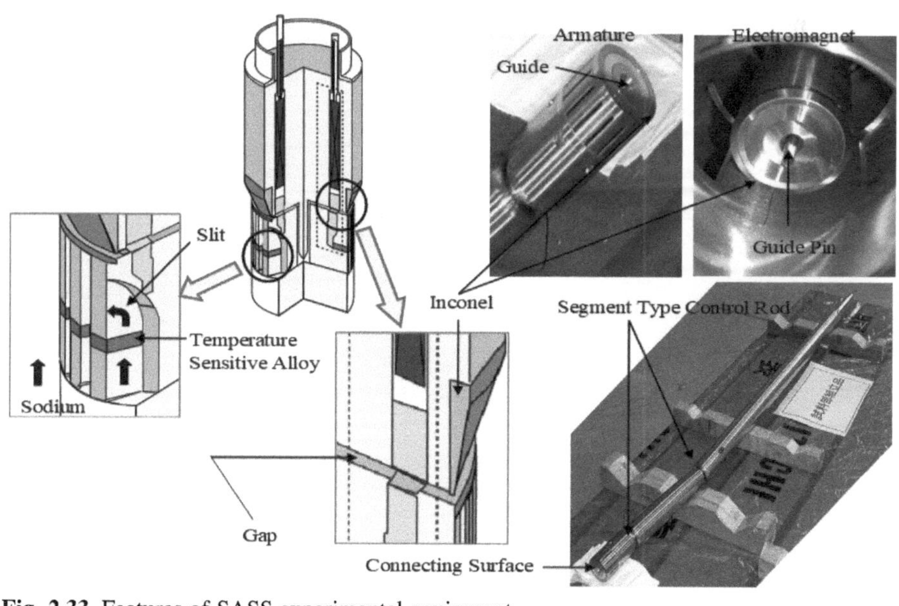

Fig. 2.33 Features of SASS experimental equipment.

In addition to the demonstration test of the holding stability, element irradiation tests were also carried out in Joyo, where the influence of the irradiation environment on the materials (temperature sensing alloy, iron core, and coil) was measured. The test specimens made from the temperature sensing alloy, iron core, and coil cable are shown in Fig. 2.34. These test specimens were put into a container, and the container was installed in the upper part of the Joyo core. The temperature sensing alloys were made from 30Ni-32Co-Fe and 23Ni-38Co-Fe, the iron core was made from soft magnetic iron and 2·1/4Cr-1Mo, and the coil cable was mineral-insulated cable.

The magnetic characteristics of the temperature sensing alloy and the iron core were measured from B-H rings. The volumetric increase of the temperature sensing alloy and the iron core was measured with swelling specimens. The tensile strength was measured with tension specimens. The insulating characteristics of the coil cable were measured with metallographic specimens. Among these data, the magnetic and insulating characteristics were measured online during irradiation. The B-H curve at 600°C for the soft magnetic iron and the temperature sensing alloy of 30Ni-32Co-Fe is shown in Fig. 2.35, as examples. Although a slight change was observed in the measured characteristics, it was confirmed that the temperature sensing alloy, iron core, and coil cable are not significantly affected by irradiation.

2.2.3 Replacement experience of large components in the cooling system

An upgrade program, designated as the MK-III project, has been conducted. In the MK-III project, modification of the cooling systems was required to support the power increase. Joyo has two main heat transport systems (HTSs). Each HTS has an IHX and a sodium pump in its primary system, as well as two DHXs and a sodium pump in its secondary system. These two IHXs and four DHXs were replaced. The IHX was

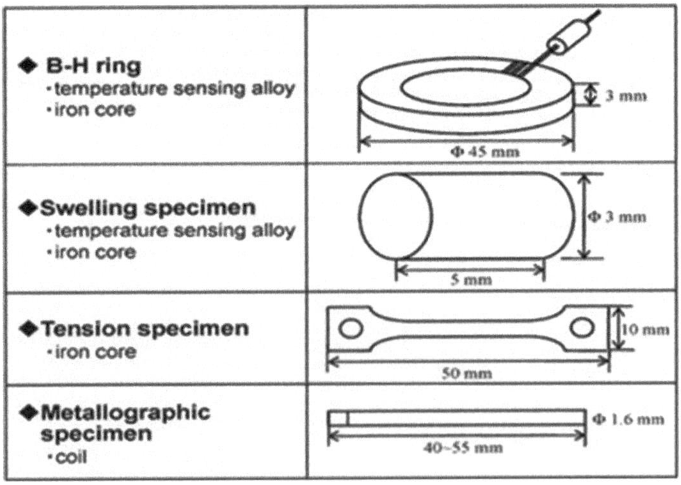

Fig. 2.34 Shape of test specimen.

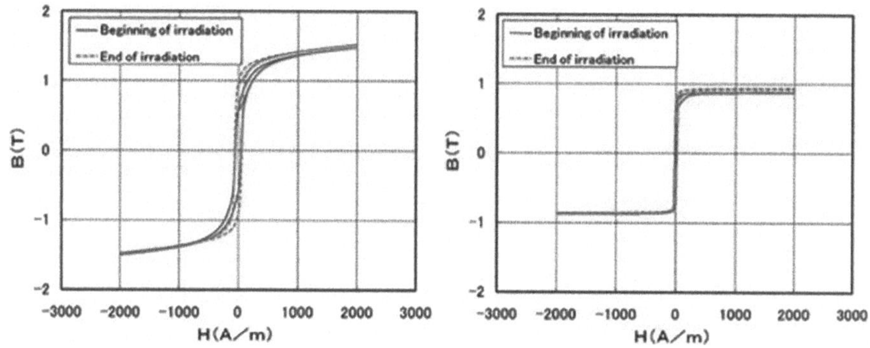

Fig. 2.35 B-H curve at 600°C for soft magnetic iron and temperature sensing alloy of 30Ni-32Co-Fe.

replaced with a new one designed to ensure heat transfer performance and reduce the pressure loss in response to the increase in power and flow rate. Additionally, as a measure against heat transients, austenitic stainless steel named 316FR (Type 316FR steel) was employed for the main material. The 316FR steel was the first domestically developed fast reactor structural material, having improved high-temperature creep-rupture properties. The design points and specifications are shown in Fig. 2.36. As for the DHX, it was necessary to increase the heat transfer area to ensure sufficient heat removal capacity. However, since there was a constraint of building size, it was determined to ensure an increased heat transfer area within the limited space by changing the type of heat transfer tube from U- to Σ-type (approximately two times the original heat transfer area) (Fig. 2.37).

The two old IHXs were removed from the HTS after cutting both primary and secondary pipes, as well as small-diameter pipes such as those of the cover gas and purification systems. During pipe cutting, to prevent the incorporation of metal fillings and in consideration of the workability, cutting was performed with turning tools up to two-thirds of the wall thickness, and afterward, a roller cutter was used in an inert gas atmosphere covered with a sealed bag. During this replacement, molten sodium and fuel were retained in the reactor vessel (RV). Consequently, the primary cooling system and cover gas boundaries had to be maintained to prevent impurity ingress to the sodium system. The following methods were used:

(1) A sealed bag filled with inert gas was applied to prevent air ingress (refer to Fig. 2.38).
(2) Oxygen, nitrogen, and hydrogen concentrations in the cover gas were monitored for indications of air ingress.
(3) Cover gas low-pressure control was used to maintain integrity at the pipe opening.

The openings of the removed IHXs were welded with stop plates to keep inert gas in the IHXs. The removed IHXs were then transferred to a solid waste storage facility in the maintenance building to await the decay of radioactivity. Fig. 2.39 shows the surface exposure rate of IHX and connecting pipes. Primary sodium contains

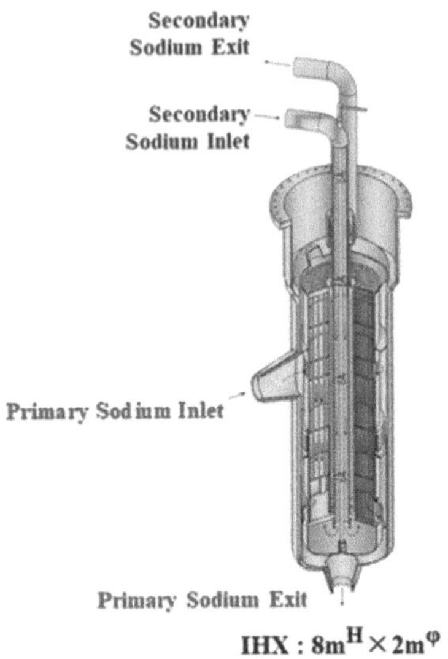

- DESIGN POINTS -
- Same size of component
- Decrease pressure loss
  → Optimization of inlet and exit nozzle shapes
- Prevent tube buckling
  → Lowering temp distribution in heat transfer tube
- Utilize newly developed 316FR

Major specs		
Material	Shell	Austenitic stainless steel (316FR/316FR F/SUS316)
	Tube	Austenitic stainless steel (316FR TB)
	Leak jacket	Austenitic stainless steel (SUS304/SUS304TP)
Flow rate		1350 t/h
Rated power		70MW
Tube No.		2088
Tube Dia./thick.		19mm/1.0mm
Heat transfer area		363m^2

Fig. 2.36 Schematic and specifications of IHX.

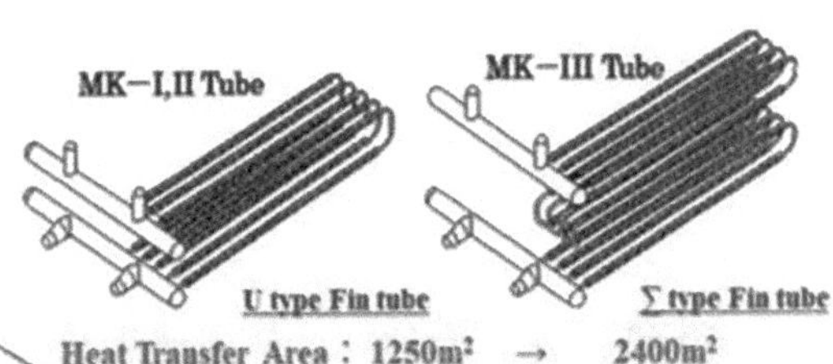

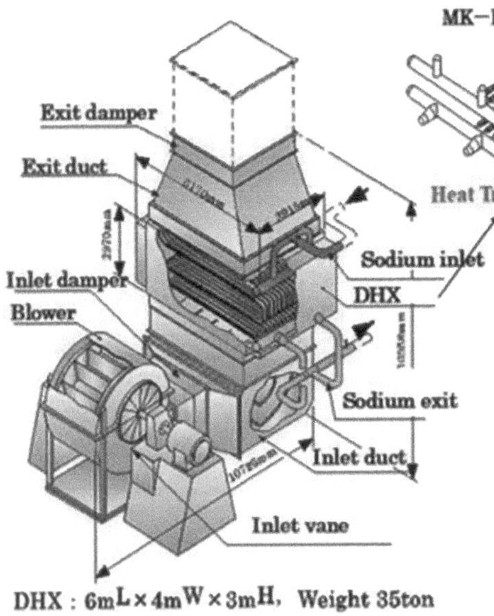

DHX major specs	
Type	Finned tube air cooler
Shell material	2 1/4 Cr-1Mo steel
Tube material	2 1/4 Cr-1Mo steel (STPA24)
Na flow rate	600t/h/unitX2unit/loop
Rated heat exchange	35MW
Tube No.	80
Tube dia./thickness	42.7mm/2.0mm
Fin dia./thickness	80.7mm/1.6mm
Heat transfer area	2400m^2
Air flow rate	7700m^3/min

Fig. 2.37 Schematic and specifications of DHX.

radioactive ^{22}Na and radioactive CPs, such as ^{60}Co and ^{54}Mn. This sodium adhered to the inner surface of the IHXs and pipes. During pipe cutting and welding, since it was required to reduce the exposure to workers as low as possible, work time was shortened by performing careful full mockup training and selecting the optimum work

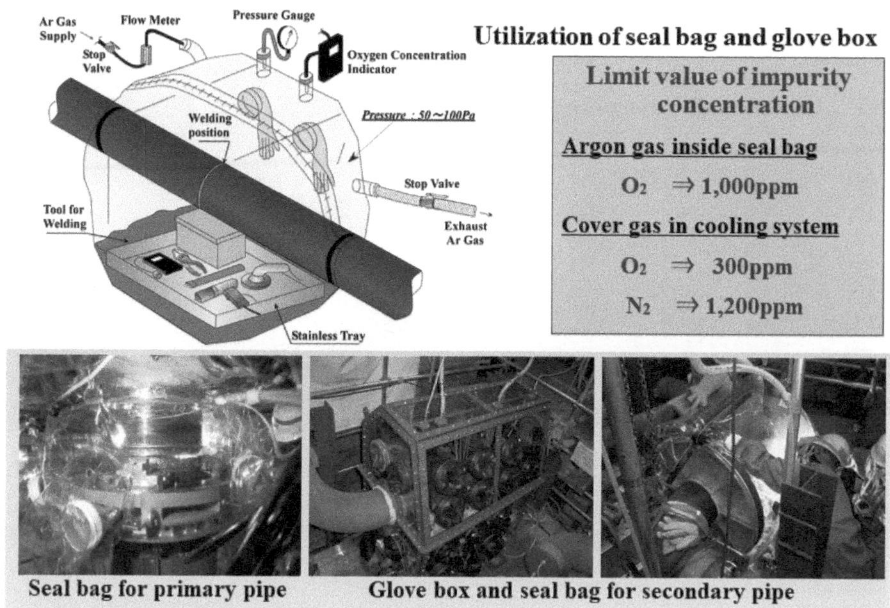

Fig. 2.38 Measures to prevent impurities from entering the system.

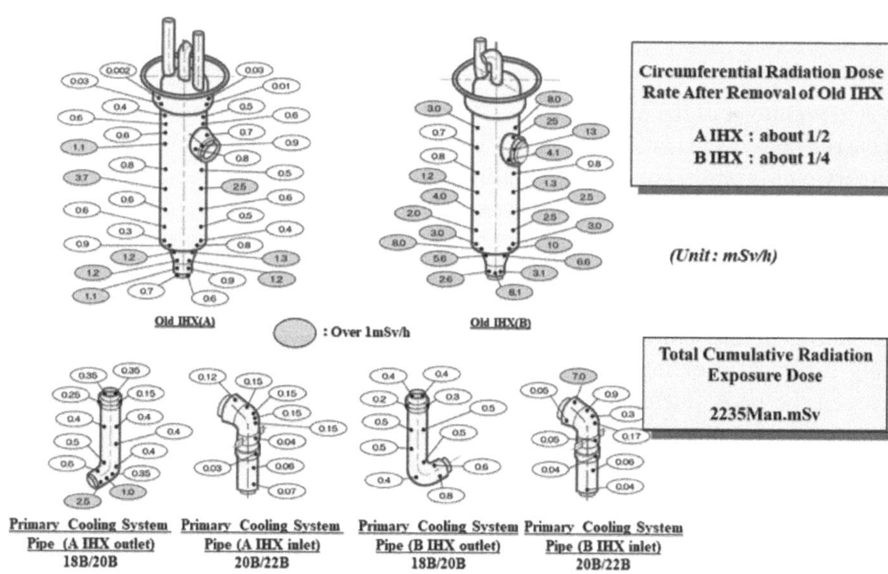

Fig. 2.39 Surface exposure rate of IHX and connecting pipes.

procedures, and additionally, temporary shields, such as lead mat, were installed at places with a high surface dose rate (Fig. 2.40). As a result of these careful preparations, work time and the ambient dose rate were reduced by approximately 7% and 20%. As a result of these radiation control efforts, the total exposure dose was reduced to 2235 man-mSv, approximately half the planned value.

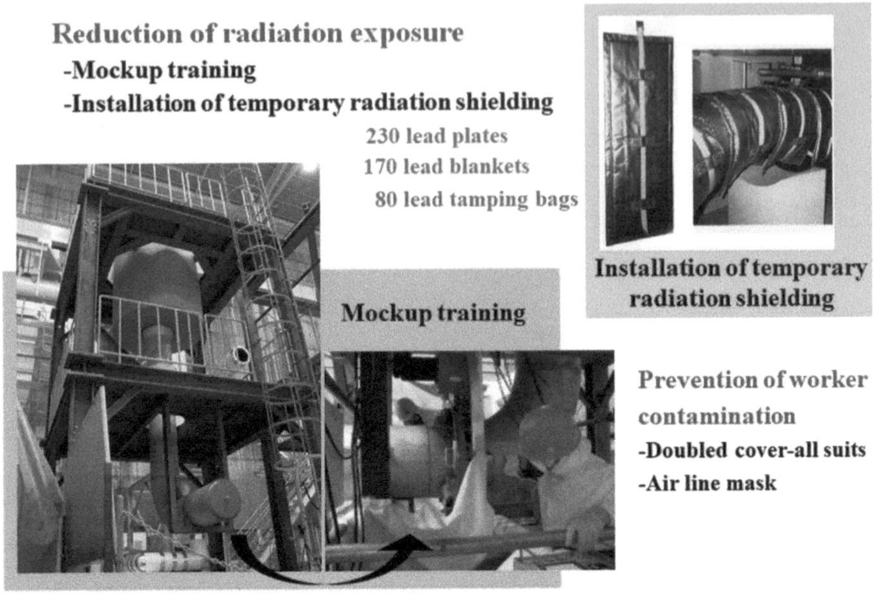

Fig. 2.40 State of mockup training.

After removal of the old IHXs, new IHXs, which had almost the same dimensions as the old ones, were installed as shown in Fig. 2.41. Sodium remaining in the existing primary piping was recovered with tools and then cleaned with alcohol water. Subsequently, the connections between the new IHX and piping were welded by covering with a seal bag to prevent the incorporation of impurities. Since the pipes to be welded were made of stainless steel, welding was done while flowing the argon gas in the system as the back shielding gas to prevent the mixing of impurities, including oxygen, nitrogen, and hydrogen, into a melting point, and the creation of faults. Before using the back shielding gas, a test was performed taking the differential pressure between the back shielding gas and the pressure in the seal bag as a parameter in the work preparation stage. After the test confirmed that no adverse impacts, such as weld defect, would occur when the above-mentioned differential pressure is equal to or less than 100 Pa, welding was performed. The cut sections and welding zones in and around the IHXs are shown in Fig. 2.42.

In the secondary system, four old DHXs and their connecting pipes were replaced. The operating procedure for these replacements was almost the same as for the primary system. Removed DHXs and pipes were dismantled, and residual sodium was removed at the sodium process facility.

The replacement of all cooling system components took approximately 1 year. After completion of the replacement, integrated functional tests were carried out to verify the basic performance of replaced components. These tests included pressure loss, vibration, and verification of satisfactory levels of impurities in the sodium at low

Fig. 2.41 New IHX installation in the reactor containment vessel.

sodium temperatures, all without reactor operation. The results of these integrated functional tests have shown that the components and system design and replacement work were successful.

2.2.4 Development of in-vessel repair techniques in sodium fast reactor [6,7]

In 2007, Joyo experienced "partial obstruction of the refueling function of the rotating plug due to its interference with an instrumented test assembly (hereinafter, MARICO-2)." Results of in-vessel observation for this trouble are shown in Fig. 2.43.

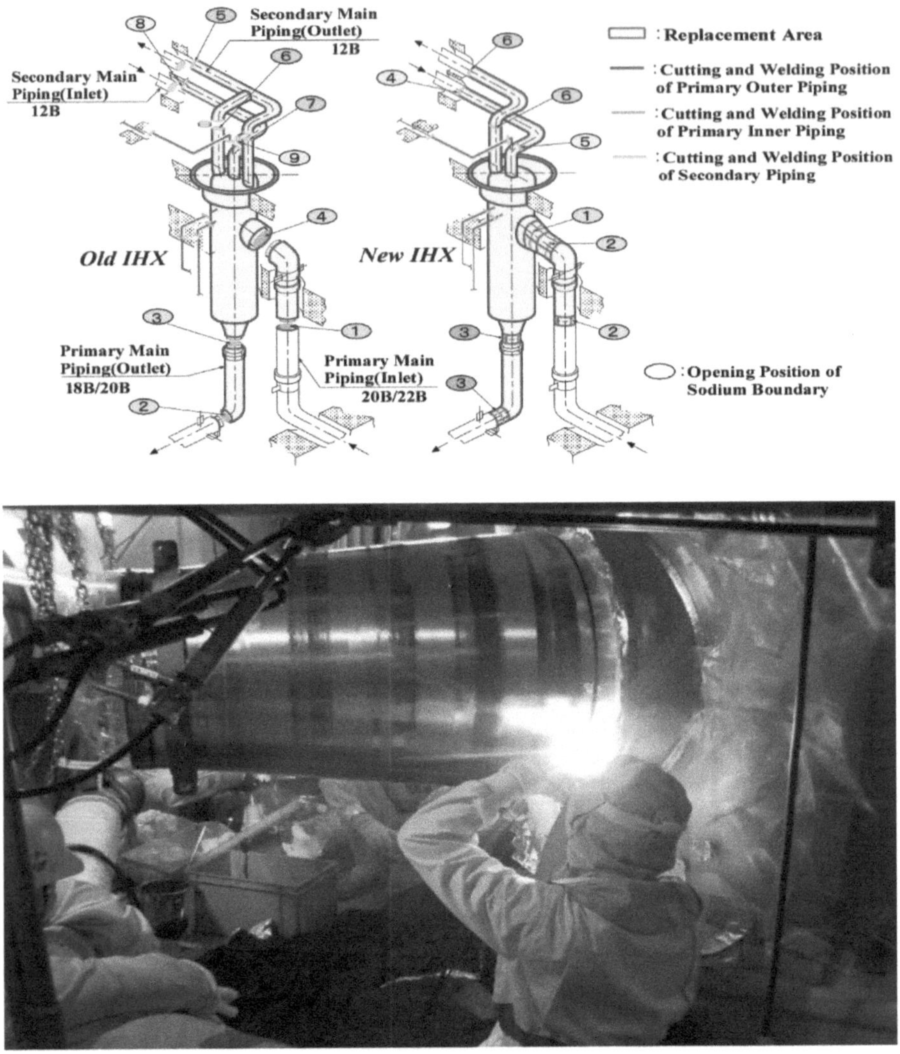

Fig. 2.42 Cut part and welded part of IHXs.

As shown in the figure, it was confirmed that a part of MARICO-2 (hereinafter, "subassembly" (S/A)) was damaged and transformed above the fuel storage rack in R/V due to this trouble, and that the lower end of the upper core structure (UCS) was partly damaged due to contact with the MARICO-2 S/A. Technologies required to retrieve the MARICO-2 S/A and replace the UCS and the related work procedures were developed, and the retrieval of the MARICO-2 S/A and the replacement of UCS were successfully performed from May to December in 2014.

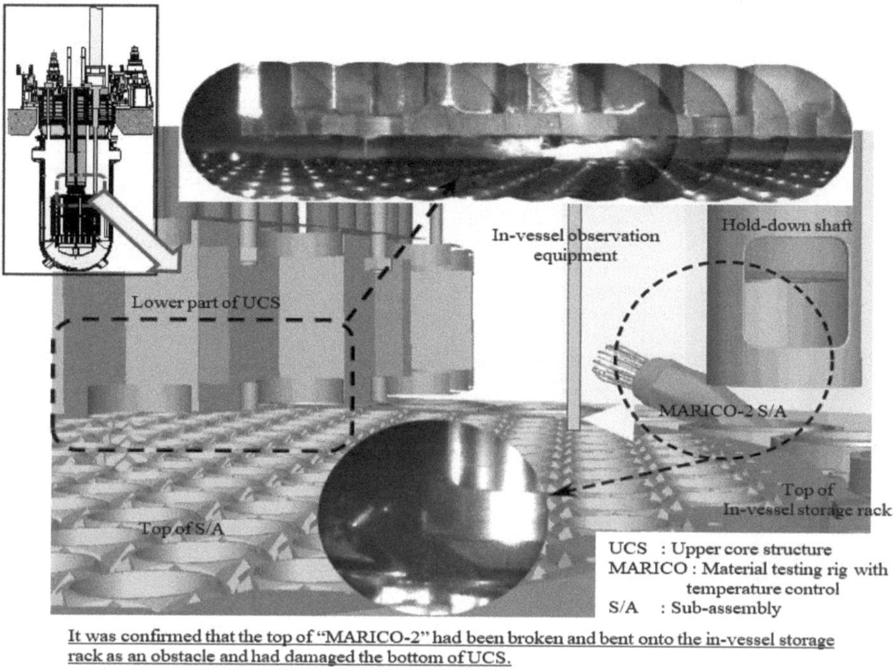

Fig. 2.43 Observation results for UCS and MARICO-2 S/A.

2.2.4.1 UCS replacement

Schematic diagrams of the R/V, rotating plug, and UCS are shown in Fig. 2.44. The R/V is a 10-m-high cylindrical vessel with an inner diameter of approximately 3.6 m and contains the core fuel subassemblies and other components, and in-core structure. Sodium is charged into the R/V to cool the core fuel subassemblies and other components therein, and is kept at approximately 200°C during reactor shutdown. The sodium in the R/V has a free surface above which an inert gas (argon gas: 140–160°C) is charged as the cover gas considering that sodium is chemically active. The rotating plug, including the UCS, installed on the R/V plays the role of the cover gas boundary, and access to the inside of R/V is available via the through-hole already provided in the rotating plug. The sodium level is normally maintained above the R/V outlet pipe (GL—6100 mm), while in maintenance conditions, the sodium level can be changed by draining the sodium in the R/V after confirming that the decay heat in the core is sufficiently low. For replacement of the UCS, it was determined to adjust the sodium level to a level 50 mm lower than the top of the subassemblies (GL—9540 mm: underfloor maintenance mode) taking into account that access to areas adjacent to the top of subassemblies is required for work to retrieve the MARICO-2 S/A that was planned in the same time period.

The UCS, a part of the rotating plug, has a cylindrical structure with a maximum diameter of approximately 1345 mm and a height of 6330 mm, as shown in Fig. 2.44, and has the flow guide and the thermocouples to measure the subassembly outlet

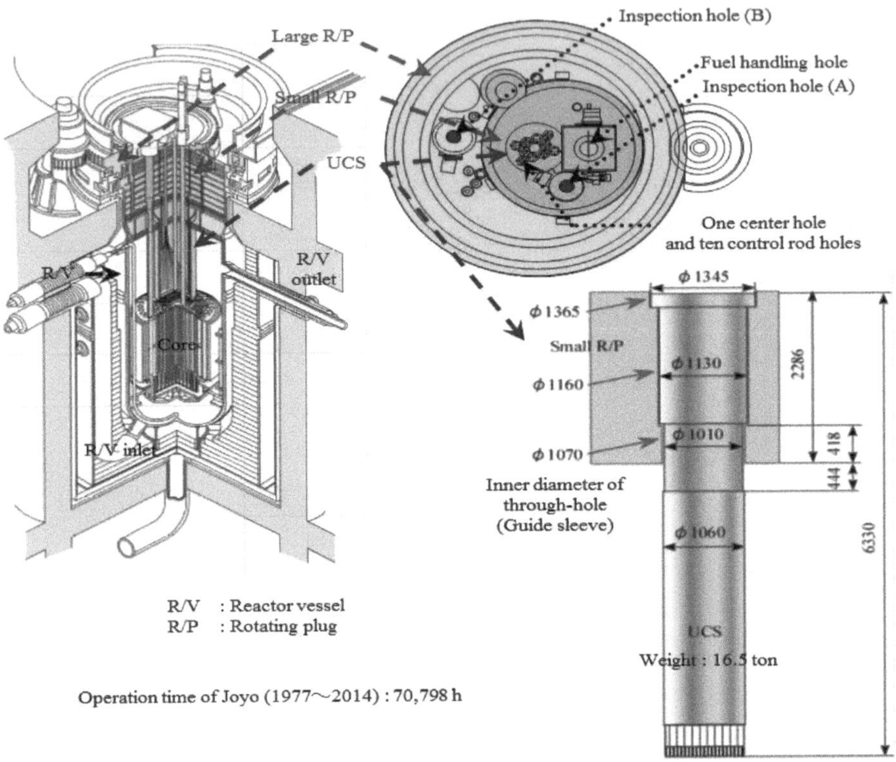

Fig. 2.44 Structure of reactor vessel, rotating plug, and UCS.

coolant temperatures in its bottom part. Outline procedures for replacing the UCS are shown in Fig. 2.45. Replacement of the UCS roughly comprises two steps: withdrawing the existing damaged (hereinafter, "old") UCS and installing the new UCS. Since the old UCS of Joyo was not designed to be replaceable and had never been replaced, there was a concern that it may become difficult to continue withdrawing the old UCS due to the following factors:

- The guide sleeve, in which the old UCS was inserted, is a 5-mm-thick thin-walled structure. If the gap between the old UCS and the guide tube is plugged with evaporated sodium due to long use, more than 30 years, the resistance (sodium shear resistance) should be caused by shearing the sodium during withdrawal of the old UCS. Excessive sodium shear resistance may cause the deformation and damage (hereinafter, "deformation, etc.") of the guide sleeve.
- While the outer diameter of the larger-diameter part of the old UCS is 1060 mm (design value), the inner diameter of the smaller-diameter part of the guide sleeve is 1070 mm (design value). Accordingly, the nominal gap on each side during withdrawing the old UCS is 5 mm. When the levelness of the old UCS is not sufficiently good during its withdrawal, the resistance due to the interference or contact (hereinafter, "interference, etc.") between the old UCS and the guide sleeve (resistance due to interference, etc.) would occur. Excessive resistance due to interference, etc., may cause the deformation, etc., of the old UCS.

To prevent these incidents, it is necessary to monitor the levelness and horizontal position of the old UCS to maintain it in an appropriate condition during withdrawal.

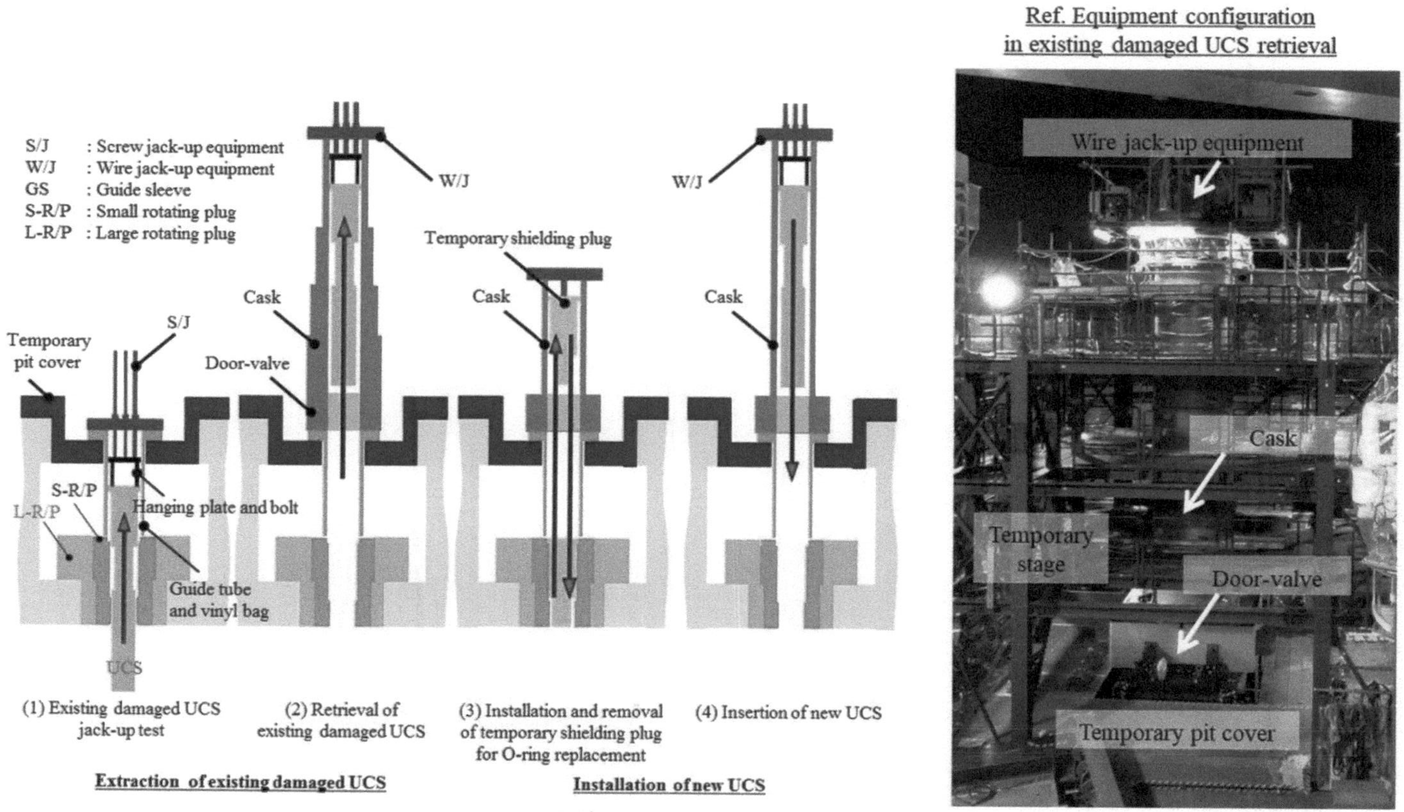

Fig. 2.45 Outline of UCS replacement.

Additionally, it is also necessary to monitor the loading to prevent the deformation of the old UCS, etc., in case of the occurrence of the sodium shear resistance or the resistance due to interference, etc. For this reason, the work to withdraw the old UCS was divided into two steps: withdrawal up to and above the 1000 mm level, at which the larger-diameter part of the old UCS goes through the smaller-diameter part of the guide sleeve. These steps were designated as "the old UCS jack-up test" and "the work to store the old UCS," respectively.

On the other hand, to install the new UCS, consideration must be given to the need to replace the double O-ring system that was placed on the installation base of the UCS. For this reason, paying attention to the work environment, the work to install the new UCS was divided into two steps: "installing the O-ring for the new UCS" and then "inserting the new UCS."

These works required accurate horizontal positioning of a hoisted object. Jack-up equipment was designed to have high stiffness to allow for accurate levelness control and manufactured as an integral structure with the old UCS. For this reason, as shown in Fig. 2.46, the jack-up equipment was installed on the top surface of the guide tube, and the interference between the old UCS and the guide sleeve was avoided by sliding the guide tube in two horizontal directions using the horizontal position adjustment bolts that were installed in the guide tube, and thereby, correcting the eccentricity of the old UCS. On the other hand, as for the wire jack-up equipment, the eccentricity correction method by sliding the guide tube was not applicable because the wire jack-up equipment has a soft structure and was installed on the door valve and cask (see Fig. 2.47), heavy objects (approximately 130 t in total). Therefore, the interference between the new UCS and the guide sleeve was avoided by correcting the eccentricity of the new UCS using the guide roller that was installed on the upper surface of the rotating plug. The old UCS jack-up test was performed on May 7, 2014, and the old UCS was withdrawn and stored on May 22. After retrieval of the MARICO-2 S/A, the new UCS was installed on November 20th and 21st, and all works including clearing up were completed on December 17.

Deposition of sodium was observed on the side surface of the old UCS, as shown in Fig. 2.47. Analysis of the recovered sodium confirmed that it consisted of approximately 97.4 wt% of metallic sodium. Additionally, it was estimated by visual inspection and tactile sensation that the recovered sodium was likely to have a porous structure consisting of aggregated and deposited fine particles, rather than a dense metallic crystal structure. It is difficult to accurately predict the properties and positions of the sodium deposited in gaps, etc., because they depend on the flow paths and temperature distribution as well as the purity control conditions. However, since there was no evidence of sheared sodium (no sodium remained on the surface of the guide sleeve), it was thought that the deposited sodium was not sheared. Instead, it was estimated that the deposited sodium slid on the surface of the guide sleeve due to a brittle fracture of the surface of the deposited sodium that was in contact with the guide sleeve and hence no sodium shear resistance was generated. The above-mentioned analysis also confirmed that the concentration of Na-22 in the deposited sodium was 5.6 Bq/g. Na-22 is produced by the (n,2n) reaction of Na-23, a stable isotope. Comparison of the analyzed concentration with the past analysis results using the inverse decay curve in consideration of the half-life (approximately 2.6 years) showed broad agreement with the Na-22 concentration in the primary coolant measured in the

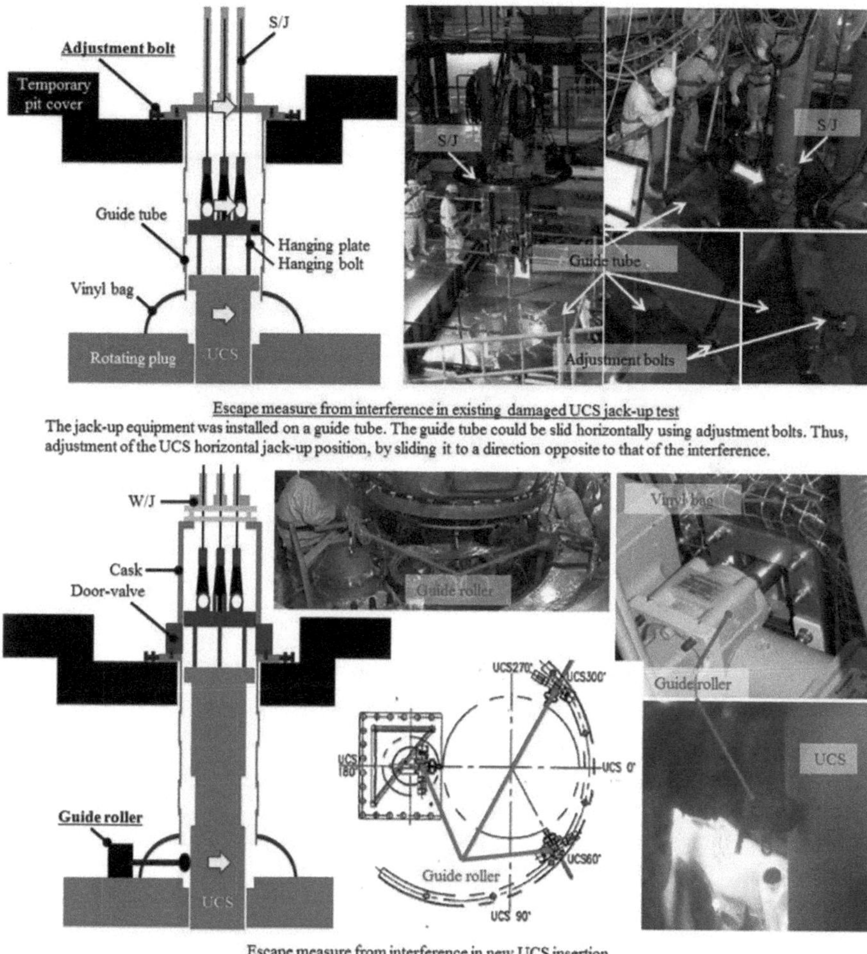

Fig. 2.46 Escape measure from interference between UCS and guide sleeve.

early phase of MK-II core operation. This result confirmed that the concerned sodium was deposited in 1986 or before (28 years ago or before) and that the deposition of sodium did not continue since then. The reason why sodium deposition did not continue includes possible blockage of routes through which sodium mist can rise.

Results of in-vessel observation during these works are shown in Fig. 2.48. These results were obtained using a radiation-resistant camera to observe the bottom surface of the old UCS and the guide sleeve in the final stage of the withdrawal of the old UCS. The condition of the bottom surface of the old UCS was observed with a radiation-resistant fiberscope in 2008, and it was confirmed that no parts fell off during withdrawing the old UCS. In addition, observation of the guide sleeve confirmed that no significant deformation, etc., of the guide sleeve was caused during withdrawing the old UCS.

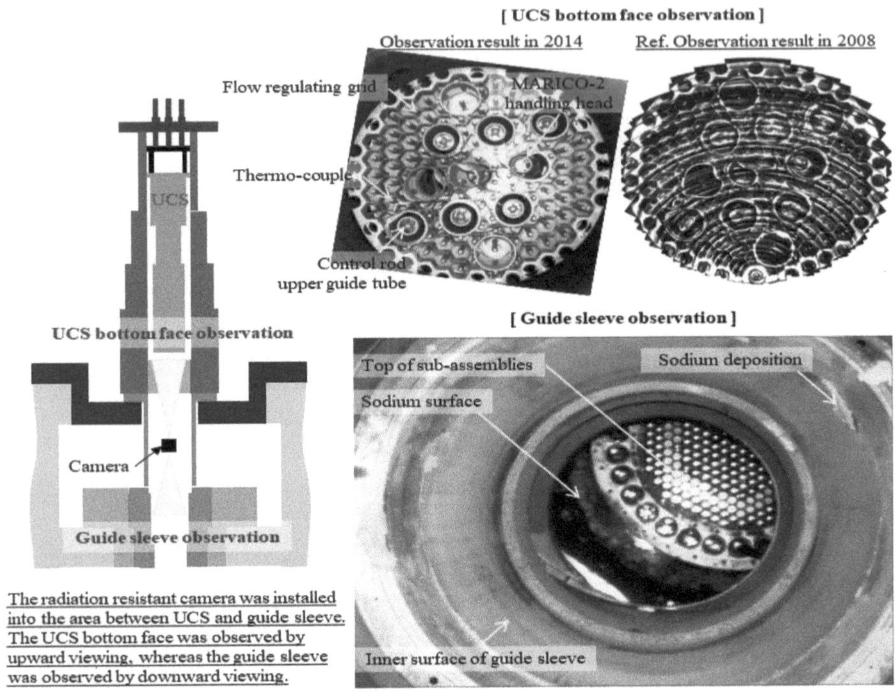

Fig. 2.47 Observation of bottom face of existing damaged UCS/guide sleeve.

2.2.4.2 MARICO-2 test subassembly retrieval

MARICO-2 is integrated-type irradiation equipment that penetrates from the upper part of R/V into the core and performs online control of the irradiation temperature. Accordingly, various material tests, including the creep rupture strength test, can be performed with high precision in MARICO-2 under irradiation environments; however, since the S/A loaded in the core is held by the holding device located at the upper part of R/V, the S/A must be separated from the holding device before discharge from the core. But the rotating plug was accidentally operated in a condition that the holding device and the S/A were not completely separated. After that, in-vessel observation revealed that the S/A was separated and bent with its top part located at a level higher than the in-core fuel storage rack (hereinafter, "rack"). The appearance of the bent MARICO-2 S/A is shown in Fig. 2.49.

Before the study of the recovery method, information on the troubles and repair experiences from foreign reactors (US EBR-II, French RAPSODIE, Russian BOR-60, etc.) was collected through meetings with foreign experts for surveying similar cases. Information on the past foreign in-vessel works and advice based on the repair experiences was useful; however, since there were quite a few differences in the in-vessel conditions, including the damage and deformation of components, and the constraints of reactors, it was determined necessary to develop an original technology to retrieve the S/A. A study of the method to retrieve the S/A based on the

[Sodium deposition on upper part of UCS (Example)]

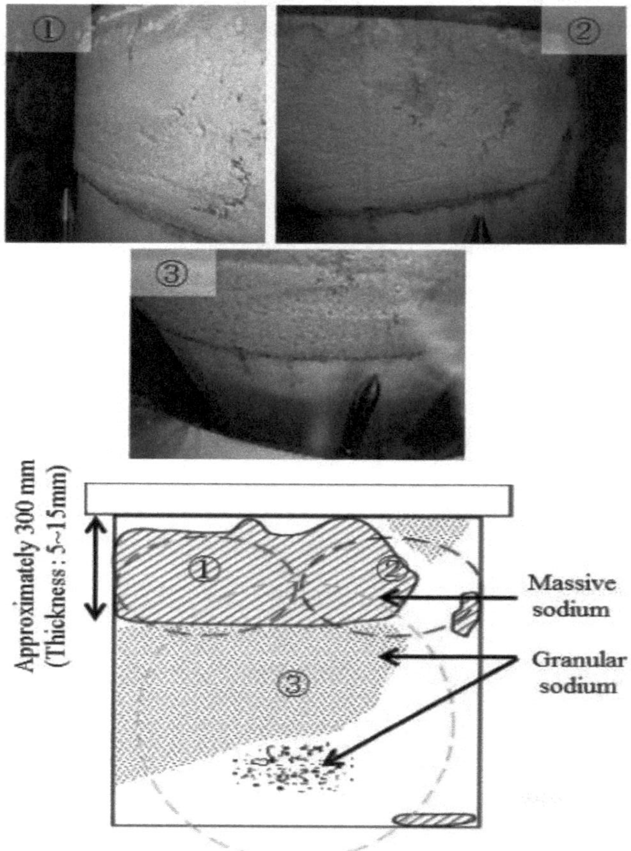

[Analysis result of sodium deposition]

Component	Measured concentration
Na (metal)	97.4 ± 2.3wt%
O	0.64 ± 0.02wt% (Na$_2$O : approximately 2.48wt%)
H	0.028 ± 0.002wt% (NaH : approximately 0.67wt%)
^{22}Na	5.65 ± 0.29Bq/g

[Comparison with past measurement (^{22}Na)]

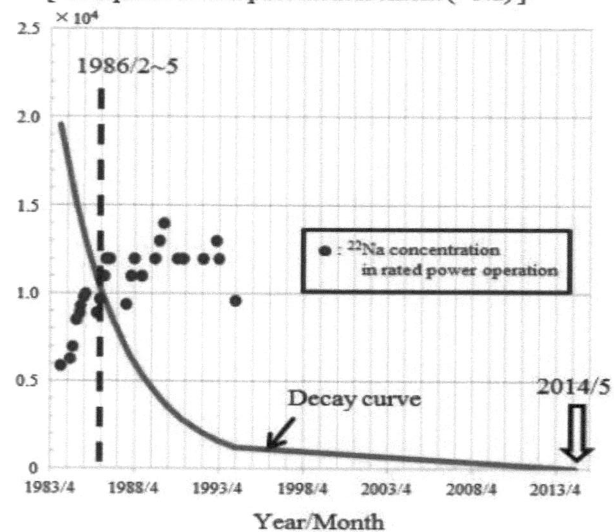

Fig. 2.48 Analysis result of sodium deposition on existing damaged UCS.

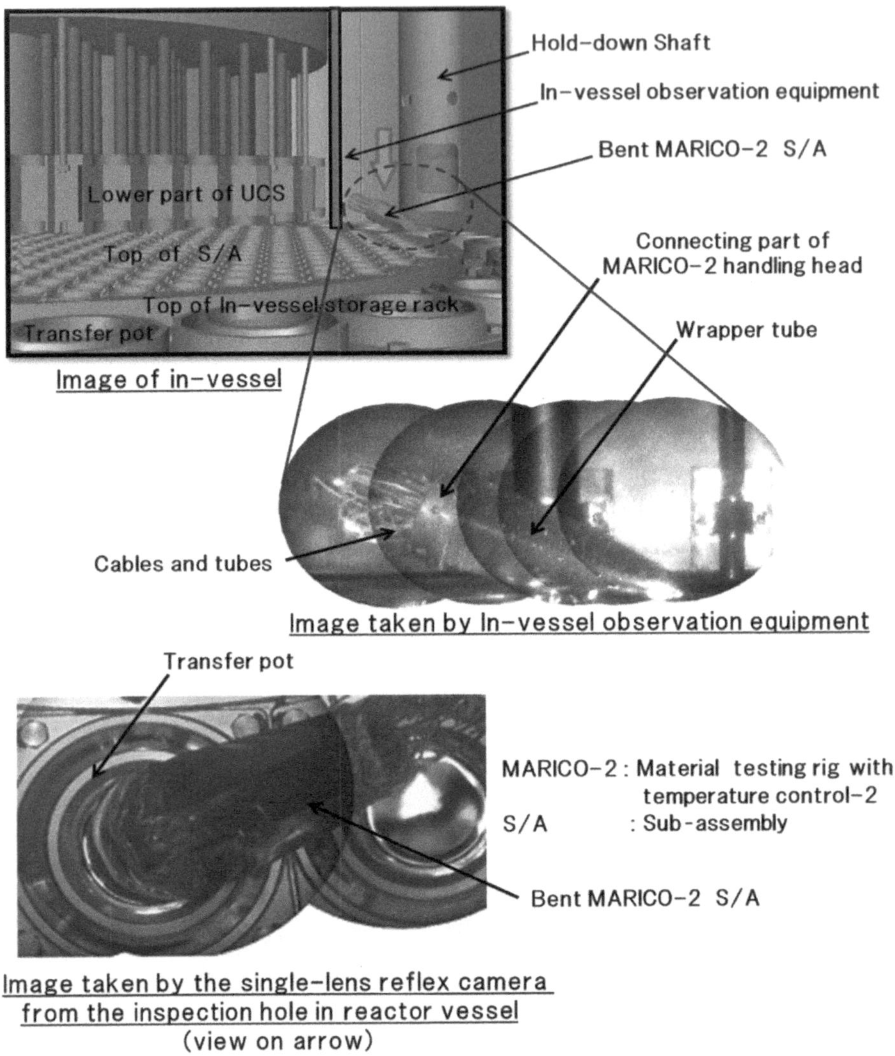

Fig. 2.49 Results of the bent MARICO-2 test assembly observation.

in-vessel investigation results was started in December 2007. After various preparations, including the design and manufacture of equipment and the mockup test, the retrieval work was performed from June to November 2014.

In an early phase of the study, a method for retrieving the S/A solely was proposed. However, since the subsequent test and observation confirmed that the S/A and the transfer pot got stuck, it was determined to adopt a method of retrieving both the S/A and the transfer pot as a single body by holding the transfer pot. In addition, since the bent S/A could not go through any existing hole penetrating the rotating plug, such as the fuel handling machine hole, it was determined to use, as the retrieval route, the hole (hereinafter, "UCS removal hole") with a diameter of approximately 1000 mm remaining after removal of the old UCS that would be separately done in the UCS replacement work.

Since the UCS removal hole has a large diameter (1000 mm), a temporary closing steel plug was inserted into the hole to reduce radiation from inside of the R/V and provide a workspace. Subsequently, the equipment forming the cover gas boundary was installed on the plug to share the door valve used for the UCS replacement work. From the above, a guide tube equipped with the holding device was inserted into the plug. Since the UCS interferes with the MARICO-2 S/A, it is not possible to position the UCS removal hole right above the S/A. Consequently, the pantograph mechanism that can horizontally move the gripper mechanism was incorporated into the retrieval cask to adjust the deviation of the central axis between the UCS removal hole and the transfer pot. In addition, the pantograph mechanism is equipped with the upper and lower stretchable arms that can be independently operated to avoid the interference between the thermal shield plate (a plate installed to prevent the occurrence of thermal stress due to rapid temperature change) installed in the R/V and a retrieved object. Interference with the thermal shield plate can be avoided by first storing the upper arm and then diagonally lifting the transfer pot in an inclined condition toward the UCS removal hole. The in-vessel holding and lifting work was performed while monitoring the condition using the in-vessel observation equipment equipped with a high radiation-resistant fiberscope. Configuration of the retrieval cask and the related components is shown in Fig. 2.50. The procedure to retrieve the MARICO-2 S/A is schematically shown in Fig. 2.51.

A noteworthy event that occurred during retrieval was that the lifting load of the transfer pot during lifting exceeded the control value although good progress was made until the holding operation of the transfer pot. Consequently, the rotational direction of the pantograph of the guide tube was adjusted to avoid excessive loading, and the lifting work resumed. The lifting of the S/A was performed in three steps: (1) lifting until the transfer pot nose passes the nose support of the pot support, (2) lifting until the S/A is completely tilted by pantographs, and (3) lifting to the upper limit.

After successful passing of the possible interference position, the S/A was smoothly lifted without generation of significant resistance from steps 1 to 3. Pictures taken during retrieval of the S/A are shown in Fig. 2.52.

2.3 Irradiation test experience and future plan

2.3.1 *Irradiation test technologies of Joyo [8]*

It is necessary to realize an economic efficiency that can compete with the light water reactor (LWR) to commercialize the SFR. From the viewpoint of fuel development, an extended fuel burnup and a higher LHR are the major points. Based on these points,

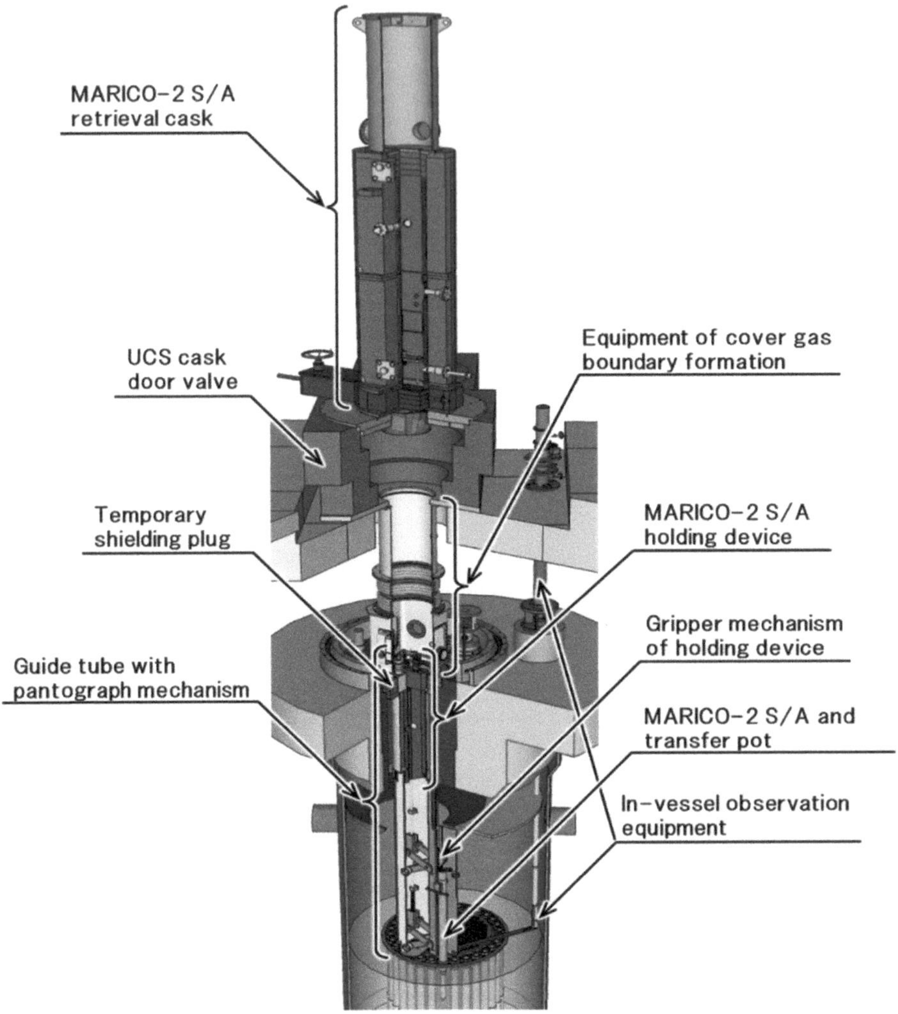

Fig. 2.50 Installation state of the MARICO-2 S/A retrieval device.

continuous efforts have been conducted to improve MOX fuel performance for large SFRs. Approximately 30 types of advanced fuel specimens installed in irradiation test rigs have been irradiated.

The material irradiation test is important because the irradiation characteristics of materials significantly influence the reactor performance and lifetime. Over 7000 test pieces, including core material such as cladding material and wrapper tube, structural material such as reactor vessel and the core support plate, and absorber material, have been irradiated in Joyo.

Fig. 2.53 shows neutron spectra that were calculated for representative irradiation fields in Joyo Mk-III core. As the position moves closer to the core, the fraction of fast-energy neutron and the neutron flux level increases.

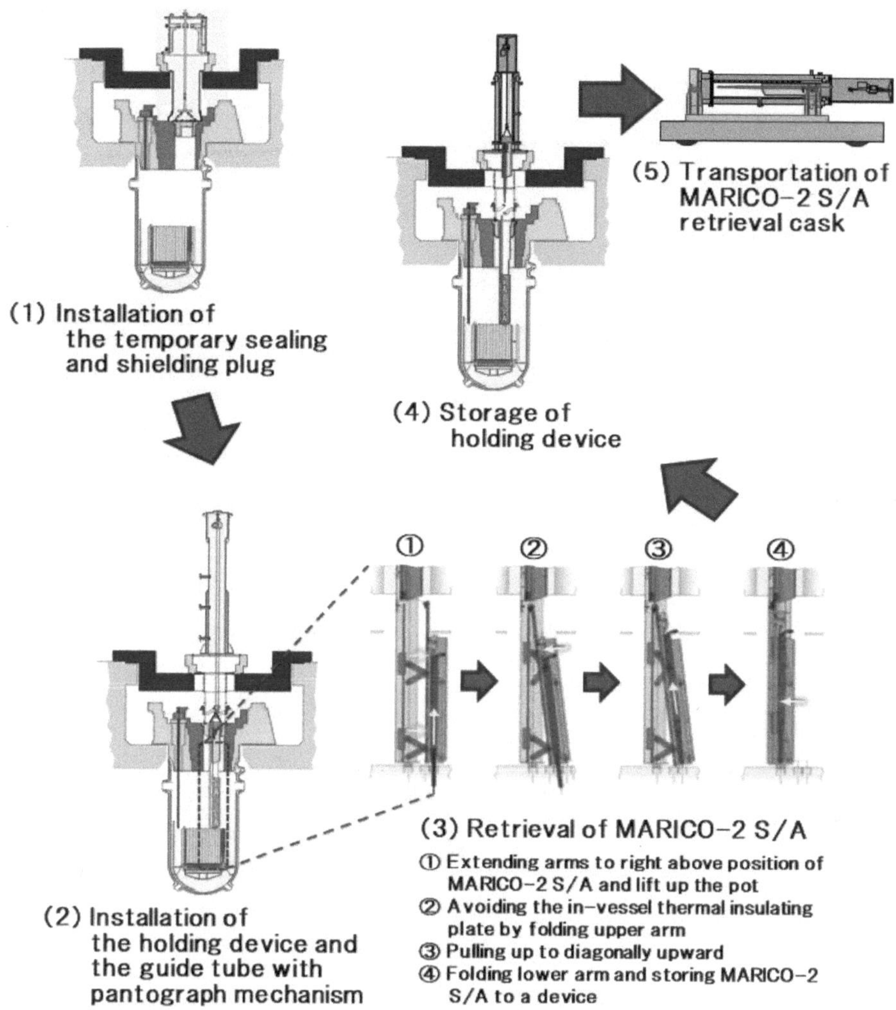

Fig. 2.51 Outline of bent MARICO-2 S/A retrieval works.

The relatively flat neutron energy distribution is available at the upper-core structure and the ex-core manhole, which are apart from the core in the axial and radial directions, respectively.

Fig. 2.54 shows an example of calculated results for axial distributions of fast neutron fluence at the core center, as well as of the temperatures of coolant and sample inside the irradiation rig in case of material irradiation. The specific irradiation conditions must be determined for each operation cycle in consideration of the location in the core, design of irradiation rig, etc.

The maximum temperature achieved in material irradiation at Joyo is approximately 750°C.

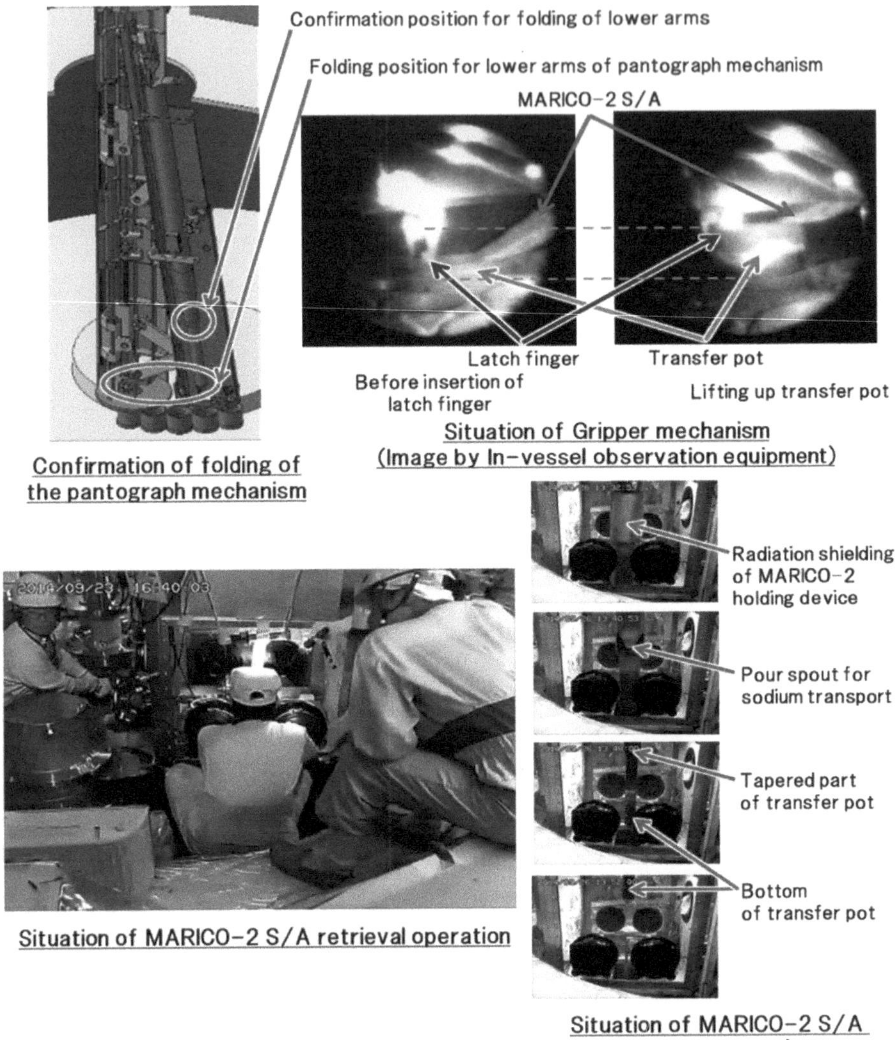

Fig. 2.52 Typical scene of MARICO-2 S/A retrieval works.

2.3.1.1 Irradiation equipment of Joyo

The irradiation rigs used in Joyo are roughly grouped into two types: the off-line irradiation rigs, which need off-line monitors to measure irradiation data, and the online irradiation rigs which are capable of online data measurement and/or temperature control.

The off-line irradiation rigs may be irradiated at any location in the core, while the installable locations for the online irradiation rigs are limited.

Typical off-line/online irradiation rigs are described in Fig. 2.55.

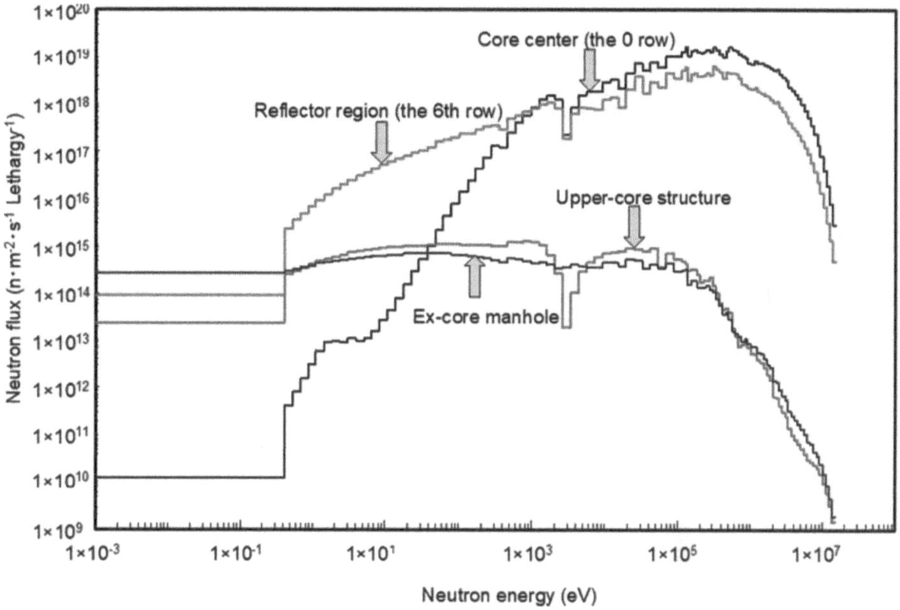

Fig. 2.53 Neutron spectra at various positions in Joyo Mk-III core [8].

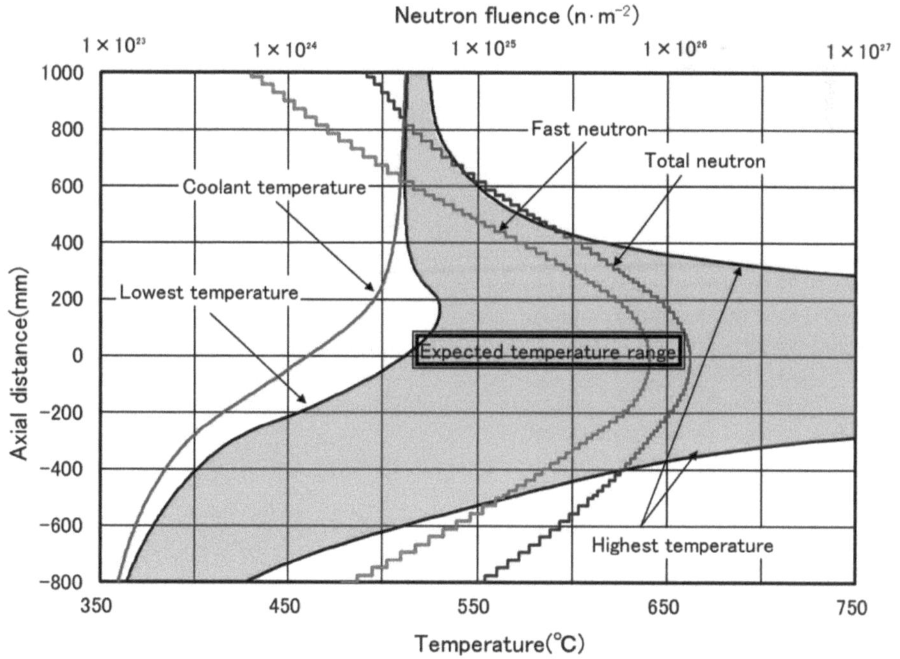

Fig. 2.54 Example of fast neutron fluence and temperature distributions at the core center (in 1 cycle [60 days]).

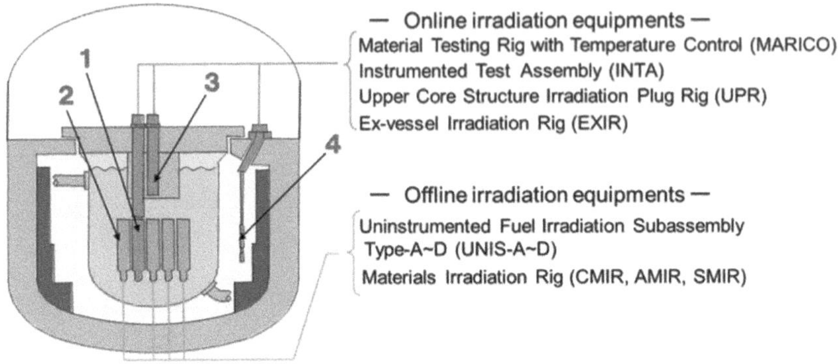

	Region	Temp. Unit:°C	Neutron flux Unit: $n/m^2 \cdot s$		Spectrum ratio
			Total	Fast (E≧0.1MeV)	
1	Fuel region	450~750	$(4\sim5)\times10^{19}$	$(3\sim4)\times10^{19}$	0.6~0.7
2	Reflector region	400~700	$10^{18}\sim3\times10^{19}$	$3\times10^{17}\sim2\times10^{19}$	0.3~0.5
3	Upper core region	500~	$10^{15}\sim10^{16}$	$10^{14}\sim10^{15}$	10^{-1}
4	Ex-core manhole	200~600	$\sim10^{16}$	$\sim10^{14}$	$\sim10^{-2}$

Fig. 2.55 Various types of irradiation rigs.

Basically, irradiation tests are carried out with irradiation rigs installed in the core and reflector regions of Joyo. The following sections introduce the representative irradiation rigs that were developed in Joyo.

(1) Uninstrumented fuel irradiation subassembly (UNIS)

To improve the economic efficiency of fast reactors, it is necessary to increase the LHR and peak burnup of fuel. In Joyo, irradiation tests for advanced fuel specimens were conducted 30 times to confirm the fuel performance. Fig. 2.56 shows the range of the maximum LHR and burnup achieved in the test. The maximum LHR was achieved by nitride fuel pin. And in the power to melt (PTM) test, the liner heat rate reached over 600 W/cm in the MOX fuel pellet. The maximum burnup of fuel irradiation in Joyo is 144 GWd/t. (It was achieved using 15-15Ti cladding material in collaboration work with CEA, France.)

Major fuel irradiation tests are performed by off-line irradiation rigs. In Joyo, four types of fuel irradiation rigs are prepared to conduct fuel irradiation tests as shown in Fig. 2.57. Their features are described below.

Type-A: A few test pins are placed within a small duct tube (flow channel) installed in the center of a Joyo driver fuel pin bundle.

Type-B: Test pins are loaded in six compartments for which different flow conditions can be set. This type of rig is also suitable for reassembling in the hot cell and reloading to the reactor, after interim examination of the test fuel pins.

Type-C: A test fuel pin bundle is wrapped by a double-walled duct to demonstrate the MOX fuel subassembly design and investigate the irradiation behavior of the MOX fuel pin bundle. The size of the inner wrapper tube is adjusted to an optional dimension for the test pin diameter and pitch.

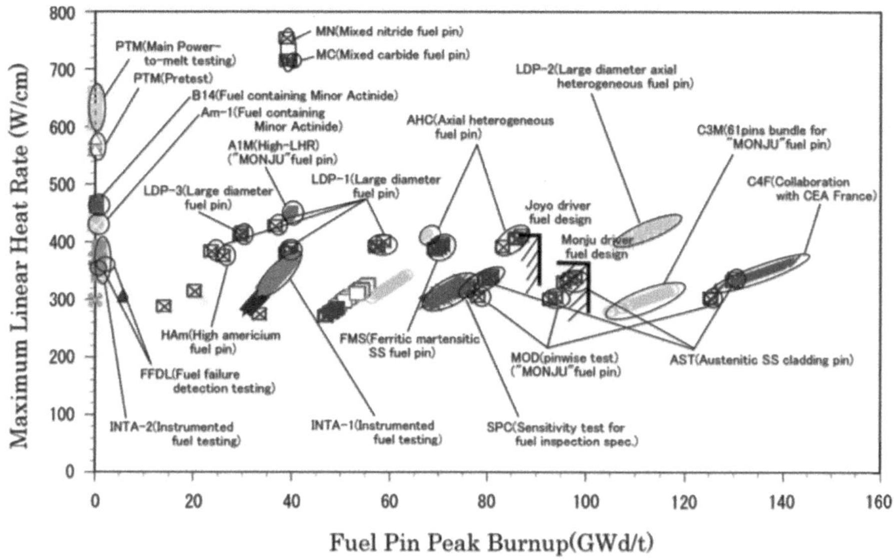

Fig. 2.56 Overview of fuel irradiation tests in Joyo.

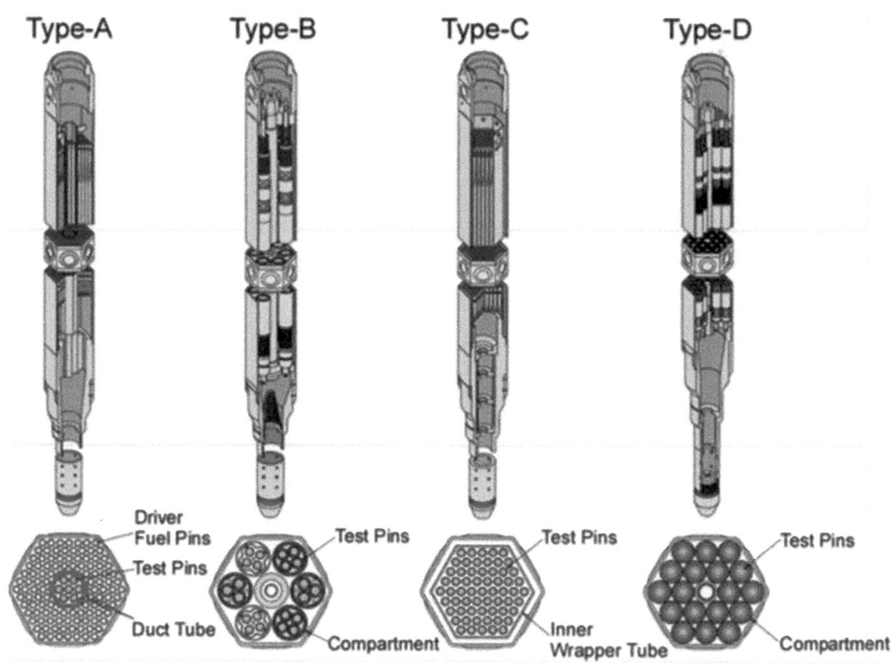

Fig. 2.57 Off-line irradiation rigs.

Type-D: A compartment type rig improved on the Type-B design. Up to 18 compartments can be placed in this rig. Each compartment holds one single test fuel pin and forms an insulated coolant flow channel for which individual pin cladding temperature parameters can be set.

One of the typical fuel irradiation tests is the PTM test. It was performed using a B-type rig called B5D [9]. The power history of the main test is shown in Fig. 2.58. The power ramp rate was increased from 0.4 to 0.5%/min in the preliminary test to reduce the structural changes before irradiation at the target temperature. Moreover, a shutdown scram was adopted to avoid additional structural changes in the molten area. The reactor power was increased to 100 MW at full power for the MK-II core based on the result of the temperature calculation code calibrated by the B5D-1 results. The larger melting area and clearer boundary of the melting area were observed by the PIE of B5D-2, different from the result of B5D-1.

Fig. 2.59 shows the capsule type rig, which is a type-B rig. The capsule-type rig is available for irradiation tests of various fuel forms including oxides, carbides, nitrides, metals, and those with minor actinides (MAs) and fission products (FPs). This rig consists of six compartments, each equipped with a capsule. The capsule can withstand the pressure increase in the event of fuel failure of the test pin. This feature allows us to irradiate materials with insufficient data. Each compartment can set the flow rate individually. And it allows interim examination and re-irradiation.

Fig. 2.60 shows the results of a short-term irradiation test for MA-bearing oxide fuel using the capsule type rig [10]. This test achieved the highest level of LHR in the world at about 430 W/cm under irradiation for 10 min, and confirmed that the fuel was not melted but structural changes had begun. Subsequently, the composition was changed under irradiation for 24 h. It is known that MOX fuel irradiated at high temperatures undergoes structural changes such as a central hole as shown in Fig. 2.60, and element movement and redistribution. It can be seen that the U concentration decreased and the Pu and Am concentrations increased in the central part during the test.

(2) Material irradiation rig (MIR)

Material irradiation tests are performed using the material irradiation rig (MIR). The uninstrumented compartment types are common material irradiation rig (MIR) in Joyo. Compartments and the internal structure of MIR are designed to set material specimens. The MIRs are classified by the type of materials, such as core material (CMIR), absorber material (AMIR), and structural material (SMIR). For instance, CMIR is loaded into the fuel region to achieve a high displacement per atom (dpa), which is required for R&D

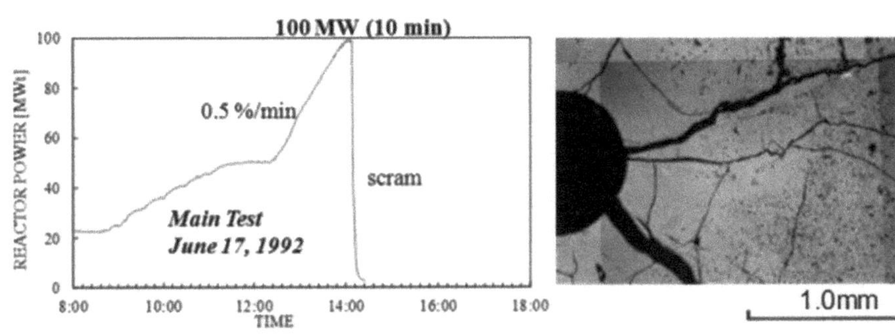

Fig. 2.58 PTM tests result for B5D-2 [9].

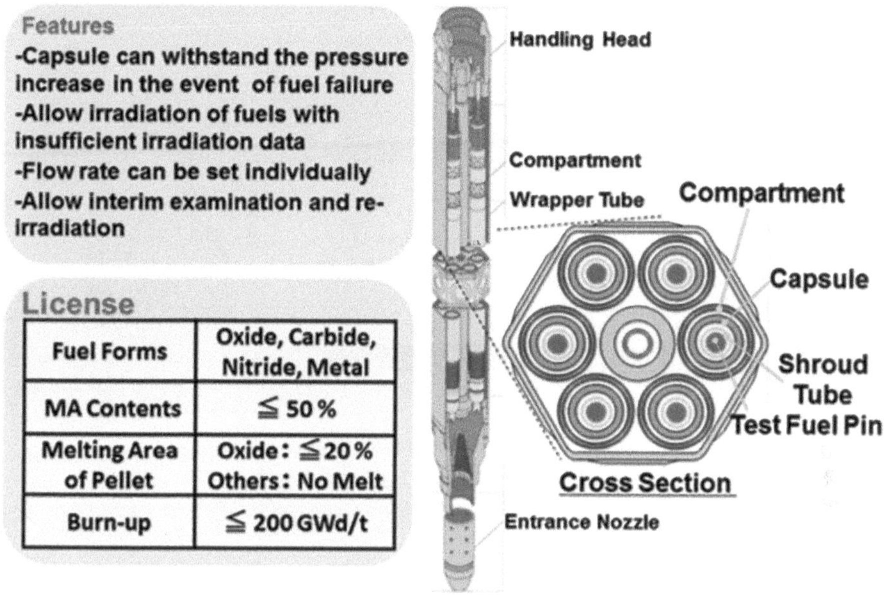

Fig. 2.59 Capsule type rig.

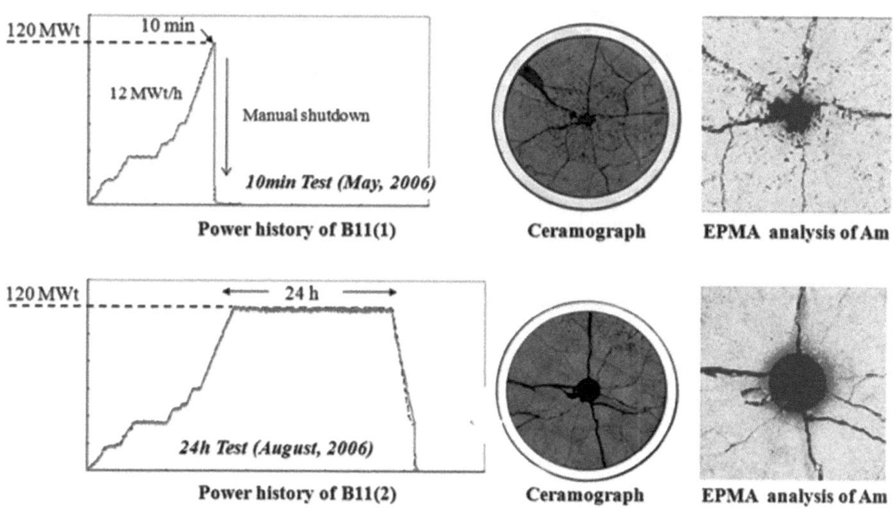

Fig. 2.60 PTM test result for B11 [10].

of cladding tubes and duct tubes to be used for future SFR core fuel subassemblies. On the other hand, AMIR and SMIR can be used in the radial reflector region. The outline of MIRs is shown in Fig. 2.61.

In CMIR, test material is usually installed in an unsealed holder to make unloading and reloading of material samples easy. Since the target fluence of the core material sample

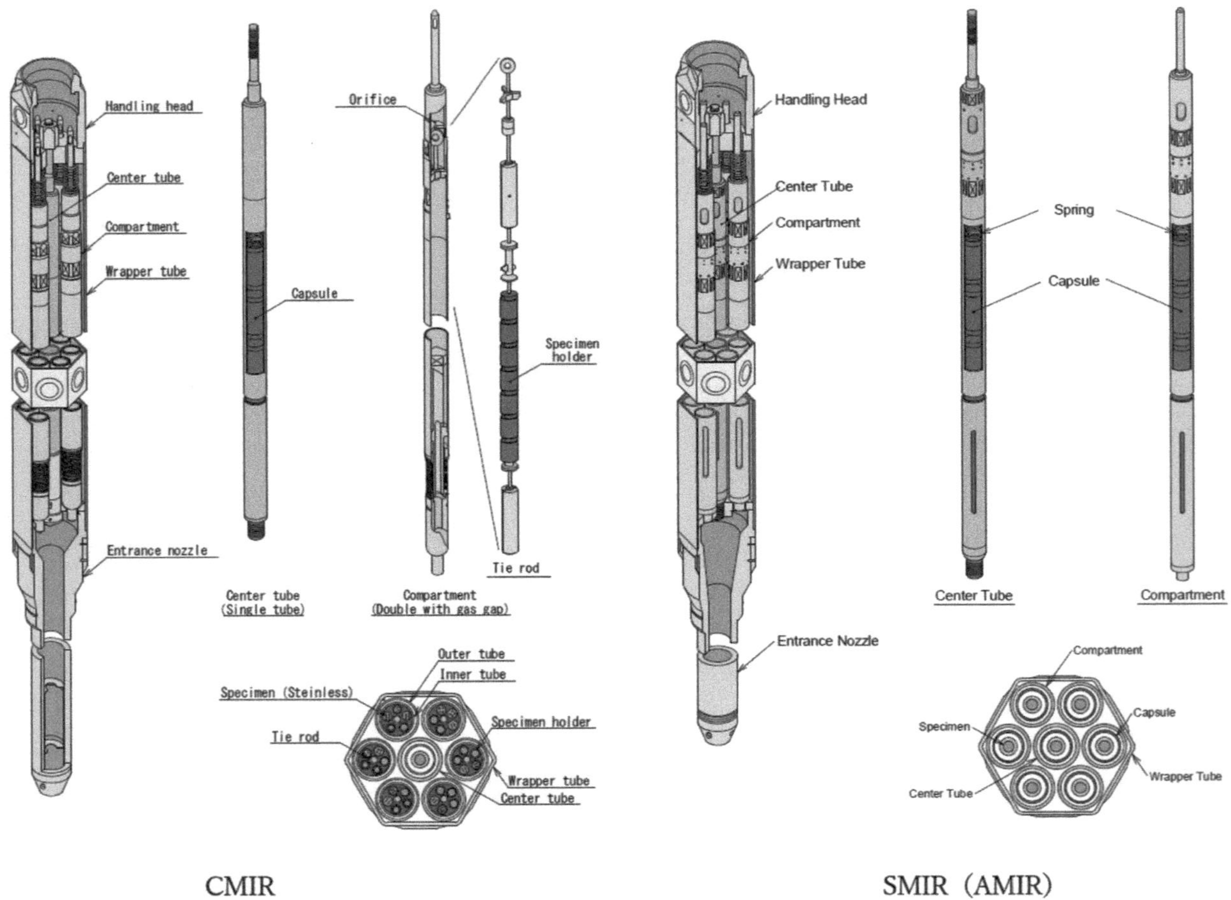

Fig. 2.61 Outline of MIRs.

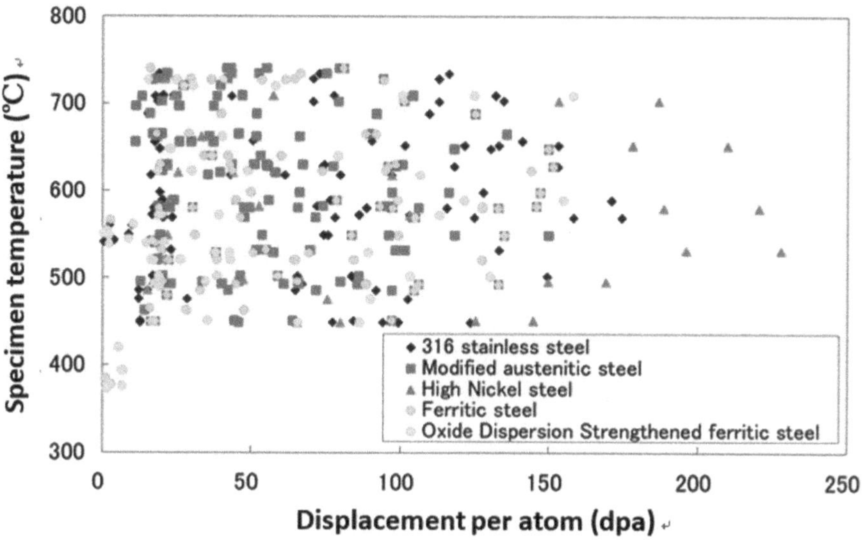

Fig. 2.62 Database for cladding and wrapper tube materials irradiated in Joyo.

(e.g., cladding tube, duct tube) is set at a very high level, the irradiated holder has to be replaced with a new one in the midterm of the irradiation test while the material sample itself is continuously used in the irradiation test. The material sample in the holder is immersed in sodium to investigate the irradiation behavior in high temperature sodium. The temperature of material samples is controlled by the following:

- the mixture ratio of argon and helium gases in the thermal insulating double tube of the compartment,
- the coolant flow rate in the compartment, and
- the layout of heat-generating elements (the additional steel rod can be installed to increase gamma heating).

In AMIR or SMIR, a test piece is normally installed in a sealed capsule. The material irradiation capsule is filled with sodium or inert gas (helium and/or argon). A unique-designed vent-type capsule, into which the primary coolant sodium may be introduced, has been irradiated in the MK-II core. For a long-term irradiation test, the material irradiation capsule is reloaded into a new irradiation rig.

Fig. 2.62 shows the achievement of irradiation tests for core materials, such as cladding tubes and wrapper tubes for fast reactors. Material irradiation tests achieve to the very high fluence region more than 100 dpa as actual results in Joyo. Several advanced reactors need research in such a high fluence region.

2.3.1.2 Online irradiation equipment

(1) Instrumented test assembly (INTA)

INTA is an online test rig equipped with thermocouples, gas pressure gauges, etc. As shown in Fig. 2.63, INTA-1 and INTA-2 were fuel bundle subassemblies. INTA has instrumentation cables, which extend from the core and penetrate the upper core structure (UCS) to transmit in-pile data. INTA-1 and INTA-2 were utilized in the MK-II core. Fig. 2.64

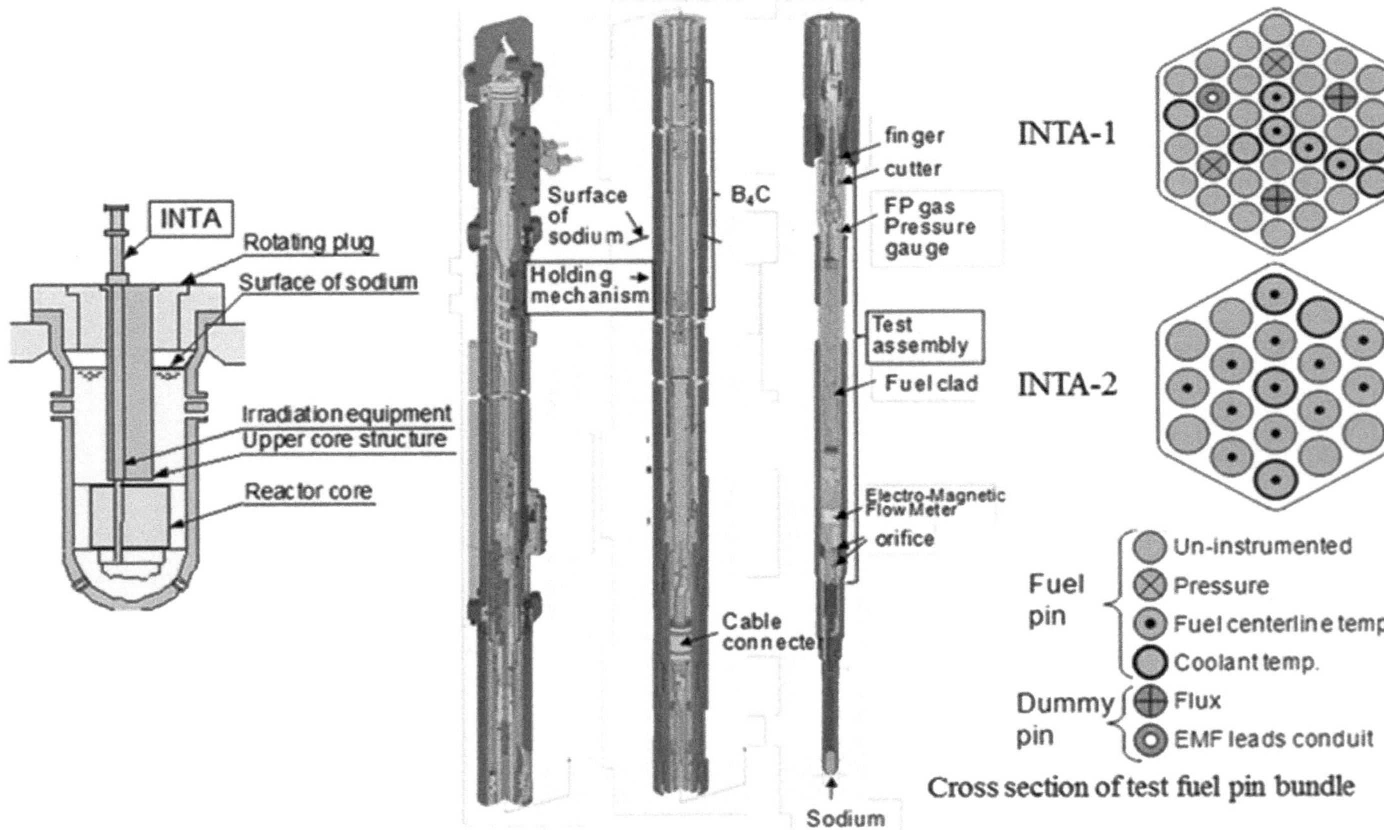

Fig. 2.63 Outline of INTA.

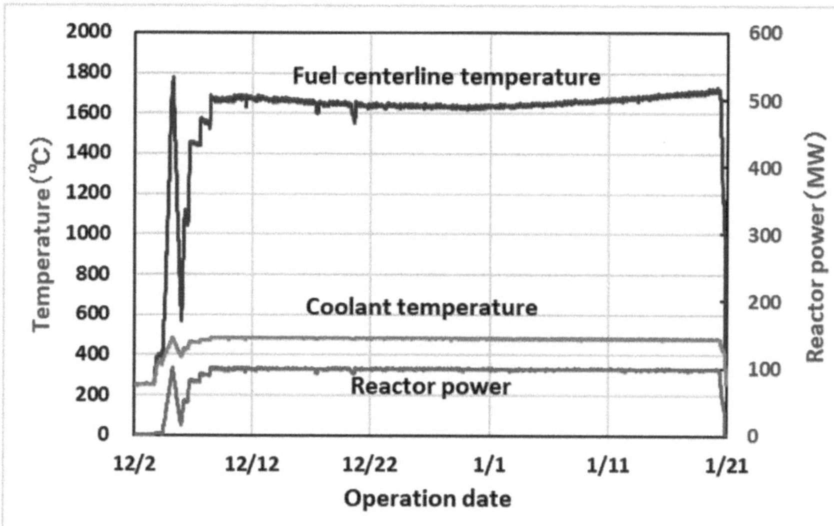

Fig. 2.64 Temperature history measured by INTA.

shows the history of a fuel centerline temperature during and after reactor start-up measured by INTA-1.

(2) Material testing rig with temperature control (MARICO) [11]

MARICO-1 was used in the MK-II core for the in-pile creep rupture test of cladding materials. MARICO can control the temperature by changing the thermal conductivity between the walls of a double-walled capsule through the adjustment of the He and Ar gas ratio. The capsule temperature was successfully controlled within ±4°C during full-power operation.

MARICO-2 was fabricated for the MK-III core. Fig. 2.65 shows the outline of MARICO-2. As an additional temperature control system, an electric heater was installed in the capsule to control the temperature during start-up and shutdown periods. Fig. 2.66 shows the result of temperature controlled by MARICO-2. In the Joyo MK-III core, creep rupture test specimens of the oxide dispersion strengthened (ODS) ferritic steel, which is the promising candidate for the cladding of long-life fuel, were irradiated with MARICO-2, and it was confirmed that the creep strength did not decrease due to irradiation.

(3) Upper core structure irradiation plug rig (UPR)

The UPR is used for irradiation testing of structure materials, including those of the reactor vessel, by installation in the upper core structure of Joyo, and allows for the measuring of specimen temperature during irradiation.

Fig. 2.67 shows the outline of UPR. Although the fast neutron flux is lower than that in the core region (10^{14}–10^{16} n/m^2 s), the specimen temperatures can be controlled by an electric heater. In the MK-III core, an UPR was used for the testing of magnetic materials and ternary alloy to measure the changes in magnetic characteristics under irradiation.

2.3.2 Postirradiation examination (PIE) technologies

The development of a high-performance fuel pin keeping its soundness under high-burnup conditions is a key for the realization of SFR having substantial economical and safety performance. The operation conditions of SFR fuel are much severe in

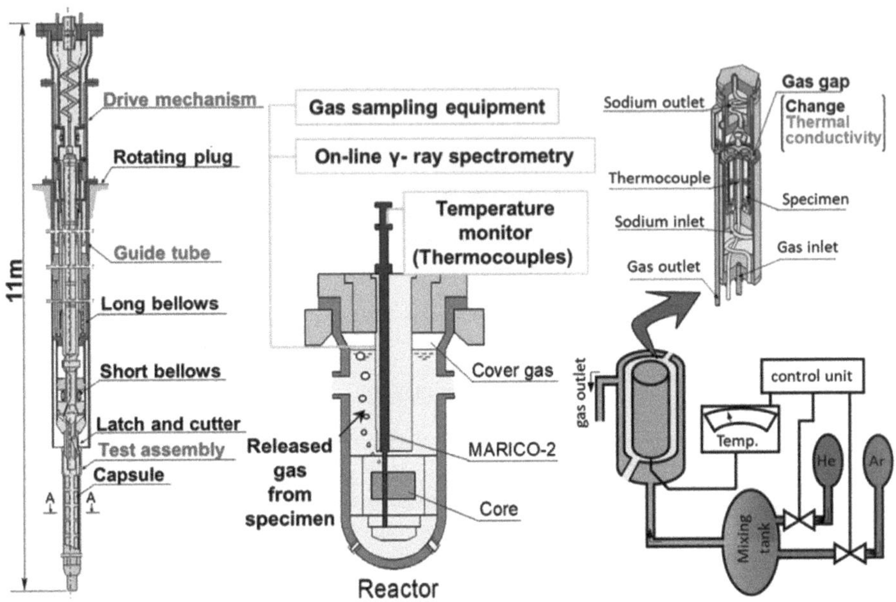

Fig. 2.65 Outline of MARICO.

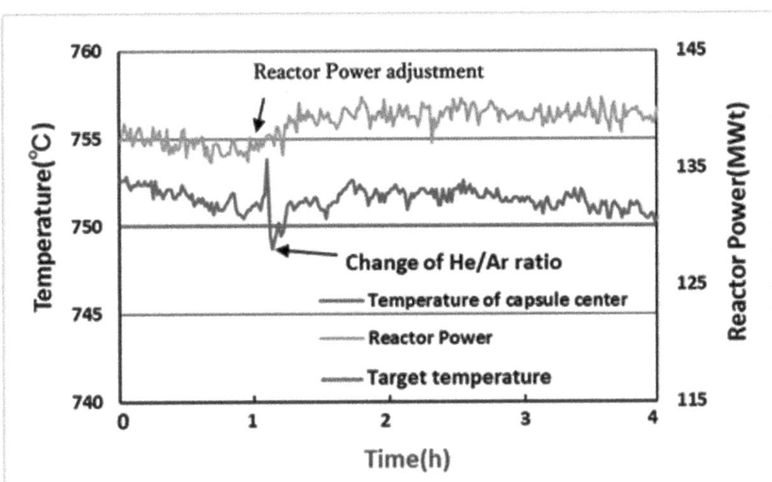

Fig. 2.66 Temperature history controlled by MARICO-2.

comparison with those for light water reactors (e.g., temperature, neutron dose). The derivation of PIE data leading to elucidation of irradiation behavior of fuels and materials is indispensable for the development of advanced fuel and materials as well as the core fuel designing.

Various types of PIE techniques and apparatuses have been installed and improved so far in the PIE facilities complex nearby the experimental fast experimental reactor

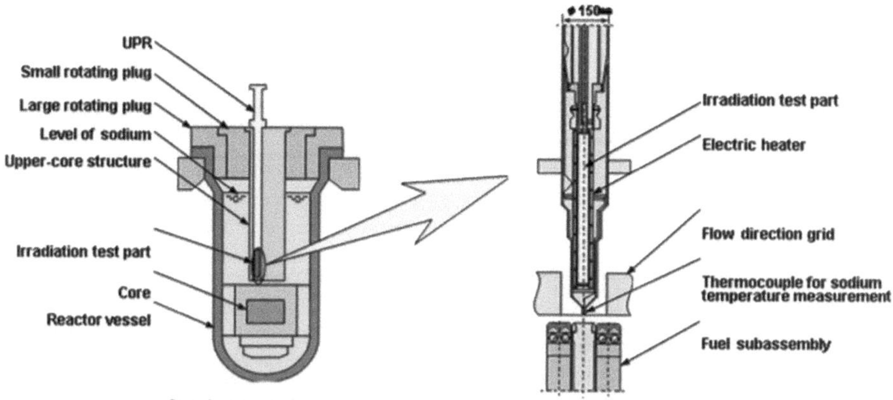

Fig. 2.67 Outline of UPR.

Joyo. They cover the basic items indispensable to characterize irradiation behavior of SFR fuels and materials (e.g., visual inspection, dimensional measurements including laser profilometry, microstructure characterizations, chemical analysis, physical and mechanical characterizations). In addition, the unique and noble technique extensively being developed is the X-ray computer tomography (CT), which has been developed specifically for nondestructively evaluating the deformation of fuel and duct (wrapper tube) as well as fuel microstructure after irradiation. Another distinctive piece of equipment is the in-cell remote fabrication system of minor actinide bearing MOX fuel. The major functions of facilities are listed in Table 2.7, and examples of outcomes are described below.

Table 2.7 Major functions of PIE facilities.

Fuels Monitoring Facility (FMF)	
Nondestructive test	Visual inspection, profilometry, Eddy current inspection, gamma scanning, X-ray radiography, X-ray CT.
Destructive test	Pin puncture test, observation by optical microscope and SEM, EPMA analysis.
Assembly handling	Disassembling and reassembling of assembly for irradiation test.
Alpha Gamma Facility (AGF)	
Fuel manufacturing test	Manufacturing of fuel pellets and pins, X-ray radiography inspection.
Evaluation of physical property of fuel	Measurement of melting point, O/M ratio and thermal conductivity, X-ray diffractometry, evaluation of fission product release.
Chemical and isotope analysis of fuel	Analysis by EPMA, ICP-AES, and ICP-MS, determination of burnup.

Continued

Table 2.7 Continued

Materials Monitoring Facility (MMF)	
Mechanical test of material	Tensile test, creep test, Charpy impact test, and hardness test.
Evaluation of physical property of material	Profilometry, measurement of density, thermal conductivity, and magnetic property.
Microstructural evaluation	Observation by optical microscope and TEM, X-ray diffractometry.

2.3.2.1 PIE of fuel assembly

Fuel subassemblies irradiated in Joyo are transferred to Fuels Monitoring Facility (FMF) connected to Joyo through an underground passage. FMF started operation in 1978; nondestructive and destructive tests have been conducted in FMF to investigate the effects of irradiation on fuel subassemblies irradiated in Joyo. The PIE data derived from here have been used for validating the design of the MOX fuel pin and assembly. In addition, FMF takes on a role for disassembling and reassembling irradiation test assemblies; the fuel and material irradiated in Joyo are taken out for disassembling and intermediate inspection at FMF to confirm their soundness after irradiation. After the inspection, the irradiated samples are reassembled to an assembly and reloaded to Joyo for further irradiation. This process has successfully enabled the stepwise increase of burnup and neutron dose. The deformation of ducts and fuel pin bundles due to irradiation is one of the most important phenomena, a major life-limiting factor for the assembly made of austenitic steel. The development of excessive deformation locally reduces the channel cross-section for sodium coolant flow in the assembly, thus overheating the fuel pins to rupture. X-ray CT test has been conducted to understand the deformation of fuel pins due to irradiation and the distribution of central voids in fuel pellets. By using the X-ray CT technique, it is possible to undestructively observe the deformation of all fuel pins in conjunction with central voids and cracks of fuel pellets in the assembly. The advantage of the X-ray CT technique is the capability of automatic mass data derivation; the data analysis technique is being studied for obtaining the information on irradiation behavior of fuels and fuel assemblies from the mass data.

The cross-section image of an assembly by X-ray CT is shown in Fig. 2.68 [12]. The fuel pin bundle and central voids in fuel pellets can be observed, clearly. Recently, the apparatus was upgraded to "High-resolution X-ray CT." The error of the dimensional measurements was especially improved by lowering the values from 0.1 to 0.03 mm [13]. This advanced apparatus gives much information to understand the structural change of irradiated fuel pellet though it is a nondestructive method [14]. The X-ray CT was also used for the verification of fast reactor fuel bundle-duct interaction analysis code by the bundle compression test [15], which simulated the fuel pin bundle and duct interaction leading to deformation of the pin bundles.

The optical microscope inspection, the electron-prove microanalysis, and scanning electron microscope examination were carried out as the major destructive PIEs. For

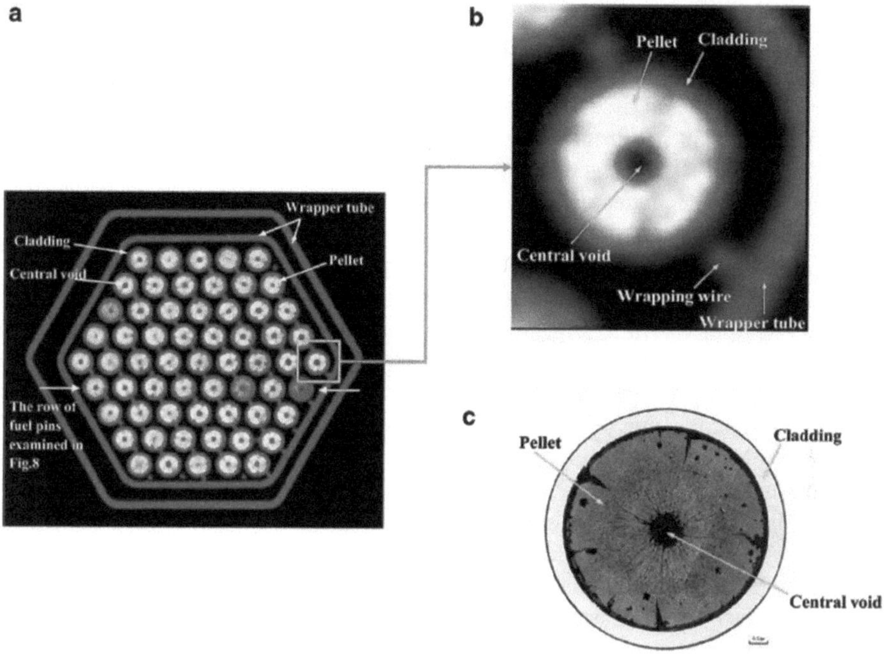

Fig. 2.68 Comparison of central void diameter by the X-ray CT and ceramography. (A) X-ray CT image obtained on the cross-section of fuel assembly-1. (B) Enlarged X-ray CT image on the cross-section of fuel pin. (C) Ceramography obtained on the cross-section examined by X-ray CT [12].

instance, the uranium (U) and plutonium (Pu) mixed oxide (MOX) fuel containing minor actinides (MA-MOX) were quantitatively analyzed. The redistributions of MA-MOX constituent elements based on experimental results were shown to agree well with the pore migration and thermal diffusion models [16,17]. One of the other destructive PIEs is the pin puncture test, which has been carried out on irradiated MOX fuels to measure the amount of gasses released from the fuel matrix. The results gave the quantitative relationship between helium release rate and fission gas (xenon and krypton) release rate [18].

Besides the above, PIE of control rod containing B_4C absorber pellets was conducted after the irradiation at Joyo to obtain the data leading to its life extension. Measurement of pin diameter by laser profilometer and surface observation was conducted to check local diameter change and surface cracking of rod caused by the relocation of B_4C absorber [19]. X-ray CT test was utilized to nondestructively investigate the irradiation behavior of control rods (i.e., deformation of control rods and B_4C pellet relocations).

2.3.2.2 PIE of fuels

Fuel specimens irradiated in Joyo are disassembled in FMF, then being transferred to Alpha Gamma Facility (AGF). AGF started operation in 1971; various characteristics of MOX fuel have been investigated in AGF. The PIE data derived from here have been used for the designing and licensing of the MOX fuel pin. To investigate the soundness of MOX fuel, which is exposed to the maximum temperature exceeding 2000°C during irradiation, the melting point of MOX fuel after irradiation has been measured. The high-temperature fuel behaviors such as fission product release behavior have been evaluated by the simulation test using irradiated fuel pin with an accident condition. In addition, various types of fuel characterizations (element distribution, phase state, crystal structure, etc.) have been comprehensively conducted to elucidate the mechanism of various phenomena that occur in the fuel under irradiation.

The melting points of MOX fuels depending on Pu content, burnup, and other parameter were correctly measured by a unique method using a rhenium inner capsule to hold the specimens while tungsten having a high melting temperature was used for the capsule in the conventional procedure. The rhenium inner capsule could prevent chemical reactions between fuels and tungsten materials, which decreases the melting temperature. The melting temperatures were successfully measured using this technique. The melting temperature was measured about 30°C higher than the previous data using tungsten capsules [20].

Determination of the burnup of irradiated fuel that indicates the fission reaction rate of uranium, plutonium, etc., is important to understand the phenomena that occurred in fuel under irradiation. The burnup is necessary for verification of neutron distribution calculations because fission products accumulate thick in the fuel, thereby disturbing the neutron distribution at high burnup. To accurately identify the burnup in the fuel, the irradiated fuel is dissolved, and the isotope fraction of uranium, plutonium, neodymium, etc., was measured. The γ-ray spectrometry focusing neodymium of FP and isotope dilution analysis technique was adopted for the determination of burnup with adequate accuracy [21].

Under severe reactor accidental conditions, FPs are released from overheated and degraded fuels. The experimental data on FP release behavior (so-called source term) from MOX fuels under SFR accidental conditions were obtained by various spectrometry. The results gave the relationships between the temperature and the source term of each element, thus leading to improvement of the safety assessment [22–24].

To confirm the effect of minor actinide additions on the irradiation behavior of MOX fuel pellets, americium (Am)-MOX fuels which were irradiated for short period in Joyo were observed by optical microscope. The redistribution behavior of actinide elements was determined by an electron probe microanalyzer (EPMA). The ceramography results showed that structural changes such as the formation of lenticular pores that led to the formation of the central void occurred quickly during the first 10 min of irradiation. The EPMA results revealed that Am, as well as Pu, migrated radially up the temperature gradient to the center of the fuel pellet as shown in Fig. 2.69 [25].

An in-cell remote fabrication and inspection technique were also developed for Am-MOX fuel pellets for the irradiation experiment of JOYO. The fuel pellet was fabricated by means of conventional powder metallurgy. A series of fuel pellet fabrication and inspection apparatuses were systematically installed in the alpha-gamma cell (hot cell) to protect workers from a strong γ-ray exposure from ^{241}Am [26].

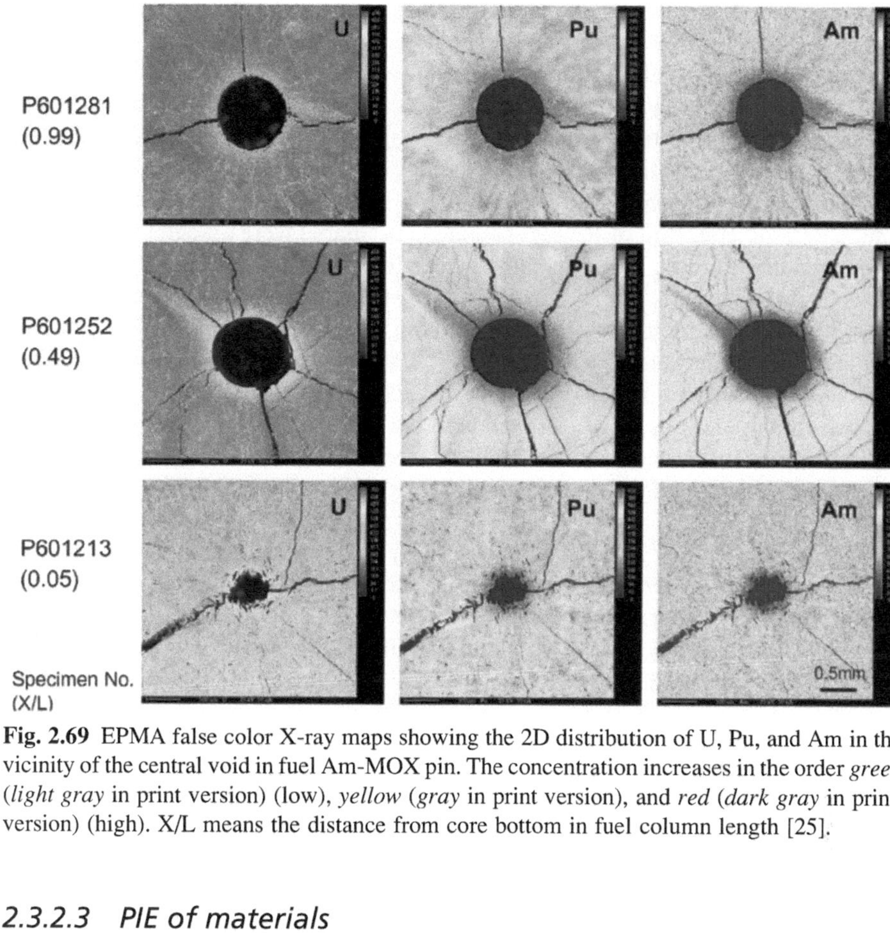

Fig. 2.69 EPMA false color X-ray maps showing the 2D distribution of U, Pu, and Am in the vicinity of the central void in fuel Am-MOX pin. The concentration increases in the order *green* (*light gray* in print version) (low), *yellow* (*gray* in print version), and *red* (*dark gray* in print version) (high). X/L means the distance from core bottom in fuel column length [25].

2.3.2.3 PIE of materials

Material specimens irradiated in Joyo are disassembled in FMF, then being transferred to the Materials Monitoring Facility (MMF). MMF started operation in 1973; various types of material tests (mechanical tests, physical property tests, and microstructural observations on core materials, structural materials, control rod materials, and so on) have been conducted in MMF. The PIE data derived from here have been used for the development of high-performance materials for SFR and the preparation of materials strength standards for core and structural materials of advanced SFR.

Core materials used in advanced SFR are assumed to be in use with much higher dose and at a higher temperature than those in light water reactors. Therefore, core materials should have superior irradiation resistance and high temperature strength. One of the important aspects of irradiation resistance is irradiation-induced swelling, which is a life-limiting factor in the case of austenitic steel. To keep the channel of sodium coolant in assembly, irradiation-induced swelling of the core material (fuel cladding tube and duct materials) should be suppressed as low as possible during irradiation. For evaluating the irradiation-induced dimensional changes of core materials, the diameter and densities of cladding tube after irradiation have been respectively measured by laser profilometer and immersion method, thus quantitatively evaluating the irradiation-induced swelling. The data obtained in these tests have given the quantitative data on irradiation-induced swelling in length and volume. The data accumulated so far for modified type 316 austenitic steel constituted the indispensable basis for the reliable formulation of the swelling equation including the swelling incubation period [27,28].

Tensile property is very fundamental while it is the most important property of core materials. The most of tensile specimens irradiated were machined in advance before irradiation tests. Some specimens were cut from either ducts or other specimens by the electrical discharge method after irradiation. The plate type tensile test is simple and easy, but it cannot test the cladding tube in the hoop direction. The ring tensile test technique has been adopted to evaluate the strength in the hoop direction. The ring tensile specimen has gauges at both sides of the ring. The test has been carried out at temperatures from room temperature to 1000°C which includes the temperatures relevant to accident conditions. Tensile data of candidate material have been steadily accumulated after irradiation, long-time thermal aging, and sodium immersion [29–31].

Impact property test is indispensable in developing ferritic steel because ferritic steels show embrittlement behavior owing to irradiation and/or thermal aging. Charpy impact test with either full or half of the standard size has been conducted by pendulum-type test machine. A miniature-sized specimen which is much suitable for irradiation tests has been tested by a hydraulic drop hammer machine with instrumentation. The Charpy test data suggest whether the candidate material maintains adequate toughness after irradiation or not [31,32].

Physical properties are important for ceramic materials such as the control element of B4C and the neutron shield element of graphite. Thermal diffusivity which is a typical and essential property for them has been measured by the laser flash method. Elastic modulus has been measured from the velocity of ultrasound waves traveling longitudinally and transversely to the direction of the cylindrical axis by ultrasonic transducers. These measurements have revealed the effects of irradiation-induced defects on the physical properties of ceramic materials [33].

The mechanical and physical property changes are related to microstructural changes. Transmission electron microscopy (TEM) observation is very important to understand the irradiation effect on the microstructure of materials. The thin plate specimen has been cut and thinned to the TEM observation sample by electropolishing or focused ion beam. TEM observation and energy dispersion spectrometry provide detailed data on microstructural change due to irradiation, such as dislocation recovery, precipitation, particle coarsening, and bubble formation on the nanometer scale. For instance, as shown in Fig. 2.70, TEM observation revealed the stability of nanometer scale structure in oxide dispersion strengthened steel which is a promising material for fuel cladding tubes of advanced SFR [34]. Orientated He bubbles and precipitations of the second phase in irradiated B_4C pellets were analyzed by diffraction and electron dispersion spectrometer equipped to TEM, and the orientation arrayed on the lattice and the effect of transmutation was discussed [35].

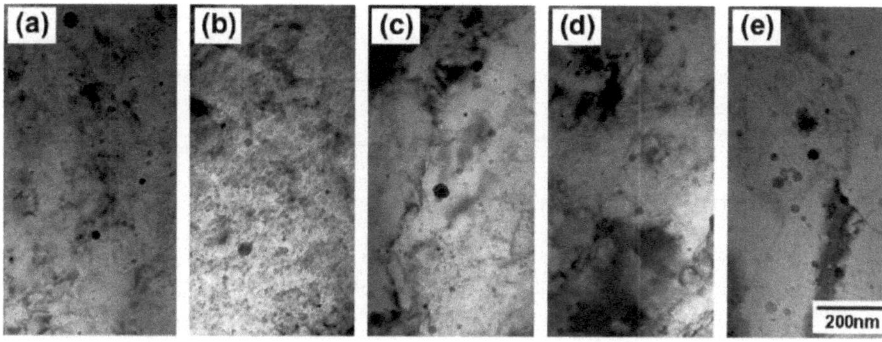

Fig. 2.70 Bright field micrographs exhibiting oxide particle distribution of 9Cr-ODS F/M steels (A) as-received and irradiated (B) at 410°C to 15 dpa, (C) at 630°C to 32 dpa, (D) at 735°C to 33 dpa, and (E) at 830°C to 28 dpa [34].

2.3.3 Future plan

As shown in Fig. 2.71, Joyo has various irradiation fields with different neutron spectra and temperatures, and can provide medium- and low-energy spectrum irradiation fields for transmutation research of MA and LLFP by loading neutron moderators. In addition, Joyo can perform a lower temperature operation to realize an irradiation temperature comparable to that of LWR. A high temperature irradiation device equipped with a high gamma heating material is also available to conduct

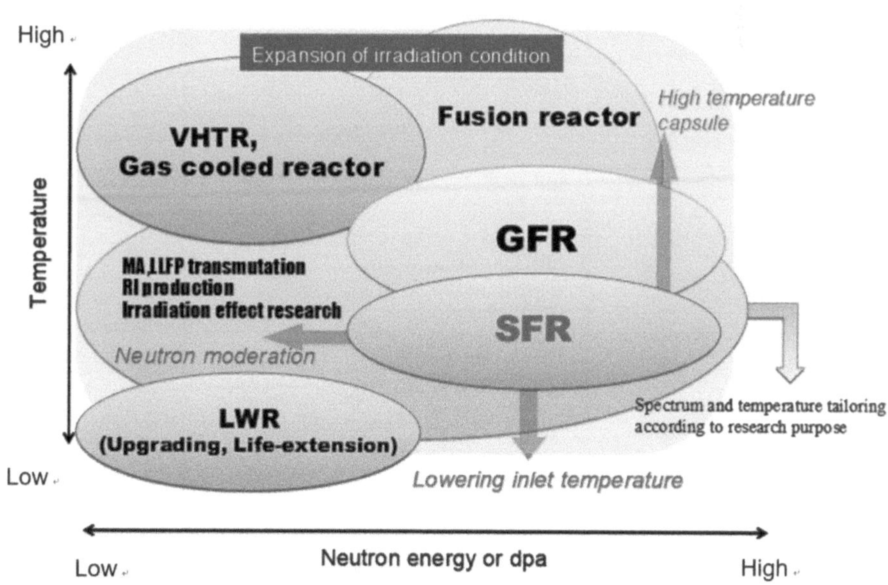

Fig. 2.71 Expansion of irradiation conditions.

high temperature irradiation at 1000°C or higher. So, it is possible to carry out research that meets a wide range of needs, such as the irradiation of fusion reactor materials and LWR materials, from various organizations including universities and research institutes.

2.3.3.1 Neutron spectrum tailoring

The cross-section of a nuclear reaction depends on the neutron energy spectrum, and generally, neutron capture reactions are more sensitive to low-energy neutrons. The fast neutron irradiation field provides a potential for tailoring the neutron spectrum individually using the neutron moderator, such as beryllium or zirconium hydride, that surrounds the target shown in Fig. 2.72. This concept enables us to provide irradiation fields suitable for various transmutation experiments for long-lived FPs or MAs and for effective radioisotope production.

2.3.3.2 Lower temperature irradiation

In conventional MK-III operation, the temperature of irradiated specimens is higher than the core inlet temperature of 350°C. Operation with lowered core inlet temperature was newly licensed based on the potential of the DHXs. This operation scheme will provide an opportunity for investigating the fundamental irradiation effect of the material and the behavior of LWR or fusion-reactor materials at low temperatures. Assuming that the inlet temperature is 290°C, a fast neutron flux of approximately 2×10^{15} n/cm^2 s can be obtained in the LWR temperature range (320–370°C).

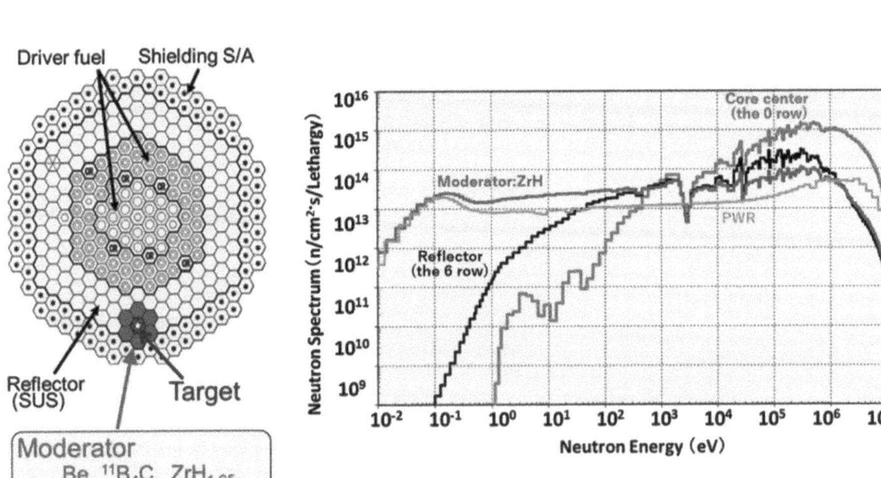

Fig. 2.72 Neutron irradiation field in Joyo.

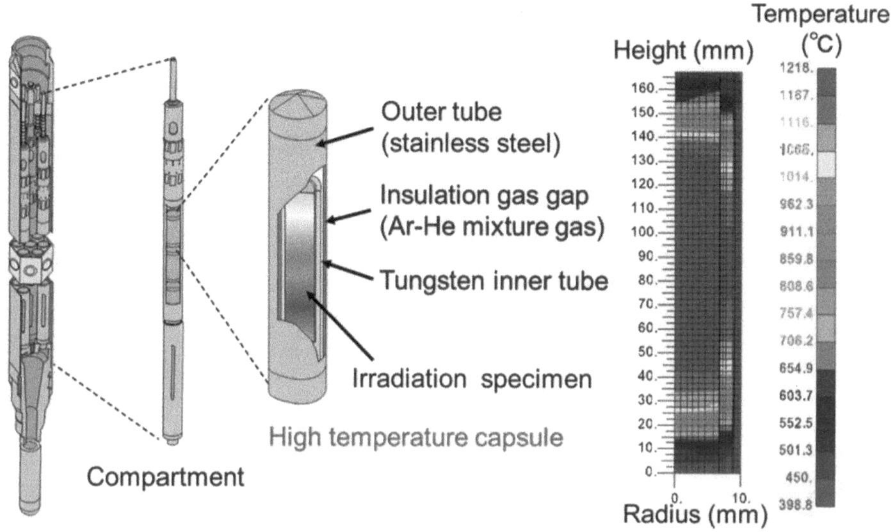

Fig. 2.73 High temperature irradiation capsule.

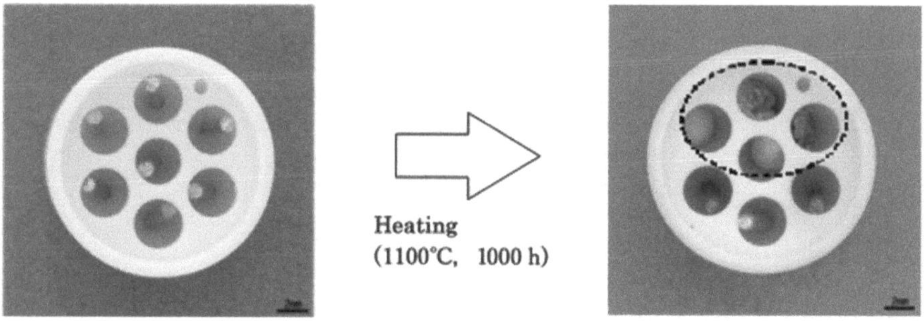

Fig. 2.74 Melt-wire-type temperature monitor.

2.3.3.3 High-temperature irradiation technique

In the R&D field of gas-cooled fast reactors (GFRs), very high-temperature reactors (VHTRs), and advanced fusion reactor systems, material irradiation tests require higher temperatures than the conventional material irradiation temperatures (~800°C) in Joyo. A high-temperature test technique has now been developed at Joyo to meet these R&D needs. An irradiation capsule will be equipped with a tungsten inner tube to produce higher gamma heating. As shown in Fig. 2.73, a preliminary temperature calculation shows that a specimen temperature of over 1000°C is feasible. As an off-line thermometer, the melt-wire-type monitor was developed to cover the high-temperature region because the available temperature range of conventional TED monitors is below 800°C. The melting wire is a device to monitor the temperature reached inside the capsule based on the molten state of multiple alloys with different melting points. Fig. 2.74 shows the result of an out-of-pile heating test at 1100°C.

References

[1] T. Aoyama, et al., Characterization of neutron field in the experimental fast reactor JOYO for fuel and structural material irradiation test, Nucl. Eng. Des. 228 (1–3) (2004) 21–34.

[2] Y. Okawachi, et al., Joyo MK-II Core Characteristics Database—Update to JFS-3-J3.2R, 2003. JNC-TN9400-2003-029 (Japanese).

[3] T. Aoyama, et al., Core performance tests for the JOYO MK-III upgrade, Nucl. Eng. Des. 237 (4) (2007) 353–368.

[4] S. Maeda, et al., Verification of JUPITER standard analysis method for upgrading Joyo MK-III core design and management, J. Power Energy Syst. 6 (2) (2012) 184–196.

[5] S. Maeda, et al., JOYO MK-III Performance Test; Criticality Test, Excess Reactivity Measurement and Burn-up Coefficient Measurement, 2005. JNC-TN9400 2005-022 (Japanese).

[6] H. Ito, et al., Inspection and Repair Techniques in the Reactor Vessel of the Experimental Fast Reactor Joyo—Replacement of Upper Core Structure, 2016. JAEA-Technology 2016-008 (Japanese).

[7] T. Ashida, et al., Inspection and Repair Techniques in Reactor Vessel of the Experimental Fast Reactor Joyo—Development of Devices for Retrieving Bent MARICO-2 Subassembly and Completion of Retrieval Work, 2017. JAEA-Technology 2017-024 (Japanese).

[8] Japan Atomic Energy Agency, Joyo User's Guide, 2008.

[9] M. Inoue, et al., Power-to-melts of uranium-plutonium oxide fuel pins at a beginning-of-life condition in the experimental fast reactor JOYO, J. Nucl. Mater. 323 (2003) 108–122.

[10] T. Soga, et al., Irradiation test of fuel containing minor actinides in the experimental fast reactor Joyo, J. Power Energy Syst. 2 (2) (2008).

[11] H. Kataoka, et al., Development of material irradiation rig with precision temperature control in experimental fast reactor JOYO, J. Nucl. Mater. 258–263 (1998) 677–681.

[12] K. Katsuyama, T. Nagamine, S. Matsumoto, S. Sato, High energy X-ray CT study on the central void formations and the fuel pin deformations of FBR fuel assemblies, Nucl. Inst. Methods Phys. Res. B 255 (2007) 365–372.

[13] A. Ishimi, K. Katsuyama, K. Maeda, T. Nagamine, T. Asaga, H. Furuya, Upgrading of X-ray CT technology for analyses of irradiated FBR MOX fuel, J. Nucl. Sci. Technol. 49 (12) (2012) 1144–1155.

[14] A. Ishimi, K. Katsuyama, H. Furuya, Restructure behavior analysis in fast breeder reactor MOX fuel by X-ray CT, J. Nucl. Sci. Technol. 56 (11) (2019) 981–987.

[15] T. Uwaba, M. Ito, J. Nemoto, S. Ichikawa, K. Katsuyama, Verification of the FBR fuel bundle–duct interaction analysis code BAMBOO by the out-of-pile bundle compression test with large diameter pins, J. Nucl. Mater. 452 (2014) 552–556.

[16] K. Maeda, S. Sasaki, M. Kato, Y. Kihara, Radial redistribution of actinides in irradiated FR-MOX fuels, J. Nucl. Mater. 389 (2009) 78–84.

[17] S. Sasaki, T. Tanno, K. Maeda, Application of FE-SEM to the measurement of U, Pu, Am in the irradiated MA-MOX fuel, in: Proc. 54th Annual Meeting on Hot Laboratories and Remote Handling (HOTLAB 2017), Mito, Japan, 2017.

[18] K. Katsuyama, A. Ishimi, K. Maeda, T. Nagamine, T. Asaga, Helium release from the uranium-plutonium mixed oxide (MOX) fuel irradiated to high burn-up in a fast breeder reactor (FBR), J. Nucl. Mater. 401 (2010) 86–90.

[19] S. Sasaki, K. Maeda, H. Furuya, Irradiation performance of sodium-bonded control rod for the fast breeder reactor, J. Nucl. Sci. Technol. 55 (3) (2018) 276–282.

[20] T. Hirosawa, I. Sato, Burnup dependence of melting temperature of FBR mixed oxide fuels irradiated to high burnup, J. Nucl. Mater. 418 (2011) 207–214.

[21] S. Koyama, M. Osaka, T. Sekine, K. Morozumi, T. Namekawa, M. Itoh, Measurement of burnup in FBR MOX fuel irradiated to high burnup, J. Nucl. Sci. Technol. 40 (12) (2003) 998–1013.
[22] I. Sato, T. Nakagiri, T. Hirosawa, S. Muyahara, T. Namekawa, Fission products release from irradiated FBR MOX fuel during transient condition, J. Nucl. Sci. Technol. 40 (2) (2003) 104–113.
[23] I. Sato, K. Katsuyama, Y. Arai, Fission gases and helium gas behavior in irradiated mixed oxide fuel pin, J. Nucl. Mater. 416 (2011) 151–157.
[24] K. Tanaka, I. Sato, T. Onishi, T. Ishikawa, T. Hirosawa, et al., Release behavior of radionuclides from MOX fuels irradiated in a fast reactor during heating tests, J. Nucl. Mater. 536 (2020) 152199.
[25] K. Tanaka, S. Miwa, S. Sekine, H. Yoshimochi, H. Obayashi, S. Koyama, Restructuring and redistribution of actinides in Am-MOX fuel during the first 24 h of irradiation, J. Nucl. Mater. 440 (2013) 480–488.
[26] H. Yoshimochi, M. Nemoto, K. Mondo, S. Koyama, T. Namekawa, Fabrication technology for MOX fuel containing AmO_2 by an in-cell remote process, J. Nucl. Sci. Technol. 41 (8) (2004) 850–856.
[27] I. Shibahara, N. Akasaka, S. Onose, H. Okada, S. Ukai, Swelling of advanced austenitic stainless steels developed for the environment of heavy neutron exposure, J. Nucl. Mater. 212–215 (1994) 487–491.
[28] T. Uwaba, S. Ukai, The Formulation of the PNC316 Swelling Design Equation, 2003. JNC TN9400 2003-007 (in Japanese).
[29] A. Uehira, S. Ukai, T. Mizuno, T. Asaga, E. Yoshida, Tensile properties of 11Cr-0.5Mo-2W, V, Nb, stainless steel in LMFBR environment, J. Nucl. Sci. Technol. 37 (9) (2000) 780–786.
[30] Y. Yano, R. Ogawa, S. Yamashita, S. Ohtsuka, T. Akasaka, et al., Effects of neutron irradiation on tensile properties of oxide dispersion strengthened (ODS) steel claddings, J. Nucl. Mater. 419 (2011) 305–309.
[31] Y. Yano, S. Ohtsuka, S. Yamashita, R. Ogawa, M. Sekine, et al., Evaluation of Irradiation Behavior on Oxide Dispersion Strengthened (ODS) Steel Claddings Irradiated in Joyo/CMIR-6, JAEA-Research 2013-030 (in Japanese), 2013.
[32] Y. Yano, T. Yoshitake, Y. Abe, The Evaluation of Irradiation Effect on Charpy Properties of PNC-FMS and the ODS Martensitic Steel Irradiation in Joyo/CMIR-5 and MARICO-1, JNC TN9400 2003-028 (in Japanese), 2002.
[33] T. Maruyama, T. Kaito, S. Onose, I. Shibahara, Change in physical properties of high-density isotropic graphites irradiated in the "JOYO" fast reactor, J. Mater. Mater. 225 (1995) 267–272.
[34] S. Yamashita, Y. Yano, S. Ohtsuka, T. Yoshitake, T. Kaito, S. Koyama, K. Tanaka, Irradiation behavior evaluation of oxide dispersion strengthened ferritic steel cladding tubes irradiated in JOYO, J. Mater. Mater. 442 (2013) 417–424.
[35] Y. You, K. Yoshida, T. Inoue, T. Yano, Helium bubbles and trace of lithium in B_4C control rod pellets used in JOYO experimental fast reactor, J. Nucl. Sci. Technol. 55 (6) (2018) 640–648.

Prototype reactor Monju

3

Taira Hazama
Sector of Fast Reactor and Advanced Reactor Research and Development,
Japan Atomic Energy Agency, Ibaraki, Japan

Chapter outline

The prototype fast breeder reactor "Monju" is a loop-type sodium-cooled fast neutron reactor with an electric power of 300 MW developed in Japan. The initial criticality was achieved in 1994, and the reactor power reached 40% of the rated power in 1995. After a long-term shutdown period, the operation of Monju restarted in 2010. In 2016, Monju was transferred to decommissioning stage without achieving the full power operation.

This chapter summarizes the design features of Monju, R&D activities to establish the design, experiences in operation and maintenance, and technological achievements through the commissioning.[1]

[1] This chapter is an abridged edition of ref. [1].

Sodium Fast Reactors. https://doi.org/10.1016/B978-0-12-824076-2.00003-1
Copyright © 2022 Elsevier Inc. All rights reserved.

(a) *Design features:* Monju is a MOX-fueled SFR with three independent cooling systems. Its major design features are described for reactor core, reactor structure, cooling systems, fuel handling systems, and instrumentations.
(b) *R&D activities for Monju design:* Various R&D activities were carried out to establish the Monju design under all-Japan cooperation since the beginning of the Monju design work in 1968. Design history and major R&D activities to develop Monju components are described.
(c) *Operation and maintenance:* A large number of events mainly on maintenance and repair have been experienced in Monju since the completion of component installation in 1991. Major events and introduced measures are described.
(d) *R&D achievements in Monju:* The commissioning of Monju was carried out to perform final adjustments and confirmation of all the plant systems, as well as validate Monju design and identify issues for future SFRs. Major achievements in commissioning are described for accomplished tests, design validation, and safety evaluation.

3.1 Design features

3.1.1 Plant overview

The major specifications of Monju are listed in Table 3.1. A bird's-eye view of the Monju plant is shown in Fig. 3.1 and its reactor system in Fig. 3.2. Monju has three independent sodium cooling systems: Each consists of a primary heat transport system (PHTS) and a secondary heat transport system (SHTS), and heat is transferred to the power-generation steam turbine of a water-steam system via the steam generators (SGs).

Table 3.1 Major specifications of Monju.

Reactor type	Sodium-cooled loop-type
Thermal power	714 MW
Electric power	~280 MW
Fuel	Plutonium-uranium mixed oxide
Burnup of fuel	80 GWd/t (average of discharged fuel)
Fuel cladding	SUS316-equivalent steel
Cladding temperature	Max 675°C (mid-wall)
Breeding ratio	1.2
Number of cooling loops	3
Location of circulation pump	Hot-leg installation
Reactor temperature	397°C/529°C (inlet/outlet)
Secondary sodium temperature	505°C/325°C (hot-leg/cold leg)
Main steam conditions	127 kg/cm^{2}G, 483°C
SG type	Helically coiled, separate type
SG layout	Concentrated arrangement
Refueling scheme	Single rotating plug with a fixed arm
Refueling interval	~6 months
Decay heat removal	Cooling in the secondary loop
Method to ensure coolant level	Pipe routing at high elevation with guard vessels

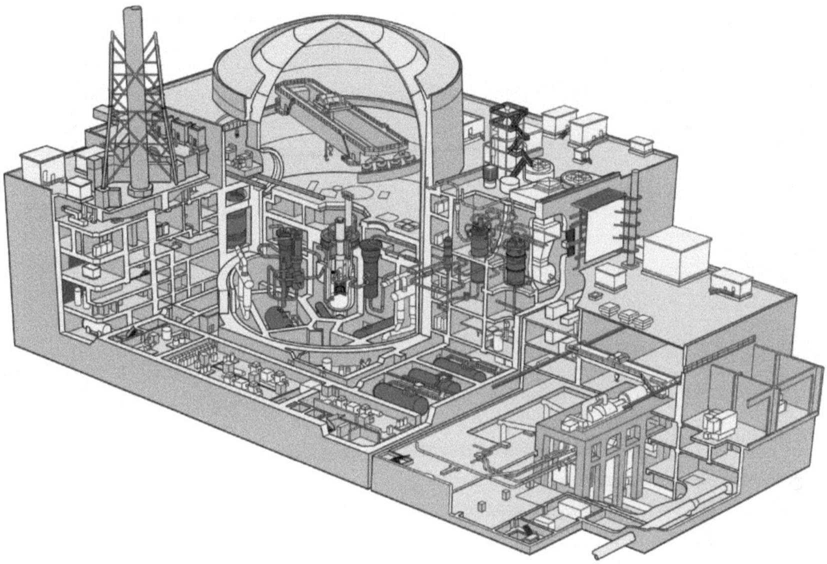

Fig. 3.1 Monju plant.

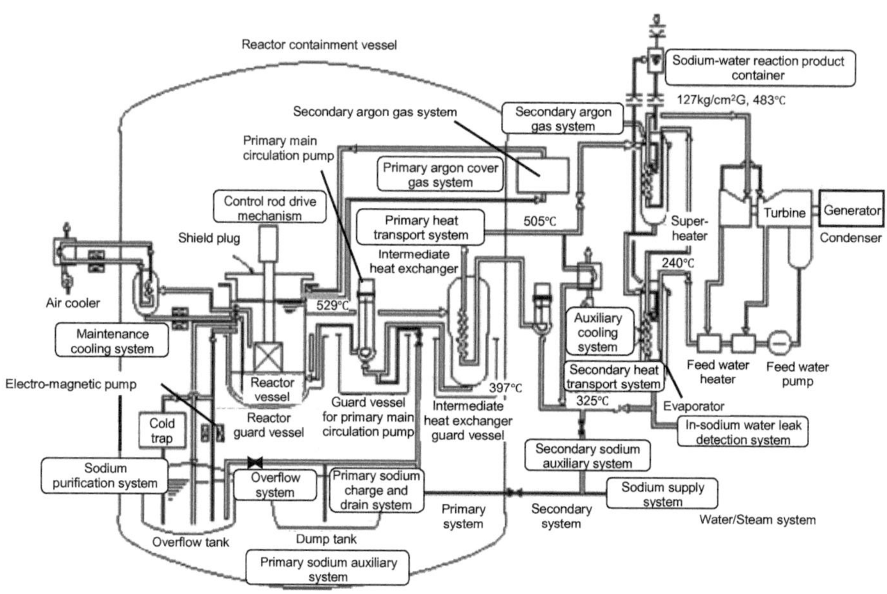

Fig. 3.2 Reactor and cooling system of Monju.

The primary sodium flows into the reactor vessel (RV) from the bottom shell at 397°C, flows out from the upper shell at 529°C after being heated in the reactor core, and then exchanges heat with the secondary sodium in the intermediate heat exchanger (IHX). The secondary sodium is heated from 325°C to 505°C and exchanges heat with water and steam in the SGs to generate superheated steam of 483°C and a pressure of 127 kg/cm^{2}G that is supplied to the turbine generator.

The reactor system consists of:

- An RV that contains a core,
- A guard vessel to maintain the coolant level for safe core cooling in case of sodium leak from the PHTS piping,
- A shield plug that shields radiation and heat from the core and maintains an argon gas atmosphere above the sodium surface,
- Core internals that support the core and provide flow distribution for each core component, and
- An upper core structure that supports control rod drive mechanisms (CRDMs) and core instrumentation wells.

The cooling system consists of the following:

- PHTS equipment (circulation pump, IHX, guard vessel, etc.),
- SHTS equipment (circulation pump, SG, etc.),
- An auxiliary cooling system (air cooler, etc.),
- A sodium-water reaction product container,
- Reactor cooling system auxiliary equipment (auxiliary primary sodium equipment, auxiliary secondary sodium equipment, maintenance cooling system equipment, primary and secondary argon gas system equipment, etc.), and
- Water-steam system equipment including steam turbines.

3.1.2 Reactor core

The reactor core consists of 198 fuel subassemblies each containing 169 pins with plutonium and uranium mixed oxide (MOX) fuel, which are surrounded by 172 radial blanket subassemblies containing depleted uranium dioxide.

3.1.2.1 Core design

(a) *Core configuration*

The Monju core subassemblies are of four kinds: core fuel, control rod, radial blanket fuel, and the neutron shield (Fig. 3.3). The core fuel region consists of inner and outer regions with different plutonium contents. Core fuel subassemblies with a higher plutonium content are loaded in the outer region to flatten the radial power distribution.

The core fuel pins contain blanket fuel pellets above and below the core fuel pellets. The radial blanket subassemblies are loaded in the surroundings of the core fuel region. With this arrangement, a breeding ratio of 1.2, a design target, is ensured, and neutron leakage from the core is reduced. Outside the blanket regions, the neutron shielding subassemblies are loaded to further reduce neutron leakage.

The maximum burnup level of the discharged core fuel subassemblies was designed to be around 80,000 MWd/t on average. However, it was tentatively set at 55,000 MWd/t until the irradiation performance of the anti-swelling properties of SUS316-equivalent stainless steel is demonstrated. The former core is named high-burnup core and the latter low-burnup core.

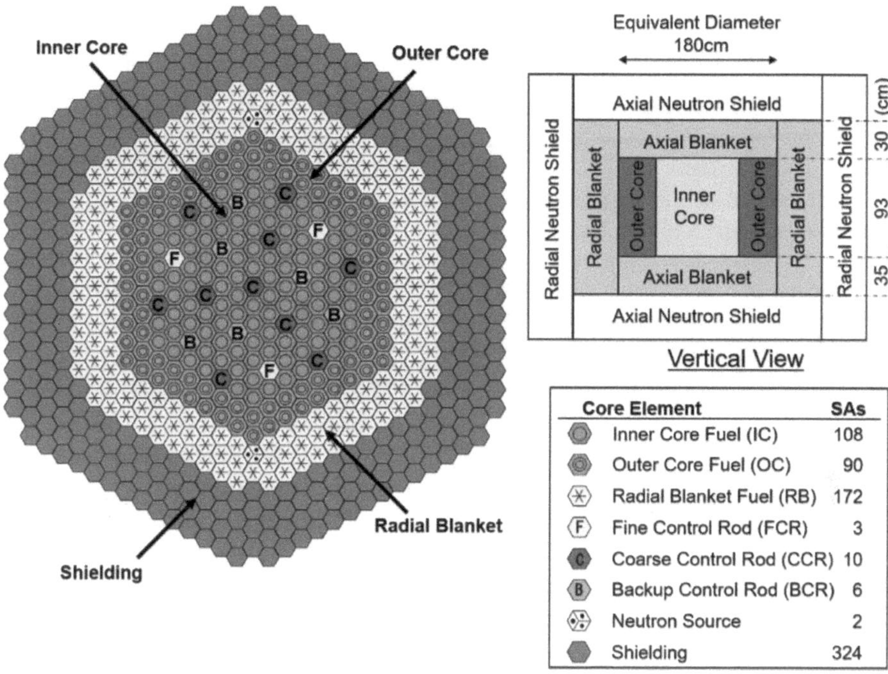

Fig. 3.3 Monju core layout.

(b) *Reactivity control*

Reactivity is exclusively controlled by control rods. The control rods are classified into regulating rods and backup shutdown rods. The regulating rods are further classified into coarse and fine control rods. A helium-bond-type control rod element is adopted, which contains B_4C pellets in a helium-gas-filled stainless steel cladding tube. A cluster of 19 control rod elements is installed in the protection tube that moves up and down in the control rod guide tube. An anti-vibration structure is adopted in the regulating rods to suppress power fluctuation. Namely, six-round bumps are circumferentially attached to the lower part of the protection tube, the design of which is based on out-of-pile hydraulic tests.

The power coefficient, an integrated value of the reactivity coefficients, such as Doppler, fuel temperature, structural material temperature, coolant temperature, and core support structure temperature, was designed to be negative to ensure safety.

(c) *Flow distribution*

The heat generation density in fuel varies with the fuel composition and location in the core. Then, the coolant flow distribution is determined to flatten the subassembly outlet temperatures and thereby suppress the peak cladding temperature at the rated reactor power. The coolant flow is distributed to 11 zones: 8 core fuel zones and 3 blanket fuel zones.

3.1.2.2 Fuel subassembly design

Fast reactor fuel generally has the following characteristics compared to light water reactors (LWRs).

- Plutonium (major fissile nuclides: ^{239}Pu and ^{241}Pu) is used as the main fuel material. The plutonium and fissile plutonium contents are much higher than those of LWR MOX fuel.
- The absorption of thermal neutrons by core material is insignificant. Therefore, cladding made of stainless steel can be used instead of Zircaloy cladding used in LWRs.
- A large volume of fission gas plenum must be provided inside a fuel pin to accommodate fission gas pressure buildup due to the high fuel burnup.
- Wire spacers are used for densely arranged fuel pins to prevent the pins from coming into contact with one another and to secure space for coolant flow.

A structural drawing of the core fuel subassembly is shown in Fig. 3.4. The core fuel subassembly is constituted of 169 core fuel elements, wrapped with wire spacers, and arranged in a regular triangular geometry, which is mounted in a regular hexagonal wrapper tube. The wrapper tube is connected with a handling head at the top and an entrance nozzle at the bottom. Each fuel element contains a stack of MOX fuel pellets (active core fuel) and UO_2 pellets (upper and lower blanket fuel) in a cladding tube.

The isotopic composition of plutonium depends on a burnup level of spent fuel reprocessed and a cooling period after reprocessing to extract plutonium. In addition, the composition varies with the period from fuel fabrication to reactor operation due to the ^{241}Pu decay. To cope with these composition changes, the "Equivalent Fissile

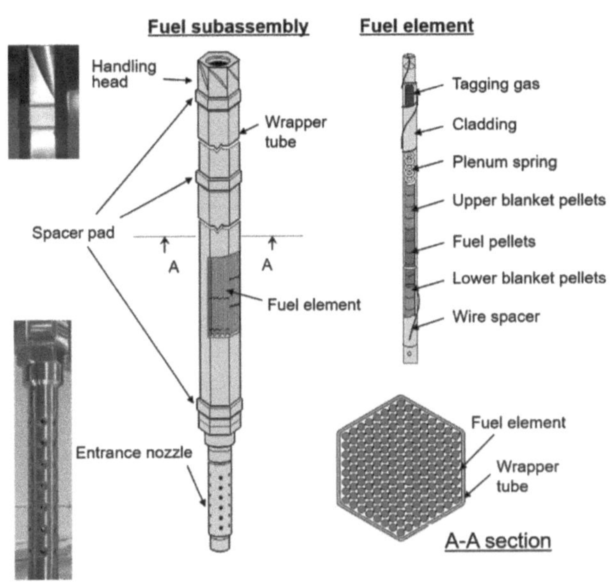

Fig. 3.4 Monju fuel subassembly.

Content Method" was introduced when setting the plutonium content in fuel fabrication. In this method, the individual reactivity effects of the isotopes of plutonium, uranium, and americium are converted to the reference reactivity effect of ^{239}Pu.

Mechanical loading due to fuel-cladding mechanical interaction (FCMI) is one of the concerns in fuel design. In the early Monju design stage, available irradiation data were limited, especially in the high burnup range. Therefore, low pellet density with 85% theoretical density (TD) was conservatively adopted to accommodate fuel swelling due to burnup and, thereby, mitigate the influence of FCMI.

The Monju fuel was designed following the five design criteria:

- Maximum fuel centerline temperature: Lower than the melting point of MOX fuel.
- Cladding strain: Less than 7% increase in the cladding outer diameter.
- Creep lifetime: Cumulative damage fraction (CDF) is less than 1, where creep deformation is caused by the tensile stress due to the cladding internal pressure.
- Cladding stress: Lower than the design allowable stress per the American Society of Mechanical Engineers (ASME) standards.
- Cumulative fatigue cycle: Less than the design fatigue lifetime, considering the CDF.

In fuel centerline temperature evaluation, fuel restructuring, which is a characteristic of fast reactor fuel, is considered. Because of the high temperatures and steep radial temperature gradient in the fuel, noticeable fuel pellet restructuring occurs through evaporation and condensation of fuel material as depicted in Fig. 3.5. Fuel restructuring has the important effect of lowering the fuel centerline temperature owing to the formation of a central hole and the increased thermal conductivity of fuel due to fuel densification in the restructured regions. This effect was incorporated in the fuel

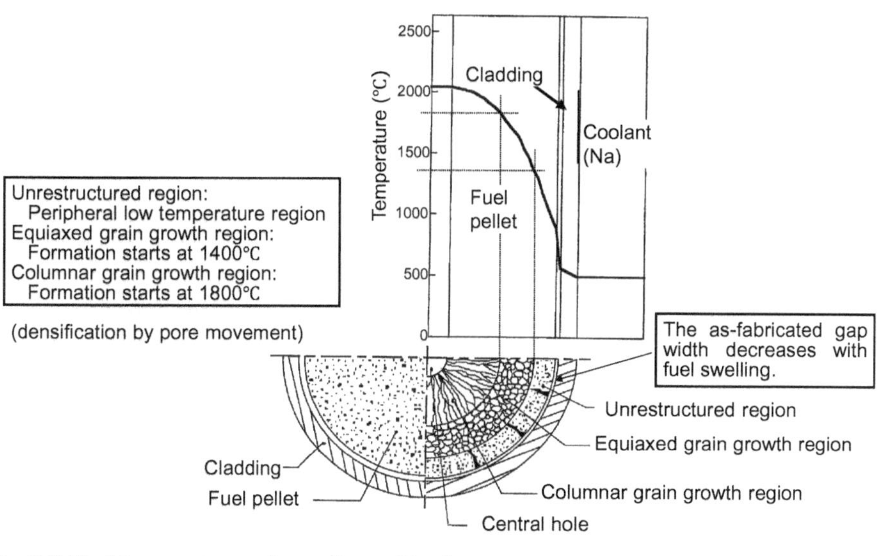

Fig. 3.5 Fuel temperature and metallographic change.

design and the reactor startup procedure. Namely, in the startup after new fuel loading, the reactor power is increased slowly such that sufficient fuel restructuring takes place.

The gap between the fuel pellets and the cladding is filled with helium gas in fuel fabrication and with fission gas generated and released during burnup. Evaluation of the heat transfer performance of the gap (i.e., gap conductance) is an important task. In the initial design, the evaluation was based on a power-to-melt (PTM) test performed at EBR-II in the United States, later based on PTM test and instrumented fuel element irradiation tests at Joyo. The gap conductance was reevaluated based on these tests, and the value used in evaluating the maximum fuel temperature in the initial design was confirmed to be sufficiently conservative.

3.1.3 Reactor structure

The reactor structure is shown in Fig. 3.6. The RV contains the core and core internals, and the reactor guard vessel encloses the RV and lower part of PHTS piping. The CRDM and a refueling system are installed on the shield plugs.

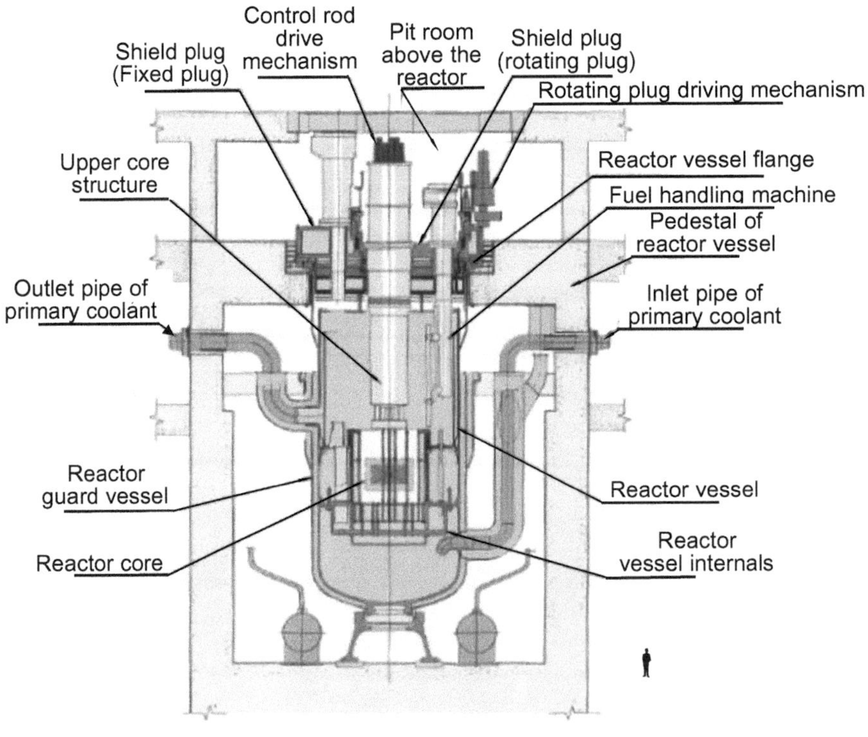

Fig. 3.6 Reactor structure.

3.1.3.1 Reactor vessel

The RV is a large, SUSF304 (forged material), a thin-walled, cylindrical vessel with an outer diameter of 7 m, a height of 18 m, and a thickness of 50 mm. The vessel is supported at the upper flange, and it has seismic support at the bottom endplate. The lower plenum forms coolant flow paths to supply coolant to the fuel subassembly and other core elements (Fig. 3.7). The coolant is supplied from the high- and low-pressure plenums below the core. The RV inlet and outlet coolant temperatures at the rated power operation are 397°C and 529°C, respectively. The space above the sodium in the RV is filled with argon gas. The sodium level is maintained within a specified range by the primary sodium overflow system.

The RV was designed with careful consideration of the characteristics of SFRs, such as lower pressures, higher temperatures, and hence more severe thermal stress during plant thermal transients. In particular, design measures such as thin-walled piping must be taken to cope with thermal stresses. On the other hand, rigid structures were requested from the perspective of the seismic design. The important task in

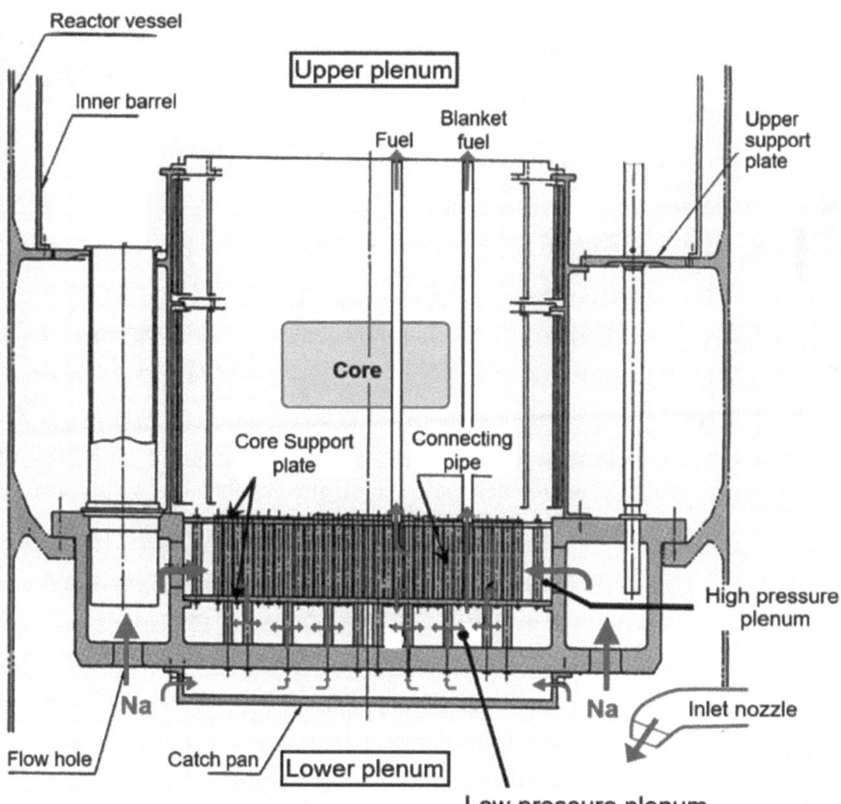

Fig. 3.7 Coolant flow paths to the reactor.

RV design, therefore, was to harmonize and optimize these two competing requirements.

Adoption of the "hot-vessel-type" structure is another unique feature of the Monju RV. It makes the RV structure simple and compact by not installing a function to cool the RV wall, which was generally adopted in tank-type SFRs.

Various R&D activities related to thermal hydraulics, seismic design, material strength, etc., were performed to establish the design methods and optimize the structure.

(a) *Selection of structural materials*

Since the main load on the RV is thermal stress, austenitic stainless steel (SUSF304) was selected to prevent excessive inelastic deformation caused by thermal stress and creep-fatigue fracture. The material's favorable properties include elevated-temperature strength (creep strength, creep-fatigue strength, etc.), corrosion resistance in sodium, and resistance to neutron irradiation.

(b) *Measures against thermal transients near sodium surface*

Upon reactor startup or shutdown, core outlet coolant temperature change is large and rapid, while that in the cover gas layer is small and slow. This difference together with the small heat-transfer rate from the gas to the RV wall would cause a steep temperature gradient in the axial direction around the sodium surface of the RV. This would generate large thermal stress on the RV wall. As a measure to prevent creep-fatigue damage due to this thermal stress, a two-sodium-level system with a protection liner was developed (Fig. 3.8). This system prevents the generation of excessive thermal stress in the RV wall by controlling the sodium level outside the protection liner. For example, in the reactor startup, the sodium level outside is initially set to the same level as inside during the first half period of the reactor startup procedure, and then, the sodium level is lowered by about 0.6 m. This operating procedure would shift the peak position of the thermal stress to adequately mitigate the overall influence on the RV.

(c) *Measures against thermal shock*

Upon a reactor trip, a large thermal shock may develop in the RV wall and outlet nozzle because of a rapid decrease in core outlet sodium temperature. To mitigate the shock, the inner barrel was installed in the upper plenum, and the RV wall was entirely covered with the thermal shield plate over the region from the top of the protection liner to the upper part of the inner barrel support. In the middle plenum, a measure was also taken to smoothen the axial temperature distribution in the RV wall by providing a bypass flow from the lower plenum through the degassing holes.

(d) *Seismic support considering thermal expansion*

To enhance the seismic capability, the lower support was attached below the bottom endplate of the RV and was fitted in the lower support structure on the RV cavity floor to constrain horizontal RV movement during an earthquake (Fig. 3.9). This seismic support structure was designed to appropriately accommodate thermal expansion associated with reactor power.

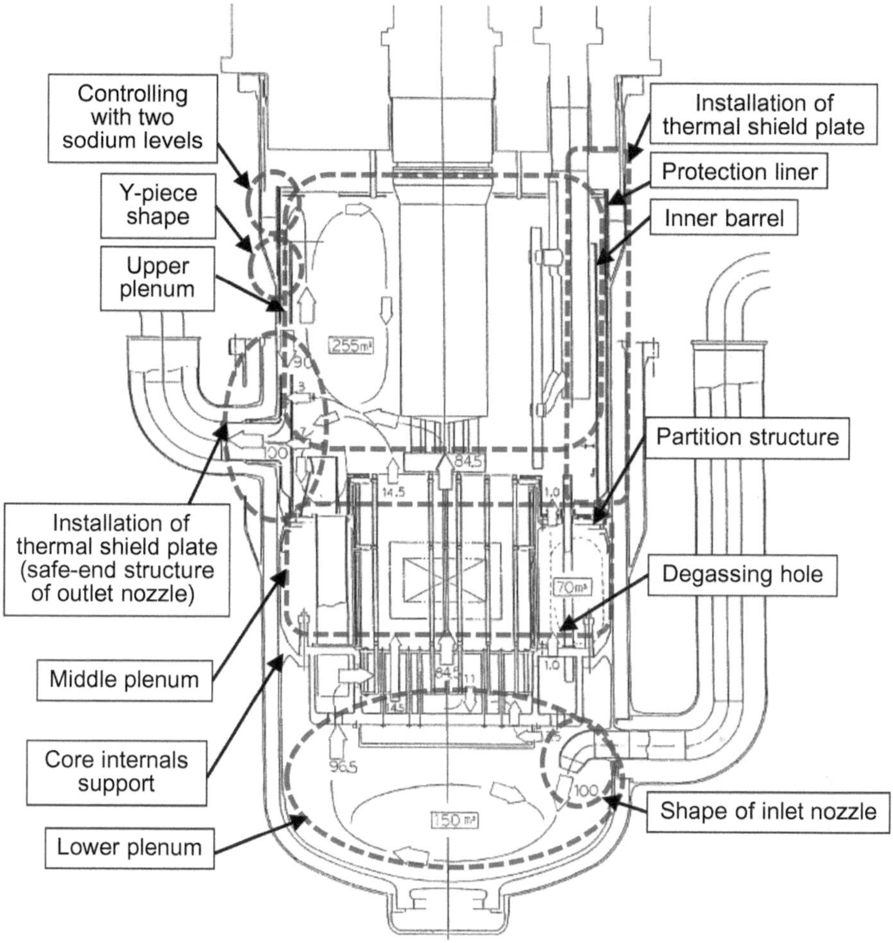

Fig. 3.8 Measures against thermal shock.

(e) *Measure to prevent fuel subassembly lifting*

To prevent the lifting of the core fuel subassemblies due to upward coolant flow, a hydraulic hold-down mechanism is provided. Namely, a downward hold-down force is ensured by supplying coolant from the high-pressure plenum to the core fuel subassemblies. A part of the coolant flows downward to the low-pressure plenum located below the subassemblies as shown in Fig. 3.7.

(f) *Feature in manufacture and installation*

Ring forging was adopted in the manufacture of the RV shell to avoid a longitudinal weld line near sodium surface where large thermal stress may occur, minimizing the

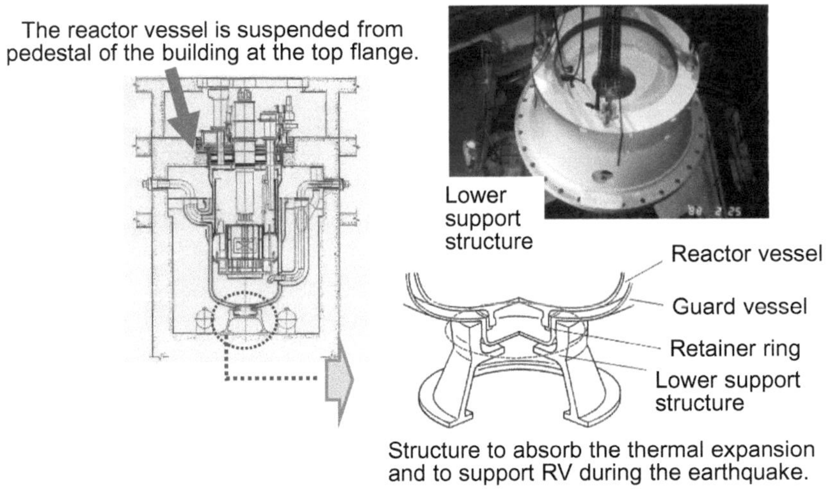

Fig. 3.9 Seismic structure of RV support.

number of weld lines from the perspectives of in-service inspection and reliability, and obtaining geometrical accuracy including roundness.

The forging of such a large-diameter thin-walled shell as the RV had never been experienced before. Furthermore, the weight of the used stainless steel ingot was about twice the largest ingot of 120 tons ever produced. Since ingot quality may significantly affect product quality, large ingot production technology with a high level of cleanliness was further advanced using a special method called "unified pouring," in which ununiformed solidification was prevented by pouring molten steel heats in multiple steps.

In assembling the ring parts of the RV, a large-scale welding machine capable of high-quality and low-strain automatic welding was developed, and the parts were assembled vertically to minimize deformation and to ensure good welding workability, realizing sufficient accuracy in the manufacturing and installation.

After the installation at the site, the straightness of the RV from the upper flange to the lower support was confirmed within 1 mm for the total height of 18 m. Concerning the installation accuracy of core internals, which is important for control rod insertability, the misalignment of the outermost control rod guide tubes was at most 1.9 mm at the upper core support frame level, sufficiently satisfying the design allowance.

3.1.3.2 Shield plug

The shield plug (Fig. 3.10) installed at the upper part of the RV is of a single rotating type and consists of the fixed plug and rotating plug. Each plug consists of the upper plate, shield shell, and dip plate. The dip plate was designed to suppress the ruffling of sodium surface in the RV during power operation. The fixed plug has a maximum diameter of 9.5 m and a total height of 2.0 m.

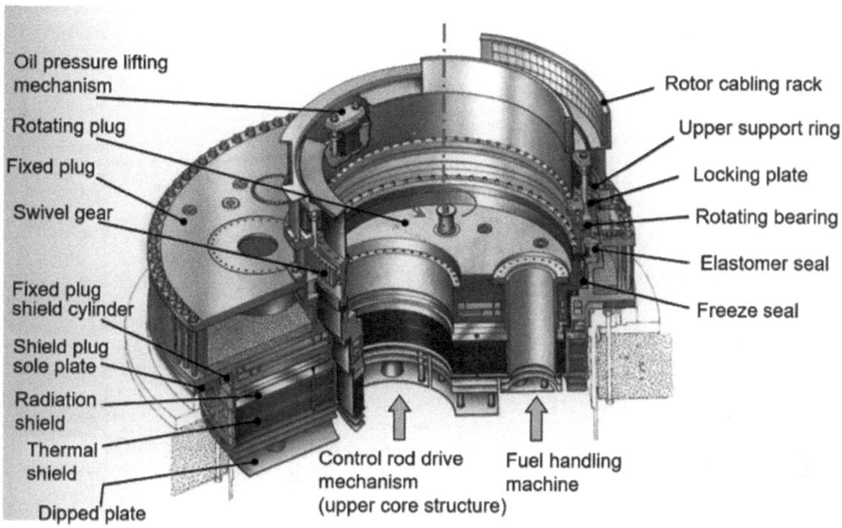

Fig. 3.10 Shield plug.

The upper core structure (Fig. 3.11) is installed at the center of the shield plug. It is a top-supported cylinder with a total height of 14 m and a diameter of 2.6 m. The structure consists of the upper plate, shield shell, middle shell, flow guide device, CRDM guide tubes, instrumentation wells, etc.

The shield plug has the following features:

(a) *Boundary function for reactor cover gas*

The boundary of the reactor cover gas around the fixed and rotating plugs was sealed by low-melting-point alloy (freeze seal metal with a melting point of 125°C) with a backup inflatable tube seal (elastomer seal). The freeze seal metal was developed to have required performances, such as resistance to oxidization, prevention of segregation upon solidification, and good wettability on and adherability with a structural surface. The freeze seal metal is kept in a solidified state when the rotating plug is not operated, such as during reactor operation. When the operation is required, the freeze seal metal is melted to allow the plug rotation, still maintaining the sealing function.

(b) *Functions to shield radiation and heat from the reactor core*

A laminated thin-plate structure consisting of stainless steel, carbon steel, polyethylene, etc., was adopted for the fixed plug to shield radiation from the core. In addition, to decrease radiation streaming, a stepped (offset) structure was adopted for the gap between the fixed and rotating plugs and between the penetrations of the installed components.

As for the thermal shield function, it is required to minimize heat radiation to above the RV and maintain the temperature of the upper surface of the shield plug at 70°C or less. Thus, a structure with multiple thin stainless steel plates was adopted taking

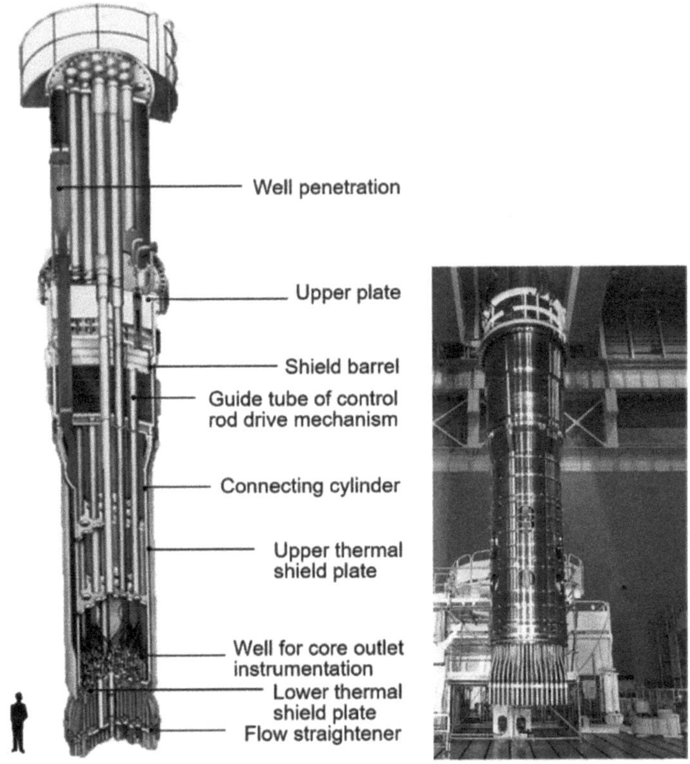

Fig. 3.11 Upper core structure.

advantage of the thermal insulation effect of gas layers. A cooling layer of forced circulation of nitrogen gas is provided as well. The pitch between the multilayered plates was optimized considering the properties: A thicker gas layer would decrease the thermal insulation effect owing to the occurrence of natural convection, while a thinner gas layer would negate the thermal insulation effect owing to the bridges formed by sodium vapor deposition.

(c) *Measures against thermal striping in the upper core structure*

The coolant temperature difference at the outlets between core fuel subassemblies and control rod assemblies would become as large as 160°C. This may induce temperature fluctuation, a phenomenon called "thermal striping." It generates high-cycle thermal stress at the near surface of the upper core structure located above the core. Therefore, an armor structure was adopted in which the components made of SUS304 as the tube plate of the flow guide device and the control rod upper guide tube are covered by Alloy 718 having excellent high-cycle fatigue strength.

(d) *Feature in manufacture and installation*

High accuracies are required in the manufacture and installation of the shield plug to ensure the performance of the fuel handling system and the insertability of control rods during an earthquake.

The misalignment between the bottom end of the control rod upper guide tube (in the upper core structure) and the control rod guide tube at the upper core support plate level was confirmed to be within 2.52 mm. This misalignment satisfies the target value of the integrated manufacturing and installation accuracy and does not exceed 5.51 mm even if the error of stop position of the rotating plug and a possible shift during jack-up and down are considered. Concerning the levelness, the tilt of the upper plate surface of the fixed plug, the largest component, was confirmed to be 0.03 mm/m, almost perfectly horizontal, and satisfied the target value of 0.12 mm/m.

3.1.3.3 Control rod drive mechanism

Structures of the CRDM of each shutdown system are shown in Fig. 3.12. The main and backup shutdown systems are mutually independent including the plant protection systems and are different in the mechanisms to accelerate and disconnect the control rods during a scram (the manufacturers were different as well) to ensure diversity. Each control rod has an independent CRDM as well: 3 CRDM units for FCRs, 10 for CCRs, and 6 for BCRs. The CRDM is comprised of the driving part, upper guide tube, and extension tube, and the sealing against the RV cover gas is formed by the bellows in the upper guide tube.

The bellows of CRDM for FCRs and CCRs must extend over a long stroke of 1100 mm within 1 s in case of reactor scram. Since sliding parts are immersed in high-temperature sodium for a long time, surface hardening was applied using Stellite and Alloy 718 to prevent self-welding and sticking of the removable parts, and seizing and galling of the sliding parts.

3.1.4 Cooling systems

The PHTS consists of three independent loops, each consisting of an IHX, PHTS circulation pump and its overflow column, piping, and valves. The PHTS piping is routed at high elevations, and the guard vessels are provided for the IHX and PHTS circulation pump, with the main piping routed at low elevations to connect these components. This arrangement ensures a sodium level in the RV required for core cooling in case of a sodium leak from the PHTS piping.

The SHTS consists of three independent loops, each consisting of the SG (evaporator and superheater), SHTS circulation pump and its overflow column, auxiliary cooling system, piping, and valves. The auxiliary cooling system removes decay heat from the core after the reactor trip.

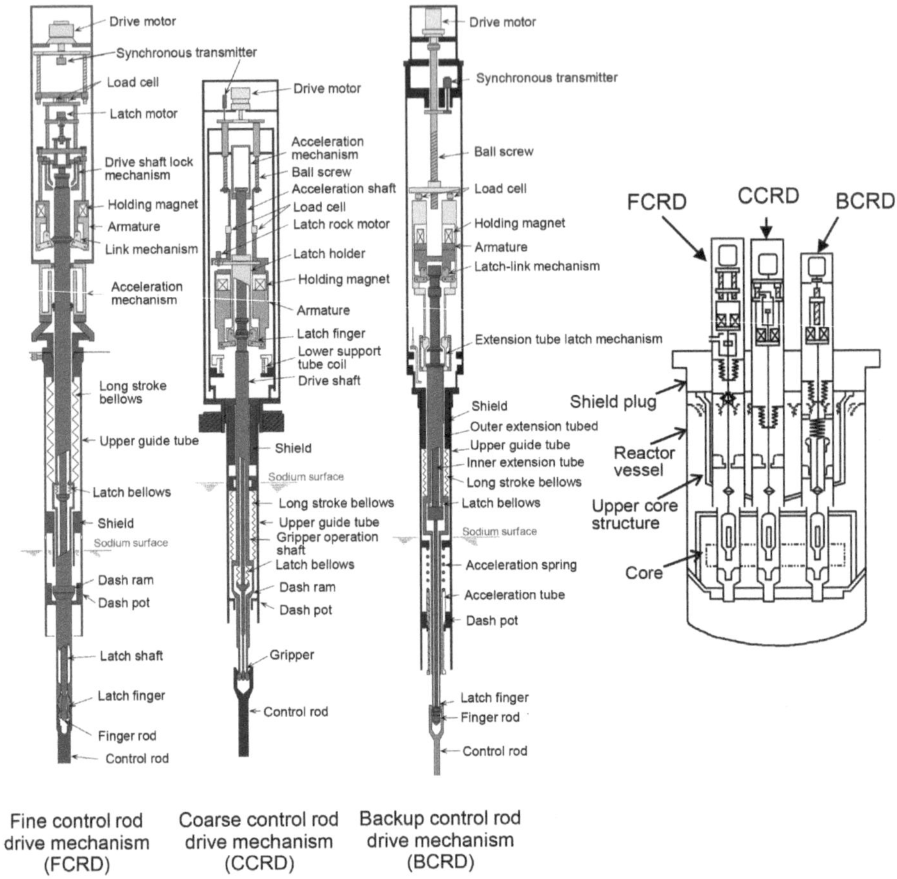

Fig. 3.12 Control rod drive mechanism.

3.1.4.1 PHTS circulation pump

The PHTS circulation pump (main circulation pump) is a safety-related active component designed to supply coolant to remove heat generated in the core, decay heat, and other residual heat (Fig. 3.13).

The pump body has a shell diameter of 1.75 m and a height of about 10 m (including the upper shaft seal). Coolant circulation is driven by the main motor (2000 kW) during the rated power operation and by the pony motor (22 kW) during decay heat removal operation after reactor shutdown. The flow rate can be controlled between 49% and 100% by the main circulation pump MG set (variable frequency power supply system of the motor-generator) to maintain the RV inlet/outlet temperature difference almost constant, regardless of reactor power.

The main circulation pump consists of a pump body and outer casing. It has a structure allowing the extraction of the pump body from the outer casing welded to the PHTS piping for easier maintenance.

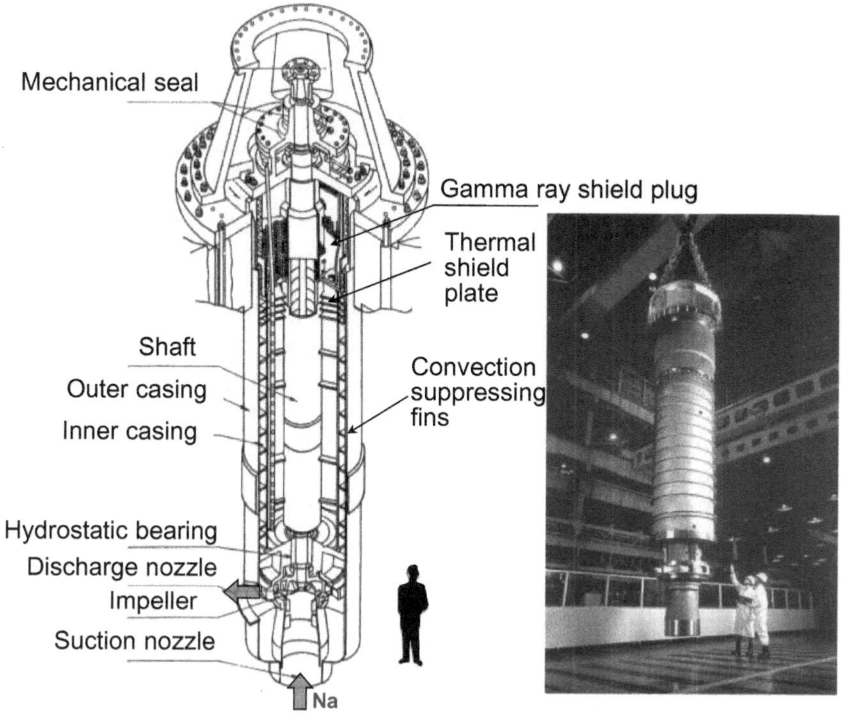

Fig. 3.13 Primary main circulation pump.

In the cover gas space above the sodium surface in the pump, vertical natural convection appears in the gap between the inner and outer casings. This causes a circumferential temperature gradient, leading to thermal deformation of the casing, and may generate excessive load on the pump drive shaft. To prevent this, natural convection suppressing fins (rectangular baffle plate) were installed in the biological and thermal shields. In addition, umbrella-like natural convection suppressing fins were provided in the shaft to prevent shaft bending due to the convection of encapsulated argon gas.

The main circulation pump is arranged in the cold leg of PHTS. Its shaft is considerably long (about 8.5 m) to accommodate a large liquid surface fluctuation and a gamma shield (about 1.4 m long). Hollow forged steel parts (diameter: 550 mm, thickness: 20 mm) were welded to the shaft at its upper, middle, and lower levels to reduce shaft weight and set the resonance rotation speed at a high value. Since a high-precision balance is required in the pump shaft, careful attention was paid to shaft bending in the manufacturing process by conducting the shaft runout test under the same high-temperature environment as the operating temperature (about 400°C). In the manufacturing process, the accuracy of the evenness of the hollow part was strictly controlled to ensure high-precision balance, and thick-walled parts were provided axially at some levels for easy correction of imbalance.

3.1.4.2 Intermediate heat exchanger

The IHX is a "vertically placed, free surface, straight-parallel countercurrent flow type" heat exchanger designed to transfer heat from the radioactive primary sodium to the non-radioactive secondary sodium (Fig. 3.14).

The primary sodium at a temperature of 529°C flows into the IHX through the inlet nozzle, rises into the space between the inlet plenum outer shell and the outer shroud, and flows into the heat transfer tube bundle through the inlet windows. In the tube bundle, sodium exchanges heat with the secondary sodium that flows inside the tubes while flowing down between the tubes and flows out from the outlet nozzle through the outlet windows at a temperature of 397°C. The secondary sodium flows into the IHX from the secondary sodium inlet at a temperature of 325°C, flows down in the downcomer, reverses direction in the lower plenum, rises in the heat transfer tubes while being heated to a temperature of 505°C by the primary sodium flowing outside the tubes, and flows out from the secondary sodium outlet nozzle through the upper plenum.

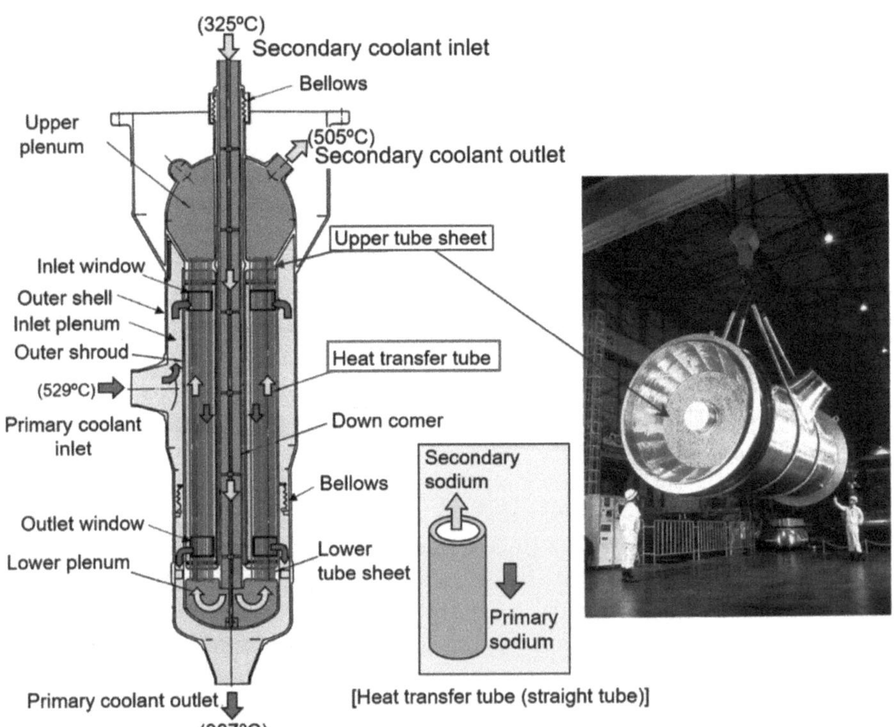

Fig. 3.14 Intermediate heat exchanger.

To prevent the radioactive primary sodium from leaking into the SHTS in case of a failure of the IHX heat transfer tube, the pressure in the IHX primary side was designed to be slightly lower than that in the IHX secondary side.

Since the IHX is operated at temperatures where significant creep occurs and may be used under severe thermal shock conditions caused by high-temperature sodium, sufficient measures were taken to mitigate thermal stress. To prevent component damage due to rapid thermal shock, thermal shield plates were installed at the primary-side inlet nozzle, support skirt, upper and lower tube plates, and secondary-side upper plenum.

The IHXs were manufactured in a factory dedicated to stainless steel products with controlled humidity and temperature under a quality control system specialized for Monju. The manufactured IHX thus generally satisfied all required dimensions within the tolerances and was then installed in Monju, so that its lower seismic support fitted the support structure of the guard vessel.

3.1.4.3 Steam generator

The SG is a component designed to generate high-pressure steam by transferring heat from the heated sodium transported from the core to the SG. Since many troubles were experienced in the SGs of earlier foreign SFRs, its performance was a key to plant availability as well as to plant safety.

To develop the Monju SG, a wide range of R&D activities from fundamental to demonstration tests were performed to accumulate test data on the thermal-hydraulic characteristics, dynamics, stable controllability, and durability, as well as operating records. The achievements were reflected in the design, manufacture, and operation of the actual component of Monju.

The Monju SG (Fig. 3.15) has the following features:

- A separated type SG consisting of an evaporator (EV) and superheater (SH) was adopted because material satisfying both anti-stress corrosion performance required in the EV and high-temperature strength required in the SH was still in the development stage and there was concern about the occurrence of flow instability on the waterside.
- A low-alloy steel (2¼Cr-1Mo steel) was selected for the EV in consideration of experience with boilers and anti-stress corrosion performance, while stainless steel (SUS321 steel) was selected for the SH for its superior high-temperature strength.
- Regarding the type of heat transfer tubes, a helical coil type was adopted for both the EV and SH, for its good heat transfer performance, good absorber for thermal expansion, and compactness.
- A once-through type (non-reheat cycle) SG was adopted. Recirculation of steam in the EV was examined in an early design phase because of its advantage in thermal efficiency; however, it was not adopted due to lower reliability and operation controllability with reference to the experience of earlier foreign plants.
- The feed-water inlet/outlet tube plate is placed above the sodium surface to mitigate the thermal shock at the tube plates and the pressure increase in case of a sodium-water reaction.
- The tube bundle is suspended at the upper plate or upper shell to allow withdrawal in inspection and repair.

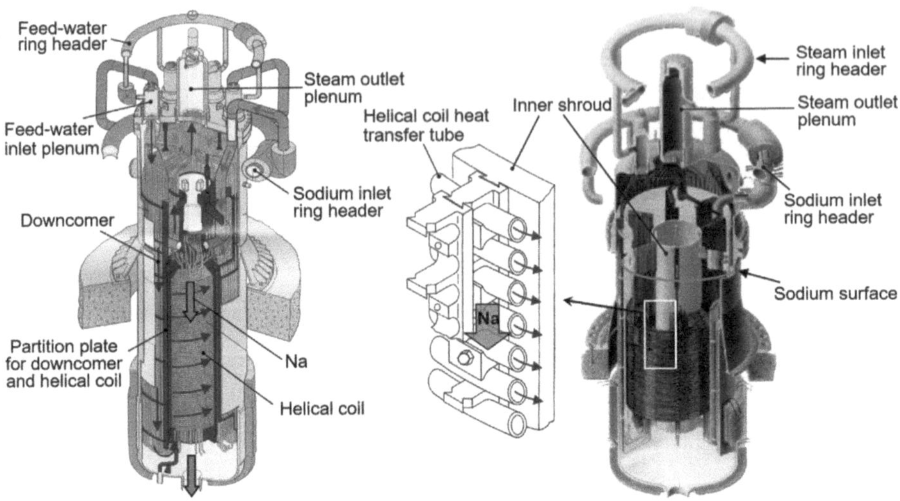

Fig. 3.15 Steam generator (left: evaporator, right: superheater).

High-temperature sodium (469°C) flows in via the inlet sodium ring header located at the upper part of the EV, flows downward between the tubes, and then flows out from the bottom end. Water heated at 240°C flows in via the feed-water ring header located at the upper part, flows downward through the downcomer, is heated while rising in the helically coiled heat transfer tubes, and then flows out via the steam outlet plenum in the form of superheated steam. The steam is further heated in the SH and then transported to the turbine.

The SG heat transfer tube is a component that forms the boundary between sodium and water steam, and hence, special attention was paid to the failure of a heat transfer tube. The main features are described below.

- No crevice (narrow gap) was provided on the inner (water-steam) side of the tubes to prevent crevice corrosion.
- Hydrogen meters were installed in sodium and cover gas to detect a small-scale water leak.
- A pressure relief plate was installed to release excessive pressure. In addition, a reaction product container was installed to suppress the emission of the reaction products to the air.
- Hydrogen generated in a sodium-water reaction is released and burnt in the air to prevent accumulation in the facility.
- The liner was installed over the inner surface of the shell to suppress wastage-type corrosion in case of a sodium-water reaction.

In manufacturing the Monju SG, special attention was paid to the welding of the tubes since many troubles in foreign SGs occurred at heat transfer tube welds. The main points to note are described below.

- Tube to tube plate welding was performed using butt welding, which allows for volume inspection.
- A long tube (EV: about 21.5 m, SH: about 32 m) was adopted for the heat transfer tube to minimize the number of welding zones.

- The heat transfer tubes for the EV were stored in full water containing hydrazine after completion of the water pressure test in a factory to suppress corrosion.

3.1.5 Fuel handling systems

The Monju fuel handling systems (Fig. 3.16) were based on those demonstrated in Joyo. Reduction of the component size around the RV was achieved by developing a compact fuel handling machine with a fixed offset arm pantograph. The use of the ex-vessel fuel storage tank (EVST) instead of the in-vessel storage adopted in Joyo contributed to the size reduction. The fuel transfer machine can run inside and outside of the containment vessel (CV) to transfer fuel subassemblies between the reactor and EVST in refueling, to transfer fresh fuel to EVST, and to transfer spent fuel from EVST.

The structure of the EVST is shown in Fig. 3.17. Fuel subassemblies stored in the EVST are cooled by the natural circulation of sodium in the EVST. Sodium is circulated both inside and outside of the rotating rack. The rotating rack has flow paths in the upper, middle, and lower support plates of the rotating rack to allow sodium to rise from the lower part to the upper part of the rotating rack.

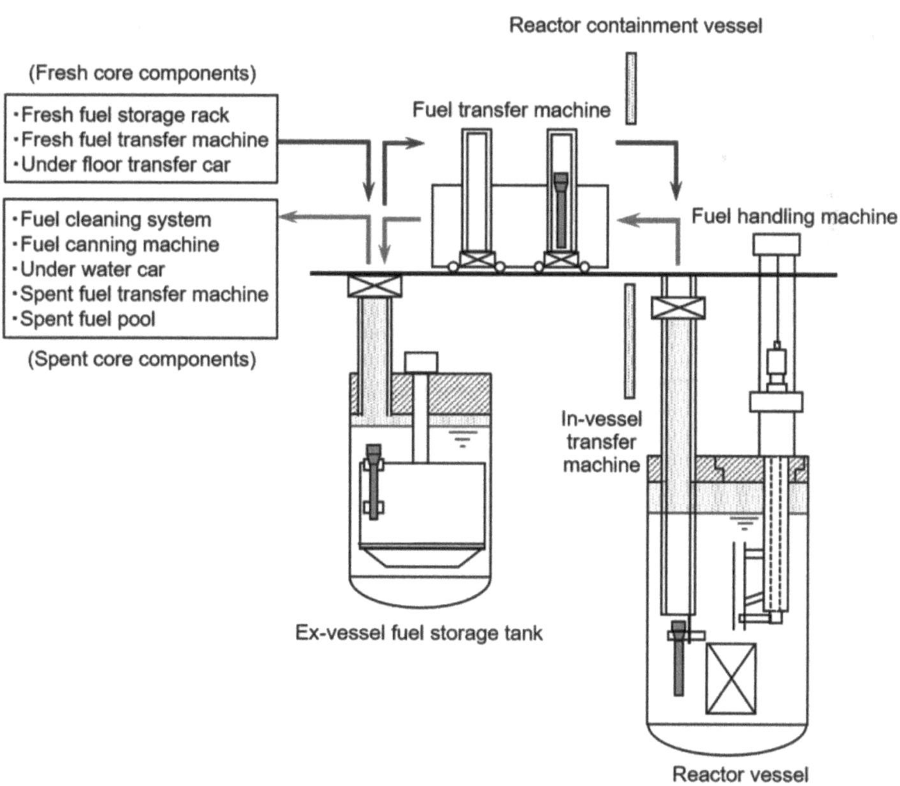

Fig. 3.16 Fuel handling systems.

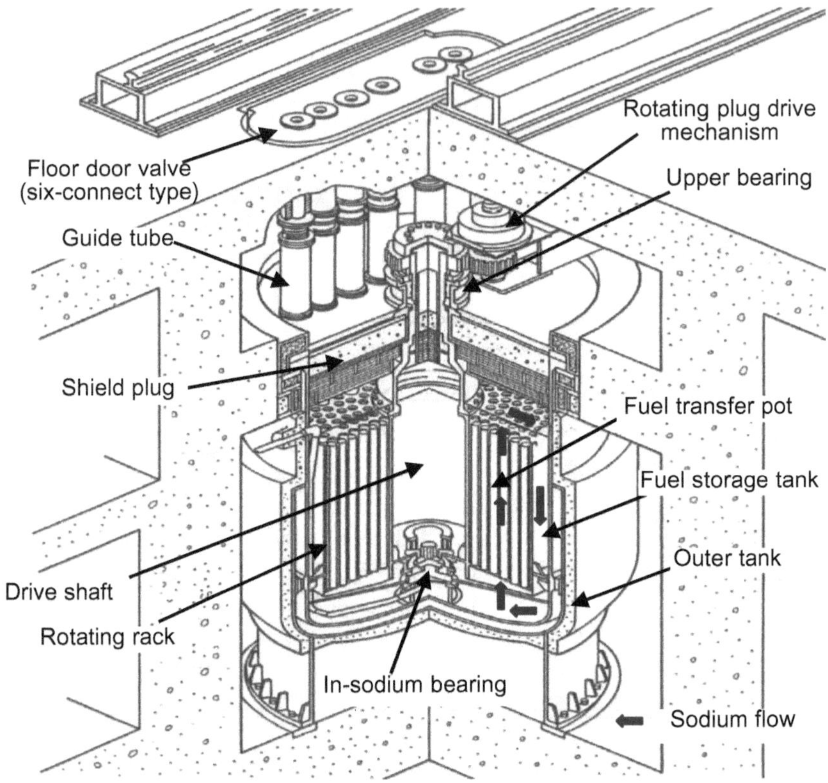

Fig. 3.17 Ex-vessel fuel storage system.

The rotating rack of the EVST is too large and heavy to be supported only by the thrust and radial bearings placed on the shield plug of the EVST. Consequently, the in-sodium radial bearing was installed as a stopper of the rotating rack at the bottom of the fuel storage tank. The bearing should be capable of being used in sodium with no lubrication (i.e., lubricated by sodium only) for 30 years and 20,000 cycles of rotation without maintenance and inspection. In case of bearing failure, a handling head is provided and an access route for replacement is secured by making the rotating rack drive shaft hollow.

3.1.6 Instrumentations

Almost all Monju instrumentation systems were developed with domestic technologies for future commercialization. The functions and performance of the developed instrumentation equipment were confirmed through the commissioning tests in Monju.

3.1.6.1 Neutron instrumentation

The neutron instrumentation consists of three monitoring systems: a neutron source range monitor (SRM) for a reactor startup and fuel loading, a wide range monitor (WRM) to cover a power range from 10^{-6} to 120% of the reactor power, and a power range monitor (PRM) to cover the range from 1% to 120%.

For the SRM, a ^{10}B-coated proportional counter and a BF_3 proportional counter were developed and their irradiation performance and durability were tested. The ^{10}B-coated proportional counter is superior in detection efficiency, but an irradiation test revealed the necessity of periodic gain adjustment of the pulse system. Then, BF_3 proportional counters were selected for the SRM by combining four counters to ensure a neutron detection sensitivity of 30 cps/nv.

For the WRM, a sensitivity of 0.3 cps/nv was required under high-temperature and radiation conditions. A fission chamber was selected, and its durability was confirmed by a high-temperature irradiation test under gamma-ray and neutron irradiation environments. Originally, a two-channel system covering the intermediate range was planned. However, to ensure the diversity of neutron detectors of the reactor trip in an event of abnormal reactivity insertion during power operation, the design was changed to use three channels and add an electric current mode.

For the PRM to be used under high gamma-ray background (BG), a gamma-ray compensated ionization chamber that can be used at high temperatures was developed. A gamma-ray compensated ionization chamber of voltage compensation type was developed and its durability was confirmed in Joyo.

3.1.6.2 Failed fuel detection system

Three failed fuel detection systems were employed in Monju: delayed neutron (DN) method, cover gas method, and tagging gas method.

(a) *Delayed neutron method*

The failed fuel detection system by the DN method was employed as a plant protection system instrumentation. Locations of the detectors were determined considering delay time due to transport and dilution by mixing of DN precursors released from failed fuel in the RV upper plenum.

The BG count rate of the DN method during the normal operation was estimated to be 45 cps in the design considering the photoneutrons from the concrete wall, the neutrons from the core, surface contamination on fuel pins, etc. However, in the 40% power test in 1995, the BG count rate at the rated power operation was estimated to be 1/50 of the design value. The count rate was too small to ensure a meaningful BG count rate (a dozen cps) to prevent false alarm and operation check of the DN method system. Consequently, a neutron source (Am-Be) was newly installed at the tip of each DN detector.

(b) *Cover gas methods*

The failed fuel detection systems using cover gas include the precipitator method and the gamma-ray detection method, both of which detect radiation from rare gases (krypton and xenon) released from failed fuel and transported to the cover gas region.

Regarding the precipitator method, "a gas replacement type precipitator" was selected because of its high detection sensitivity, high reliability, and easy maintenance and inspection. A prototype precipitator was developed, and its performance was confirmed by a durability test in the research reactor JRR-3.

Regarding the gamma-ray detection method, a NaI scintillator was installed in the cover gas system to measure the gamma rays emitted from rare gases, by which a larger-scale fuel failure not covered by the precipitator method could be detected.

(c) *Tagging gas method*

Two methods were examined to identify the location of failed fuel: the selector valve and tagging gas methods. In Monju, the tagging gas method was adopted in consideration of the configuration of the upper core structure and experience in foreign SFRs.

The tagging gas method identifies the location of failed fuel by detection and analysis of the tag gases encapsulated in each fuel element. Different compositions of stable krypton (Kr) and xenon (Xe) isotopes are assigned to each of the fuel subassemblies. The United States and France were leading the development of this method, but there remained certain challenges. These included the understanding of gas transport behavior, the development of a compact tag gas concentrator/separator, an analysis computer code to model tag gas generation and incineration, and the development of a tag gas capsule.

Concerning the development of the gas concentrator/separator, basic data using a small-scale activated charcoal bed were obtained on temperature, time, and carrier gas flow rate in the gas recovery and desorption processes. Based on the data, it was foreseen that nearly 100% of gas could be recovered. However, a gas recovery rate of 100% could not be achieved in experiments, especially for Kr. The cause of the small recovery rate was thought to be slow rare gas desorption. The temperature was not sufficiently increased in the interior of a large-diameter adsorption bed because of the small thermal conductivity of the activated charcoal. Consequently, a U-shaped adsorption bed with twice the length was adopted in the actual plant.

3.1.6.3 Sodium purity control devices

Sodium purity must be appropriately controlled to prevent corrosion of the structural materials of the components and piping, and the fuel cladding. The impurity concentrations in the sodium must be kept low to reduce exposure caused by radioactive corrosion products in the PHTS and to improve the detectability of water leaks from an SG tube in the SHTS.

Sodium purity is maintained and controlled within a prescribed range using a cold trap (CT) for the removal of impurities in sodium, and a plugging (PL) meter for measuring the impurity concentrations.

(a) *Cold trap*

The CT is a device to purify sodium. It supersaturates and precipitates oxygen and hydrogen in sodium by lowering the sodium temperature (generally to 120°C–150°C) and traps the precipitates with a stainless steel mesh. It is installed in a purification system branching from the main sodium circulation systems.

(b) *Plugging meter*

The PL meter is an instrument designed to continuously measure impurities in sodium. The measurement principle uses the precipitation of an impurity in sodium at a temperature equal to or lower than its saturation temperature. The impurity concentration in sodium is estimated based on the relationship between a measured saturation temperature and the saturated solubility of the impurity. The determination of impurity becomes difficult when various types of impurities with different saturated solubilities are present in sodium (called the multiple breaking phenomenon).

3.2 R&D activities for Monju design

3.2.1 Design history

The design study of Monju started in 1968, under the basic policy to bring a sodium-cooled, MOX-fueled prototype FBR with an electrical power of 200–300 MW into criticality by around 1976.

Since there was little engineering experience with FBRs in Japan, the first goal was to select a basic concept of the ideal image of Japan's commercial FBRs. Basic design concepts such as loop-type, cold-leg pump, reheating cycle, integrated once-through helically coiled SG, and refueling with a single rotating plug with a fixed offset arm were extracted through evaluation and review of a preliminary design study. The major specifications of Monju were selected based on Japan's own philosophy by evaluating the design study and the related R&D as well as the information on the preceding foreign prototype FBRs, expecting future possible technological advancement.

After completion of the Preliminary Design, Monju design progressed to Phases 1 through 3, followed by the Adjustment Design started in 1974. In 1977, the Manufacturing Preparation Design started with a focus on the preparation of the Application for the Reactor Installation Permit, finalizing the design to confirm that the Permit could be obtained and the construction cost would fall within an appropriate range.

In the design elaboration process, the layout of components, piping, wiring, and ducts was investigated in further detail and some specifications were changed as follows:

- Elimination of spent fuel in-core storage,
- Installation of the auxiliary cooling system in the secondary heat transport system,
- Adoption of horizontal pipe routing at high elevation for the primary heat transport system hot-leg piping,

- Elimination of core clamping structure,
- Change from the selector valve to tagging gas method for failed fuel detection, and
- Elimination of SG reheater.

R&D activities for Monju were carried out under all-Japan cooperation, gathering experts from industry, academia, and government. In addition, a series of safety technology guidelines for SFRs were developed.

3.2.2 Sodium components

There was little experience in handling a large amount of high-temperature liquid sodium in Japan when Monju design work started. Thus, facilities to develop sodium technologies were constructed at the Oarai Engineering Center (OEC, currently Oarai Research and Development Institute of JAEA), and various research activities ranging from fundamental to engineering studies were conducted.

3.2.2.1 Reactor vessel

Upon a reactor trip, the core flow rate rapidly decreases to 10% of the rated value, and low-temperature sodium from the core flows into the stagnant high-temperature sodium region in the upper plenum. This creates the low-temperature sodium layer below a high-temperature layer, and the boundary between the two layers gradually rises. This phenomenon is called "thermal stratification" and may cause large thermal stress caused by the temperature gradient along the stratification boundary (Fig. 3.18).

A similar thermal stratification phenomenon is seen at the RV outlet nozzle, where a stagnant high-temperature sodium region appears in the upper layer and a low-temperature sodium flow through the inner barrel flow holes appears in the lower layer.

These phenomena were experimentally simulated by water and sodium thermal-hydraulic tests. The obtained data were used in design improvement to mitigate the influence of the thermal stratification as follows:

- The stress at the Y-piece part of the RV is mitigated by increasing the inner barrel height by 1 m,
- The thermal stratification in the upper plenum is eliminated in a short time by generating a circulating flow using the sodium pumped up from the overflow system, and
- The local thermal stratification at the outlet nozzle is mitigated by providing the flow holes in the upper and lower parts of the inner barrel.

3.2.2.2 Shield plug

The thermal insulation performance of the shield plug, RV, and the pedestal was confirmed using a mockup device with an actual axial size and scale of 1/2.5 in the radial direction. In a thermal insulation test to investigate the behavior of the RV cover gas, the non-uniform circumferential temperature was observed in the narrow annulus region of the shield plug, which may cause unfavorable structure deformation. This was caused by the natural convection of cover gas in the annulus. To mitigate this,

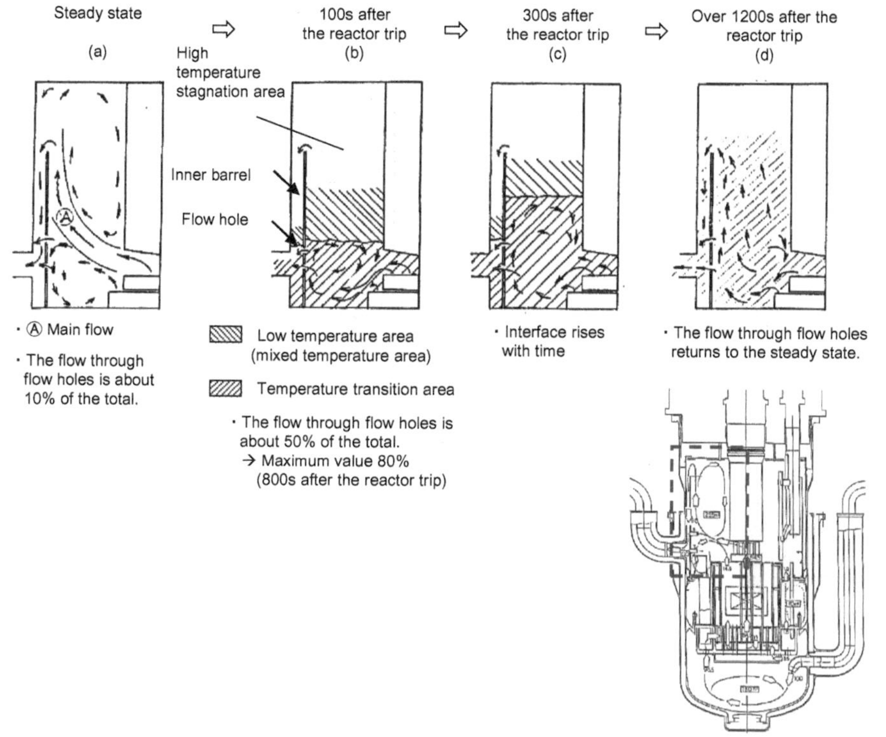

Fig. 3.18 Schematic description of thermal stratification.

convection suppressing fins were inserted into the annulus region. Furthermore, various thermal-hydraulic tests using water and sodium were performed for a detailed understanding and evaluation of the thermal striping phenomenon, and the results were reflected in the design of the lower part of the upper core structure.

3.2.2.3 Control rod drive mechanism

The CRDM is an active component having the crucial functions of reactivity control during reactor operation and emergency reactor shutdown in abnormal conditions. To ensure the reliability of these functions, a significant R&D effort has been made using the prototypes (trial manufacturing) from the early design stage.

The trial manufacturing of the mockup was performed in three phases. In each phase, tests were conducted to check various performances, such as:

- The under-water functioning, under-sodium functioning, under-sodium durability for three times the number of activations assumed during the service period of 30 years, and
- Under-water vibration performance (the general performance, structural integrity, and scram performance during earthquake).

Scram tests were performed more than 16,200 times using the prototype, and no failure of control rod insertion was observed. In the under-sodium durability test, intermittent CRDM operation in high-temperature sodium after the long-term shutdown was simulated to confirm the presence or absence of self-welding, sticking, and galling of movable parts, the presence or absence of sodium sticking on the interface with the cover gas, and the integrity of bellows.

3.2.2.4 Circulation pump

The main circulation pump is the only rotating component that directly handles high-temperature radioactive sodium. Thus, R&D activities on the pump were conducted step by step aiming at high reliability. A prototype pump, the first domestically developed unit (capacity of 1 m^3/min), was manufactured in 1966 and was followed by the design, manufacture, and testing of a sodium pump (5 m^3/min) and then a mockup of Joyo main circulation pump (21 m^3/min). In the development of the main circulation pump for Monju, which is about five times as large as that of Joyo, development was focused on the shaft bearing and seal, and a full-scale mockup test. The test results were reflected in the design and manufacturing of the actual Monju pump as shown in Fig. 3.19.

(a) *Development of the shaft bearing and seal*

Mockup parts tests of the shaft bearing were conducted to obtain design data extrapolatable to actual components. These included tests in water and sodium, and low-speed tests, which are more severe for the bearings. For the surface hardener, Colmonoy alloy (Ni-Cr alloy with addition of boron and carbon) having a low cobalt content, a source of radioactive corrosion product, was adopted to reduce radiation exposure.

Double mechanical seals were employed in the pump shaft to seal the argon cover gas in the pump. The space between the two seals is filled with high-pressure lubricating oil to prevent the leak of radioactive cover gas to the outside. Furthermore, an argon gas diffusion seal is also used, by which the cover gas is purged and shaft sticking due to the deposition of sodium vapor is effectively prevented. Since the operating conditions become more severe with increased shaft diameter, a series of parts tests were performed using a test component of actual size. Particularly for the argon gas diffusion seal, the effects of the flow rate and pressure of purge gas were investigated in detail.

(b) *Full-scale mockup test*

In the series of full-scale mockup tests, the pump hydraulic characteristics were confirmed by water tests, and then, the mechanical and structural reliability and durability were verified by sodium tests with a partial flow rate using a special impeller.

As an example, the axial distribution of the circumferential temperature difference is shown in Fig. 3.20. The convection suppressing fins installed in the relevant space were found to effectively decrease the peak circumferential temperature difference in the outer casing from 74°C to 10°C and thereby significantly reduce the amount of thermal deformation of the casing.

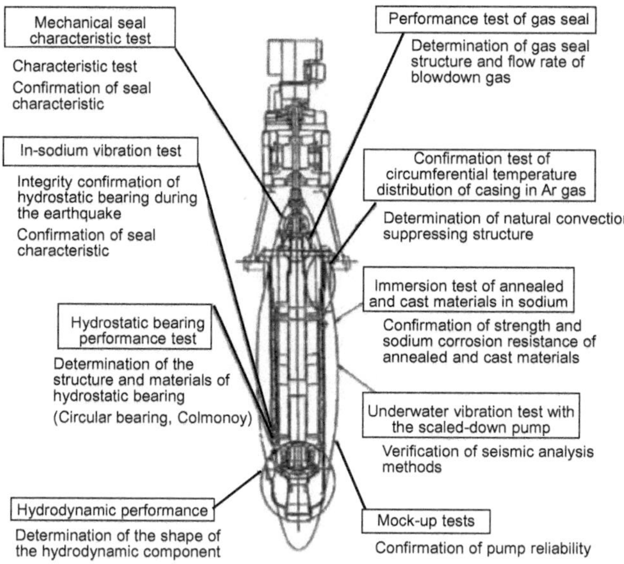

Fig. 3.19 Improvement of the main circulation pump.

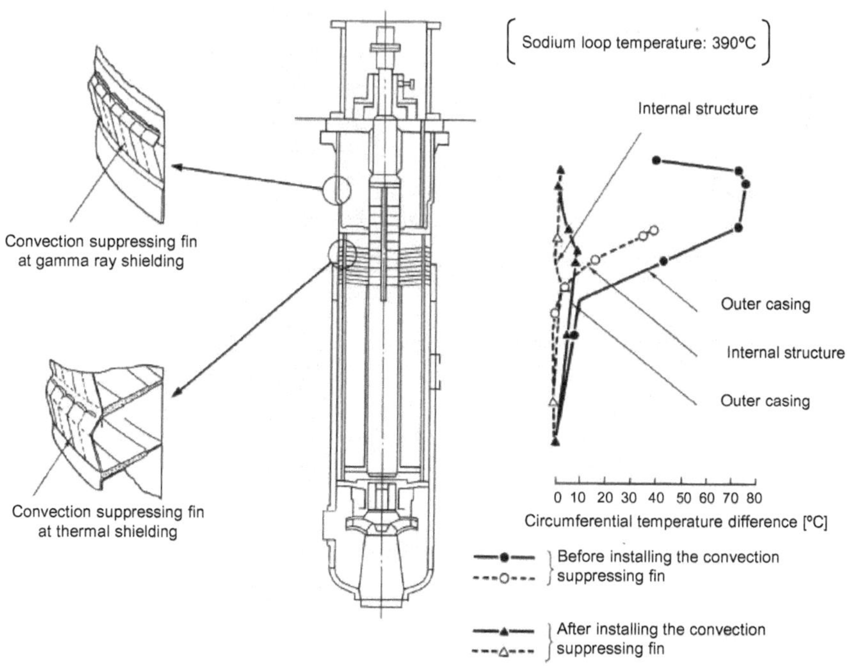

Fig. 3.20 Distribution of circumferential temperature difference measured in mockup pump.

Deformation of the casing caused by natural convection in the annulus space became evident in the scale-up from Joyo to Monju. This is a good example showing the importance of full mockup tests simulating the actual temperature conditions and size to identify problems due to scale-up.

3.2.2.5 Intermediate heat exchanger

The IHX of Monju has a heat exchange capacity of 238 MW with a primary-side inlet temperature of 529°C and 3294 heat transfer tubes. These specifications differ from the IHX of Joyo in the larger heat exchange capacity (about five times), higher operating temperature, and the adoption of a free surface type.

R&D activities were conducted in various fields, including thermal hydraulics, material, structural integrity, and fabrication technologies. Technologies were developed to mitigate thermal stress and to achieve uniform flow distribution in the component.

(a) *Flow characteristics*

Uniform flow distribution is desired to attain good IHX performance and to ensure structural reliability by limiting the temperature difference among heat transfer tubes. Thus, water flow tests using a full-scale partial model and a small-scale whole model were conducted. It was confirmed that uneven flow could be suppressed by adjusting the shape of flow paths in the IHX, the flow guide plates, and the baffle shape.

(b) *Material*

Austenitic stainless steel (SUS304) was selected as the main material of the IHX for its superior corrosion resistance and high-temperature strength. For the material used in the discontinuous part at high temperatures, the creep rupture strength of SUS304 was increased by limiting carbon and nitrogen to no greater than 0.1%. Concerning the weld material, the contents of trace elements as niobium and vanadium in the base material were optimized by experimentally confirming their effects on high-temperature strength.

(c) *High-temperature strength and integrity of tube plate*

Since the IHX is directly connected to the hot-leg piping of the RV outlet, significant thermal shock is applied to the IHX during transients upon reactor trip, PHTS pump trip, and SHTS pump trip. In particular, serious effects are anticipated at the joint between the shrouds and the upper tube plate, on the tube-tube plate joint, and the primary-side nozzle. A thermal shock test was performed using a 1/2.5 reduced size test unit to understand thermal-hydraulic behaviors at various parts of the IHX.

(d) *Integrity of heat transfer tubes against compressive buckling*

The IHX tube bundle has a configuration in which about 3000 straight heat transfer tubes are welded at the upper and lower tube plates. When a significant temperature difference between the tubes arises due to non-uniform sodium flow in the tube bundle, a compressive load is axially applied to the tubes at relatively higher temperatures, possibly causing buckling. Therefore, a test simulating the actual configuration was conducted to confirm that the strain and horizontal deflection are sufficiently small under the load conditions anticipated in the actual plant.

(e) *Structural integrity of bellows*

Bellows in the IHX is intended to accommodate the thermal expansion difference between the secondary-side downcomer and tube bundle and between the outer shell and tube bundle. The former bellows forms a part of the reactor coolant boundary. The integrity of these bellows was confirmed through various tests and stress analysis including fundamental tests for determining the bellows crest, stress measurement, fatigue durability tests, and immersion in sodium.

(f) *Tube-tube plate welding method*

The complete socket welding method was selected for tube-tube plate welding to ensure the manufacturability and reliability of the heat transfer tube bundles. The crevice corrosion, which may occur in a crevice (the gap between tube and tube plate), was judged unlikely to occur based on operation experiences.

3.2.2.6 Steam generator

The Monju SG is a separated once-through type consisting of an evaporator and a superheater, equipped with helically coiled heat transfer tubes. The SG technology was developed at the OEC from a basic research phase with a small-scale SG of 1 MW to one of the world's largest SG test facilities of 50 MW (50-MW SG), equivalent to about one-fifth of the Monju SG (Fig. 3.21). R&D activities and test operations for a cumulative time of 30,000 h efficiently yielded demonstration data related to structure and material, operation and handling, maintenance and repair, and safety and reliability. Concerning a possible water leak, a wide variety of tests and analyses were conducted and measures against water leak accidents in the Monju SG were developed.

(a) *Development of a thermal-hydraulic evaluation formula*

The development of a thermal-hydraulic evaluation formula was the most basic and important item in designing the SG. The performance tests using the 1-MW and 50-MW SG were conducted by taking the pressure, temperature, and load level as parameters to evaluate and validate thermal-hydraulic evaluation formulae (for the water-steam-side and the sodium-side heat transfer coefficients) and thereby to establish the evaluation formulae for the Monju SG.

(b) *Measures against water-side flow instability*

Flow instability in the water side may arise from the fluctuation of the pressure drop of a steam-liquid two-phase flow when the water boils in the heat transfer tubes. An analysis code to evaluate the flow instability of the two-phase flow was developed and validated, and a range of operating conditions that can securely eliminate the occurrence of flow instability has been established as a stability map. This map was used to determine the operating range of the actual component.

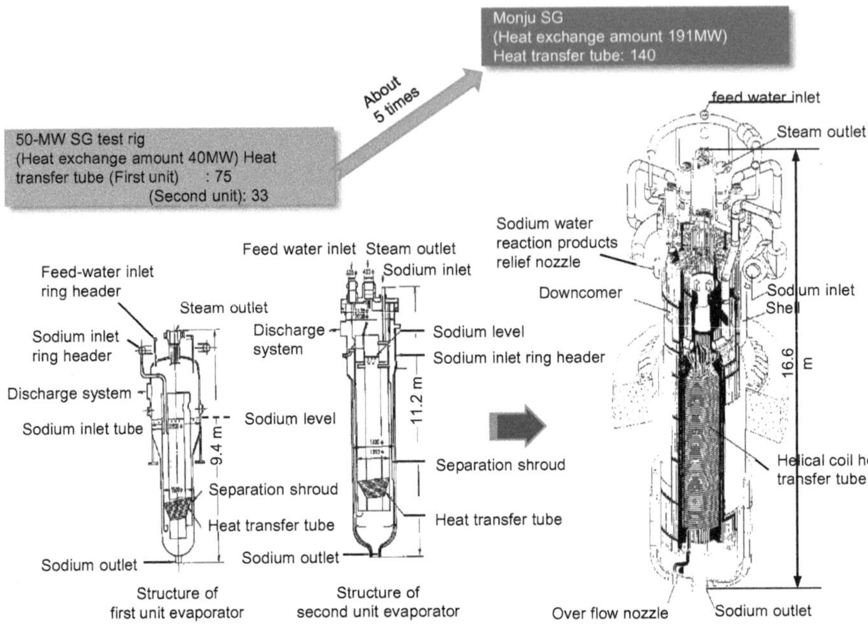

Fig. 3.21 Comparison of SG models between 50-MW SG and Monju.

(c) *Understanding of dry-out phenomenon in tubes*

Water flowing in heat transfer tubes is heated by sodium and boils in a two-phase flow. As the ratio of steam (quality) increases, a state in which the transfer tube inner surface becomes dry (dry-out state) arises at a certain point. The start of dry-out triggers the change in the boiling state from the nucleate to film boiling, followed by a rapid decrease in the heat transfer coefficient. Therefore, it is necessary to evaluate the quality with high accuracy to control heat transfer performance. In addition, the fluctuation of tube wall temperature near the dry-out point is important from the perspective of thermal stress exerted on tubes.

To resolve these issues, a test focusing on dry-out was performed to clarify the dry-out phenomenon in a helical coil-type sodium-heated SG. Accordingly, a method to evaluate the dry-out quality and thermal fatigue was developed.

(d) *Confirmation of material properties and strength*

Austenitic stainless steel, a candidate material for the SH, has good high-temperature strength, but the stress corrosion cracking was a concern. Therefore, several types of austenitic stainless steel were tested under actual plant conditions and accelerated conditions assuming the inflow of wet steam and the increase in dissolved oxygen. Test results confirmed that stress corrosion cracking is unlikely to occur for all tested materials under the assumed operating conditions.

To confirm the aging of material properties, a portion of the heat transfer tubes used at 1-MW and 50-MW SG were cut out for detailed material tests to investigate decarburization and carburization, corrosion on the sodium side, generation of scale on the waterside, and the change in material strength. Consequently, no anomaly was observed on the heat transfer tube material during extended operation of up to 16,000 h, and it was confirmed that deterioration due to aging is negligible.

3.2.2.7 Sodium handling technologies

(a) *Sodium purity control*

When the CT is used over an extended period, the mesh part of the CT would be clogged with impurities. Thus, uniform trapping of the impurities in the mesh was a major R&D issue.

Through various tests in the sodium test facilities at the OEC and operating experience in Joyo, findings were obtained on flattened temperature distribution in the CT, an increase in the inflow area of the mesh part to prevent local blockage formation, and optimization of the mesh packing density.

CTs in the secondary cooling system are subjected to a heavy load caused by impurities due to the permeation of hydrogen from SG tubes during plant operation and thus need to be replaced during the plant lifetime. For this reason, R&D activities on the regeneration of the CTs were conducted to reduce waste, and the applicability to an actual plant was examined.

(b) *Sodium leak detector*

The sodium leak detectors employed in Monju are listed in Table 3.2. In their development, the focus was placed on a gas sampling type capable of detecting fine leaks. R&D necessary to demonstrate the feasibility of the gas sampling type included: the detector itself, the generation and attenuation characteristics of aerosol, self-plugging behavior, corrosion caused by leaked sodium, and a method of gas sampling. It was confirmed that a sodium leak with a rate of 100 g/h can be detected within 24 h in both the primary and secondary systems. For the leak detectors in the EVST and its cooling system, the amount of aerosol generated is less due to the low system temperature, and a sodium leak with a rate of 100 g/h can be detected within 150 h.

Reflecting also experience in the commissioning tests, the following improvements were made:

- Platinum filament with a purity of 99.999% is used in the sodium ionization detector to suppress BG increase due to impurities.
- The alarm setting of the radiative ionization detector (RID) was originally based on the absolute value of the output signal. However, it was revealed that the RID signal depends on the atmospheric temperature (i.e., the output decreases with increasing temperature) and notable signal fluctuation associated with the ionization of ^{241}Am was observed. Accordingly, a digital signal processing system using the moving average method was added.
- Deteriorated insulation or false activation of leak alarm occurred in a silver-soldered closed contact-type sodium leak detector (CLD) at high temperatures. The ion migration of silver solder causes the deterioration and short circuit (a phenomenon that causes migration of ionized metal atoms between electrodes, resulting in short circuit). Therefore, a gold-soldered closed CLD in which ion migration is insignificant was introduced.

Table 3.2 Sodium leak detectors used in Monju.

	Detector type	Schematic drawing
Gas sampling type sodium leak detector	Sodium ionization detector (SID) Sodium aerosol in sampling gas introduced to detector is dissociated and ionized by filament kept at a high temperature, and the ion current flows between filament collectors. By detecting this ion current, a fine sodium leak can be detected	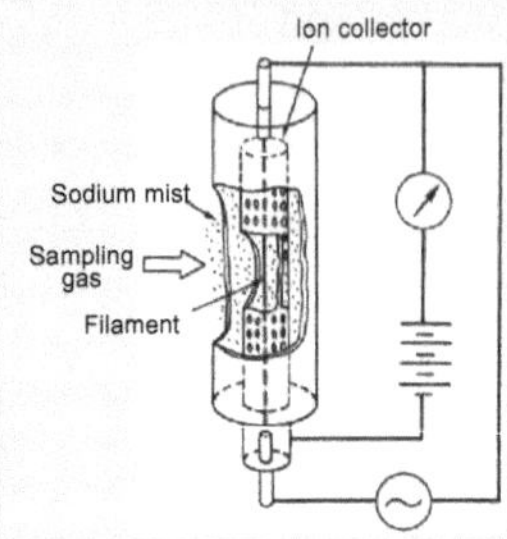
	Radiative ionization detector (RID) Sodium aerosol in sampling gas introduced to detector adheres to gas ionized by Am-241, decreasing the electric current flowing between electrodes to which external electric field is applied. By detecting this change in electric current with the voltage difference from the reference ion chamber, a fine sodium leak can be detected	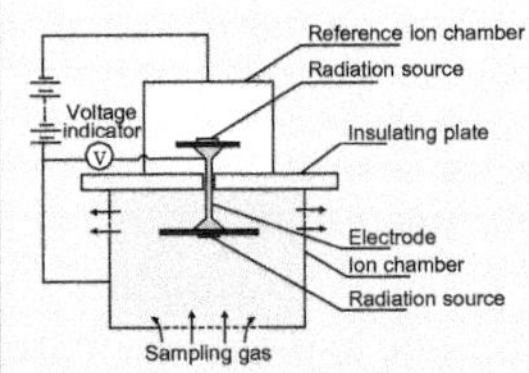
	Contact-type sodium leak detector (CLD) The electric conductivity of sodium is utilized. When leaking sodium adheres between detector electrodes or electrode and the earth, an electric current flows. By detecting this electrical short circuit, a sodium leak is detected	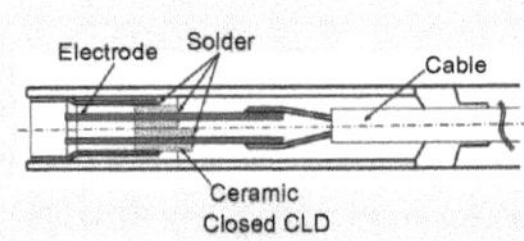

3.2.3 *Development of elevated temperature structural design guide*

Monju is a prototype FBR operated at higher operating temperatures with a plant operation life longer than Joyo. Therefore, it was essential to develop a technological basis that could be applied to its component and structure design and integrity evaluation. Furthermore, a reasonable elevated-temperature structural design method had to be developed to extend the plant operation life for future commercialization.

To meet these demands, the Elevated Temperature Structural Design Guide for Class 1 Components of the Prototype FBR was developed based on the ASME design standards through various types of structural material and strength tests.

The evaluation method of the creep damage was improved by incorporating such effects as stress relaxation with time.

3.2.3.1 Structural tests

Around 1970, virtually no structural tests directly concerning the design standards were conducted in Japan or abroad. Therefore, the early structural tests began with low-cycle fatigue tests and thermal ratchet tests in the area of ASME Section III using simple piping elements. Later, tests using a piping element model, tests of a piping system under a mechanical load that substituted a thermal load in the creep range, and thermal creep ratchet tests were performed to reflect the test results to the Elevated Temperature Structural Design Guide. In particular, creep-fatigue tests were performed for a wide variety of structures and under various thermal stress conditions to clarify the various types of thermal stress to be considered in the design. When designing the specimens for the tests, the established strength evaluation method and the accumulated basic testing experience were proven to be effective. These structural test data and the related analytical evaluations were used to formulate and validate the Elevated Temperature Structural Design Guide. Representative three structural tests are explained below.

(a) *Elastic follow-up test of PHTS hot-leg piping*

In a high-stress section of piping kept at a high temperature, the strain increases associated with the reduction in elastic deformation in the low-stress section during a stress relaxation process caused by creep. This phenomenon is called "elastic follow-up." The Elevated Temperature Structural Design Guide for Monju stipulates a specific method to evaluate the elastic follow-up strain. This method was developed using simplified inelastic analysis of piping validated by high-temperature tests of many piping elements. Thus, actual behavior had to be confirmed in the entire piping system like the PHTS hot-leg piping of Monju, a new concept of the horizontal pipe routing at high elevation. Then, an elastic follow-up test at a high temperature (600°C) was conducted using a quarter-scale model of the PHTS hot-leg piping (piping from the RV outlet to the IHX inlet), and the following results were obtained:

- The elastic follow-up strain caused by the thermal expansion stress is small in the PHTS hot-leg piping, and the thermal expansion stress could be classified as secondary stress.
- The "elastic follow-up determination method related to the thermal expansion stress" in the Elevated Temperature Structural Design Guide was confirmed to be conservative.
- The developed simplified inelastic (elastic creep analysis) code was validated.
- The entire piping system was confirmed to exhibit the behavior expected in the design.

(b) *Thermal fatigue test of SUS304 steel pipe weld joint*

To evaluate the fatigue or creep-fatigue strength of a weld joint, it is essential to consider the discontinuity of a geometrical structure caused by welding deformation and weld crown (buildup due to welding heat) and the material discontinuity caused by differences in materials between the base material, the heat-affected zone, and the weld metal. Moreover, the influence of residual stress should be considered.

Therefore, a thermal fatigue test of a SUS304 steel pipe weld joint was conducted by alternately pouring high- and low-temperature sodium onto a testing device to apply thermal cycles.

Cracks generated in the inner surface of the specimen are shown in Fig. 3.22. The crack in (A) was generated from a gentle crevice formed by a counter-boring process and a weld crown on the inner surface of the piping in a heat-affected zone, and the crack in (B) originated from a crevice generated by heat shrinkage of the weld metal. It was concluded that a safety factor (stress concentration factor) taking into account such shape irregularity should be incorporated in the evaluation of thermal fatigue failure of a weld joint in piping.

(c) *Integrated structural test*

The Elevated Temperature Structural Design Guide should prevent creep-fatigue damage when actual components undergo cyclic thermal transients. To confirm the overall validity of the damage prevention criteria, a failure test applying a thermal transient load with the magnitude and number of cycles exceeding those assumed

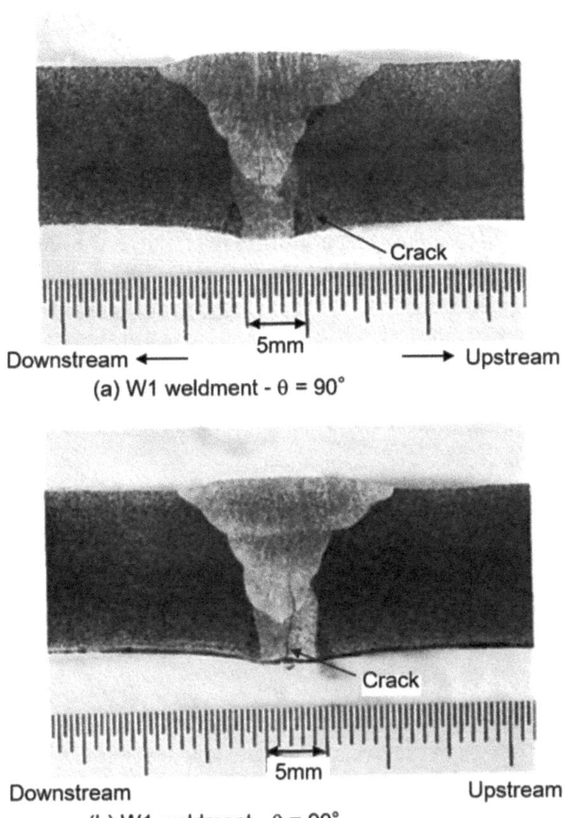

Fig. 3.22 Crack in SUS304 welded joint after thermal fatigue test.

in an actual plant, in other words, a test to monitor the generation of a significant crack, was conducted at the Thermal Transient Test Facility for Structures at the OEC.

The facility constituted two (high- and low-temperature) sodium loops, through which high- or low-temperature sodium is alternately supplied to a test model to conduct a creep-fatigue test until failure and strength data can be obtained. The test model incorporates typical structural parts to be focused on in terms of structural strength. The same manufacturing process as those used in the actual plant was applied to the model.

As a result of the test, it was confirmed that the creep-fatigue strength method based on the elastic analysis of the Elevated Temperature Structural Design Guide has a large safety margin (a factor of 50 in the gas tungsten arc welded area of the RV outlet nozzle and larger than 50 in other areas). This implies that further reduction in the safety margin is possible through the application of inelastic analysis. Furthermore, it became clear that the thermal shield plates can significantly reduce transient thermal stress because sodium present between multilayered shield plates mitigates the change in temperature of metal surface in contact with sodium.

3.2.3.2 Elevated temperature structural design guide

A comparison between the Elevated Temperature Structural Design Guide for Monju and ASME Code Case N-47 is shown in Table 3.3. Using the developed Elevated Temperature Structural Design Guide, Monju design was simplified. A typical example is the simplification of the PHTS hot-leg piping layout shown in Fig. 3.23. The plant volume was successfully reduced with the new piping layout.

The Guide and the Material Strength Standards were eventually developed as national guidelines ("Structural Design Guide for Class 1 Components of Prototype Fast Breeder Reactor for Elevated Temperature Service" and "Structural Design Guide for Prototype Fast Breeder Reactor for Elevated Temperature Service – Material Strength Standards"). Later, the standards are published as the "Standards for Nuclear Power Generation Equipment: Design and Construction Standards Edition II – Rules for Fast Reactors" by the Japan Society of Mechanical Engineers. Furthermore, some of the concepts developed in Japan were incorporated in the RCC-MR regulations, the French standards for SFRs.

3.2.4 Reactor core and fuel

3.2.4.1 Research on MOX fuel subassembly

Research on the physical properties of MOX fuel in Japan included the development of MOX fuel phase diagrams and acquisition of basic physical properties, such as O/M ratio, X-ray diffraction, metallographic observation, as well as thermophysical properties on the melting point, thermal conductivity, and thermal expansion coefficient. Subsequently, research was performed on the measurement of the physical properties of MOX fuel during irradiation and detailed analyses of irradiation behaviors.

Table 3.3 Features of elevated temperature structural design guide for Monju

	Elevated temperature structural design guide for Monju	**ASME code case N-47**	**Remarks**
Scope	Prototype FBR class 1 components	Class 1 components used at temperatures beyond those to which ASME Sec. III NB is applicable	
Materials covered	SUS304, SUS316, SUS321, 2¼Cr-1Mo	Type 304SS, Type316 SS, Alloy 800H, 2¼Cr-1Mo, (Alloy 718[a])	
Upper temperature limit	SUS304, SUS316, SUS321: 650°C 2¼Cr-1Mo: 550°C	304SS, 316SS: 1500°F (816°C) Alloy 800H: 1400°F (760°C) 2¼Cr-1Mo: 1200°F (649°C)	Temperature limits specified by N-47 depend on the allowable values
Environmental effect	Evaluation methods of the sodium environment and neutron irradiation effects are provided	Not provided	
Evaluation method	• Long-term and short-term loads are classified • A method of judging elastic follow-up is provided • The strain and creep fatigue damage limits are provided for the following cases: (1) General provisions, (2) Case of low long-term primary stress, and (3) Case of insignificant creep effect. • Design fatigue curve including the strain rate effect (the holding time effect is evaluated as the creep damage.)	• Long-term and short-term loads are partially classified (creep damage evaluation) • No evaluation method for elastic follow-up is provided • The strain and creep fatigue damage limits are provided for the following cases: (1) General provisions, and (2) Case of insignificant creep effect. (Test No. 4) • Design fatigue curve for elastic analysis including the holding time effect (the holding time effect is evaluated as the fatigue damage.)	

[a] Applicable only to a bolt member.

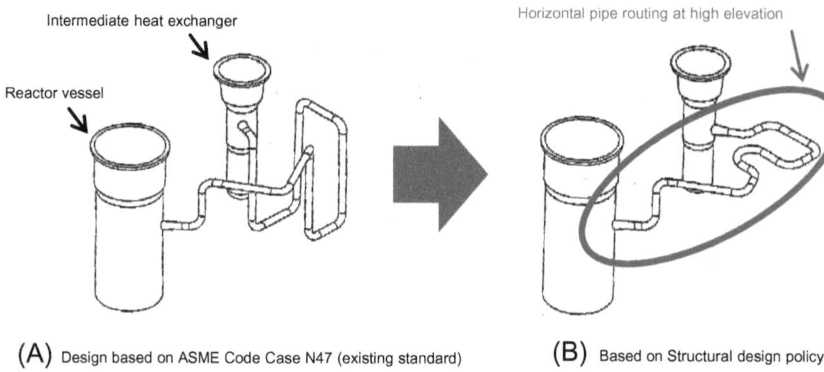

Fig. 3.23 Simplification of primary hot-leg pipe routing.

A variety of irradiation tests were conducted to understand the irradiation behavior and to confirm the integrity of the SFR fuel. Initially, irradiation tests were conducted at foreign reactors, including the DFR in the United Kingdom and Rapsodie in France. After the commissioning of the Joyo MK-II core, irradiation tests were performed primarily at Joyo. In addition, the data from the Transient Overpower (TOP) tests obtained through the US-Japan Operational Reliability Testing (ORT) program at EBR-II were used in the Safety Review for the restart after the Secondary Sodium Leak Accident in Monju.

Through this research, a wide range of physical property data on MOX fuel was obtained domestically and from abroad. A comprehensive collection of data on physical properties was compiled and utilized to develop the design evaluation formulae of Monju fuel.

3.2.4.2 Development of PNC316 for fuel cladding material

The cladding requires high-temperature strength and good compatibility with sodium and fuel material. The former requirement is to endure the internal pressure due to fission gas at a high temperature of 650°C–700°C for extended periods, and the latter includes the effects of excess oxygen generation and fission product (FP) accumulation with burnup. Under a high fast neutron fluence condition, it is also important to limit the deformation of cladding and wrapper tube caused by swelling and irradiation creep to ensure the heat removal from the fuel subassembly and smooth withdrawal of the fuel subassembly from the core.

Austenitic type-316 stainless steel (JIS SUS316), which had demonstrated satisfactory performance as a high-temperature structural material, was selected as the cladding and wrapper tube materials. Before the application of SUS316 to Monju, the material underwent improvements through collaboration with steel manufacturers, research institutions, and universities. Developed SUS316-equivalent stainless steel (named "316-equivalent steel" or PNC316) has superior high-temperature strength, anti-swelling properties, creep rupture strength, and short-time strength required as cladding material.

The maximum fuel subassembly burnup of Monju core fuel was designed to be 94 GWd/t (80 GWd/t in the core average of discharged subassemblies), and the maximum design fast neutron fluence was 2.3×10^{23} n/cm^2 (E > 0.1 MeV). Since only limited irradiation data on the cladding material were available at the time of the initial Reactor Installation Permit of Monju in 1983, it was decided to initially limit the maximum fuel subassembly burnup to 64 GWd/t (to 55 GWd/t in the core average), until irradiation data on anti-swelling properties of the 316-equivalent steel cladding could be obtained up to fast neutron fluence of 2.3×10^{23} n/cm^2.

3.2.4.3 Reactor physics

Since the analysis accuracy was insufficient in the early design stage of Monju, it was essential to understand and improve the accuracy of nuclear characteristics analysis. Various research topics were addressed in a cooperative framework with domestic institutions, universities, and manufacturers, including research using the FCA critical assembly.

International cooperation was also used effectively for technical exchanges with countries advanced in SFR development and for reactor experiments that were difficult to conduct in Japan. Such cooperation includes fuel irradiation in foreign reactors, Operational Reliability Testing at the US EBR-II, and in-pile safety tests at the French CABRI reactor. In addition, a series of Monju core simulation tests (the MOZART project) using the UK critical assembly ZEBRA was conducted for one and a half years, and the results were reflected in the Monju design. The MOZART experiment data are currently available in the OECD/NEA benchmark data collection.

3.2.5 Safety

The technological bases for safety design and evaluation of Monju were provided by various research activities on sodium leak and fire, sodium-water reaction, thermal-hydraulic characteristics of sodium during severe accidents, reactor safety research, including in-pile tests performed through international cooperation, development, and improvement of the safety analysis computer codes, and probabilistic risk assessment study.

3.2.5.1 Sodium leak and fire

When high-temperature sodium leaks in a room of air atmosphere, sodium reacts (burns) with oxygen and moisture in the air to produce heat resulting from the chemical reaction and smoke (sodium aerosol as a reaction product).

Since the 1970s, experimental studies on sodium fire of various scales and modes (pool fire, spray fire, etc.) had been carried out using the sodium fire test facilities at the OEC. The experimental database was developed to quantitatively understand the phenomena related to sodium leak and fire and to reflect the findings to the validation and improvement of safety analysis codes and the design of equipment to mitigate the effects of sodium leak.

More than 200 pieces of test data, including small-scale elementary experiments, were accumulated as a database, which was effectively used for the quantitative understanding of sodium fire behavior as well as validation and improvement of the analysis codes.

3.2.5.2 Sodium-water reaction

When a tube rupture accident occurs in the SG, high-pressure water leaks into sodium to cause a sodium-water reaction that generates heat and pressure, and hydrogen and corrosive reaction products are produced.

Similar to the experimental studies on sodium leak and fire, experimental studies of various scales and conditions were carried out using sodium-water reaction test facilities at the OEC. The accumulated data were used to quantitatively understand the phenomena associated with sodium-water reactions and to reflect the findings to the validation and improvement of safety analysis codes and the design of equipment to cope with sodium-water reactions.

Fig. 3.24 shows a schematic drawing and a photograph of the release and combustion of hydrogen from the largest test facility, SWAT-3.

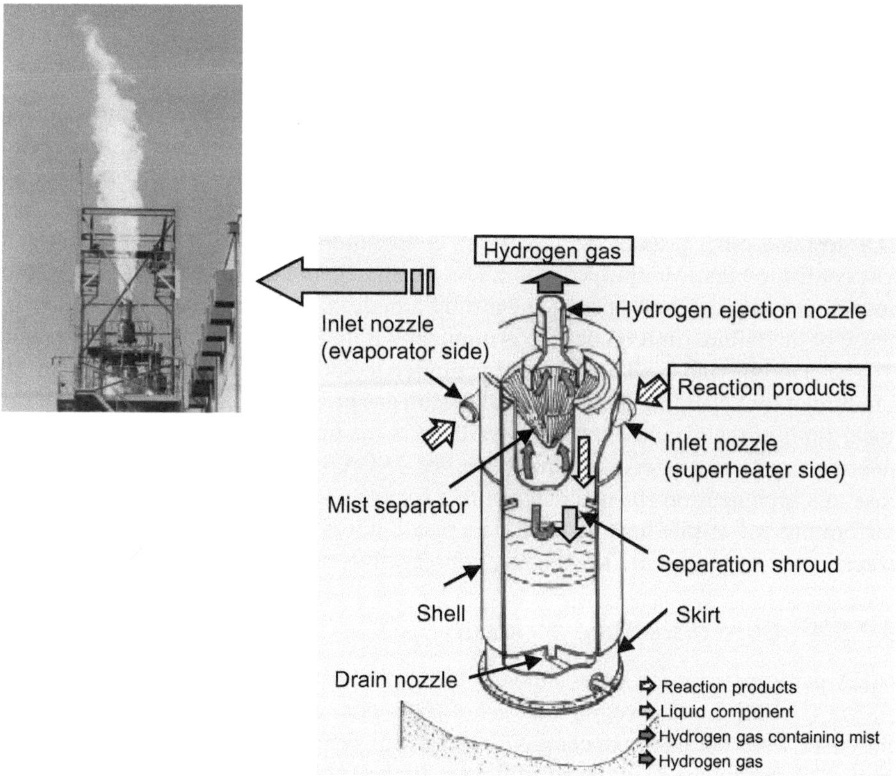

Fig. 3.24 Release and burning of hydrogen gas from SWAT-3 test facility.

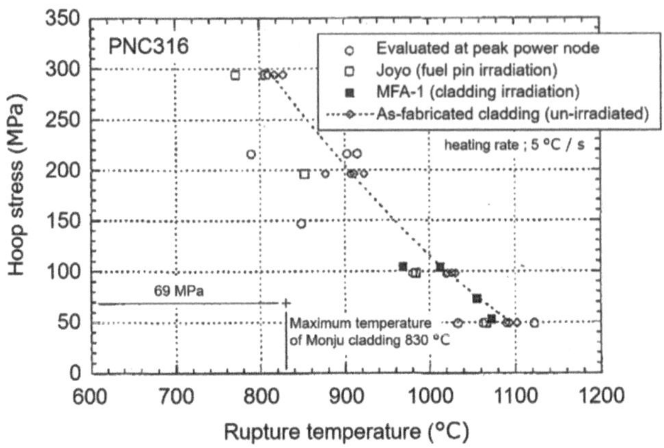

Fig. 3.25 Rapid heating rupture strength characteristics of cladding.

It was confirmed that the wastage on the adjacent tubes by jet impingement of sodium-water reaction products generated in a tube rupture event is the dominant mechanism of failure propagation, and the relevant phenomena were quantitatively understood through a series of SWAT tests.

3.2.5.3 Fuel failure criterion

Concerning the fuel failure criterion under overpower conditions, comprehensive analysis and evaluation were performed for the in-pile test data obtained from a series of the Transient Overpower (TOP) tests in the Operational Reliability Test program conducted at EBR-II in cooperation with the United States, and a slow-heating-rate TOP test and other tests of the CABRI in cooperation with France and Germany. It was confirmed that Monju fuel with a low density (pellet density of 85%TD) would not fail even if fuel melt occurs, suggesting a high failure limit, and that the dependence of the failure limit on burnup is minimal within the range of available test data.

The cladding mid-wall temperature is limited to 830°C or less in Monju to prevent a failure of fuel cladding due to internal fission gas pressure. This value was set at the lower limit of the database that was available in the initial Monju design phase, considering a sufficient safety margin. Thereafter, JAEA made efforts to expand the database to a high neutron fluence range with a focus on slow heating rate conditions by performing out-of-pile heating tests. As a result, it was confirmed that the acceptance criteria used in Monju are sufficiently conservative, as shown in Fig. 3.25.

3.2.5.4 Core disruptive accident

JAEA participated in an international joint in-pile test project using the French test reactor CABRI. In this program, a total of 63 in-pile tests were conducted, and valuable data were obtained concerning the transient behavior, failure, and post-failure motion of fresh or pre-irradiated SFR fuel. In the CABRI tests, accurate test instrumentation using the neutron hodoscope (capable of measuring the change of fuel distribution by selectively measuring the fast neutrons emitted from test fuel) was developed.

Concerning the stable cooling and retention of damaged core materials in the RV, experimental research was conducted at the molten core material behavior test facility at the OEC using various types of simulants. It was confirmed that high-temperature molten material was effectively solidified by the excellent heat transfer characteristics of sodium and was relocated in a form (particulate debris bed) that can be readily cooled.

3.3 Operation and maintenance

3.3.1 Operation

The operation of Monju has the following features compared with that of LWRs.

- Operation is simpler because the reactor is controlled solely by the control rod system, and sodium systems are controlled mainly by temperature, not by pressure.
- Plant power from 40 to 100% can be controlled by automatic operation.
- Plant startup/shutdown operation takes longer to moderate thermal transient load on components associated with high-temperature operation.
- Systems for heating sodium and air ventilation are kept in operation even during reactor shutdown to maintain plant integrity.

After completion of the Monju component installation in 1991, sodium was charged into the cooling systems and the commissioning of Monju started.

In 1994, the initial criticality was achieved, followed by the first connection to the power grid and 40% power test.

Major operation records during the 40% power test in 1995 are shown in Fig. 3.26. The 40% power test was interrupted due to the Secondary Sodium Leak Accident on December 8, and Monju was forced into the long-term shutdown until the restart in May 2010.

In December 2016, the Government announced its "Fast Reactor Development Policy," under which it was decided that Monju would be transferred to decommissioning without achieving the full power operation.

The total reactor operation hours and the power production are as follows:

- Reactor operation hours: 5300 h and 45 min (from initial criticality to the end of the Core Performance Confirmation Tests) and
- Power production: 102,325 MWh (883 h).

3.3.2 Maintenance

3.3.2.1 Maintenance program

In Japan, the regulatory ordinance concerning maintenance management was amended in August 2008 for commercial nuclear power reactors. At the same time,

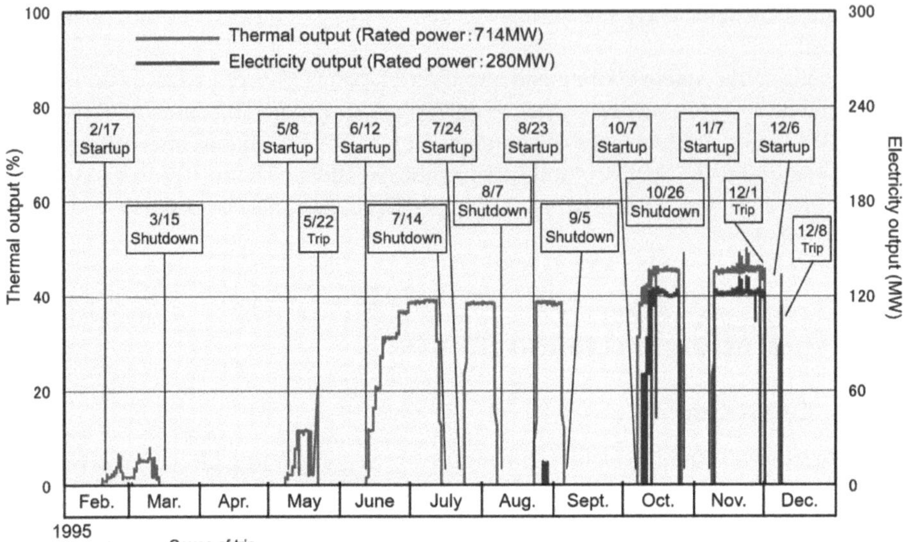

Cause of trip
5/22: Due to pump trip in the secondary heat transport system
12/1: Reactor trip test from 40% rated power
12/8: Due to sodium leak in the secondary heat transport system

Fig. 3.26 Plant operation records in the 40% power operation.

the ordinance for the power reactors in the R&D stage was also amended, and thereby, a maintenance program was introduced in Monju.

Since there had been no experience in operation and maintenance through an in-service operation cycle at Monju, a program was developed based on preventive maintenance using the available information from Joyo, LWRs, etc., with reference to the Code for Maintenance at Nuclear Power Plants and its Guide (JEAC4209/JEAG4210). Because deterioration of sodium components can be largely ignored in sodium environments, function check and visual inspection were basically performed without open inspection or overhaul.

Based on the maintenance program, the pre-service maintenance plan was developed for each of the three pre-service maintenance cycles: the Core Performance Confirmation Tests at zero power, the Plant System Confirmation Tests at 40%, and those at 100% power.

The first pre-service maintenance cycle (from January 2009 to July 2010) was completed on schedule. The activities included system check required for the Core Performance Confirmation Tests and integrity confirmation of the water-steam system equipment.

The second pre-service maintenance cycle (from July 2010) was prepared by reflecting the results of the evaluation of the effectiveness of maintenance for the first pre-service maintenance cycle. The primary and secondary cooling system equipment were inspected according to the inspection plan to confirm the integrity of equipment required for the tests at 40% power. The water-steam system equipment, which had been kept in a long-term storage state, was also inspected to confirm integrity.

The second pre-service maintenance cycle, however, was never completed due to several reasons. First, the overall plant schedule was extended several times due to failures of the equipment important to safety, including the drop of the In-Vessel Transfer Machine (IVTM) in August 2010 and the cracking in a cylinder liner of one of the diesel generators in December 2010. Second, urgent safety measures were required following the accident at the Fukushima Daiichi Nuclear Power Station of Tokyo Electric Power Company (1F Accident) in 2011. Consequently, the plant shutdown state was further prolonged and the end date of the second maintenance cycle became uncertain. Under these circumstances, inadequate maintenance management problem was revealed where many components were found uninspected within their time limits. JAEA attempted to overcome the problem by introducing an all-Japan framework in collaboration with utilities and manufacturers. However, while improvement activities were underway, the decommissioning of Monju was decided by the Government in December 2016 for significant uncertainties in the schedule and economic prospects toward the restart.

The Monju maintenance program was first introduced based on regulatory requirements to achieve a level of maintenance management equivalent to that for LWR plants, but the attempt was unsuccessful due to several reasons. First, the maintenance program was introduced too mechanically, conservatively, and hastily without sufficient time and preparation work. Second, Monju, still in the commissioning stage, had never gone through a cycle of plant operation and periodic maintenance, with staff clearly lacking experience in implementing and reviewing the maintenance program.

3.3.2.2 Maintenance and repair experiences

A large number of events with respect to maintenance and repair have been experienced since the completion of component installation in May 1991. The representative cases are outlined below:

(a) *Load increase in fine CRDM*

During the commissioning, an alarm indicating an abnormal load to fine CRDM unit 2 was set off. A disassembling investigation revealed that sodium compounds had become stuck in the gap between the shield and driveshaft in the upper guide tube (Fig. 3.27). A possible cause of the increased load was thought to be the dirt (a film of impurities floating on the surface of the sodium) left during the initial sodium charge in combination with the configuration in which sodium becomes stagnant. As a countermeasure, flow holes were added to the upper guide tube at the surface wetted by sodium to facilitate sodium flow. This event might occur not only during the initial sodium charge but also during sodium charge after work involving the opening of a sodium boundary.

(b) *Damage to superheater pressure relief plate*

In March 1998, during the planned replacement of a pressure relief plate for the SH, a small amount of sodium was found adhered to the secondary side (in a nitrogen atmosphere) of the pressure relief plate. Stress corrosion cracking was thought to be a

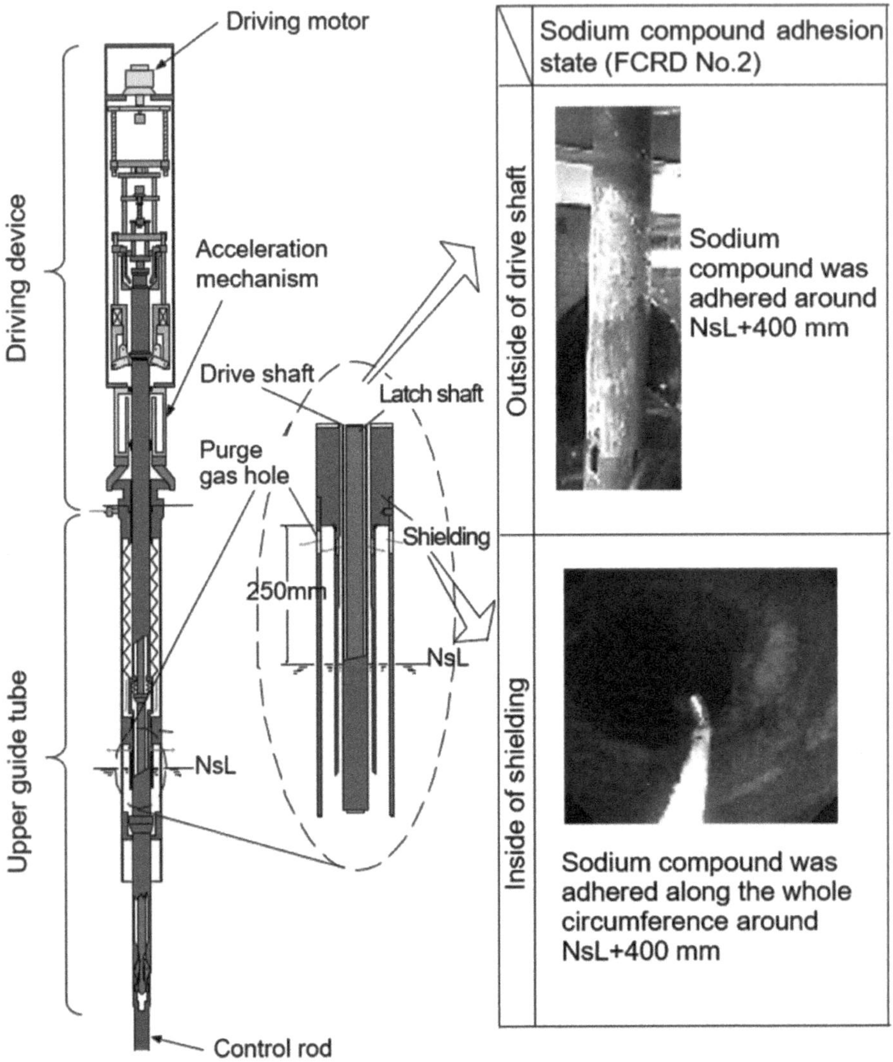

Fig. 3.27 Sodium compound adhesion state in CRDM.

reason, which was caused by melted caustic soda (sodium hydroxide) generated by a reaction of sodium with moisture absorbed in the pressure relief plate. In general, stress corrosion cracking would not occur in an environment with a large amount of sodium vapor since caustic soda should change to sodium oxide by reaction with sodium. However, in this case, caustic soda remained as it was because sodium vapor was not supplied in a narrow space between the vacuum support and the plate. As a countermeasure, the method of polishing the vacuum support was improved.

(c) *Increased pressure loss in the primary argon system*

The pressure loss between the RV vapor trap outlet and compressor surge tank increased with sodium temperatures during the reactor startup in October 1995. The cause was a blockage formed due to the deposition of sodium vapor to the valves downstream from the vapor trap. Sodium vapor in the cover gas should be effectively removed by the two-step process consisting of mist capture by a mist trap and condensation capture using a vapor trap. However, the performance of the vapor trap was found not at a satisfactory level. As countermeasures, two sintered filter lines and one HEPA filter line were added to the vapor trap outlet.

(d) *Tests of an alternative dew-point meter*

A lithium chloride dew-point meter is used for the containment vessel leak rate test (CV-LRT) in Monju. A problem with the meter is that the recommended validity of the lithium chloride solution used in the humidity sensing is limited to from 3 to 6 months depending on the service environment. In Monju, the period of validity is conservatively set at 3 months. This short validity period severely restricts the schedule of the CV-LRT, because access to the PHTS rooms to reapply the lithium chloride solution is allowed only when sodium is drained and the nitrogen atmosphere is replaced by air.

A new capacitive dew-point meter was tested in Monju as an alternative that did not employ lithium chloride solution. Two tests were conducted: verification in the CV-LRT and 2-year long-term verification. Test results confirmed the absence of significant difference from the lithium chloride dew-point meter in the CV-LRT. The 2-year long-term verification test satisfied the instrument accuracy required by the Code for Reactor Containment Vessel Leak Rate Test (JEAC4203-2008). It was concluded that the capacitive dew-point meter can be used in the CV-LRT.

(e) *Oxygen concentration control in sodium*

The history of oxygen concentration in sodium, measured by the PL meter, is shown in Fig. 3.28. The CT in the PHTS has been operated at 130°C to keep the oxygen concentration at 3 ppm or lower (except for the initial purification period). At the beginning of the operation, a short increase in the oxygen concentration due to the elution of oxygen attached to the surface of the structure was observed. Thereafter, the oxygen concentration was appropriately controlled, demonstrating the satisfactory performance of both the CT and the PL meter. The oxygen concentration in the SHTS was also confirmed at a satisfactory level.

(f) *Performance of cold trap*

In the SHTS, it is important to keep the hydrogen concentration low to ensure early detection of water leak from an SG tube. Fig. 3.29 shows purification performance of the CT when the operating temperature of the CT in the SHTS were changed in a ramp-wise manner from 140°C to 120°C. Good agreement is observed in the response curve of hydrogen concentration between the calculation and the measurement when the purification efficiency is set at 0.85, which is better than the design value of 0.7.

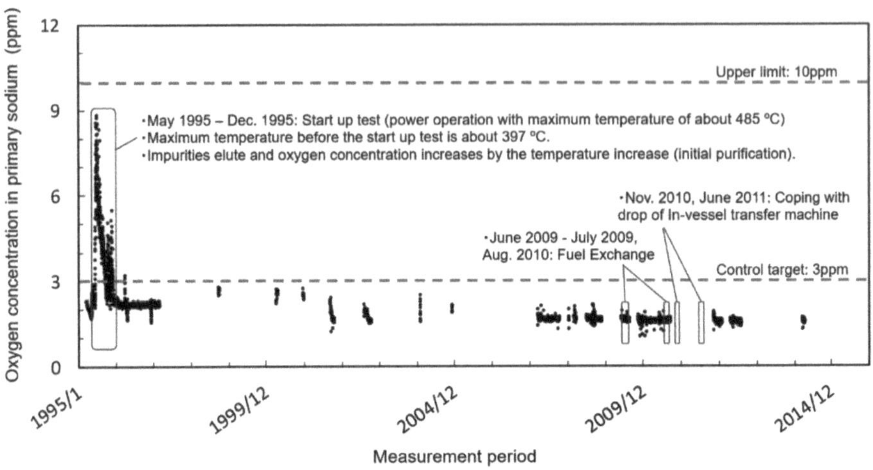

Fig. 3.28 Oxygen concentration in sodium in the primary cooling system.

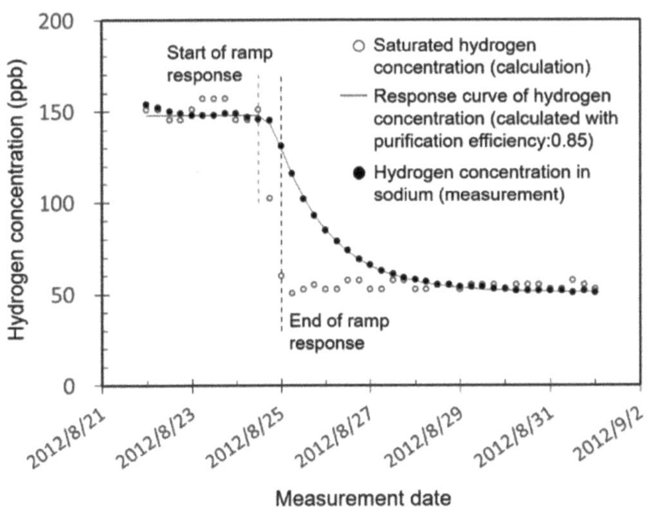

Fig. 3.29 Purification efficiency of cold trap in the secondary cooling system.

3.3.3 *Accidents and failures*

Representative failures Monju experienced are given below:

3.3.3.1 *Thermal displacement of SHTS piping*

In June 1991, the temperature of the SHTS piping was raised in a preliminary test. When the system temperature reached 120°C, the pipe adjacent to the IHX of loop C was displaced toward the RV side (by 36 mm). A part of the pipe was displaced toward the IHX side (by 5.5 mm), which differed from the prediction.

It was identified that the spring constant of the bellows at the CV penetration was three times larger than the design value and that the spring constant of the nitrogen gas seal bellows was somewhat larger than the design value, both of which constrained the thermal displacement of piping.

To cope with the problem, the CV penetration bellows was replaced with softer ones by changing from a two- to one-layer configuration, and the nitrogen gas seal bellows was changed to a reinforced rubber seal. Furthermore, additional pipe supports were installed.

Since the relevant CV penetration bellows had a larger ratio between the crest and pitch compared to the conventional type, the two layers of the bellows interfered mutually (moved as if they were one-layer bellows), resulting in the spring constant three times larger than the design value.

3.3.3.2 Pressure drop of water-steam system flash tank

In February 1995, during the reactor startup test, in the process of increasing the reactor power from 2% to 10%, the outlet pressure of the flash tank in the startup bypass line of the water-steam system dropped temporarily.

The pressure drop was caused by the following mechanism: Flow resistance was increased by steam entrainment due to a circulating flow of hot water in the flash tank, and then, the heat to be recovered in the deaerator was decreased. The increased amount of steam consumed then caused the pressure drop.

As countermeasures, the flash tank structure was modified to prevent steam entrainment, and the drainpipe to the deaerator was replaced with a larger one.

The startup bypass line of the water-steam system was introduced as an advanced system to collectively recover the waste heat produced in the evaporators of the three loops to heat feedwater effectively before aeration into the superheaters. However, the steam entrainment could not be predicted in the design since the configuration of the three loops merged into a single tank from different angles was too complicated. This experience highlighted the importance of design verification work when an advanced system aiming at streamlining/improving efficiency is adopted, even if it is an extension of a general-purpose technology already proven.

3.3.3.3 Secondary sodium leak accident

In December 1995, during the 40% power test, the "IHX C SHTS outlet sodium temperature high" alarm and fire alarms were simultaneously set off, followed by the "SHTS sodium leak" alarm. A reactor operator went immediately to the SHTS loop C piping room and observed white smoke. The reactor was manually shut down after reducing the reactor power.

A detailed investigation confirmed that the sheath of the thermocouple installed in the IHX outlet piping was broken and about 640 kg of sodium leaked to the room atmosphere from the bottom end of the connector through the opening created by the break.

Holes were formed on the air duct and grating floor located under the leaked piping area, and the leaked sodium was piled up on the floor in an oval shape with a diameter of about 3 m (Fig. 3.30). In addition, the sodium compound (sodium aerosol) generated during the burning of sodium in air diffused to the other rooms of SHTS loop C and outside through the heating, ventilation, and air-conditioning system.

The cause of the thermocouple sheath failure was investigated by various tests and analyses. It was clarified that the narrow part of the thermocouple sheath of a stepped structure was oscillated by the hydrodynamic force of the symmetric vortexes generated in flowing sodium. The resultant high-cycle fatigue concentrated at the stepped part of the sheath caused thc breakage of the sheath. A numerical simulation of a coupled structural and hydrodynamic analysis successfully reproduced the generation of symmetric vortexes and thermocouple oscillation (i.e., side-to-side displacement of the thermocouple centerline) as shown in Fig. 3.31. When the thermocouple is stationary, the alternating vortexes are generated, while when assuming the coupled oscillation with the flow field, the symmetric vortexes are generated as they are.

Based on the above results, the thermocouples were replaced with shorter, tapered ones with no stepped structure to prevent the breakage of the sheath due to hydraulic oscillation (Fig. 3.32).

The findings related to the flow-induced oscillation were generalized to develop the Evaluation Guides for Hydraulic Oscillation of Cylindrical Structure inside Piping (Japan Society of Mechanical Engineers standards: JSME S 012-1998).

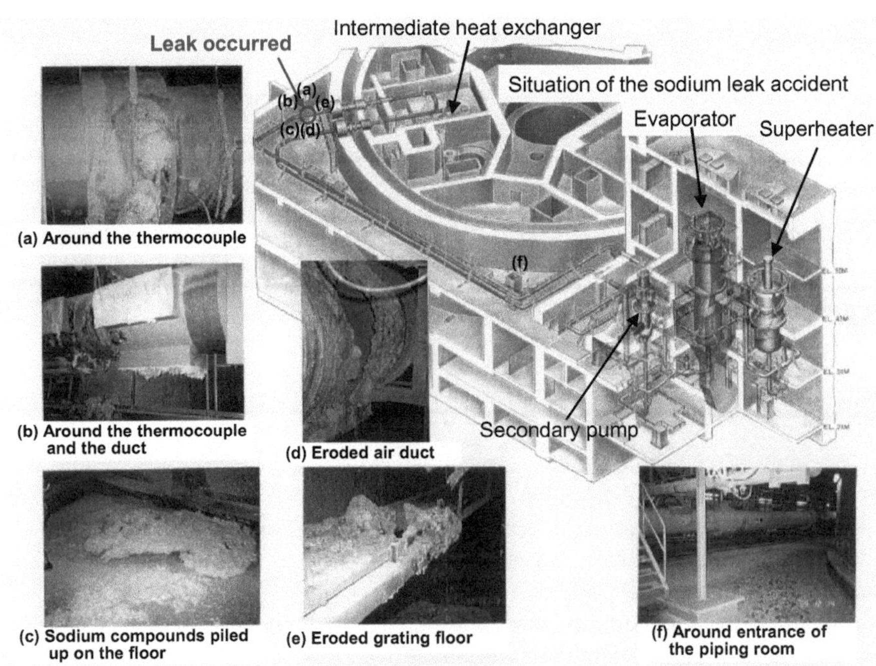

Fig. 3.30 Secondary sodium leak accident.

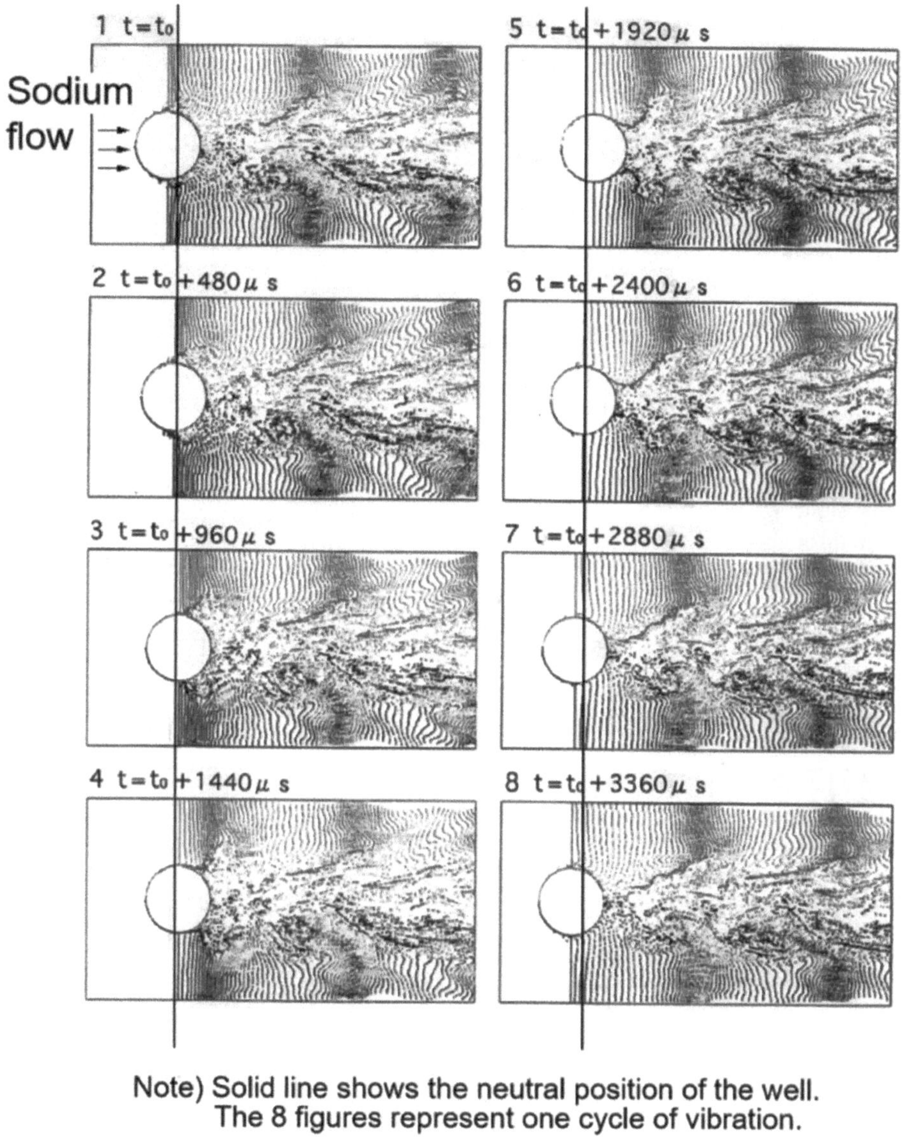

Fig. 3.31 Symmetrical vortex generated at the backside of the thermocouple well.

Concerning the corrosion behavior of steel structures, two types of steel corrosion mechanisms, Na-Fe complex oxidation-type corrosion and molten salt-type corrosion, were identified based on various simulation tests. Na-Fe complex oxidation-type corrosion is induced by a mechanism in which complex oxides are produced by the reaction between sodium oxide and iron. This corrosion occurred on the floor liner during the Secondary Sodium Leak Accident. On the other hand, the molten salt-type

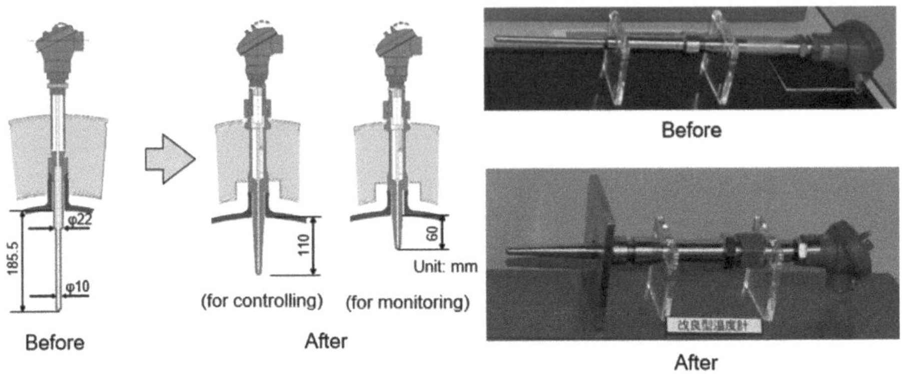

Fig. 3.32 Improvement of thermocouple reflecting the secondary sodium leak accident.

corrosion, which occurred in the reproduction experiments, is induced by a mechanism through which peroxide ions melted in sodium hydroxide under the specific condition of the experiments. The corrosion rate of the molten salt type is about five times larger than that of the Na-Fe complex oxidation type. The model used to predict the corrosion thinning rate of steel floor liner was revised based on the findings.

3.3.3.4 In-vessel transfer machine dropping

In August 2010, the IVTM was dropped inside the RV from 2 m above the installed position when being removed after refueling. It was being lifted using the auxiliary handling machine (AHM) for removal from the RV (Fig. 3.33).

The drop occurred due to the unstable holding of the IVTM with AHM gripper fingers, which were not fully opened to hold the IVTM. The AHM gripper finger drive rod is plane-shaped and must be carefully designed, manufactured, and maintained to prevent rotation. However, the care to prevent rotation was not sufficiently taken in replacing the power cylinder of AHM, resulting in the screws gradually becoming loose to induce the finger drive rod to rotate. As a countermeasure, the finger drive rod was changed to one with a welded structure that does not rotate.

In the recovery work, an in-vessel observation device was newly developed to observe the IVTM in the RV. The device was tested in a full-scale mockup simulating the actual RV conditions. It was then inserted into the RV and the damaged state of the IVTM was observed; the IVTM inner guide tube was deformed and interfering with the inner surface of the fuel transfer hole sleeve. The observation confirmed that the damaged IVTM could not be withdrawn by the normal method, and then, it was removed together with the sleeve. Finally, the new IVTM was manufactured and installed.

The developed in-vessel observation technique is useful for future sodium handling in a high-temperature argon gas and sodium vapor atmosphere.

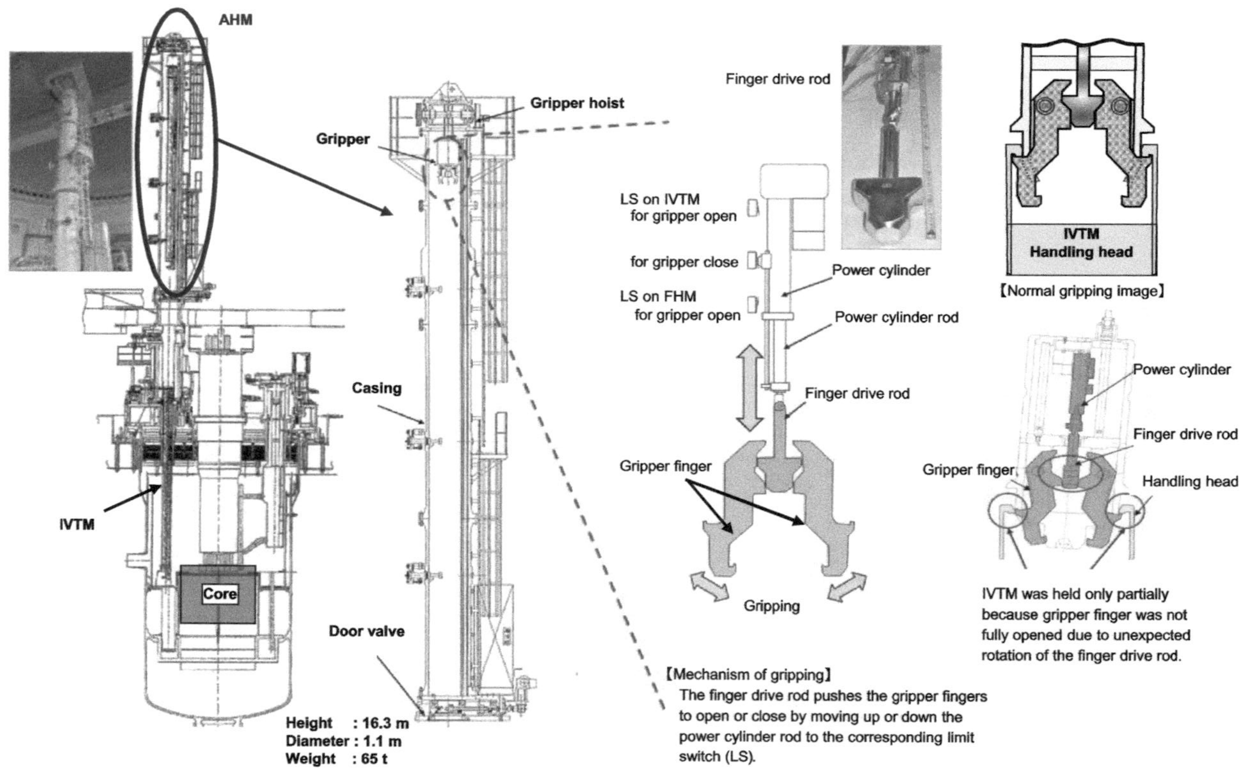

Fig. 3.33 In-vessel transfer machine (IVTM) and auxiliary handling machine (AHM).

3.4 R&D achievements in Monju

3.4.1 Achievements in commissioning

Monju is an FBR plant that was designed and constructed as a reactor in the R&D stage toward the commercialization of SFRs; thus, relatively large design margins are considered. Accordingly, it was expected in the commissioning to validate the design including appropriateness of the margins.

3.4.1.1 Major commissioning steps

The commissioning (i.e., pre-operational test) was divided into two phases: the Comprehensive System Function Tests before loading core fuel and the System Startup Tests (SST) from the core fuel loading to the start of rated power operation to confirm plant performances. The SST was started in October 1993 and suspended in December 1995 due to the Secondary Sodium Leak Accident. After a long-term shutdown state, the SST was resumed in 2010 with a revised plan.

3.4.1.2 Comprehensive system function tests

The core with dummy fuel subassemblies was constructed in May 1991, and then, the functional tests on 125 items were carried out for the primary and secondary cooling systems, fuel handling and storage systems, and other systems in in-air tests at room temperature, in-argon gas tests, and in-sodium tests.

In the in-air tests at room temperature, the operability and controllability of the fuel handling machine and the IVTM in the RV were confirmed by visual observation. A preheating test performed before sodium charge confirmed that the relevant systems were preheated uniformly as designed.

Sodium was procured in France and shipped to Japan in tank containers. As much as 1700 tons of sodium was received at the Monju site and charged into the relevant systems through temporarily prepared storage tanks. After the sodium charge, flushing operation, sodium circulation, and purification tests in the cooling systems were performed. Subsequently, scram tests were performed in air and sodium, and it was confirmed that the control rods were successfully and rapidly inserted within the time assumed in the design.

The preliminary plant performance tests were scheduled before core fuel loading. For the evaluation of natural circulation performance in the primary cooling system, the primary and secondary cooling systems were heated up to 325°C by heat input from the pump, and the core flow rate was measured after the activation of the auxiliary cooling system by a reactor trip signal and the stopping of primary pump pony motor operation. As a result, a core flow rate of 80 m^3/h (larger than 1% of rated flow rate) by natural circulation was measured as predicted (Fig. 3.34).

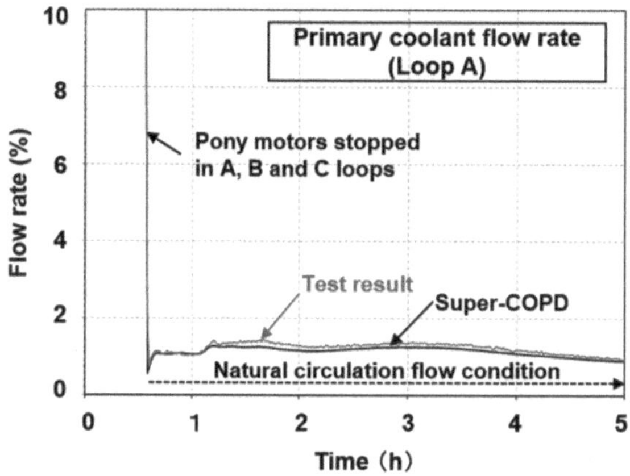

Fig. 3.34 Preliminary test of primary system natural circulation with flow rate of 1%.

3.4.1.3 System startup test

Monju used core fuel consisting of degraded plutonium (i.e., plutonium with high contents of ^{240}Pu and heavier isotopes relative to ^{239}Pu) obtained from the reprocessing of LWR spent fuel. Consequently, it was expected that the reactor characteristics, which cannot be simulated by critical assemblies, are acquired through a variety of reactor physics tests.

(a) *Initial criticality achievement*

A total of 108 inner core fuel subassemblies were loaded into the core in two batches from October 13, 1993. Loading of outer core fuel subassemblies (90 in total) into the core started on January 27. When 60 outer core fuel subassemblies were loaded in 7 batches, initial criticality was achieved on April 5, 1994, with the core consisting of 168 core fuel subassemblies.

In the fuel loading steps, the number of fuel subassemblies to be loaded in the subsequent step was estimated using the inverse multiplication method. The estimated range was narrowed by the source multiplication method with the central control rod inserted or withdrawn to change the level of subcriticality.

The criticality approach was monitored using two different neutron instrumentation systems: two neutron-source-range detectors and two fuel-loading-range detectors that were installed outside and inside the RV, respectively. During the approach, neutron source subassemblies were arranged differently from those in normal operation to improve the efficiency in monitoring neutron multiplication factor.

The equipment hatch of the CV is supposed to be open during fuel loading and is closed during reactor operation. In early fuel loading steps, control rod operation for criticality approach was performed with the hatch open. After predicted subcriticality became less than 1%Δk, the criticality approach operation was performed with the hatch closed. Since about 1 week was required for the opening/closing of the hatch and fuel loading step, new fuel subassemblies were temporarily loaded in the in-vessel rack for efficient fuel loading.

Before the final step of the criticality approach, it was predicted that criticality could be reached by loading the 167-th fuel subassembly. Since Monju's initial criticality was becoming a public concern, two fuel subassemblies were added in the final loading step to ensure that criticality was reached. Initial criticality was achieved with 168 fuel subassemblies.

(b) *Power (reaction rate) distribution measurements*

After completion of the initial core configuration, the experimental fuel subassemblies were loaded in the core, irradiated, and discharged from the core for the evaluation of power distribution (Fig. 3.35). The procedure was repeated six times changing loading patterns. It took considerably longer time for preparation and measurement than any other tests on reactor physics. Experimental fuel subassemblies containing activation foils (Pu, U, Ni, Au, etc.) were loaded and irradiated in the core to measure the reaction rates of the foils. The experimental fuel subassembly is a special subassembly in which a neutron detection element containing the foils is inserted in the central area of a normal fuel subassembly (equivalent to 7 fuel pins). Five core fuel subassemblies, three blanket fuel subassemblies, and four neutron shielding subassemblies were prepared. The effectiveness of an anti-floating mechanism of the detection element in the experimental subassembly was demonstrated in the water hydraulic test loop at the OEC.

The neutron detection elements irradiated and activated in the core were connected to a glove box and cut inside the greenhouse, and the foils were discharged. The radioactivity of the activated foils was determined by gamma-ray measurement using a Ge solid-state detector and then converted to the reaction rates.

(c) *Reactivity worth measurements*

In parallel with the power distribution measurements, reactivity worth was measured for the control rod, sodium temperature, coolant, fixed absorber, and fuel subassembly.

In the measurement of control rod worth, the reference control rod worth was measured at the core center by the asymptotic period method, and then, control rod worth at the other positions was measured by the replacement method. The reactivity meter based on inverse kinetics analysis was effectively used in the test. The modified neutron source method was also tentatively employed to measure reactivity under subcritical conditions; however, it was not successful due to the restriction of neutron detector positions and insufficient count rates.

The temperature reactivity was measured from 190 to 300°C by changing the core temperature by heat input from the primary pump. In the measurement process, the flow reactivity was also measured by changing the flow rate from 49% to 100%.

Coolant reactivity was measured by the difference in reactivity with and without sodium around the core center, where void reactivity is positive. Six experimental fuel subassemblies each with sodium or void (helium gas) regions around the axial midplane were loaded around the core center for the measurement.

《Equipment to handle neutron detection element》

(a) Replacement of neutron detection elements

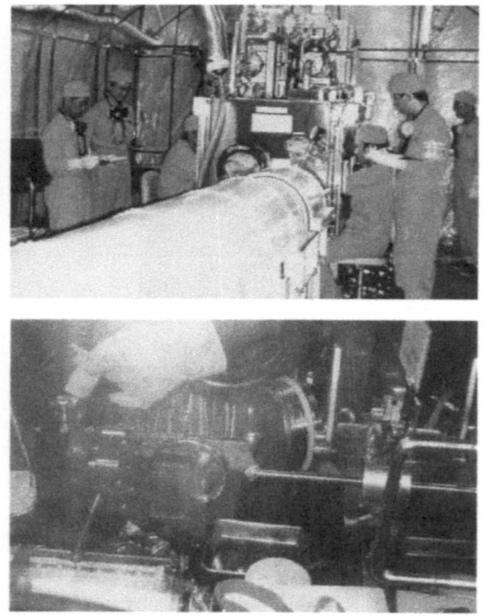

《Equipment to handle neutron detection foil》

(b) Take-out of neutron detection foil (using globe box)

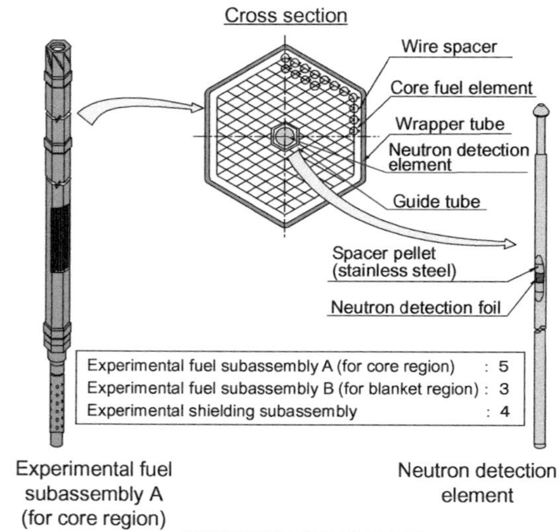

Neutron detection foil

Total 2,000 pieces

(1) fission foil
Pu-239, U-235, U-238, Np-237

(2) activation foil
a) (n, p) reaction : for high energy neutrons
Ni, Ti, Fe (Fe-54)
b) (n, γ) reaction : for low energy neutrons
Au, Fe (Fe-58), Na, Sc, Co

Fig. 3.35 Devices for reaction rate measurement.

(d) *Core flow rate*

The core flow rate distribution was measured under reactor shutdown conditions by sequentially placing the flow rate measurement device on the top of each core element. The device has a configuration similar to the fuel handling machine that connects the gripper to the top of a fuel subassembly. The device is equipped with an electromagnetic flowmeter in place of the gripper.

3.4.1.4 Power operation

(a) *Connection to the power grid*

The nuclear heating test started in September 1995 following the startup procedure of the reactor including reactor power increase, preheating and startup of the water-steam system, adjustment of the plant control systems, and confirmation of plant system performance.

When the evaporator outlet temperature reached the saturation temperature (330°C) at the operating pressure (127 kg/cm^{2}G) with the increase in reactor power, water started to boil and steam was generated in the evaporator.

On August 29, 1995, Monju succeeded in generating electricity and connecting to a power grid as Japan's first FBR, marking its first step as a prototype reactor. Subsequently, the reactor power was increased step by step to 40% rated power.

With the increase in reactor power, characteristics of the turbine system were confirmed, and operation conditions, such as the steam control valves and turning equipment, were successfully adjusted. In addition, the performance of the individual systems was confirmed at various reactor power levels.

Data on steam blow characteristics of the SGs were obtained in a plant (turbine) trip test at an electric power of 40%. It was confirmed that the performances of the feedwater stop valve and superheater drain valve as well as the depressurization characteristics are appropriate.

(b) *Temperature distribution in shield plug*

The temperature distribution in the shield plug was measured to confirm the function of the nitrogen gas cooling system and to collect information about the temperature rise behavior in case of the loss of the function. The circumferential temperature distribution of the rotating plug was confirmed to be uniform regardless of axial positions.

(c) *Pump flow coast-down*

In the primary cooling system characteristics test, the coolant flow rate was increased to 50% under a power operation, and various behaviors of the main circulation pumps and systems were investigated. In the plant trip test at 40% rated power, the flow coast-down characteristics of the primary main circulation pumps were confirmed to be as expected and consistent in the three loops as shown in Fig. 3.36. It was also confirmed that the coolant temperature has a negligible effect on the pump coast-down characteristics.

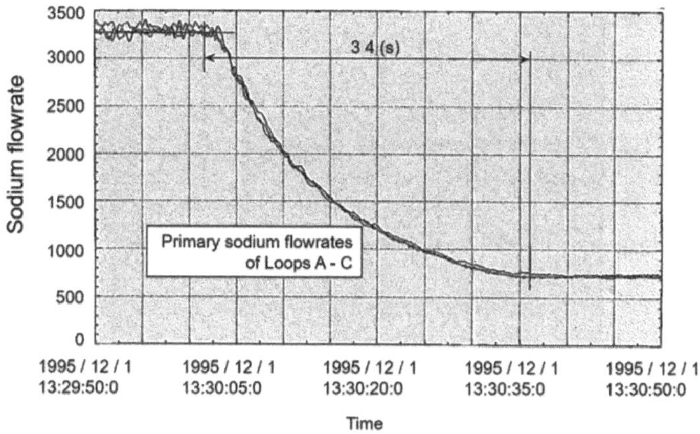

Fig. 3.36 Flow coast-down characteristics of primary circulation pump.

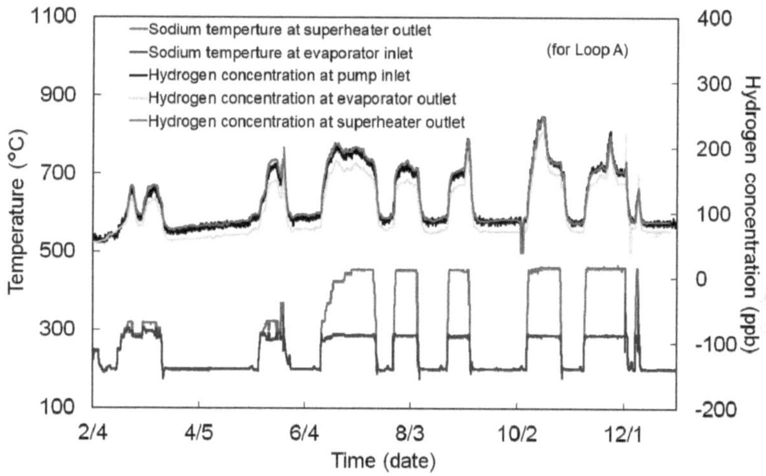

Fig. 3.37 Hydrogen concentration in secondary cooling system.

(d) *Hydrogen concentration in the secondary cooling system*

The hydrogen concentration in the system is important for monitoring a water leak in the SGs and evaluating the removal performance of cold traps. The histories of hydrogen concentrations in sodium and cover gas were measured during SST when the reactor power was increased to 40% (Fig. 3.37). Based on the measurements, the hydrogen concentration was confirmed to vary with operating conditions such as coolant temperature, and the permeability of hydrogen from the SGs was evaluated. These results were then used to optimize the alarm setting for abnormal hydrogen concentration to improve the reliability of a water leak monitoring device.

3.4.1.5 Resumed SST

(a) *Revision of original SST*

The SST was resumed after the Secondary Sodium Leak Accident. "Core Performance Confirmation Tests" at zero power and "40% Power Performance Confirmation Tests" were added to the original SST because plant systems and components were kept in a long-term standby state. The latter Tests were specifically aimed at confirming the startup and operation of the water-steam and turbine systems.

Before resuming SST, tests of 141 items were performed as "Plant System Confirmation Tests" in the 2 years from August 2007 to August 2009 to confirm that the planned SST could be safely resumed.

(b) *Characteristics of americium-containing core (Core Performance Confirmation Tests)*

As the first part of the resumed SST, Core Performance Confirmation Tests were performed for two and a half months from May 2010.

Since about 1.5% of ^{241}Am accumulated in the core fuel due to the decay of ^{241}Pu during the long shutdown period, it was expected that useful data for future study on the minor actinide burning in SFRs would be obtained. Since the refueling was not sufficient to recover the reactivity loss due to the ^{241}Pu decay, core reactivity was expected to be fairly small and predicted carefully by maximum use of the results of the past tests performed in 1994 and the nuclear data uncertainties. As a result, criticality was achieved within the prediction range.

Other items performed in the Confirmation Tests included the feedback reactivity confirmation and the advanced sodium thermometer performance evaluation.

The feedback reactivity confirmation test demonstrated that the plant state is stabilized owing to the effect of inherent negative reactivity feedback mechanisms including the Doppler Effect, after the addition of a positive reactivity (2–6 cents) at the critical condition (Fig. 3.38).

The advanced sodium thermometer performance evaluation test was intended to confirm the performance of a newly developed ultrasonic thermometer, which can measure the sodium temperature outside the pipe. An ultrasonic thermometer was installed in the SHTS (loop C), and temperature data were compared with those of

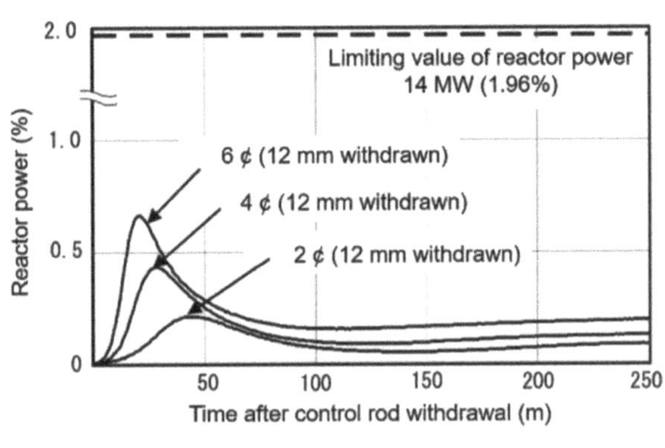

Fig. 3.38 Change in reactor power after positive reactivity insertion.

conventional thermocouples equipped in the loop. Data processing methods were developed to extract temperature data efficiently from the signal waveforms. It was confirmed that the temperature measured by the ultrasonic thermometer agrees well with those by the conventional thermocouple.

(c) *Water-steam system confirmation tests*

To resume SST, the water-steam system was recovered from the long-term storage state and inspected from April to December 2010. Subsequently, in February 2011, the Water-Steam System Confirmation Tests started. Water supply and flushing operations were performed for the condensate system, feed-water systems, and condensate demineralizer to confirm the conditions of pumps and the control system. As a result, no anomaly, such as a water leak or abnormal vibration, was identified. In addition, operation testing of the steam turbine (gland exhaust fan) and the generator system (gas, cooling, and oil systems) was performed, while water chemistry was carefully controlled to minimize the impurities. Through these tests, it was confirmed that long-term storage was appropriate, and the relevant systems were operable without a problem.

However, considering the social environment in the wake of the 1F Accident that occurred during these test series, it was decided to place priority on the so-called stress test against severe accidents and other urgent safety measures to ensure the safety of Monju. Therefore, the Water-Steam System Confirmation Tests were interrupted before supplying water to evaporators. In October 2011, the water-steam system was brought back to a storage state again.

(d) *SST not accomplished*

SST was terminated at 40% rated power without accomplishing the goals of a prototype FBR in the R&D stage.

Regarding the core characteristics, valuable data of an actual FBR core fueled by degraded plutonium fuel were obtained in the reactor physics tests. On the other hand, it was not possible to obtain sufficient data associated with power operation, such as the changes in reactivity, power coefficient, feedback, and Doppler reactivity coefficients. The data were not sufficiently obtained on either burnup reactivity, the change in fuel composition, or the burnup behavior of fuel associated with reactor power operation.

Although sodium-related components and equipment comprising the primary and secondary cooling systems were developed, designed, and manufactured through mockup tests and design studies to be operable at the rated power, data related to performance and behavior at the rated power were not obtained. Other data that were not obtained were data over the entire plant system at the rated power, such as the cooling capability by natural circulation, the transport behavior of radioactive materials, the combined operational controllability of water-steam and sodium systems, and the transport behavior of hydrogen.

Although these test items remained unfinished, it is believed that the data obtained in Monju are extremely valuable and could be effectively used for future SFR development in Japan.

3.4.2 *Design validation through commissioning*

Monju achieved initial criticality with 168 core fuel subassemblies in April 1994. In the initial criticality approach during commissioning, as many as six fixed absorbers were prepared for cases with excess reactivity much larger than predicted. As a result, the critical mass was as predicted, demonstrating high design accuracy.

In the successively performed reactor physics tests and Core Performance Confirmation Tests, data for validation of reactor core characteristics were acquired.

In particular, due to the decay of ^{241}Pu, the core tested in May 2010 (Core2010) became a globally unprecedented one with an ^{241}Am content of about 1.5%, three times larger than that of the core in 1994 (Core1994) (Fig. 3.39). Valuable data, including the effect of ^{241}Am accumulation on design accuracy, were obtained through a comparison of the data obtained in Core2010 and Core1994.

3.4.2.1 *Neutronic design validation*

As for the validity of the neutronic design of Monju, it was demonstrated that the nuclear characteristics data obtained in Monju satisfied the neutronic limits.

(a) *Excess reactivity and control rod worth*

It was confirmed that the excess reactivity (a margin for reactor operation) and the control rod worth (capability for reactor shutdown) both satisfied the neutronic limits.

(b) *Power coefficient measurement*

Measurement of the reactivity change during the power increase from zero to a thermal power of 45% confirmed that the power coefficient is negative over the entire power range.

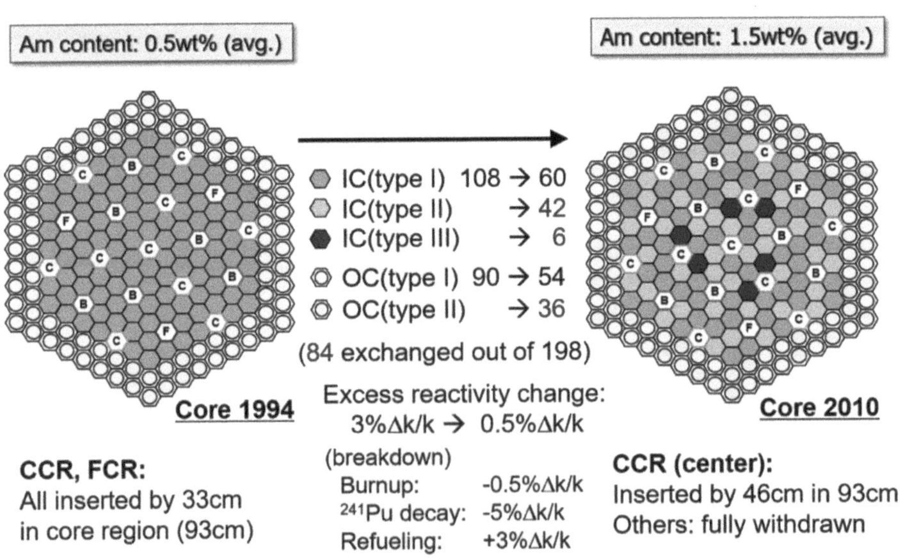

Fig. 3.39 Core configurations of Core1994 and Core2010.

(c) *Isothermal temperature coefficient*

The isothermal temperature coefficient is defined as the change in reactivity when the core temperature is uniformly increased. The temperature was gradually increased by heat input from the coolant circulation pumps, and the change in reactivity was measured by the change in control rod positions for the temperature range from 190°C to 300°C, confirming that the isothermal temperature coefficient was negative.

(d) *Power distribution*

The power distribution measured with activation foils showed that the ratios of calculation to measurement (C/E value) of the ^{239}Pu fission reaction rates, which are closely related to the reactor power, deviated from unity by a maximum of 3% and 5%, respectively, in the core and blanket fuel regions. These values are within the design margins for power distribution: ±5% for the core fuel region and ±10% for the blanket fuel region.

Using the fission reaction data, the C/E values for the maximum linear heat rate in the core region of the initial core were estimated to be from 1.003 to 1.009. The breeding ratio was also estimated to be 1.185, which agreed well with the design target of 1.2.

(e) *Design method validation*

Fig. 3.40 compares the measured and analyzed values of the major nuclear characteristics. The measured and calculated values agreed well within the design margins, confirming the validity of the core design method.

Fig. 3.41 shows the C/E values of the control rod worth obtained in Core1994. The C/E values are almost 1.0 with measurement uncertainties of 2% after correcting the interference effect, which is the change in worth of a control rod caused by the insertion of other control rods.

Fig. 3.42 shows the validation of the criticality obtained in Core1994 and Core2010. There is a 0.2% difference in the analyses between the two cores when using the nuclear data library JENDL-3.3, which is mainly due to the difference in the content of ^{241}Am. The difference becomes negligible when using the revised library JENDL-4.0, suggesting high analysis accuracy of its ^{241}Am nuclear data.

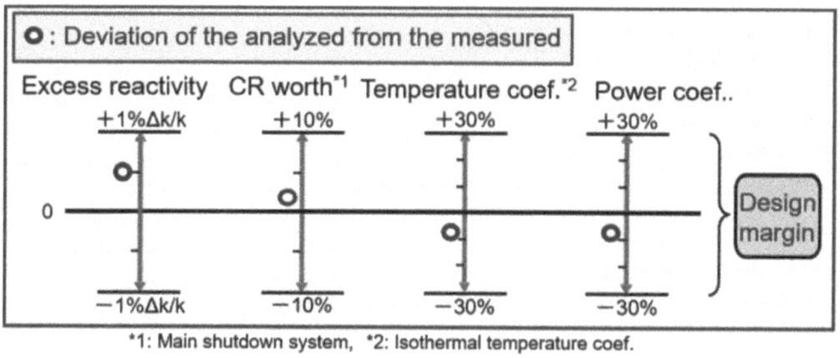

Fig. 3.40 Validation of core design margin.

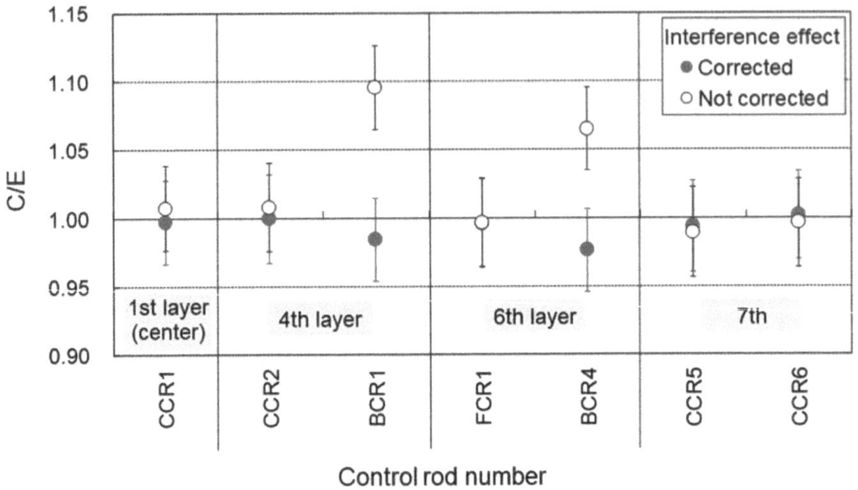

Fig. 3.41 C/E values of control rod worth.

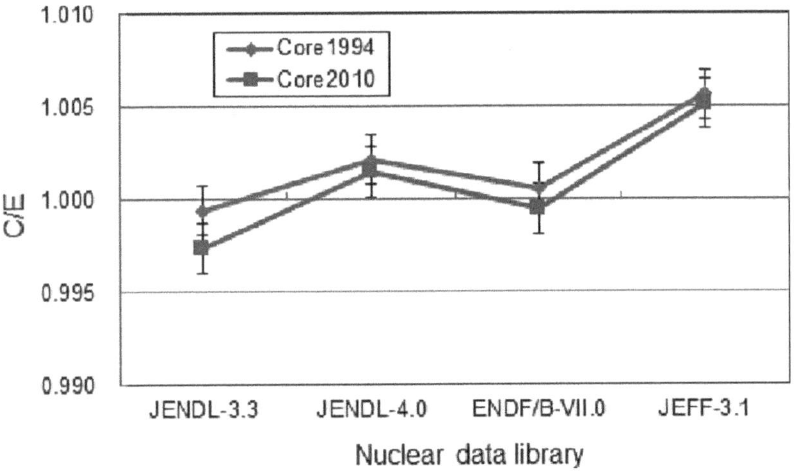

Fig. 3.42 C/E values of criticality.

The calculation method was validated for other major nuclear characteristics measured in Monju; the calculated values almost agree with the measured within the measurement uncertainties. They included reaction rate distribution (power distribution), temperature coefficients, and power coefficient. It is noteworthy that all results were obtained without the use of "correction based on critical experiment data" that was introduced in the original Monju core design.

The measurement of temperature and power coefficients, corresponding to the loss of reactivity associated with the increase in core temperature and power, was performed only at thermal powers up to 45% and might be insufficient for actual reactor data. However, since calculation accuracy was confirmed to be about 5% for both coefficients, it is possible to reduce and rationalize the design safety margin (±30% in the original Monju core design) for the temperature and power coefficients. This is also a significant research achievement to be used in future SFR design.

3.4.2.2 Thermal hydraulic design validation

(a) *Core flow distribution*

The core subassembly flow rates were directly measured using a specially prepared measurement device to confirm the appropriateness of the core flow distribution.

A comparison between measured and calculated values showed good agreement within 2% (Fig. 3.43), indicating that the flow distribution design based on water mockup tests was valid and effective.

(b) *Fuel subassembly outlet temperatures*

The thermocouples are installed at the outlet of all the core fuel subassemblies and 16 blanket fuel subassemblies.

The temperature changes following a reactor trip from 40% power level are compared between measured and calculated values as shown in Fig. 3.44. The subassembly outlet temperatures in the core (inner and outer) are almost flat. The outlet temperatures in the blanket region are initially about 50°C lower than the core and became slightly higher after the trip. The calculation reproduced the measured characteristics very well.

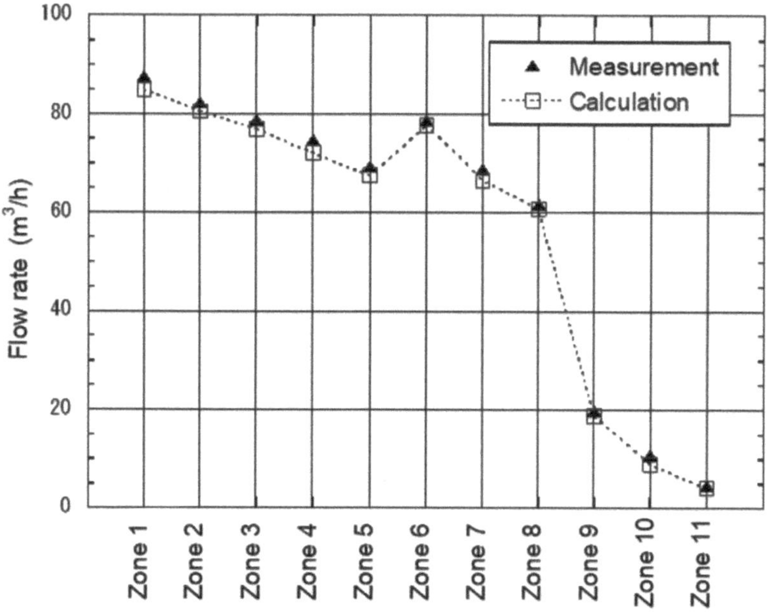

Fig. 3.43 Measurement of flow distribution.

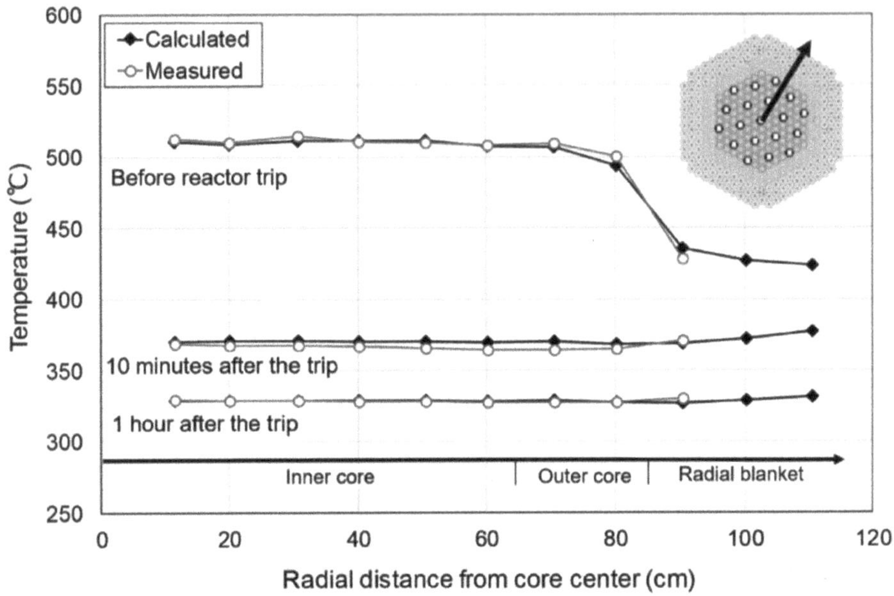

Fig. 3.44 Fuel subassembly outlet temperature change after reactor trip at 40% power.

(c) *Temperature distribution in the upper plenum*

Axial temperature distribution in the upper plenum was measured in the reactor trip test from a 40% power level. It was observed that temperature dropped more sharply in the lower elevations. The change in the temperature distribution is successfully simulated by calculation (Fig. 3.45). In addition, this result on the thermal stratification is well within the range postulated in the design. The data obtained have been internationally utilized in the IAEA benchmark analysis.

3.4.2.3 Component design validation

(a) *Control rod insertion speed*

The control rod latch/delatch test, normal drive test, scram performance test, and a test to confirm elongation of the drive shaft were performed for each of the three CRDM types under the condition that sodium was charged and a dummy core was configured in the RV. Regarding the safety-related scram function of the CRDM, a test performed at the rated coolant flow rate confirmed that control rods can be inserted into the core within the required scram time (1.2 s).

(b) *Function of natural convection suppressing in the main circulation pump*

The main circulation pump has been operated during reactor shutdown, and operational data were accumulated for more than 20 years. Using temperature data from the gas layer in the main circulation pump, the relationship between sodium temperature and the maximum circumferential temperature difference (at the middle level of

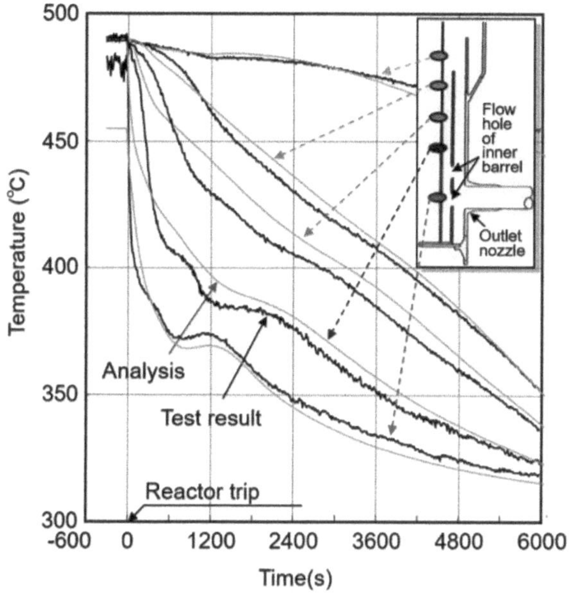

Fig. 3.45 Temperature change in the upper plenum after reactor trip.

the gas layer) was investigated (Fig. 3.46). The maximum temperature difference at the upper level in the gas layer was 6°C for design allowance 10°C, and those at the middle and lower levels in the gas layer were 12°C for design allowance: 15°C. The results show the effectiveness of the natural convection suppressing fins.

3.4.2.4 Development of ISI technology

The Monju in-service inspection (ISI) program was formulated following the basic ISI policies:

- The reactor coolant boundary is inspected by visual observation of welds and by sodium leak monitoring of the main components and piping throughout the in-service period.
- The reactor cover gas boundary is subjected to leak monitoring using radioactive cover gas monitors.
- The important structural materials undergo tests using a material surveillance piece. The environmental effects on the materials are assessed, as needed.

The program was based on R&D of inspection devices for RV visual testing and for the volumetric examination of the PHTS piping as well as the inspection programs of Joyo and LWRs along with their achievements.

(a) *Inspection of RV*

The visual inspection of the main RV parts and inlet piping is to be performed under the unprecedented conditions of a narrow space inside the guard vessel, high temperature, and radiation environment. Therefore, a remote-controlled inspection device was newly developed adopting an environment-resistant remote visual sensor. For

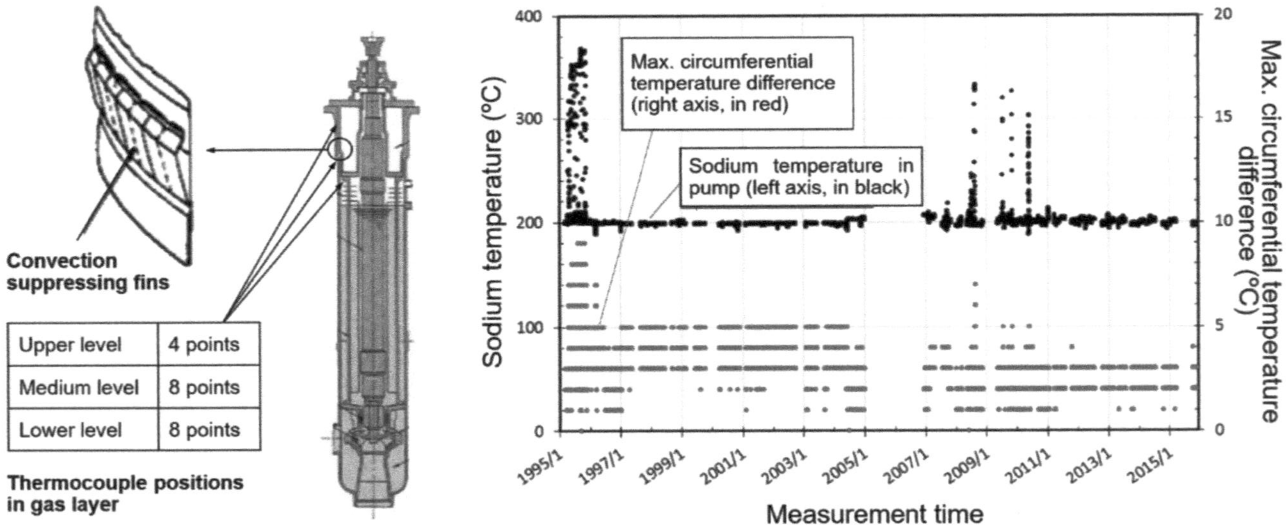

(a) Circumferential measurement points

(b) Measured circumferential temperature differences (at medium level)

Fig. 3.46 Maximum circumferential temperature difference and its temperature dependence.

the RV, a device equipped with an electromagnetic acoustic transducer (EMAT) sensor for volumetric inspection was also developed to detect small defects before they grow to penetrate the RV wall.

The newly developed inspection devices were applied to the visual tests of a preservice inspection (PSI) of Monju. The absence of a significant defect in the RV wall was confirmed, and several points of improvement were identified on inspection device operation.

To reduce the inspection time and improve reliability and durability, the RV inspection device was modified by improving accessibility to narrow spaces, widening the field of view using a CCD camera, and reducing the maintenance time by improving the durability of mounted components. The improvements in accessibility are illustrated in Fig. 3.47. The improved functions were checked using a mockup system, and the performance was confirmed to be satisfactory.

The EMAT was further improved by enhancing signal strength and reducing size and weight by adopting the technologies developed in the field of accelerators. As a result, defects with a depth-to-plate thickness ratio of 20% were successfully detected with the new sensor. However, the problem of electrical noise arose when the sensor was mounted on the inspection device. Further improvement is needed for practical application.

(b) *Inspection of PHTS piping*

The ISI for the PHTS piping is based on visual inspection and volumetric inspection. Since the visual inspection device was already available, the development was focused on a volumetric examination device considering thin-walled, large-diameter piping specific to SFRs.

One of the key features of the device is the use of a tire-type ultrasonic probe. To reduce worker exposure, non-couplant probes, not requiring a contact medium, were employed for continuous inspections in vertical and oblique angles.

In the function tests and PSI using the PHTS piping volumetric examination device, the inspection device demonstrated satisfactory performance in terms of driving, flaw detection, and data storing functions, while identifying remaining issues, such as the complexity of the inspection procedure and the deterioration of sensitivity due to aging.

The tire-type probes and the scanner were further improved, and the control device and the data storage device were updated. The functions of the new system were confirmed in a mockup system and in the piping of Monju. The sensitivity and other probe performances were further improved based on the obtained data.

(c) *Inspection of steam generator tube*

The SG tubes have a complex structure with a long, helically coiled shape (maximum length: 85 m). The evaporator tubes are made of a magnetic material, for which a conventional eddy current test (ECT) is not suitable. In addition, they have a thick wall, which is difficult for the eddy currents and magnetic fields to penetrate. Because of the above, the inspection technologies successfully used for LWRs could not be applied without modification. New inspection technologies were developed with a focus on the ECT for ferromagnetic and thick-wall piping.

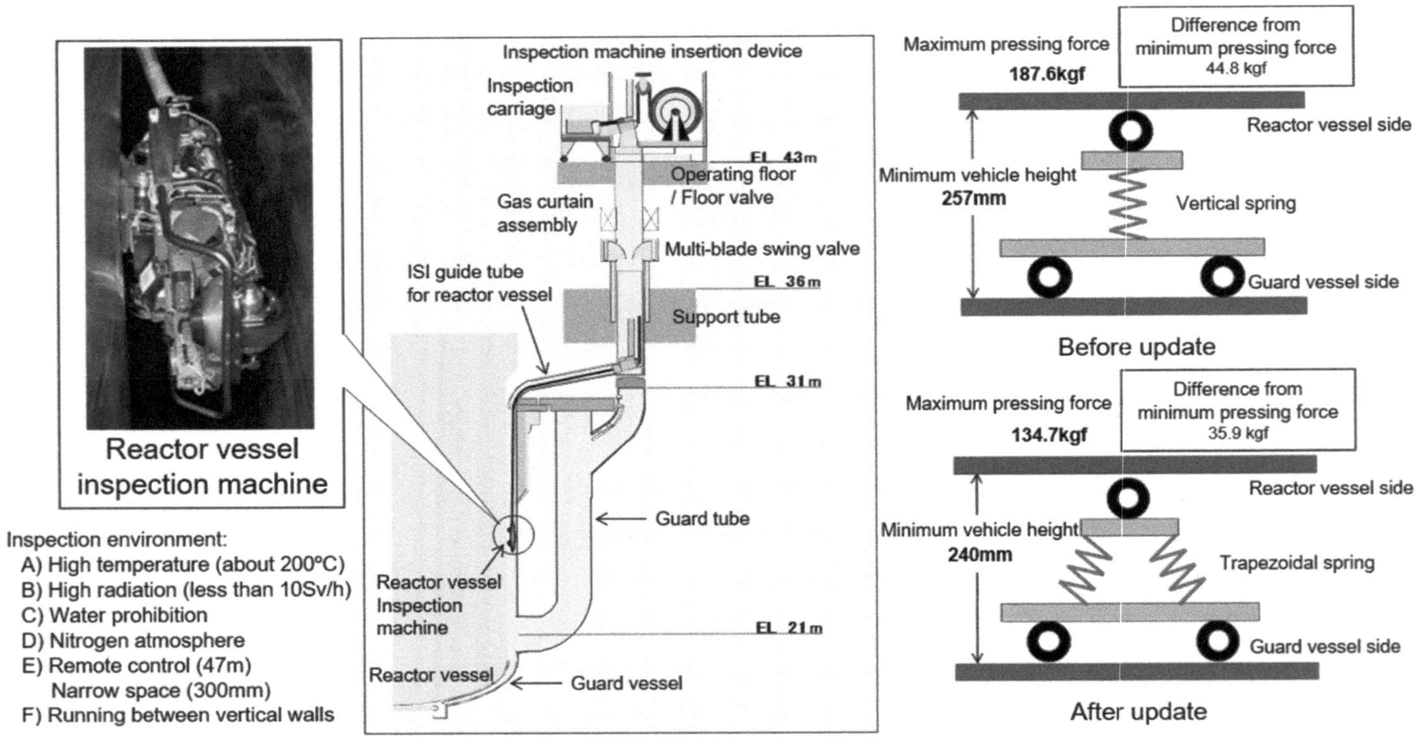

Fig. 3.47 Update of RV inspection machine for use in narrow space.

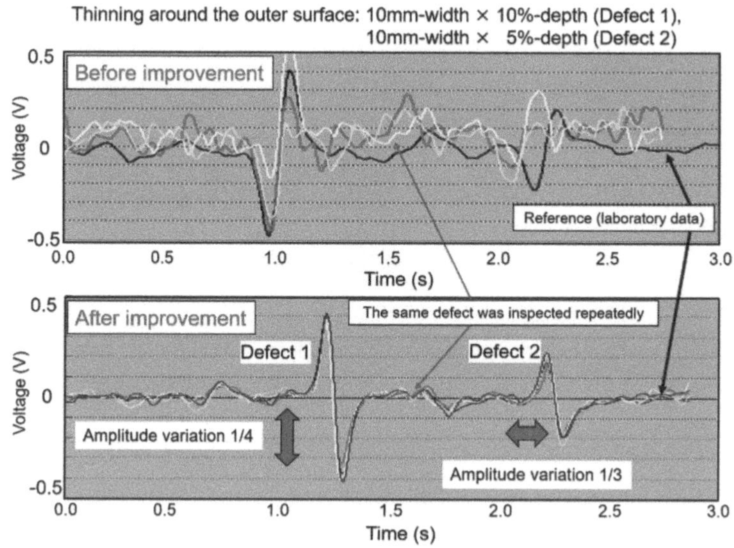

Fig. 3.48 Improvement in defect detectability by vibration countermeasures.

In the function tests, data were accumulated through a series of operational performances related to transfer, assembly, installation, adjustment, and flaw detection using the device. In addition, flaw detection data on all the heat transfer tubes in the evaporators and superheaters were collected for ISI.

The defect detection performance and the aligning mechanism were improved, while noise reduction was achieved (Fig. 3.48). Performance was further improved to detect defects on the support structure with sodium attached to the outer surface of SG tubes. Evaporator flaw detection data were obtained in the Plant System Confirmation Tests, and stable and improved detectability was confirmed.

3.4.3 Safety evaluation

3.4.3.1 Safety margins

Safety margins in the reactor design were evaluated based on the data obtained in the commissioning. In the design stage, large safety margins were adopted in the initial and analytical conditions conservatively to obtain more severe results. The results of safety assessment in the design stage were compared with those evaluated using the data measured in Monju for the evaluated accident events including the PHTS circulation pump seizure (the pump stick accident) (Fig. 3.49). It was confirmed that there were significant margins in the design, such as an evaluated cladding temperature of 702°C, much lower than the safety assessment result of 796°C.

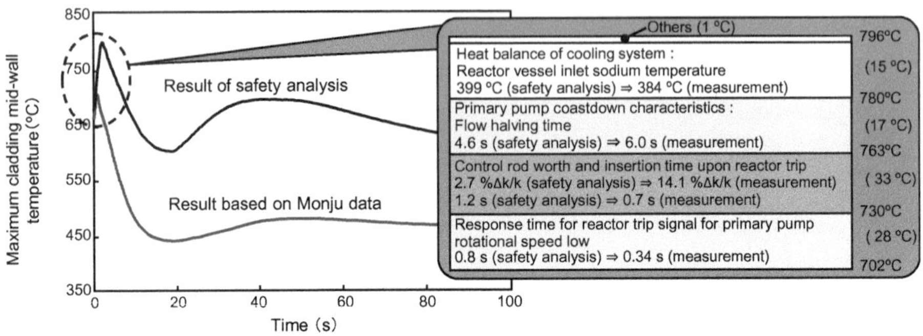

Fig. 3.49 Safety margin for primary pump stick accident in design basis accident analysis.

3.4.3.2 Probabilistic risk assessment

The probabilistic risk assessment (PRA) methods developed in the United States in the 1970s were applied to Monju. A reliability database for sodium components specific to SFRs was developed, and the data required for PRA, such as the component failure rate, have been continuously expanded.

In the PRA, potential initiating events are systematically identified first, and then, the accident sequences from each initiating event are analyzed by successively quantifying the success/failure response of safety systems and components (equipment responsible for safety functions) to assess the occurrence frequency of core damage (Level 1 PRA). Next, the containment failure frequency and the source terms of radioactive materials are evaluated by analyzing the in-vessel and ex-vessel accident progression for each core damage accident sequence (Level 2 PRA). Major achievements are as follows:

- A PRA technological base applicable to SFRs was established through the implementation of the detailed PRA for Monju.
- For Level 2 PRA that evaluates the core damage progression processes, the latest safety analysis codes and in-pile and out-of-pile test data obtained through safety research activities and international cooperation were used as much as practically possible.
- It was estimated that the core damage frequency (CDF) and the loss of containment function (including the containment isolation failure) were, respectively, in the orders of 10^{-7} and 10^{-9} per reactor year, and hence, the risk level of Monju would be at a sufficiently low level compared with that of LWRs.
- The estimated probability of the occurrence of an early and large-scale release of radioactive material that is most notable from the risk perspective was much lower: less than 1/10,000 per core damage.
- The PRA results were effectively used for design improvement of the plant protection system, etc., in the detailed design stage and the preparation of accident management measures after the start of operation.

The above results confirmed that the potential risks of Monju are maintained at an extremely low level and that the application of PRA methods is extremely useful in evaluating the comprehensive appropriateness of safety design and in examining measures to further improve the safety of equipment and operating procedures.

3.4.3.3 Seismic back-check

The seismic back-check review was performed based on the Regulatory Guide for Reviewing Seismic Design revised in 2006 considering new findings from seismology and seismic research. As for Monju, the seismic safety of building/structure and component/piping systems was reevaluated by increasing the design basis earthquake ground motion (horizontal direction) from the original design acceleration of 466 Gal (cm/s^2) to 760 Gal (other simulated seismic waves were also considered). It was confirmed that the acceptance criteria were satisfied and seismic safety would be ensured.

To ensure an appropriate seismic safety margin, the 100-m-tall stack was reinforced since its seismic margin was reduced by increased design-basis earthquake. The stack strength was increased by installing a damper at the top of the stack to suppress vibration during an earthquake and by reducing the number of fixed positions of the support towers. Concerning the seismic stability of the slope on the backside of the plant, a large amount of soil was removed from the slope surface.

In addition, the possibility of heat removal by natural circulation was confirmed even if the seawater pump intake limit is exceeded by the postulated tsunami wave.

3.4.3.4 Safety improvement following the 1F Accident

(a) *Safety measures in consideration of the 1F Accident*

In consideration of the occurrence of the 1F Accident following the Tohoku Region Pacific Coast Earthquake on March 11, 2011, various safety measures assuming emergency situations including station blackout (SBO) were immediately taken.

Since most facilities of Monju are placed 21 m or higher above sea level, they are tolerant against tsunami. However, measures to stop water were taken around the seawater pump at the Monju port and the seawater intake piping penetration to ensure safety.

During SBO, the core cooling by natural circulation of sodium is possible without an electric power source or urgent operator actions such as depressurization of the system and water injection as required in LWRs.

(b) *Safety evaluation of natural circulation cooling*

The direct cause of the 1F Accident is the complete loss of all cooling functions caused by SBO and the loss of heat sinks due to the flooding of seawater system equipment and power supply system resulting from a tsunami far beyond the design basis. Postulating a hypothetical large-scale tsunami at the Monju site, the coolability of the core and spent fuel was evaluated.

Concerning the removal of core decay heat, an examination of the conditions required for maintaining natural circulation confirmed that the core can be stably cooled down to a cold shutdown state and the integrity of the coolant boundary is ensured even when considering the uncertainties in the cooling capability and the availability of coolant flow paths. As shown in Fig. 3.50, the coolant temperatures naturally decrease without relying on the power supply or operator actions. The coolability of spent fuel and the integrity of the EVST are ensured as well.

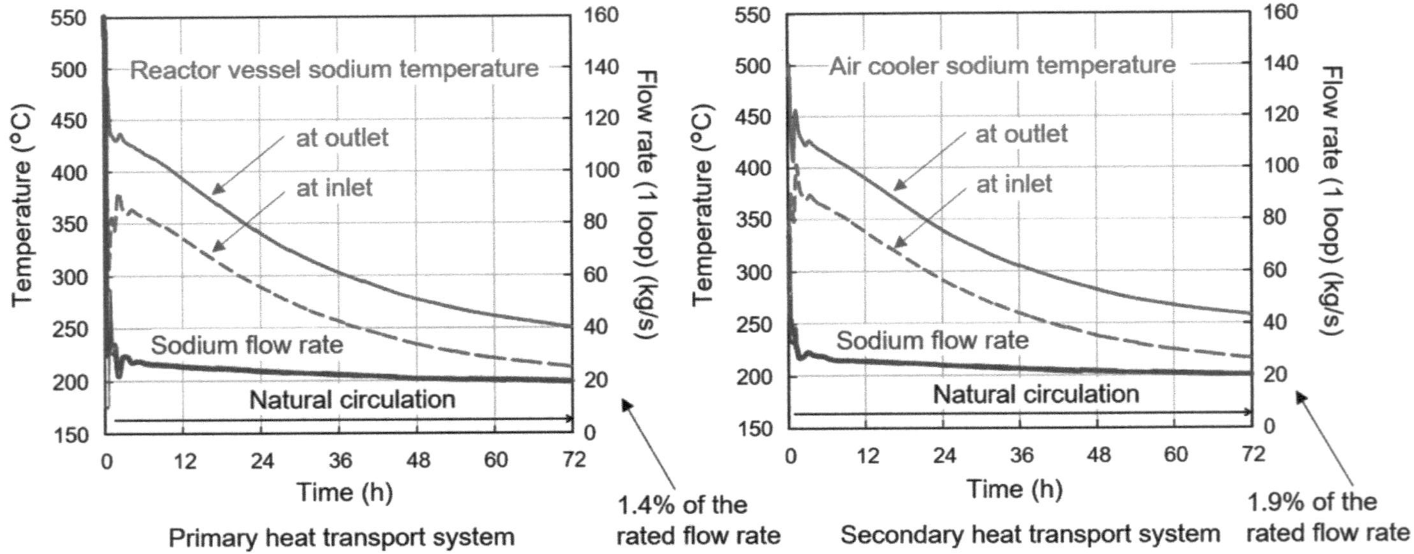

➢Emergency diesel generators stopped at 13 s after the earthquake
(1ry/2ry main cooling system circulation pumps and air coolers stopped)
⇒Transition to natural circulation and natural ventilation
⇒Cold shutdown condition* after 3 days

* All control rods inserted (except for failed rods) and primary coolant temperature stays between 180 and 250°C.

Fig. 3.50 Analyses of natural circulation cooling during station blackout.

Concerning the spent fuel water pool, it was confirmed that there would be a large time margin of longer than 2 months before the top of cans containing spent fuel is exposed to air by water evaporation and that the water temperature would not increase over 70°C.

(c) *Comprehensive safety evaluation*

In consideration of the 1F Accident, a comprehensive evaluation of the safety of Monju, a so-called "stress test," was performed, similarly to the LWR plants. The events evaluated were an earthquake, tsunami, SBO, and the loss of ultimate heat sink. The margin to significant fuel damage was quantitatively evaluated for each of these events. As a result, the safety margins for the reactor and EVST were confirmed to be 1.86 and 2.2 times the design basis earthquake, respectively. Against tsunami, safety was confirmed up to a tsunami reaching the plant installation level, 21 m above sea level, much higher than the design tsunami height of 5.2 m. In addition, in the case of the SBO and the loss of the ultimate heat sink, it was confirmed that continuous cooling would be possible by natural circulation of sodium and natural air ventilation.

(d) *Addressing the new regulatory standards*

Toward the resumption of operation of Monju after the 1F Accident, JAEA developed a policy to ensure safety with full consideration of safety features of SFRs and based on the trend of international safety standards. The policy was peer-reviewed by domestic and foreign experts. It is particularly important to consider the characteristics of a low-pressure system that can virtually eliminate the possibility of containment overpressure failure, different from a high-pressure LWR system, in which containment overpressure failure is a dominant failure mode. It is also important to take advantage of the capability of in-vessel retention of a damaged core and passive decay heat removal by natural circulation.

To undergo a conformity review based on the New Regulatory Standards enacted in 2013 by the Nuclear Regulation Agency, the preparatory effort was made in Monju until the project was terminated at the end of 2016.

When a new SFR development project is studied in Japan in the future, much of the knowledge, including the achievements and experience in the licensing of Monju, safety analysis codes advanced through applications to Monju, and the supporting experimental database, should be applicable.

Reference

[1] F. Nakashima, et al. (Eds.), Prototype Fast Breeder Reactor Monju—Its History and Achievements, JAEA-Technology, 2020. 2019-020.

Demonstration and commercial plant design study

4

Yoshitaka Chikazawa[a], *Atsushi Kato*[a], *Masatoshi Kameyama*[a,b], *Shoji Kotake*[a,c], *Shigenobu Kubo*[a], *Shigeo Ohki*[a], *Shuji Ohno*[a], *Kazuya Takano*[a], *and Tomohiko Yamamoto*[a]
[a]Sector of Fast Reactor and Advanced Reactor Research and Development, Japan Atomic Energy Agency, Ibaraki, Japan, [b]Projects Development Department, Japan Atomic Power Company, Tokyo, Japan, [c]Japan Atomic Power Company, Tokyo, Japan

Chapter outline

Following the development of the experimental reactor Joyo and the prototype reactorh Monju, several design studies of large-scale SFR plants were conducted in Japan, which includes a demonstration plant design by the utilities and a 1500 MWe SFR that is named Japan sodium-cooled fast reactor (JSFR) as a commercial plant jointly by JAEA, the utilities, and the industry.

Sodium Fast Reactors. https://doi.org/10.1016/B978-0-12-824076-2.00004-3
Copyright © 2022 Elsevier Inc. All rights reserved.

1. ***Demonstration reactor project***: A demonstration fast breeder reactor "DFBR" project was carried out primarily by industries, namely utilities and manufacturers. However, due to both domestic and international circumstances, the DFBR project was canceled.
2. ***JSFR in FaCT project:*** As a commercial plant concept in the FaCT project, JSFR with a capacity of 1500 MW was developed. MOX fuel was studied as a reference for the core, but metal fuel was also considered as an option. JSFR employs several innovative key technologies to achieve high safety and economic targets, and these innovative technologies were evaluated in terms of maturity before moving forward with further development steps such as conducting large-scale demonstration tests, basic design, and detailed design of a demonstration reactor. For the demonstration reactor, a comparison of reactor output between 500 MWe and 750 MWe was conducted and 750 MWe was selected. The AtheNa sodium test facility was constructed to conduct component and system demonstration tests for the large-scale demonstration test.
3. ***Safety improvement:*** Since 2011, to contribute to the development of safety design criteria and safety design guidelines, in the framework of the Generation IV International Forum (GIF), a design study on the 750 MWe demonstration JSFR enhancing safety and maintainability was conducted, which included lessons learned from the TEPCO's Fukushima Daiichi nuclear power plants accident (the 1F accident).
4. ***Pool-type SFR:*** International collaboration on ASTRID was initiated in 2014, and a design study of a pool-type concept accumulating knowledge from the ASTRID collaboration was conducted showing feasibility even against Japanese severe seismic conditions.

4.1 DFBR: Demonstration reactor project lead by utilities

4.1.1 Background of DFBR design study

Since 1978, the Federation of Electric Power Companies of Japan conducted a conceptual design study of a demonstration reactor based on the technology adopted for the preceding reactor, namely Monju. The construction cost of the demonstration reactor was found to be several times that of the light water reactors (LWRs). Therefore, to improve the efficiency of the demonstration reactor, it was decided to proceed with the design study by anticipating innovative technologies that could be incorporated into the demonstration reactor in the next 10 years of development.

In the design of the loop-type reactor, which was adopted as the first reactor, the lengths of the primary and secondary piping were shortened, and the main components were arranged close to each other to reduce building volume. In the design of the pool-type reactor, which was developed in Europe, the design applicable to the severe seismic design conditions in Japan was studied.

In the mid-1980s, when the construction work of the prototype fast breeder reactor Monju was in full swing, fast breeder reactors (FBRs) were being promoted in Japan according to the "Long-term plan for the development and utilization of nuclear energy" of the Atomic Energy Commission in 1987 with the aim of commercialization of FBRs in 2020–2030.

Since 1985, Japan Atomic Power Company has been the main developer of the demonstration reactor, and the rationalization of the plant design, systematization of innovative technology, and comparative evaluation of reactor types have been conducted. The loop-type reactors such as Monju have a disadvantage in that the amount of construction material increases because of the piping installed for thermal stress mitigation. On the other hand, pool-type reactors, such as Superphénix, which was being constructed in France at that time, have less piping outside the reactor vessel and require less containment space; however, compared to the loop-type reactors, the pool-type reactor vessel has a larger diameter and inferior maintenance and repairability capabilities.

In Japan, the top-entry loop-type reactor was selected as a suitable type for the demonstration reactor in 1990. This selection was made based on the considerations that inspection and maintenance capabilities are important for stable operation and it is important to be able to take prompt action even if main components need repairs. The basic specifications of the demonstration reactor were selected in January 1994 after estimating the construction cost. As the power scale of the commercial reactors is equivalent to that of LWRs, i.e., 1 million kW class, the power of the demonstration reactor was set to be approximately 660,000 kW with three loops. Following the selection of the basic specifications, the conceptual design of the DFBR plant was continued, and the development of the elemental technologies necessary for improving economic efficiency and safety was promoted.

4.1.2 Outline of DFBR design

The main purposes of the demonstration reactor are to determine the economic prospect of the commercial stage and to demonstrate the technical feasibility of a large power plant in the commercial stage. These purposes are achieved by constructing and operating the demonstration reactor with innovative technologies that improve economic efficiency and safety. In terms of economics, an important goal of the design study was to reduce the construction cost of a demonstration reactor with an electrical output of 660 MW to less than 1.5 times the cost of an LWR with the same output as the cost of a demonstration reactor is converted to the cost of a commercial reactor of 1 MW class.

4.1.2.1 Study to improve economic efficiency

The concept of the reactor and main cooling system is shown in Fig. 4.1. The main specifications are listed in Table 4.1. The design of the demonstration reactor was greatly rationalized by incorporating the following new design concepts that differ from those of Monju.

- The length of the primary piping and its installation space was significantly reduced by adopting the top-entry loop-type reactor, in which the primary piping enters and exits from the top of the reactor vessel. (The length of the primary piping was reduced to about 1.5 times that of Monju.)
- In Monju, SGs were separated into an evaporator and a superheater because of the restrictions imposed by the material used in the high-temperature water environment. However, in the demonstration reactor, a new material (Mod. 9Cr—1Mo steel), which has excellent high-temperature strength and stress corrosion cracking resistance, was adopted, and the evaporator and superheater were integrated. (The amount of materials required for the SGs was reduced to about one-third that of Monju.)

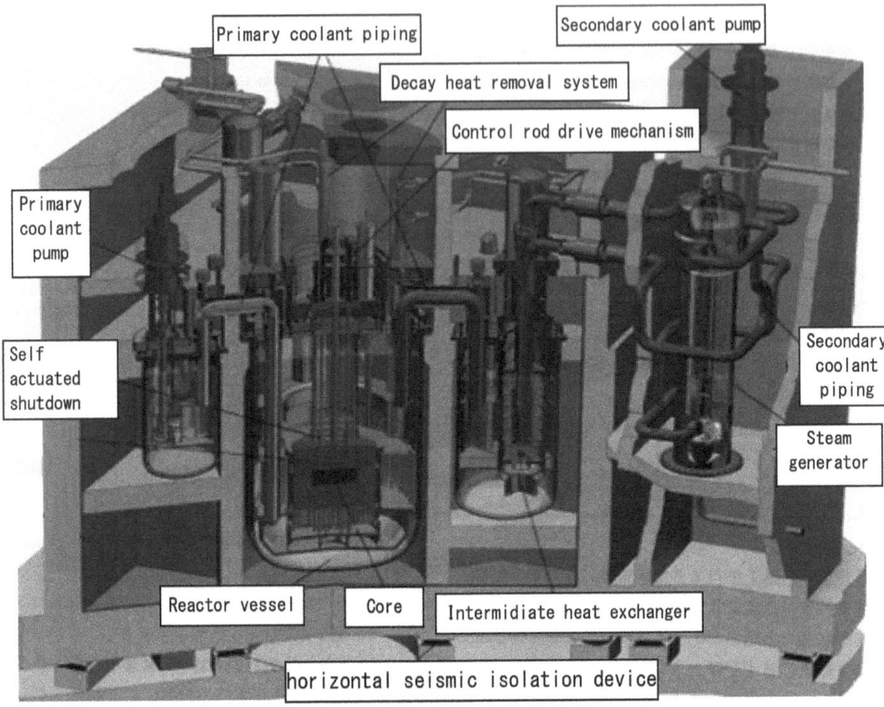

Fig. 4.1 Concept of reactor and main cooling system for DFBR [1,2].

Table 4.1 Main specifications of the DFBR [1,3].

Item	Specifications
Thermal output	1600 MW (electrical output about 672 MW)
Reactor type	Top-entry loop-type reactor
Reactor outlet temperature	550°C
Core	Homogeneous core
Fuel	U/Pu mixed oxide fuel
Burnup	About 150,000 MWD/t (high burnup stage core)
Breeding ratio	About 1.2 (with blanket)
Reactor shutdown system	Two independent systems, one of which is equipped with a static reactor shutdown mechanism
Decay heat removal system	Four systems, direct core cooling system
Containment facility	Liner-lined rectangular concrete containment facility
Fuel handling system	Rotary plug and manipulator system
Reactor building	Horizontal seismic isolation design

- In Monju, a steel cylindrical containment vessel was adopted as the containment vessel, but in the demonstration reactor, a rectangular, reinforced concrete containment vessel integrated with the building was adopted to reduce the building volume significantly. (The volume of the reactor building was reduced to about a quarter that of Monju.)

In the demonstration reactor, the core, primary coolant pump, and intermediate heat exchanger are placed in independent vessels and connected by inverted U-shaped piping, considering economy and ease of maintenance and repair. The reactor structure was simplified, and the reactor vessel diameter was reduced; the fuel handling equipment was rationalized by reducing the number of fuel handling machines, and a building design with a horizontal seismic isolation device was adopted to make the structure thinner and reduce the amount of materials. The construction cost was evaluated to be less than 1.5 times that of the design-target LWR.

4.1.2.2 Study to improve safety [4,5]

Similar to the LWRs, the safety design of FBRs is also based on three safety functions: "stop (reactor shutdown)," "cool (core cooling)," and "contain (containment)."

For reactor shutdown, two reactor shutdown systems were designed using different detectors, logic circuits, and absorber holding mechanisms and absorber shapes to ensure their independence and high reliability. This active safety function was designed to detect the abnormal events that occur in a minor stage and to terminate the abnormal events at an early stage.

In addition to the design of the active shutdown system, a self-actuated shutdown mechanism was developed and adopted. In this mechanism, the control rods are naturally detached and inserted by their own weight when the core coolant temperature rises during an abnormal event. By test and analysis, it was confirmed that the reactor can be safely shut down in consideration of the assumed uncertainties.

About the core cooling function, a direct core cooling system with a heat exchanger immersed in the reactor vessel was designed to suppress the temperature rise of the core fuel quickly and to reduce the heat load on the cladding and structural materials caused by forced circulation with the main circulation pump and the direct core cooling system after reactor scram. In addition to this active safety function, the core was designed to be cooled for a long period by the natural circulation driven by the density difference of the coolant even in the event that the power supply to drive the circulation pump is lost. In particular, to enhance the natural circulation capability, a structure capable of promoting the interwrapper flow (IWF) and a code for analyzing the natural circulation behavior very precisely were verified in cooperation with the then JAEA and the Central Research Institute of Electric Power Industry.

By a probabilistic safety assessment, it was confirmed that the core damage frequency can be significantly reduced by "shutdown" and "core cooling," which have both active safety functions with high reliability and passive safety functions based on natural phenomena. Furthermore, considering the core safety features of FBRs, countermeasures against core disruption were also studied extensively with the cooperation of the Power Reactor and Nuclear Fuel Development Corporation (PNC).

The core of an FBR is capable of fuel breeding, high fuel burnup, and long-term operation, thereby enabling the effective use of uranium resources and reducing the fuel cost. In addition, the reactor outlet temperature can be 550°C, and the thermal efficiency of the plant can exceed 40%. Moreover, the sodium coolant is kept in a low-pressure system, and the main cooling pipes do not rupture, and the design does not lead to any loss-of-coolant accident (LOCA), which was a safety problem in the case of LWRs. Therefore, the emergency core cooling system (ECCS) is not required in the FBR, and the core coolant level can be maintained by means of the guard vessel that covers the reactor vessel. The main cooling system piping is installed at a higher location than the sodium level in the reactor vessel. Because of these features, the effects of accident events assumed in the design-basis events are mostly minor, and the core fuel is not being damaged in such events.

However, the coolant temperature coefficient becomes more positive as the core of a fast reactor (FR) becomes larger, and the concentration of the molten fuel may lead to recriticality. Fuel failure and fuel melting are assumed in situations that exceed the design-basis events, and in such cases, the consequences of core damage are expected to be severe. Therefore, research on hypothetical core disruptive accidents (HCDAs) in FRs commenced in the United States, and this was followed by active research in the United Kingdom, Germany, France, and Japan. Research on HCDA included fuel failure tests and failed fuel behavior tests using an experimental reactor, out-of-pile tests using simulant materials, and the development of safety analysis codes that incorporate the physical models obtained from these tests. In Japan, based on the experience of the safety review of Monju and the progress of safety research at that time, more rational and appropriate core damage event transition and countermeasures against this transition in demonstration reactors were studied with the cooperation of domestic and foreign research institutes.

As shown in Fig. 4.2, the probabilistic safety assessment method was applied to the FR design in the latter half of the 1990s, and active and passive safety functions were combined to significantly reduce core damage frequency. In addition, HCDAs were also evaluated to terminate the consequences of core disruptions inside the reactor vessel and to prevent the impacts of core disruptions from spreading beyond the site.

4.1.2.3 Influence of the Monju accident and further rationalization of the design

A sodium leak occurred in the secondary piping of Monju in December 1995. Although this accident did not affect the core safety at all, the recovery tasks, such as searching for a part of a damaged thermometer and removing the leaked sodium aerosol, took a long time. Therefore, the importance of countermeasures against sodium leakage, which is not related to core safety, to secure the operation rate of the FBR was recognized. Owing to the increased liberalization of the electricity market at that time, there was a greater need to reduce the construction and operation costs of nuclear power plants, including LWRs.

Under these circumstances, a design study to establish a drastic countermeasure against sodium leakage and to further reduce the cost was conducted since 1997.

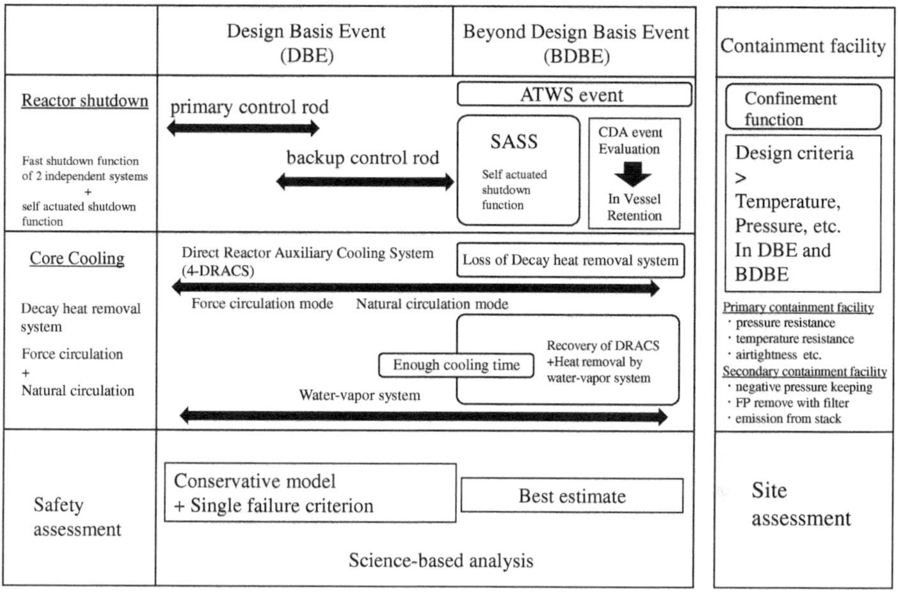

Fig. 4.2 Concept of safety design for DFBR [6].

4.1.3 Feasibility study on practical application strategies

As described in Sections 4.1.1–4.1.2, various measures to improve economic efficiency and safety were adopted in the demonstration reactor design. However, the design study was terminated at the end of 1999 because of changes in circumstances, such as the easing of the natural uranium supply-demand balance, demand for further economic improvement of FBRs, and development of FRs in Europe and America.

Many valuable research results in the DFBR design study were passed down to the "Feasibility Study on Practical Application Strategies," which was started in 1999 with the participation of manufacturers and research institutes, in addition to cooperation between the government and utilities.

4.2 JSFR developed in the FaCT project

4.2.1 Design requirement on commercial concepts

4.2.1.1 Introduction

Japan Atomic Energy Agency conducted a "Fast Reactor Cycle Technology Development (FaCT)" project in cooperation with the Japanese electric utilities [7,8]. Mitsubishi FBR Systems, Inc. (MFBR) contributes significantly to the design and engineering of Japan Sodium-cooled Fast Reactor (JSFR). The FaCT project was constructed on the findings of the previous project, the Feasibility Study on Commercialized Fast Reactor Cycle Systems (FS), which was completed over seven years from

1999 to 2005 [9,10]. In the FS, a combination of the sodium-cooled fast reactor with oxide fuel, the advanced aqueous reprocessing, and the simplified pelletizing fuel fabrication was preferred as the main concept to be developed primarily because this combination would meet all of the development targets as the future nuclear energy system, namely sustainability, economy, proliferation resistance, and safety, as well as a reasonable prospect for technical feasibility. The advanced loop-type reactor named "Japan sodium-cooled fast reactor" was investigated as the most promising concept for commercialization, which had attractive features by adopting innovative technologies for ensuring safety, increasing the economic competitiveness, and long-term viability of nuclear energy utilization. Both the conceptual design study for JSFR and the development of innovative technologies were carried out in the FaCT project, with special attention paid to the consistency between the design and the necessary technologies.

4.2.1.2 Development targets and design requirements

Fast reactor cycle technology needs to possess enough economic advantage on the premise of ensuring safety to deploy commercial facilities. Furthermore, fast reactor cycle technology must be flexible to meet the needs of society in the future. The development targets and their breakdowns, i.e., the design requirements, were established at the start of the FaCT project by revising ones from FS and taking into account recent changes in Japanese society and the global situation. The development targets are shown in Table 4.2, and the major design requirements are shown in Table 4.3 [11].

Table 4.2 Development targets in the FaCT project [11].

Safety and reliability	
SR-1	Safety level shall be equal to future LWRs and related fuel cycle system.
SR-2	Reliability level shall be equal to future LWRs and related fuel cycle system.
Sustainability	
Environmental protection	
EP-1	Radiation effect on the environment shall be lower than future LWR cycle system.
EP-2	Materials emission to the environment shall be controlled within the acceptable level.
Waste management	
WM-1	Amount of radioactive waste shall be reduced compared with future LWR cycle.
WM-2	Manageability of radioactive waste shall be improved compared with future LWR cycle.
WM-3	Radiotoxicity of radioactive waste shall be reduced compared with future LWR cycle.
Efficient utilization of nuclear fuel resources	
UR-1	Sufficient breeding ratio shall be achieved for smooth transition to fast reactor cycle, and flexible core performance shall be kept for wider purposes of utilization.

Table 4.2 Continued

Economic competitiveness	
EC-1	Electric generation cost shall be competitive with other power generation systems.
EC-2	Investment risk shall be competitive with future LWR cycle.
EC-3	External cost shall be competitive with future LWR cycle.
Nuclear nonproliferation	
NP-1	Institutional measures and technological features shall be adopted for enhancement of nonproliferation.
NP-2	Physical protection shall be considered in system design.

Copyright 2008 by the American Nuclear Society, La Grange Park, Illinois.

Table 4.3 Major design requirements for SFR in the FaCT project [11].

Safety and reliability	
SR-1.1	Fundamental safety principles shall be observed. Safety standards and guidelines for former SFRs shall be kept, while specific features of new reactors shall be considered.
SR-1.2	Prevention and mitigation against severe accident initiators shall be considered so as to avoid exercise of off-site emergency plan.
SR-1.3	Total CDF shall be less than 10^{-6}/reactor year considering multiple units in a site, and total frequency of loss of containment function in core damage conditions shall be less than 10^{-7}/reactor year.
SR-2.1	Adequate maintenance/repair rule shall be developed, and design concept shall be well fitted with this rule. Inspection devices shall be suitably developed.
Sustainability	
Environment protection	
EP-1.1	Effective radiation doses under normal operation conditions shall be less than future LWR cycle system.
EP-2.1	Emission of greenhouse effect gas shall be less than future LWR cycle system.
Waste management	
WM-3.1	Contents of MA in the fuel shall be from 1 to 5%.
Efficient utilization of nuclear fuel resources	
UR-1.1	BR: ca.1.2 for high-breeding core (TRU fuel with FP content, compound system doubling time (CSDT): ca. 30 years), BR: ca. 1.1 for low-breeding core (do., CSDT: 60–70 years)
Economic competitiveness	
EC-1.1	Commercial reactor (1500 MWe, twin plant): [Construction cost] 120,000 JPY/kWe (NOAK, overnight cost), Construction period: 30–35 months, Plant life: more than 60 years.

Continued

Table 4.3 Continued

[Operation cost] Operation period: more than 24 months, Periodical inspection term: 40 days, Energy availability factor: more than 90%. [Fuel cost] Average burnup for driver fuel: 150,000 MWd/t, for whole core including blanket: 60,000 MWd/t for high-breeding core, 80,000 MWd/t for low-breeding core.	
Nuclear nonproliferation	
NP-1.1&2	Institutional measures and technological features shall be at least of the same level with that of future LWR cycle system.
NP-2.1	Physical protection system shall be adopted by considering characteristic features of fast reactor cycle.

Copyright 2008 by the American Nuclear Society, La Grange Park, Illinois.

As the principle of specifications for fast reactor cycle technology at the full-scale deployment stage in around 2050, new development targets and design requirements were investigated.

4.2.1.3 Design approach to meet design targets and requirements

JSFR is an advanced loop-type sodium-cooled fast reactor with mixed oxide (MOX)-fueled core. Fig. 4.3 shows the schematic view of JSFR, and the major specifications for the 1500 MWe plant are shown in Table 4.4 [11,12].

To meet the design requirements for safety and reliability, environmental protection, waste management, efficient utilization of nuclear fuel resources, economic competitiveness, and nuclear nonproliferation, several innovative technologies are used in the JSFR design. The JSFR approaches to meet those requirements are as follows:

(a) Core design

One of the most important requirements for the core design is a high core average burnup of around 150 GWd/t. The application of oxide-dispersion-strengthened steel to the fuel cladding material is a key technology for completing this. Another design goal for the breeding ratio is to have a flexible range between 1.0 and 1.2 while maintaining the same subassembly and core dimensions. Low-decontamination transuranium (TRU) MOX fuel is applied, and the fuel contains a certain amount of minor actinides, which shall be from 1 to about 5 wt% (percent in heavy metal weight). Fuel cycle carryover fission products (FPs) corresponding to the decontamination factor for the used reprocessing system are also present in the fresh fuel. The volume of carryover FPs in the future fast reactor multirecycle stage was assumed to be 0.2% MOX fuel volume. The fuel composition is considered to have some variation based on the wide variety of the spent fuel composition of LWRs and FRs expected to be treated during the LWR-to-FR transition period.

From the safety point of view, the core design limits core coolant void reactivity 6 $ or less, core-specific power 40 kW/kg-MOX or more, and core height 100 cm or less. These

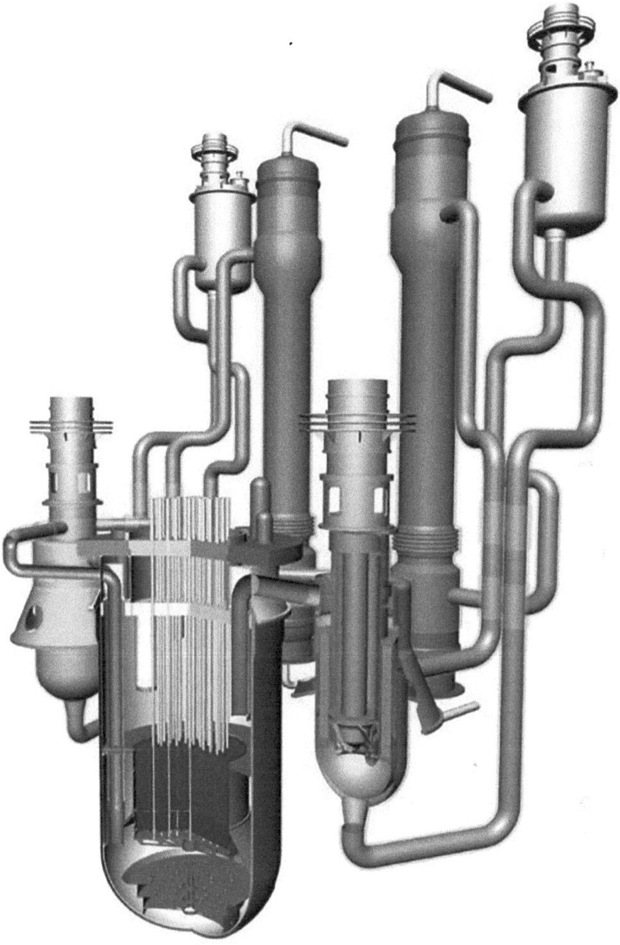

Fig. 4.3 JSFR system configuration [11].
Copyright 2008 by the American Nuclear Society, La Grange Park, Illinois.

are preliminary design requirements values to avoid excessive positive reactivity insertion in the initiating phase of unprotected flow loss (ULOF). The JSFR core implements a modified fuel assembly with an inner duct structure to mitigate core disruptive accidents (CDAs) (FAIDUS). In the case of CDA, the FAIDUS is expected to perform better for molten fuel discharge. Such molten fuel discharge from the core region is a key to preventing severe recriticality in the course of core damage progression.

(b) Safety design

The safety design for JSFR is based on "defense in depth." The reactor shutdown system includes a self-actuated shutdown system (SASS) as well as two separate active reactor shutdown systems. SASS is developed to maximize reactor shutdown capability and has been developed to shut down the reactor even if the reactor scram fails for anticipated transients due to a massive drop of the backup control rod.

The primary cooling piping and secondary one are equipped with guard pipes and enclosures, respectively. Guard vessels surround both the reactor vessel (RV) and the

Table 4.4 Design specifications of JSFR [11].

Items	Specifications
Electricity output	1500MWe
Thermal output	3570MWt
Number of loops	2
Primary sodium temperature and flow rate	550/395°C 3.24×10^7 kg/h /loop
Secondary sodium temperature and flow rate	520/335°C 2.70×10^7 kg/h /loop
Main steam temperature and pressure	497°C 19.2 Mpa
Feed water temperature and flow rate	240°C 5.77×10^6 kg/h
Plant efficiency	Approximately 42%
Fuel type	TRU-MOX
Burnup (average) for core fuel	Approximately 150GWd/t
Breeding ratio	Breakeven (1.03), 1.1, 1.2
Cycle length	26 months or less, 4 batches

Copyright 2008 by the American Nuclear Society, La Grange Park, Illinois.

intermediate heat exchanger (IHX) integrated with the primary main circulation pump. These design features allow for the maintenance of the liquid level required for cooling the core even if a coolant leak due to a coolant boundary failure is assumed to occur. The decay heat removal system is designed to have redundancy and diversity, and the capability of removing the decay heat only by natural circulation of sodium, taking advantage of its property of easier generation of difference in density owing to good heat conductivity.

The ability of the RV to contain radioactive materials is ensured by two key measures: recriticality-free core and in-vessel retention (IVR). The recriticality-free core concept is based on the successful discharge of molten fuel from the core whenever core melt occurs, with an effective core and fuel design consideration, such as FAIDUS, to promote an early discharge of molten fuel. IVR is to retain and cool the fuel debris generated by the core disruption in the RV, with a reactor system design effort to provide a specific plate (core catcher) to support the fuel debris in the bottom area of the RV. The IVR approach, as well as design measures such as enclosing the coolant boundaries in the containment vessel (CV) with guard pipes and guard vessels, can eliminate the factor of excessive pressure increase caused by sodium combustion leaking in the CV.

(c) Reactor vessel and reactor internal structure

To reduce the number of structural materials, the diameter of the RV shall be minimized and the reactor internal structures shall be simplified.

The compact RV diameter is achieved through the use of a compact core design in conjunction with a single rotating plug structure. The JSFR method describes a shell-less column-type slit upper internal structure (UIS). The fuel handling machine enters the UIS through the slit where the control rod guide tubes are not present. As this type of UIS makes it possible to conduct fuel handling with the UIS remaining just above the core, the diameter of a single rotating plug can be set smaller, and this leads to the compact design of the closure head structure. The wall thickness for the RV shall be minimized. This can be done by introducing seismic isolation technology, which can mitigate seismic load, as well as recriticality-free core technology, which can mitigate mechanical energy release caused by a hypothetical CDA.

In addition, a simple hot vessel design has been employed. The establishment of an advanced elevated temperature structural design standard, which allows for the use of an inelastic analysis, and a direct assumption method of the thermal stress are used to ensure structural integrity against thermal stress caused by an axial temperature profile around the liquid surface of the RV for the RV structure without a wall cooling system [13]. Adoption of a simple skirt-type core support structure and elimination of a liquid surface-level control also contribute to the simplification of the reactor structure. A dipped plate attached beneath the liquid surface effectively suppresses cover gas entrainment in the coolant from the liquid surface, which is caused by increased sodium velocity due to the RV's compact design.

(d) Heat transport system

The most advantageous cooling system concept, the loop number, primary piping system, and applicability of component integration have all been thoroughly examined in terms of construction cost reduction, safety, maintainability, and manufacturability. As a result, shortening of piping, two-loop cooling system, integrated design of an intermediate heat exchanger (IHX), and a primary pump are introduced in the design.

The significantly shortened primary and secondary piping layout results in a compact plant configuration due to the close arrangement of components and the reduction of piping materials. Consideration was given to the primary hot-leg piping. It is made up of simple L-shaped piping with a structure that absorbs thermal expansion through only one elbow. Such a shortened piping layout with one elbow is realized with the aid of the adoption of high chromium steel, which has high strength in the elevated temperature and low thermal expansion coefficient.

The approach to reducing the loop number is summarized in Fig. 4.4 [11].

The sodium-cooled system has some advantageous properties, such as a high boiling point and high heat conductivity, which can increase the heat transfer capacity per unit volume of heat exchangers under low-pressure conditions. As a result, larger capacity heat exchangers are easier to design. The adoption of high chromium steel with high strength and high heat conductivity makes it possible to enlarge the heat transfer capacity further than

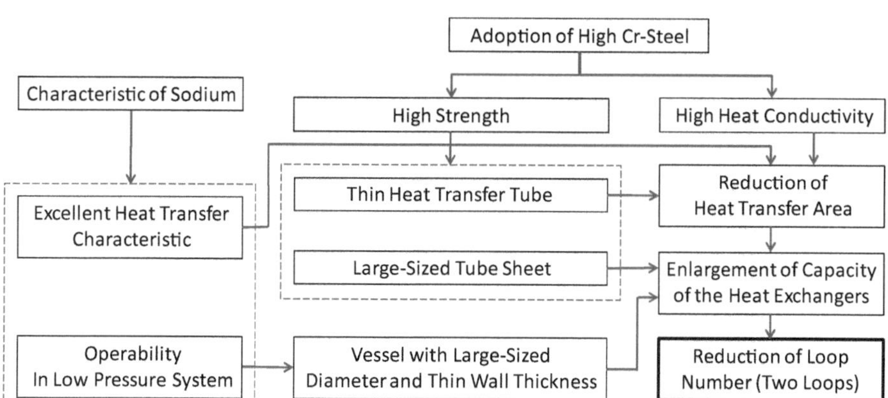

Fig. 4.4 Approach to reduction of loop number [11].
Copyright 2008 by the American Nuclear Society, La Grange Park, Illinois.

that of the conventional design because a thinner heat transfer tube is available and it is easier to ensure the structural integrity against thermal stress by a large tube sheet. FS included a comparison of Mod. 9Cr—1Mo steel and 12Cr steel. In terms of data accumulation for standards, Mod. 9Cr—1Mo steel has been chosen as a reference. The capacity of the heat exchangers, i.e., the heat transfer capacity per one-cooling loop, has been increased in this design to ensure manufacturability and structural integrity of the components, and the number of loops has been reduced to two. Although the fuel integrity under a primary pump stick accident and the structural integrity of the large tube sheets of the heat exchangers against thermal stress become critical issues for the two-loop cooling system, prospects for a solution have been obtained through some design countermeasures and safety analyses.

The integrated design of an IHX and a primary pump allows for the elimination of middle-leg piping. Installing the primary pump in the center of the IHX tube bundle constructs the integrated component. The IHX is a no-liquid-surface type that allows for high sodium coolant velocity in the primary hot-leg piping without increasing cover gas pressure in the RV. The primary sodium flows within straight heat transfer tubes, and the secondary sodium flows zigzag outside of the tube. With the adoption of thinner heat transfer tubes made of high chromium steel, the heat transfer performance has been improved and the tube bundle has become compact. The integrated component has a diameter of 4.3 m and a height of 14.2 m. Although the integration of the IHX and the primary pump results in a larger tube sheet diameter of the IHX, structural integrity is ensured by the use of high chromium steel with high strength and low thermal expansion.

(e) Containment vessel

Reduction of the materials of the nuclear steam supply system (NSSS) leads to reducing the reactor building volume, thus the amount of the building materials. The CV is drastically compacted due to the double-walled structure for the sodium boundary and the elimination of mechanical energy release from recriticality events in severe accidents. When compared to LWRs, the pressure load on the CV is not as high. As a result, the CV would be shaped like a rectangle and would be designed to be a part of the reactor building. The rectangle shape reactor building is suitable for compact component arrangement, i.e., less dead space in the building. Steel plate-reinforced concrete (SC) is applied to all parts of the building instead of conventional reinforced concrete by steel rods. In Japan, the SC structure is already being used to support buildings for LWRs. Because the steel part of the SC structure can be manufactured in a factory as large modules, using the SC structure results in a shorter on-site construction period. SC structure design technology for JSFR is being developed.

(f) Seismic isolation

In Japan, severe seismic condition is requested, and the primary components are exposed to severe seismic loading. Severe earthquakes have a strong impact on the thin-walled structure of SFR. As a result, JSFR employs a seismic isolation system to reduce seismic force. A combination of laminated rubber bearings and oil dampers is used in the advanced seismic isolation system for SFR. Using the proposed seismic isolation system, acceleration in the natural period range of the major components such as the RV could be decreased for both horizontal and vertical directions.

4.2.2 Core

4.2.2.1 Basic concept

A reference core concept with mixed oxide (MOX) fuel for the JSFR is called the high-internal-conversion-type core [14–19]. In this core concept, large diameter fuel pins are used to increase the internal conversion ratio and thereby to reduce the amount of blanket fuels to the greatest extent possible. This concept has the economic advantage that in addition to sufficient breeding capability with a small amount of blanket, a high average discharged burnup can be achieved for the entire core and long operation cycle length.

The core performance requirements and design conditions are summarized in Table 4.5. Two development phases, which correspond to the demonstration and commercial core design, are assumed.

(a) Plant conditions

The power outputs in the demonstration and commercial phases are 750 and 1500 MWe, respectively. The core outlet and inlet temperatures are 550°C and 395°C, respectively. The primary coolant flow rate is 9000 kg/s in the demonstration phase and 18,000 kg/s in the commercial phase.

(b) Fuel burnup

In the commercial phase, the important target from the cost viewpoint is to achieve a high average discharged burnup. The total average discharged burnup (i.e., the whole core discharged burnup, including the blanket), which is related to fuel cycle cost, must be more than 80 GWd/t for a low-breeding core. For a high-breeding core, some mitigation of the total average discharged burnup can be accepted. In both cases, the target core average discharged burnup is around 150 GWd/t. To realize such a high value of core burnup under high temperatures, oxide-dispersion-strengthened (ODS) steel is being developed for application as the cladding material.

In the demonstration and early commercial phases, only one condition has to be fulfilled from the viewpoint of competitiveness against generation III reactors: The total average discharged burnup should be higher than 60 GWd/t for a low-breeding core. The target value of the core average discharged burnup is not specified. Rather, it is expected that a demonstration fuel irradiation test will be performed to achieve the peak burnup needed for commercial core development.

(c) Operation cycle length

The operation cycle length of the demonstration phase should be longer than 13 months to ensure competitiveness against generation III reactors. The commercial phase requires an operation cycle longer than that of the demonstration phase.

(d) Breeding ratio

The breeding ratio (BR) of the commercial phase is intended to have some flexibility and is set to 1.0–1.2 according to the design philosophy of using compatible fuel assemblies. A BR of 1.1 (low-breeding core) is targeted for achieving a smooth transition from LWRs to fast reactors (FRs). Moreover, a higher BR of 1.2 (high-breeding core) is considered to cater to any possible high Pu demand. Given that the Pu demand will decrease as FR deployment is saturated, a breakeven BR of around 1.0 is adequate to sustain the FR fuel cycle in the equilibrium recycle state.

In the demonstration phase, only a low-breeding core with BR of 1.1 is required. The purpose of this phase is to demonstrate the breeding capability of the reactor, which can then be extrapolated to evaluate the design accuracy of the commercial-phase BR.

(e) Safety requirements

Safety requirements are common to the demonstration and commercial phases. The requirements to prevent super-prompt criticality in the initiating phase of core-disruptive accidents (CDAs) are as follows: a sodium void reactivity of about 6 $ or less, a core height of 100 cm or less, and an average core-specific power of about 40 kW/kg-MOX or more [20].

To prevent recriticality in the transition phase of CDA, an innovative fuel assembly concept called FAIDUS (fuel assembly with inner duct structure) was introduced. The FAIDUS

Table 4.5 Core performance conditions for JSFR MOX fuel core [18].

Item	Demonstration phase	Commercial phase
Plant conditions		
Power output	750 MW_e	1500 MW_e
Coolant temperature (outlet/inlet)	550/395°C	Same
Primary coolant flow rate	9000 kg/s	18,000 kg/s
Core performance requirements		
Total average discharged burnup (low-breeding core)	>60 GWd/t	>80/>60[a] GWd/t
Core average discharged burnup	–	150/–[a] GWd/t
Operation cycle length	> 13 months	Longer than that of demonstration phase
BR		
(low-breeding core)	1.1	1.1–1.0
(high-breeding core)	–	1.2
MA content rate	Same as that of commercial phase (step-by-step MA introduction could be assumed.)	
(core average)		<3 wt%
(in harmless loading positions)		<5 wt%
Safety requirements		
Sodium void reactivity	<6 $	Same
Core height	<100 cm	Same
Average core-specific power	>40 kW/kg-MOX	Same
Fuel assembly concept for recriticality-free core	FAIDUS	Same
Thermal hydraulic conditions		
Cladding mid-wall temperature	<700°C	Same
Bundle pressure drop	<0.2 MPa	Same
Fuel integrity limits		
Linear heat rate	<430 W/cm	Same
Fast neutron fluence (E_n > 0.1 MeV)	$<5 \times 10^{23}$ n/cm^2	Same
Cladding CDF (steady state)	<0.5	Same

[a] For the early commercial phase.

concept was designed as a measure for facilitating the early discharge of molten fuel. Molten fuel enters the inner duct channel and exits the core through the upper shielding [14].

In addition, the use of hydrogen-releasable materials such as zirconium hydride as shielding is not permitted to eliminate the potential risk of a hydrogen explosion.

(f) Thermal hydraulic conditions

For both demonstration and commercial phases, the maximum cladding mid-wall temperature was limited to 700°C in accordance with the development target of the ODS steel. The fuel pin bundle pressure drop was reduced to under 0.2 MPa to facilitate the natural convection flow in the event of a primary pump seizure accident in the loop-type cooling system.

(g) Fuel integrity limits

The design limit of the linear heat rate is about 430 W/cm. The development target of fast neutron fluence (neutron energy $E_n > 0.1$ MeV) is about 5×10^{23} n/cm^2. The cumulative damage fraction (CDF) of the cladding under steady operation should be less than 0.5. Those limits are applied to the core design in both the commercial and demonstration phases.

(h) Fuel composition

The fuel composition (i.e., heavy metal nuclide composition) for FR has a broad range depending on the variety of spent fuels of LWR and the recycling of them in the FR fuel cycle. According to studies [21–23] on the changes in fuel composition and the effects of such changes on FR core design, there are remarkable correlations between the main core characteristics such as burnup reactivity, sodium void reactivity, and Doppler coefficient. Using these correlations, a design envelope that includes any fuel compositions appearing in the FR operation phase can be defined.

For the demonstration and early commercial phases, two design envelopes were introduced for uranium-plutonium (U—Pu) and transuranium (TRU) cores using representative FR-deployment scenario simulations. The start-up U—Pu core (fuel form: (U, Pu)O_x) reflects the prospective situation of the domestic nuclear fuel cycle in which Pu is expected to be supplied from a reprocessing plant based on the PUREX technology. The successive TRU core (fuel form: (U, TRU)O_x) is fed with low-decontaminated TRU fuel that contains a small amount of minor actinoids (MAs) and residual fission products, which are treated in an FR fuel cycle facility based on the advanced aqueous reprocessing with MA extraction process. Table 4.6A and B summarize the design envelopes of the fuel composition. Each design envelope consists of high-fissile, standard, and low-fissile compositions. The standard composition exhibits the most probable composition in the corresponding FR operation phase.

Table 4.6C provides the specifications for the design envelope for the commercial phase. Assuming that the development after FR deployment proceeded successfully, the high-fissile composition is defined as the average of the high-fissile TRU composition in the early commercial phase and the fuel composition after one round of recycling in the commercial FR cycle. The low-fissile composition is defined similarly. The standard composition is a multiple recycled equilibrium composition to which all fuel compositions approach upon recycling in a commercial FR fuel cycle.

Nevertheless, some rationalizations of the design envelopes of fuel composition have been performed so far from the perspectives of the fuel-composition trend in the representative FR-deployment scenarios [18]. First, the U—Pu high-fissile side boundary can be shifted to the U—Pu standard level since the U—Pu high-fissile composition originating from spent fuel with relatively short cooling time appears rarely in the scenario

Table 4.6 Design envelopes of the fuel composition for the JSFR MOX fuel core [18].

(a) Pu composition in the demonstration and early commercial phases (in wt%)

Nuclide	High-fissile composition	Standard composition	Low-fissile composition
Pu-238	2.5	2.0	2.3
Pu-239	52.4	54.0	51.6
Pu-240	26.2	28.0	25.8
Pu-241	10.3	7.0	6.3
Pu-242	7.6	8.0	7.4
Am-241	1.0	1.0	6.6
Origin	Average of PWR and BWR spent fuel		
	– Time for cooling and storage		
	6 y	15 y	3 y
	– Time after reprocessing		
	2 y	2 y	15 y

(b) Transuranium (TRU) composition in the demonstration and early commercial phases (in wt%)

Nuclide	High-fissile composition	Standard composition	Low-fissile composition
Pu-238	2.2	2.0	1.7
Pu-239	46.4	46.0	46.7
Pu-240	23.3	23.0	23.6
Pu-241	8.3	4.0	2.0
Pu-242	6.7	7.0	6.7
Np-237	5.7	6.0	6.2
Am-241	5.6	10.0	11.5
Am-243	1.4	1.0	1.4
Cm-244	0.4	1.0	0.2
Origin	Average of PWR and BWR spent fuel		
	– Ex-core time for cooling, storage, reprocessing, and fuel fabrication		
	10 y	25 y	40 y

(c) TRU composition in the commercial phase (in wt%)

Nuclide	High-fissile composition	Low-fissile composition	Standard composition
Pu-238	3.1	3.3	1.7
Pu-239	49.9	49.1	55.9
Pu-240	23.7	23.9	30.5
Pu-241	5.8	2.2	3.4
Pu-242	6.3	6.3	3.3
Np-237	4.0	4.4	0.3
Am-241	4.9	8.5	3.0
Am-242m	0.1	0.2	0.1
Am-243	1.5	1.5	1.0
Cm-244	0.6	0.5	0.6
Cm-245	0.1	0.1	0.2

Table 4.6 Continued

(c) TRU composition in the commercial phase (in wt%)			
Origin	Average of high-fissile composition of (b) and that recycled in FR cycle	Average of low-fissile composition of (b) and that recycled in FR cycle	FR multiple recycled composition
	– Ex-core time for cooling, storage, reprocessing, and fuel fabrication: 5 y		

calculations. This allows the reduction in burnup reactivity, maximum linear heat rate, and necessary coolant flow rate.

In contrast, for the low-fissile TRU composition, as the recent scenario excluded the management of intensive reprocessing of LWR-MOX spent fuel within a short period, the maximum core average MA content rate in the FR core fuel (defined as the weight percentage in heavy metal) can be reduced from 5 wt% to 3 wt%. The MA content rate can be adjusted by changing the mixing fraction of the U-Np-Pu and Am—Cm feed. Therefore, it is possible to decrease the sodium void reactivity. MA content rates of up to 5 wt% are acceptable if the contents are loaded to harmless positions where the linear heat rate and CDF of cladding do not become high.

4.2.2.2 Reference core specification

The design results of 750 MWe demonstration [19] and 1500 MWe commercial [24] core with BR of 1.1 (low-breeding core) are summarized in Table 4.7. The core configurations are shown in Fig. 4.5. The core height is 100 cm with upper and lower axial blankets of 20 or 25 cm each. Considering the high-internal-conversion-type core concept, the fuel pin diameter is set as 10.4 mm. There are 255 fuel pins in a fuel subassembly, and the space corresponding to 16 fuel pins is occupied by the inner duct structure. The core design is a homogeneous two-region type that consists of inner and outer core fuel subassemblies. There is one layer of radial blanket subassemblies. These core and blanket regions are surrounded by shielding regions as shown in Fig. 4.5. A lower gas plenum-type pin is adopted to decrease the internal pressure created by the released fission product gas. The thickness of the upper and lower gas plenum is 10 and 110 cm, respectively.

Considering the core design parameters, the BR is set as 1.1. The average discharge burnup is 145–148 GWd/t for the core and 83–89 GWd/t for the whole core, including the blanket. The operation cycle length is 18 and 26 months for the demonstration and commercial cores, respectively. The sodium void reactivity is 5.2–6.0 $, and it is less than the upper limit of about 6 $. The average core-specific heat is 40 kW/kg-MOX. The maximum cladding mid-wall temperature is 700°C. The maximum linear heat rate is 376–392 W/cm for the demonstration core and 401–404 W/cm for the commercial core; the core configuration of the demonstration core was optimized to suppress the fluctuations in the maximum linear heating rate that occur during control rod

Table 4.7 JSFR demonstration and commercial MOX fuel core designs (low-breeding core) [19,24].

Item	Demonstration core [19]	Commercial core [24]
Core and fuel specification		
Core height (cm)	100	100
Axial blanket thickness (cm) (upper/lower)	20/25	20/20
Fuel pin diameter (mm)	10.4	10.4
Number of fuel pins per subassembly	255	255
Subassembly pitch (mm)	206	206
Number of fuel subassemblies (inner core/outer core/radial blanket)	184/94/66	288/274/96
Operation cycle length (month)	18	26
Number of fuel exchange batches	6	4
Fuel composition	Pu and TRU composition in Table. 4.6A and B	TRU composition in Table. 4.6C
Core characteristics		
Breeding ratio	1.1	1.1
Average discharged burnup [GWd/t] (total/core)	83/148	89/145–146
Maximum sodium void reactivity [$]	5.2–6.0	5.2–5.7
Average core-specific power [kW/kg-MOX]	40	40
Maximum cladding mid-wall temperature (°C)	700	700
Maximum bundle pressure drop (MPa)	0.2	0.2
Maximum linear heat rate (W/cm)	376–392	402–414
Maximum fast neutron fluence (n/cm^2) ($E_n > 0.1$ MeV)	5×10^{23}	5.5×10^{23}
Maximum cladding CDF (steady state)	0.5	0.5

insertion. The maximum fast neutron fluence is 5.0–5.5 $\times$ 10^{23} n/cm^2. The remaining parameters also satisfy the design requirements.

The design requirements allow some flexibility in fixing the BR in the range from 1.0 to 1.2. Table 4.8 summarizes the design results to meet these requirements. For breakeven breeding, it is possible to change the BR of a low-breeding core from 1.1 to 1.03 easily by replacing the radial blanket with steel shielding and by adjusting the axial blanket thickness. On the other hand, for high breeding, an effective method to increase the BR is to use a thicker blanket. Hence, the chosen method was to increase the axial blanket thickness and reduce the core height so that the high-breeding core in a reactor plant can be replaced with a low-breeding core. Meanwhile, it was necessary to increase the number of fuel pins per subassembly using pins with a smaller diameter because of the limitation imposed by the maximum linear heating

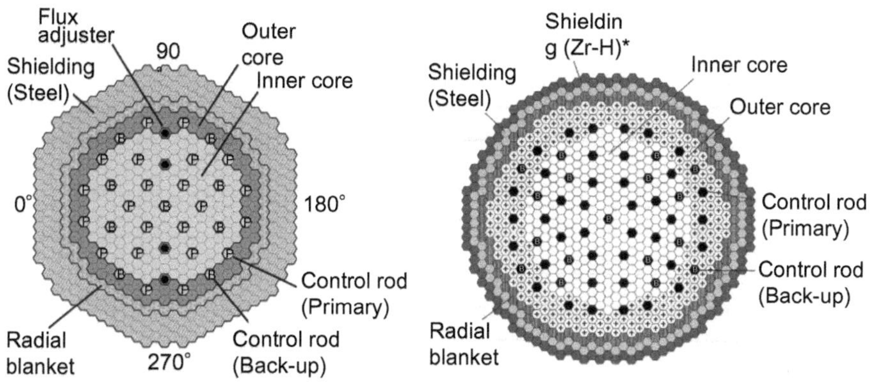

Fig. 4.5 Configurations of demonstration and commercial cores for JSFR. *Zirconium hydride shielding will be altered to eliminate the potential risk of hydrogen explosion.
(A) Demonstration core [19], (B) Commercial core [24].

Table 4.8 Breeding capability of the JSFR commercial MOX fuel core [15,24].

Item	Commercial core		
	Low breeding (breakeven) [15]	**Low breeding [24]**	**High breeding [24]**
Core and fuel specification			
Core height (cm)	100	100	75
Axial blanket thickness (cm) (upper/lower)	15/20	20/20	40/50
Fuel pin diameter (mm)	10.4	10.4	9.3
Number of fuel pins per subassembly	255	255	315
Subassembly pitch (mm)	206	206	206
Number of fuel subassemblies (inner core/outer core/radial blanket)	288/274/0	288/274/96	288/274/96
	26	26	21
Operation cycle length (month)	4	4	4
Number of fuel exchange batches	TRU standard composition	TRU standard composition	TRU standard composition
Core characteristics			
Breeding ratio	1.03	1.1	1.2
Pu fissile inventory [t/GW$_e$]	5.8	5.8	5.1
Average discharged burnup [GWd/t] (total/core)	115/150	89/145	60/149

Continued

Table 4.8 Continued

Item	Commercial core		
	Low breeding (breakeven)	**Low breeding**	**High breeding**
Maximum sodium void reactivity [$]	5.3	5.2	4.6
Average core-specific power [kW/kg-MOX]	41	40	52
Maximum linear heat rate (W/cm)	411	406	424

rate. A BR of 1.2 (Table 4.8) can be obtained with the following specifications: an axial blanket thickness of 90 cm, a core height of 75 cm, and a fuel pin diameter of 9.3 mm and 315 fuel pins per subassembly. In the high-breeding core, the total average discharge burnup decreased by 33%, and the operation cycle length decreased by 20% compared to those of the low-breeding core, respectively. These results indicate the economic disadvantage of the high-breeding core. Nevertheless, this core has the merits of a high BR of 1.2 and a smaller Pu fissile inventory, which may contribute to the reduction of the FR introduction period. Therefore, the high-internal-conversion-type core can flexibly comply with the changes in Pu breeding requirements ranging from those caused by the LWR-to-FR transition to the FR equilibrium status.

4.2.2.3 Detailed core design highlights

The design study of the JSFR demonstration core proceeded extensively in the FaCT project and the succeeding activities. Some technical highlights of the core design are described below.

(a) Multiple fuel-composition loading method

As mentioned in Section 4.2.2.1, there exists a wide variety in fuel compositions, especially in the LWR-to-FR transition phase. An FR core should be designed to accept various fuel compositions simultaneously. Based on a comparative study on the multiple fuel-composition loading methods for an FR core [25], we selected a method to determine the fuel enrichment so that the end-of-cycle (EOC) core reactivity is raised by the same value α for any fuel composition; that is,

$$\rho'_{EOC} = \rho_{EOC} + \alpha \tag{4.1}$$

where α is the lack of reactivity arising from a combination of burnt high-fissile composition fuels and fresher low-fissile composition fuels in the state of multiple fuel-composition loading. In terms of excess reactivity, we should add the values α and β to the conventional excess reactivity as

$$\rho'_{Excess} = \rho_{Excess} + \alpha + \beta \tag{4.2}$$

The value β represents the increment in excess reactivity due to a combination of fresher high-fissile composition fuels and burnt low-fissile composition fuels. This method was applied to the demonstration core design [18,19]. Although multiple fuel-composition loading also affects the maximum linear heating rate, the impact is the least in the present method so that it can be treated as a part of the power distribution uncertainty.

(b) Optimization of the core configuration

In the conventional JSFR core designs, the maximum linear heating rate increases with the insertion of the control rods, which also results in an increase in the necessary primary coolant flow rate. The suppression of this increment in the linear heating rate was one of the challenges in developing advanced JSFR core design. To tackle this challenge, an optimization study was performed for the demonstration core configuration [19].

The optimized core configuration is shown in Fig. 4.5A. Based on the demand arising from the upper internal structure design, a zone is created such that control rods cannot be arranged in the direction of 90–270° other than at the core center because of the tendency that neutron flux increases in four positions that are distant from the primary control rods in this zone. Careful arrangement of the control rods and introduction of flux adjusters at the four high-flux positions can greatly reduce the distortions in the neutron flux distribution. As shown in Fig. 4.6, the horizontal distributions of the linear heating rate are quite stable even under changes in the primary-control-rod insertion depth.

(c) Evaluation of power distortion due to the inner duct

The FAIDUS-type fuel subassembly has asymmetrical geometry. In this subassembly, an inner duct is placed in one corner, and it is predicted that a power distortion arises horizontally. To clarify the power distribution deformation according to the inner duct directions, analyses were performed for an extreme case in which many inner ducts are directed

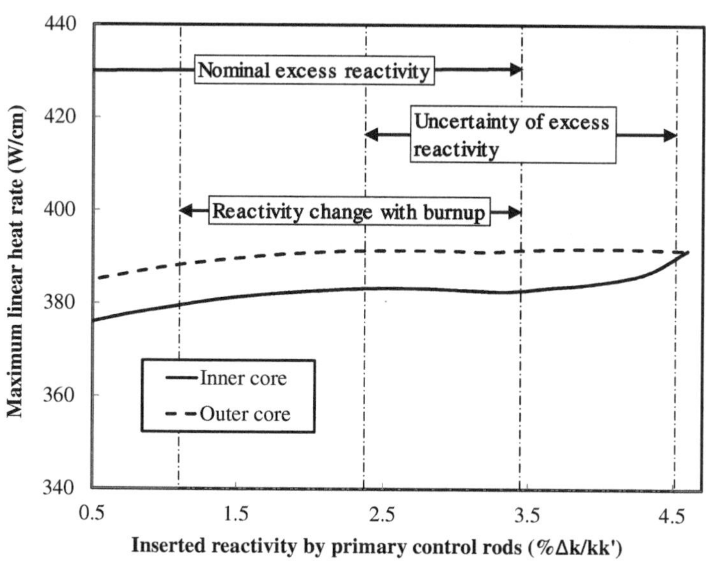

Fig. 4.6 Fluctuations in maximum linear heating rate that occur during control rod insertion in demonstration core [19].

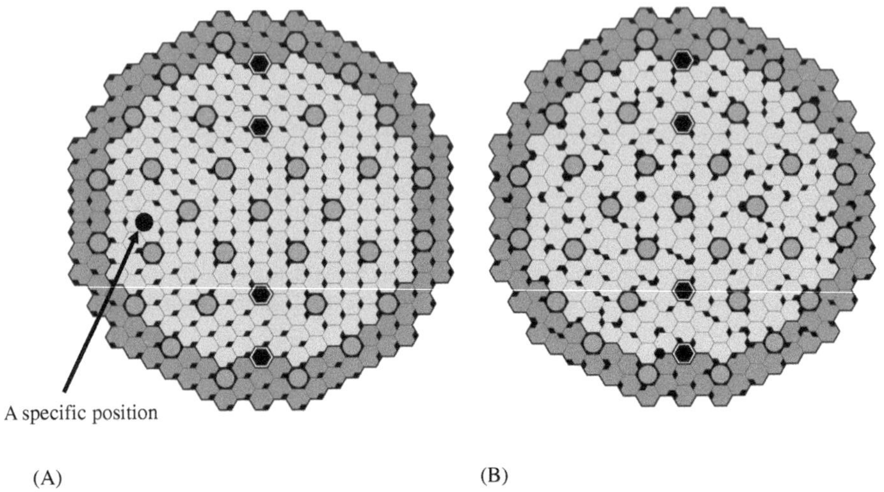

Fig. 4.7 Sample loading patterns for inner duct directions [19]. (A) Extreme case (B) Random case.

such that they focus on a specific position in the core, as shown in Fig. 4.7A, and for realistic cases in which the inner duct directions are randomly arranged for 200 operation cycles, as shown in Fig. 4.7B. The results show that in the extreme case, the power deformation rate may greatly exceed 1.1. On the other hand, in the realistic cases, the power deformation rate is only 1.01–1.02. Assuming random duct directions, the inner duct effect can be treated as a part of the power distribution uncertainty [19,26].

4.2.2.4 Metal fuel core alternative

The U-Pu-Zr base metal fuel is considered as an attractive fuel type owing to its high heavy metal density. Nevertheless, the maximum cladding inner-wall temperature is limited to less than 650°C to prevent the formation of a liquid phase in the part of metal fuel that is in contact with the cladding. The conventional metal fuel core designs with two Pu-enriched regions could not easily attain a high reactor outlet (coolant) temperature of 550°C, which is identical to that of the MOX fuel core.

In the JSFR development framework, a new metal fuel core concept was proposed [27]. This core can achieve high reactor outlet temperatures while satisfying the cladding temperature limitation. The core concept is as follows. The local conversion ratio at any point of the core is made close to unity by commonly applying the Pu enrichment of 12% in the core; then, the power distribution of the core becomes stable during the operation cycle as shown in Fig. 4.8. For flattening the radial power distribution, the inner and outer cores have different Zr contents and smear densities. The stable radial power distribution can minimize the necessary coolant flow rate, and hence, the core can achieve a high reactor outlet temperature.

Design studies were conducted for the commercial and demonstration cores [28–30]. One of the design efforts was to reduce sodium void reactivity to mitigate its influence on anticipated transient without scram events. The fuel pin diameter

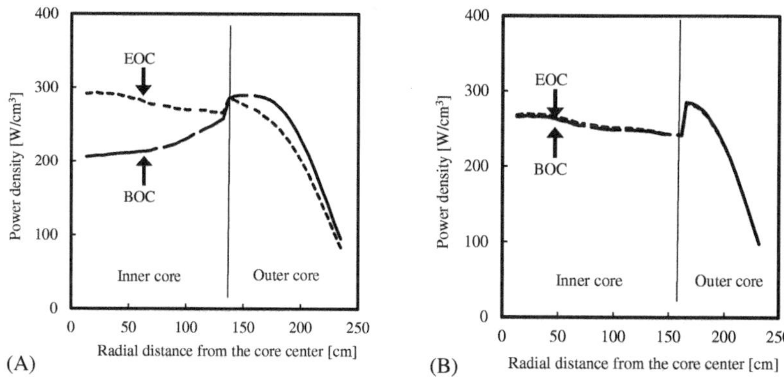

Fig. 4.8 Comparison of radial power distributions between conventional and "high reactor outlet temperature" metal fuel core [27]. (A) Conventional core (B) High reactor outlet temperature core.

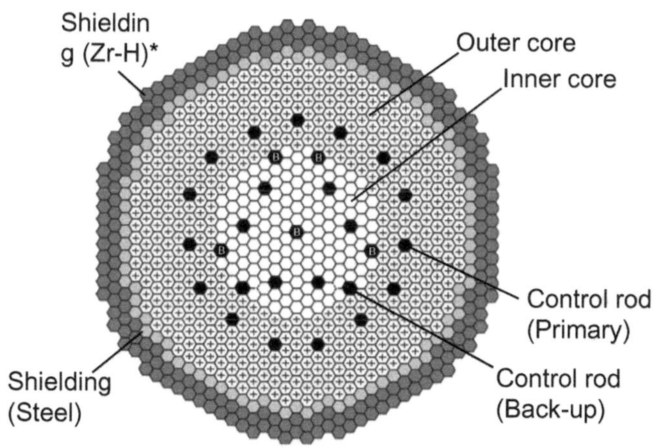

Fig. 4.9 Configuration of commercial metal fuel core [28]. *Zirconium hydride shielding will be altered to eliminate the potential risk of hydrogen explosion.

was increased from 7.2 mm to 8.5 mm to reduce the coolant volume fraction and core height. Additionally, the boundaries of the inner and outer core regions were revised. By adopting these design modifications, the sodium void reactivity decreased by about 20%. The reference commercial core configuration and its main design results are shown in Fig. 4.9 and Table 4.9, respectively. Because of the superior neutronic economy, the BR reaches 1.03 without a blanket. The core also achieves a high total average discharged burnup comparable to that of the MOX fuel core. Because of these favorable characteristics, this advanced metal fuel core is an alternative for JSFR.

Table 4.9 Main specifications and the characteristics of commercial metal fuel core [28,29].

Item	Commercial core (breakeven) [28,29]
Plant conditions	
Power output [MW_e]	1500
Coolant temperature (outlet/inlet) (°C)	550/395
Primary coolant flow rate (kg/s)	18,000
Core and fuel specification	
Core height (cm)	75
Axial blanket thickness (upper) (cm)	0
Fuel pin diameter (mm)	8.5
Zr content [wt%] (inner core/outer core)	10.0/6.0
Smear density [%TD] (idem)	75/75
Number of fuel pins per subassembly	331
Subassembly pitch (mm)	191.8
Number of fuel subassemblies (inner core/outer core)	150/495
Operation cycle length (month)	24.5
Number of fuel exchange batches	3
Fuel assembly concept for recriticality-free core	Lower axial blanket is not installed.
Fuel composition	Metal fuel FR multiple recycled composition[a]
Core characteristics	
Pu enrichment (wt%)	12.2
Breeding ratio	1.03
Average discharged burnup [GWd/t] (total/core)	97/97
Maximum sodium void reactivity [$]	6.2
Maximum cladding inner-wall temperature (°C)	649
Maximum bundle pressure drop (MPa)	0.22
Maximum linear heat rate (W/cm)	347
Maximum fast neutron fluence (n/cm^2) ($E_n > 0.1$ MeV)	5.4×10^{23}

[a] $^{238}Pu/^{239}Pu/^{240}Pu/^{241}Pu/^{242}Pu/^{237}Np/^{241}Am/^{243}Am/^{244}Cm/^{245}Cm$ = 1.1/66.0/25.2/2.4/2.4/0.4/1.6/0.5/0.4/0.1 wt%

4.2.3 Safety design

4.2.3.1 Development goals

JSFR has been developed with the goal of achieving a high level of safety as a generation IV nuclear system. Therefore, the following development goals for the safety and reliability of GIF were applied [31].

- SR-1: Excel in Operational Safety and Reliability.

  Safety and reliability during normal operation, and likely operational events that assume forced outage rate.

- SR-2: Very low likelihood and degree of reactor core damage.
  Minimizing frequency of initiating events, and design features for controlling and mitigating any initiating events w/o causing core damage.
- SR-3: Eliminate the need for off-site emergency response.
  Safety architecture to manage and mitigate severe plant conditions, for minimizing the possibility, and the amount of releases of radiation.

In order to achieve the above, design measures based on the concept of defense in depth were taken, especially the design measures of prevention of core damage and mitigation of core damage consequences for design extension conditions were adopted as the fourth level of defense in depth, as described in Chapter 5.1.2.

Additionally, the core damage frequency (CDF) was set at 10^{-6} per reactor year or less, and the containment failure frequency (CFF) was set at 10^{-7} per reactor year or less as probabilistic goals.

On the other hand, it was required to meet the regulatory safety requirements in Japan at that time, considering the existing nuclear site conditions in Japan.

4.2.3.2 Safety design concept

Fig. 4.10 shows the safety design concept of JSFR [32]. The following is a list of characteristics.

(a) Providing passive reactor shutdown capability
  The self-actuated shutdown system (SASS) in the delatch mechanism of the backup reactor shutdown system is installed to provide an effective passive reactor shutdown capability for typical anticipated transient without scram (ATWS) events (unprotected loss of flow (ULOF), unprotected transient overpower (UTOP), and unprotected loss of heat sink

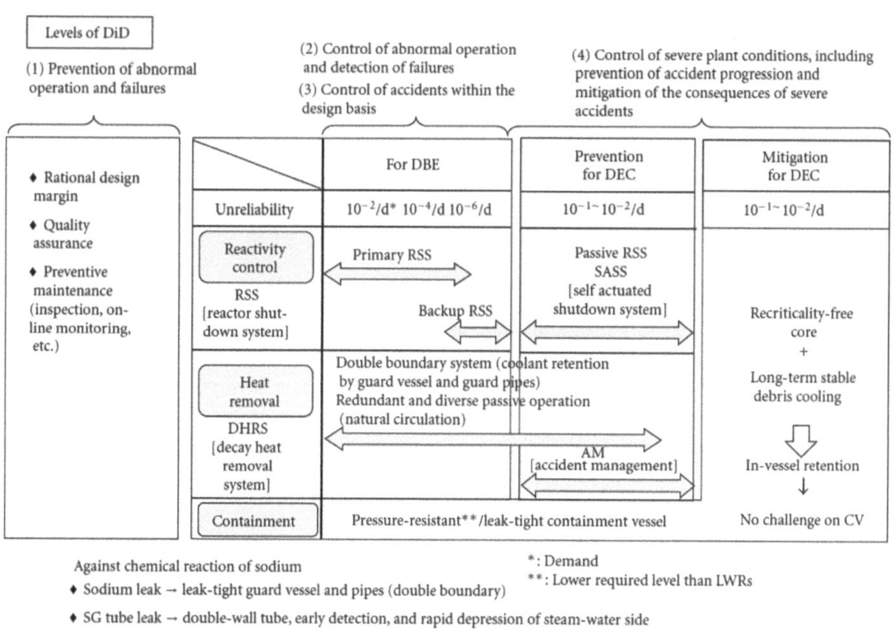

Fig. 4.10 Basic framework of safety design in JSFR [32].

(ULOHS)) that may lead to core damage in a short time in addition to the two independent reactor shutdown systems of control rods. The control rods of the backup reactor shutdown system are designed to fall passively by responding the surrounding coolant temperature increase to surely prevent core damage even if the reactor fails to shut down due to failure of the safety protection system signal. SASS development is described in Chapter 5.1.3.

(b) Ensuring core cooling capability in case of reactor coolant boundary failure

The primary pipes are designed to have guard pipes, and the reactor vessel and integrated primary pump-intermediate heat exchanger are designed to have guard vessels. This design ensures that the liquid level required for core cooling is ensured even in the case of a coolant leak due to the reactor coolant boundary failure.

(c) Decay heat removal

The decay heat removal systems of fully natural circulation type, that take advantage of the liquid metal cooling feature, are designed to have sufficient vertical distance between the core and heat exchangers and to reduce the pressure loss in the cooling circuits. Forced convection of the coolant by pumps or blowers is not required, and heat removal capacity required only by natural circulation is ensured for all assumed operational modes. The decay heat removal systems are designed to have redundancy or diversity not to loose their functions even in accidental conditions with a single failure. The redundancy and diversity are considered in both system level and component level. As the system level, decay heat removal systems are designed to consist of multiple subsystems or various types of subsystems, which have different configurations, for instance, different location of heat exchangers. As the component level, air cooler dampers are designed to be doubled and different types of dampers are designed for an example.

(d) Elimination of severe recriticality [33]

Sodium void worth is appropriately limited in the core design to prevent prompt criticality due to positive sodium void reactivity induced by rapid coolant boiling expansion of the coolant in core damage sequences. As a reference, it should be less than six dollars for an oxide-fueled core. In order to avoid recriticality in the transition phase, design measures for molten fuel discharge to the outside of the core region are introduced in the core design. Moreover, a core catcher for the long-term retention and cooling of degraded core debris is installed at the bottom of the reactor vessel in order to achieve in-vessel retention (IVR).

(e) Reactor containment

The pressure increase associated with leaked sodium combustion is significantly reduced by covering all reactor coolant boundaries in the containment vessel with closed guard vessels and guard pipes. Also, the pressure rise factor is eliminated by achieving IVR of core debris during core damage, and radioactive release to the environment is limited by the steel plate concrete containment vessel (SCCV).

(f) Design measures against sodium chemical reactions

As shown in Fig. 4.11, double-walled structure is introduced as the design measure against sodium leak. The thermal consequences associated with sodium combustion and the effects of sodium aerosols are limited only inside guard vessels, guard pipes, and enclosures. In order to practically prevent the sodium-water reaction by a water-steam leak in steam generator heat transfer tubes during the plant lifetime, double-walled tubes are used in the steam generators.

4.2.3.3 Safety evaluations

The following are the safety evaluation results for JSFR with 1500 MWe output power, which core design is shown in Section 4.2.2, the cooling system design shown in Section 4.2.4, and the design of SCCV shown in Section 4.2.5.

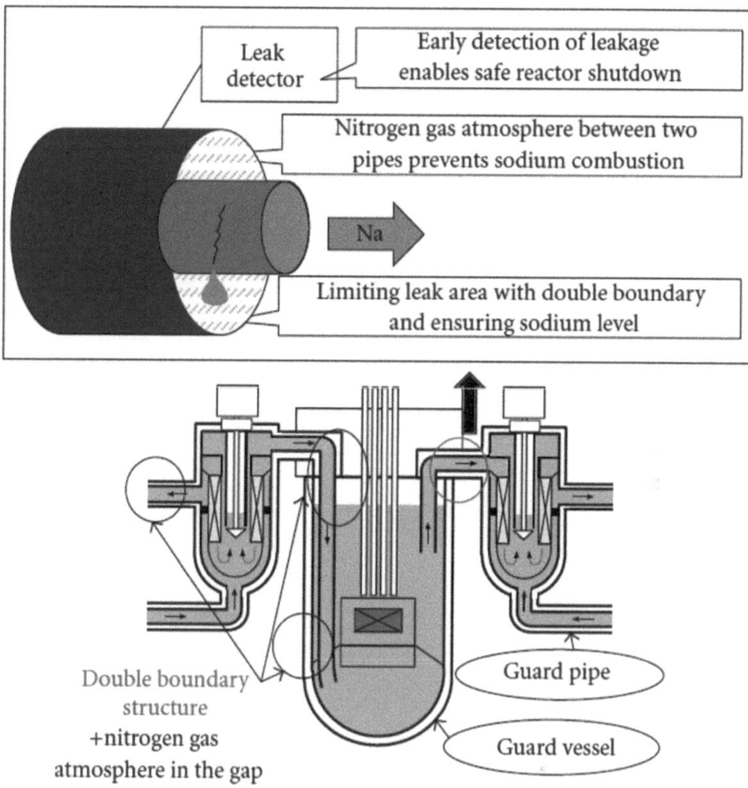

Fig. 4.11 Schematic diagram of double-walled structure [32].

(a) Evaluation of anticipated transients and design-basis accidents

Transient analyses on representative events were conducted. Loss of flow and transient overpower type events are thought to be critical from the view of ensuring core fuel integrity. The primary pump seizure with the most rapid loss of flow was targeted among the selected events. This is because JSFR adopts two-loop configuration [34]. The primary pipes in this concept are covered with all guard pipes, and the primary pipe failure event does not become severe because both the leakage amount and flow rate decrease be controlled even if the primary pipes are damaged. Transient overpower-type events are targeted at control rod withdrawal events. In addition, decay heat removal capability by natural circulation, the water-steam leak in steam generators, sodium leak, and evaluation of public radiation dose are mentioned.

(i) Loss of flow-type events (primary pump seizure)

In this event, the cladding tube temperature and coolant temperature of the core fuel led to an increase by seizing the primary pump shaft in the one loop, resulting in reducing the core coolant flow steeply during reactor power operation. Fig. 4.12 shows the result of the reactor shutdown case by the main shutdown system, and Fig. 4.13 shows the reactor shutdown case by the backup shutdown system. Both results satisfied the safety criteria.

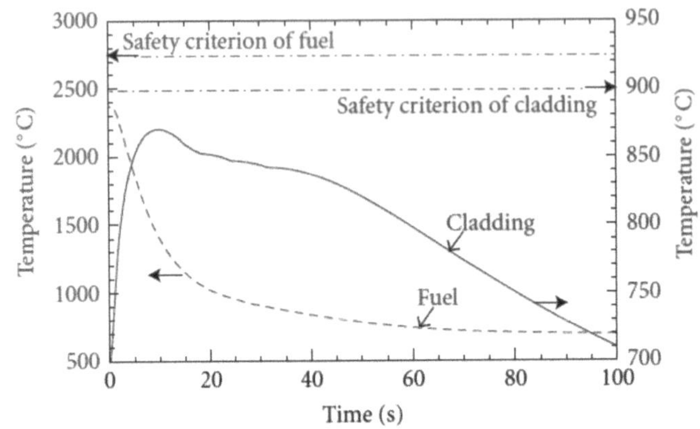

Fig. 4.12 Primary pump seizure accident with primary RSS [32].

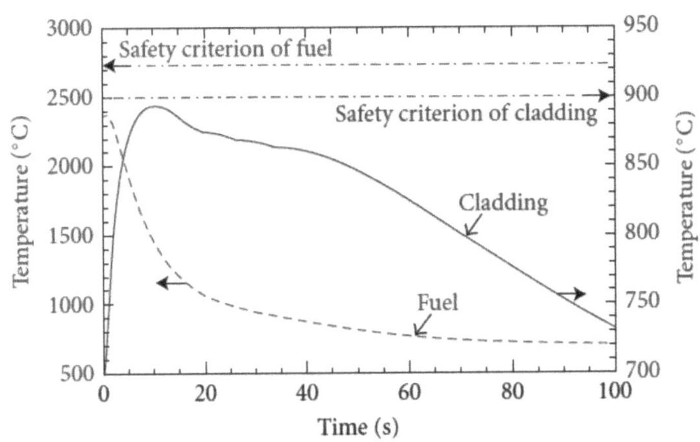

Fig. 4.13 Primary pump seizure accident with backup RSS [32].

(ii) Transient overpower-type events (control rod withdrawal)

During power operation in this event, positive reactivity is abnormally inserted, and the reactor power and fuel temperature increased due to continuous withdrawal by the malfunction of the control rod drive mechanisms or misoperation by the reactor operating staff. The results for both the control rods withdrawal corresponding to the anticipated transient and the rapid control rods withdrawal corresponding to the design basis accident have shown that those transient could meet safety criteria.

(iii) Evaluation of decay heat removal capability

The decay heat removal systems performing the function of core cooling in the event of an abnormal condition in JSFR need to ensure the necessary circulation and heat removal, although the coolant injection function is unnecessary because the closed guard vessels and guard pipes ensure the coolant needed for core cooling even in boundary failure. It is designed to have redundancy or diversity, and the heat removal capacity is set by systematical consideration of the expected operating conditions, including abnormal conditions. Long-term core cooling is possible even if

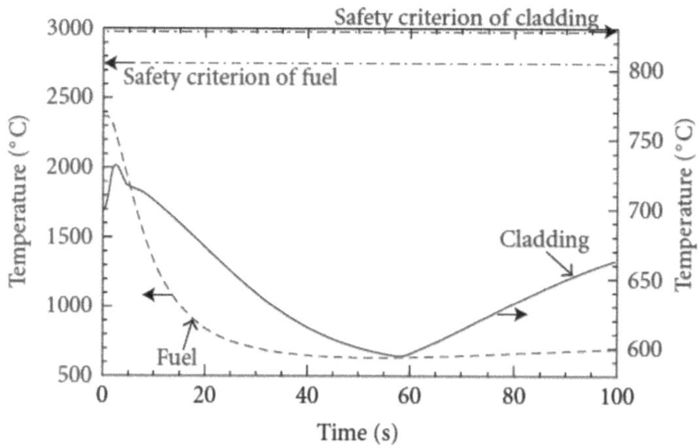

Fig. 4.14 Loss of off-site power event (short term) [32].

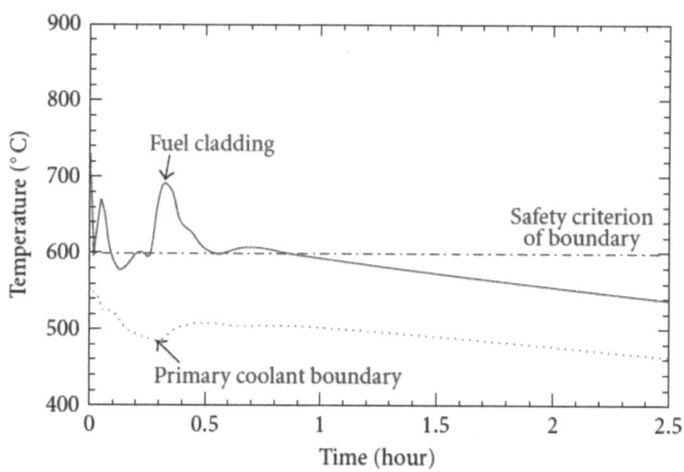

Fig. 4.15 Loss of off-site power event (long term) [32].

assuming a total loss of alternative current electric power source because this system does not rely on active components. Figs. 4.14 and 4.15 show the analysis results of decay heat removal at the loss of off-site power. A scram due to loss of off-site power caused reactor shutdown, and the maximum temperature of the fuel cladding tube drops. In the process of establishing the natural circulation of the primary coolant during the next 30 min, cladding maximum temperature peaks appear, but the cumulative damage of the cladding satisfied the criteria.

(iv) Water-steam leak events in steam generator

During power operation in this event, steam generator heat transfer tubes are damaged, and a water-steam leak occurs into sodium. The pressure generated by the sodium-water reaction may damage the structures, such as the steam generator shell, components and pipes in the secondary cooling system, and the intermediate heat exchanger tubes.

When double wall tubes are adopted for steam generator heat transfer tubes, failure modes to cause simultaneous failures of the inner tube and outer tube could be excluded by design evaluations. Such failure modes are common causes and dependent failures (inner tube failure caused by outer tube failure or outer tube failure caused by inner tube failure). Even if considering independent failures of the inner tube and outer tube, the possibility of a water-steam leak is extremely low, and even if a leak occurs, the leak rate may be limited to a range of small leaks, which leak rate is order of several gram/second. Therefore, this event does not cause a safety problem in practice. Even postulating the double-ended break of a heat transfer tube, the system is equipped with a leak detection system such as a hydrogen detector, isolation and water-steam blow inside the steam generator, and a pressure release and reaction product treatment system and can safely terminate the accident.

(v) Sodium leak events

In this event, the sodium boundary of the primary or secondary cooling systems is failed during the reactor power operation and the coolant leaks. As a result of a sodium combustion analysis conducted in an enclosure assuming a nitrogen gas atmosphere with an oxygen concentration of 2%, the maximum temperature of the enclosure was approximately 10°C higher than the leaked sodium temperature, and the ambient pressure increased slightly with combustion, and the sodium retention function in the enclosure can be maintained.

(vi) Evaluation of public radiation dose

The results on the evaluation of public radiation dose can be severe in case of simultaneous failure of fuel and reactor coolant boundary. However, fuel integrity is maintained even during primary pump seizure and control rod withdrawal, which are considered to have severe consequences for fuel integrity. For this reason, events leading to significant radioactive materials released into the containment vessel are difficult to predict within the range of design-basis accidents. Therefore, evaluation of public radiation dose was conducted assuming "primary pipe failure" and "primary argon gas leak," considering the possibility of a statistical failure of fuel pins in normal operating conditions and assuming the source term inside the primary coolant. Both cases satisfied the criteria. As a hypothetical assumption, the criteria are also satisfied even when the source term in the containment vessel was set the condition of noble gases 100% (fraction of core inventory) and iodine 10% as in Monju.

(b) Evaluation of design extension conditions

(i) Prevention of core damage

From the viewpoint of confirming the feasibility of SASS as a passive reactor shutdown mechanism, each of three types of ATWS, TOP-type ATWS, LOF-type ATWS, and LOHS-type ATWS, was analyzed assuming typical abnormal conditions. This evaluation was on a nominal basis because it is positioned as design extension conditions. Fig. 4.16 shows the results of the LOF-type ATWS analysis. The SASS activation temperature was set at 660°C (delatch temperature), and the coolant transportation and temperature response from the core outlet to the SASS temperature sensing alloy location were evaluated. After reaching the cutoff temperature, the insertion by the gravity of the backup control rods resulted in reactor shutdown without exceeding the fuel and coolant temperature specified in the safety criteria.

(ii) Mitigation of core damage sequences

The core damage sequences and consequences were evaluated assuming ULOF not considering SASS.

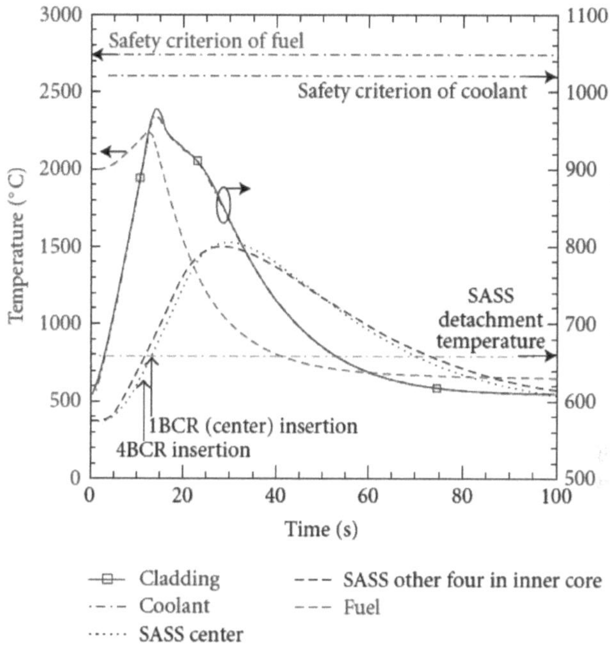

Fig. 4.16 Loss of flow without scram accident with passive shutdown [32].

Tools and knowledge used for this evaluation are described in Chapter 5.1.4. In the initiating phase, the results of analysis using the SAS4A code, which was validated by the CABRI in-pile experiments through international collaboration with France and Germany and other countries, confirmed that the net reactivity during the transients does not excess prompt criticality by setting the maximum sodium void worth to approximately 6 dollars or less considering the uncertainty. In the early fuel discharge phase, SIMMER-III calculation results showed that more than 15% of the total core fuel was confirmed to be discharged through an inner duct installed in each fuel assembly before the molten fuel pool extension in the core. SIMMER-III code was reflected the findings of EAGLE in-pile and out-of-pile experiments conducted in international collaboration with the Republic of Kazakhstan. Fig. 4.17 shows the concept of fuel assembly with inner duct structure (FAIDUS), and Fig. 4.18 shows the results of the analysis of core power and reactivity changes with fuel discharge from the inner ducts. These results confirmed the effectiveness of the fuel assembly with inner duct structure in the evaluation of early fuel discharge.

In the material relocation phase, a static neutronics calculation was conducted taking into account core materials fall down, melt progression by decay heat and neutron absorber material behavior. The results confirmed that a sub-critical state could be maintained throughout the all along the material relocation phase. In the long-term cooling phase, the degraded core materials moved to the lower plenum in the reactor vessel are retained in the core catcher. The fragmentation of the damaged core materials in the lower plenum was estimated to be possible to fragment in a range of approximately 3 m to the debris tray, referring to the findings of out-of-pile molten material

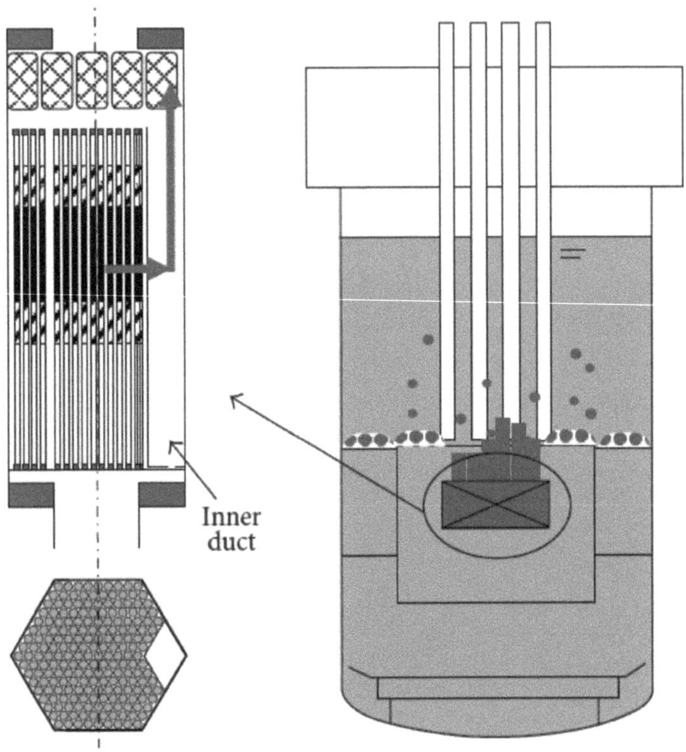

Fig. 4.17 Fuel assembly with inner duct for enhancing molten fuel discharge [32].

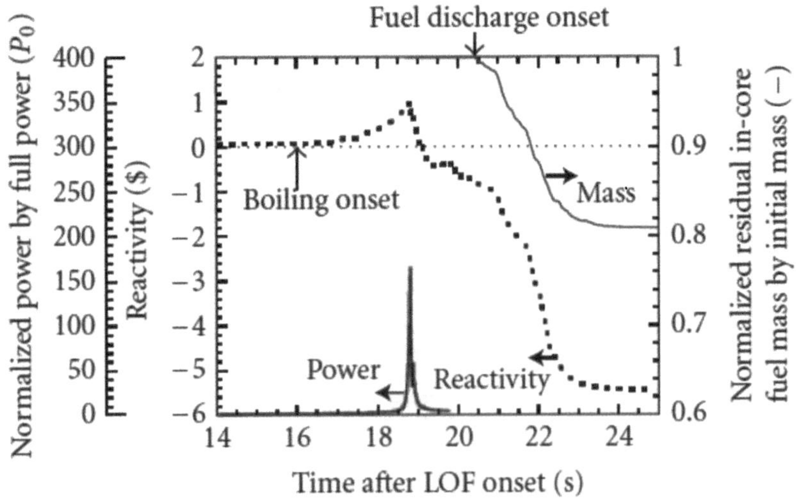

Fig. 4.18 Initiating and early discharge phases of unprotected loss of flow accident [32].

injection experiments. In addition, the cooling path by natural circulation could be secured even if 100% of the core fuel was located on the core catcher and formed a debris bed, and these results confirmed that the debris bed could be cooled and maintained. These results showed the prospect of IVR in the evaluation of the relocation behavior and the stability and coolability of degraded core materials in the reactor vessel.

(c) Probabilistic analysis

The core damage frequency was evaluated for internal events during power operation in order to have the prospect of achieving the probability target shown in Fig. 4.10. A set of event sequences leading to core damage was classified into three categories, ATWS, loss of reactor level (LORL), and protected loss of heat sink (PLOHS), and the point estimation value of each core damage frequency was calculated [35]. ATWS was 1×10^{-8}/reactor year, LORL was 4×10^{-9}/reactor year, and PLOHS was 9×10^{-9}/reactor year, and total CDF became 2×10^{-8}/reactor year.

4.2.4 Reactor cooling system

4.2.4.1 Overview

(a) Concept and specifications

The major specifications of JSFR are shown in Table 4.10. A schematic of the reactor cooling system is shown in Fig. 4.19. The reactor inlet and outlet temperatures are 395°C and 550°C, respectively. A two-loop configuration is used for the reactor cooling system to

Table 4.10 Major specifications for the JSFR cooling system [36].

Items	Specifications
Reactor type	Sodium-cooled loop-type
Electricity output	1500MWe
Thermal output	3530MWt
Number of loops	2
Primary sodium temperature	550°C/395°C
Primary sodium flow rate	3.24×10^4 t/h/loop
Secondary sodium temperature	520°C/335°C
Secondary sodium flow rate	2.70×10^4 t/h/loop
Main steam temperature/pressure	495°C/18.7 MPa
Feed water temperature/flow rate	240°C/2.88×10^3 t/h/loop
Main material of reactor vessel	316FR
Main material of reactor cooling system	Mod.9Cr-1Mo steel
Primary piping type	L-shaped piping
IHX and primary pump type	Integrated IHX with pump
Steam generator type	Once-through type with straight double-walled tubes
Decay heat removal system	PRACS x 2 + DRACS x 1

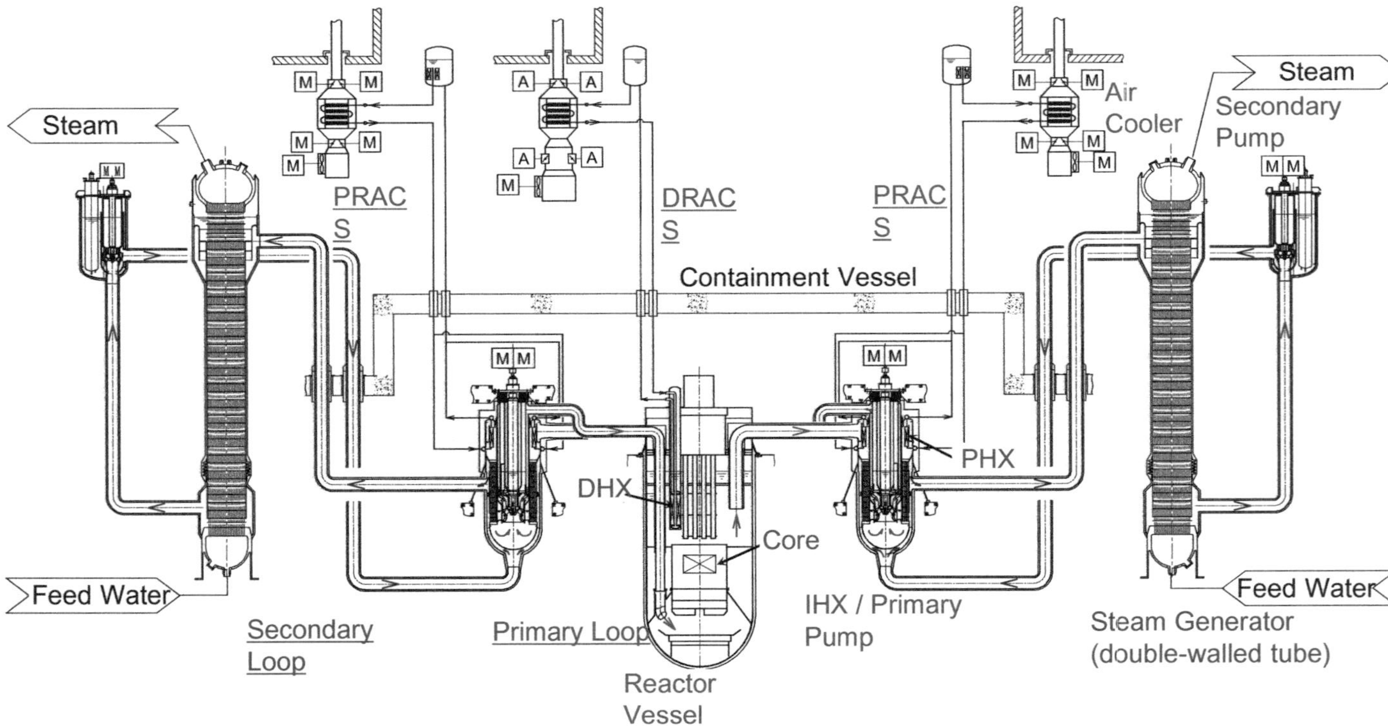

Fig. 4.19 JSFR primary cooling system [36].

enhance its economic performance. The reactor cooling system consists of a primary system and a secondary system to prevent the possibility of contact between radioactive sodium in the primary system and the water in a steam generator (SG) in the secondary system. An intermediate heat exchanger (IHX) and an SG are installed per loop [36].

JSFR achieves the FaCT development targets and generation IV reactor goals by adopting innovative technologies. For the reactor cooling system, the following key technologies are installed:

- A compact reactor system adopting a hot vessel and in-vessel fuel handling with a combination of an upper internal structure (UIS) with a slit and advanced fuel handling machine (FHM)
- A two-loop cooling system with large diameter piping made of Mod. 9Cr—1Mo steel
- An integrated IHX-pump component
- A reliable SG with a double-walled straight heat transfer tube
- A natural circulation decay heat removal system (DHRS)

In FaCT phase 1, these technologies were evaluated as suitable for installation to the demonstration JSFR and large-scale demonstration experiments [37].

4.2.4.2 Reactor structure

(a) Concept and specifications

Sketches of the reactor structure are shown in Fig. 4.20 [38]. The reactor vessel (RV) diameter is 10.7 m, and the height is 21.2 m. There are two hot-leg pipes and four cold-leg pipes. The hot- and cold-leg pipes are installed into the top of the RV (called the "top-entry system"). This top-entry system can reduce not only the reactor cavity diameter but also the displacement by thermal expansion of the piping, which contributes to the reduction of the mass amount of the piping system.

The flow path of the coolant is shown in Fig. 4.20. The 395°C primary coolant flows into the reactor lower plenum from the four cold-leg pipes at a flow rate of 6.48×10^4 t/h. The coolant mixed in the reactor lower plenum flows into the core. The coolant absorbs the core's heat, and the 550°C coolant is suctioned into the hot-leg pipes.

The diameter of the rotating plug is decreased by adopting an advanced refueling system, which consists of a UIS with a slit and an FHM with an extension arm. The FHM installed on the rotating plug can insert the arm into the center of the UIS through a slit (see Fig. 4.21). By extending the arm through the slit of the UIS, FHM's gripper can access any location of the fuel assemblies. The refueling route in the reactor system is shown in Fig. 4.21. The new fuels are put into the in-vessel relay tank by the ex-vessel fuel transfer machine. The FHM puts out spent fuels from the core and sets new fuels instead.

The hot RV can reduce the vessel diameter by eliminating the wall cooling system. The prototype reactor Monju uses an overflow and makeup system instead of a wall cooling system, which controls the coolant surface level in the RV to reduce the thermal stress during start-up. Conversely, JSFR eliminates the overflow system to decrease the amount of plant material. Because thermal loading is larger and the operating temperature is higher in JSFR than those in Monju, the RV material adopts 316FR stainless steel, which has greater strength at elevated temperatures.

Fig. 4.20 shows the configuration of the components on the roof deck and the rotating plug. The main components except the hot- and cold-leg pipes are listed below. The direct

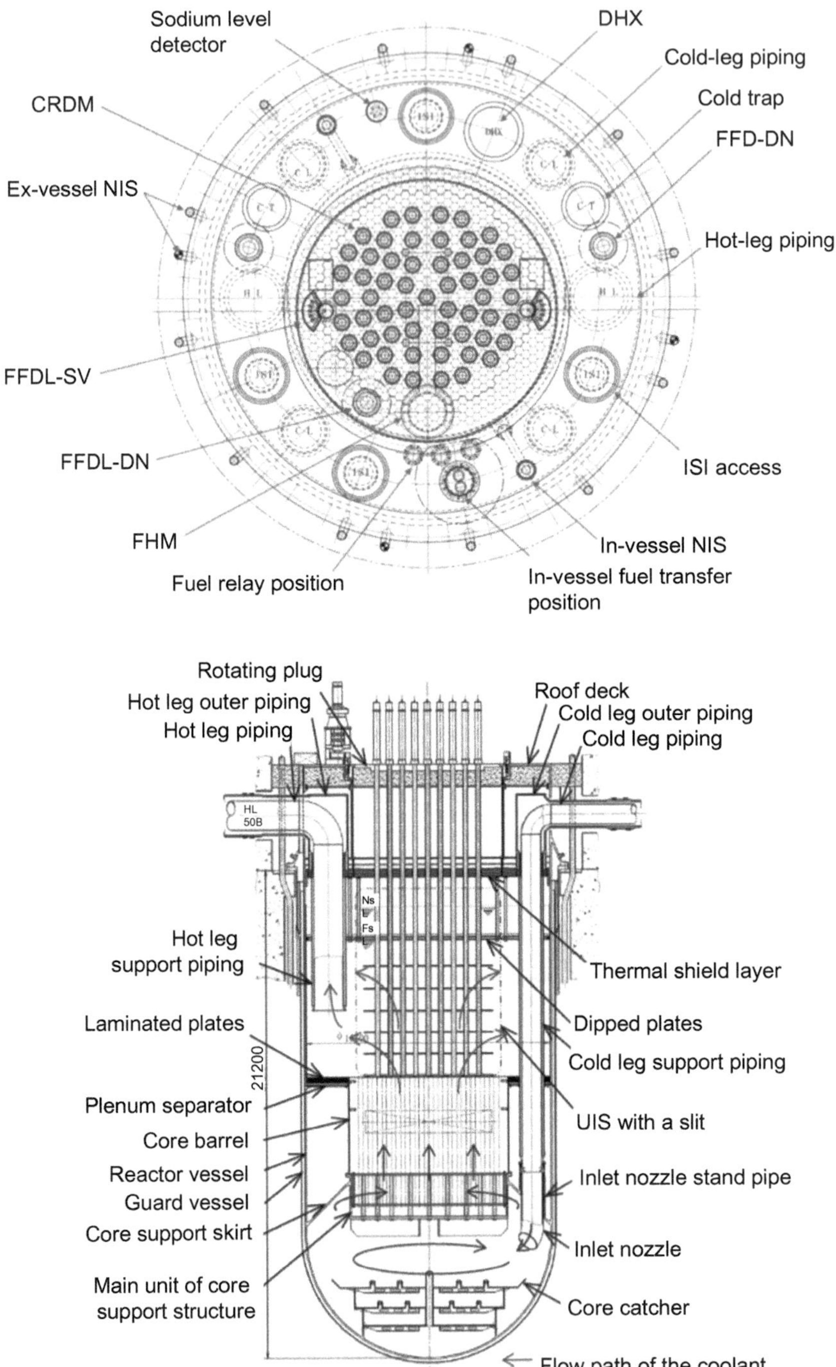

Fig. 4.20 Sketches of the JSFR reactor structure [38].

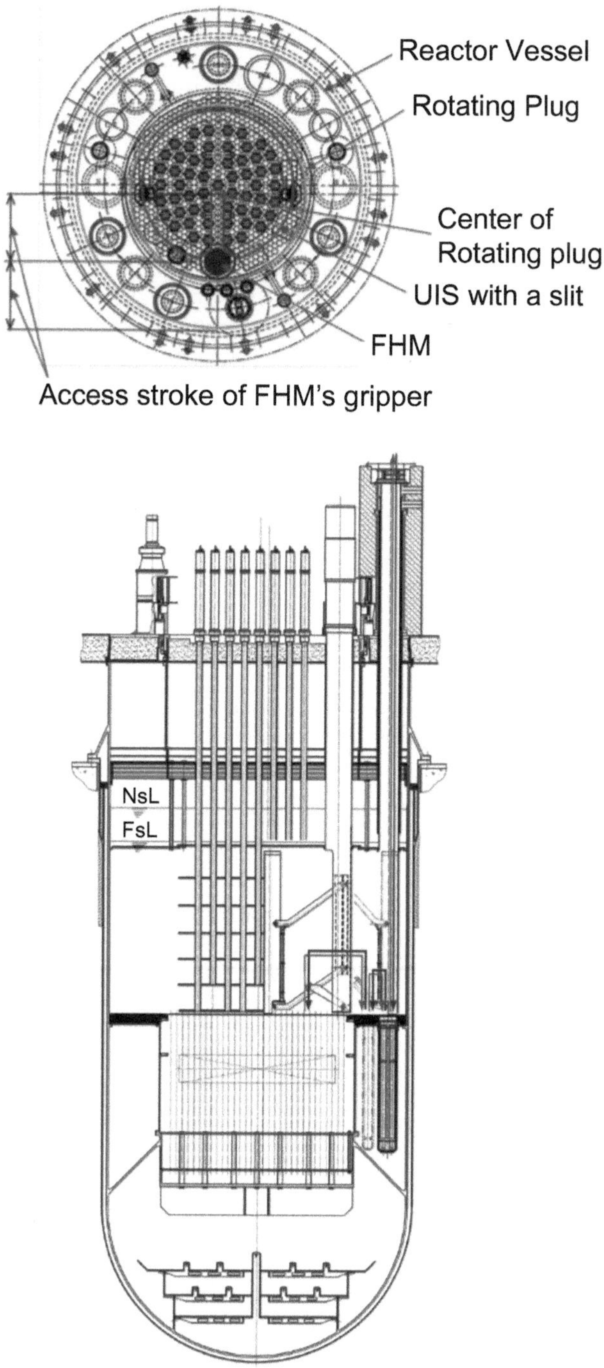

Fig. 4.21 Sketch of the refueling rout [38].

heat exchanger (DHX) is a component of the DHRS, and it cools the primary coolant after the reactor scram. The cold traps (CTs) remove impurities from the primary sodium. The control rod drive mechanism (CRDM) drives the control rods for power control and reactor shutdown during emergencies.

- UIS
- DHX: 1
- CTs: 2
- CRDM: 57
- FHM
- Guide tube of the in-vessel fuel transfer machine

In addition, the measurement components below are installed on the roof deck and the rotating plug. The core exit thermocouples measure the fuel outlet temperature, and they detect extraordinary temperature elevations. The failed fuel detection (FFD) system detects the delayed neutrons emitted out when the fuels fail. In the case of an FFD system detection, the failed fuel detection and location (FFDL) system is activated, and it locates the failed fuel. The sodium level detector can detect various liquid levels from normal sodium level (NsL) to emergency sodium level (EsL). The neutron instrumentation system (NIS) monitors the neutrons emitted from the core and detects extraordinary neutron levels during accidents.

- Core exit thermocouple
- FFD
- FFDL
- Sodium level detector
- NIS

Four in-service inspection access holes, through which an inspection device gets into the RV, are set at the roof deck. Dummy plugs are installed at the holes during normal operations.

(b) Flow optimization in the reactor upper plenum

The flow velocity in the reactor upper plenum is increased by reducing the RV diameter. Consequently, cover gas entrainment can occur. Dipped plates (DPs) are installed below the sodium free surface as a countermeasure for the cover gas entrainment, reducing the flow velocity just below the free surface and preventing cover gas entrainment (see Fig. 4.22).

Three-dimensional thermal-hydraulic analyses were conducted to confirm the flow pattern in the reactor upper plenum. Fig. 4.23 shows the calculated velocity vector in a horizontal cross section below the sodium free surface level. Whereas the free surface flow velocity without the countermeasure is very high, the velocity with the countermeasure is low [39].

In addition to the above analysis, water experiments were also conducted. A 1/11-scale model water experiment by the Japan Atomic Energy Agency (JAEA) for the reactor upper plenum confirmed that the cover gas entrainment phenomena did not occur [40].

The pressure of the fluent at the center of a vortex gets lower than that without a vortex. When the pressure gets lower than the saturated steam pressure, cavitation can occur.

The coolant flows out from the core outlet into the reactor upper plenum. Finally, the coolant is suctioned into the hot-leg pipes. The discharged flow from the core outlet passes through the core instrumentations plate (CIP) and enters the UIS. Most of the discharge flow is distributed radially by the lower three baffle plates (BP-1 to BP-3), and the rest goes up directly to the DP through the inside of the UIS and expands radially to the RV wall. A part of the radially distorted flow directly enters the hot-leg piping intakes. The rest of the distorted flow goes to the RV wall and ascends toward the DP, forming recirculation flows

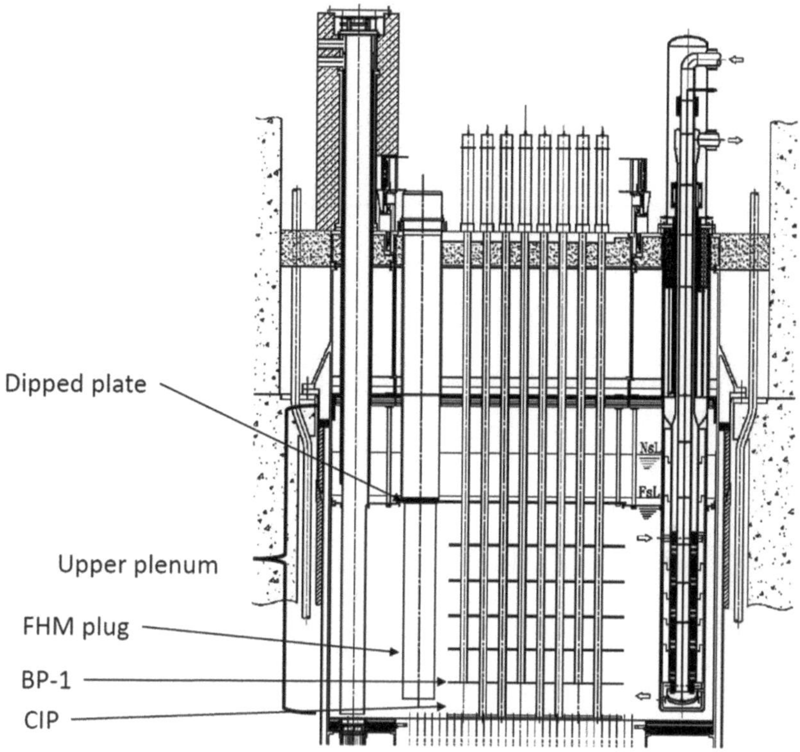

Fig. 4.22 Countermeasures for flow optimization in the upper plenum [38].

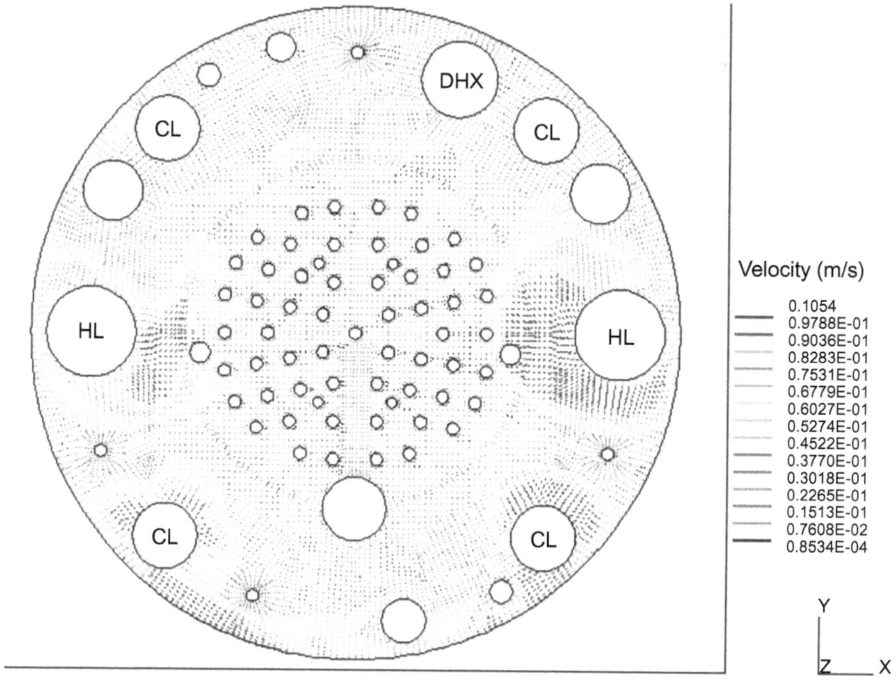

Fig. 4.23 Sodium velocity at the free surface [38].

between the components installed in the reactor upper plenum. Additionally, the flow at the radial slit is considerably fast and causes an asymmetric flow in the reactor upper plenum. Accordingly, the flow velocity along the RV wall is high, and several vortex tubes stretch into the hot-leg pipes. Significant depressurizations might occur in the vortex tubes near the intakes of the hot-leg pipes and might induce cavitation [39].

The FHM plug is installed at the slit above the middle height between the CIP and the BP-1 as a countermeasure to balance the flow rate discharged from the slit side of the UIS and the other side. The flow holes at the BPs and the CIP as the other countermeasures reduce the flow velocity along the RV wall (see Fig. 4.22).

Fig. 4.24A shows the flow distribution in the vertical cross section near the hot-leg intake of the reactor upper plenum with the countermeasures. The vortexes due to the radial

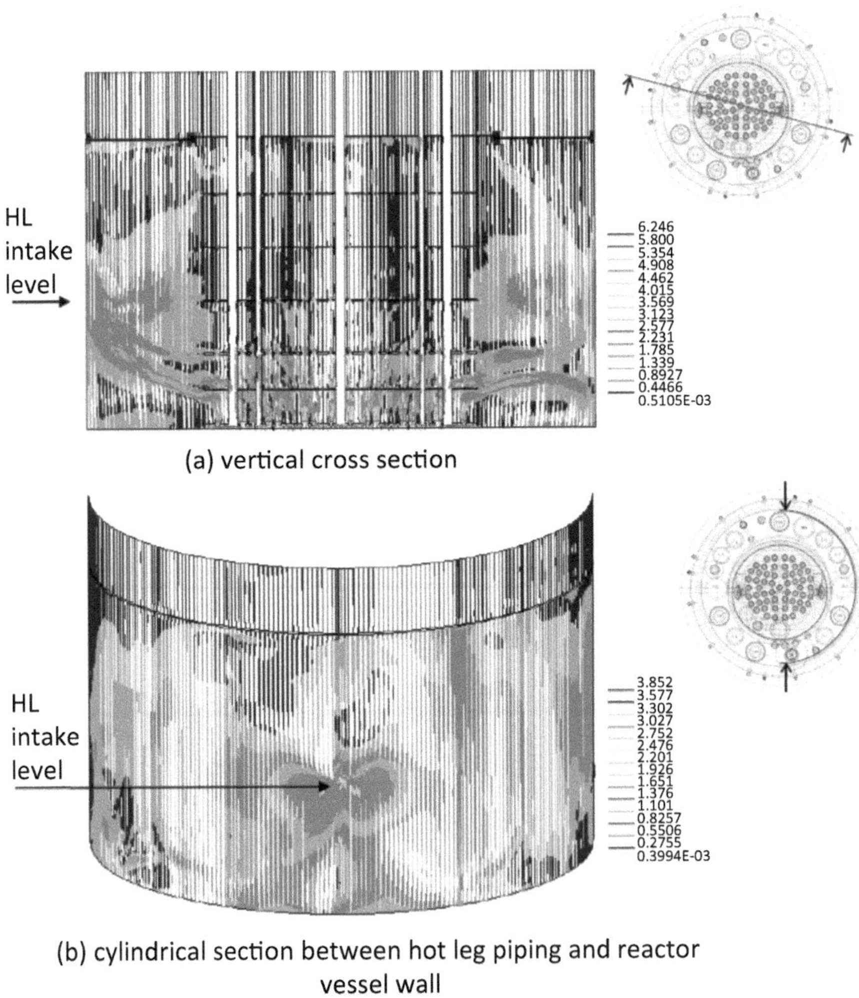

Fig. 4.24 Flow velocity in the upper plenum [38].

flow from the lower part of UIS are more weakened than those without the countermeasures. Fig. 4.24B shows the flow distribution in the cylindrical section between the hot-leg piping and the RV wall with the countermeasures. The velocity near the hot-leg intake is decreased compared with that without the countermeasures.

The validity of the proposed measure was confirmed from the three-dimensional thermal-hydraulic analyses and the 1/11-scale model water experiment for the reactor upper plenum of the RV [40].

(c) Seismic design

Considering past severe earthquakes, the Nuclear Safety Commission of Japan (NSC) revised the regulatory guideline for reviewing the seismic designs of nuclear power reactor facilities in 2006 [41]. Yet another massive earthquake, the so-called Chuetsu-oki Earthquake, occurred in 2007. Under those influences, the earthquake conditions for the JSFR design were reviewed and modified to be severer, as shown in Fig. 4.25. A seismic evaluation for the reactor system was performed using the severer earthquake condition [42,43]. The buckling of the RV and jump-up amount of fuel subassemblies were found to be crucial factors in terms of the integrity of the reactor system. Two measures were adopted to solve this problem: One is improving the stiffness of the RV by thickening its wall, and the other is mitigating the input seismic motion by optimizing the natural periods of seismic isolation system considering the reactor system components, as shown in Fig. 4.26. These two measures obtained good results on the integrity of the reactor system when the RV wall thickness was 60 mm or more and the seismic isolation system had longer natural periods, that is, 0.29 Hz of natural

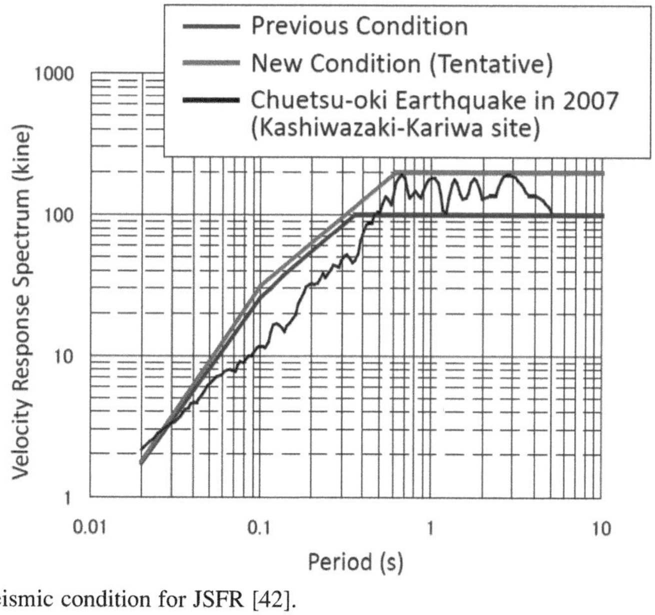

Fig. 4.25 Seismic condition for JSFR [42].

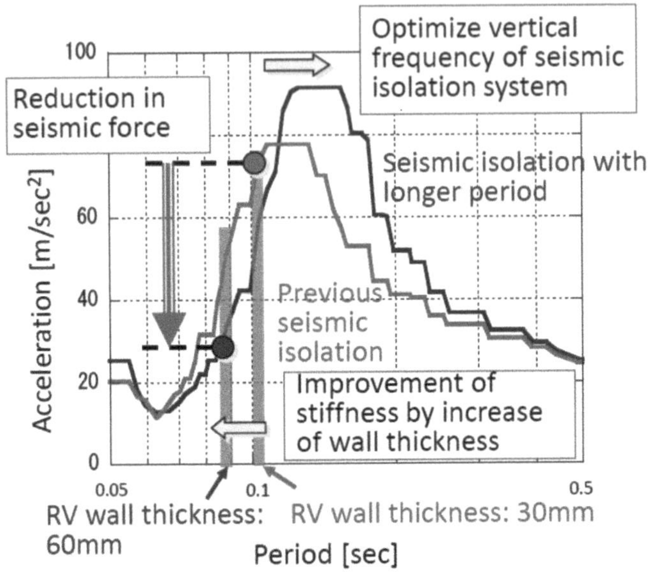

Fig. 4.26 Reduction of the vertical seismic force [42].

frequency in a horizontal direction and 8.0 Hz in a vertical direction. Thicker laminated rubber bearings were introduced to the aforementioned advanced seismic isolation system to optimize the natural periods. Oil dampers were also adopted instead of steel bar dampers to improve the damping performance.

(d) Thermal stress

The vertical temperature distribution of the RV wall during start-up is shown in Fig. 4.27. The temperature of the roof deck bottom plate is about 50–100°C because the plate is cooled. The temperature of the RV wall below the sodium surface level is similar to the coolant temperature. Accordingly, the temperature gradient of the RV gets steep between the coolant surface and the roof deck bottom plate.

Because JSFR eliminates the overflow system, the sodium surface level moves during start-up and shutdown. In particular, the temperature distribution is sharp during start-up because the sodium surface level rises, while the coolant is heated up. Hence, evaluating the thermal stress generated around the sodium surface level is important.

Fig. 4.28 shows the relation between the start-up durations and the thermal stress (in the range of primary-plus-secondary stress intensity). It can be confirmed that the longer the start-up durations, the lower the thermal stress. In addition, it is deduced that more than 4 days are needed to satisfy the thermal rachet evaluation.

After a reactor trip, thermal stratification is generated in the upper plenum. Fig. 4.29 shows the vertical distribution of the coolant temperature when thermal stratification occurs, where the temperature distribution profile goes up slowly thereafter. In this case,

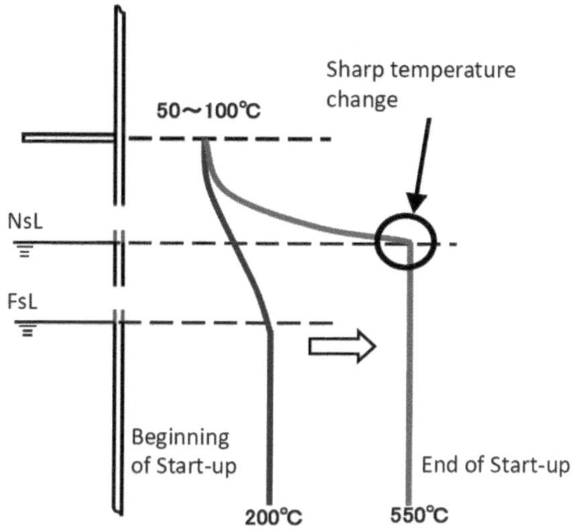

Fig. 4.27 Temperature distribution during start-up [38].

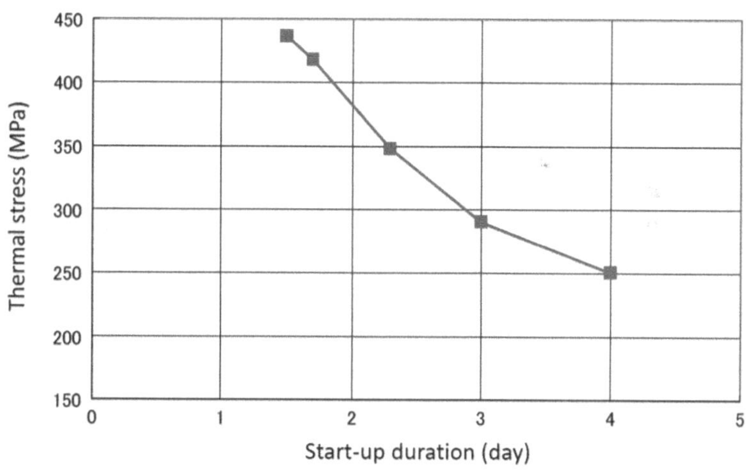

Fig. 4.28 Relation between the start-up durations and the thermal stress [38].

the most critical evaluation item is the thermal rachet. The stress analysis results show that the thermal rachet evaluation is satisfied because the thermal membrane stress is low.

4.2.4.3 Two-loop primary cooling system

(a) Concept

A two-loop cooling system can be realized by introducing innovative technologies [34]. One of such innovative technologies is high-chromium ferrite steel, which has excellent

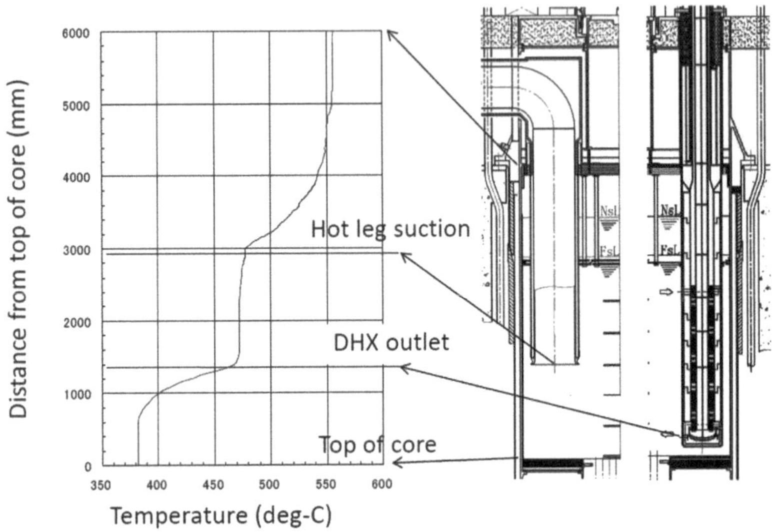

Fig. 4.29 Vertical temperature distribution during thermal stratification [38].

physical properties: high strength, high heat conductivity, and low thermal expansion. This high strength feature allows the use of a thin heat transfer tube and a large tube sheet, by which the structural integrity could be ensured against thermal stress. With such a heat transfer tube, the heat transfer capacity of heat exchangers could be increased compared with those of conventional plant designs by utilizing the high heat conductivity of the steel. For this capacity enlargement, adequately considering the desirable characteristics of sodium-cooled fast reactors is critical: the excellent heat transfer characteristic of the sodium coolant and a low-pressure system. The increase in heat exchanger capacity is reflected in the increase in the heat transport capacity per cooling loop. Thus, the number of loops can be reduced to two in the cooling system design.

In addition to the reduced number of loops, drastically shortened primary and secondary piping systems could be realized by utilizing the high strength and low thermal expansion characteristics of high-chromium steel. The shortened piping design results in a compact plant configuration because of the close arrangement of the plant's components. The scale merit can be achieved by increasing the electricity output per reactor to 1500 MWe, which is comparable with that of a state-of-the-art commercial light water reactor.

In the early stages of the conceptual design phase, the plant construction cost is compared for the two-loop and four-loop systems. The comparative study of these systems found that the amount of commodities of the nuclear steam supply system in the two-loop system was about 16% less than that in the four-loop system. The two-loop system design was also estimated to reduce about 14% of the reactor building volume compared with that with the four-loop system. A further cost analysis indicated that the two-loop system design significantly contributed to reducing the construction cost compared with other design improvements such as a compacted RV or component integration. However, the above economic advantage of the two-loop cooling system in JSFR raises the following technological issues:

- Hydraulic and structural integrity issues of the components
  - Flow-induced vibration due to the higher flow rate
  - Gas entrainment and vortex cavitation in the upper plenum
- Safety and reliability issues of the system
  - Safety design against primary pump seizure accident
  - Safety reliability assurance of DHRS

The gas entrainment and vortex cavitation are discussed in the reactor structure subsection. The other issues are described in the following paragraphs.

(b) Flow-induced vibration

A flow-induced vibration test with water was conducted using a 1/3-scale L-shaped elbow test section, which simulated the JSFR hot-leg piping, to clarify the hydraulic and vibration behaviors in the primary cooling piping [44,45]. The experimental facility for this test is depicted in Fig. 4.30. The test section was positioned in the upper region of an inlet tank. A rectifier with a honeycomb structure was installed below the intake of the test section within the inlet tank to obtain a uniform velocity distribution at the test-section intake. Two kinds of test sections were prepared. One was an acrylic pipe to observe flow patterns and measure pressure fluctuations on the piping wall, corresponding to the sources of the flow-induced vibration excitation force. The other was a stainless steel pipe, the stiffness of which agreed well with that of the JSFR hot-leg piping. The steel-pipe test aimed to obtain a piping vibration response to verify the flow-induced vibration analytical methodology.

Because isothermal fluid forces on the wall were necessary for the flow-induced vibration evaluation, water was used instead of sodium as the working fluid considering the

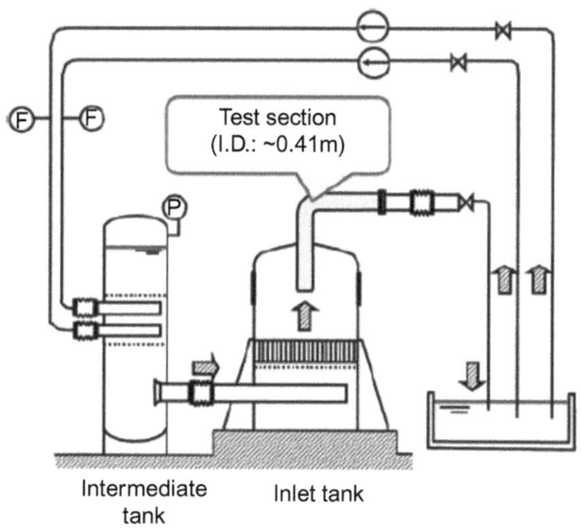

Fig. 4.30 Flow-induced vibration experimental facility [34].

reasonable implementation of the experiment. The test parameters were the mean velocity (from 0.8 to 9.2 m/s) and water temperature (~15°C, corresponding to room temperature, and 60°C). The mean velocity of 9.2 m/s is equivalent to the JSFR hot-leg piping flow velocity. Our experiments covered a wide range of Reynolds numbers: $Re = 3 \times 10^5$ to 8×10^6. The maximum Reynolds number in the experiments was one-fifth of the JSFR plant condition $\sim 4 \times 10^7$.

Flow visualization experiments using the acrylic pipe were performed using an ink injection method as the first step. Fig. 4.31 shows a typical flow visualization result in the model elbow at the mean velocity of 9.2 m/s (Reynolds number: $\sim 3.7 \times 10^6$). The main flow in the piping separated from the intrados wall near the elbow exit and then reattached at ~290 mm downstream from the separation point. Further, vortex shedding was observed intermittently from the flow separation region to the downstream region. The flow pattern was also examined in lower-velocity conditions, and a similar flow pattern was observed. This showed that the flow pattern did not depend on the Reynolds number.

The radial distribution of the main flow direction velocity was measured along the centerline in the piping using laser Doppler velocimetry. The nondimensional time-averaged velocities in the main flow direction were compared with regard to the non-dimensional radial position in Fig. 4.32. The nondimensional distance was defined as the distance from the inner surface of elbow pipe z normalized by the pipe diameter D. The nondimensional velocity was defined as the local velocity U divided by the mean velocity U_m. They were measured at 75 mm downstream from the elbow exit. Fig. 4.32 compares five experimental results with different coolant velocities and temperatures. No clear difference could be seen in the range of our experimental conditions ($Re = 3 \times 10^5$ to 8×10^6). Therefore, we concluded that the velocity distributions in the piping are independent of the Reynolds number.

The pressure loss coefficients of a single elbow are summarized in terms of Reynolds number in Fig. 4.33. In the past, the pressure loss coefficients of an elbow pipe were measured up to 3.5×10^5 by Idelchik [46]. According to Idelchik, the pressure loss coefficient was defined as a differential pressure divided by the kinetic pressure of the main flow, as

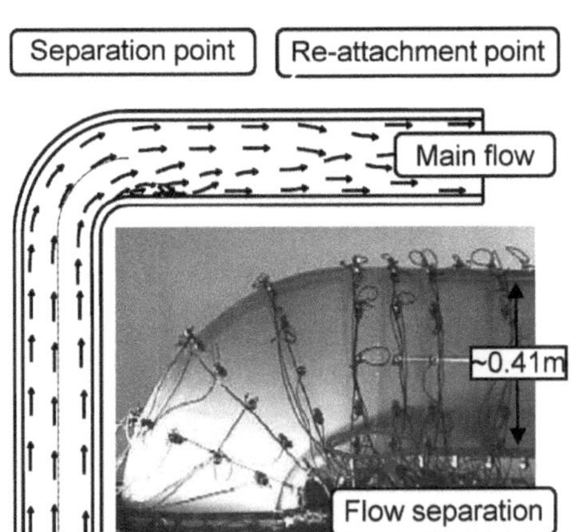

Fig. 4.31 Flow pattern in the model elbow [34].

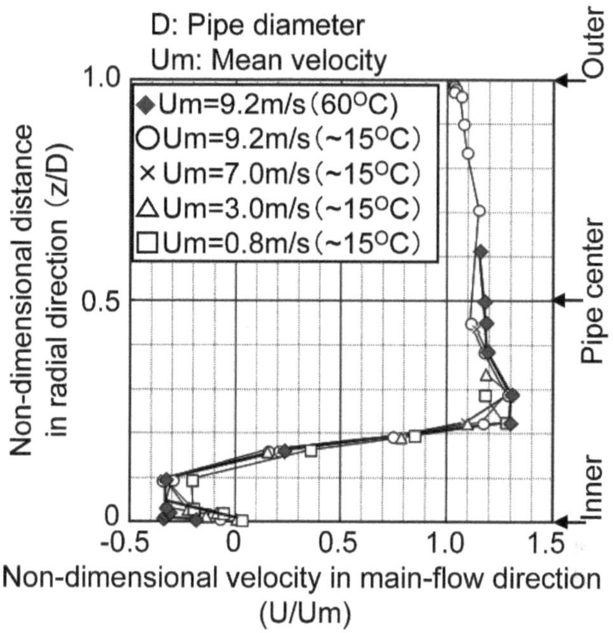

Fig. 4.32 Radial distribution of nondimensional time-averaged velocities. (main flow direction at 75 mm downstream from the elbow exit) [34].

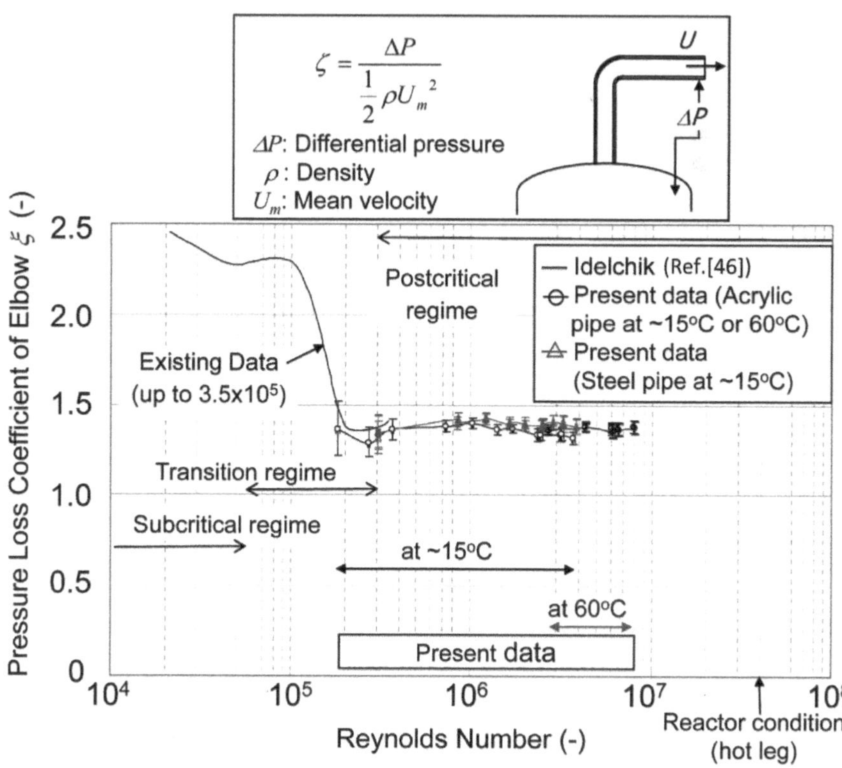

Fig. 4.33 Pressure loss coefficients of a single elbow [34].

shown in Fig. 4.33. The experiments for JSFR give new experimental data in a higher Reynolds number range than those of existing data. All experimental data up to 8×10^6 are plotted in this figure.

Similar to the flow around a cylinder, the pressure loss coefficient can be divided into three regimes: subcritical, transition, and postcritical. Almost all the present data are in the postcritical regime. The pressure loss coefficient becoming minimal in the transition regime is consistent with the existing data. In the postcritical regime, the coefficients are nearly constant. In other words, the dependency on the Reynolds number is relatively small in the postcritical regime. This experimental evidence suggests that the pressure loss coefficient can be extrapolated to the hot-leg piping condition.

Pressure fluctuations on the piping wall that correspond to the excitation forces of the piping vibration were measured using 124 pressure transducers in the axial and circumferential directions of the test section. Fig. 4.34 shows representative nondimensional power spectrum densities (PSDs) of the pressure fluctuations at a coolant mean velocity of 9.2 m/s at 60°C. Given that the PSDs become input data for the flow-induced vibration analysis, they should be summarized to simplify the vibration analysis. In Fig. 4.34, the piping is divided into five sections, where the measured PSDs were similar in each section at different positions in the experiment. The PSDs in this figure were averaged with the measured PSDs in the respective regions. As expected, the pressure-fluctuation PSD in Fig. 4.34C, including the flow separation region, became dominant. The peak in the PSD in Fig. 4.34C appears at ~0.45 of the nondimensional frequency (i.e., the Strouhal number). According to the flow visualization, this peak frequency corresponds to the periodicity of vortex shedding from the flow separation region. This Strouhal number agrees well with separately measured velocity data by Kawamura et al., who indicated periodic turbulence at ~0.5 of the Strouhal number in the area downstream from the elbow [47]. The nondimensional PSDs at the flow separation region are compared to investigate their dependency on velocity and water temperature (viscosity) in Figs. 4.35 and 4.36, respectively. No clear dependency on velocity and viscosity can be seen in these figures. Kawamura et al. also reported

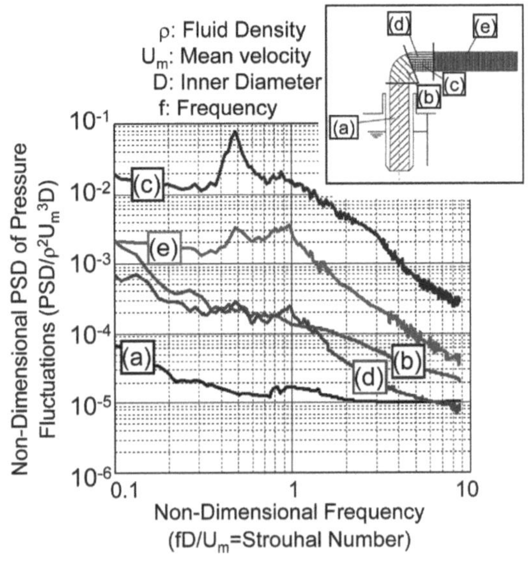

Fig. 4.34 Nondimensional PSDs of pressure fluctuations in representative regions [34].

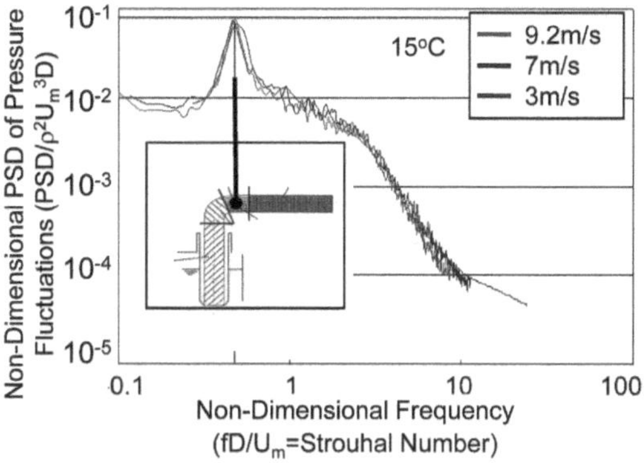

Fig. 4.35 Comparison of nondimensional PSDs of pressure fluctuations among different flow velocities [34].

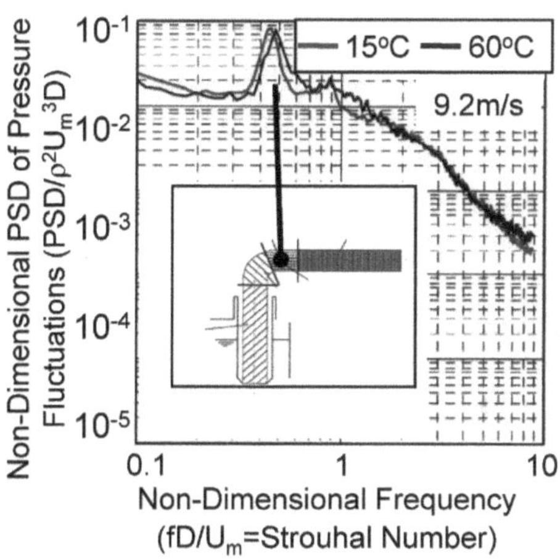

Fig. 4.36 Comparison of nondimensional PSDs of pressure fluctuations between different temperatures [34].

no effect of the Reynolds number on the power spectra of turbulence. This experimental evidence agrees with our experimental results described earlier.

Steel-pipe experiments were also conducted to simultaneously measure the pressure-fluctuation PSDs and the vibration response of pipings. In these experiments, the measured PSDs of the pressure fluctuations were confirmed to be equivalent to those of the acrylic pipe experiments under the same experimental conditions. By specifying the flow-induced vibration force with the experimental PSDs, the flow-induced vibration analysis methodology calculated the mechanical stress on the piping. This methodology was verified by comparing its results to the stress obtained in the steel-pipe experiments [48].

The analysis methodology was then applied to the calculation of the maximum stress on the JSFR hot-leg piping. This calculation specified simplified PSDs of the pressure fluctuations in the five sections (Fig. 4.34) to roughly evaluate the stress in the conceptual design phase. Because no clear dependency on the Reynolds number can be seen in Figs. 4.35 and 4.36, the reactor evaluation was possible by applying the assumption of analogy. For a conservative evaluation, each of the simplified PSDs enveloped all the PSDs measured in the respective regions. Using this analogy, we extended the simplified PSDs to the JSFR hot-leg

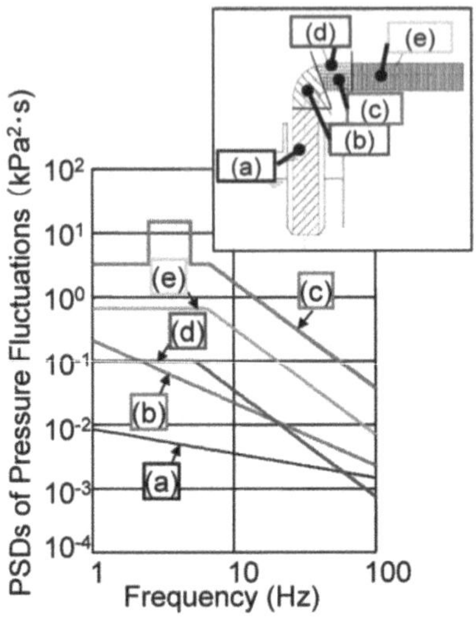

Fig. 4.37 Simplified PSDs of pressure fluctuations for the JSFR hot-leg piping [34].

Fig. 4.38 Calculated maximum stress for the JSFR hot-leg piping [34].

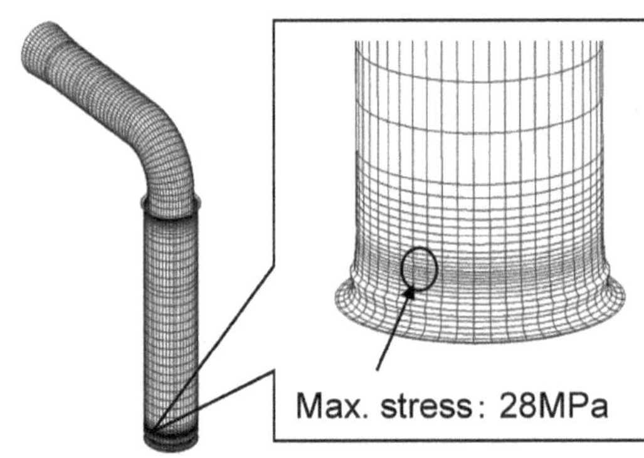

piping, the diameter of which is three times larger than that of the experimental one. Fig. 4.37 shows the pressure-fluctuation PSDs under the JSFR condition, which were applied to the flow-induced vibration evaluation for JSFR using the verified methodology mentioned earlier. As shown in Fig. 4.38, the predicted maximum stress (28 MPa) fulfilled the tentative design criterion (49 MPa at 550°C) of the high-cycle structural fatigue of the piping material (2.25Cr-1Mo steel). We plan to obtain the high-cycle fatigue data of high-chromium ferrite steel, the strength of which is generally higher than that of 2.25Cr-1Mo steel.

4.2.4.4 Integrated IHX-pump

(a) Concept and specifications

The integrated IHX-pump concept is among the technologies used to reduce the mass of a reactor cooling system. The integration of the IHX and the primary pump also contributes to the reduction of the middle-leg pipe connecting the two components and the volume of the reactor building, as shown in Fig. 4.39.

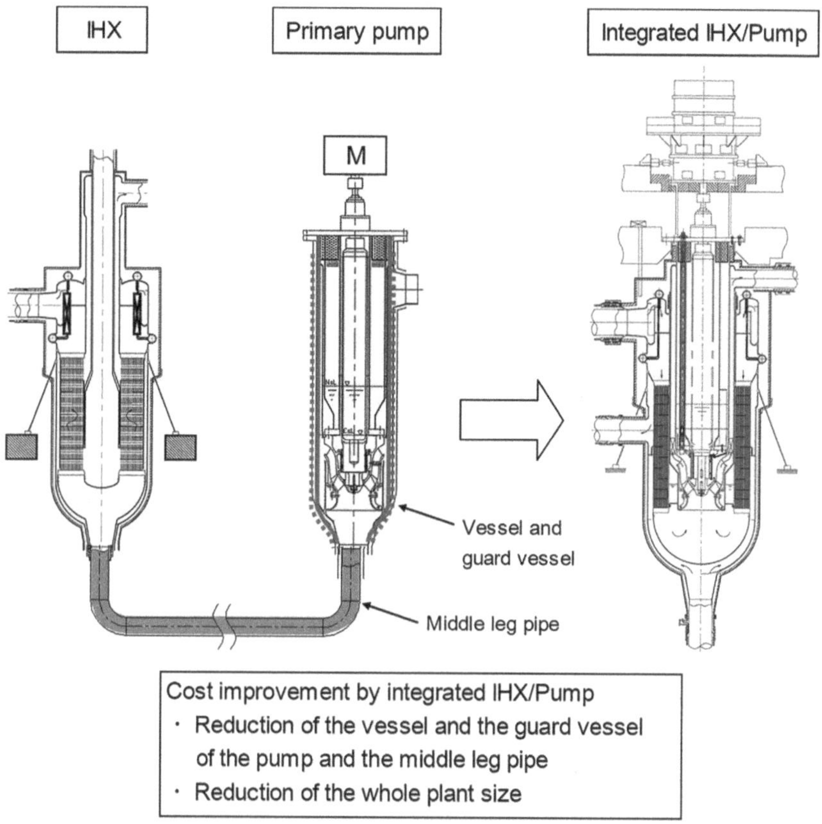

Fig. 4.39 Cost improvement by the integrated IHX-pump [49].

The primary pump is located in the center of the combined component, and the IHX surrounds the primary pump. The primary pump is installed into the IHX and fastened on the upper flange of the IHX. The primary pump is removed when the primary pump and the IHX are to be maintained.

The integrated IHX-pump has one inlet and two outlets for the primary coolant. The primary coolant through the inlet is distributed by a submerged circular weir and a perforated plate at the upper plenum to make the flow uniform and avoid the temperature distribution in the IHX tubes. The temperature distribution causes tube bundle deformation, tube buckling, tube dent due to contact with the baffle plates, and pump function disturbance. In addition, a primary heat exchanger (PHX) for decay heat removal is located in the upper plenum to eliminate the need for a separate vessel for PHX. While passing through the heat exchanging tubes, the primary coolant is cooled from 550°C to 395°C. After that, the primary coolant flows into the lower plenum, that is, the pump suction. The integrated IHX-pump has the gas layer between the IHX and the pump discharge line to insulate the heat transfer between them and a ring plate at the lower plenum to prevent gas entrainment from the gas layer. In addition, the gas layer is closed, and the surface level is not controlled. The pump discharge line from an impeller to the two outlets of the primary coolant is annular. The two outlets connecting two RV inlets with the cold-leg pipes are located on the outer surface of the pump discharge line at a 120° angle. A secondary coolant inlet is located at the bottom of the integrated IHX-pump. Secondary coolant flows zigzag in the tube bundle area with the baffle plates to increase the heat transfer efficiency.

The support points of the IHX and the primary pump are different. The IHX has a support point on the lower floor, whereas the primary pump has the other support point on the upper floor. The different locations of the support points require a convoluted expansion joint to absorb the thermal expansion difference of the IHX and the primary pump.

The other main features are as follows:

- Prevention plates are located in the pump to prevent the heat from coming upward and keep the upper bearing and the pump base at low temperatures.
- A biological shield wall is arranged beneath the upper flange.
- Ring plates against the gas entrainment around the shaft are installed at the inner face of the pump casing.
- A squeeze film damper (SFD) is located behind the hydrostatic bearing to increase the stability of the long shaft vibration. Therefore, the pump design must consider the damage of the rods that support the hydrostatic bearing to make appropriate clearance for the SFD.

The structure of the integrated IHX-pump is shown in Fig. 4.40, and its major specifications are shown in Table 4.11.

The primary coolant flows into the integrated IHX-pump from its upper nozzle, flows down to the heat transfer tubes, making a turn at the lower plenum, and finally returns to the RV from the upper nozzle after being pressurized by the primary pump. The secondary coolant flows into the component from the bottom nozzle, flows around the tubes as a cross flow, exchanges heat with the primary coolant, and finally flows out from the side nozzle.

Important considerations whether or not to integrate a primary pump into an IHX are the resonance of components and the wear of heat transfer tubes caused by vibration during pump rotation. The natural frequency of each part should be adjusted to avoid the resonance of components with pump rotation. This effect is confirmed by structural analyses [49].

(b) Wear-out of IHX tubes

The wear-out of IHX tubes caused by vibration can be evaluated by the combination of vibration analysis, work rate (integration of the product of sliding length and frictional force over a second) analysis, and wear property of tube materials [50–52]. Flow-induced vibration by zigzag flow outside of the tubes and pump vibration, which propagates through the structure and fluid, are considered the causes of tube vibrations (Fig. 4.41).

The vibration responses of the baffle plates and tubes by mechanical and hydraulic excitation forces of the pump are calculated using the whole analysis model of the integrated IHX-pump. The collision and frictional sliding behavior of each tube at the baffle plates should be simulated to calculate the work rate. The boundary conditions are given from the analysis of the whole model. The wear property is a coefficient measured by a wear test in high-temperature liquid sodium. It is the value that divides the wear volume by the work rate and test time. The vibration and wear-out evaluation steps of IHX tubes are as follows:

(1) The vibration responses of the baffle plates and tubes are calculated using the whole analysis model.
(2) The work rate is simulated using the boundary conditions given from analysis (1).
(3) The wear volume is calculated by multiplying the work rate by the wear property measured by the wear test.

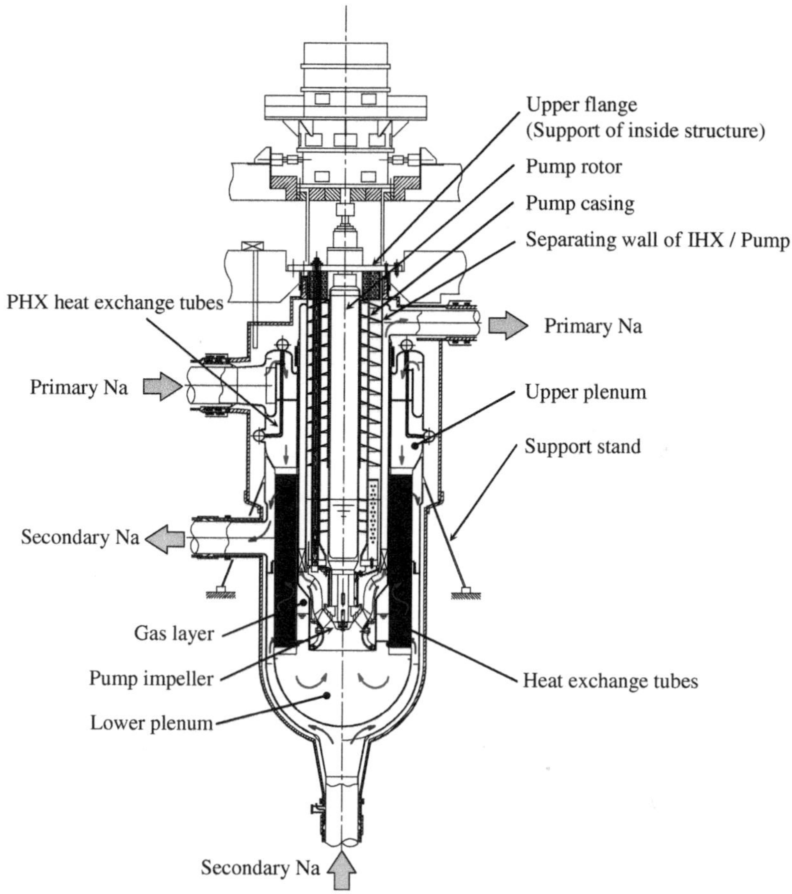

Fig. 4.40 Sketch of the integrated IHX-pump [36].

Table 4.11 Specifications for the integrated IHX-pump [36].

Items	Specifications
IHX type	Cross flow of sodium straight tube type
Heat transfer capacity	1765 MWt
Heat transfer area	4480 m^2
Outer diameter of heat transfer tube	25.4 mm
Heat transfer length	6.0 m
Number of heat transfer tubes	9360
Pump type	Single-stage, single-suction type
Pump head	79 mNa
Pump rated speed	550 min^{-1}

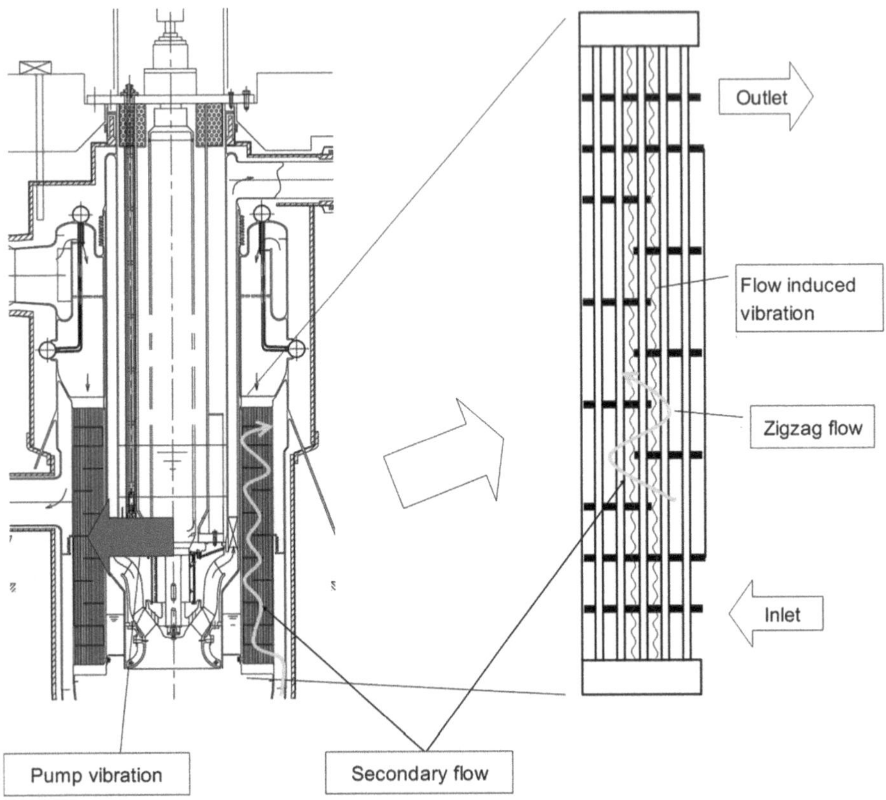

Fig. 4.41 IHX tube vibration due to pump vibration and flow [49].

The propagation of pump vibration to IHX was pointed out as the first problem related to the concept of the integrated IHX-pump. The vibration responses were simulated using the finite element nonlinear structural analysis system (FINAS), which can calculate fluid-structure interactions (FSIs). FINAS was developed by JAEA in 1976 and is still being updated and improved [53]. FINAS is a general-purpose nonlinear structural analysis code using the finite element method. Static stress, dynamic response, heat transfer, and other complex analyses are available. FINAS can use many materials with inelastic behavior, a common characteristic for fast reactor materials, and solve FSI problems such as the sloshing problem of RVs with a lot of liquid sodium and free surface. FSI analysis results were verified by comparing them with theoretical solutions and model test results [54]. Because the integrated IHX-pump consists of multiple cylinders with liquids, FINAS, which can analyze FSI problems, was selected.

(c) 1/4-Scale vibration tests

The following is an overview of the 1/4-scale vibration test model. The considered causes of the excitation force generated by the pump are shaft imbalance and fluid imbalance. Both are determined by manufacturing accuracy, and the test model does not have a scaling law for the excitation force. Therefore, vibration analysis is necessary to evaluate the vibration responses of actual components.

We tried to scale down the changes to actual components to as little as possible. Still, some parts, such as the IHX tubes, were modified structures that had equivalent stiffness because of manufacturing difficulties. Because the eigenvalues of the structures of the 1/4-scale test model were about four times bigger than those of the actual components, the test pump rotational speed was four times as high as that of the actual components to obtain the same ratio of the pump rotational speed to the structures' eigenvalues. The specifications for the test pump are as follows: (1) The rated rotational speed is 2960 min^{-1}, (2) the rated flow is 20 m^3/min, and (3) the rated head is 70 m. The specific speed of the test pump conformed to that of the actual one. Secondary flow was not modeled, but the region of secondary sodium was filled with water because flow-induced vibration would be evaluated at another test. Fig. 4.42 shows a vertical cross section and a photograph of the 1/4-scale test model. Fig. 4.43 shows the analytical model, including the pedestal and the motor.

The vibration responses of some parts of the test model were measured. The eigenmode and response were checked through the frequency analysis of the pump test data.

Consequently, a rotational frequency component (*N* component) and rotational frequency component multiplied by the number of blades (*NZ* component) were found as excitation forces by the pump. The *N* component was presumably caused by shaft imbalance, whereas the *NZ* component was caused by the interference between the impeller and the diffuser vane. Also, large peaks were found around 240 Hz for most parts at rotational speeds over 2700 min^{-1} and flow rates over 18.2 m^3/min. We found that the cause of these peaks was either the *NZ* component or the vibration by overflow at a submerged circular weir near the primary inlet.

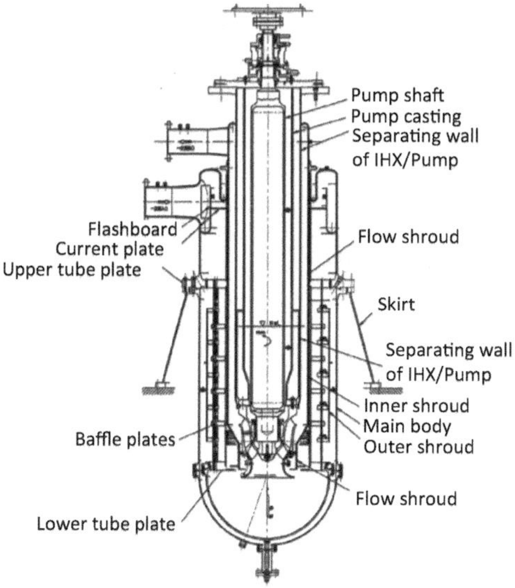

Fig. 4.42 1/4-Scale integrated IHX-pump test model [49].

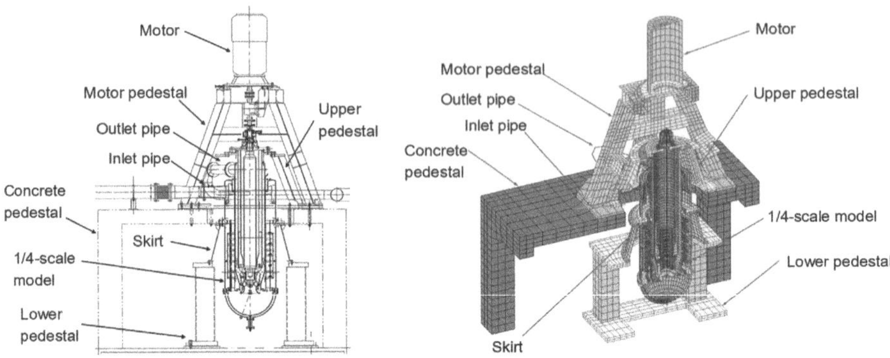

Fig. 4.43 Analytical model of the 1/4-scale test model [49].

Two additional pump tests were conducted to find the cause of these peaks. One was the test range of rotational speed under 2650 min^{-1} and rated flow (flow increase test). The other was the test range of flow under 17.8 m^3/min and rated rotational speed (rotational speed increase test). If peak responses appeared in the flow increase test, the overflow at the submerged circular weir would have been thought to be the cause. If peak responses appeared in the rotational speed increase test, the NZ component would have been thought to be the cause.

Fig. 4.44 shows the test results. We found that peak responses increased not in the flow increase test but in the rotational speed increase test, and the NZ component caused peak responses around 240 Hz or higher. A hydraulic design of the impeller and diffuser is being modified to decrease the interference of the impeller and diffuser, such as the vaneless diffuser, to reduce the NZ component vibrations.

The vibration response was evaluated by comparing the results of a vibration test for an exciter and the vibration analysis results using the FINAS code. Fig. 4.45 shows a comparison result of the shaking test with the analysis of IHX No. 4 baffle plate vibration response, which affects the wear-out of tubes. Many peak positions matched between the test and analysis results except for peaks around 200 Hz observed in the tests.

(d) Gas entrainment around the pump shaft

The shaft rotation of the primary pump in the integrated IHX-pump possibly enables gas entrainment from the free surface in the primary pump. Thus, a countermeasure to prevent gas entrainment is required. A gas entrainment measurement test with a 1/4-scale test model with a prevention structure was conducted to confirm the prevention effect.

Fig. 4.46 shows the gas entrainment route and the analysis area for the fluid analysis using Star-CD for the 1/4-scale model. Fig. 4.47 shows the analysis results. Case 0 is the result for the case without any prevention structures, and the other cases are the results for those with various designed prevention structures. The basic concept of the prevention structure is the parting of the space around the shaft horizontally by some plates in the pump casing. The differences among the cases are the (a) number of plates, (b) thickness, (c) angle of the plates, and (d) clearance between the plates and shaft. The following summarizes the analyses.

(a) Number of plates

Both four and six plates had some effect (cases 1 and 6).

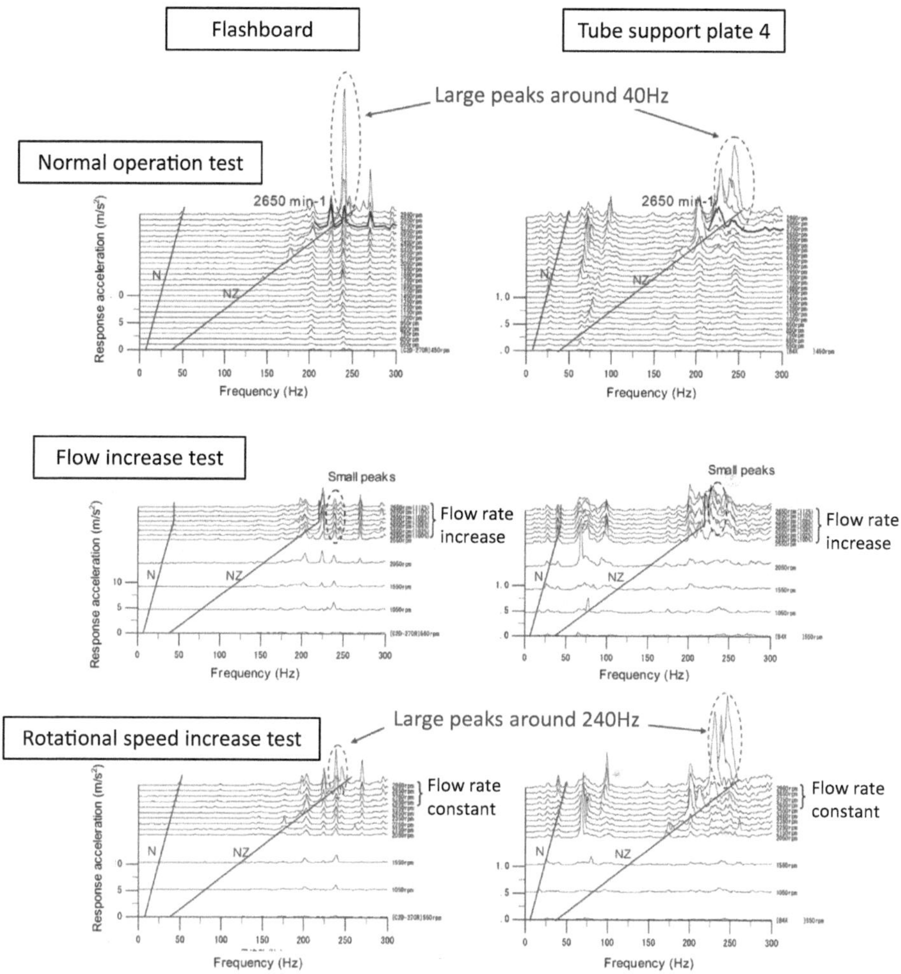

Fig. 4.44 Relation between the frequency spectrum and the flow increase-pump rotational speed [49].

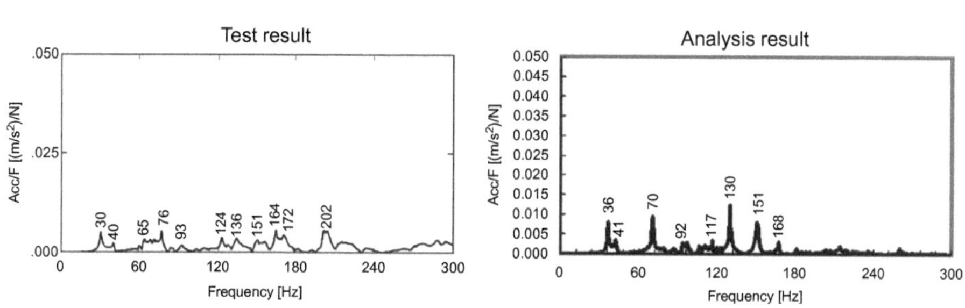

Fig. 4.45 Comparison of test and analysis results on the frequency spectrum [49].

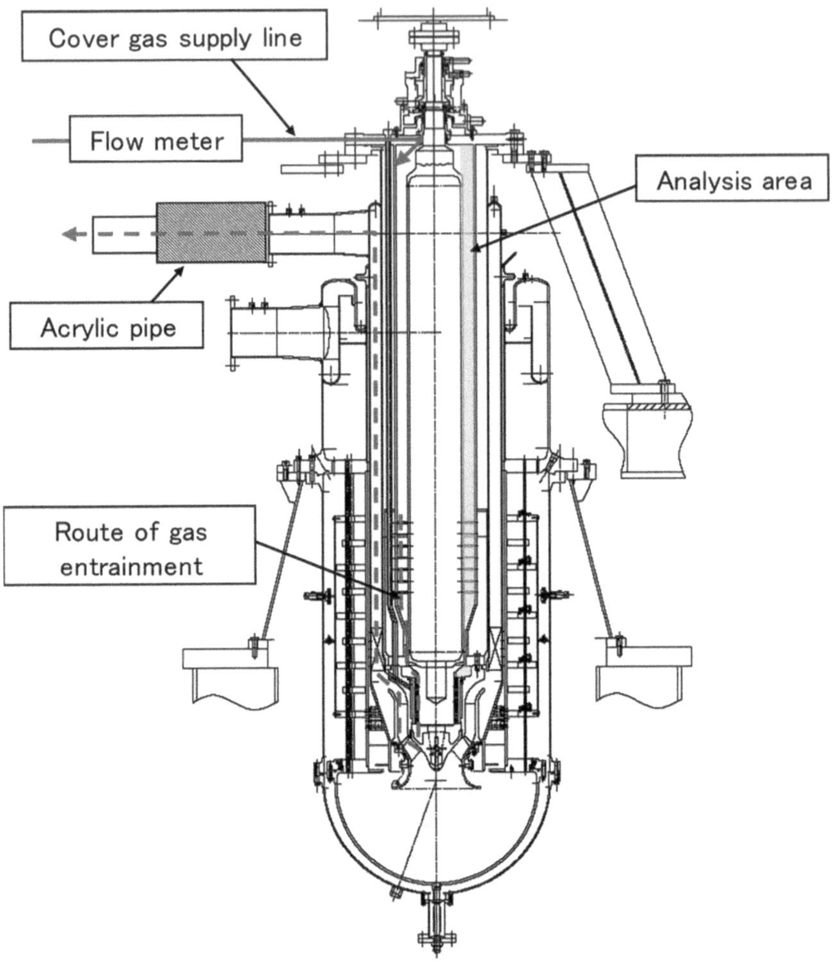

Fig. 4.46 Sketch of the gas entrainment route and analysis area [49].

(b) Thickness

Both the 10- and 20-mm thicknesses had some effect (cases 1 and 3). However, the thicker plates caused greater friction loss of the rotating shaft because the surface area close to the shaft is greater than that of the thinner plates. Consequently, the thinner plates are preferred as far as they can prevent gas entrainment.

(c) Angle of the plates

Horizontally mounted plates worked better than did the inclined ones (cases 1 and 4).

(d) Clearance between the plates and the shaft

Plates of 5 mm or smaller clearance (2.5 mm) had good effects, but the 10-mm plates did not work well (cases 1, 2, and 5).

Following these results, we designed ring plates for the gas entrainment prevention test and attached them to the 1/4-scale model. In this test, the acrylic pipe was located as a part

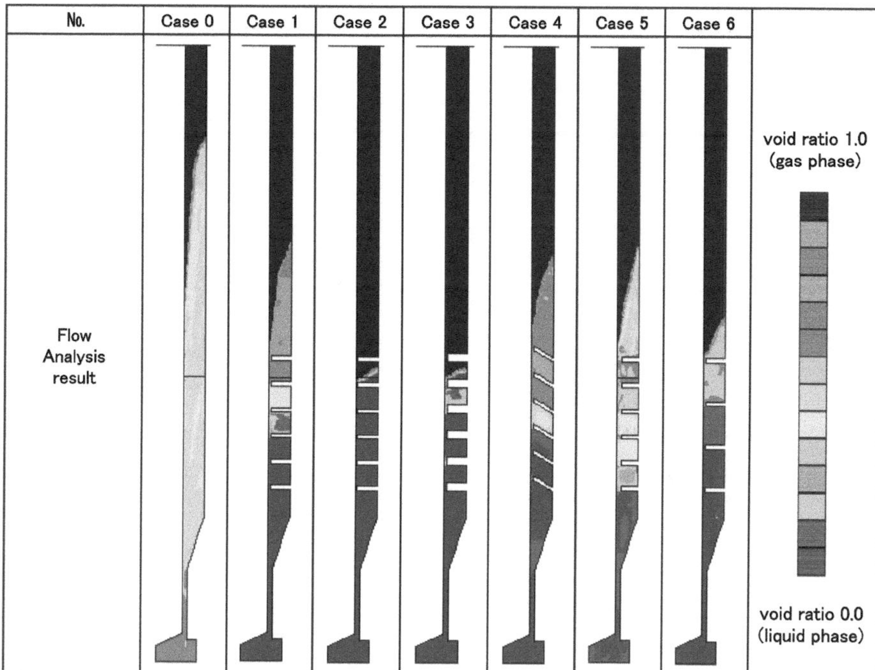

No.	Case 0	Case 1	Case 2	Case 3	Case 4	Case 5	Case 6
N-plates	0	6	6	6	6	6	4
thickness (mm)	–	10	10	20	10	10	10
angle (deg)	–	flat	flat	flat	60	flat	flat
clearance (mm)	–	5	2.5	5	5	10	5

Fig. 4.47 Analysis results on the gas entrainment around the pump shaft [49].

of the outlet pipe, and a flow meter of the cover gas supply line was equipped to measure the entrained gas volume. Fig. 4.48 shows the test results. For the test without the ring plates, gas entrainment occurred even at a low rotational speed (1150 min^{-1}). Conversely, gas entrainment did not occur for the test with the ring plates. Moreover, the ring plates could prevent gas entrainment at three times higher rotational speed if the free surface was assumed to be at the same level in both tests. We found that the ring plates effectively prevent gas entrainment.

(e) Gas convection around the pump shaft

When the pump is halted for plant standby mode operation, thermal natural convection occurs in the cover gas around the pump shaft, inducing asymmetric temperature distribution on the shaft periphery. In this case, the shaft is bent by the thermal effect. Then, the shaft contacts the lower hydrostatic bearing, which places the bearing at risk for damage. This thermal bending effect strongly appears in the long shaft of the integrated IHX-pump.

Therefore, convection restriction plates are placed in the cover gas area to prevent thermal bending. Thermal shield shrouds are also installed here to block the heat from the sodium flow annulus. Because the thermal convection of the interior shaft space influences the temperature distribution, this region is vacuumed.

Three-dimensional thermal-hydraulic analyses were executed to reveal the asymmetrical temperature distributions around the shaft [55]. The calculation areas are the pump shaft, the cover gas region, the pump casing, the sodium flow shroud, and the support flange. The

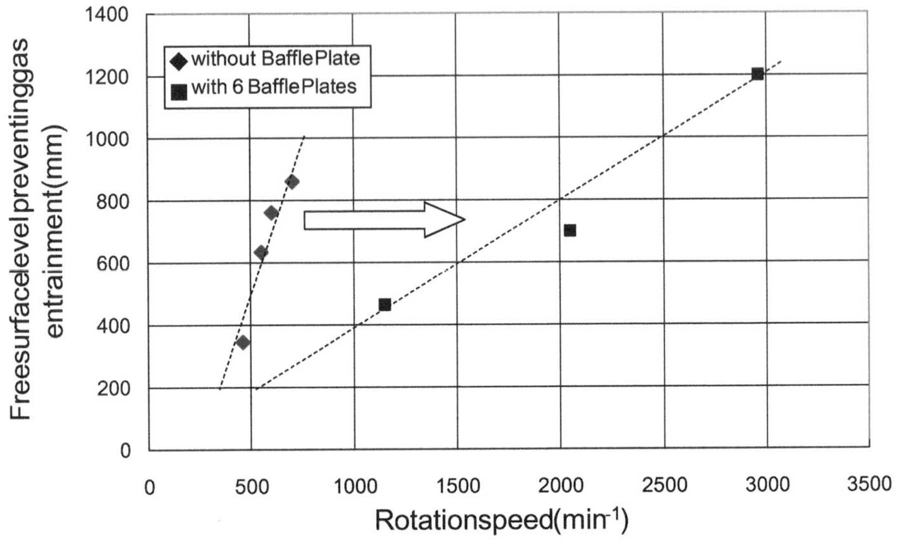

Fig. 4.48 Test results for the gas entrainment measurement [49].

Fig. 4.49 Cover gas temperature profile around the pump shaft by the 3D analysis (halt mode) [55].

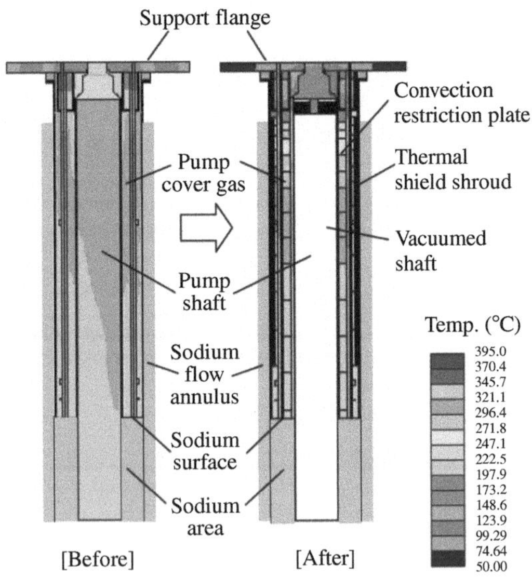

analytical model considered the radiation heat from the sodium flow annulus and the pump sodium surface, the thermal convection of the gas regions, and the thermal conduction of internal structures such as shrouds.

Fig. 4.49 shows the temperature contour around the shaft both before and after installing these measures. Without the measures, asymmetrical temperature distribution occurred around the shaft through gas thermal convection, as shown in the left contour of Fig. 4.49. The adoption of these measures reduced the gas thermal convection, and the

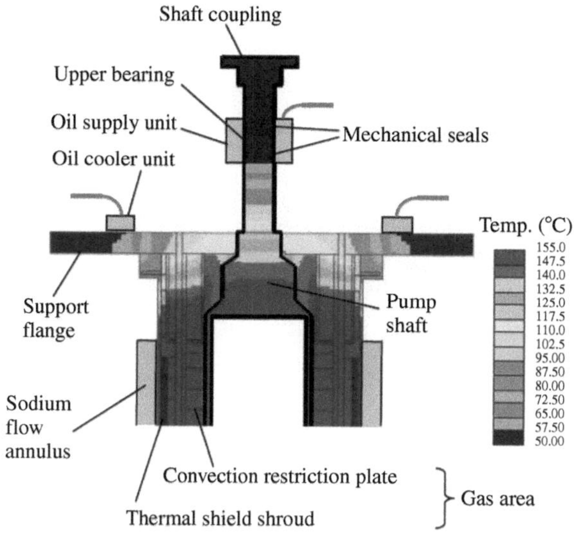

Fig. 4.50 Temperature profile of the upper pump area by the 3D analysis (100% operation) [55].

temperature distribution became symmetrical, as shown in the right contour. The maximum circumferential temperature difference on the shaft decreased from 30°C to 2°C. The thermal bending calculation of the shaft shell model indicated that no contact arose between the shaft and the lower hydrostatic bearing.

During the plant system's 100% operation, the temperature of the upper pump areas increases because the sodium in the primary pump is under the hottest condition. The top of the primary pump is set with the support floor and connected to the motor. This region must be cooled below its allowance temperature to protect the support floor concrete and the motor from the heat.

Fig. 4.50 illustrates the upper pump structures and the heat transfer calculation result [55]. Oil cooler units are installed on the support flange. The upper bearing and the mechanical seals are equipped at the top of the pump shaft, and oil is supplied to these devices for lubrication and seal. The oil supply rate is increased to enhance the cooling capacity. The thermal shield shrouds and the convection restriction plates, which are installed to prevent gas thermal convection, also contribute to the decrease in temperature.

The heat transfer calculation model includes the thermal conduction, convection, and radiation of the pump's internal structures such as the shaft, the support flange, the thermal shield plates, and the gas area. The heat sources are the sodium flow annuls and the sodium around the lower part of the shaft. The heat is removed by the oil cooler units and the oil supply unit. Some of the heat is transmitted to the support floor concrete and the motor. The cooling performances of the oil cooler unit and the oil supply unit are given in the calculation. By adopting these methods, the temperatures of the support flange and the shaft coupling remained below 50°C, protecting these areas from the heat.

(f) Pump rotation stability

The high aspect ratio of the pump shaft leads to lower shaft stiffness and lower natural frequency. Therefore, the natural frequency of the shaft overlaps the pump rotation frequency region, and sympathetic vibration might occur between the natural frequency and the rotation frequency. Thus, the natural frequency of the shaft should be shifted out

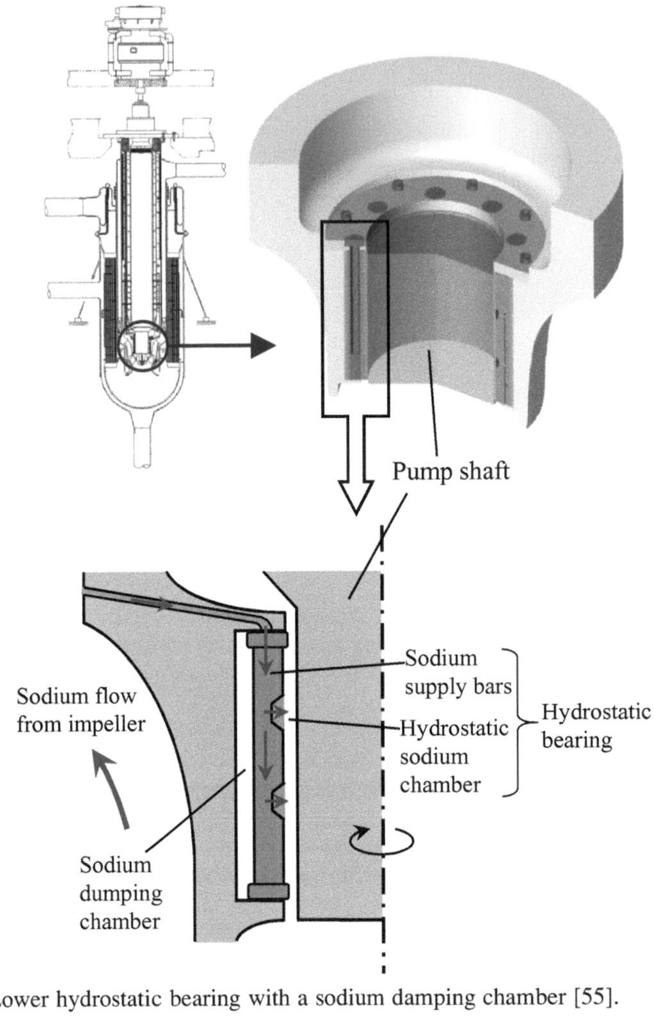

Fig. 4.51 Lower hydrostatic bearing with a sodium damping chamber [55].

from the pump rotation frequency area by increasing the shaft stiffness to avoid sympathetic vibration. In addition, a vibration damping device is necessary to reduce the shaft vibration.

Therefore, a shaft with a large diameter is adopted to raise the stiffness and heighten the natural frequency. A chamber filled with sodium is installed at the lower hydrostatic bearing to reduce shaft vibration using its damping effect. Fig. 4.51 illustrates the mechanics of the lower hydrostatic bearing with the sodium damping chamber. A portion of sodium pressurized by the pump impeller is supplied to the hydrostatic sodium chamber. The sodium flow to the hydrostatic chamber keeps a clearance between the shaft and its bearing during pump rotation. The sodium damping chamber is located at the outer side of the hydrostatic bearing.

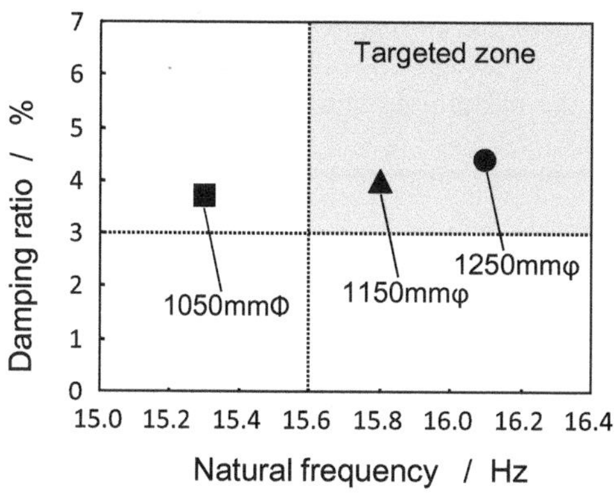

Fig. 4.52 Pump shaft natural frequency and damping ratio [55].

Vibration analysis was performed to clarify the natural frequency and damping ratio of the shaft, as shown in Fig. 4.52. In this analysis, the internal structures such as the shaft and the shrouds were treated as beam models, and the mass and damping effect of the sodium in the pump were taken into consideration.

The shaft diameter was surveyed from 1050 to 1250 mm to determine the appropriate stiffness. The natural frequency of the shaft must be allocated to more than 15.6 Hz, which is the pump rotation frequency with 20% redundancy, to prevent sympathetic vibration. The 20% redundancy is required considering errors in the design analysis and natural frequency distribution range. A computer calculation revealed that the 1150- and 1250-mm shaft diameters met the targeted natural frequency. A diameter of 1150 mm was chosen to minimize the pump size. The graph also shows the shaft damping ratio. The damping ratio must be greater than 3% for stable rotation, and this requirement was also satisfied by the adoption of the sodium damping chamber.

4.2.4.5 Secondary system

The secondary heat transport system consists of an SG, an IHX with pump, a secondary pump, and secondary piping, among others. One secondary system is connected to one primary system.

Secondary sodium, which comes out from the secondary pump, is directed to the outer side of the heat transfer tubes in IHX and receives the heat from the primary sodium. The secondary sodium flows out of IHX, enters the SG, and returns to the secondary pump after heat exchange with the water and steam in the heat transfer tubes.

The sodium level in the SG is set higher than that in the RV to prevent the flow-out of the primary sodium to the secondary system even when the primary-secondary system boundary fails and the secondary system cover gas pressure is lost in IHX. The cover gas in the SG and the secondary pump is linked with an isobaric line.

The secondary sodium main piping is designed to have fewer elbows and a reduced piping length using high Cr-based steel, which has high creep strength, high yield stress, and a small thermal expansion coefficient. The space between the secondary piping and the outer piping is filled with an inert gas atmosphere, similar to that in the primary piping. This prevents the combustion of the leaked sodium. The flow rates are 9.7 m/s for the HL piping and 9.2 m/s for the middle-leg and the CL piping.

4.2.4.6 SG

(a) Concept and specifications

The structure of the JSFR SG is shown in Fig. 4.53, and its major specifications are shown in Table 4.12 [36].

Sodium-water reaction accidents in the event of an SG tube failure are considerably reduced by adopting a double-walled tube type. In this regard, a shell and tube type with a straight tube bundle is adopted to enhance the manufacturability of double-walled tubes. The inner tube directly contacts with the outer tube in this design concept, which aims to enhance the inspection capability for the inner and outer tubes during periodic outages for plant maintenance and refueling, as described in the next part of this section.

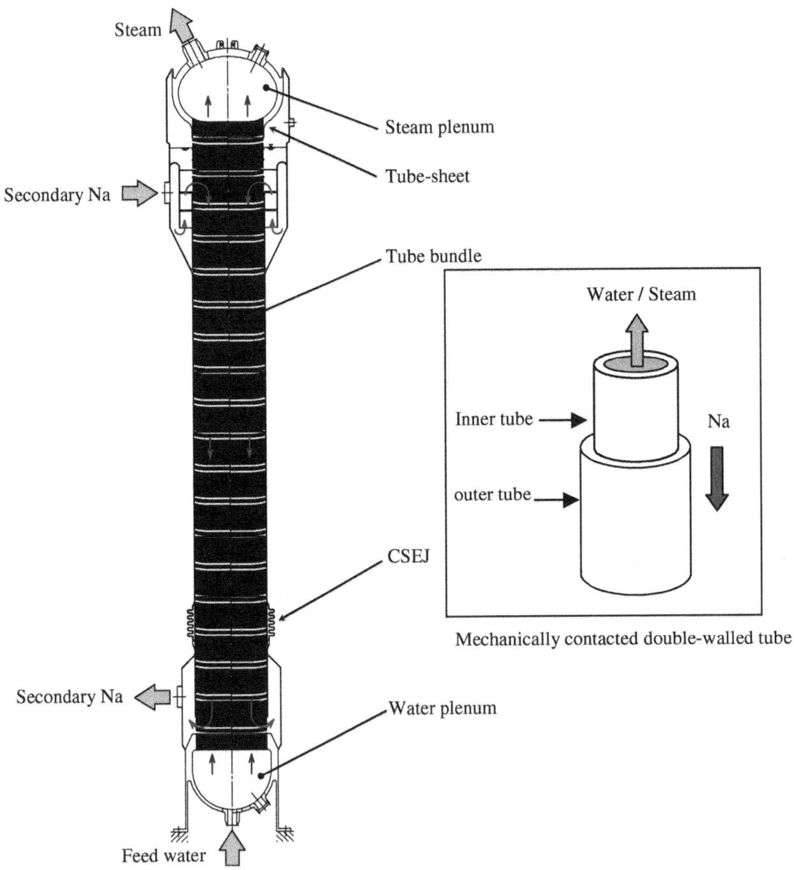

Fig. 4.53 Sketch of the JSFR SG [36].

Table 4.12 Specifications for the JSFR SG [36].

Items	Specifications
SG type	Double-walled straight tube type
Thermal capacity	1765 MWt
Heat transfer area	13,850 m^2
Inner tube diameter	15.6 mm
Outer tube diameter	19.0 mm
Heat transfer length	29.0 m
Number of tubes	8000

The use of high-Cr steel, which is resistant to high temperatures and has high thermal conductivity, for heat exchanger tubes reduces the heat transfer area. By adopting high Cr-based steel, which is also free from stress corrosion cracking (SCC), the SG is designed as a once-through-type generator, enabling a reduction of the component numbers.

The secondary sodium flows into the SG from its upper part, is flow-adjusted in the flow-uniforming structure, flows down in the outer side of the heat transfer tubes while exchanging heat, and finally flows out from the bottom of the SG. The feed water is distributed in the heat exchanger tubes by the feed water tube sheet at the bottom, flows upward while exchanging heat, and flows out from the steam outlet nozzle as superheated steam. The thermal-mechanical behavior of the heat exchanger tube bundle is dominated by the tube-to-shell temperature differences and the difference between the comparatively low water temperatures on the tube side and the high sodium temperatures on the shell side. Therefore, the thermal expansion of the shell is larger than that of the heat transfer tubes, causing a significant difference in the thermal expansion. Thus, a convoluted shell expansion joint is installed on the shell to absorb this expansion.

(b) Prevention of tube failure propagation

The JSFR design adopts the straight double-walled tube SG for both safety and investment protection. Periodical inspections of the inner and outer tubes are required to keep reliable sodium-water boundaries. The development target of SG tube inspection devices is the detection of thickness defects of 10% for inner tubes and 20% for outer tubes. The JSFR straight double-walled tube SG can eliminate tube failure propagation in design-basis events (DBEs) considering the above inspection capabilities. The prevention of tube failure propagation has been confirmed, covering the following double-boundary failure modes [37]:

1. Common-mode failure (inner and outer tube failures due to a common cause): Initiators of common-mode failures are fretting, flow-induced vibrations, and departure from nucleate boiling (DNB) oscillation. The fretting speed has been evaluated to be small enough. For flow-induced vibration and DNB, crack propagation analyses have been conducted. Assuming that the initial crack depths for the inner and outer tubes are 10% and 20%, respectively, on the basis of the inspection capability, the cracks are evaluated to take 7 and 8 years to penetrate the tube wall. Comparing the penetration time and the operation period of 26 months, the cracks can be found before penetration in the next inspection.
2. Dependent double failure (inner tube failure caused by outer tube failure or outer tube failure caused by inner tube failure): Creep damage accumulation under single-walled failure has been evaluated. The results show that either a single inner tube or a single outer tube can withstand operating conditions until the next inspection.

3. Independent double failures (inner and outer tube failures happen at the same tube coincidently): For independent double failures, leak positions and rate have been evaluated probabilistically. The result shows that the frequency of double leaks within a few meters is $<10^{-6}$/reactor year. Also, experimental results with double failures with a certain distance between the outer and the inner tube failures showed that the water leak is self-plugged when the annulus gap is <10 μm. Then, it can be said that the leak caused by independent double leaks will be self-plugged in the design base range and can be detected in the next inspection.
4. Tube-to-tube sheet weld failure (leak at the tube-to-tube sheet weld): Because the double tube expands after welding, the annulus between the tube and the tube sheet is <10 μm. Even assuming a total tube-to-tube sheet welding failure, the leak rate is evaluated to be <1 g/s. For this leak rate, the hydrogen detection system can detect the leak before any failure propagation happens.

   The above discussion shows that there is no failure propagation in the range of DBEs. Even though large leaks are eliminated in the DBE, a double-ended guillotine rupture of one double-walled tube is assumed as the maximum leak rate for a bounding event to confirm a certain design margin. SG tube failure propagation analyses using the leak enlargement and propagation code (LEAP) [56,57] have been conducted with initial leak rates from small to 1DEG or DBE with hydrogen monitoring failure. The results show that the maximum tube failure propagation is within the range of 5DEG. The spike pressure on the primary-secondary and secondary sodium boundaries due to this range of sodium-water reaction has been evaluated using the large leak sodium-water reaction analysis code (SWACS) [58] and found to be within the design limits.

(c) Double-walled tube fabrication

For the double-walled tube, there are three requirements from the viewpoint of failure propagation prevention and performance as follows [59]:

- Crack propagation resistance

  The SG tubes of fast breeder reactors (FBRs) must keep high reliability during operations. In the case of single-walled tubes, cracks on the surface are considered extended by fatigue load and grow up to penetration. Thus, we expect the cracks to stop at the boundary (of the inner and outer tubes) in case of crack growth from the surface of the double-walled tube under low stress levels. Thus, the inner and outer tubes must be separated, and their boundary must not be bonded.
- Clearance between the inner and outer tubes

  The clearance between the inner and outer tubes must be less than 10 μm. This is a requirement considering the leakage rate. The penetration leakage of pressurized water (or steam) must be limited to significantly low levels because of the high flow resistance in the clearance in an event in which the crack location is longitudinally different in the inner and outer tubes.

- High contact pressure

  The inner and outer tubes must be separated because the outer tube is exposed to sodium (operated under high-temperature conditions), whereas the inner tube is exposed to water (operated under low-temperature conditions). However, the double-walled tube must be manufactured with high contact pressure in the drawing process of the outer tube to reduce contact thermal resistance. Although a design requirement for the contact pressure has not been concretely decided, we set the target value to approximately 30 MPa.

Single tubes for the inner and outer tubes are manufactured in domestic mills. The manufacturing process is shown in Fig. 4.54. Annealing and normalizing and tempering (N.T.) heat treatments are conducted in inert gas furnaces to prevent the tubes from carburizing.

Because the diameter and thickness of the tube are relatively small, several passes of cold drawing are required.

In the first step, the inner tube was inserted into the outer tube. As the clearance of the inner and outer tubes was controlled, this process went well. Next, we coupled the end of both tubes on one side to prevent their separation during the cold drawing. After coupling both tubes, they were drawn together through a die with the diameter of a double-walled tube by a drawing machine, as shown in Fig. 4.55.

After the double-walled tube was drawn, it partly had a slight bend, as shown in Fig. 4.56A. Thus, it was straightened using a straightening machine with five-pair rollers. The five-pair rollers were adjusted to compress the double-walled tube, and the bend was removed (shown in Fig. 4.56C). In the case of a badly bent tube, straightening had to be done several times. Thus, keeping the straightness after drawing to avoid repeated straightening is important.

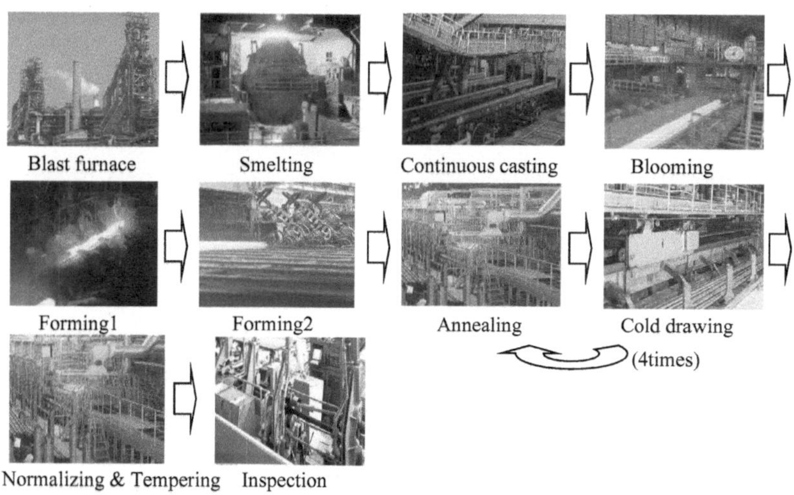

Fig. 4.54 SG tube manufacturing process [59].

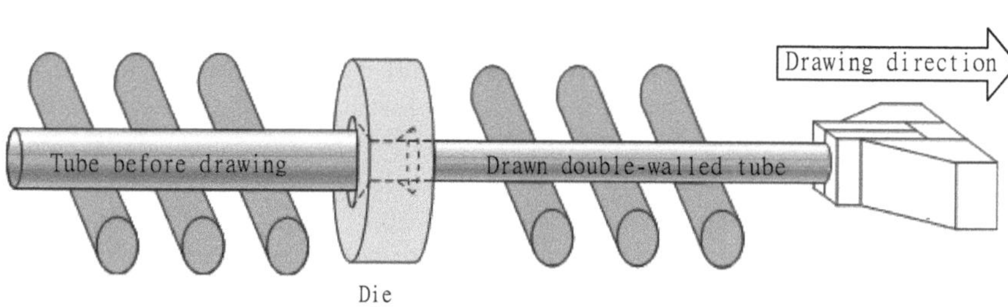

Fig. 4.55 Concept of the drawing machine [59].

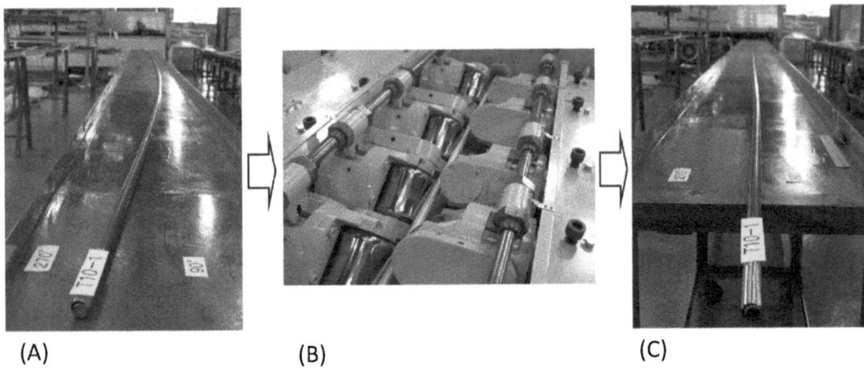

Fig. 4.56 Straightening process of the double-walled tube [59]. (A) Before straightening, (B) straightening machine, (C) after straightening.

Fig. 4.57 Trial outer and inner tube products [59]. (A) Outer tube (ϕ 20.4 mm × t 1.7 mm), (B) inner tube (ϕ 15.7 mm × t 1.4 mm).

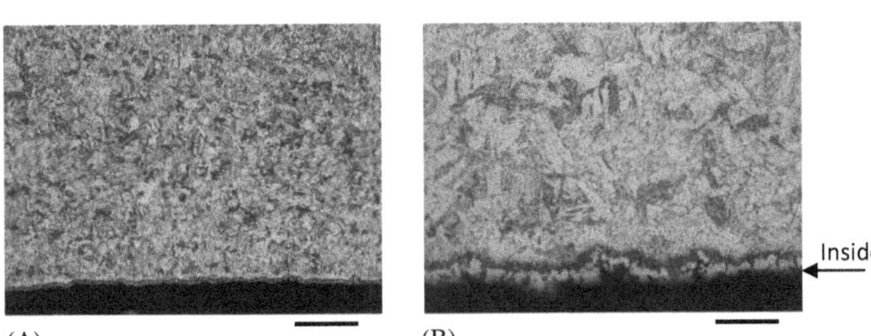

Fig. 4.58 Optical photographs of the tube after N.T. [59]. (A) Low-power field, (B) high-power field.

The outer and inner tubes manufactured in this study are shown in Fig. 4.57. They were manufactured accurately (the outer diameter was ±0.1 mm, and the thickness was −0/+0.4 mm), and the production yields reached more than 98% in this trial.

The metallographic structures of the outer tube after N.T. are shown in Fig. 4.58. Carburization occurred inside of both tubes, and this issue should be cleared by commercialization.

Table 4.13 Tensile properties of the single tubes [59].

Item	Size	Temperature	Test results (S_y, S_u^*)	
			0.2%Y.S.	T.S.
Inner tube	ϕ 15.7 mm × t1.4 mm	R.T.	496 (440)	653 (620)
		500°C	367	464
		550°C	327	400
Outer tube	ϕ 20.4 mm × t1.7 mm	R.T.	498 (440)	656 (620)
		500°C	375	469
		550°C	332	404

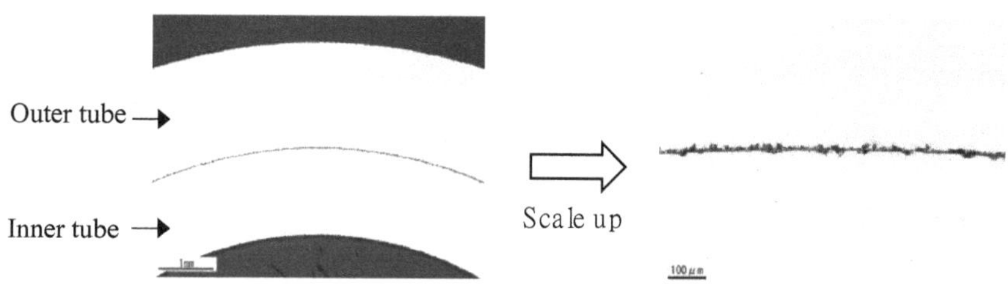

Fig. 4.59 Photograph of the double-walled tube [59].

The tensile properties of the inner and outer tubes are shown in Table 4.13. These data satisfy the requirements of STBA28.

The clearance between the outer and inner tubes shown in Fig. 4.59 was measured using a microscope, and the results are shown in Table 4.14. This shows that almost all samples were manufactured stably, although microscopic asperities on the surface were observed. These asperities were caused by the scale inside of the outer tube. Then, we had to decide whether it must be descaled or not after the measurement of the thermal conductivity of the double-walled tubes.

The contact pressure between the outer and inner tubes was measured using the sectioning method, calculated by Eq. (4.3) as

$$P = \frac{E}{2(1-\nu^2)} \frac{r_2^2 - r_1^2}{r_2^2} (\varepsilon_\theta + \nu\varepsilon_z) \tag{4.3}$$

Table 4.14 Clearance between outer tube and inner tube (10 m) [59].

T.P. No.	Clearance (μm: average)		
	Top	Center	Bottom
T10-1	1.9	1.5	1.1
T10-2	2.3	Not done	1.3
T10-3	2.8	Not done	3.5
T10-4	2.4	Not done	1.5
T10-5	3.9	Not done	2.9
T10-6	2.6	Not done	1.9

Table 4.15 Contact pressure of the double-walled tube [59].

T.P.	Contact pressure (MPa)	Target (MPa)
Top	34, 36	Over 30
Center	36, 38	
Bottom	29, 32	

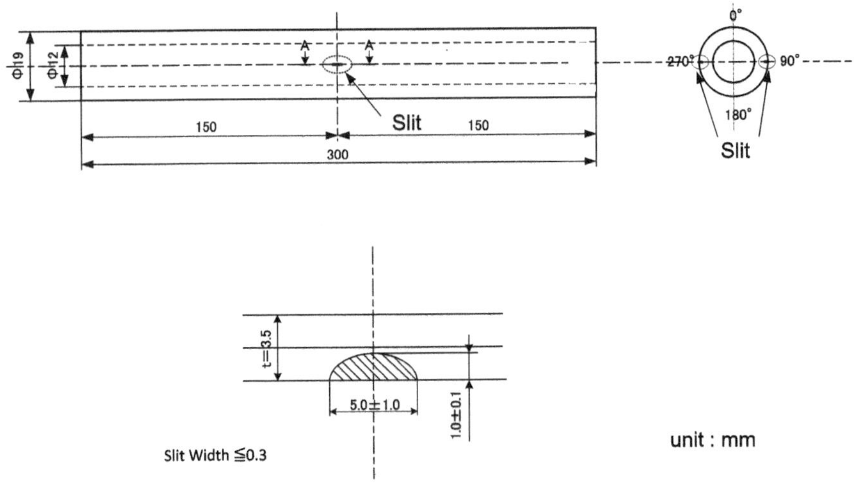

Fig. 4.60 Test piece for the crack propagation test [59].

where P: contact pressure (MPa), E: Young's modulus (MPa), ν: Poisson's ratio, r_1: inner radius of the inner tube (mm), r_2: outer radius of the inner tube (mm), ε_z: axial strain (mm/mm), and ε_θ: hoop strain (mm/mm).

The results of the contact pressure measurement between the outer and inner tubes are shown in Table 4.15. The target value was determined as 30 MPa and tentatively over that of the relaxation analysis. These results satisfy relevant requirements. The dispersion of the contact pressure value appeared because of the accuracy of the sectioning method. On another note, tubes with extreme bends after drawing showed contact pressures of less than 30 MPa. The repeated straightening process applied to improve the bending can cause this low contact pressure.

Crack propagation tests of the double-walled tube were conducted. Half elliptic precracks were introduced on both sides of the outer tube, as shown in Fig. 4.60. Then, a cyclic compression load was applied in the 0–180° direction of the test piece (Fig. 4.61) to extend the precrack by fatigue. The cracks were extended until they penetrated the outer tube as preparation at 350 MPa. The penetration of the outer tube was detected using the potential difference method. After the penetration of the outer tube, the compression load was reduced to below the fatigue limit (200 MPa). Then, the cyclic compression loading was continued over 10^7 times. Fig. 4.62 shows the test

Fig. 4.61 Test equipment for crack propagation [59].

	Outer Tube	Inner Tube
Outer surface		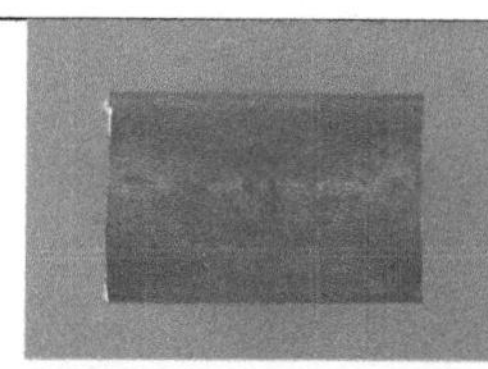
Surface after liquid penetrant test		
Fractured surface	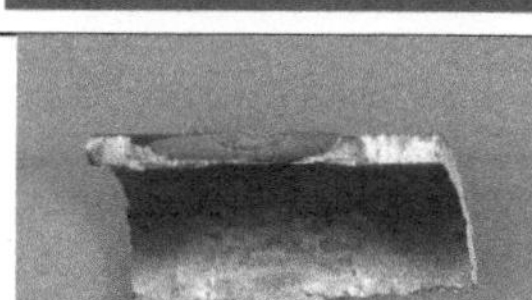	Not fractured

Fig. 4.62 Test piece after the crack propagation test [59].

piece after the crack propagation test. It shows crack penetration for the outer tube but no crack indication for the inner tube. From this result, we confirmed that the surface crack would stop at the boundary of the double-walled tube.

The welding joints of the tubes and tube sheets must have high reliability because of the boundary between sodium and water and the large pressure difference. As regards performance, a small tube pitch is desirable. However, the smaller the tube pitch, the more difficult the welding in such a narrow space. Thus, we decided to develop a compact welder that enables welding in such a narrow space. A prototyped welding torch for narrow spaces was developed, and tubes and tube sheet mockups were welded to check its ability. The tube is expanded from inside prior to welding to reduce the clearance between the tube and the tube sheet.

The welding joint of the tube and the tube sheet is shown in Fig. 4.63. This form was designed to improve the reliability of welding joints so that nondestructive inspection can be conducted.

Trial welding was performed from outside using a prototype torch, which can be inserted in the spacing between two tubes (Fig. 4.64). We confirmed the ability of the prototype welder in the simulated spacing and dependence of groove accuracy.

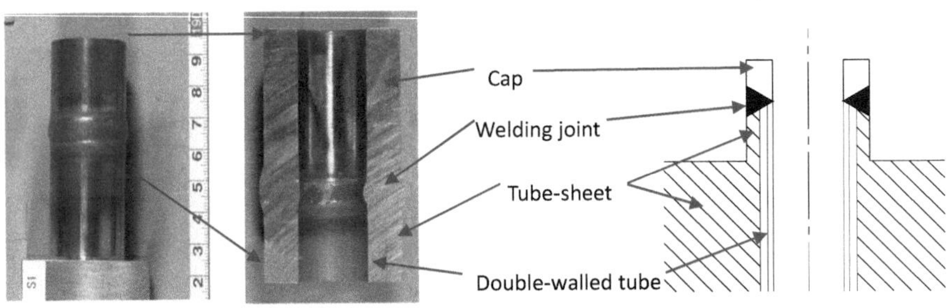

Fig. 4.63 Form of the welding joint between the tube and the tube sheet [59].

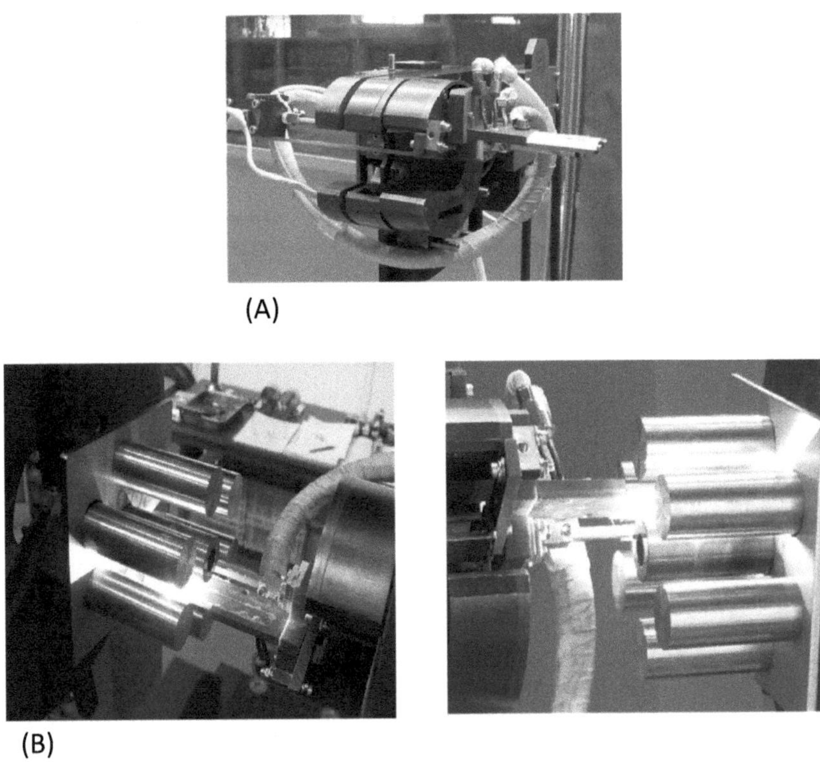

Fig. 4.64 Prototype torch for welding the tube and tube sheet [59]. (A) Prototyped welder, (B) view of trial welding of simulated tube and tube-sheet.

Expansion tests were conducted using a mechanical torque control tube expander and a hydraulic drive torque control tube expander. Figs. 4.65A and B show the mechanical tube expander unit and mandrel, respectively. Figs. 4.65C and D show the hydraulic tube expander unit and mandrel, respectively. Fig. 4.65E shows a mockup of the tube sheet and double-walled tube after expansion. After expansion, the holding force between the tube and tube sheet was evaluated using tensile tests.

Tensile tests of the welding joints between the tube and tube sheet were conducted at room temperature. They ruptured at the base metal of the double-walled tube. Therefore, these joints were considered to have enough strength against tensile load.

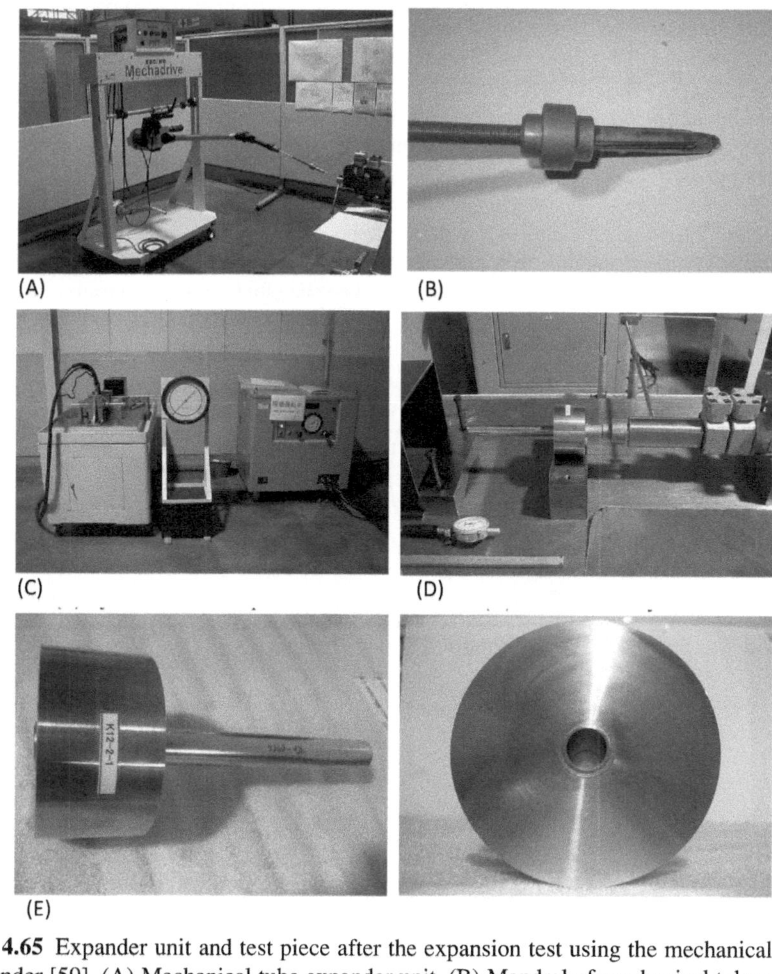

Fig. 4.65 Expander unit and test piece after the expansion test using the mechanical tube expander [59]. (A) Mechanical tube expander unit, (B) Mandrel of mechanical tube expander, (C) hydraulic tube expander unit, (D) Mandrel of hydraulic tube expander, and (E) test piece after expanding by mechanical tube expander.

The holding forces of the expanding joints made by the mechanical expander were 31–35 kN on the tensile side and about 38 kN on the compression side. These results satisfied relevant design requirements.

(d) Temperature distribution in the tube bundle

One of the significant requirements for heat exchangers is making the flow and the temperature distribution uniform. If nonuniform flow or temperature distribution occurs in the straight double-walled tube SGs, there is a risk of thermal expansion mismatch among tubes, and tube buckling might occur. Thus, sodium flow distributors have been evaluated using thermal-hydraulic calculations to avoid tube buckling [55]. The sodium flow calculation area was divided into two regions: the sodium inlet plenum and the tube bundle. These were individually analyzed, as stated in the following sections [60].

Fig. 4.66 shows the internal structure of the sodium inlet plenum, including the upper part of the bundle. The inlet plenum contains a cylindrical barrage, a porous plate, and a porous shroud that function as flow distributors. The cylindrical barrage separates the inlet plenum into two areas: an outer annulus and an inner annulus. The porous plate and the porous shroud are installed at the bottom of the outer annulus and at the inner side of the inner annulus, respectively. The location of the porous shroud is also the entrance of the tube bundle region.

Sodium enters the single entrance nozzle of the inlet plenum horizontally. Then, it flows downward in the outer annulus with a circumferential flow component, and then, it flows upward in the inner annulus after passing under the barrage. The sodium permeates the porous plate and the porous shroud and enters the tube bundle region laterally.

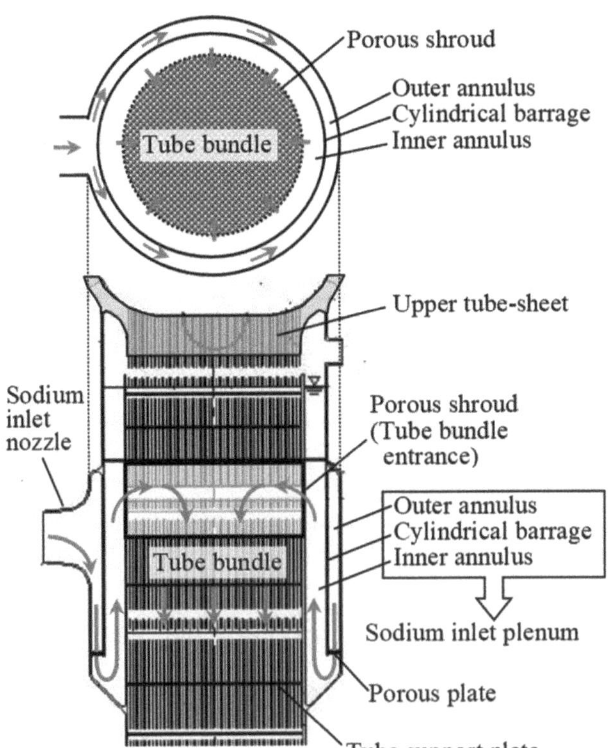

Fig. 4.66 SG inlet plenum [55].

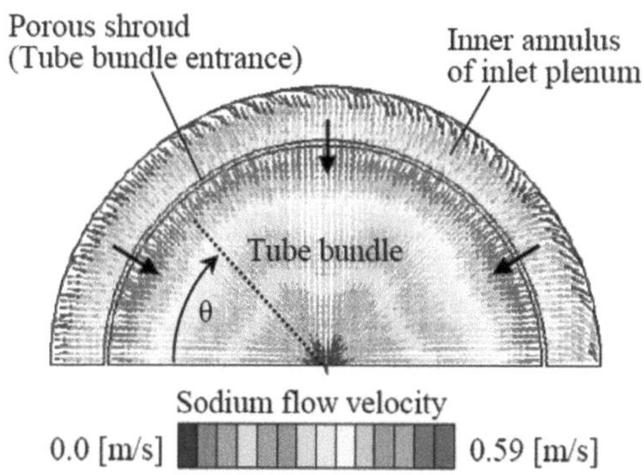

Fig. 4.67 Sodium flow velocity distribution near the tube bundle entrance by the 3D model [55].

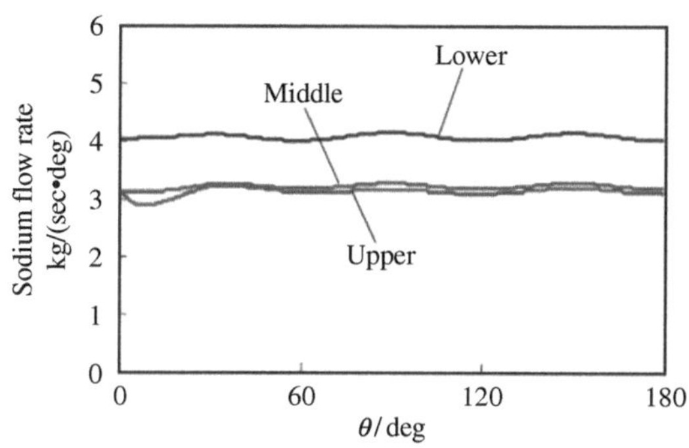

Fig. 4.68 Sodium flow circumferential distribution at the tube bundle entrance by the 3D model [55].

Uniform sodium flows are expected because of the pressure drop of the porous structures and the dual flow path of the annulus. A three-dimensional flow calculation was conducted to confirm the function of the flow distributors.

Half of the plenum was modeled using its symmetry structure. Because there was no heat exchanging region, the analysis model covered sodium flow only. The upper tube bundle region was incorporated into the calculation mesh to account for the effect of the bundle pressure drop on the inlet plenum flow.

The results of the sodium flow calculation are shown in Figs. 4.67 and 4.68. Fig. 4.67 is an upper view of the sodium flow velocity from the inner annulus to the tube bundle. In contrast, Fig. 4.68 depicts the circumference distribution of the sodium flow rate at each elevation of the porous shroud (the tube bundle entrance). The sodium flow rate was almost flattened in Fig. 4.68, and it is uniformly distributed at the bundle entrance.

Because the sodium flow was uniform enough in the circumferential direction, as described in the previous section, a two-dimensional model with cylindrical coordinates was employed in the bundle region calculation.

Sodium laterally enters the tube bundle from the inlet plenum and gradually shifts downward in the upper tube bundle region. The tube support plates contribute to the uniformity of the sodium downward flow through their pressure drops. The tube volume and its pressure drop were considered in modeling the porosity mesh on the sodium side. A one-dimensional multichannel homogeneous model was used for the water side. A thermal-hydraulic calculation was performed, and the heat transfer convection and the sodium-water flow were correlated to reveal the temperature distribution.

The sodium temperature contour when the SG was operated at 100% capacity is illustrated in Fig. 4.69. The contour is laterally magnified to make it easier to see. A low-temperature area appeared at the center of the bundle. This was because the hot inlet sodium laterally flows from the bundle, whereas the cooled sodium flows toward the center.

Fig. 4.70 shows the tube temperature deviation distribution in the radial direction. The left and right sides correspond to the center and the shell side of the bundle, respectively. The tube temperatures were the averages of values taken at various positions in the longitudinal directions of each tube, and they were relative temperatures to the averages of all the tubes. As the tubes were connected to the upper and lower tube sheets, compressive stress and tensile stress occurred in the hotter and colder tubes, respectively, due to axial thermal expansion differences. Tube buckling is caused by the compressive stress of the hotter tubes. The tube buckling load is determined by the tube size (diameter and thickness), Young's modulus, and the interval of the support plates. Tube thermal expansion analysis indicates that the permissive temperature range to avoid tube buckling is below +20°C of the average temperature of all the tubes. As the hottest tube temperature was approximately 3°C above the average, no tube buckling occurred.

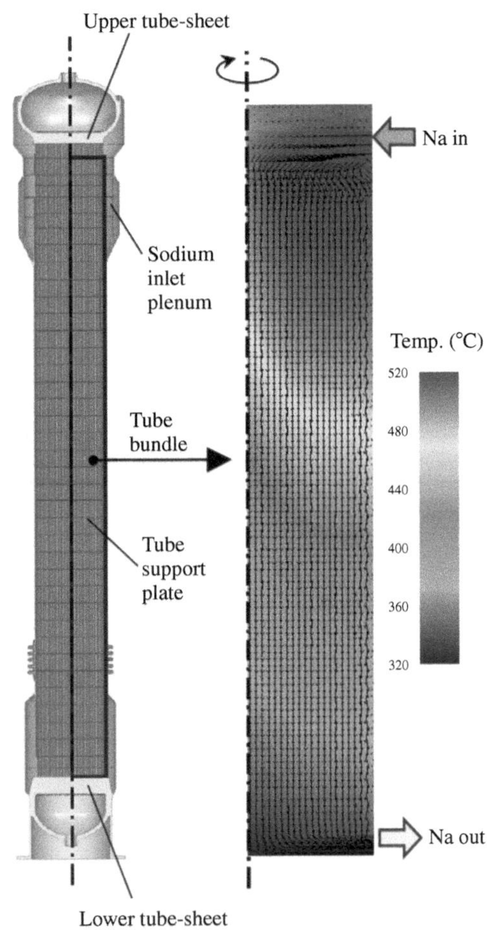

Fig. 4.69 Sodium temperature contour of the tube bundle region by the 2D model [55].

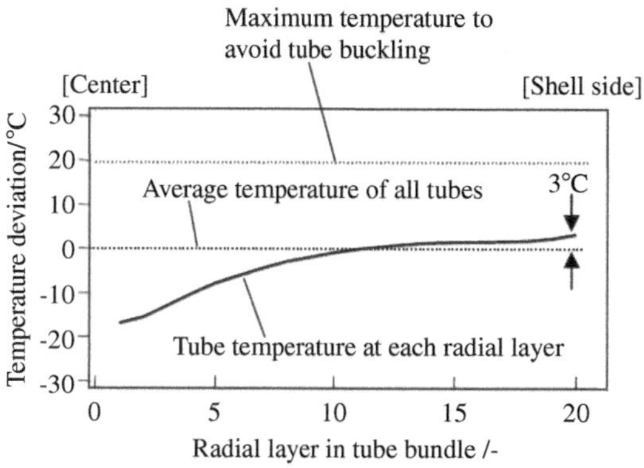

Fig. 4.70 Tube temperature deviation radial distribution by the 2D model [55].

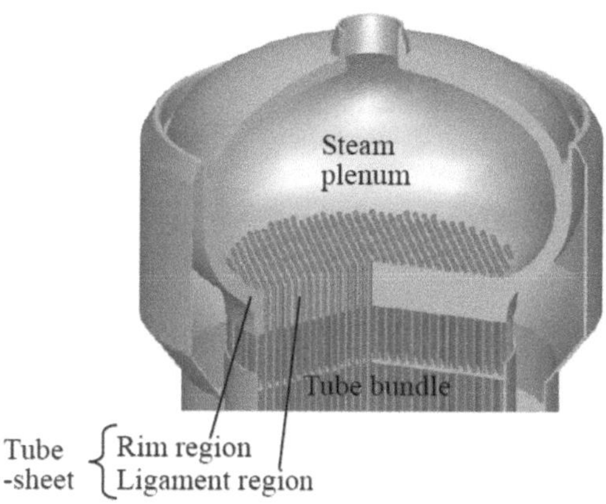

Fig. 4.71 Sketch of the JSFR SG tube sheet [55].

(e) Thermal stress on the tube sheet

The structural soundness of the tube sheet is an important consideration in the case of transient phenomena. Fig. 4.71 shows the tube sheet structure of the JSFR SG. The tube sheet consists of two regions: a ligament region, which incorporates all the tubes, and a rim region. The rapid temperature change of the steam through the tubes causes a large temperature difference between the ligament and rim regions. As the ligament area includes the tubes, its temperature follows the steam temperature quickly. Conversely, the rim temperature follows the steam temperature change with a delay because the rim area includes no tubes. This temperature difference between the two regions causes high thermal stress at their boundary. We marked this issue because the tube sheet of the JSFR SG is an especially thick structure because of the high thermal capacity and the high pressure on the water side.

Fig. 4.72 illustrates the valve constitution of the water-steam system in JSFR. The water-steam system stops water-steam flow through the SG by closing a water inlet valve and a steam outlet valve when transient events occur to avoid high thermal stresses on the tube sheet.

However, if a water release valve opens because of a malfunction, water-steam rapidly flows out from the SG. The steam-water pressure in the SG decreases quickly. Consequently, its temperature is also largely decreased by adiabatic expansion. The low-temperature steam flow produces high thermal stress on the tube sheet, as stated

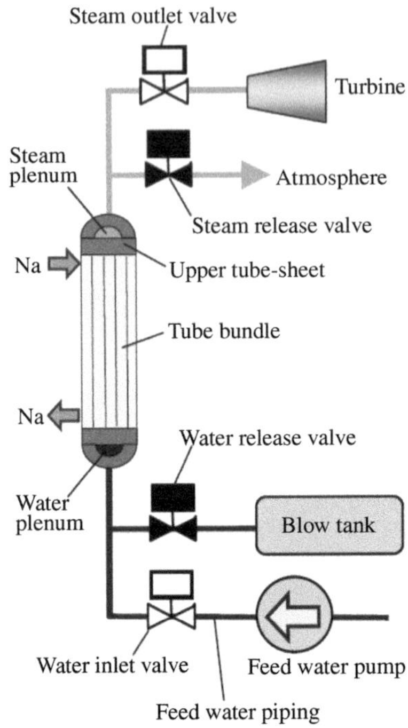

Fig. 4.72 Valve configuration of the steam-water system [55].

previously. This scenario must be considered as a DBE. Thus, the valve operation of the water-steam system was determined using an SG transient computer code to mitigate the rapid temperature decrease and the thermal stress. When a water release valve opens during a malfunction, the water-steam system opens a steam release valve to depressurize the steam plenum. Also, the system closes both the steam outlet valve and the water inlet valve to isolate the SG from the turbine system to restrict steam flow through the tube sheet.

The SG computer code calculated the behaviors of the temperature, flow rate, and pressure of the steam-water in the SG during this event. The results are indicated in Fig. 4.73. The water release valve opens because of a malfunction at 0.0 s. Because the water flows out from the water plenum, the steam flow rate through the upper tube sheet and the steam plenum pressure begin to decrease. The system detects this failure because of the unusual water flow rate signal, and it opens the steam release valve and closes the water inlet valve-the steam outlet valve with a 3.6-s delay. This time delay is caused by the time lags of a pump trip signal and a reactor trip signal. The steam flow rate through the tube sheet begins to increase with the depression of the steam plenum.

Fig. 4.73 also shows that the steam temperature in the plenum falls down by adiabatic expansion at first but recovers thereafter. This is because the steam, which is heated up by the sodium in the bundle region, flows out to the steam plenum. Although the steam temperature decreases by approximately 150°C, the upper tube sheet temperature is hardly affected. As the fall and rise time of the steam temperature by the valve operation is short enough (~15 s), the temperate of the tube sheet does not follow that of the steam.

The transient calculation clarified that the valve operation effectively functioned to avoid large temperature changes at the tube sheet.

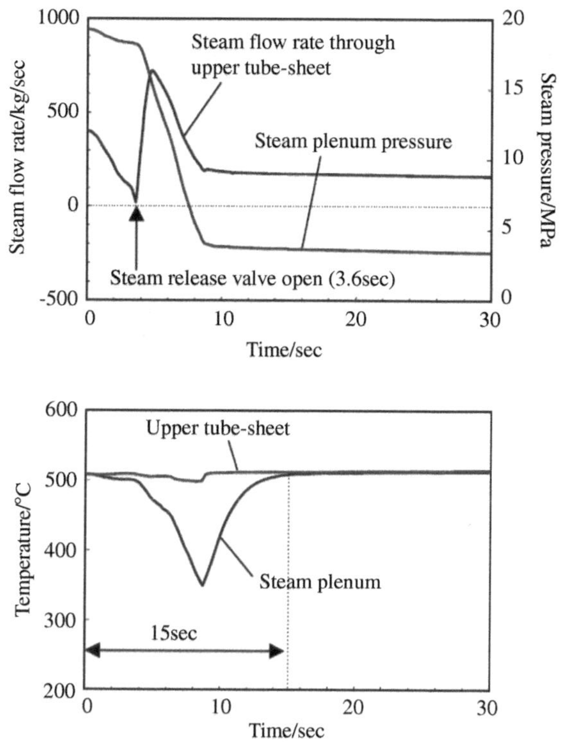

Fig. 4.73 Transient phenomena after a water release valve opened because of a malfunction [55].

4.2.4.7 *Decay heat removal system*

(a) Concept and specifications

The DHRS is a cooling system utilizing natural circulation. DHRS equipment can be operated under passive natural circulation without the active operation of pumps and blowers to maintain the core cooling function after a reactor shutdown. The large temperature difference between the core inlet and outlet can ensure the core flow rate through the buoyancy force. The natural circulation system increases the system's reliability and reduces its construction cost.

The DHRS consists of two units of primary reactor auxiliary cooling systems (PRACSs) and one unit of a direct reactor auxiliary cooling system (DRACS). The heat exchanger of the PRACS (PHX) is located in the integrated IHX with pump, and that of the DRACS (DHX) is located in the RV [36].

As the DHRS can be operated under passive natural circulation, its systems only require direct current power supplied by battery backup systems to operate the dampers of the air coolers. The dampers also have redundancies built into their designs so that their cooling function is retained even if a single failure criterion is applied. This is achieved by each air cooler having redundant dampers placed in parallel airflow paths; an opening failure of a single damper causes a less than 50% reduction in the airflow rate. Furthermore, component diversity is incorporated by adopting different mechanical designs of the air cooler dampers for the DRACS and PRACSs to prevent common-mode failures.

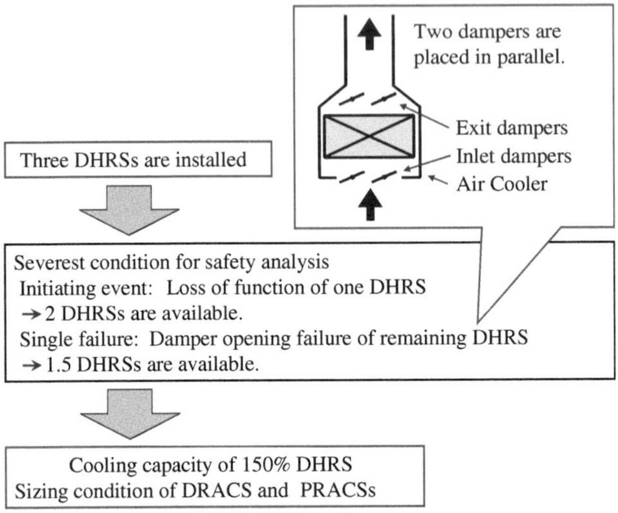

Fig. 4.74 Outline of the DHRS sizing conditions [36].

Fig. 4.74 shows the outline of the DHRS sizing conditions. The function loss of the DRACS or one of the two PRACSs is the most severe initiating event for evaluating the cooling capability after the reactor shutdown. In addition, the loss of off-site power and a single active failure should be assumed for a conservative evaluation for the DBE. In this case, the most severe single active failure is a damper not opening in the remaining DRACS or PRACS. Considering the design features of the DHRS system, maintaining 150% of the cooling capacity against the most severe DBE is possible. This ensures that enough capacity to cool the core is maintained by either the DRACS or one of the two PRACSs, with an additional 50% capacity from either the DRACS or one of the two PRACSs, where a single failure is assumed. Thus, the equipment of DRACS and PRACS is accordingly sized to meet the safety criterion of 150% cooling capacity.

Fig. 4.75 shows the height difference of the DHRS components. The height difference between the core and the heat exchangers, such as the air coolers, is adjusted to ensure the passive decay heat removal capability by natural circulation. The design specifications for the major components of DRACS and PRACS are as follows:

- Countercurrent flow straight-tube-type heat exchangers are utilized for both DRACS and PRACS.
- Double-walled tubes with cooling fins are adopted for the air cooler.
- The diameter of the secondary circuit piping of about 400 mm is selected to obtain the necessary natural circulation flow rate.
- Blower lines driven by a nonsafety power source are installed in each DHRS to operate the cooling process. The blowers are also used for accident management.

(b) Performance evaluation

The performance of the DHRS under accident conditions was evaluated. The calculated results indicate that the design is sufficiently robust to perform the required cooling function for the reactor.

For evaluating the cooling capability after the reactor shutdown, the most severe initiating event is the function loss of the DRACS or one of the PRACSs. In addition to the

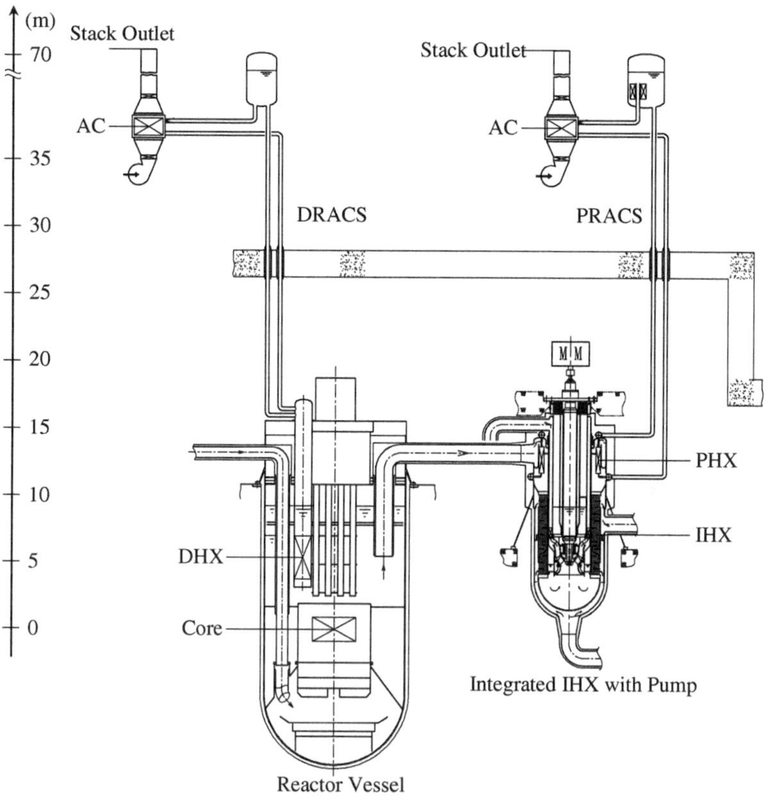

Fig. 4.75 Height difference of the DHRS components [36].

function loss of one of the DHRS components, the loss of off-site power and a single failure of the damper system should be assumed for a conservative evaluation for the DBE. In this case, the most severe single failure is a damper opening failure in the remaining DRACS or PRACS. Fig. 4.76 shows the results from the analysis of the function loss of DRACS along with a single damper opening failure of PRACS, which confirmed the 150% DHRS cooling capacity. The analytical results show that a stable coolant flow is established by natural circulation and that the peak temperatures of the fuel cladding and the primary coolant boundary are below the safety criteria.

(c) 1/10 Water tests

Water tests were performed using a 1/10-scale model, which physically simulates the reactor core and vessel, the two sets of the primary loop, and the DHRS consisting of one set of DRACS and PRACS [61]. Because the water tests were concerned with the whole primary system, they were named "system water tests." Fig. 4.77 shows the overview of the water test apparatus, which was set up at the Central Research Institute of Electric Power Industry. The upper plenum of the RV, the upper plenum of the IHXs, including the heat transfer tubes for PRACS, and the hot- and the cold-leg piping were made of acrylic

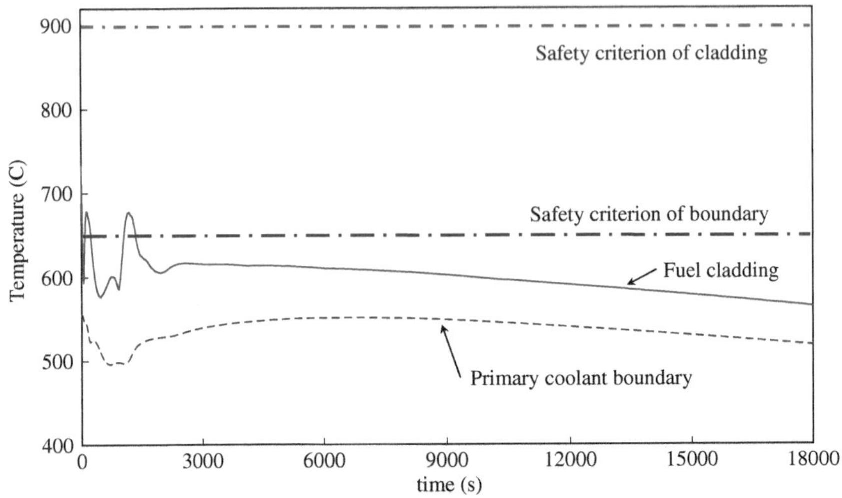

Fig. 4.76 Transient evaluation on loss of off-site power with PRACS cooling [36].

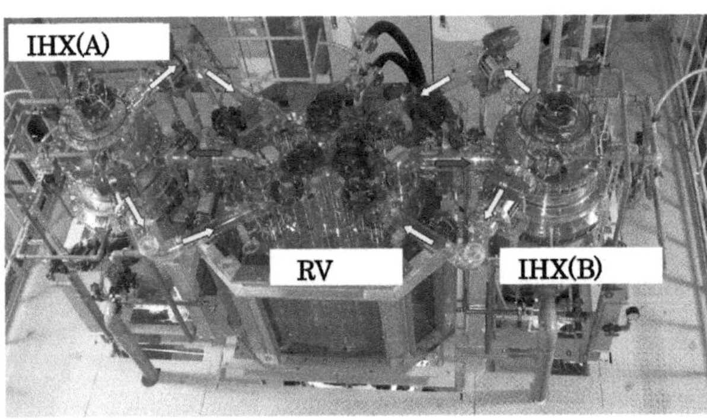

Fig. 4.77 1/10-scale water test apparatus [61].

material to enable the direct observation of the thermal-hydraulic phenomena. The core region was divided into three zones: the inner core, the outer core, and the radial blanket. Each zone consisted of several vertical channels with electric heaters simulating the heat generation of the JSFR subassemblies. The temperature rise from the core inlet to the outlet was set at 9.3°C, corresponding to 155°C for JSFR at full power. The rated power of the apparatus was about 120 kWt, enabling the simulation of the transient phenomena from the forced circulation to the natural circulation condition. The core power of the natural circulation mode was about 17 kWt, according to the similarity rule.

The similarity rule of the natural circulation tests has been discussed in a previous study for the top-entry loop-type FBR [62], indicating that three major dimensionless parameters to be considered: the Euler number (Eu), modified Boussinesq number (Bo), and modified Grashof number (Gr). Eu stands for the ratio of the pressure loss to the inertial force, equivalent to half the value of the pressure loss coefficient in a continuous flow conduit.

Bo stands for the ratio of the thermal inertia with natural convection to molecular thermal diffusion, and Bo1/2 is equivalent to the Peclet number (Pe) in the case of forced convection. Gr stands for the ratio of the inertial force with natural convection to molecular momentum diffusion, and Gr1/2 is equivalent to the Reynolds number (Re) in the case of forced convection. Bo1/2 and Gr1/2 are convenient for identifying the natural circulation test conditions because the representative flow velocity depends on the heating power. Conducting the scale model thermal-hydraulic test based on the similarity rule by matching the above dimensionless numbers to those of JSFR is ideal. However, it is practically impossible to match the above three dimensionless numbers at the same time. Therefore, the water test was performed with matching *Eu*, whereas the Bo1/2 of the water test was set at the same order as that of JSFR (by about a factor three times larger than that of JSFR). Under this test condition, although Gr1/2 equivalent to *Re* was reduced to about 1/350 of that of JSFR, its absolute value was as high as several thousands in the primary piping under natural circulation. Therefore, the flows during the water test were likely to stay in turbulent states in major parts of the apparatus.

The water tests were conducted for five typical transient events selected from the DBEs of JSFR: (1) loss of off-site power, (2) sodium leakage in the secondary loop, (3) sodium leakage in the secondary loop of DRACS, (4) sodium leakage in the secondary loop of PRACS, and (5) one pump stick in the primary loop [61].

The progress in the case of the loss of off-site power is as follows. In this case, two primary loops (loops A and B) behave symmetrically. The flow rates in the primary loops decrease just after the loss of forced circulation, and the cold-leg temperature increases due to the decrease in the secondary flow rate in IHX. At 500 s after the onset of the transient, the PRACS comes into operation with the natural circulation mode. Afterward, a stable and sufficient natural circulation flow rate (corresponding to about 2%–3% of the flow rate of JSFR) is established in the primary loops, as shown in Fig. 4.78. It is confirmed that a stable heat removal is also established in the other tests (events for sodium leakage in the secondary loop of DRACS and one pump stick in the primary loop) [61]. These experimental results demonstrate that decay heat removal with natural circulation can be used as a robust cooling method for JSFR by employing the proposed PRACS and DRACS design.

(d) Sodium tests

The sodium tests aim to evaluate the heat transfer characteristics of the PHX under the natural circulation condition and to confirm the transition behavior from the stationary forced circulation conditions to the natural circulation conditions using an integrated system consisting of the primary sodium system, secondary sodium system of the PRACS, and tertiary air cooling system.

The PHX consists of straight heat transfer tubes arranged along the inner shroud installed in the upper plenum of the IHX, as shown in Fig. 4.79. Under the high flow rate conditions of the rated power operation, the primary coolant flows down uniformly through the annular region between the outer and the inner shrouds with a low-pressure loss. Under the low flow rate conditions of the natural circulation regime, the coolant flows around the PHX heat transfer tubes because of the buoyancy effect and then flows down, transferring heat on the tube walls. The sodium tests were required to employ heat transfer tubes of the same scale as those of JSFR to verify the heat transfer characteristics of PHX under *Pe* conditions equivalent to those of JSFR.

The sodium tests were conducted using the Plant Dynamics Test loop (PLANDTL) [63] of the JAEA Oarai Research and Development Center. A new IHX partially simulating PHX was manufactured and installed in PLANDTL, as shown in Fig. 4.80. The primary system consisted of the core with seven mockup fuel subassemblies, an electromagnetic

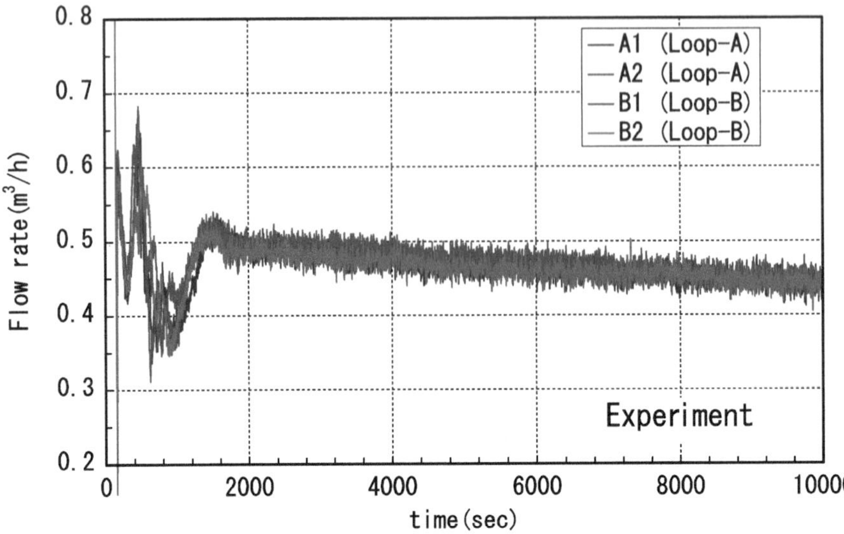

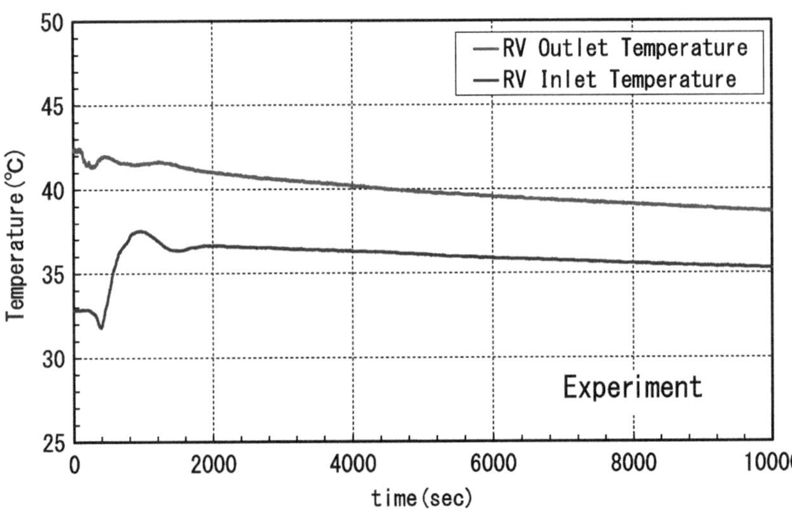

Fig. 4.78 Test results on the loss of off-site power [61].

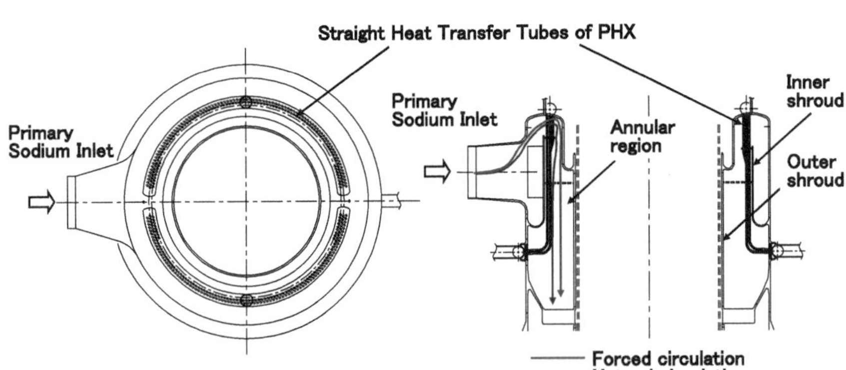

Fig. 4.79 Schematic of the PRACS tubes in the IHX plenum [61].

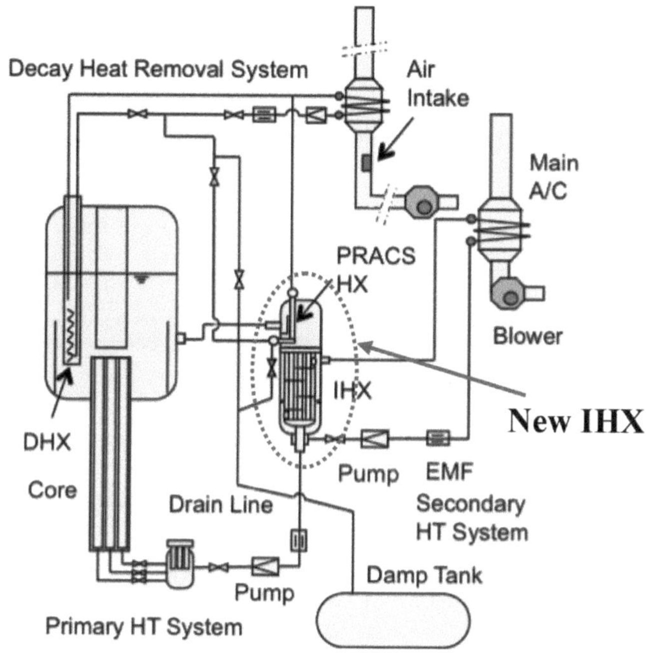

Fig. 4.80 Schematic of the sodium test loop [61].

pump, RV simulating the upper plenum and lower plenum, a couple of hot- and cold-leg pipings, and the new IHX. The secondary system consisted of the main air cooler and a couple of hot- and cold-leg pipings. The auxiliary cooling system for the decay heat removal consisted of the PHX, a couple of hot- and cold-leg pipings, and the air cooler with a blower and an air damper in the inlet duct. A dipped heat exchanger (DHX) installed in the upper plenum was not operated during the test. The geometry of the primary system of PLANDTL was not so similar to that of the JSFR design compared with that in the 1/10-scale water test apparatus. However, the sodium tests could be conducted under equivalent *Eu* conditions of the DHRS of JSFR. The heat removal capability of the PRACS was determined as 100 kWt, accounting for the Richardson number (Ri) similarity and a model scale of nearly 1/8 of JSFR.

First, the heat transfer characteristics of PHX were tested. Then, the start-up characteristics of PRACS were tested, and the integrated test simulating the DBE "loss of off-site power" was conducted. Figs. 4.81 and 4.82 show the vertical coolant temperature profile in the annular region and the *Pe* dependency of the heat transfer coefficient on the tube outer wall under natural circulation conditions, respectively, as PHX characteristic test results. A temperature decrease was locally found at the lower part around the tubes. Such temperature drop near the PHX tubes strongly depended on the flow velocity distribution in the IHX annular region. The obtained heat transfer coefficients exceeded that evaluated by the Lubarsky-Kaufman empirical correlation employed in the PHX design of JSFR.

A transient experiment was conducted from forced to natural circulation. Figs. 4.83 and 4.84 show the transient of the primary flow rate and the core inlet-outlet temperatures, respectively. Just after the onset of the transient, the primary flow rate steeply decreased because of the loss of the forced circulation. The core outlet temperature also dropped

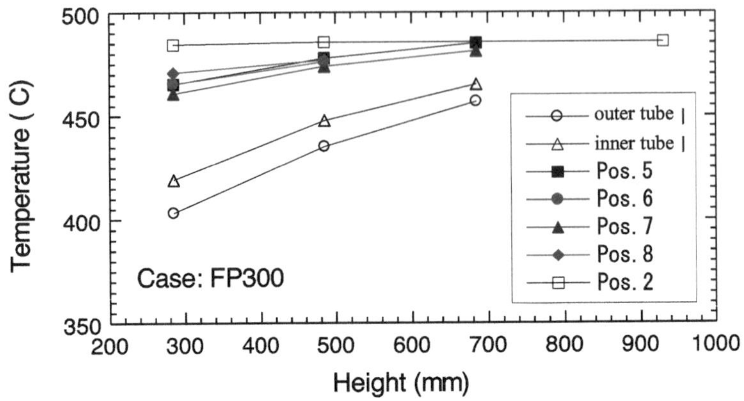

Fig. 4.81 Vertical coolant temperature profile in the annular region [61].

Fig. 4.82 Pe dependency of the heat transfer coefficient [61].

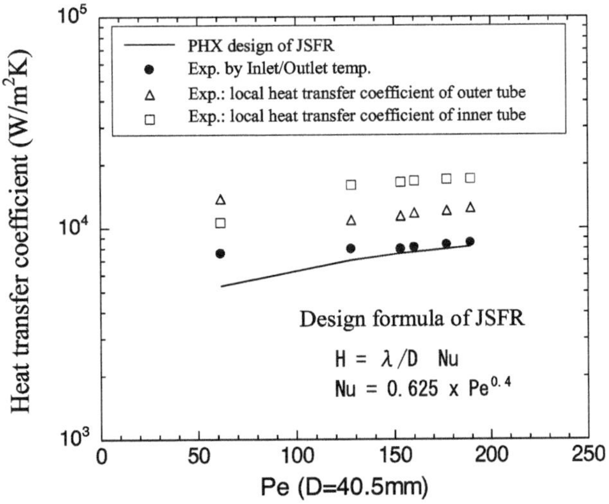

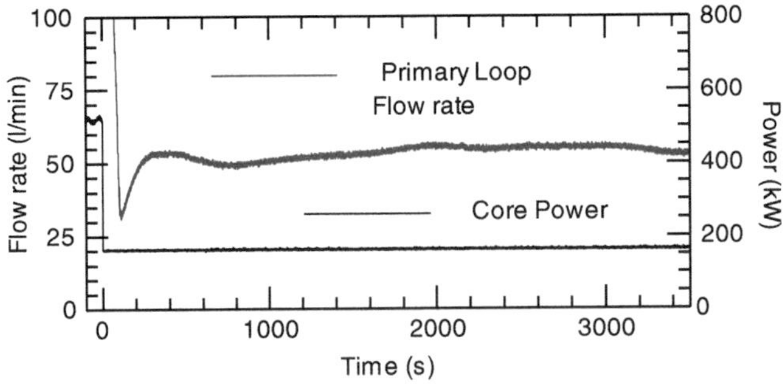

Fig. 4.83 Transient of the primary flow rate in the PRACS sodium test [61].

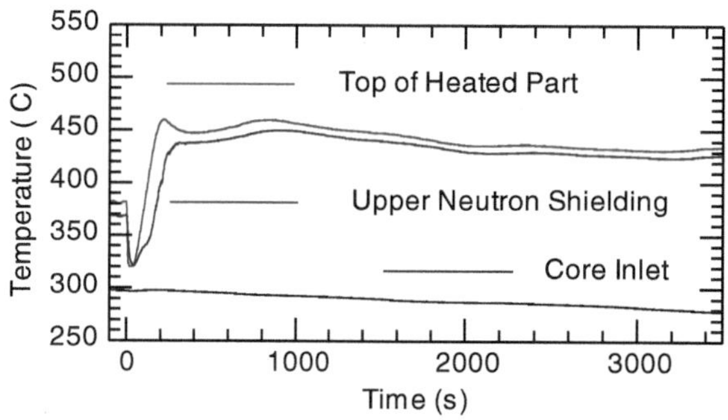

Fig. 4.84 Transient of the core inlet-outlet temperatures in the PRACS sodium test [61].

because of the steep core heating power decrease. Then, the core outlet temperature rose to the peak value called "secondary peak temperature" because of the core decay heat, and the primary flow rate started increasing because of the natural circulation head corresponding to the temperature rise in the core. After 800 s from the onset of the transient, the core outlet temperature slightly increased to the peak value called "tertiary peak temperature," corresponding to the cold-leg temperature rise due to the loss of the IHX heat removal. Afterward, the PRACS came into operation with the natural circulation mode, as shown in Figs. 4.85 and 4.86. A stable and sufficient natural circulation flow rate was established in the primary system, and the core outlet temperature was controlled stably. We confirmed that the natural circulation flow in all systems was generated smoothly by opening the entrance damper of the PRACS air cooler.

(e) Development of evaluation tools

The one-dimensional flow network method is widely applied to safety analyses for SFR plants because of its simplicity and ease of handling. This method was developed following the Super-COPD, which is an improved version of the plant dynamics code applied to the safety analysis for the licensing of the Japanese prototype SFR called Monju [64]. Therefore, the one-dimensional flow network method is expected to be applicable for the future licensing of JSFR. However, this method needs to be verified because the flow network model is constructed differently for each plant. In this study, the one-dimensional flow

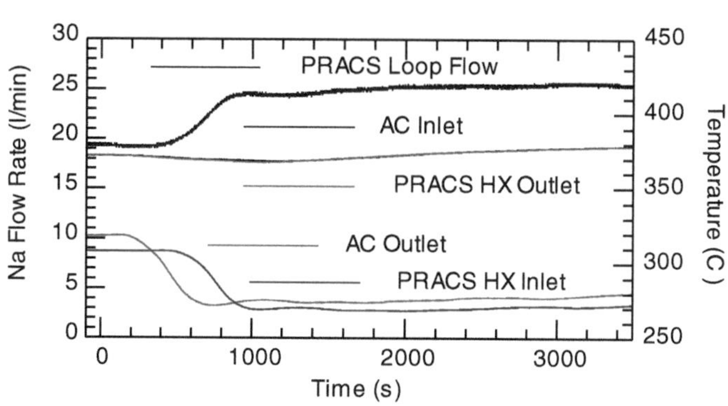

Fig. 4.85 Transient of the sodium flow rate-temperature at PRACS IHX in the PRACS sodium test [61].

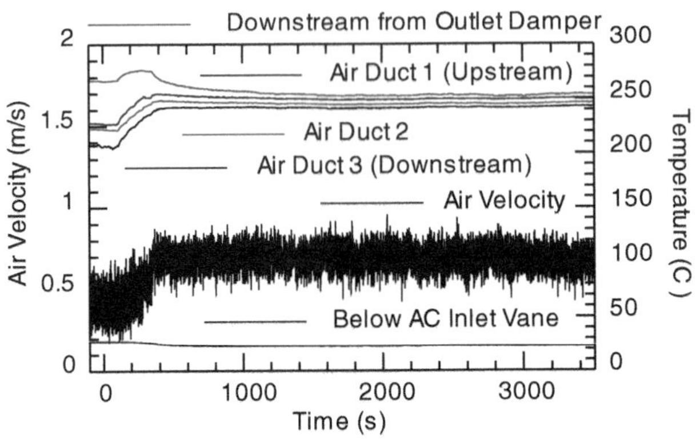

Fig. 4.86 Transient of the airflow rate-temperature air cooler in the PRACS sodium test [61].

network method was improved and validated using a 1/10-scale water test and a 1/8-scale sodium loop test. Improvements were made mainly for the large plenum, such as the reactor upper plenum large diameter primary pipings, where three-dimensional complex flows due to buoyancy were expected.

A three-dimensional fluid flow analysis method was developed to deal with such phenomena as local natural convections and thermal stratifications of the primary cooling system under decay heat removal conditions. The computational fluid dynamic analysis code "STAR-CD (Vr4.14)" was employed as the main frame of the three-dimensional fluid flow analysis. Moreover, a set of options such as the high Reynolds number-type standard k-ε turbulent model [65,66] and the second-order advection scheme named Monotone Advection and Reconstruction Scheme were also applied to the three-dimensional fluid flow analysis. PRACS, DRACS, and IHX secondary model were newly developed and assembled as outer subroutines. A three-dimensional analysis model for JSFR was made, as shown in Fig. 4.87. The total number of meshes was about 6,500,000. It currently takes about 4 days with 32 nodes (Xeon dual-core) for parallel computing to simulate the transient of a loss of off-site power event for 3600 s.

The flow network model was reconstructed for the water test apparatus to validate the one-dimensional flow network method using the 1/10-scale water test results. Then, the apparatus characteristics (pressure loss in the primary cooling system and heat transfer performance of IHX, PHX, and DHX) were incorporated into the one-dimensional flow network method as input data. The core heating power decrease, the primary pump coast-down characteristic, the secondary side inlet flow rates, and the temperature changes of IHX, PRACS, and DRACS were incorporated into the one-dimensional flow network method as the boundary condition data. The three-dimensional fluid flow analysis model was also constructed, as shown in Fig. 4.88. The total number of meshes was about 5,600,000. Characteristic data and boundary condition data were incorporated into the three-dimensional fluid flow analysis method and in the one-dimensional flow network method.

A validation analysis of the one-dimensional flow network method was conducted for the water test results for the loss of off-site power. Fig. 4.89 compares the test and analyses results for the primary flow rate and the inlet and outlet temperatures of RV. The one-dimensional flow network method simulated the water test result well.

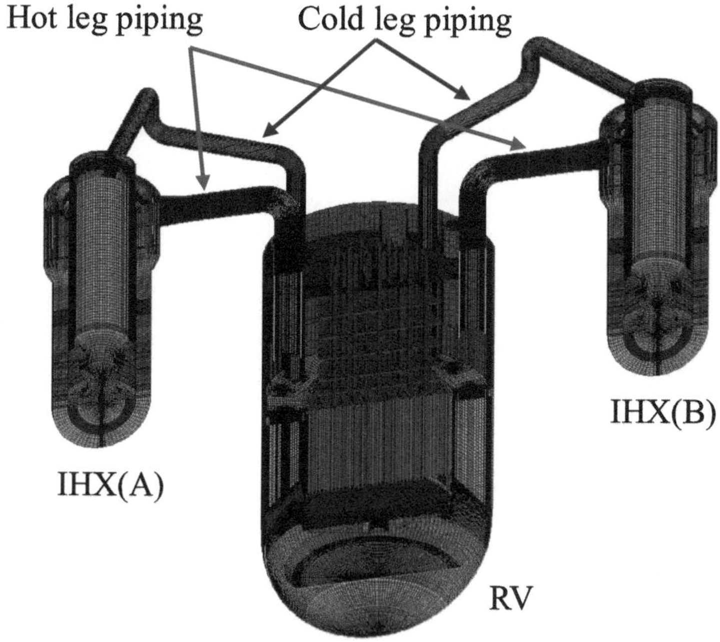

Fig. 4.87 3D analysis model for JSFR DHRS [61].

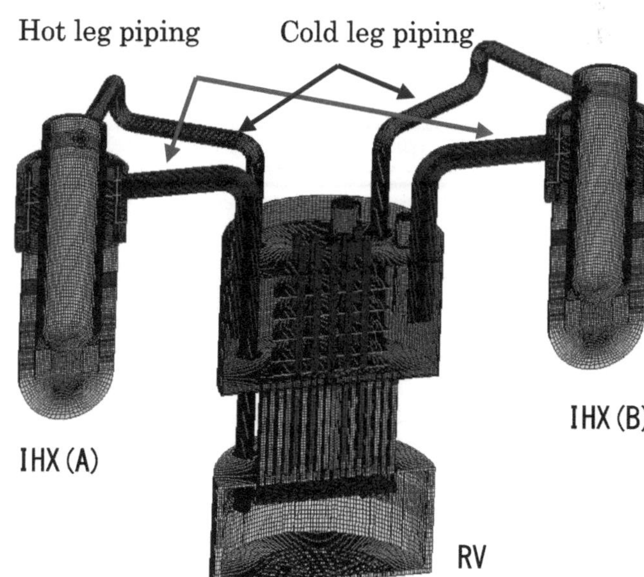

Fig. 4.88 3D analysis model for the 1/10-scale water test [61].

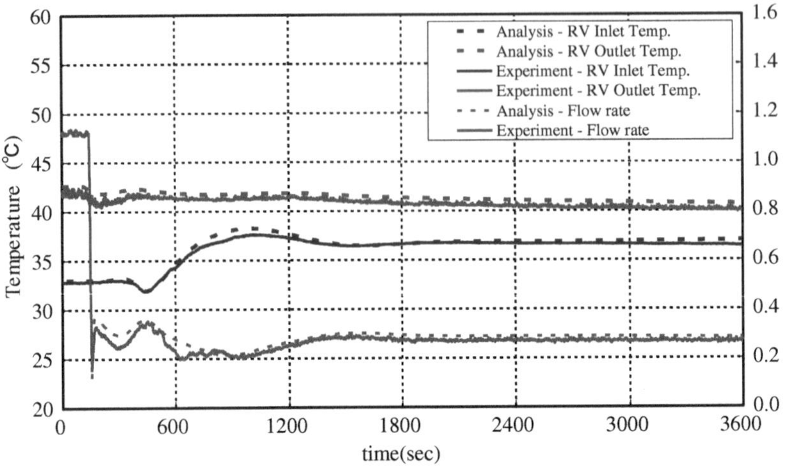

Fig. 4.89 Comparison between the experiment and 1D analysis in the 1/10-scale water test [61].

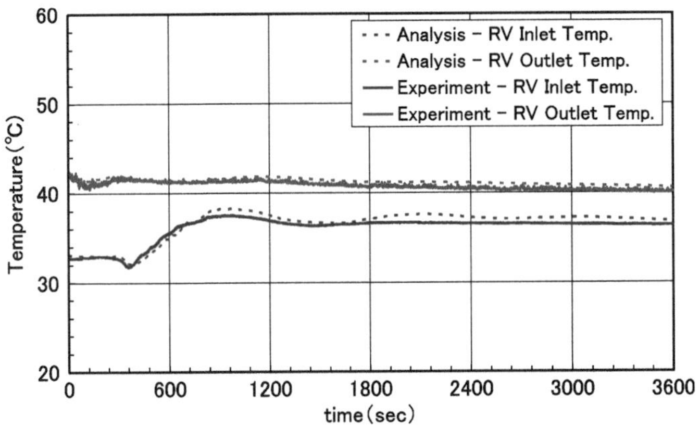

Fig. 4.90 Temperature comparison between the experiment and 3D analysis in the 1/10-scale water test [61].

A validation analysis of the three-dimensional fluid flow analysis method was conducted, as has been done for the one-dimensional flow network method. Fig. 4.90 compares the test and the analyses results. A couple of cold-leg piping could be observed in the primary loop. Fig. 4.91 compares the flow rates in the two cold-leg piping of Loop A in a short period. The three-dimensional fluid flow analysis method simulated the water test results well, including the fluctuation of the flow rate in the cold-leg piping. These phenomena are caused by unstable upward flows in the annular region of IHX, and their durations are limited to 800 s from the onset of the transient because the unstable flow is terminated when PHX cooling is established.

Both the one- and three-dimensional methods are expected to be applicable to JSFR. The three-dimensional fluid flow analysis method can simulate more precisely than can the one-dimensional flow network method.

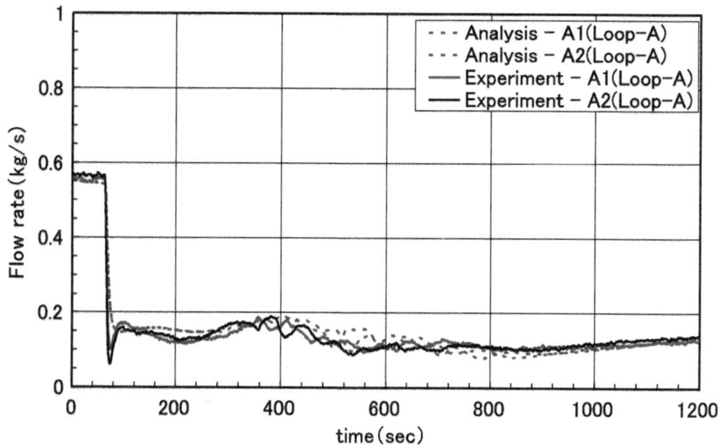

Fig. 4.91 Flow rate comparison between the experiment and analysis in the 1/10-scale water test [61].

4.2.5 Balance of plant and reactor building

To complete a nuclear power plant design, the required functions of various structures, systems, and components must be clarified, and their design concepts must be materialized. For balance of plant, fuel handling system, reactor building, including containment, and seismic isolation system have been investigated. A twin plant with two JSFR reactor units with 1500 MWe each has been designed for the commercial reactors. Meanwhile, a single-plant JSFR with 750 MWe has been designed for the demonstration reactor. In addition, the conceptual design or estimation of the capacity for the steam-water power conversion system and various auxiliary systems, such as the instrumentation and control system, electric power supply system, ventilation system, and waste treatment system, has been conducted for the JSFR in the FaCT project. This section describes the design concepts of the fuel handling system and the reactor building, including containment and seismic isolation system.

4.2.5.1 Fuel handling system

(a) Concept and specifications

The JSFR is a loop-type plant with 1500 MWe of electricity output. The power station consists of two main reactor units and is provided a set of FHS as common facilities. The core of the JSFR has 562 fuel assemblies, 96 blanket assemblies, and 57 control rods (CRs). The cycle length of the commercial concept is 26 months. The fuel and blanket assemblies are separated into four batches. One-fourth of the fuel and blanket assemblies and all of CRs, total 222 assemblies, are handled in one refueling operation [67].

The pre-conceptual design of the FHS has been conducted based on the 750 MW demonstration JSFR [68]. The dimensions of the fuel subassembly are basically the same between the commercial 1500 MWe and the demonstration 750 MWe concepts. For the fuel handling system, the system configurations and component specifications are basically common or similar between the commercial and demonstration concepts, except for the

Table 4.16 Design conditions of the fuel handling system for the demonstration JSFR [68].

Items	Value
Electric/thermal output	750/1765 MW
Number of driver fuel SAs	274
Number of blanket fuel SAs	66
Number of control rod SAs	27
Number of batches	6
SA pitch	206 mm
SA length	Approximately 4.5 m
SA weight	690 kg
Operation cycle	18 months
SA decay heat after 17 days	35 kW
SA decay heat after EVST	10 kW
Fresh SA decay heat	2.6 kW

storage capacity. Table 4.16 presents the design conditions for the 750 MWe concept. The JSFR core installs minor actinide (MA) and low decontamination fuel considering the reduction of environmental burden [69]. For the demonstration JSFR, fuel will be reprocessed from the light water reactor (LWR) spent fuel. Various LWR-to-FBR transition fuels have been investigated considering various LWRs, burnup, and cooling conditions [70–72]. The FHS design conditions are identified to cover the wide range of LWR-to-FBR transition fuels. The dimensions of the fuel subassembly (SA) for the demonstration JSFR are similar to those of the commercialized JSFR. The batch number of 6 is slightly different from that of the FBR multirecycle equilibrium core of 4. The operation cycle is 18 months, which is shorter than 26 months of the FBR multirecycle equilibrium core. The maximum decay heat of the spent fuel subassembly after 17 days from the reactor shutdown, which means decay heat during refueling, is evaluated as 35 kW with a design margin. Spent fuel subaasemblies are stored in the EVST during one operation cycle and the reduced decay heat is less than 10 kW to handle in the gaseous area. The JSFR handles the MA-bearing fuel; hence, fresh fuel SA also has certain decay heat, while conventional fresh fuel does not. In the FaCT design conditions with an MA content up to 5% in the homogeneous core, the maximum decay heat of the JSFR fresh fuel is 2.6 kW.

The JSFR adopted a simple fuel handling system with advanced technologies (Fig. 4.92). The JSFR in-vessel fuel handling system consists of a combination of an upper inner structure with a slit (slit UIS) and a flexible arm-type pantograph fuel handling machine (FHM) to dramatically reduce the reactor vessel diameter. The FHM is removed during power operation from the reactor vessel. From the reactor vessel to the EVST, a spent SA, which is accommodated by a sodium pot, is transported by an ex-vessel transfer machine (EVTM) (Fig. 4.93). A two-position sodium pot is installed for the transportation of the subassemblies from the reactor vessel to the EVST to reduce the refueling time and increase the plant availability. Active cooling is not necessary during the transportation from the reactor vessel to the EVST because the sodium pot provides enough heat capacity. Only when the transportation has malfunctioned and stuck, the sodium pot cooling system, which consists of a combination of direct and indirect cooling, is activated. The EVST has

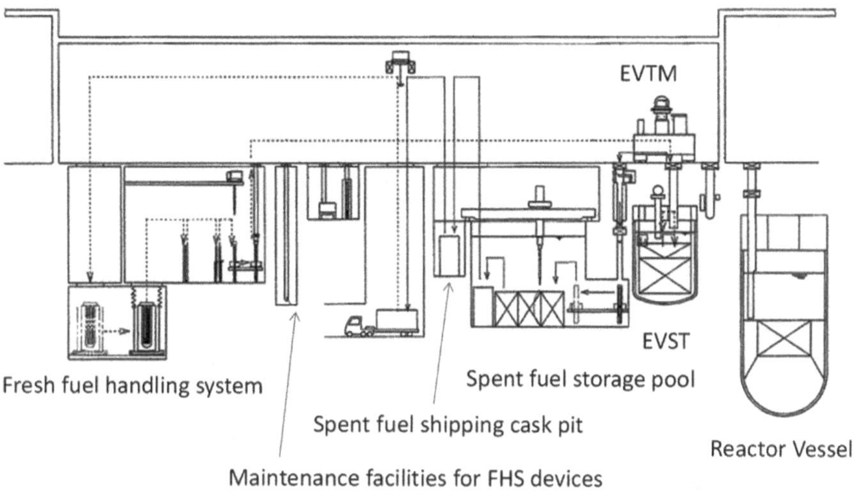

Fig. 4.92 JSFR fuel handling system [68].

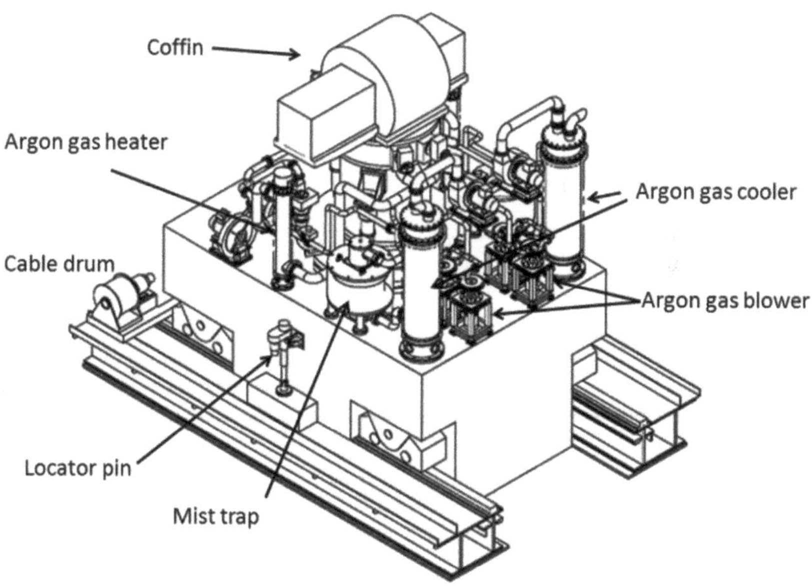

Fig. 4.93 Sketch of the ex-vessel transfer machine [68].

enough capacity for full core evacuation to enhance the plant's in-service inspection and repair capability (Fig. 4.94). It has 431 SA positions, including 84 positions for one refueling campaign, 340 for whole core evacuation, and seven for failed fuel. The EVST diameter is 10.5 m, with a heat removal capacity of 9.8 MW. After an approximately one-cycle operation, the spent fuel SAs are transported to the spent fuel storage pool by EVTM with argon gas cooling. During the transportation to the storage pool, the spent fuel SA is cleaned to remove the residual sodium from the SA surfaces. The spent fuel cleaning

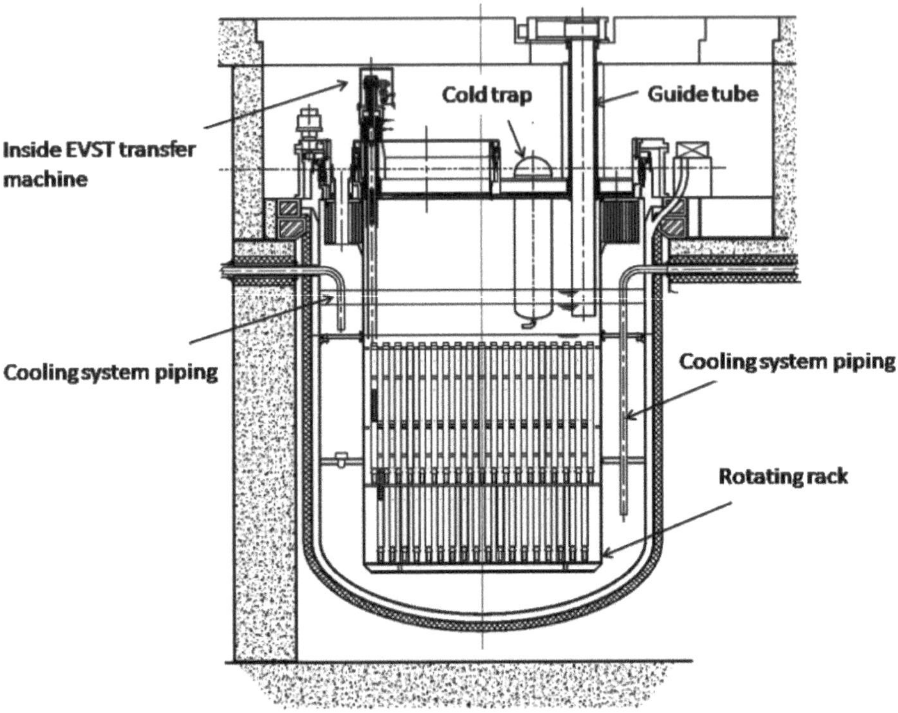

Fig. 4.94 Sketch of the ex-vessel storage tank [68].

method has changed from conventional water rinse to argon gas dry cleaning to eliminate a facility for subassembly cleaning and reduce radioactive liquid waste. After argon gas dry cleaning at the SA guide tube of the EVST, the SA is transported by EVTM, and the residual sodium is made inert by using moist argon gas at the SA lifter at the entrance of the spent fuel storage pool. The storage capacity of the water pool is 227 to maintain the total SA storage capacity of the nuclear building of 658, including the storage capacity for an 8-year operation, one batch of CR, a whole core evacuation, and failed fuel. A fresh fuel handling system has been designed to accommodate decay heat and radiation from the minor actinide-bearing fuel in accordance with JSFR's core concept: transmuting MAs from the LWR and containing transuranium elements inside the fast reactor fuel cycle system.

Table 4.17 lists the key technologies for the proposed FHS. The four new technologies must be evaluated: (1) flexible arm-type pantograph FHM; (2) sodium pot with two core component positions; (3) dry cleaning of the spent fuel subassembly; and (4) fresh fuel shipping cask. The JSFR installed the flexible arm-type pantograph FHM instead of an offset arm-type pantograph FHM in Monju. The flexible arm positioning and the interaction between the FHM-slit UIS in the seismic conditions are the major issues in the new FHM. The cooling capability for the two spent fuel subassemblies and various conditions are the key issues for the sodium pot for two core component positions. Meanwhile, the performance of the dry cleaning process, sodium inert process after dry cleaning, and cladding integrity in the spent fuel pool are the major issues for the dry cleaning of the spent fuel subassembly because the cleaning process is completely different from the conventional

Table 4.17 Key technologies for the fuel handling system [68].

Items	Main issues
Flexible arm-type pantograph FHM	– Positioning of valuable arm – Seismic tolerance
Sodium pot with two core component positions	– Cooling capability of the sodium pot (two core component positions and subassembly with inner duct)
Dry cleaning of spent fuel subassembly	– Performance of dry cleaning – Sodium inert process – Integrity of cladding in the spent fuel pool
Fresh fuel shipping cask for MA-bearing fuel	– Cooling capability of the fresh fuel shipping cask

wet process with water rinse, and the cladding material is changed to oxide dispersion strengthened (ODS) steel. Lastly, for the fresh fuel shipping cask, the cleaning performance of the shipping cask is the key issue. The JSFR adopted the MA-bearing fuel; thus, the cooling and shielding of the fresh fuel cask must be confirmed. On the contrary, the conventional fresh fuel does not have to mind these points. The evaluation results of these key points are summarized in the subsequent chapters.

(b) Flexible arm-type pantograph FHM

A full-scale mockup of the pantograph FHM was manufactured to evaluate basic feasibility [73]. The major requirements on the FHM are as follows:

(1) slim structure suitable for the slit UIS (400 mm width);
(2) prevention of interaction with the UIS;
(3) positioning accuracy within a few mm order;
(4) stiffness against SA insertion-extraction (25 kN);
(5) short refueling time (30 min per SA);
(6) recovery from abnormal events; and.
(7) ease of maintenance and repair.

Fig. 4.95 displays a sketch of the FHM mockup and photographs. The experiment was conducted in the air with a dummy SA. The gripper mechanism was installed with a self-positioning adjustment mechanism that can cover a 30 mm offset. However, considering the offsets caused by the rotating plug motion, FHM rotation, SA deformation due to irradiation-thermal expansion, UIS displacement, and offset of UIS-FHM positions, the horizontal positioning accuracy for the mockup FHM in the air must be within 3 mm. For the vertical direction, the limit positioning accuracy was set to avoid the interaction of the FHM and the SA. The gap between the SA top and the UIS bottom was 35 mm; hence, the vertical positioning accuracy requirement was 10 mm considering the FHM-SA offsets and SA swelling. The mockup test showed that the FHM can meet the positioning accuracy requirements by keeping required speed of 2.3 m/min.

A seismic analysis was performed using Femap with NX Nastran Ver. 9.3.1 to prevent the interaction between the FHM and the UIS. The limit of the seismic displacement of the FHM was 20 mm in the horizontal direction considering the UIS displacement, offsets, and thermal expansion. The vertical limit displacement was 15 mm, preventing the interaction between SA and FHM. The seismic analysis result with solid hexagonal meshes on the

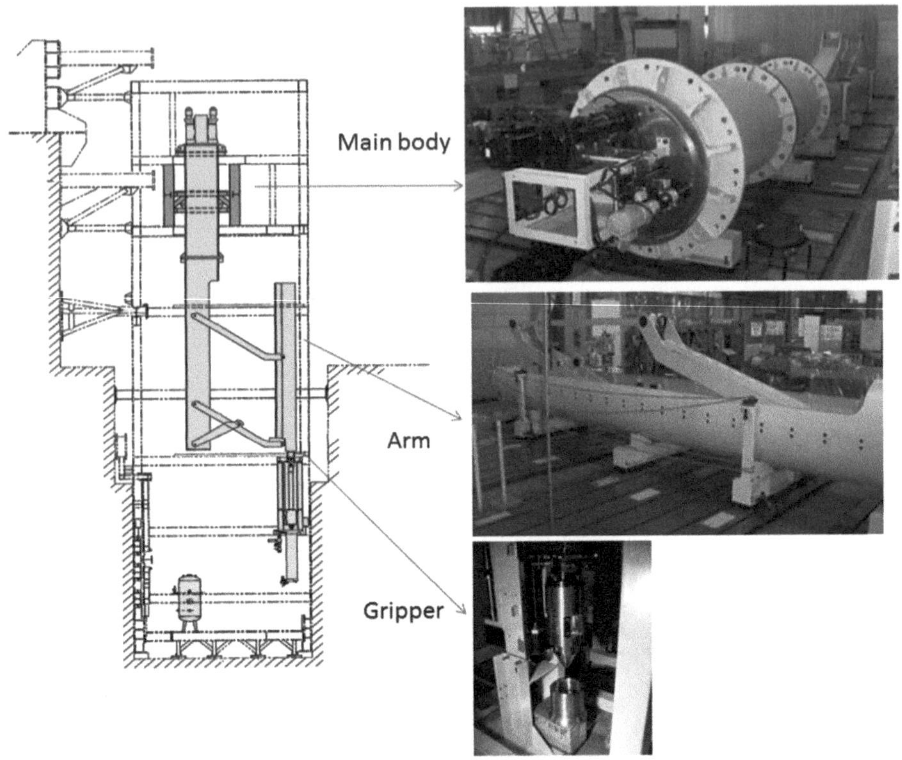

Fig. 4.95 Full-scale mockup of the pantograph fuel handling machine [68].

design base seismic condition showed that the maximum displacements of 19.6 mm and 5 mm in the horizontal and vertical directions, respectively, met the requirements. Fig. 4.96 shows the stress distribution. The maximum stress was 131 MPa, which was lower than the design limit of the 304SS yield stress at 200°C of 144 MPa.

For recovery from abnormal events, such as FHM arm stuck, a comprehensive survey on initiating malfunction was conducted, and proper recovery measures were listed. The most severe event is gripper malfunction holding SA because the FHM must delatch SA before extraction. In these cases, a device to delatch the SA forcedly from the FHM gripper will be inserted from the top of the reactor vessel plug [74].

(c) Sodium pot with two core component positions

During refueling, the spent fuel SAs accommodated in the sodium pot were transported from the reactor vessel to the EVST. The JSFR sodium pot had two core component positions to reduce the refueling time. Active cooling is not necessary in the normal refueling operation because the sodium pot has enough heat capacity. The active cooling system is activated when the sodium pot is stuck at the reactor vessel guide tube. The cooling capacity of the sodium pot for the demonstration JSFR was 35 kW. A mockup was fabricated, and the cooling performance was tested to evaluate the cooling performance of the sodium pot [75]. Fig. 4.97 depicts the fabricated mockup pot. As a result of the fabrication tests, fin machining showed no technical issues, and the high-velocity oxygen fuel thermal process was selected because it was found to be more suitable for coating on the rather complicated

Fig. 4.96 Calculated stress in the design base seismic conditions [68].

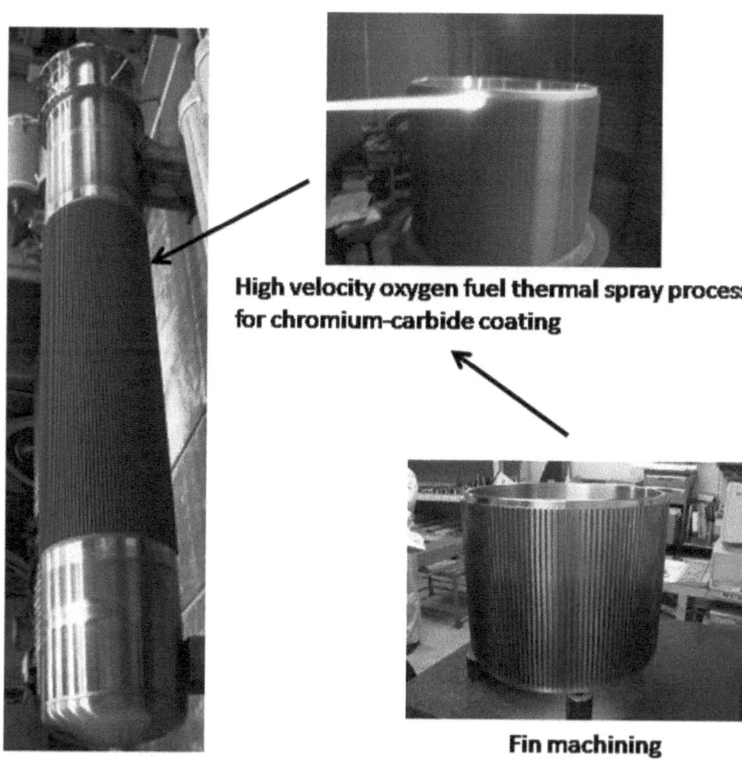

Fig. 4.97 Mockup fabrication on the sodium pot [68].

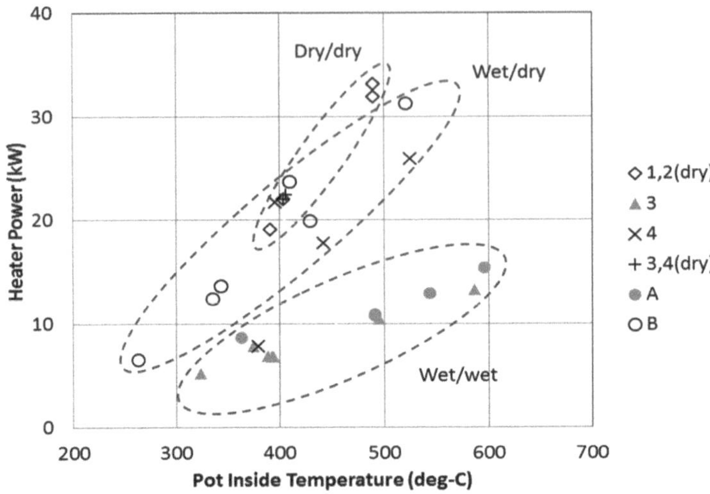

Fig. 4.98 Experimental results of emission cooling [68].

shaped surface (e.g., fin) than the detonation process. The properties of the chromium carbide coating, such as hardness, thermal conductivity, and thermal expansion coefficient, were obtained.

Heat transfer performance tests using the mockup transfer pot were performed. The experimental apparatus simulating the most severe condition of transfer pot stoppage in the guide tube on the way from the RV was composed of a test pot and a guide tube, whose cross section, including fin configurations, was full scale to confirm the effect of the fin and coating performances in the actual condition. Fig. 4.98 shows the experimental results of the emission cooling performance [75]. The experimental results were categorized into three groups as follows from the viewpoint of the surface conditions of the pot and the guide tube:

Case 1: Dry pot and dry guide tube.

Case 2: Sodium-immersed pot and dry guide tube.

Case 3: Sodium-immersed pot and guide tube (hypothetical).

For Case 1, the emissivities of the pot and the guide tube were evaluated as 0.5 and 0.5, respectively. Case 2 shows the results after the sodium immersion. In Case 2, the cooling performance was similar to that in Case 1 because sodium on the pot surface was first evaporated (Case 2-1). As the thermal output of the pot increased, the emission from the pot reduced due to the deposition on the guide tube surface. The evaluated emissivities of the pot and the guide tube were 0.5 and 0.3, respectively, in this case. Case 3 is a hypothetical case with sodium on the guide tube because the guide tube in the reactor vessel did not have to change to make direct contact with sodium. The guide tube emissivity was 0.1 in Case 1. During this stage, the sodium pot cooling system was designed with the conservative emissivity of the hypothetical Case 3.

A three-dimensional (3D) analysis model using the CFX code for the heat transfer evaluation of the JSFR fuel transfer pot was developed [76]. The heat transfer model inside the pot was validated using the CCTL-CFR experiments [77], while that outside the pot was validated using the mockup pot experimental results. The cooling performance of the pot was evaluated using the developed 3D model. The calculation results showed that the fuel transfer pot can be cooled using a combination of indirect cooling with air and direct cooling with argon gas, even with conservative emissivity. Fig. 4.99 illustrates one of calculation cases with two subassemblies (with 22.4 kW each) [76]. For a simpler design tool, a two-

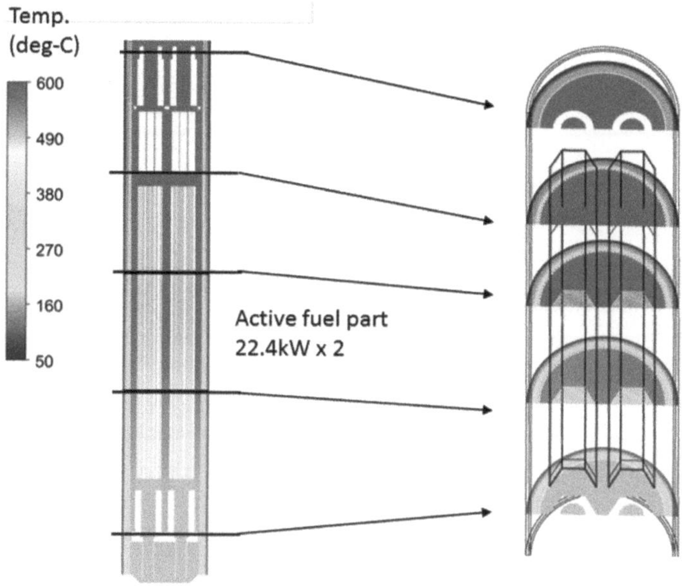

(A)

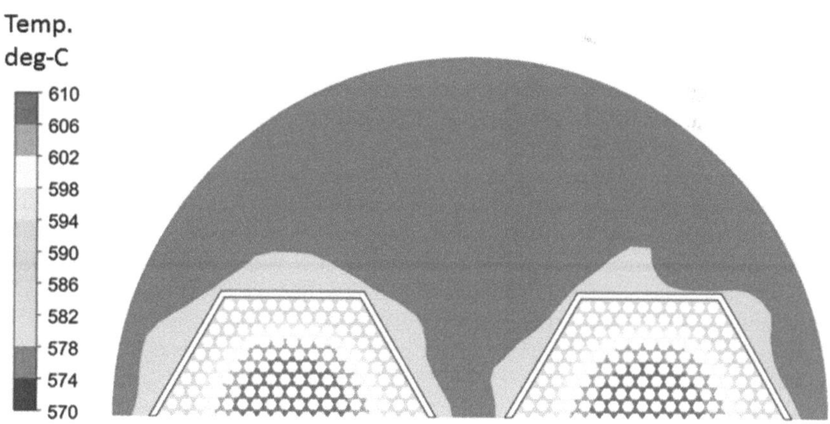

(B)

Fig. 4.99 3D evaluation on direct-indirect combination cooling [68]. (A) Temperature distribution. (B) Section at the top of active fuel.

dimensional (2D) analysis model using ZEPHYRUS code was developed and validated by a 3D analysis model [76]. The sodium pot for the demonstration JSFR was evaluated. In the calculation, the air flow rate at 60°C for indirect cooling was 20 m^3/min normal, while the argon gas flow rate at 150°C for direct cooling was 8 m^3/min normal. The results showed that the sodium temperature at the center of the subassemblies was lower than the limit 600°C (Fig. 4.100).

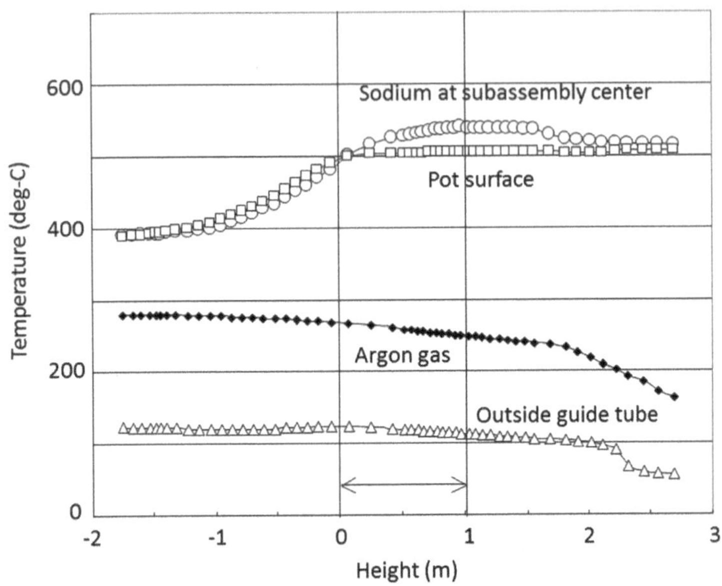

Fig. 4.100 Evaluation on the sodium pot for the demonstration JSFR [68].

(d) Dry cleaning of the spent fuel subassembly

In the JSFR, an advanced spent fuel cleaning process was adopted, and a spent fuel subassembly was directly stored in the water pool without canning. Fig. 4.101 presents the concept of the advanced spent fuel cleaning, called dry cleaning. In the dry cleaning method, a special facility for spent fuel cleaning like the fuel cleaning facility in Monju is unnecessary. The spent fuel discharged from the EVST is held at the EVST guide tube, a tube through which the EVTM accesses the EVST, and argon gas circulation by the argon gas cooling system is provided during a certain period to blow away sodium on the spent fuel subassembly. After the gas blow, the subassembly is transported to the water pool guide tube by the EVTM. At the guide tube, the residual sodium on the spent fuel subassembly is made inert by moisture argon gas and immersed into the water pool. The JSFR spent fuel subassemblies are stored in the water pool without canning; thus, a certain amount of residual sodium is acceptable as long as the water pool purification system can handle it. The water conditions of the JSFR water pool are managed by the purification system, and sodium carried into the water pool would be accumulated at the purification system. One major feature of the new cleaning method is that the radioactive liquid waste caused by spent fuel cleaning is reduced because the new process is free from spent fuel rinse, which cannot avoid a certain amount of liquid waste.

The JSFR SA has an inner duct for enhancing the molten fuel release in case of severe accidents; thus, the sodium drain from the inner duct is an issue. In a previous study, inner duct drain tests using inner duct mockups were conducted, and the drain performance of the inner duct was confirmed. A full-scale mockup of the JSFR fuel subassembly, which involves a fuel pin bundle, a wrapper tube, and an inner duct, was manufactured to demonstrate the proposed dry cleaning process. Dry cleaning performance tests after sodium immersion were then conducted, as shown in Fig. 4.101. The experimental results showed that the residual sodium on the mockup SA was less than 200 g after the cleaning process. Based on the experimental results of the residual sodium density of each part of the mockup SA, the amount of residual sodium on the JSFR SA was evaluated (Table 4.18) considering

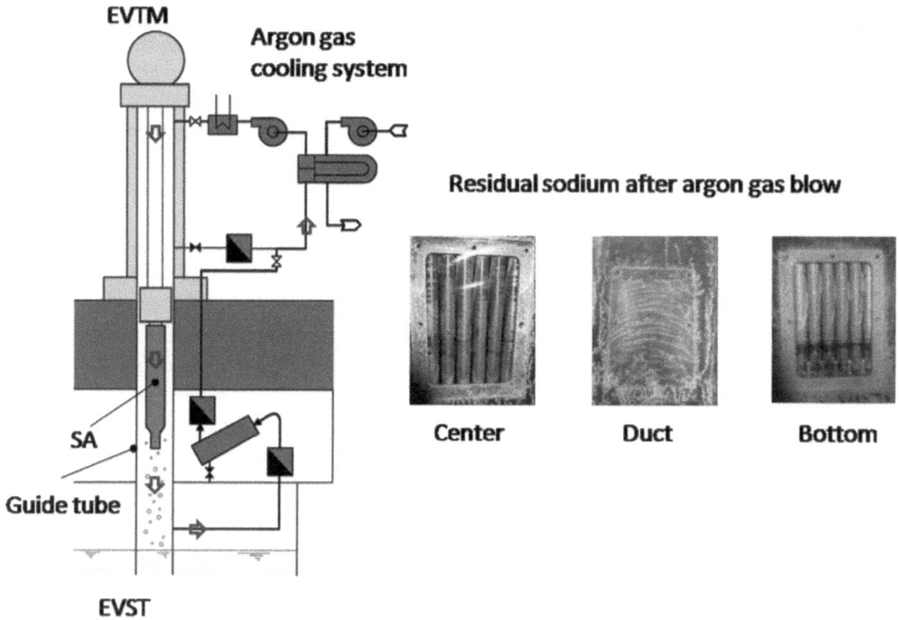

Fig. 4.101 Dry cleaning and residual sodium on a mockup fuel subassembly [68].

Table 4.18 Evaluation of residual sodium after cleaning [68].

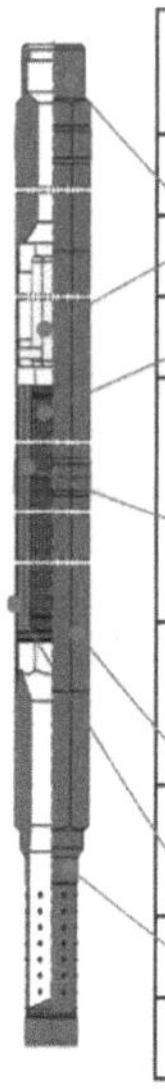

Item		Density (g/m^2)	Weight (g)
Handling head		8	5.7
Shielding		8	5.4
Pin bundle		11	270.1
Duct	Out	8	3.3
	In	8	5.5
	Bottom	40/1000	1.8
Wrapper tube	Out	8	15.2
	In	8	12.3
Duct-Wrapper tube		40	11.1
Entrance nozzle		110	68.0
Total		-	399

the uncertainty and the design margin [78]. The residual sodium of the SA was evaluated as approximately 400 g. After dry cleaning, the spent fuel subassembly was transported to the water pool guide tube, and the residual sodium was made inert by humid argon gas and steam. The experiences in Joyo and Monju showed that the duration of the sodium inert process would depend on the sodium density on the spent fuel subassemblies, and the inert process duration would be approximately 25 min. For the JSFR fuel handling design, the inert process duration was conservatively designed to be 30 min based on the Joyo and Monju experiences [78]. From the viewpoint of the FHS operation, two SAs in the EVST were transported to the spent fuel storage pool. The pool water can be maintained at less than pH 10 with a proper pool purification system. The ODS cladding is thought to be tolerable based on the corrosion experiments [79].

(e) Fresh fuel shipping cask

Fresh fuel shipping is always an issue when a sodium-cooled reactor is going to handle the MA-bearing fuel, which requires shielding and cooling. In the JSFR design range, the fresh fuel SA might have a 2.6 kW decay heat per SA. The cladding temperature limits are set to 395°C for the inert gas environment (core inlet temperature) and 300°C in air [80] to prevent creep damage and oxidation. The JSFR fresh fuel shipping cask adopts a helium gas coolant to avoid water contact before core loading and provide a better heat removal performance than other gaseous coolants. Fig. 4.102 illustrates the designed helium gas cask with five SA positions.

Two design evaluation models were developed [81]. The cooling performance was roughly evaluated using the ABAQUS code with a 3D whole cask model. The conservativeness of the whole cask calculation was confirmed by the fact that the cladding temperature evaluated by the detailed SA calculation by Star-CD showing the detailed SA calculation provided a lower cladding temperature compared to the whole cask analysis. Fig. 4.103 shows the calculation results of a cask with two subassemblies of 2.6 kW decay heat each. In the 3D whole cask model, the maximum cladding temperature was evaluated

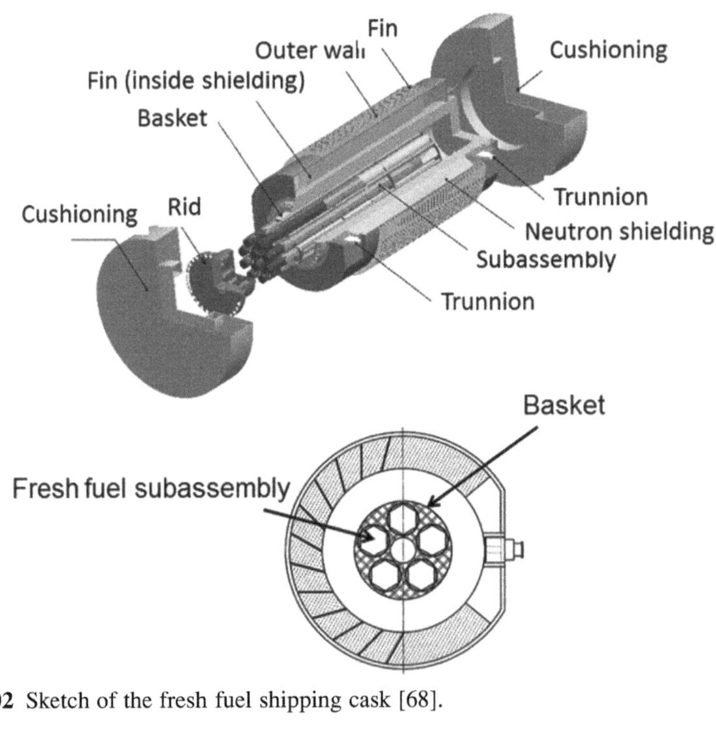

Fig. 4.102 Sketch of the fresh fuel shipping cask [68].

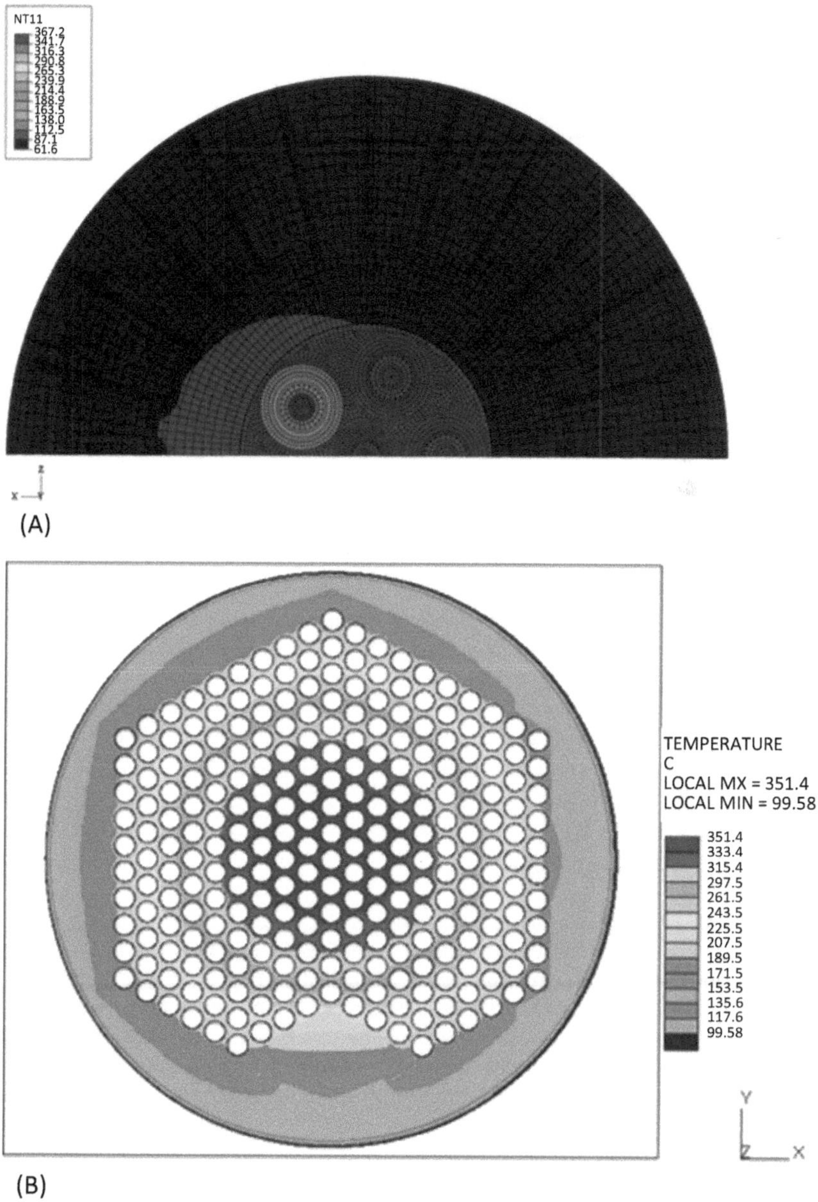

Fig. 4.103 Fresh fuel temperature evaluation [68]. (A) 3D whole cask calculation. (B) 2D detailed subassembly calculation.

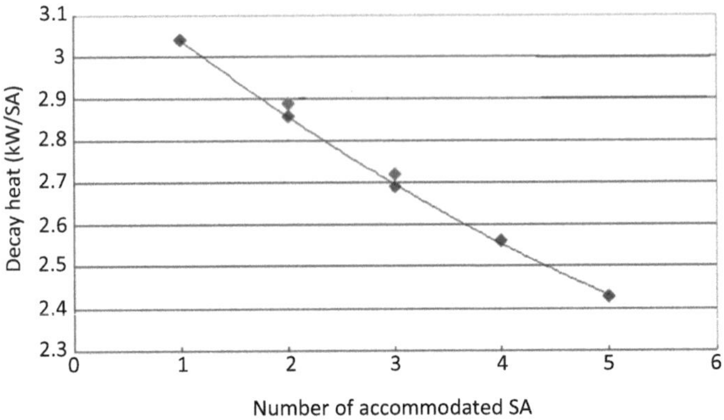

Fig. 4.104 Cooling performance of the fresh fuel shipping cask [68].

to be 367°C, while the maximum temperature from the detailed 2D calculation was 351°C. The analyses showed that the 3D whole cask model can provide a conservative evaluation of the maximum cladding temperature. The relation between the accommodated SAs and the maximum decay heat was evaluated as shown in Fig. 4.104. The results showed that the proposed helium gas cask can handle SAs with decay heat of up to 2.2 kW/SA. The cask can even manage higher decay heat SAs by reducing the SA accommodations from five to a few. In short, the MA-bearing fresh fuel in the FaCT deign range can be handled using the proposed helium gas cask.

4.2.5.2 Reactor building layout

(a) Concept and specifications

Fig. 4.105 depicts the bird's-eye view of the entire plant of the JSFR planned at present [82]. In many cases in Japan, reactor sites are located along the bay-shore close to the mountain area. This is related to where the hard base rocks can be found. Moreover, the nuclear power plant in Japan uses seawater as the cooling water; hence, the JSFR is assumed to be installed along the bay-shore, where the turbine building is located closer to the ocean. The JSFR plant consists of a reactor building, a turbine building, a freshwater treatment building, a wastewater treatment building, an incinerator building, a solid waste storage building, an office building, a switch yard, and a ship yard.

Fig. 4.106 presents the reactor building and turbine building layout. The layout design aims to shorten the length of the piping and the cables.

Fig. 4.107 displays the reactor building layout plans. Volume reduction is one of the most important items in the layout design study of a reactor building. To examine the minimization of a reactor building, the volume is reduced by rationalizing the component arrangement by sharing facilities taking advantage of the twin-unit plant or by adopting a rectangle containment vessel. The low-pressure primary coolant and the acceptance of a two-loop configuration make it possible to enlarge the unit size of the cooling components (i.e., RV, IHX, and steam generator (SG)) for an FBR plant compared with an LWR plant.

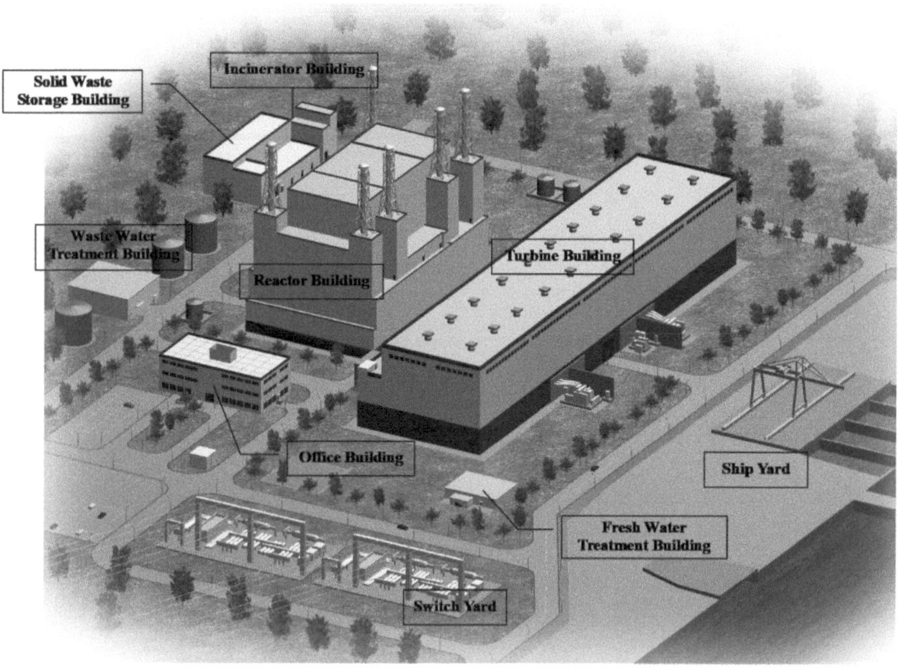

Fig. 4.105 Bird's-eye view of the JSFR plant [82].

A double-walled piping system is also adopted in JSFR's design to prevent the sodium-air interaction in the case of a sodium leak. This double-walled piping system lowers the containment vessel (CV) design pressure. These considerations lead to the adoption of a rectangle-shaped CV with sufficient arrangement efficiency. The vessel contributed to the reduction of the volume of the whole building. Under the advantage of a rectangular CV, the reactor building volume can be reduced compared to the conventional cylindrical CV.

The adoption of a steel plate-reinforced concrete structure (SC structure) is essential in adopting a rectangle CV. The SC structure is composed of steel plates partly assembled at a factory, which contributes to the structure quality improvement. Hence, the amount of on-site construction is expected to be reduced, and the construction cost reduction (construction term shortening) can be attained.

In this reduction examination, nonsafety electric and control panels conventionally installed in the reactor building are relocated in the turbine building, enabling the reduction of the reactor building volume.

Installation and maintenance works are considered in this layout design to secure the area. The corridor width is determined to maintain spaces to install pipes, ducts, and cable ladder and handle the parts whose components are needed for maintenance. Moreover, pipe shafts are placed in several areas to avoid pipe interventions.

For the component layout, heavy components (e.g., tanks or diesel generators) are located in the lower floor to improve the balance of the reactor building and bear earthquake force. To keep the natural circulation performance, DHRS is located in the upper floor for natural cooling. Heating, ventilation, and air conditioning systems are also located considering pipe and duct shortening.

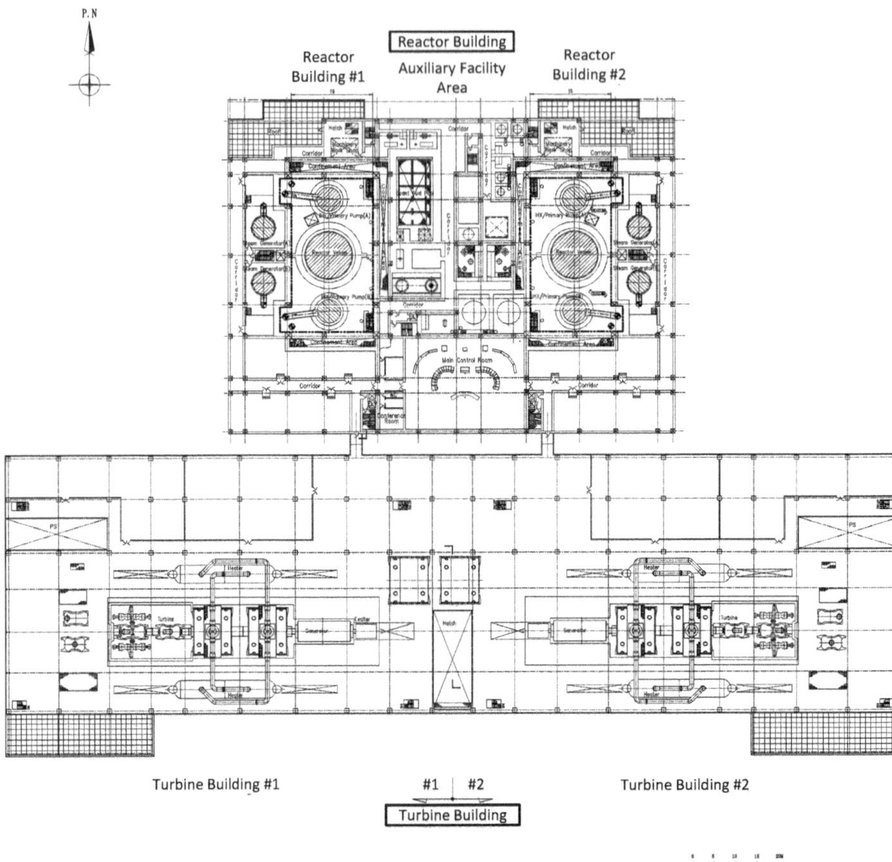

Fig. 4.106 Reactor building-turbine building layout [82].

Consequently, the building size became a 104 m × 77 m × height of 70 m, and the whole building capacity became 186,000 m^3 for one single plant.

(b) Reduction of the construction period

The JSFR adopts the modular system from the viewpoint of shortening the construction period and quality assurance of steel structure products [83]. In the outline of this method, the steel structure parts of the module are first assembled in a factory. Some of the components, including ducts, pipes, and cable trays, are then installed in advance. Weld inspections are also performed in the factory to assure high quality for the CV boundary. Meanwhile, large-size or heavy components, such as the RV and the SG, are separately delivered to the construction site after assembly and inspection. The reactor building is divided into several modular units depending on the component and equipment configuration in the building layout. The assembled steel modules are shipped by barge from the factory to the site dock. After lying alongside the site dock, they are transported by several dollies and put on the top of seismic isolators, which have been installed beforehand at the site. The installation of the aforementioned large-size heavy components is started after the placing and hardening of concrete.

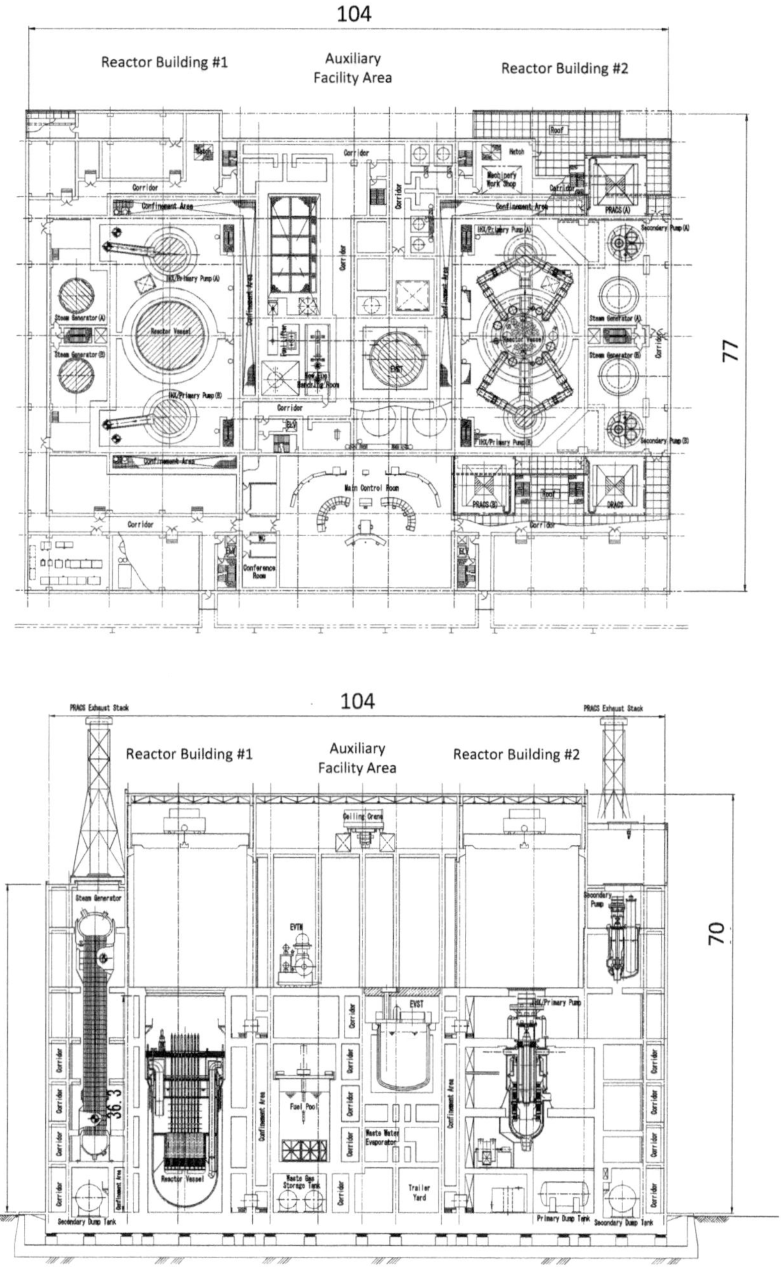

Fig. 4.107 Reactor building layout [82].

However, the reactor building is installed on a bed rock from the viewpoint of earthquake resistance; therefore, the surface stratum from the ground surface level to the bed rock should be excavated. The level difference between the ground surface and the bed rock, which depends on a site, influences the modular unit installation. The inclination of the installation road is within 3% based on the practical performance of the dollies; thus, if the level difference is large, the slope area, including the installation road and slopes on either side of it, for the reactor building installation interferes the surrounding buildings (e.g., buildings of the turbine and waste storage). This causes the possibility of interference of construction or an additional increase in the amount of excavation, which is not preferable from the economical point of view. The construction period study on the modular unit installation by cranes was also performed. Fig. 4.108 shows the installation concept of the construction method. If the dollies are used for the installation, the reactor building of the demonstration reactor of the JSFR (750 MWe) is divided into nine units (Fig. 4.109) considering the equipment composition and the transportability restriction. In contrast, if the cranes are used, the modular units are to be divided into 48 units.

Fig. 4.110 shows the construction periods of the JSFR demonstration reactor. The construction period consisted of the base-mat work, installation of seismic isolators, architecture work of the operation floor, architecture work of the reactor building, equipment

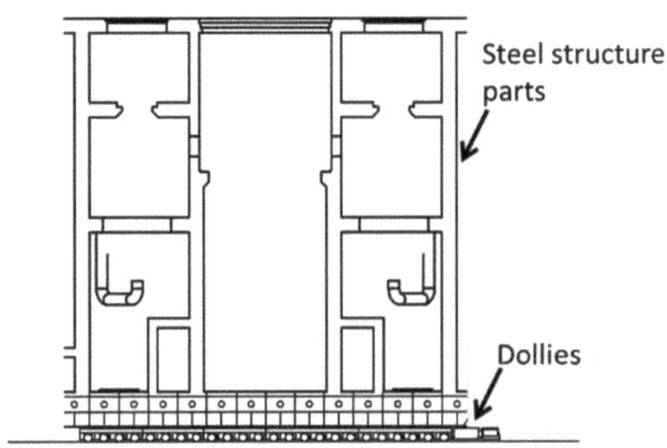

(A)

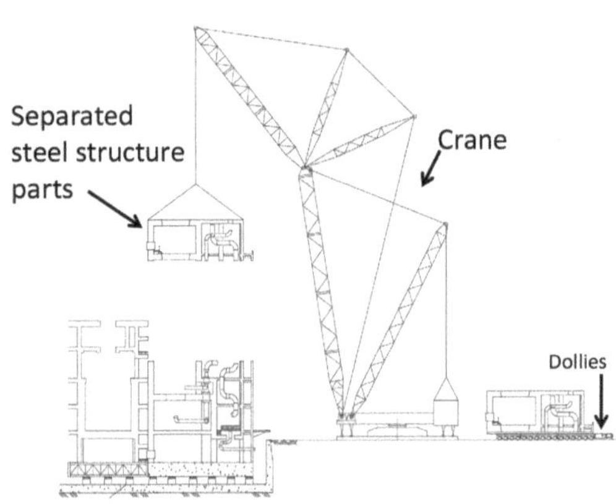

(B)

Fig. 4.108 Image of modular construction [83]. (A) Shallow rock level (transport by dollies) (B) Deep rock level (transport by cranes).

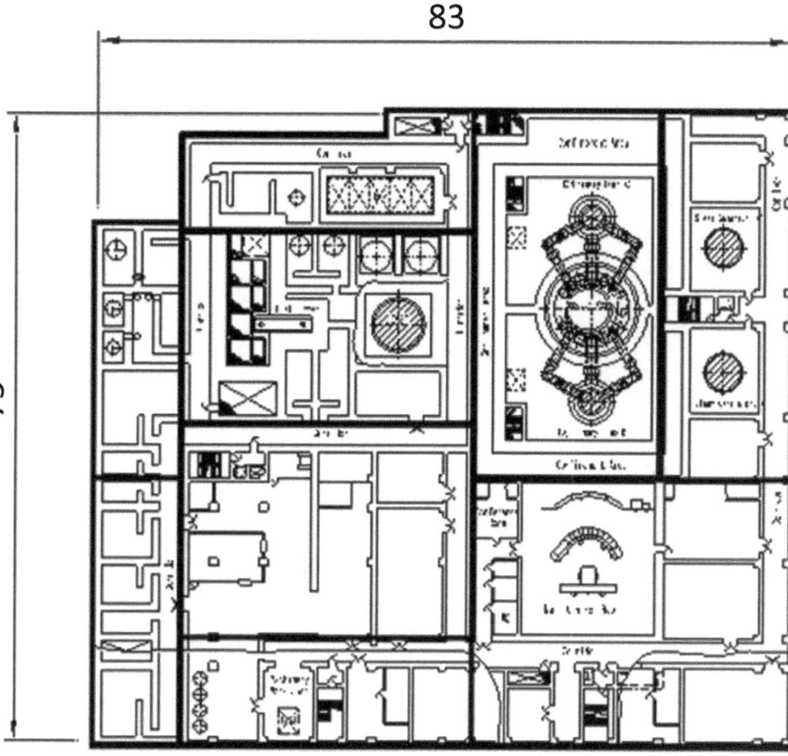

Fig. 4.109 Example of the unit layout for the demonstration reactor [83].

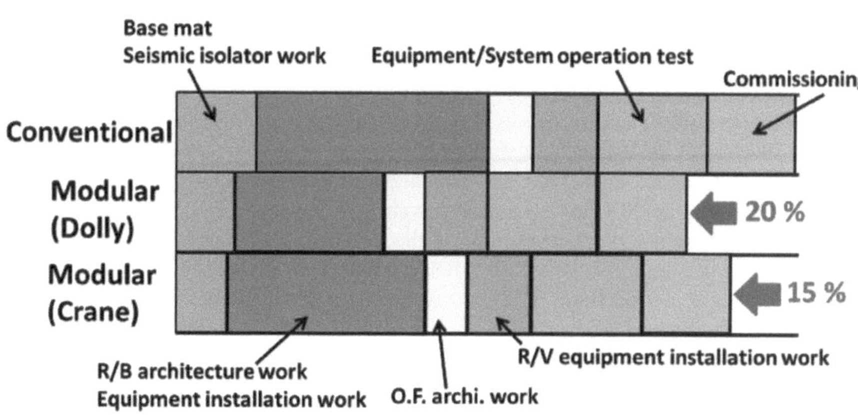

Fig. 4.110 Reduction of the construction period for the demonstration plant [83].

installation, equipment and system operation test, and commissioning. In the conventional construction method, the reactor building and the CV are constructed by the reinforced concrete (RC) structure, and the SC structure is used only for walls around the RV. The works of reinforcing bar installation are necessary at the construction site. Regarding the effect of the SC structure in the assembly work in the factory, the architecture work may be reduced

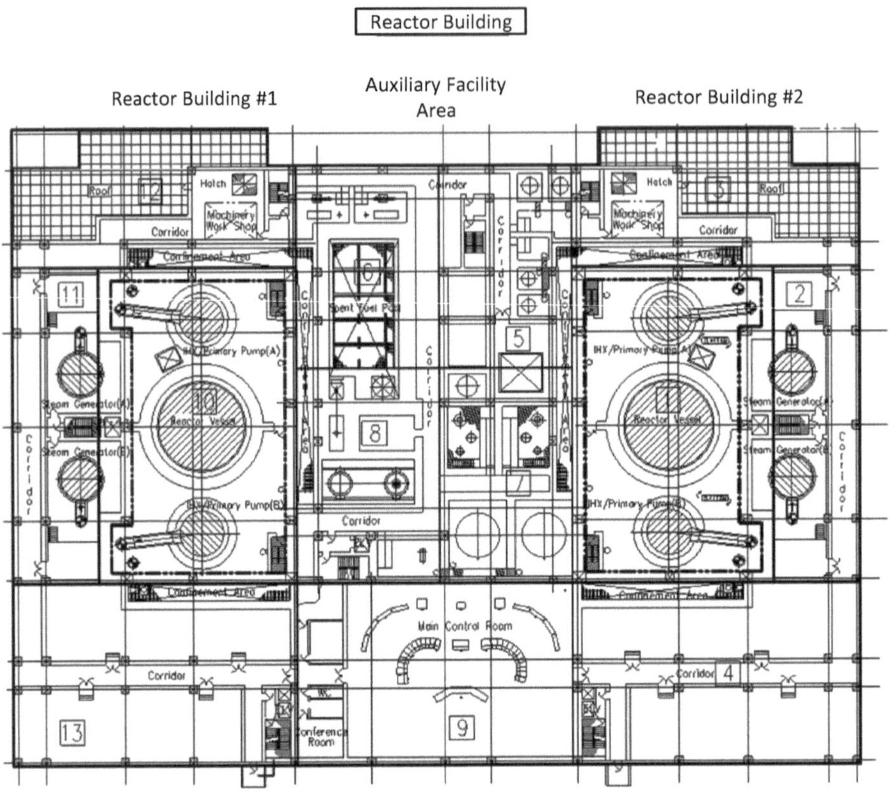

Fig. 4.111 Example of the unit layout for the commercial twin plant [82].

to about half of its work in the conventional RC construction method. Furthermore, it is practicable to start the architecture work partly because of the division of the modular units. Based on the module design study, the late 30 months can be achieved. Compared with the conventional construction method, the modular construction schedule by dollies and cranes could be reduced by 20% and 10%, respectively.

For the commercial plant, an example of the modular unit division of the reactor building is shown in Fig. 4.111, where the whole buildings are divided into 13 modular units [82]. The heaviest unit is the CV part, weighing approximately 6000 tons after prefabrication in a factory.

In this layout design, the JSFR is set beside the bay-shore, which enables the handling of heavy modular units from the factory to the JSFR site by ships. This factory assembly contributes to quality improvement.

Fig. 4.112 shows the plant construction term necessary for the completion of the JSFR.

The target schedule trial calculation is conducted, and the latter half of the 30 months can be attainable based on the examination of the construction method of a reactor building.

The modular structure of the reactor building enables the reduction of the construction term. By adopting the SC structure, the RC structure's steel rod-reinforced bar work and installation-withdrawal of the mold work at site become unnecessary. This effect reduces the construction term.

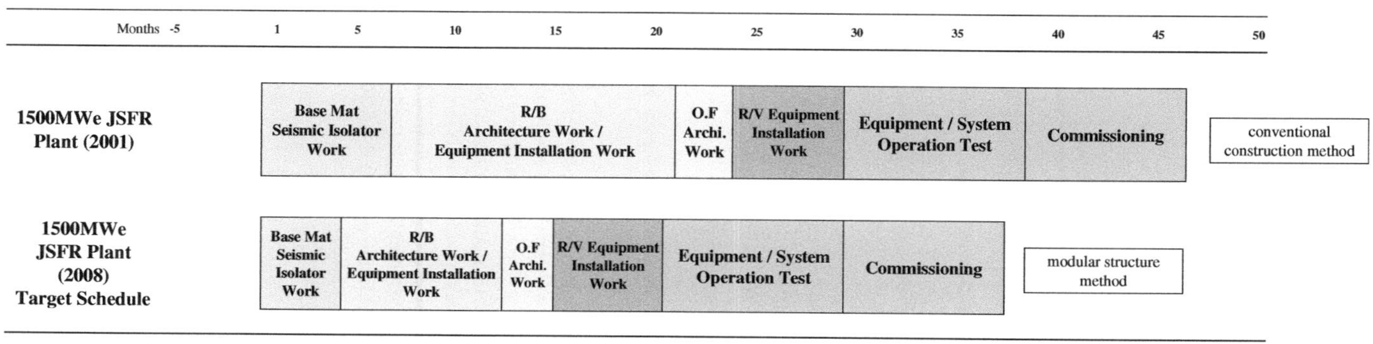

Fig. 4.112 Reduction of the construction period for the commercial plant [82].

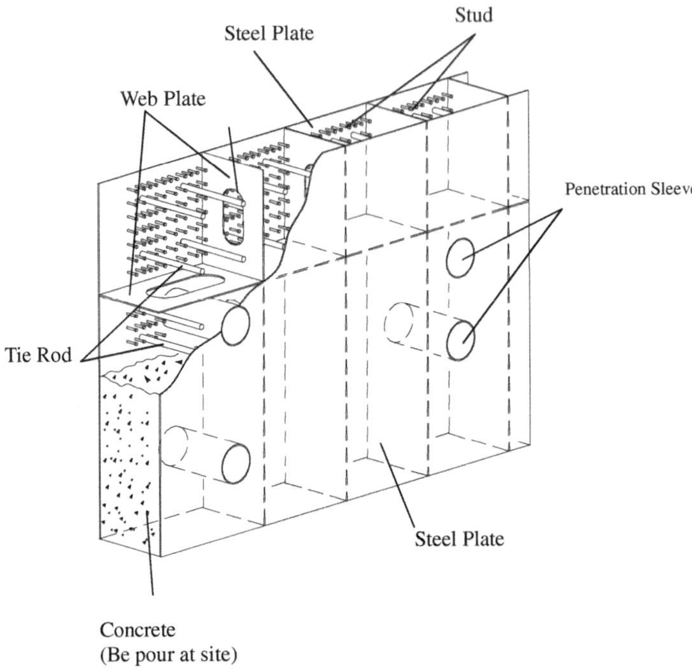

Fig. 4.113 Concept of the SC structure [82].

4.2.5.3 Steel plate-reinforced concrete structure

(a) Concept and specifications

The JSFR is planning to adopt an SC structure (Fig. 4.113) not only for a reactor building, but also for a CV (Fig. 4.114) from the viewpoint of shortening a construction period by adopting modular construction and quality assurance of steel structure products [82,83]. As for conventional buildings, the SC structure has been used for the radioactive waste disposal facility of a LWR and the shielding wall of MONJU, the prototype FR in Japan. Although the bearing strength of the SC structure has been well investigated at ambient temperatures [84], the SCCV has two main problems to be solved: (1) The SCCV shall withstand high temperatures and pressure conditions; and (2) the SCCV shall prevent radioactive nuclide leakage under severe conditions.

The SCCV is also investigated in the next-generation LWR. The major design conditions are as follows: maximum temperature: 171°C and pressure: 0.31 MPa [85]. The characteristics of the SFR under high temperature must be investigated to adopt the SC structure to the CV because the environmental temperature of the SFR is much higher than that of the LWR in the coolant leakage accident.

This paper describes the construction of a design evaluation method from design (construction period shortening) and accident management in both experimental and analytical points of view.

A CV is the final barrier against radioactive nuclide leakage. As for MONJU, its CV is a steel vessel and consists of a top hemisphere, a central cylinder, and a bottom shallow dish with an inside diameter of approximately 50 m and an overall height of approximately 80 m (Fig. 4.115). The CV of MONJU can maintain the structural integrity and the boundary

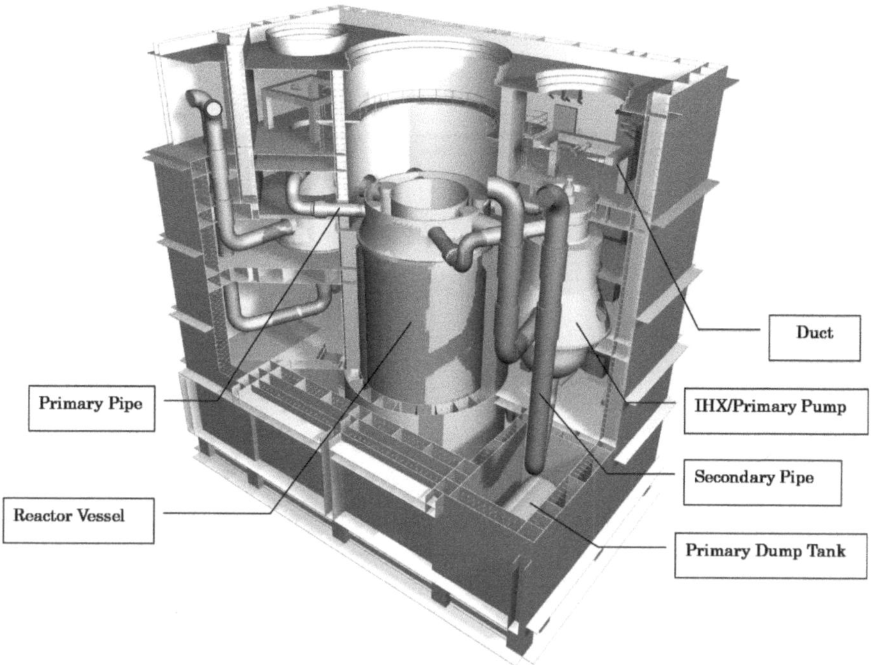

Fig. 4.114 Bird's-eye view of the SC structure containment vessel [82].

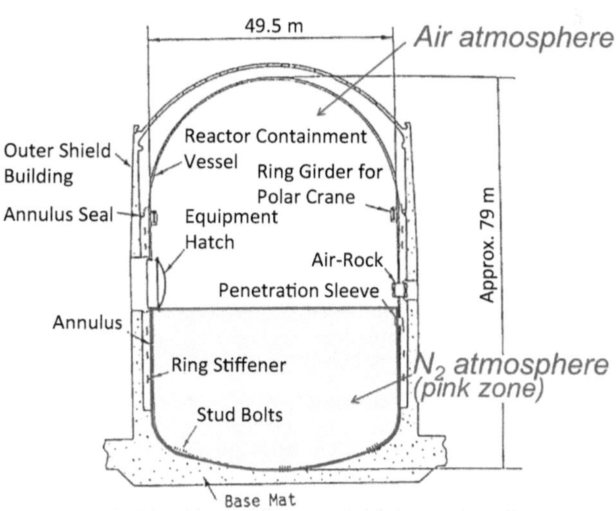

Fig. 4.115 Monju containment vessel [83].

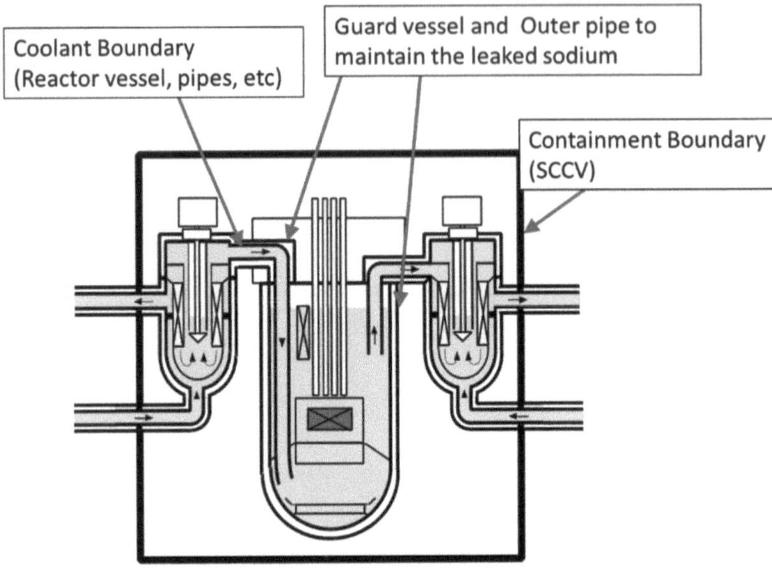

Fig. 4.116 JSFR containment vessel [83].

function against temperatures and pressures in case of accidents, such as primary coolant leakage. Furthermore, the atmosphere around the sodium-contained equipment is nitrogen gas, which prevents sodium combustion. The large space filled with air atmosphere mitigates the influence of pressure in case of accidents.

In contrast, in the JSFR safety design, a complete double-walled structure for the sodium-contained components, such as double-walled pipes of the coolant and the guard vessel for the RV, is adopted to prevent the combustion of leaked sodium in the CV (Fig. 4.116). Furthermore, the severe mechanical energy release due to recriticality events can be eliminated from core disruptive accident (CDA) scenarios by some safety designs, such as the special fuel assembly features for molten fuel discharge from the core in the case of the CDA and the limitation of the core sodium void worth, which is a multilayered structure at the bottom of the RV for core debris retention within the RV. The design features significantly reduce the loads on the CV and allow a compact containment design. These design approaches prevent sodium combustion in the CV; thus, the influence of the pressure caused by the sodium combustion can be excluded in the JSFR CV design. The design pressure of the CV is lower than that of the LWR. Accordingly, the JSFR adopts the rectangular-shaped CV consisting of the SC structure to achieve economies by eliminating the layout dead space and shortening the construction period.

Table 4.19 shows the tentative design specifications of the SCCV. Although they were set in accordance with the JEAC4618, which is the standard for the SC structure in Japan, the inner plate specifications are different from those in JEAC4618. The inner plate is exposed to high temperatures in the case of the hypothetical sodium leakage; hence, the carbon steel plate for the pressure vessels for the intermediate and moderate temperature services was used in the inner plate material. This full SC structure will be adopted as the bearing walls and floor deck, and the half SC structure will be adopted as the top floor, which has less influence on the sodium combustion.

Table 4.19 SCCV specifications [83].

Item	Value
SC wall thickness	1500 mm
Concrete	
Design strength	$21 \leq F \leq 36$ N/mm^2
Surface plate	
Inner plate material (TS[a])	Carbon steel plate for pressure vessels for intermediate and moderate temperature services (410 N/mm^2)
Outer plate material (TS[a])	Rolled carbon steel for welded structure (400–490 N/mm^2)
Thickness	12 mm
Stud	
Material (TS[a])	Hot rolled carbon steel deformed bar (400 N/mm^2)
Diameter/pitch/length	22 mm/400 mm/176 mm
Partition plate	
Material (TS[a])	Rolled carbon steel for general structure (400–490 N/mm^2)
Thickness	12 mm
Tie bar	
Material (TS[a])	Hot rolled carbon steel deformed bar (345 N/mm^2)

[a] TS, tensile stress.

(b) Requirements against sodium combustion

The JSFR adopts the complete double-walled structure for sodium-contained equipment to prevent the combustion of leaked sodium in the CV; thus, the sodium leakage accident can be excluded from design-basis events (DBEs). The following conditions are selected to assume loads for the SCCV under the DBE condition: (1) heat generation of gaseous and volatile radioactive materials inside the containment (100% inventory of noble gas and 10% iodine, etc., are assumed); (2) heat radiated from the sodium-containing components and piping without ventilation and cooling of the containment atmosphere in the case of containment isolation; and (3) low ambient pressure due to conservative meteorological condition. The analysis results showed that the design loads for the SCCV at the DBEs become less than 200°C and 50 kPaG.

In addition to the design-basis approach, the SCCV must explicitly provide the containment (boundary) function against severe accidents. Such conditions for extended safety design are defined as the design extension conditions (DECs). However, the JSFR safety design, including the double-walled structure against sodium leakage, and some countermeasures against severe recriticality events could achieve prevention and mitigation against typical DECs. Therefore, the SCCV loads at the DECs are not dominant in the SCCV design. In spite of the technical prevention of the sodium combustion in the SCCV at the DBEs and DECs, the hypothetical sodium leakage into the SCCV should be considered because it is the final barrier against radioactive nuclide leakage. The hypothetical

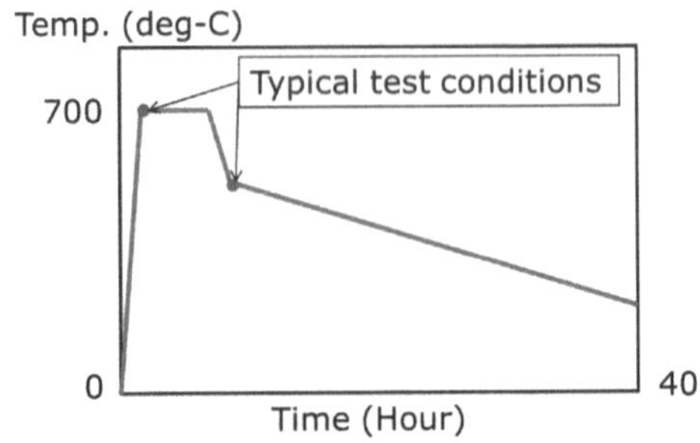

Fig. 4.117 Temperature at the inner plate [83].

sodium leakage and combustions by a sodium spray and a pool fire are considered under this condition. The leakage rate of the sodium spray from a branch pipe on an outer pipe is assumed to be up to a few kg/s, and the maximum amount of sodium is privately 160 tons, which is assumed concerning the configuration of the sodium-containing equipment. As a result of the hypothetical sodium combustion evaluation by the SPHINCS code [86], the inner steel plate temperature reaches 700°C, where it continues for 20 min (Fig. 4.117), and the maximum pressure reaches 120 kPaG for a moment.

(c) Experiments on the SCCV integrity

The structural characteristic tests of the SCCV under the hypothetical sodium leakage and the combustion condition were investigated to determine the potential characteristics of the SCCV. Table 4.20 shows the experiment outline. The data obtained by these tests were planned to be used for investigating the SC structure characteristics under high temperature,

Table 4.20 Experiments outline [83].

Purpose	Main Parameter of experiment	Use* ①	②	③	④
1.Structural integrity of the SCCV					
(1)Material property					
1)Steel property (pull-out, bending and creep strength)	Temperature	○	○	○	○
2)Concrete property (tensile/compressive strength)	Temperature	○	○	○	○
(2)Anchor strength of attached stud on concrete					
1)Tensile strength	Temperature, Diameter of stud, Embedded depth of stud	○	○	○	○
2)Shear strength	Temperature, Diameter of stud	○	○	○	○
(3)Load-displacement relation of the SC beam					
1)Out-of-plane shear strength	Temperature, Reinforcement type (tie bar, partition plate), Reinforcement ratio	○	○		○
2)Out-of-plane bending strength	Structural type, Temperature, Number of partitioning plate (single partition / multi partition), Restriction / non-restriction of initial thermal deformation	○	○		
3)In-plane behavior (strength against seismic load)	Temperature	○	○		○
(4)Strength of specific parts					
1)Strength of beam which has penetration part	Temperature	○	○		○
2)Strength of corner part	Temperature	○	○		
(5)Function of steam releasing structure in concrete	Temperature	○			
2.Boundary function of the SCCV					
(1)Behavior of heated plate (damage mode)	Temperature	○		○	
(2)Critical strain at attaching portion of stud on plate	Temperature, Diameter of stud	○		○	

*Test results are used mainly for;
① research the property of the SC structure under high temp.
② development of the analytical model to evaluate the structural integrity of the SCCV
③ development of the analytical model to evaluate the boundary function of the SCCV
④ construction of the bearing force evaluation formula in the design code

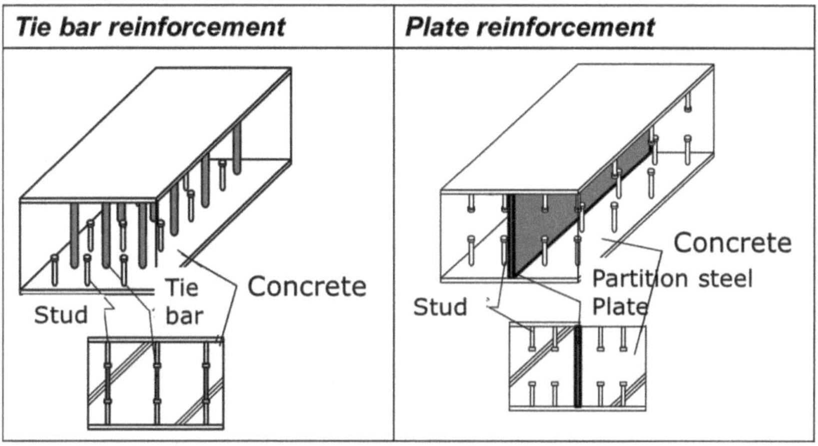

Fig. 4.118 Reinforcement type of specimens [83].

developing the analytical model to evaluate the structural integrity and the boundary function of the SCCV, and proposing the bearing strength formula in the design standards.

The out-of-plane force on the SC structure became dominant because of the adoption of the rectangular-shaped CV in the JSFR. Shear failure causes a drastic decrease of the bearing strength of the SC structure; thus, shear failure must be avoided in the SCCV design to assure its structural integrity. On the contrary, bending failure is acceptable due to the gradual decrease of the bearing strength. To assure that the bending failure precedes shear failure, the shear strength must be larger than the bending strength. Accordingly, the out-of-plane shear strength and bending characteristics were experimentally investigated. In addition, the in-plane characteristics should be considered against an earthquake. The characteristics of the corner part of the rectangle CV under high temperature must also be investigated because the local stress concentration due to the thermal expansion of the steel plates and concrete would occur at the corner part. The results of the load-displacement relation of the SC beams are outlined in this section.

The out-of-plane shear tests under high temperatures were performed to verify the influence of high temperature on the SC structure strength and propose an estimation method of the bearing strength. The test parameters were the type and amount of reinforcing (Fig. 4.118). These tests were conducted at several points of the transient temperatures as shown in Fig. 4.117. The outline of the tie bar reinforcement type is introduced here.

This specimen was about in half the scale of the tentative design of the SC structure. Fig. 4.119 shows the failure mode of the specimen loaded under 700°C. Diagonal shear cracks between the load and the support point were observed, and the brittle failure by the diagonal tensile force occurred. Tie bars near the heating surface were yielded. Fig. 4.120 shows the load-displacement curves of the SC beams with the same reinforcement ratio. The shear strength decreased as the temperature increased. The displacements at the maximum load were almost the same.

A series of tests for the out-of-plane bending were performed to verify (1) the influence of high temperature on the bending strength, (2) the influence of confined concrete between the partition steel plates on the bearing strength (Fig. 4.121), and (3) the influence of the restriction of the initial thermal deformation on the specimen (Fig. 4.122). The tests were conducted at several points of the transient temperatures (Fig. 4.117).

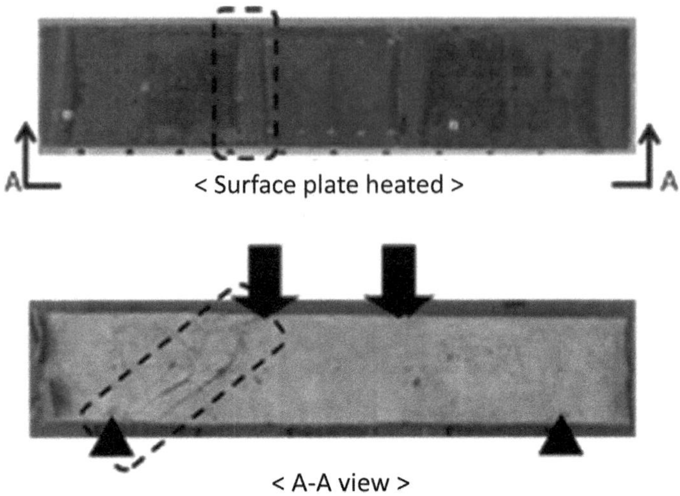

Fig. 4.119 Failure mode (tie bar type, 700°C) [83].

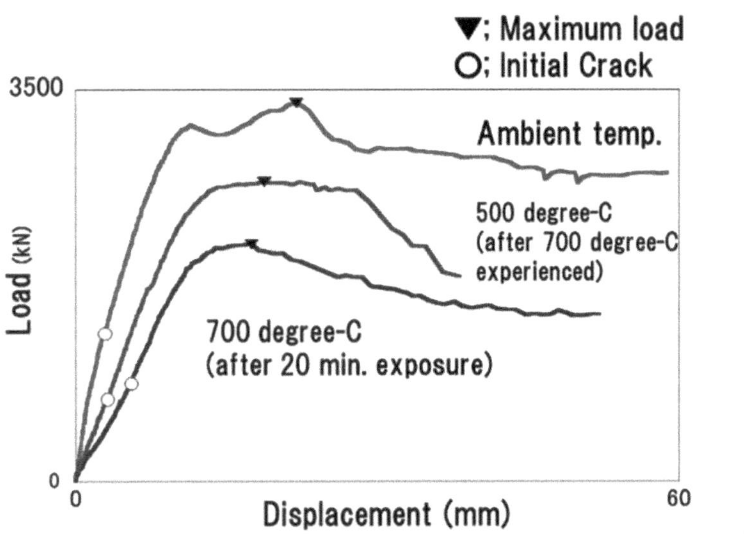

Fig. 4.120 Load-displacement curves of the SC beams [83] (tie bar reinforcement type)

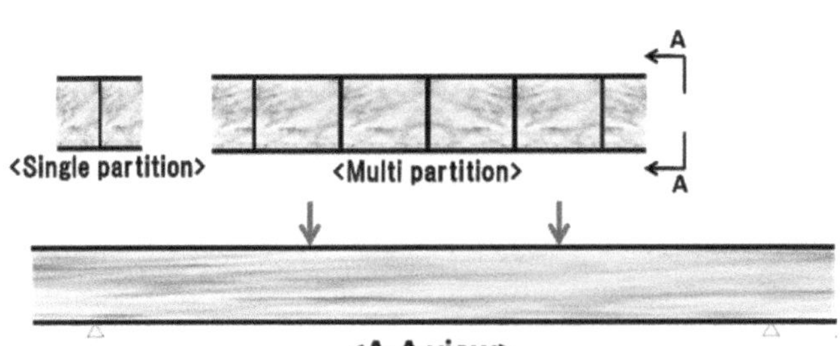

Fig. 4.121 Specimens for verifying the bounding effect of concrete [83].

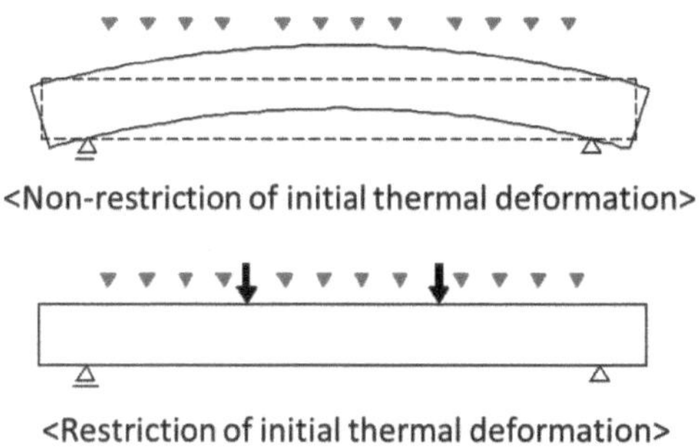

Fig. 4.122 Restriction of the initial thermal deformation [83].

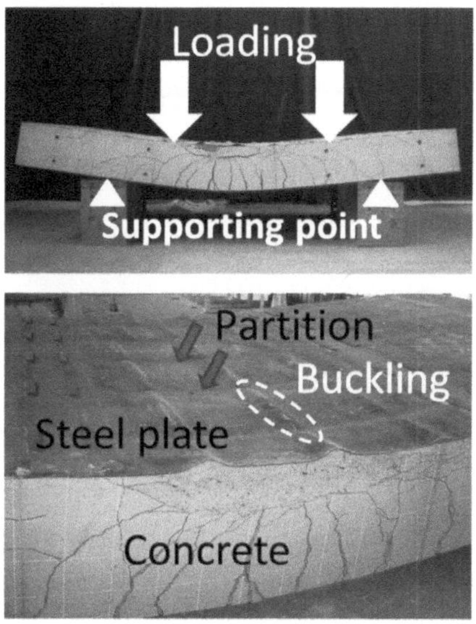

Fig. 4.123 Failure mode of the multipartition specimen [83].

Fig. 4.123 shows the failure mode of the multipartition-type specimen loaded under 500°C after the 700°C experience. Bending cracks near the tensile plate and crush of concrete near the compression plate were observed. Although the compression plate yielded and buckled, the surface crack of the compression plate for the SCCV boundary was not observed. Fig. 4.124 illustrates the bending load-displacement relationships. In this figure, "heated" means the loaded case under 500°C after the 700°C experience, and "restrict" means the restriction of the initial thermal deformation (Fig. 4.121). The loads of the single partition were modified to be five times to directly compare the load-displacement relationships of the single partition with those of the multipartition, which had five partition plates. From this figure, a significant effect of confined concrete between the partition plates was observed to raise the bearing strength (comparison of Case 1 with Case 2). The load-displacement relationships of the multipartition were stable, even in the large-displacement region, compared with the single partition. The decrease of the bearing strength by the high temperature was not clearly observed (comparison of Case 2 with Case 5). The restriction of the initial thermal deformation during heating has little influence on the bearing strength compared with the nonrestrict case (comparison of Case 1 with Case 3).

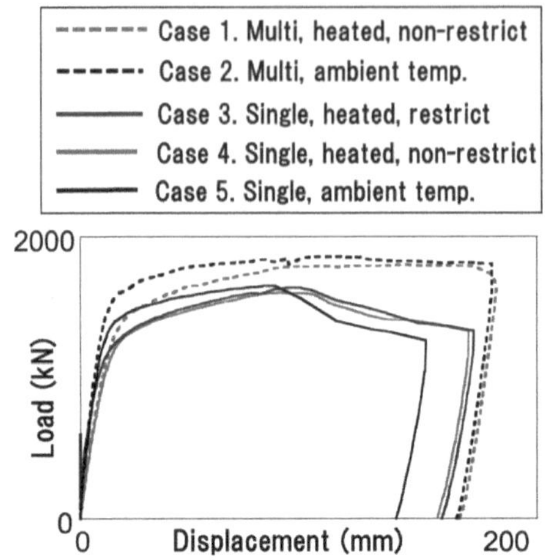

Fig. 4.124 Load-displacement curves [83].

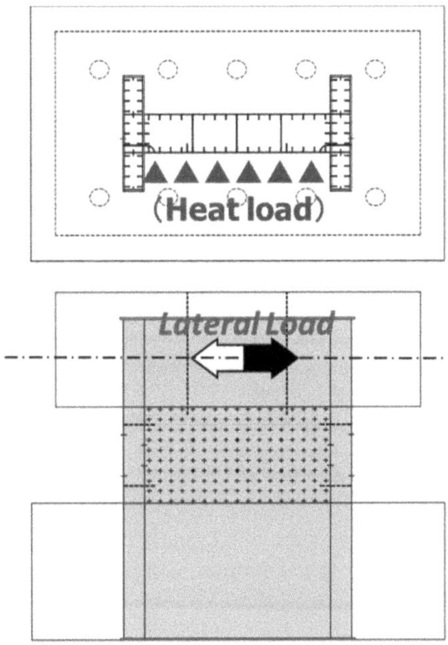

Fig. 4.125 Specimen of the in-plane strength test [83].

The lateral loading test shown in Fig. 4.125 was performed to estimate the seismic in-plane shear strength after the high-temperature experience due to the hypothetical sodium leakage combustion and verify the heated surface boundary condition. This specimen was approximately in one-fifth the scale of the tentative design of the SC structure. The static and cyclic in-plane forces at the top of the specimen were gradually loaded when the specimen was cooled down to the ambient temperature after the 700°C experience. The strain data on the surface plates were also logged.

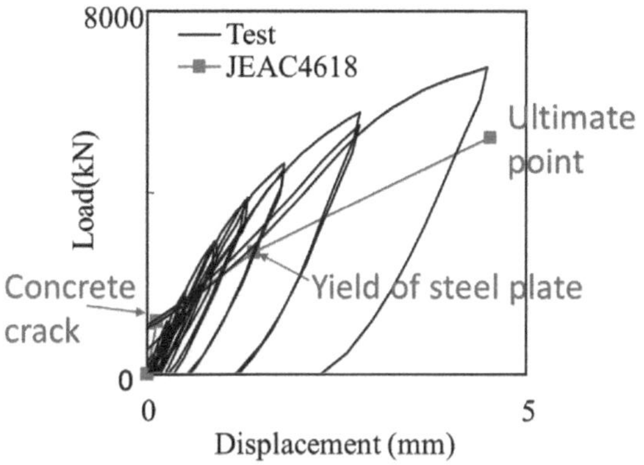

Fig. 4.126 Load-shear strain hysteresis curve [83].

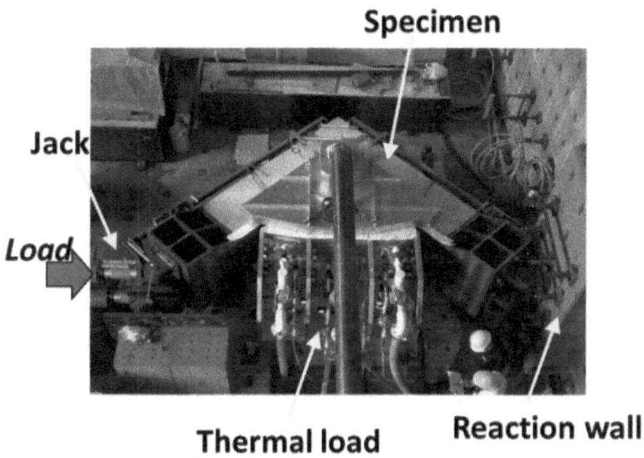

Fig. 4.127 Test specimen of the corner part [83].

Fig. 4.126 shows the load-shear strain curves compared with the skeleton curve under the ambient temperature described in JEAC4618 (ultimate shear strain: 6×10^{-3} rad.). The drastic load decrease was not observed, and the stable seismic characteristics were expected. Even after the high-temperature experience, the specimen maintained a higher strength than the JEAC4618 shear strength formula under ambient temperature. Although buckling in the surface plate occurred due to the heat increase as the lateral load increased, surface cracks were not observed. In other words, high temperatures have little influence on the in-plane shear characteristics and boundary function.

The bending tests were performed at the temperatures of 700°C and 500°C after the 700°C experience to verify the condition of the boundary surface and the load-deformation capability (Fig. 4.127). The corner part of the specimen was reinforced by tie bars and a thick plate. This specimen was about in half a scale of the tentative design of the SC structure.

Fig. 4.128 illustrates the failure condition of the specimen loaded after the 700°C experience. Buckling occurred on the heated plate. As a result of the magnetic particle testing, no crack was found around the buckling plate; hence, the boundary function was thought to be maintained.

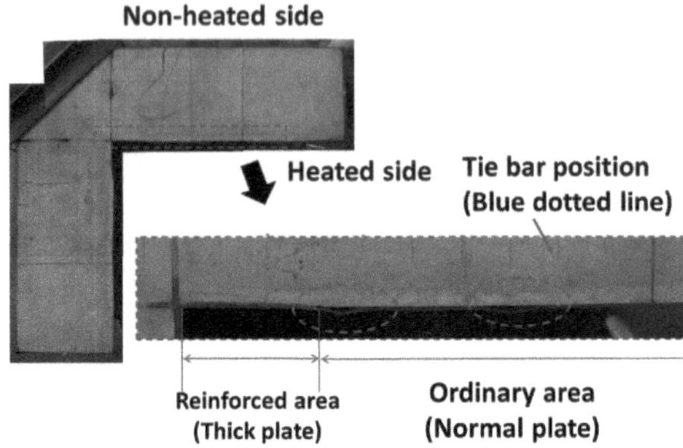

Fig. 4.128 Failure condition of the corner part [83].

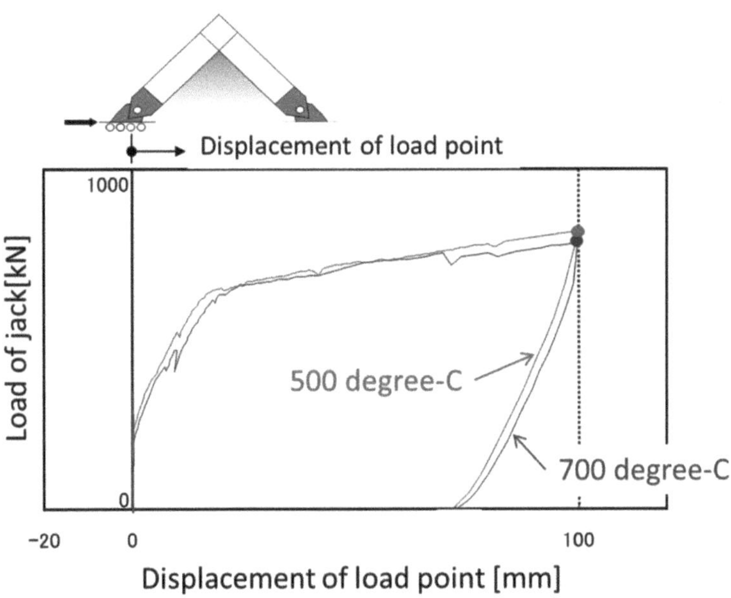

Fig. 4.129 Load-displacement curves of the corner part [83].

Fig. 4.129 shows the load-displacement curves. The stable deformation characteristics were observed without a drastic decrease of the load to end. The specific difference of the deformation depending on the temperature was not obtained in these tests. Based on the experimental test results, the SCCV is thought to be practical, even under high temperature: hypothetical sodium combustion. The experimental results were also used to make the analytical models for the evaluation of the structural and boundary integrities of the SCCV under high temperature.

(d) Numerical evaluation on the SCCV integrity

The SCCV is the final barrier against the leakage of radioactive nuclides. In the case of the extensive material deterioration and crack penetration caused by sodium combustion in

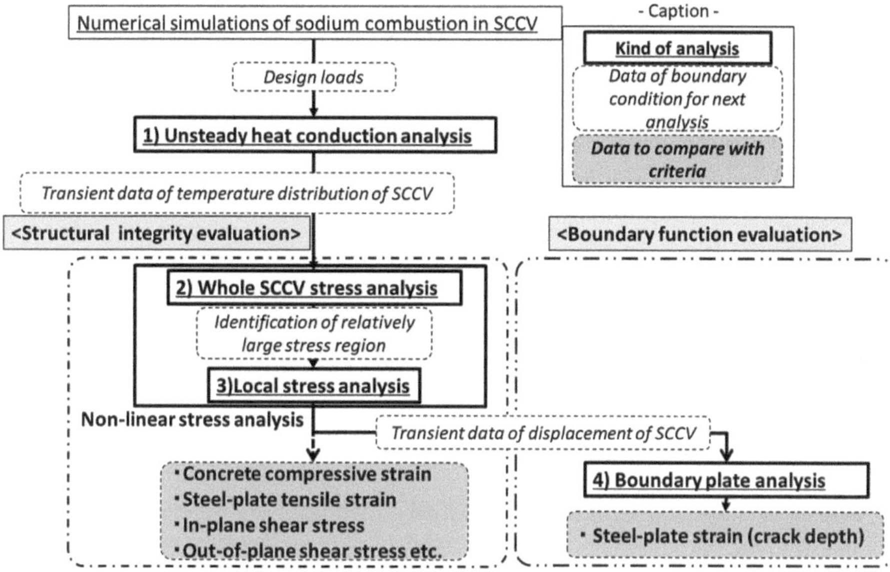

Fig. 4.130 Analytical evaluation flow [83].

the CV, a significant amount of hydrogen and steam could be generated by the sodium-concrete reaction, and the radioactive nuclides are finally released due to the CV boundary failure. Therefore, the structural and boundary integrities of the SCCV must be ensured against sodium leakage, even if all sodium boundaries are doubled in the JSFR design.

Under the high-temperature conditions caused by hypothetical sodium combustion, material deterioration occurs in the concrete and the steel plate, and buckling happens in the boundary steel plate. Therefore, a new SCCV design evaluation method is required to evaluate the structural integrity of the SC structure and the boundary function of the SCCV under the condition. The method consists of four analyses: (1) unsteady heat conduction analysis; (2) whole SCCV stress analysis; (3) local stress analysis; and (4) boundary plate analysis. Fig. 4.130 presents the analytical evaluation flow.

The SC structure is a composite structure consisting of a steel plate, concrete, studs, etc., and it is especially important to evaluate the temperature transient at the connecting part of the studs. The thermal resistance between the surface plate and concrete is set up based on the heating test results for these parts and considered in the unsteady heat conduction analysis. The model is validated by comparing the temperature distribution in the SC structure with that in the experimental results. Fig. 4.131 shows the experimental and analytical results of the concrete temperature transient in the out-of-plane bending tests, in which the specimen is heated up to 700°C. The analyses are expected to reconstruct the temperature transient with sufficient accuracy at each concrete depth from heating surface. The boundary conditions of the nonlinear stress analysis are based on the transient temperature distribution.

Through the nonlinear stress analysis of the SC structure, the bearing force and the deformation of the SC structure under high temperature are analyzed by evaluating the concrete compressive stress, steel plate tensile strain, in-plane shear stress, and out-of-plane shear strain to confirm the structural integrity. The nonlinear stress analysis of the SC structure consists of a whole SCCV analysis and a local stress analysis. As mentioned above, the

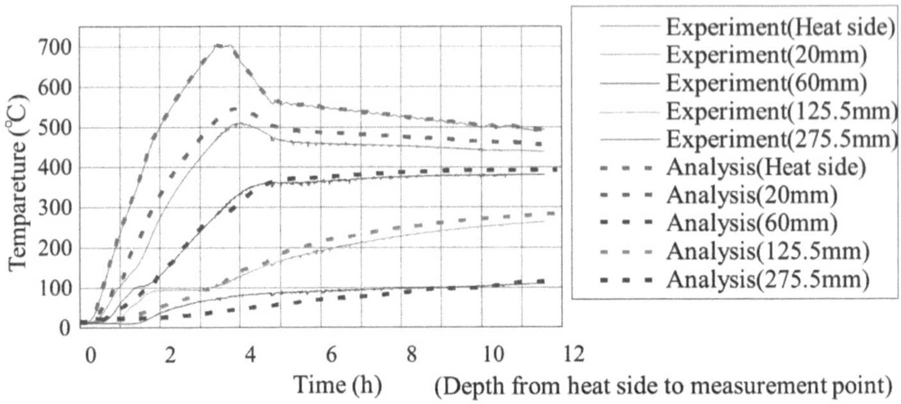

Fig. 4.131 Transient temperature distribution on the out-of-plane bending tests [83].

boundary conditions of the nonlinear stress analysis are based on the unsteady heat conduction analysis. The whole SCCV analysis aims to identify the dominant stress region of the steel plate and the concrete.

The whole SCCV stress analysis adopts shell elements for the SC wall. In the SC structure, local strain and stress are expected to be redistributed and mitigated by causing a local crack on concrete. Hence, the influence of the local strain and stress can be neglected at the whole plant analysis, and the selection of the adequate mesh size is rather important. The mesh size is set up based on the existing relating standards of the RC structure, etc.

The out-of-plane shear force under the hypothetical sodium combustion condition is analyzed by modeling a whole SCCV. As shown in Fig. 4.132, the shear force is relatively large at the region near the hypothetical sodium leakage portion. The other region only has a slight influence on this condition [87].

In the response of the whole SCCV result, the local analysis area is set considering the shear stress distribution. The local analysis adopts solid elements for the SC wall, shell elements for the steel plate, and bond elements for the stud, enabling the performance of a detailed analysis of the SCCV. The stud load-displacement is modeled based on the experimental results with temperature dependency.

The other temperature dependency of the material property is based on Eurocode 4 [88] and the material test data acquired in this R&D. For the constitutive modeling of concrete, the compressive and tensile stress-strain relations are based on the conventional models proposed in the past. The stress-strain relation of the steel plate is also based on the conventional bi-linear model. Regarding these modeling, their applicability is mainly verified through studies for the RC structure, etc. The nonlinear thermal stress analysis method is validated. For the out-of-plane bending strength, the analytical tolerance for the maximum bearing strength of various specimens under temperatures is evaluated in the range of 15% (Fig. 4.133).

A boundary plate (inner steel plate) analysis model based on the FEM is made for the boundary region: the stud, steel plate, and concrete. The boundary condition is the transient data of the SCCV displacement and temperature based on the unsteady heat conducting and nonlinear stress analyses. Fig. 4.134 shows the boundary plate analysis model that evaluates the plastic deformation with buckling and the creep deformation under high temperature.

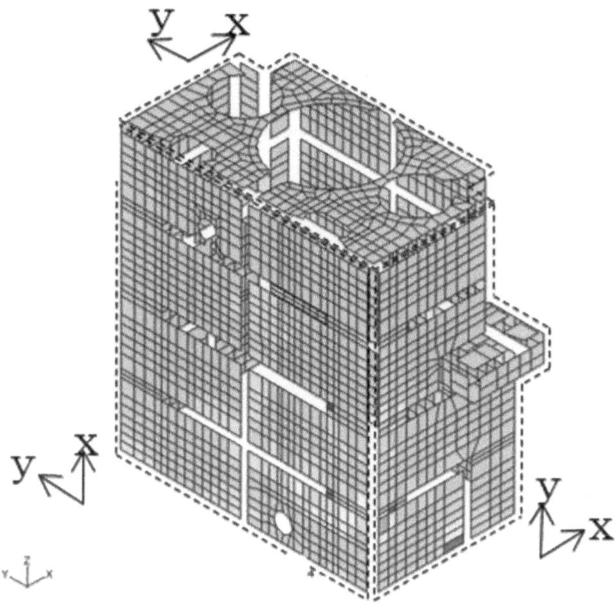

Fig. 4.132 Shear force in the hypothetical sodium combustion condition [83].

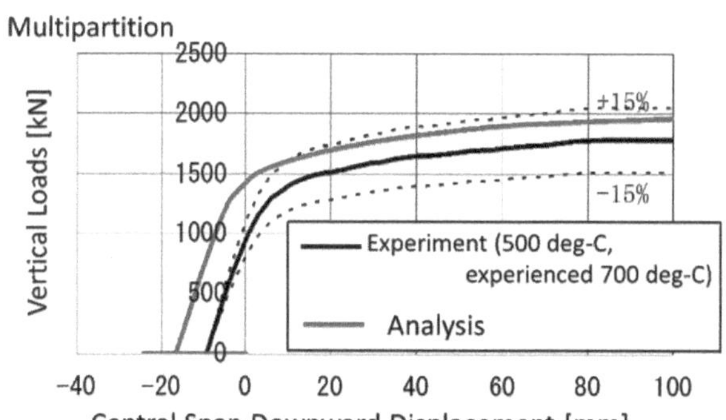

Fig. 4.133 Load-displacement relation of the out-of-plane bending strength [83].

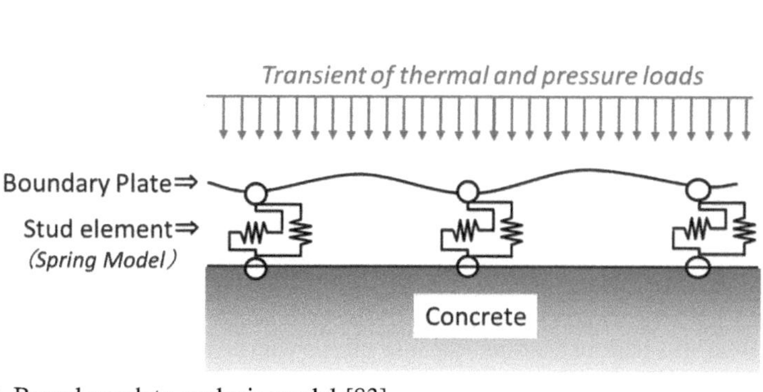

Fig. 4.134 Boundary plate analysis model [83].

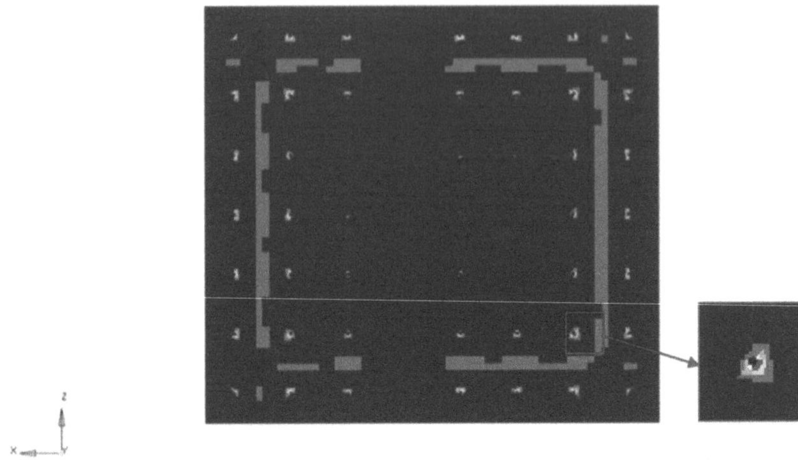

Fig. 4.135 Plastic strain of boundary plate in hypothetical sodium combustion condition [83].

One of the typical damage modes under the hypothetical sodium combustion condition is the crack from the stud joint part caused by the thermal buckling deformation of the boundary plate.

The boundary plate under hypothetical sodium combustion is analyzed by evaluating the steel plate strain and crack depth. The crack penetration on the steel plate must be prevented because the boundary plate shall prevent the sodium-concrete reaction.

A boundary evaluation result shows the inner steel plate plastic strain in the hypothetical sodium combustion condition (Fig. 4.135). The analytical result depicts that the strain is relatively large between studs due to thermal buckling and especially concentrated at the connecting parts of the stud. Consequently, a crack will occur at the inner steel plate, but this does not penetrate through the steel plate, preventing the sodium-concrete reaction. The steel plate integrity experiment also shows thermal buckling, but not a crack occurrence under the hypothetical sodium combustion condition [89]. Compared with the past experimental result, the analytical method could conservatively evaluate the crack depth.

Applying the combination of the four analysis methods, the SCCV integrity can be systematically evaluated under the hypothetical sodium combustion conditions. Each analysis method is validated by experimental results. The JSFR SCCV design is evaluated by the developed method, and the feasibility is confirmed to be introduced into the conceptual design of the demonstration reactor.

4.2.5.4 Seismic isolation design

(a) Concept and specifications

In Japan, the severe seismic condition is requested, and the primary components are exposed to severe seismic loading. The thin-walled structure of the SFR tends to be strongly affected by severe earthquakes. Therefore, the JSFR employs a seismic isolation system to mitigate the seismic force [90].

Fig. 4.136 shows a comparison of the floor response spectra with and without an isolation system. The peak of the floor response spectra with an isolation system shifts to a longer period and becomes lower than that without an isolation system. Consequently, the floor

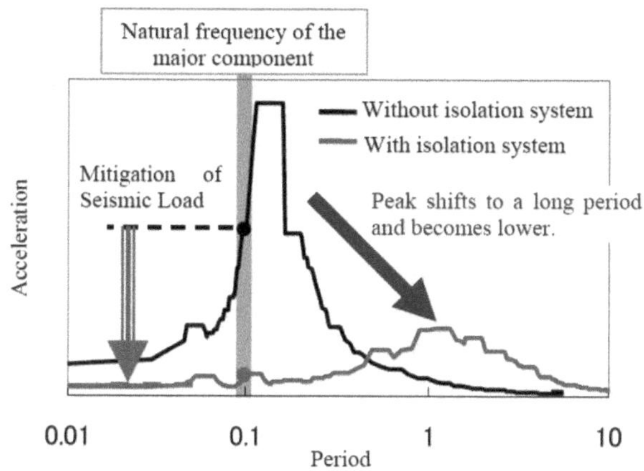

Fig. 4.136 Comparison of the floor response spectra with and without seismic isolation [90].

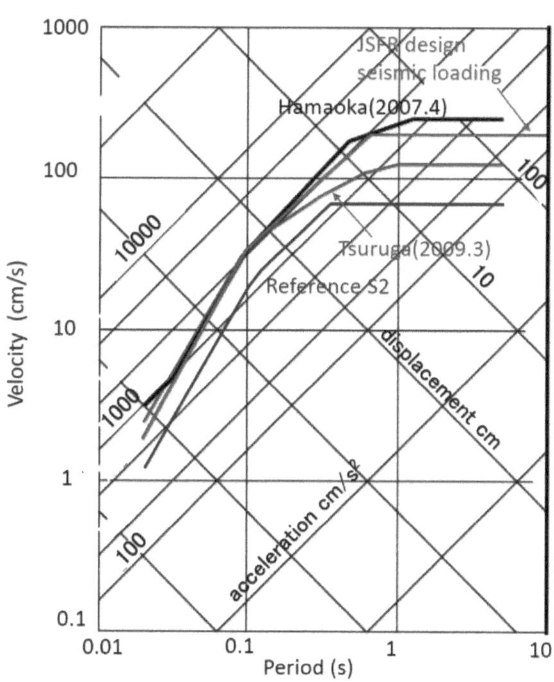

Fig. 4.137 Design seismic loading of the JSFR [90].

response spectra with a seismic isolation system in the range of the natural frequency of the primary component decrease compared to that without an isolation system. Consequently, the seismic loading to act on the primary component reduces.

Fig. 4.137 illustrates the design seismic loading of the JSFR and of the typical sites of the LWRs [91,92].

The Reference S2 wave is a design seismic loading of the initial design stage of the SFR. The design seismic loading is modified based on the Niigata-ken Chuetsu-oki Earthquake in

2007. The modified design seismic loading is approximately twice as large as Reference S2 in the natural period area (approximately 10 Hz, 0.1 s) of the principal equipment. In the natural period area (approximately 0.5 Hz, 2.0 s) of the seismic isolation system, the design seismic loading is approximately thrice as large as Reference S2. The ground condition is assumed to be on the hard rock ground (Vs = 1500 m/s), where it is difficult to expect the damping of the scattering attenuation and so on.

The design seismic loading of the JSFR cannot cover that of the typical site of LWRs in the shortest period (0.2 s). The seismic loading in the shortest period is not in the natural period area of the principal equipment, although that loading influences the amplitude of the acceleration of the seismic ground motion. Therefore, the seismic load in the shortest period (0.2 s) does not influence the equipment. The design seismic loading of the JSFR covers the seismic loadings of the typical site of the LWR in the natural period area (approximately 10 Hz, 0.1 s) of the primary equipment. Moreover, the ground condition is the hard rock ground; hence, the design seismic loading is not inferior to the seismic loadings in the LWR.

The dynamic seismic analysis evaluation in the vertical direction is demanded in the "Regulatory Guide for Reviewing Seismic Design of Nuclear Power Reactor Facilities" published by the Nuclear Safety Commission of Japan. In this design, the vertical seismic loads are assumed to be 2/3 times that of the horizontal seismic loads.

Fig. 4.138 displays the reactor building and the seismic isolation system of the JSFR. The seismic isolation system is installed between an upper base mat and a lower base mat. Fig. 4.139 depicts the floor response spectra at the reactor system setting position. The broken line in the figure shows that of the previous seismic isolation system. The response of the natural period area of the reactor system in the vertical direction is approximately 8 G because the peak is near the natural period area of the reactor system. Therefore, the peak of the floor response spectra must be removed from the natural period area of the reactor system. The response in the horizontal direction must also be decreased as much as possible because the seismic loads are made more severe.

Accordingly, the advanced seismic isolation system for the SFR, which can decrease the response in the horizontal and vertical directions, is examined. The features of the advanced seismic isolation system for the SFR are as follows:

- mitigation of the horizontal seismic force by thicker laminated rubber bearings with a longer period and improvement of the damping performance by adopting an oil dumper instead of steel bar dumpers;
- mitigation of the vertical seismic force by thicker laminated rubber bearings with a longer period.

Here, the important concept of the advanced seismic isolation system for the SFR is to make a practicable modification of the previous technologies.

As a result, the advanced seismic isolation system for the SFR has the following specifications:

- The natural frequency (period) in the horizontal direction is 0.29 Hz (3.4 s).
- The natural frequency (period) in the vertical direction is 8.0 Hz (0.125 s).

The diameter of a laminated rubber bearing can be set to become 1600 mm. The thicker laminated rubber bearing for satisfying the abovementioned specifications has a layer thickness of approximately 30 mm and approximately 10 layers. The first shape factor of this laminated rubber bearing is approximately 10. The damping ratio of the oil dumper is approximately 40% because the relative displacement is suppressed to 350 mm. Fig. 4.138 shows the advanced seismic isolation system for the SFR. The solid line in

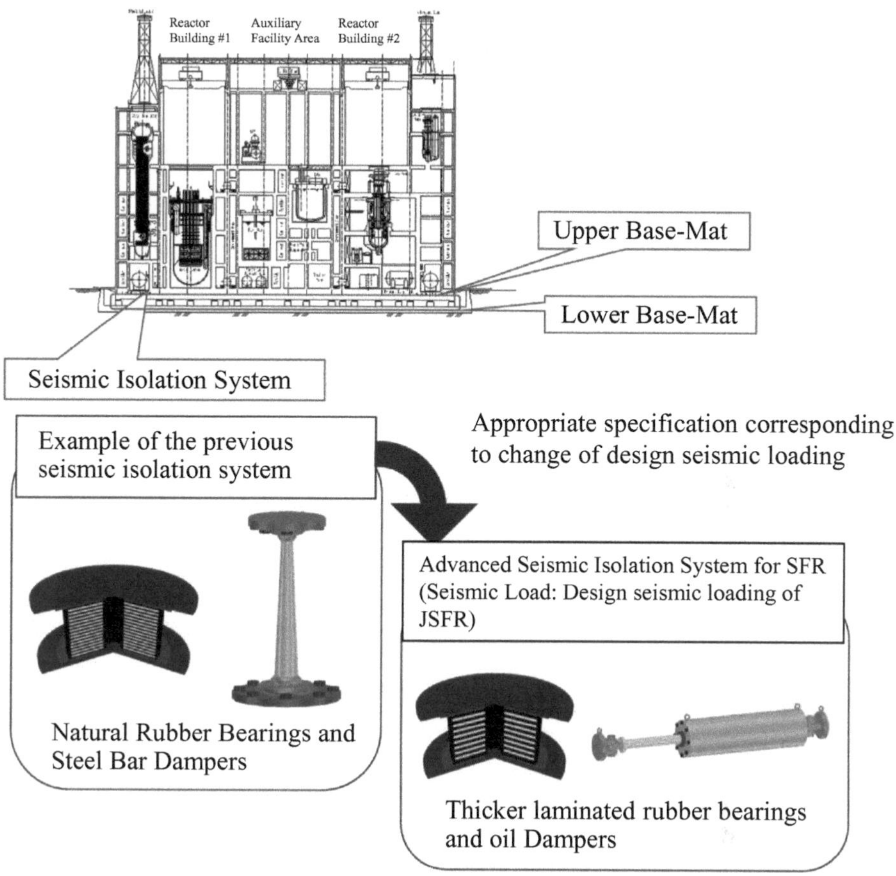

Fig. 4.138 JSFR seismic isolation system [90].

Fig. 4.139 depicts the floor response spectra of the advanced seismic isolation system for the SFR. In the natural period area of the nuclear reactor system, the response of the horizontal direction has been decreased to approximately 0.75 times, while that of the vertical direction has been decreased to approximately 0.4 times.

(b) Evaluation on the reactor structure

Table 4.21 shows the rough estimate evaluations of the seismic loads in a reactor system. The reactor vessel thickness and diameter are 60 mm and 10.7 m, respectively [93]. The evaluation items are the reactor vessel buckling, jumping amount of a fuel assembly, and reactivity input, which are critical in the previous seismic evaluation. The evaluation values of the advanced seismic isolation system for the SFR are roughly estimated from the ratio of the response values of the advanced seismic isolation system for the SFR and the previous seismic isolation system in the first natural period of the reactor system. The evaluation of the jumping amount of a fuel assembly is satisfied by adopting the advanced seismic isolation system for the SFR. The margin of the evaluations of the advanced seismic

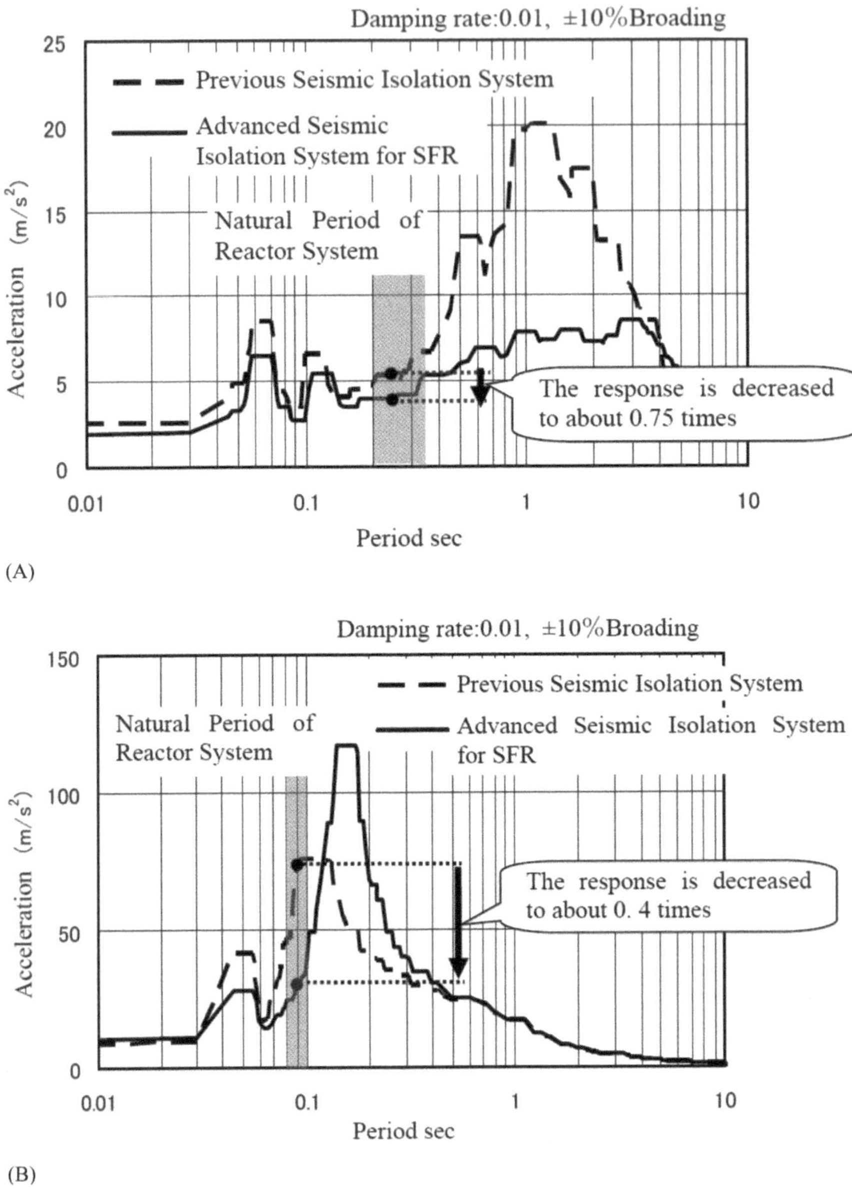

Fig. 4.139 Floor response spectra at the reactor system location [90]. (A) Horizontal direction. (B) Vertical direction.

isolation system for the SFR increases more than that of the previous seismic isolation system. The change of the values of the seismic evaluation is expected in advancing the reactor system design. However, the design margin will certainly increase if the advanced seismic isolation system for the SFR is adopted.

Table 4.21 Seismic evaluation on the reactor structure [90].

Isolation system			A-SIS[b]	P-SIS[c]
Thickness of reactor vessel		mm	60	
Buckling	① Evaluation value		3.0[d]	1.5
	② Design safety rate		1.5	
	Design margin(①/②)		2.0	1.0
	Evaluation[a]		◎	○
Jumping amount of fuel assembly	③ Evaluation value (Velocity of reactor core support structure)	m/s	0.42[d]	0.76
	④ Design limitation value	m/s	0.58	
	Design margin(④/③)		1.4	0.76
	Evaluation[a]		◎	×
Reactivity input	⑤ Reactivity	¢	39[d]	72
	⑥ Design limitation value	¢	100	
	Design margin(⑥/⑤)		2.5	1.4
	Evaluation[a]		◎	◎

[a] Margin ≧ 1.3:◎, 1.3 > Margin ≧ 1.0:○, 1.0 > Margin ≧ 0.8:△, 0.8 > Margin:×.
[b] A-SIS: advanced seismic isolation system for SFR.
[c] P-SIS: previous seismic isolation system.
[d] The evaluation values of advanced seismic isolation system for SFR are estimated roughly from the ratio of the response values of advanced seismic isolation system for SFR and previous seismic isolation system in the first proper period of RV.

(c) Experiments on the laminated rubber bearings

A basic characteristic test with a 1/8 reduced scale model (diameter: 200 mm) is performed to confirm the feasibility of the thicker laminated rubber bearings. Table 4.22 presents the shape of the characteristic test model. The horizontal and vertical direction static loads are added in the examination.

Fig. 4.140 illustrates one example of the hysteresis curve in the horizontal direction, which is confirmed to be a steady hysteresis curve. Fig. 4.141 shows the relation between the first shape factor and the shear modulus. The shear modulus has the same characteristics expected in the design (i.e., approximately 0.54 N/mm^2). Therefore, the horizontal characteristics (rigidity) prescribed in the design are expected to be obtained in a real machine.

Fig. 4.142 shows the relation between the first shape factor and the damping rate. The damping rate of the previous laminated rubber bearings is approximately 2%, while that of the thicker laminated rubber bearings is approximately 3% or 4%. The thicker laminated rubber bearings are expected to have higher damping than the previous laminated rubber bearings.

Fig. 4.143 depicts the relation between the first shape factor and the longitudinal modulus. The longitudinal modulus rules the vertical characteristic of the rubber bearing. The longitudinal modulus in the static test cannot be reproduced when the formula introduced by the "Technical Guidelines on Seismic Base Isolation System for Structural Safety and Design of Nuclear Power Plants JEAG 4614-2000" [94] is applied. On the contrary, the longitudinal modulus in the static test can be reproduced when the formula that considers the hole in the rubber bearing [95] is applied, confirming the use of this formula for the rubber bearing design.

Fig. 4.144 presents the relation between the first shape factor and the horizontal linear characteristic limit. The strain γ of the horizontal linear characteristic limit is approximately 220%

Table 4.22 Shape of the characteristic test model [90].

Outside diameter	Inside diameter	First shape factor
D_o	D_i	S1
mm	mm	–
200	40	8.0
		10.0
		12.0
		14.0
	20	11.3
		13.5
	60	8.8
		10.5

Outside diameter D_o	Inside diameter D_i	First shape factor S1
mm	mm	-
200	40	8.0
		10.0
		12.0
		14.0
	20	11.3
		13.5
	60	8.8
		10.5

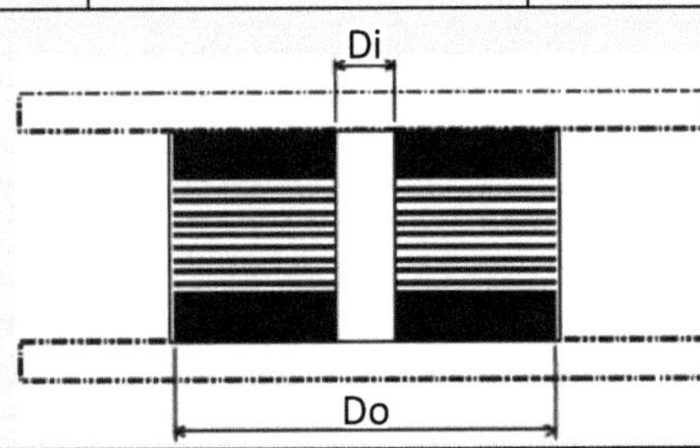

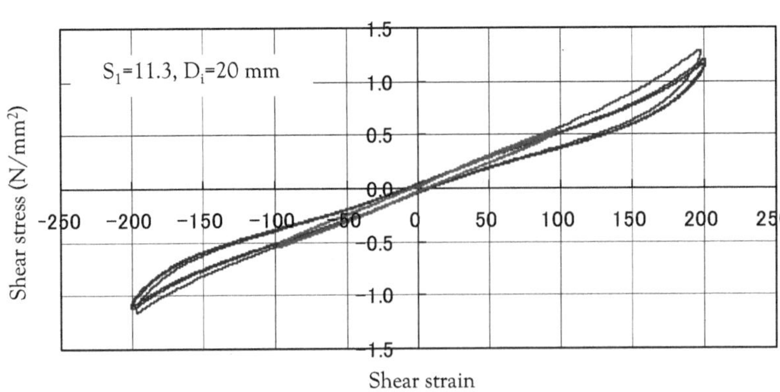

Fig. 4.140 Hysteresis curve of examination shape A-5 (shear strain γ = 100%, 200%, pressure of vertical σ = 6 N/mm^2) [90].

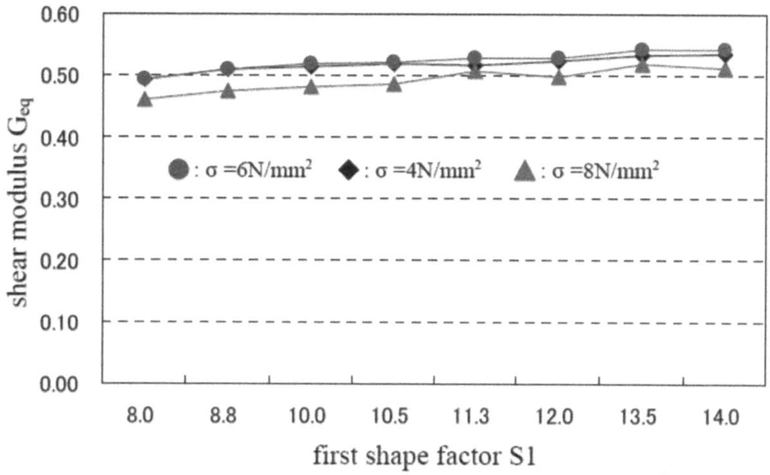

Fig. 4.141 Relation between the first shape factor and the shear modulus (shear strain $\gamma = 200\%$) [90].

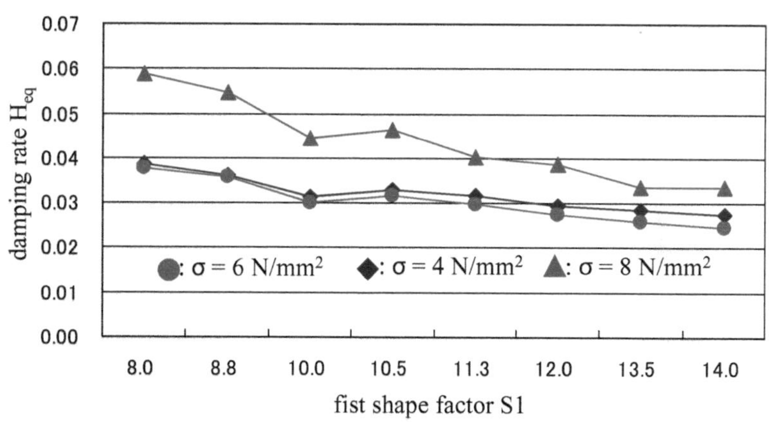

Fig. 4.142 Relation between the first shape factor and the damping rate (shear strain $\gamma = 200\%$) [90].

or 230% for the first shape factor $S1 = 12$ (i.e., full-scale model). The strain of the design-basis earthquake of this rubber bearing is approximately 110% (i.e., the quotient of 350 mm divided by the total thickness of the rubber layer, which is 325 mm). Therefore, the strain of the horizontal linear characteristic limit becomes 1.5 times more than that of the design-basis earthquake. The buckling was not caused, and the characteristic is stable in all the examinations. As a result, the possibility of the application of thicker laminated rubber bearings is confirmed.

4.2.6 In-service inspection and repair

Maintainability must be considered in the conceptual design stage as the maintenance program may significantly impact component design. In the design study of a commercial JSFR, an in-service inspection (ISI) program was tentatively defined

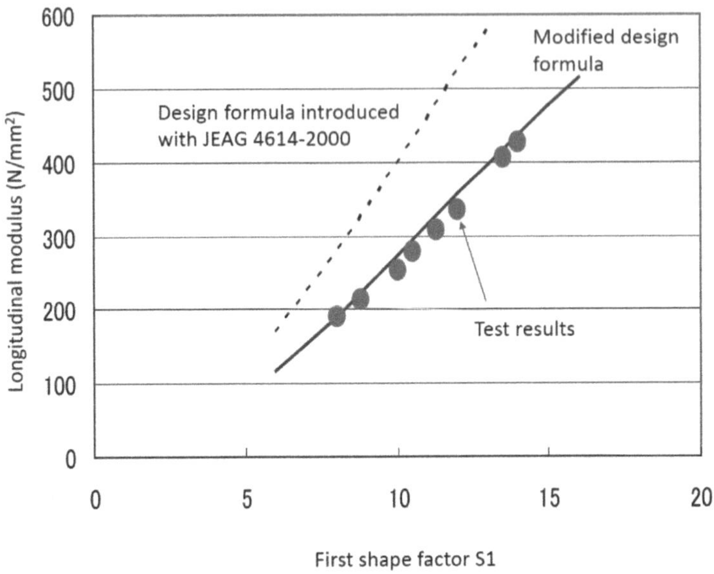

Fig. 4.143 Relation between the first shape factor and the longitudinal modulus [90].

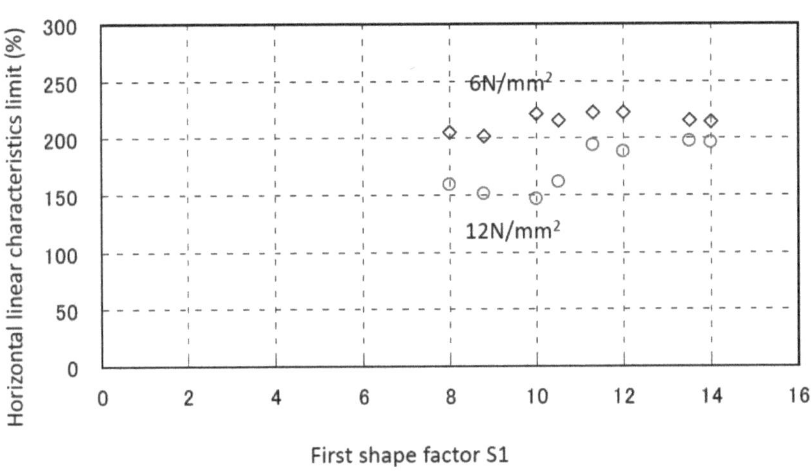

Fig. 4.144 Relation between the first shape factor and the horizontal linear characteristic limit [90].

considering the characteristics of the sodium-cooled reactor, the applicable inspection technology, and LWR practice, among other factors. Moreover, as a part of the repair program, anticipated failures were listed and categorized by the expected frequency and required repair level. The JSFR design study was conducted to determine the feasibility of these ISI and repair programs, in which access routes for inspection using remote sensing devices and dismantling and reassembling procedures for repair were investigated [96]. The development of inspection devices applicable to SFRs is undergoing.

4.2.6.1 ISI program

At the beginning of the 21st century, inspection was classified into regulatory and voluntary ISI in LWR practice, in Japan. The regulatory ISI program has been developed from a safety perspective, whereas the voluntary ISI program primarily concerns property protection. Sodium-cooled reactors have certain maintenance drawbacks, such as chemical activity, opacity, and high temperature (200°C) even during reactor shutdown. Alternatively, low coolant pressure is advantageous for continuously monitoring sodium leakage, and sodium has good compatibility with the structure's material. The ISI program was tentatively defined for JSFR's design study.

(a) Regulatory ISI

Because Japan lacked an ISI standard for general sodium-cooled reactors, JSFR created the following tentative regulatory ISI program. To ensure effective maintenance, maintenance importance for each component has been determined based on the light water reactor's safety classification. Components that should be inspected have been selected based on maintenance importance. Inspection methods and frequency have been set based on the ISI program of foregoing "Monju," the ISI code for the light water reactor (Japan Society of Mechanical Engineers, JSME S NA1) [97] and that for the liquid metal cooled reactor (American Society of Mechanical Engineers, ASME section XI division 3). ISI's basic concept is as follows. In the event of a coolant leak accident, a coolant leak in sodium-cooled reactors does not affect core cooling as long as the RV coolant level is maintained. JSFR has complete double-walled sodium boundaries (RV, piping, and IHX), and the sodium level could be secured even in primary sodium failure. Therefore, from the safety perspective, continuous leakage monitoring (CM) is applied to the primary sodium boundary as ISI. Leak-before-break (LBB) is the condition of applying CM for ISI of the sodium boundary, and the LBB applicability of piping and vessels is examined. Regulatory ISI items of the main parts are listed as follows.

- Primary sodium boundary: CM
- Guard vessel and outer tube: leak rate test
- Reactor core support structure: visual test
- Sodium-water boundary (steam generator heat exchanger tube): volumetric test and CM

(b) Voluntary ISI

The voluntary inspection program has also been developed from the perspective of property protection and the short-time operating experience of sodium-cooled reactors. For instance, a visual test is applied to the RV surface as voluntary inspection to detect minor leakage below the lower limit of the sodium leakage detectors used for CM. The ISI program for major components based on the foregoing is shown in Table 4.23.

4.2.6.2 Repair program

There is no standard for repair requirements on general sodium-cooled reactor components at the conceptual design stage. In this study, anticipated failures have been listed and categorized according to their frequency and required repair level.

Possible events require Repair Level 1 as follows. Each component is required to be easily repaired in a short period. Access and maintenance spaces are considered in the component structural design. For some minor components, complete component

Table 4.23 ISI program for major components [96].

Component		Part	Maintenance importance	Classification	Method
Reactor vessel	Reactor vessel	Primary boundary	A(PS-1)	Regulatory Voluntary	CM[a] VTM-2[b] Volumetric test[c]
		Cover gas boundary	B(PS-2)	Regulatory	CM
	Reactor core support structure	Support cone	A (PS-1)	Regulatory	VTM-3[d]
	UIS	Instrumentation well	A (MS-1)	Voluntary	VTM-3
	Guard vessel	Guard vessel	A (MS-1)	Regulatory	Leak rate test
Pump-integrated IHX	Upper plenum	Primary boundary	A (PS-1)	Regulatory Voluntary	CM VTM-2
	Tube bundle	Heat exchanger tube	A (PS-1/ MS-1)	Regulatory	CM
	Pump	Upper bearing	(PS-3)	Voluntary	Overhaul
Primary piping		Primary boundary	A (PS-1)	Regulatory Voluntary	CM Volumetric test[c]
		Outer piping	A (MS-1)	Regulatory	Leak rate test
Steam generator	Upper body	Secondary boundary	B (PS-2)	Regulatory	CM
	Tube bundle	Heat exchanger tube	B (PS-2/ MS-2)	Regulatory	CM Volumetric test

[a] CM: continuous monitoring.
[b] VTM-2: visual test (signs of leakage).
[c] For welded joint (high stress parts).
[d] VTM-3: visual test (deformation of equipment).

removal and overhaul in the reactor building are considered to be applied (examples: SG tube plugging and refueling machine difficulty).

Unlikely incidents that occur about once or more over the plant's lifetime require Repair Level 2 as follows. Repair processes are roughly investigated. This category does not require an easy repair process or short plant outage (examples: not through the RV crack and through the IHX tube crack).

Hypothetical events require Repair Level 3 as follows. Repair is not investigated in this design stage (example: through the RV crack).

Examples of the repair program based on the foregoing are shown in Table 4.24.

Table 4.24 Example of repair program [96].

Component	Part	Event	Repair level	Method
Reactor vessel	Primary boundary	Crack (large)	3	Replacement
	Instrumentation well	Crack (large)	2	Replacement
	Fuel handling machine	Failure	1	Replacement
Pump-integrated IHX	Heat exchanger tube	Crack/ denting (through)	2	Plugging (in-place)
Primary piping	Primary boundary	Crack (large)	3	Replacement
Steam generator	Heat exchanger tube	Crack/ denting (not through)	1	Plugging (in-place)

4.2.6.3 *Design accommodation*

According to the proposed ISI&R program, the JSFR component design has been improved. JSFR adopts a simple configuration of in-vessel structures, such as core support structures. This feature reduces the points to be inspected, and it is relatively easy to access inspection targets using remote inspection equipment. For example, the support skirt, which supports the core support structure, is designed to be simple to make access to welding lines easier (Fig. 4.145). A pair of "Y junctions" made of single-piece forgings minimizes the number of welding lines and provides sufficient space for the inspection equipment around the welding lines [98].

Accessibility for inspecting the in-vessel structure is also a critical aspect of the JSFR design. A loop-type reactor has essentially fewer inspection targets in the RV than a pool-type reactor. In terms of taking advantage of the loop-type reactor, JSFR considers installing access holes in the roof deck to easily insert inspection equipment into the RV. The access holes are also provided in the partition plate of the plena and support cone to allow access to the structures in the middle and lower plena. Fig. 4.146 shows the access holes.

The under-sodium area monitor (USAM) is being developed to reduce the size and weight of the transducer and for the pursuit of higher resolution. Until recently, the resolution of the sensor required to conduct VTM-3 in sodium has been obtained through the element test. To apply the USAM system for in-vessel structure inspection, a transport system is needed to deliver the USAM sensor to each structure submerged in sodium. Thus, an under-sodium vehicle, which is driven by electromagnetic pumps and can freely move in the RV, was developed. Experiments of inspection techniques under sodium both underwater and under-sodium were conducted.

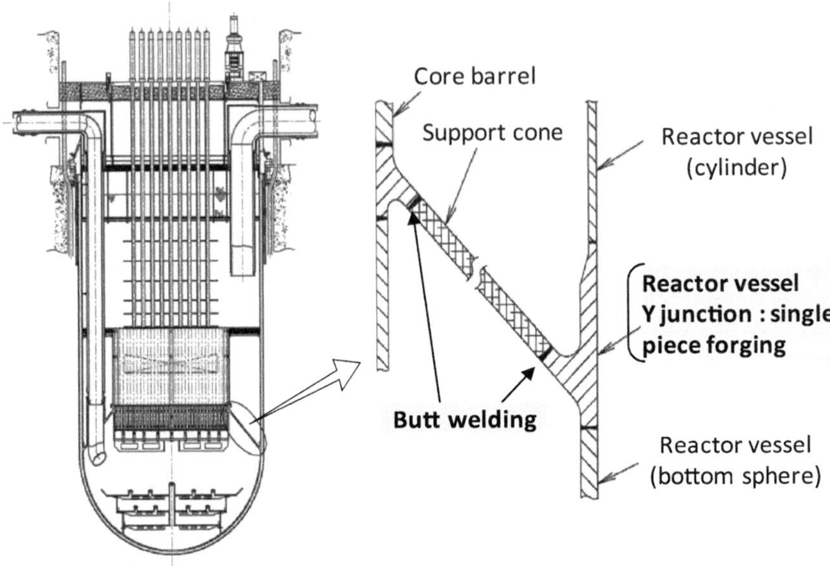

Fig. 4.145 Sketch of core support skirt [96].

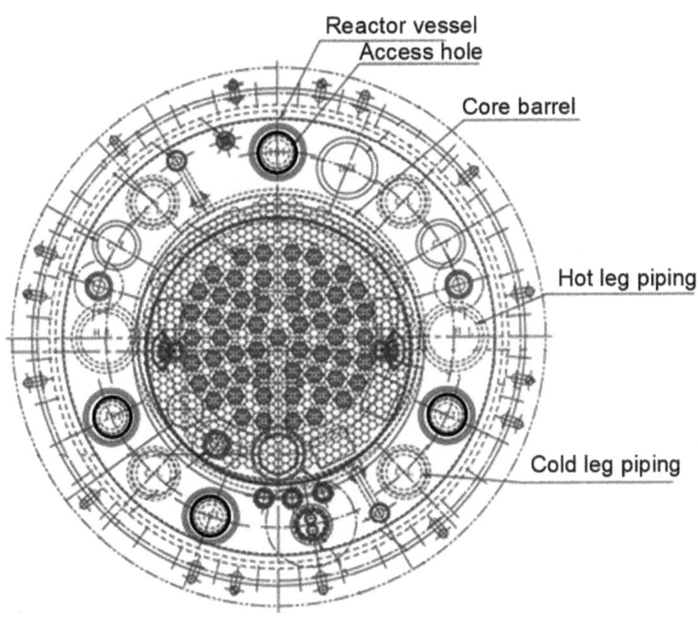

Fig. 4.146 ISI access holes [96].

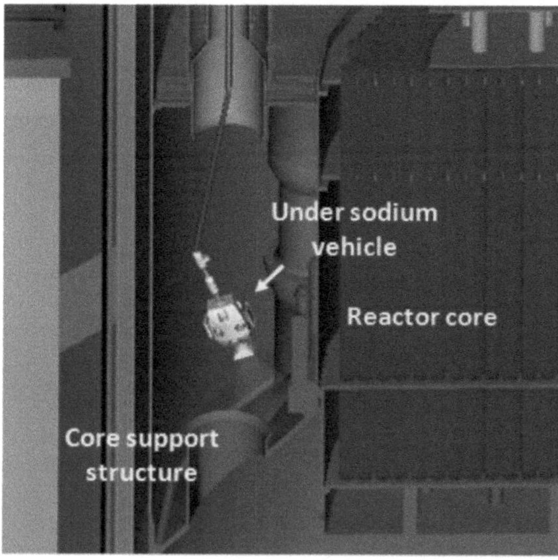

Fig. 4.147 Image of under-sodium viewing system [96].

Fig. 4.147 shows the under-sodium viewing system in the RV [99]. Continuous monitoring will be accomplished via the use of an advanced laser sodium leak detector (LLD). LLD could be more sensitive and avoid false alarms than the conventional ones [100]. Several sampling points are installed between the RV and guard vessel for the continuous monitoring of sodium leakage.

Straight double-walled tubes with a length of 30 m will be used as the heat exchanger tubes of the steam generator (SG). Remote field eddy current test, ultrasonic test (UT), and other volumetric inspection techniques are being developed for double-walled tubes. Experimental results have demonstrated that these inspection methods can detect damage to the outer and inner tubes. However, the inspection speed of the UT sensor requires improvement.

The JSFR IHX is combined with a primary main pump, with the pump shaft located at the component's center. The pump shaft can be removed to allow for the inspection and repair of the primary pump, IHX tubes, and other components. Fig. 4.148 shows the cask of the IHX-pump shaft. Devices for inspecting the IHX tubes and primary sodium boundary are planned to be developed.

Workspace for inspection and maintenance has been factored into component and building design. For instance, the operating floor has been designed to allow for simultaneous primary pump maintenance, RV ISI, and refueling. Moreover, the reactor building houses the inspection devices that are stored in special casks during normal plant operation.

ISI&R on sodium-cooled reactors is conducted using remote control devices considering high coolant temperature and radiation dose. For instance, the radiation dose at the time of removing the primary pump shaft has been roughly evaluated and it has been confirmed that the evaluated radiation dose satisfies the work environment requirements.

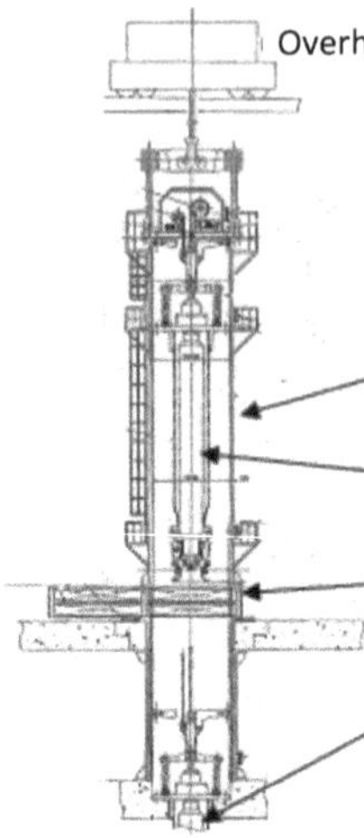

Fig. 4.148 Cask for pump shaft [96].

Table 4.25 Requirements on ISI device development [96].

	Device	Contents
RV	Under-sodium viewing system[a]	VT-3, UT
	Volumetric test for RV	EMAT[b], CCD[c]
	Visual test in cover gas atmosphere	CCD
SG	Volumetric test for double-walled tube	RF-ECT, UT, Guide Wave
IHX	Visual test for IHX tube	CCD

[a] Under development.
[b] EMAT: electromagnetic acoustic transducer.
[c] CCD: charge-coupled device camera.

The development of inspection and repair devices is critical for realizing the ISI&R program, as previously stated. The requirements for developing ISI&R devices are provided in Table 4.25.

4.2.7 Selection of demonstration reactor specifications

4.2.7.1 Requirements on demonstration reactor

A conceptual design study for the demonstration reactor was conducted for two power plants with outputs of 500 and 750 MWe, respectively [101]. The design study for the demonstration reactor in Japan includes various specific requirements that should be considered.

- The commercial reactor should be a 1500 MWe-class advanced loop-type sodium-cooled fast reactor (JSFR).
- The demonstration reactor should involve all innovative technologies that will be adopted for the commercial reactor.
- The demonstration reactor's output power should be sufficiently high to validate the performance of the fuel, components, structure, and system of the commercial reactor.
- The demonstration reactor's output power should be sufficiently high to obtain prospects for fabricating a commercial reactor.

These requirements stem from discussions with utilities that will be responsible for commercialization. To satisfy these conditions, the demonstration reactor concept must be similar to that of the commercial JSFR but with a smaller size.

4.2.7.2 Reactor structure

A 10.7-m reactor vessel diameter is realized for the commercial reactor using a compact design of the reactor core and a single rotating plug structure. The reactor diameter of the 750 MWe reactor was estimated at 8.9 m (Fig. 4.149). The coolant volume per power of the 750 MWe reactor is much larger than that of the commercial reactor

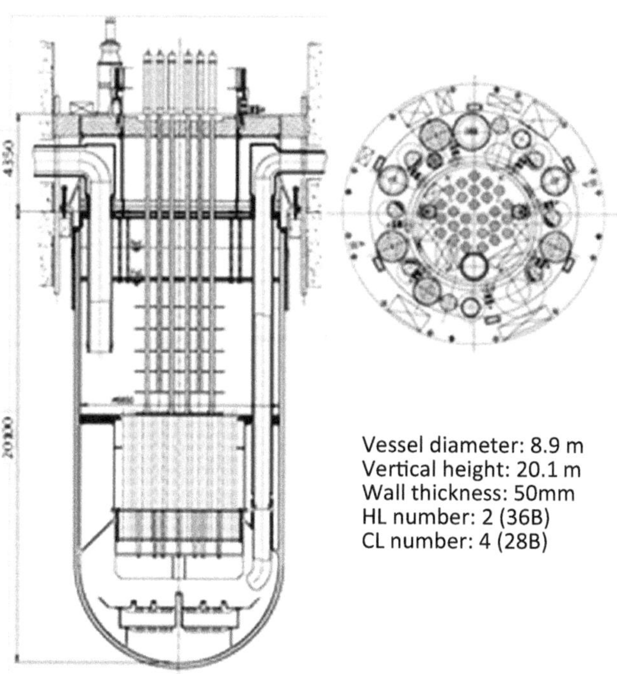

Fig. 4.149 Sketch of reactor structure of 750 MW [101].

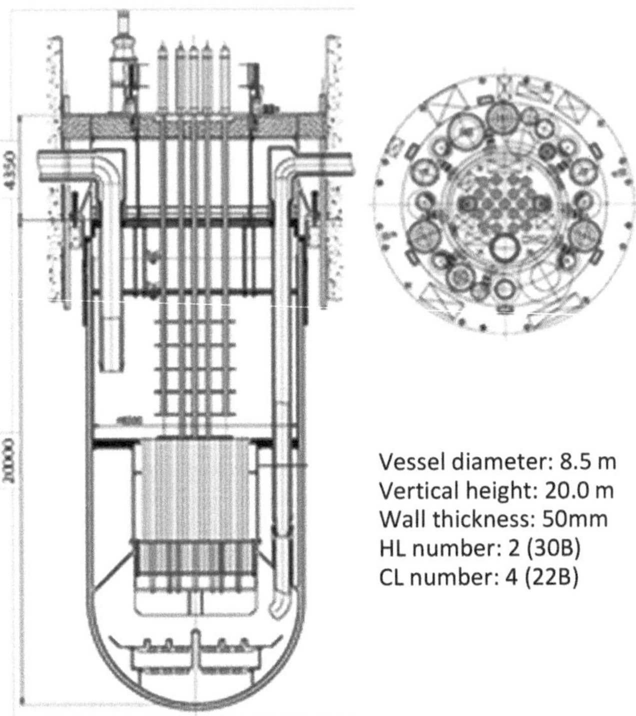

Fig. 4.150 Sketch of reactor structure of 500 MW [101].

because of the limitations of component arrangement at the reactor's upper deck. Thus, the coolant velocity in the vessel is smaller than that of the commercial reactor. For the 500 MWe reactor with an 8.5-m diameter, the coolant volume per power increases as shown in Fig. 4.150. Several design checkpoints were compared, such as gas entrainment characteristics at the liquid surface, thermal-stratification tendencies during transients, and flow-induced vibration of inner structures. The lower velocity condition makes it rather easy to establish the design in a reactor vessel. This implies that the commercial reactor is better demonstrated using a higher velocity. Therefore, in terms of technological demonstration for the commercial plant, the 750 MWe reactor outperforms the 500 MWe reactor.

Fig. 4.151 shows a CFD result of the velocity profile near the liquid surface of the 750 MWe reactor. The maximum velocity is reasonably suppressed below 0.2 m/s using a dipped plate in the reactor vessel. Therefore, it is expected to avoid gas entrainment at the liquid surface. Fig. 4.152 shows the temperature profile after scram. A thermal-stratification phenomenon was observed at the hot-leg inlet level in the upper plenum. However, the transient vertical temperature gradient is expected to be within the acceptable range for vessel design.

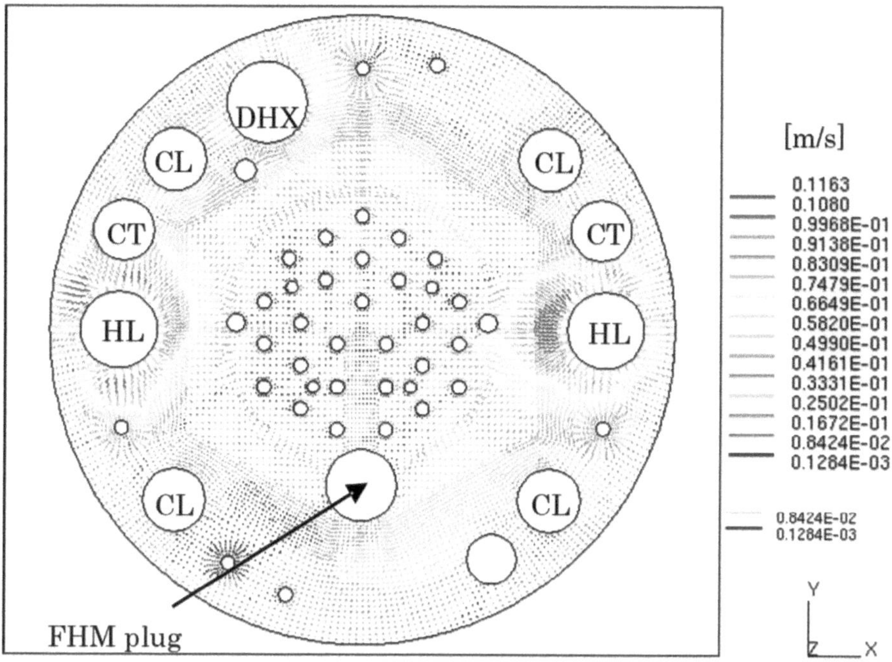

Fig. 4.151 Velocity profile near the surface of the 750 MWe reactor (100% power condition) [101].

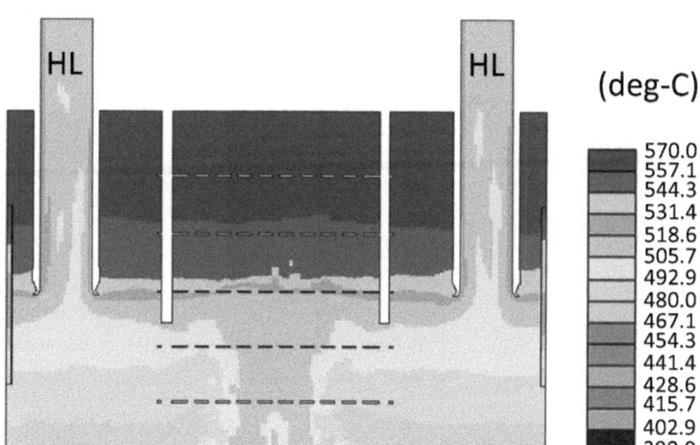

Fig. 4.152 Temperature profile in the upper plenum of the 750 MWe reactor (300 s after scram) [101].

4.2.7.3 Cooling system

(a) Piping system and transient

JSFR adopts a two-loop system to minimize the capital cost by introducing Mod. 9Cr—1Mo steel large diameter piping. Particularly, the primary hot-leg has simple L-shaped piping with a structure that absorbs thermal expansion via a single elbow. It is enabled by the adoption of Mod. 9Cr—1Mo steel, which has high strength and low thermal expansion coefficient.

Table 4.26 shows a comparison of heat transport system specifications between the 750 and 500 MWe designs for the demonstration reactor. In this comparison, the velocity differencces appear to be insignificant from the technological demonstration perspective. However, the fabrication experience of the larger diameter piping is a critical demonstration point for the commercial reactor. Therefore, the 750 MWe reactor also has an advantage in this regard.

Safety evaluation for representative events was performed tentatively for the 750 MWe reactor. For the design-basis event evaluation, transient overpower (TOP) and loss of flow (LOF) events were estimated. Fig. 4.153 shows an example of the analysis result for TOP with 10 ¢/s reactivity insertion due to one control rod malfunction. It reaches the trip level after 0.84 s in this case. Notably, the maximum temperature satisfies the safety criteria for the accident. For the LOF condition, the primary pump seizure accident was evaluated as a representative example since it is assumed to be important for the two-loop system.

As a safety feature of JSFR, natural circulation is adopted for the decay heat removal system to enhance the cooling system's reliability. Fig. 4.154 shows the safety evaluation result of the decay heat removal by natural circulation for the loss of off-site electric power event. The result satisfied the safety criterion of the peak cladding temperature during transients. Because the demonstration reactor has the same core inlet and outlet temperatures as the commercial reactor, the former also has the same component elevation in the plant as the latter. Therefore, the cooling system's performance by natural circulation of the commercial reactor would be almost similar to that of the demonstration reactor. The difference between 750 and 500 MWe was also negligible from a safety perspective.

Table 4.26 Specifications of cooling system [101].

Items	750MWe/1765 MWt	500MWe/1177MWt
Number of loops	2	2
Primary sodium temperature	550/395C	550/395C
Primary flow rate	1.62×10^7 kg/h/loop	1.08×10^7 kg/h/loop
HL diameter	914.4 mm	762 mm
HL flow velocity	8.8 m/s	8.6 m/s
CL diameter	711.2 mm	558.8 mm
CL flow velocity	7.3 m/s	8.1 m/s
Piping material	Mod.9Cr-1Mo steel	Mod.9Cr-1Mo steel
Second sodium temp.	520/335C	520/335C
Second flow rate	1.35×10^7 kg/h/loop	9.0×10^6 kg/h/loop
Steam temperature	497C	497C
Steam pressure	19.2 MPa	19.2 MPa
Steam flow rate	1.44×10^6 kg/h/loop	9.61 x10^5kg/h/loop

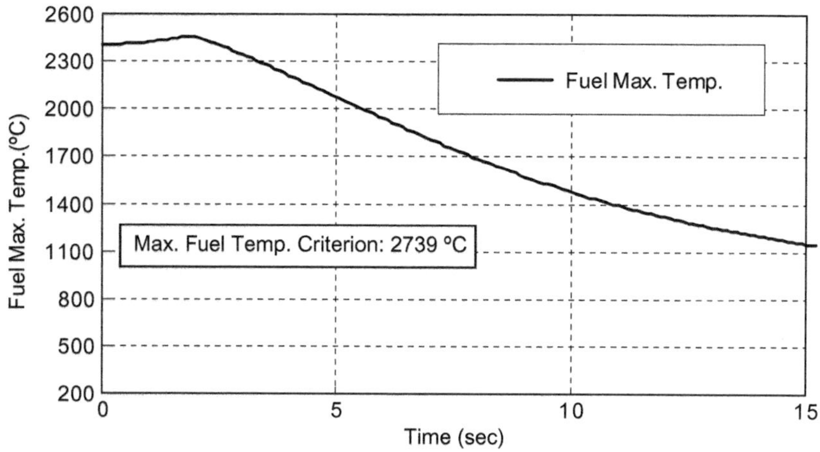

Fig. 4.153 Transient analysis for TOP (10 ¢/s reactivity insertion) [101].

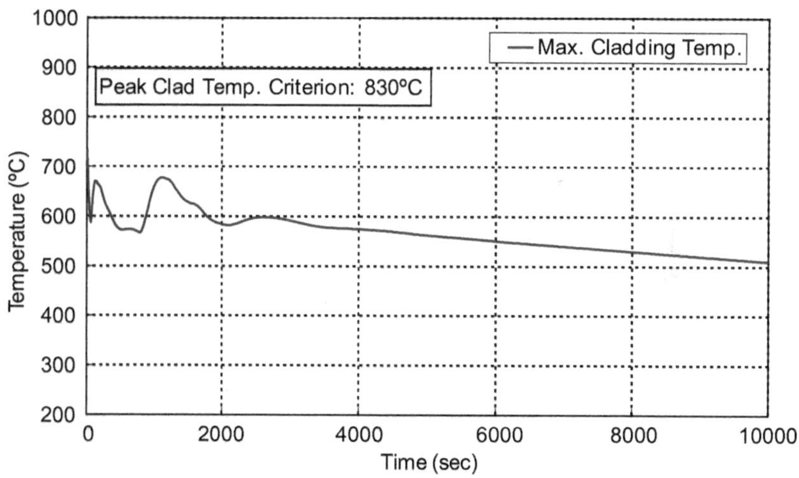

Fig. 4.154 Transient analysis for decay heat removal by natural circulation (loss of off-site electric power event) [101].

(b) Integrated IHX-pump

An integrated intermediate heat exchanger (IHX) with a primary pump is introduced in JSFR. The integrated design of an IHX with a primary pump enables the elimination of middle-leg piping. The integrated component is formed by installing the primary pump at the center of the IHX tube bundle. The diameter and height of the integrated component in the commercial reactor are 5.3 and 16.3 m, respectively. Fig. 4.155 shows the configurations of IHX-pumps due to the design study on the demonstration reactor. The heights of the IHX-pump were virtually the same for the 750 and 500 MWe reactors to maintain

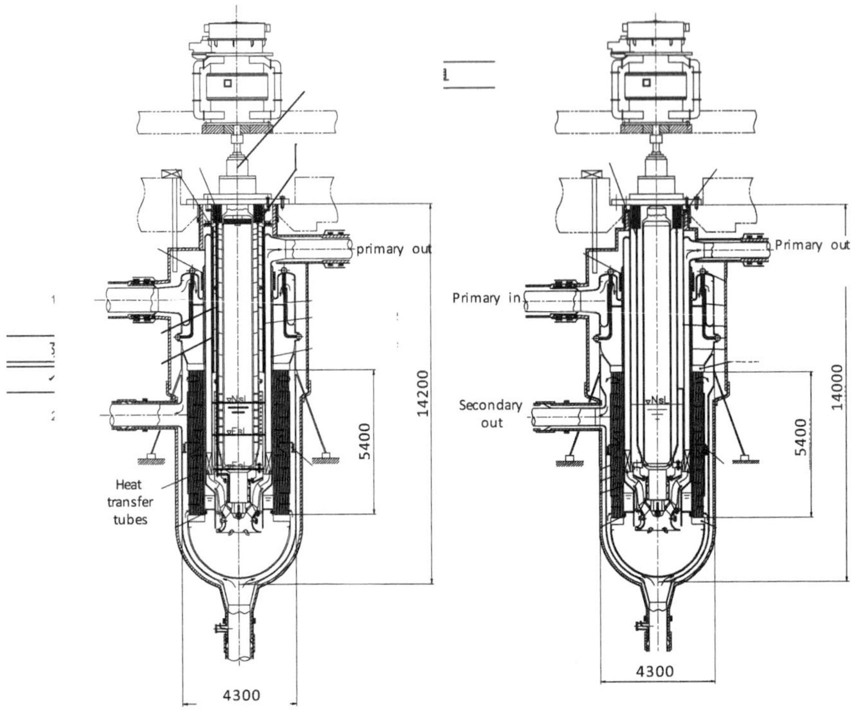

Height : 14.2m
Casing diameter (upper) : 4.3m
Casing diameter (lower) : 4.3m
Heat transfer tube number : 6140

750MWe IHX-pump

Height : 14.0m
Casing diameter (upper) : 4.3m
Casing diameter (lower) : 3.9m
Heat transfer tube number : 4100

500MWe IHX-pump

Fig. 4.155 Sketch of integrated IHX-pump for demonstration reactors [101].

balance with other components. The number of heat transfer tubes in an IHX is reduced to accommodate the 500 MWe reactor's heat exchange capacity. However, the effect on the component diameter is not as remarkable. The difference in the casing diameter was only 0.4 m at the lower part of the component. Alternatively, the 750 and 500 MWe reactors demonstrated similar characteristics of pump rotating stability of the long axle with a flexible structure. Notably, the 750 MWe reactor's rotating blades have a configuration similar to those of the commercial reactor. However, the 500 MWe reactor requires a different configuration because of the limitation on the pump system's natural frequency.

Thermal-hydraulic analysis was performed for 750 MWe to evaluate the temperature profile around the long axle. If vertical gas convection occurs in the annulus around the axle, the temperature profile at the periphery of the axle will be asymmetric. To prevent the thermal deformation of the axle, the placement of convection restriction plates in the annulus zone was considered. Fig. 4.156 shows that the temperature profile became symmetrical around the axle due to the convection restriction plates.

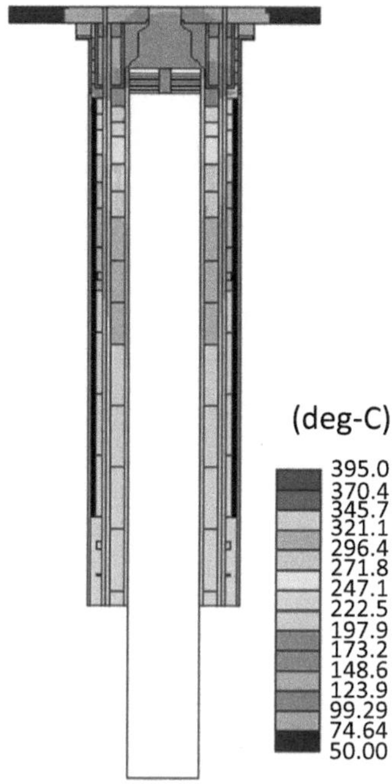

Fig. 4.156 Temperature distribution around pump shaft [101].

Flow velocities at the rated condition and temperature profile during transient were evaluated for the IHX upper plenum using CFD code for the 750 MWe reactor (Fig. 4.157). The velocity profile at the inlet of the IHX was sufficiently uniform in a circumferential direction. During the transient after scram, the flow is cooled using a primary reactor auxiliary cooling system (PRACS) tube in the upper plenum. Fig. 4.157 shows a low-temperature flow along the PRACS tube. However, the low-temperature coolant is mixed in the upper plenum until the entrance of the IHX tube bundle.

(c) Steam generator

The steam generator (SG) in JSFR is intended to use double-walled tubes for its heat transfer tubes to enhance reliability against sodium-water reaction hazards in the event of SG tube failure. In addition, the SG uses a straight tube bundle due to the advantages of using double-walled tubes. SG tubes are made of Mod.9Cr-1Mo steel, which is resistant to high temperatures and has low thermal expansion properties. The SG length is determined by the balance of the heat exchange area, SG diameter, and water-steam flow stability in the heat transfer tubes.

Fig. 4.158 compares the SG configurations for the 750 and 500 MWe reactors. The demonstration reactor should use the same diameter and length of heat transfer tubes as the commercial reactor to validate a key technology. Thus, the total SG heights of both reactor powers are almost similar. The heat exchange capacity was designed by adjusting the number of tubes. Therefore, the SG system performance of the commercial reactor will be

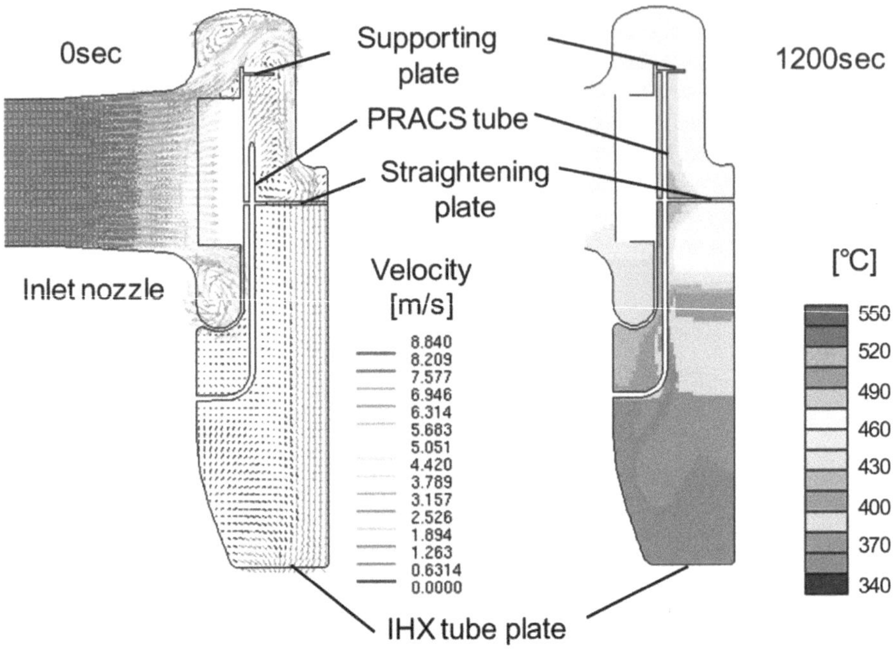

Fig. 4.157 Velocity and temperature profile at the upper plenum of IHX-pump [101].

reasonably demonstrated by the demonstration reactor. When the 750 and 500 MWe reactors are compared in terms of technological achievement, the latter's tube plate size and convoluted shell expansion joint diameter are lesser than those of the former. However, it should be possible to compensate for the difference by conducting test fabrication and design evaluation in stages.

4.2.7.4 Reactor building

The JSFR building is characterized by the use of seismic isolation and steel plate-reinforced concrete (SC) rather than conventional concrete reinforced with steel rods. Sodium-cooled fast reactors have relatively thin-walled structures due to their low-pressure system and for minimizing thermal stresses due to large temperature differences.

Protecting the thin-walled structure from severe earthquakes is critical. JSFR employs a seismic isolation system to mitigate earthquake force. For the SC structure, because its steel part can be manufactured in a factory, the application of the SC structure minimizes the on-site construction period.

Fig. 4.159 compares the reactor building layouts for the 750 and 500 MWe reactors. The effect of power difference was negligible in the common utility area. In addition,

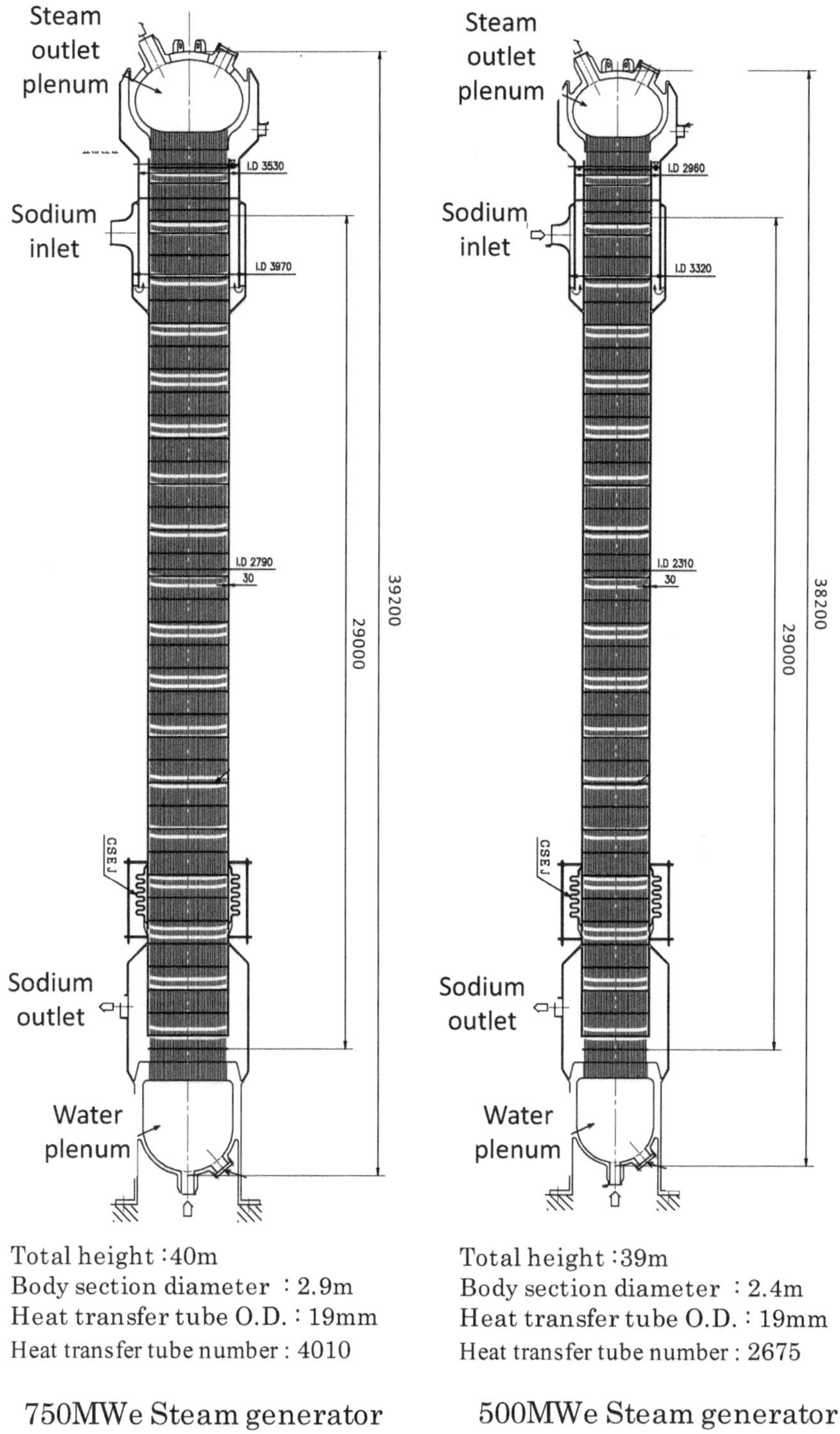

Fig. 4.158 Sketch of steam generators for demonstration reactors [101].

the difference in containment vessel size was only 0.5 m each in the horizontal width. Therefore, the reactor building sizes of both the reactor powers became almost the same. This implies that the building material requirements for the 750 and 500 MWe reactors are almost similar. The 750 MWe reactor appears to have an advantage in construction when cost benefits for the demonstration reactor are considered.

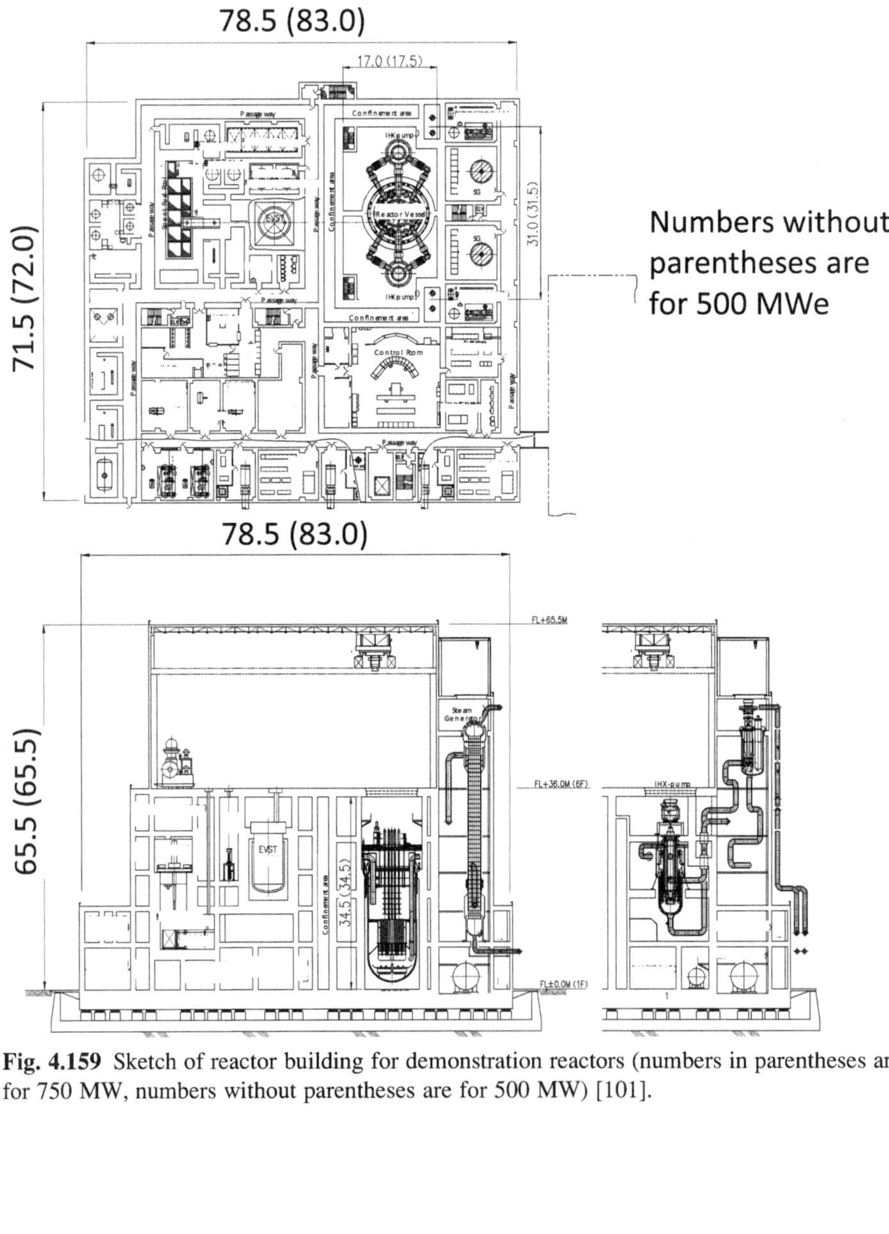

Fig. 4.159 Sketch of reactor building for demonstration reactors (numbers in parentheses are for 750 MW, numbers without parentheses are for 500 MW) [101].

4.2.7.5 Comparison of demonstration reactor output

In conclusion, the following results were obtained and a demonstration reactor with a power output of 750 MWe was selected.

- The reactor vessel diameters were evaluated at 10.7, 8.9, and 8.5 m for 1500 MWe, 750 MWe, and 500 MWe, respectively. The ratio of the coolant volume per power increases as the reactor power decreases. This implies that the coolant velocity in the reactor vessel decreases as the reactor power decreases. Therefore, 750 MWe has an advantage if we intend to demonstrate the phenomena in the vessel.
- Concerning the heat transport system, depending on the piping diameter selected, the cooling system's velocity can be controlled at both reactor powers. The 750 MWe reactor is preferred if we consider the importance of the fabrication experience of larger diameter pipes.
- From a safety perspective, several representative occurrences were examined to evaluate the system's safety performance. The difference between 750 and 500 MWe was negligible due to the same component elevation in the plant.
- The sizes of the IHX-pump components of the 500 and 750 MWe reactors are almost similar. However, the configuration of the rotating blades must be changed in the case of 500 MWe as a countermeasure against vibration.
- For the SG, the same heat transfer tube height and diameter are applied to 500 MWe, 750 MWe, and 1500 MWe. The SG's technology demonstration would be reasonably achieved by 500 and 750 MWe.
- The difference in building size between 500 and 750 MWe reactors was determined to be negligible. However, the 750 MWe reactor appears to have an advantage in construction when cost benefits for the demonstration reactor are considered.

4.3 Design improvement of JSFR

In the framework of GIF, the design study focuses on safety design improvement to contribute to the development of safety design criteria (SDC) and safety design guidelines (SDG), which include the lesson learned from the TEPCO's Fukushima Daiichi nuclear power plant accident (1F accident) [102]. Design measures should be enhanced against design extension conditions (DECs), including those for external events considering the lessons learned from the 1F accident, to achieve the safety goals for the generation IV reactor. DECs and related design measures are identified and selected with due consideration of the safety features of SFR. Several situations to be practically eliminated are discussed with possible design measures [103].

From the viewpoint of a plant operator, maintenance and repair capabilities are crucial. The design requirements for maintenance and repair were updated to enhance accessibility to the sodium-contained structures such as the reactor vessel and its internal structures, the integrated primary pump-intermediate heat exchanger, and the primary piping. The design conditions regarding external hazards, including earthquakes, were also updated to enhance plant robustness against external hazards.

A design improvement study was conducted on the basis of these updated requirements and conditions for the demonstration JSFR with 750-Mwe output power. The updated requirements and conditions are described in Section 4.3.1. The safety design improvement is described in Section 4.3.2. The design updates of the major components and piping of the reactor cooling system are described in Section 4.3.3.

4.3.1 Update on requirements and conditions

4.3.1.1 Safety requirements

(a) SDC for generation IV SFR

The SDC for generation IV SFR (GIF-SFR-SDC) developed in the framework of GIF was used as the basis for the safety design of this study [104]. GIF-SFR-SDC refers to the structure of IAEA SSR 2/1 and is formulated consistent with GIF's basic safety approach and to achieve the safety and reliability goals of GIF. Specific technical features of SFRs, the latest knowledge such as R&D results for innovative technologies, and lessons learned from the Fukushima Daiichi accident were incorporated. It should be noted that the following criteria, in which general characteristics of SFR are reflected, were introduced as design measures for DECs corresponding to the fourth level of defense in depth. Establishing design concepts to meet these criteria is thus important.

- Passive or inherent reactor shutdown capabilities
- Mitigation provision to avoid large mechanical energy release during a core degradation progression and means for decay heat removal of a degraded core
- Decay heat removal system with passive mechanism and diversity, which can be available even under extreme external hazards and their consequences such as long-term loss of all AC power supplies

(b) Lessons learned from the 1F accident

The Fukushima Daiichi accident was caused by a design-basis level earthquake accompanied by a beyond-design-basis tsunami during the Great East Japan Earthquake on March 11, 2011. Generation IV SFR was designed to avoid the release of significant radioactive materials to the environment even under such severe conditions. Any sort of severe external events beyond design basis selected as a result of siting evaluation (e.g., aircraft crash, volcano ashfall, strong wind, heavy snowfall, extreme temperature, external explosion, fire, flood, earthquakes, and tsunamis) were considered in the design and accident management for both the reactor core and the fuel handling and storage systems.

JSFR adopted a seismic isolation technology to ensure the integrity of the thin-walled structures of its sodium-contained components under severe earthquake conditions [105]. Because the reactor building is installed on seismic isolation devices (i.e., laminated rubber bearings), seismic loads on the sodium-contained components (those of primary and secondary coolant systems, fuel handling and storage systems, and sodium auxiliary systems) in the reactor building can be reduced. The safety systems, which include those for reactor shutdown, decay heat removal, and containment, are also seismically isolated. Thus, JSFR can feasibly save a comparable structural margin to those of the current LWRs in Japan.

Basic countermeasures against tsunamis or floods entail sufficient site elevation and require dikes to prevent the submersion of the reactor building. In addition, waterproof measures must be applied for rooms for safety-grade electricity facilities and sodium-contained components such as dump tanks, which are on the lower levels of the building. Moreover, natural circulation heat removal by air cooling is effective under loss of electric power and water splash situations because it does not require driving power for motors and/or pumps and seawater or groundwater cooling systems, which are located on lower levels. The main decay heat removal systems (DHRSs) for JSFR, one direct reactor auxiliary cooling system (DRACS) and two primary reactor auxiliary cooling systems (PRACSs), were designed to maintain core cooling for long times even under long-term loss of AC power after severe earthquakes and tsunamis.

Because the sodium-air heat exchangers of DHRSs are directly connected to the surrounding environment of the plant, they might affect the influences of severe external events such as aircraft crashes, volcano ashfalls, strong winds, heavy snowfalls, extreme temperatures, external explosions, and fires. Design measures (e.g., reinforcement of air stacks for strong winds, physical barriers, and separation for external missiles) are thus required to protect the sodium-air heat exchangers against severe external events.

As mentioned above, taking the lessons learned from the 1F accident into account, preventing core damage and the release of significant radioactive materials even for severe external events beyond the design basis is a must. Reinforcing decay heat removal measures is especially important to cope with these challenges [106].

(c) Measures against DECs

For DECs, the design measures are those for the "prevention of core damage" and "mitigation to ensure containment function," as shown in Fig. 4.160. The time margin to core

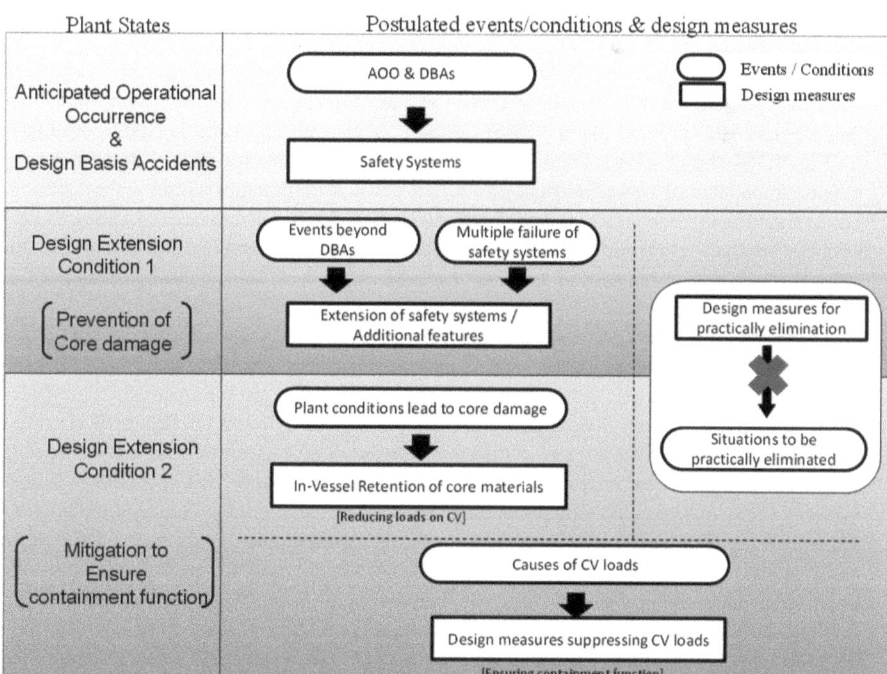

Fig. 4.160 Plant states and design measures for JSFR [103].

damage and the number of applicable measures must be considered to determine the plant conditions postulated as a DEC (e.g., core degraded or not). Built-in design measures must be incorporated for DEC. However, the application of possible accident management measures is supplementally considered in the plant design in advance.

SFR DEC events can be grouped into the following two categories based on the characteristics of SFR and PSA studies [107,108].

(1) Failure to shut down the reactor following an off-normal initiating event (ATWS type), loss of flow with failure to scram, overpower transient with failure to scram, and loss of main heat removal with failure to scram.

(2) Failure to remove heat from the core following an initiating event with scram (LOHRS type), loss of primary coolant flow (flow path becomes disrupted), loss of primary coolant level (core becomes uncovered), and loss of heat sink (LOHS).

As an SFR is operated under low-pressure conditions, close to atmospheric pressure, and at temperatures far below the boiling point, coolant leakage or pipe break does not lead to the type of coolant loss accident experienced in an LWR with depressurization, coolant boiling, and loss of cooling capability. Therefore, emergency core cooling systems for coolant injection under high- and low-pressure conditions, as used in the LWR, are not necessary in an SFR. The only requirements for SFR core cooling are the maintenance of the sodium coolant level above the reactor core in the reactor vessel and sufficient heat removal capability. Primary coolant leakage does not cause pressure buildup inside the containment vessel with adequate design for mitigating leaked sodium combustion, that is, holding leaked sodium by guard vessels and guard pipes with inertized atmosphere.

In case of ATWS-type events, degraded core materials can be retained inside a reactor vessel with an adequate design (i.e., prevention of severe mechanical energy release and in-vessel core catcher). Thus, SFR can be designed to have less impact on the containment with measures for in-vessel retention of four degraded core materials for ATWS-type events, maintenance of core cooling for LOHRS-type events, and mitigation of sodium combustion (sodium-concrete reaction should also be prevented). In these cases, the main role of the containment is to prevent the significant release of gaseous and volatile fission products.

Fig. 4.161 shows a basic idea of the JSFR design measures for DECs. In ATWS-type events, unbalance of power and cooling might cause core damage within a short period. Passive shutdown mechanisms such as the self-actuated shutdown system (SASS) can prevent core damage even under such conditions. In addition, mitigating core damage is considered in the design because of the short time period to core damage and of the potential mechanical energy release, which might occur in core damage situations.

On another note, in LOHS-type events of LOHRS, because the decay heat is at a few percent of the nominal power, the temperature of the reactor coolant system, including the core, coolant, and reactor coolant boundary, increases slowly for long periods. Therefore, there is a sufficient time margin to do recovery actions for failed DHRSs and/or implement backup cooling measures. If no heat sink is available, coolant boundary failure might occur because of creep damage anywhere in the primary and secondary coolant systems. This might cause containment bypass. Then, the system temperature reaches the sodium boiling point, and sodium vapor is released into the containment. This might cause significant thermal loads that the containment could not cope with. Such situations must be practically eliminated using design measures to enhance core cooling. Concerning the loss of primary coolant flow events of LOHRS (the flow path is disrupted), the total loss of flow paths to the heat sink has consequences similar to those of the LOHS-type events. Therefore, these consequences can also be practically eliminated using design measures.

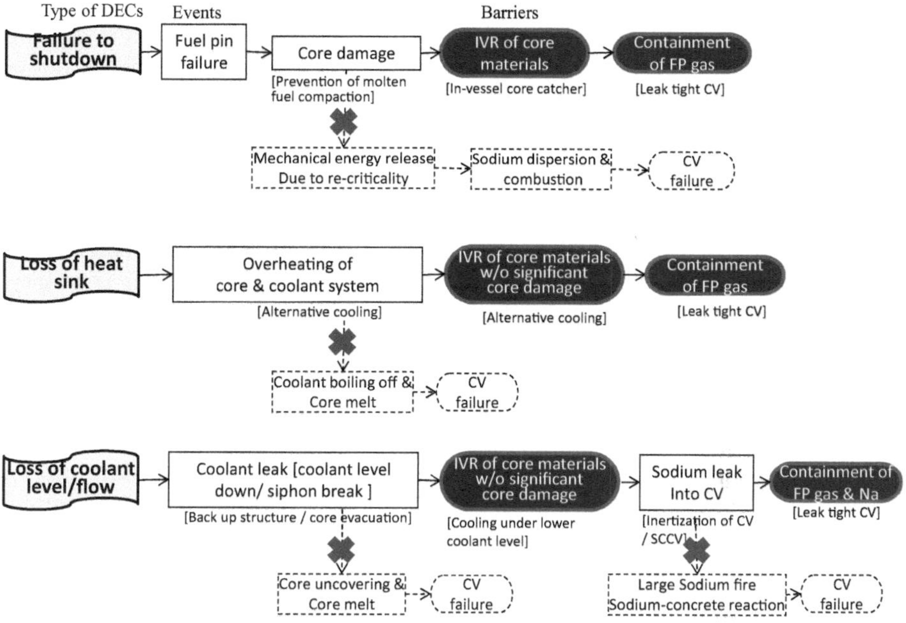

Fig. 4.161 JSFR design approach to ensure the containment function [103].

In the loss of primary coolant-level events of LOHRS, avoiding core melting and the release of significant radioactive materials into the containment atmosphere is impossible if the core is uncovered. The containment function under such situations would not suffice in mitigating the release of radioactive materials to the environment below emergency response levels. Thus, design measures should practically eliminate such uncovering of the core.

(d) Situations to be practically eliminated

Mitigating the consequences of some accident situations should be excluded by design where feasible. This is because the implementation of additional mitigation devices or the R&D necessary for demonstrating their effectiveness may be prohibitively expensive or difficult to prove effective under DECs. However, for situations that are physically possible, the design process must consider, within economic and physical constraints, all situations independent of their probability. Several situations to be practically eliminated, tentatively selected on the basis of the design characteristics of SFR, are as follows.

- Design measures are investigated and provided to prevent the occurrence of the following situations in the JSFR design study
- Abnormal reactivity insertion leading to prompt criticality (large bubble ingress into the core, core configuration change due to beyond-design-basis earthquakes, and core displacement due to significant failure of the core support structure)
- Severe mechanical energy release due to coherent sodium boiling, molten fuel compaction, or failure of decay heat removal from the degraded core in ATWS-type events
- Significant core damage in LOHRS-type events
- Large-scale sodium spray combustion inside the containment

- Hydrogen accumulation and deflagration-detonation due to sodium-concrete reaction inside the containment
- Containment bypass due to the large-scale rupture of steam generator tubes
- Fuel melt in the fuel storage system

4.3.1.2 Maintenance requirements

Maintainability and repairability (M&R) are among the major design requirements for nuclear reactors. Investigating the improvement of M&R and presenting more practical reactor concepts from the viewpoint of future plant users are thus necessary. Although JSFR already took into account M&R, a further investigation including design modifications was conducted, aiming at enabling the maintenance and repair of almost all components to a level equivalent to that attained for LWRs [109].

Maintenance and repair works are classified into three categories: inspection, checking, and repair. Inspection means the mandatory periodic inspection for high-importance vessels and piping or structures in the Japanese regulatory framework and voluntary inspection to assure the reliability of important components and structures. Checking means examining portions or recovering parts in case of unforeseen troubles (including the generation of lost parts [breaking away from the original position]). Repair means, in case of unforeseen troubles, repairing in place or replacing (including repairing after extraction). The respective criteria for these categories were investigated. For inspection, continuous monitoring to ensure that there are no indications of small leaks, which do not immediately affect core cooling and do not entail reactor trip, shall be applied as a major inspection method. For checking, access to all parts and volumetric tests for welding lines should be feasible. For repair, the original criteria are feasible, and in case in-place repair cannot be realized, repairing after extraction must be feasible.

4.3.1.3 Updates on design conditions

(a) Conditions for external hazards other than earthquakes

Because of the magnitude of the Tohoku-Pacific earthquake and the severity of the accident response by Fukushima I NPP, JSFR must be updated for it to have robustness in its design to withstand severe earthquakes, tsunamis, and other external hazards compared with the JSFR design prior to the Fukushima I NPP accident [110]. JSFR has to protect important safety components and remove decay heat when it experiences these external hazards. The evaluated external hazards are listed in Table 4.27 [111].

Table 4.27 Selected external hazard conditions [110].

External hazards	DBE design condition	Proposed extremely condition
Earthquake	Ss	2 x Ss
Tsunami	Height 15 m	Height 25 m
Strong wind	70 m/s(average)	130 m/s
Snowpack	Height 4 m	–

For screening the design conditions, a statistical approach based on historical worst-case values for the postulated plant site or in Japan in general and a deterministic approach in the case with no statistical data were used.

For screening the proposed extreme condition, this study used the probabilistic and statistical approaches based on historical worst-case values in Japan. For the probabilistic approach, the extreme condition is defined such that severe consequences will not appear in the risk region immediately exceeding the design conditions to prevent a cliff edge on the risk curve.

For example, the fastest wind speed ever recorded in Japan is ~70 m/s at Cape Muroto in Kochi Prefecture due to a typhoon, according to Japanese meteorological records. The tornado proposed in the extreme condition is specified as 130 m/s considering US severe conditions. The greatest snowfall ever recorded in the Japanese meteorological record is ~4 m, except for mountainous areas. In the proposed extreme condition, the snowpack must be higher than the design condition. However, the snowpack height can be reduced through the manual removal of snow during accident management. The following chapters describe the evaluation of external hazards. The JSFR building design for 2010 was evaluated.

(b) Seismic conditions

After the Fukushima Daiichi nuclear power plant accident, the design seismic condition for SFR was updated [112]. Because countermeasures for external hazards were installed, the acceleration level of the floor response spectra to be used for the design seismic conditions of the components has increased. This is because the design seismic condition is larger and the weight of the reactor building is increased according to the improved external hazards.

JSFR adopted a steel plate-reinforced concrete (SC) structure not only for the reactor building but also for the containment vessel (SCCV). For the floor response analysis, the multiple degrees of freedom system (mass-spring model) was used. Fig. 4.162 shows the mass-spring model of the demonstration JSFR. After the Tohoku earthquake and the Fukushima Daiichi nuclear power plant accident, the design seismic condition was changed, as shown in Fig. 4.163. Japanese nuclear power plants were divided into two groups in this analysis: hard rock and soft rock. Hypothetical seismic conditions enveloping site conditions for both hard rock and soft rock groups were evaluated.

4.3.2 Safety improvement

SDC and SDG for generation IV SFRs were developed in the framework of the GIF. The SDC and SDG, which incorporate lessons learned from the 1F accident, require the enhancement of design measures against severe accidents and the provision of design measures against severe external events using inherent or passive safety features based on general SFR characteristics. This section describes the improvements of the decay heat removal system (DHRS) and the reactor building considering severe conditions. Design studies were conducted on the demonstration JSFR with 750-MW electric output. For CDA, both prevention and mitigation measures are already discussed in Section 4.2.3.

In coping with severe external events, the reactor building and the plant ray-out have important roles as barriers protecting the reactor cooling system, fuel handling and storage systems, and other facilities important to safety. The seismic isolation

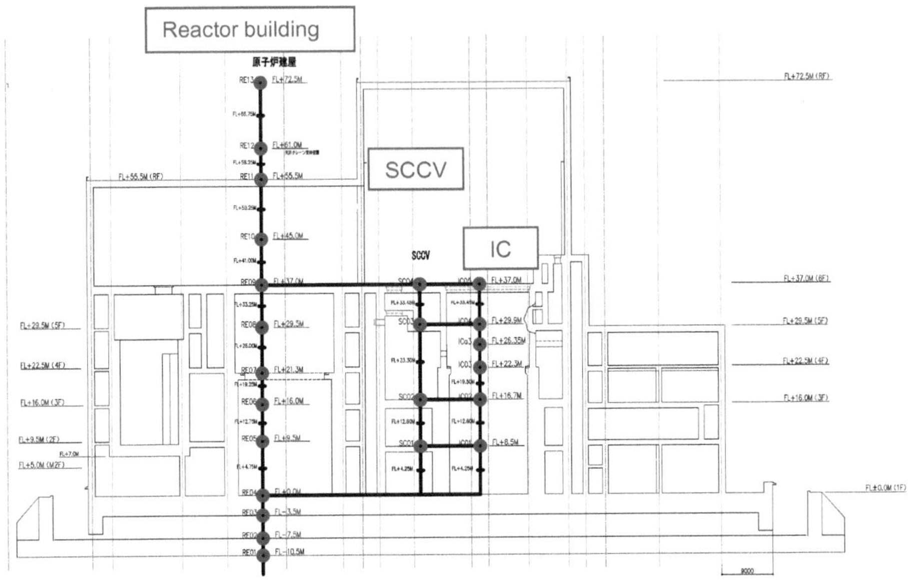

Fig. 4.162 Mass-spring model of the JSFR reactor building [112].
Copyright 2016 by the American Nuclear Society, La Grange Park, Illinois.

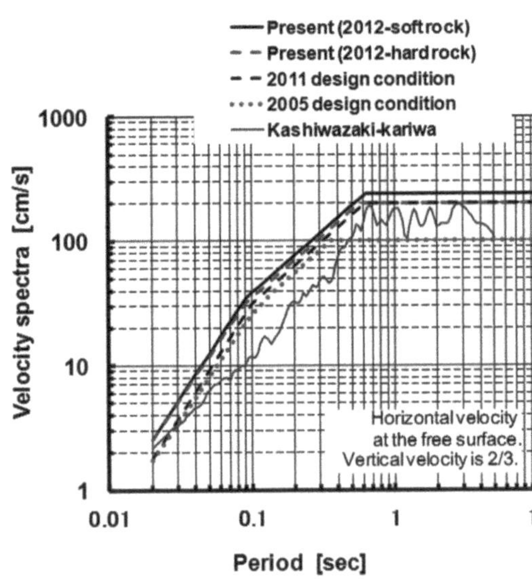

Fig. 4.163 JSFR design seismic condition in 2012 [112].
Copyright 2016 by the American Nuclear Society, La Grange Park, Illinois.

system and the structure of the reactor building are also important in mitigating seismic loadings on the structures, systems, and components important to safety, which are installed inside the reactor building. In the latter part of this section, the reactor building tolerance against external hazards (i.e., earthquakes, tsunamis, strong winds, and snowpacks) is evaluated. Regarding earthquakes, design measures for the reinforcement of the reactor building are investigated to reduce the acceleration on the floor on which the components of the reactor cooling system are installed.

4.3.2.1 DHRS

(a) Update requirements

A safety design study of the demonstration version of JSFR to meet SDC and SDG was conducted. For the decay heat removal function, design measures against internal events exceeding design-basis accidents (DBAs) and severe external events related to loss of heat removal events are concretized according to the following policies [113].

- Design measures against LORL-type events
  - Reliability should be secured by preventing dependent failure and common cause failure between the reactor vessel (RV) and the guard vessel (GV) while securing the design margin.
  - The cooling function should be secured to prevent core damage even if leakage on the primary cooling circuit is postulated.
- Design measures against LOHS-type events
  - Core damage and overheating failure of the reactor coolant boundary should be prevented through functional extension, alternative measures, and accident management measures against the function loss of PRACS and DRACS to secure the cooling function.
  - Core damage and overheating failure of the reactor coolant boundary should be prevented using independent measures to secure the cooling and containment functions, considering cooling failure through the above measures.

(b) Countermeasure against LORL-type events

Abnormal conditions from a wide variety of internal and external events were extracted, and the necessity of design modifications was examined. The results are summarized in Table 4.28. The double failure of RV and GV can be prevented by securing the high reliability of GV as regards preventing dependent failure and common cause failure. The reactor coolant boundary failure resulting from severe earthquakes can be avoided by ensuring sufficient margin against design-basis earthquake. On another note, "siphon break" on all the primary cooling circuits or the remarkable drop of the sodium surface level in RV, which is caused by coolant boundary failure severer than that in design-basis events, has been shown to result in loss of the primary coolant circulation. In the case of multiple coolant boundary failures for which the breakage of gas-vent lines placed at the upper section of the intermediate heat exchanger (IHX) in all of the main circuits is assumed as the representative case, siphon break might happen in a few hours. In the case of double failure of the reactor coolant boundary and the guard pipe in the primary circuits, sodium leakage from the failure locations lowers the sodium surface level in RV near the emergency sodium level (EsL), which is set at the elevation of intake of the DRACS heat exchanger (DHX), and leads to siphon break. Thus, the DRACS was functionally extended to improve the cooling capability by adjusting the EsL and DHX design to address the postulated siphon break conditions.

Table 4.28 Abnormal conditions related to LORL-type events [113].

Factor of function loss		**Abnormal condition**	**Event**	**Necessity of measure**
Heat removal path failure between reactor core and decay heat removal system (all system siphon break, remarkable drop of sodium surface level)	Large-scale damage occurred on coolant boundary	Large-scale failure of primary pipe (double-ended pipe break)	Internal	–
		Large-scale failure of coolant boundary resulting from an earthquake exceeding a design base	External	–
	Small-scale damage on more than two places of coolant boundary	Breakage of gas-vent line in both of main loops	Internal	✓
		Small-scale failure on more than two places of coolant boundary resulting from an earthquake exceeding a design base	External	–
	Double damage with coolant boundary and GV or GP	Double failure with RV and GV	Internal	–
		Double failure of coolant system	Internal	✓
		Double failure with coolant boundary and GV or GP resulting from an earthquake exceeding a design base	External	–
	Large-scale cover gas mixing	Large-scale failure of dip plate resulting from an earthquake exceeding a design base	External	–
		Sodium surface sloshing resulting from an earthquake exceeding a design base	External	–

(c) Countermeasure against LOHS-type events

(1) Identification of conditions

Abnormal conditions from a wide variety of internal and external events were extracted, and the necessity of design modifications was examined.

Secondary sodium boundary failure on all DHRSs (DRACS and PRACS) resulting from severe earthquakes can be avoided by ensuring sufficient margin against design-basis earthquake. Because the primary pipes are located inside the stiff roof deck, their coolant paths are maintained in case of severe events on the top of the roof deck such as sodium fire and heavy load drop. Thus, the simultaneous failure of DRACS and PRACS in the primary and the secondary circuits can be prevented. The collapse of the air stack resulting from severe earthquakes and strong wind loads can also be avoided with sufficient margin. However, filters installed at the air cooler intake might be clogged with volcanic ash. Countermeasures such as cleanups or replacements may be provided.

Abnormalities in air cooler dampers require design and operational measures to prevent loss of heat removal. "Stuck of all air cooler dampers at full close" might interrupt air inflow, and "stuck of all air cooler dampers at full open" might cause the freezing of secondary sodium. These events might be induced by common causes such as common control system failure and runout of DC power in case of long-term station blackouts.

(2) Enhancement of the safety design

Concerning abnormalities in air cooler dampers, accident management measures are effective. This is because there is a sufficient time margin to reach severe plant conditions such as during reactor coolant boundary failure due to undercooling and secondary sodium freezing due to overcooling. Manual operation of dampers and preparation of additional control power sources can be implemented considering the necessary time for operator actions such as plant state diagnosis, movement to places where damper movers are installed, and execution of operations [114].

Considering the cooling failure of the above management measures, core damage and overheating failure of the reactor coolant boundary should be prevented using independent measures. Therefore, alternative cooling measures independent of DRACS and PRACS are considered. The candidate systems are as follows and listed in Table 4.29.

(a) Auxiliary core cooling system (ACS)

This system consists of a sodium-sodium heat exchanger allocated outside the RV. For the UHS in the ACS, both seawater and air are applicable. When air is adopted as the UHS, heat removal with natural circulation is possible by keeping an adequate height difference between the heat exchangers and the reactor core (see Fig. 4.164). In contrast, seawater is not suitable for natural circulation because the heat sink is located at a lower position below the reactor core, and a high flow rate is necessary for the intermediate nitrogen circuit.

(b) Nitrogen gas cooling system with steam generator (SG) tubes.

JSFR already has this system for maintenance and inspection. After blowing water held in tubes, a circulation pump injects and circulates nitrogen gas into the tubes to remove heat possessed in secondary sodium. The UHS is seawater.

(c) Air cooling system with SG tubes.

This system is similar to the above (b) system. However, it uses air directly instead of nitrogen gas. Its heat removal capacity is smaller than that of the above (b) system.

(d) RV external cooling system.

This concept involves the removal of decay heat generated from the core by indirect means from outside of the RV. Because it is an indirect cooling system and only some parts of the outer surface of the RV contribute to heat transfer, the heat removal capacity seems small. Both air and water are applicable as UHS (see Fig. 4.165).

Table 4.29 Candidate systems for the enhancement of the decay heat removal function [113].

		Auxiliary core cooling system (ACS)		Nitrogen gas cooling system with SG tubes	Air cooling system with SG tubes	RV external cooling system		Sodium circulating system utilizing EVST
		Air cooling	**Water cooling**			**Air cooling**	**Water cooling**	
Heat removal capacity		Large		Large	Medium	Medium or large		Large
Secondary system	Coolant	Sodium	Sodium + nitrogen	Sodium + nitrogen	Sodium	Water or liquid metal	Water or liquid metal	Sodium
	Circulating mode	Natural	Forced	Forced	Natural	Forced	Forced	Forced
	Driving electric power	Not required	Required	Required	Not required	Required	Required	Required
Ultimate coolant system	UHS	Air	Seawater	Seawater	Air	Air	Seawater	Air
	Circulating mode	Natural	Forced	Forced	Forced	Forced	Forced	Forced
	Driving electric Power	Not required	Required	Required	Required	Required	Required	Required

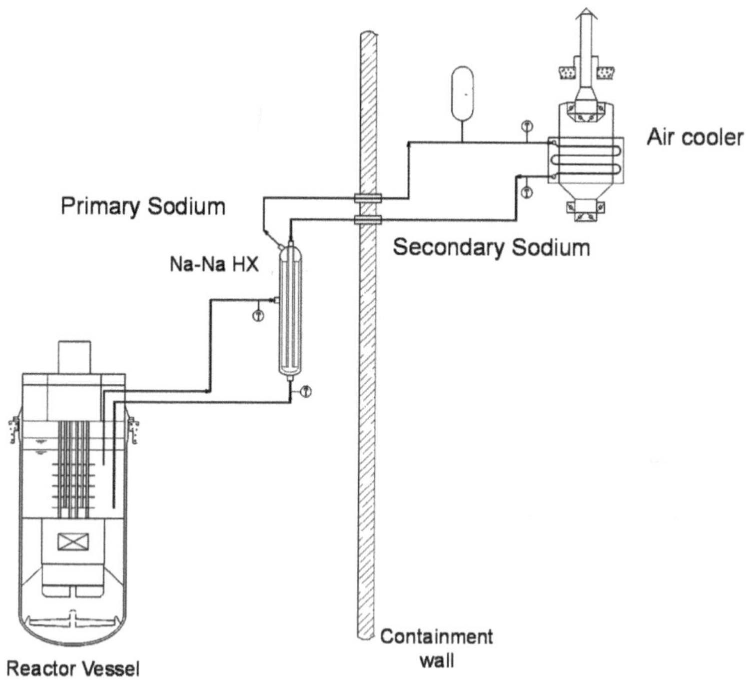

Fig. 4.164 ACS with air as the UHS [113].

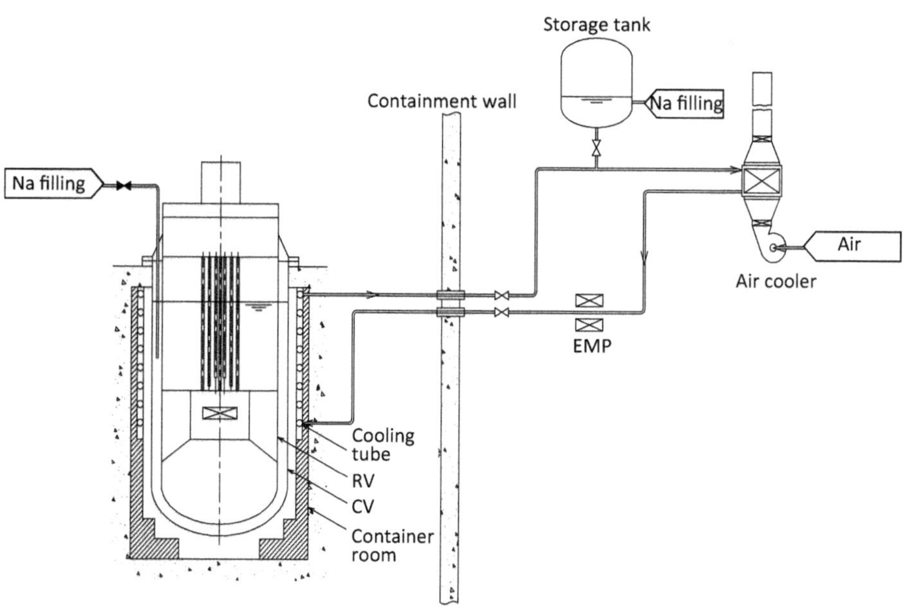

Fig. 4.165 RV external cooling system (with air as the UHS) [113].

(e) Sodium circulating system utilizing the external vessel storage tank (EVST) system.

This concept utilizes the EVST cooling system effectively and removes decay heat by circulating sodium between the RV and the EVST. Connection pipes and electromagnetic pumps are additionally needed. The system is expected to have sufficient heat removal capability.

The ACS with air as the UHS was selected from the systems described above as the most effective cooling means. The reasons are as follows.

The heat removal capacity required to limit the maximum coolant temperature below the allowable value is 17 MW. The ACS, the nitrogen gas cooling system with SG tubes, and the sodium circulating system utilizing EVST have large heat removal capacities enough to achieve 17 MW.

Regarding the RV external cooling system, filling sodium into the GV from the outside to increase the heat removal capability is necessary because the gas layer inside the GV has a large heat transfer resistance. Once sodium is injected into the GV, recovery (i.e., removing sodium and cleaning) is difficult. Furthermore, installing and maintaining the cooling tube is challenging.

The sodium circulating system utilizing EVST takes a long time for preparatory processes such as preheating the connection pipe between EVST and RV and filling sodium into the pipe. Another problem is making a penetration on the containment boundary directly connected to the reactor coolant boundary.

About the nitrogen gas cooling system with SG tubes, active components, a large-capacity power supply, and complicated operation are required because this is a forced circulation system connected to seawater as the UHS. The reliability of this system is not so high because of its many possible causes of failure, such as the failure of active components, loss of driving power, and human error. Furthermore, this system requires much additional plant equipment.

When seawater is adopted as the UHS, ACS would have the same problems on reliability and plant material amount as those of the nitrogen gas cooling system with SG tubes. In contrast, ACS with air as the UHS has sufficient heat removal capacity and higher reliability because it can be operated through natural circulation. Moreover, it requires less additional plant equipment. Measures were provided to prevent the common cause failure of the air cooler with DRACS and PRACS.

Fig. 4.166 shows a plant configuration image to adopt candidate measures against loss of heat removal events. A combination of selected measures can be expected to satisfy SDC requirements for LORL- and LOHS-type events.

4.3.2.2 Reactor building

(a) Earthquake

Seismic inputs at the reactor building base mat for the 1F accident and JSFR design-basis earthquake conditions are shown in Section 4.3.1. Seismic analysis of the original JSFR building was conducted to prove the safety tolerance of JSFR against seismic conditions. The JSFR building analytical model is shown in Fig. 4.167. This model is a multimass model of the reactor building connecting the base mat using the damper model. As a prerequisite, the lower base mat must be solid and the base mat itself must not collapse and slide with the soil. A criterion for the evaluations is such that all the evaluated items have margins twice that of the Ss condition.

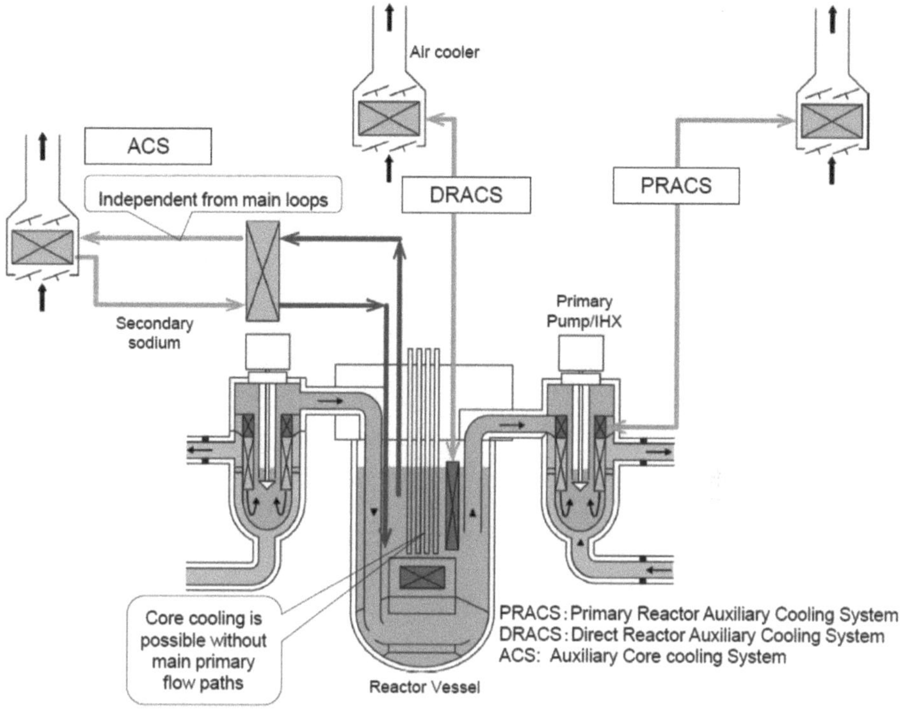

Fig. 4.166 Measures against LORL- and LOHS-type events [113].

The evaluated items and results are shown in Table 4.30. These items for the reactor building were evaluated using the following viewpoints:

- Structural integrity: margin to tolerance
- Bending moments: bending crack and/or stress to steel plate
- Shear force: to building walls
- Stability: margin to tolerance
- Overturning: nonoverturning of the entire building such that the action point of ensuring force at the lower base mat is within 1/3 at the center point
- Sliding: no sliding of the entire building such that the horizontal force acting at the lower base mat is less than the shear resistant force
- Vertical bearing capacity: to lower the base mat such that the vertical ground reacting force at the lower base mat is less than the vertical bearing capacity of the ground

The items for the seismic isolation system were evaluated using the following viewpoints:

- Rubber bearing: margin to linear limitation
- Damper: margin to design stroke.

As the results of the evaluation, Table 4.30 shows that all the evaluated items have margins twice that of the Ss seismic condition. Therefore, the JSFR reactor building and advanced seismic isolation system exclude the viewpoint of structural integrity and stability.

However, as discussed at the end of this section, design measures for the reinforcement of the reactor building were required to ensure the seismic margin for the components of the reactor cooling system.

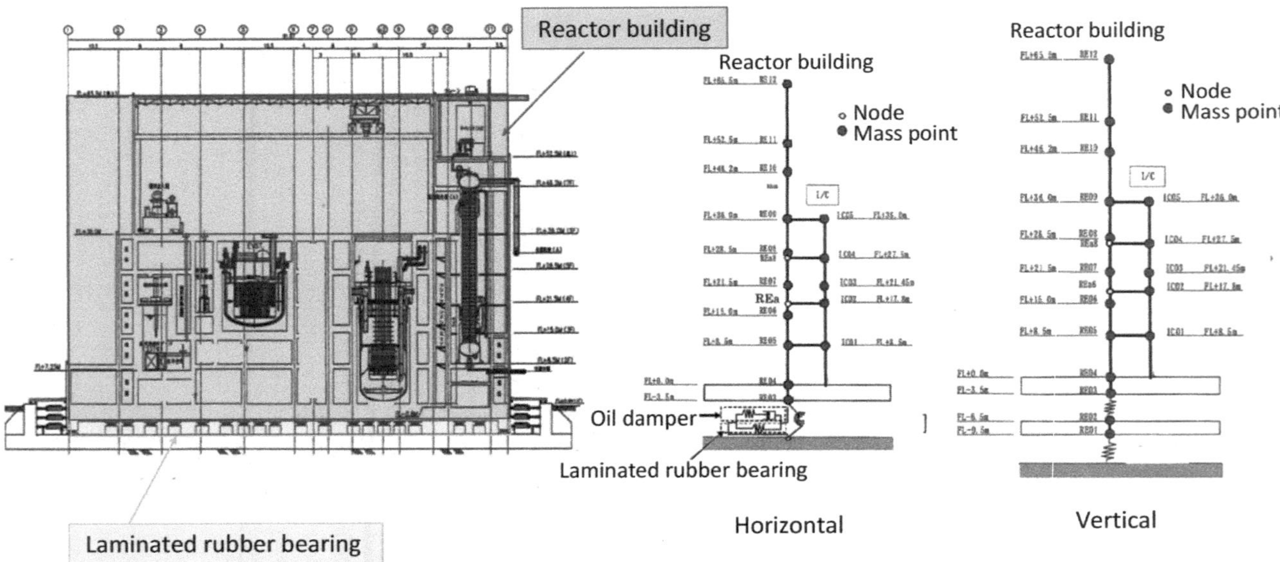

Fig. 4.167 Seismic evaluation model for the reactor building [110].

Table 4.30 Seismic evaluation results on the Ss condition [110].

<table>
<tr><th></th><th colspan="2">Evaluated items</th><th>Margin</th></tr>
<tr><td rowspan="3">Reactor building</td><td rowspan="2">Structural integrity</td><td>Bending moment (NS/EW)</td><td>2.8/2.9</td></tr>
<tr><td>Shear force (NS/EW)</td><td>2.2/3.6</td></tr>
<tr><td>Stability</td><td>Overturning
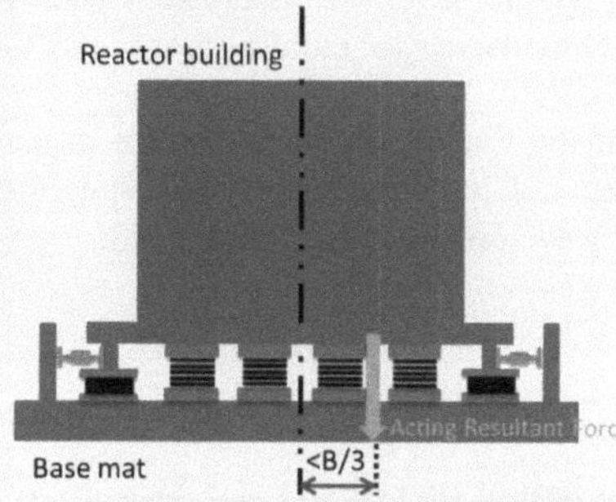

Sliding
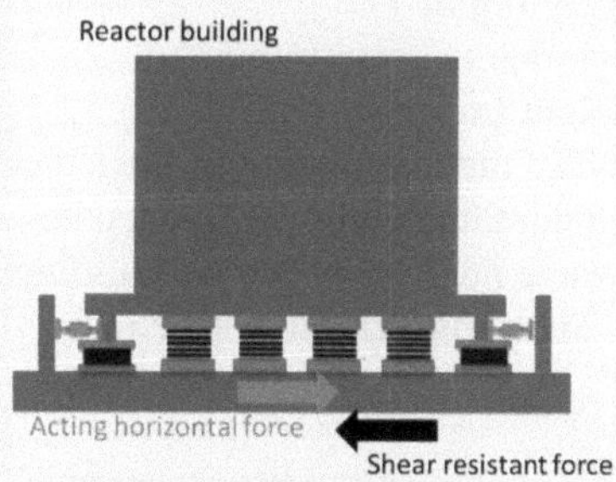

Vertical bearing capacity
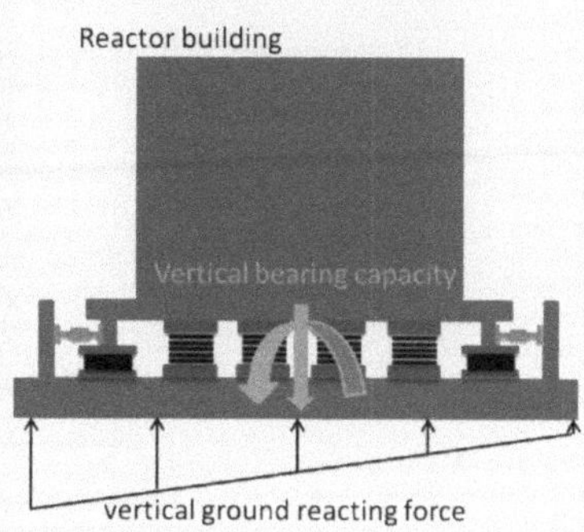

</td><td>7.7
6.5
7.8</td></tr>
<tr><td rowspan="4">Seismic isolation system</td><td rowspan="3">Rubber bearing</td><td>Shear strain</td><td>2.1</td></tr>
<tr><td>Tensile strain</td><td>9.7</td></tr>
<tr><td>Compressed strain</td><td>2.4</td></tr>
<tr><td>Damper</td><td>Displacement</td><td>2.3</td></tr>
</table>

(b) Tsunami

Notably, during the 1F accident, the reactor building and the emergency power supply were flooded. If a huge tsunami attacks the JSFR reactor building, the outer wall can be damaged. Thus, countermeasures for huge tsunamis were evaluated to prevent floods and damage to the reactor building. In this evaluation, tsunami heights were assumed to be 15 m for the design and 25 m for the proposed extreme condition. For the analysis, the building ground level was assumed to be at EL 15 m with reference to the sea level of EL 0 m. Thus, the reactor building is at a level that cannot be flooded by design condition tsunamis.

The Tsunami Evaluation Subcommittee of the Nuclear Civil Engineering Committee, Japan Society of Civil Engineers (JSCE), suggests an evaluation formula for the wall load by tsunami wave pressure as follows [115]:

$$q(z) = \rho g(3h - z) \tag{4.4}$$

where

z: pressured height (m) ($0 < z < 3h$).
$q(z)$: water pressure at z (N/m^2).
ρ: water density (kg/m^3).
g: acceleration due to gravity (m/s^2).
h: flooded height (m).

Fig. 4.168 shows the distribution of the tsunami wave pressure when the building height is higher than three times the flooded height. Fig. 4.169 shows the distribution of the tsunami wave pressure when the building height is lower than three times the flooded height. These figures show schematic views of the evaluation formula.

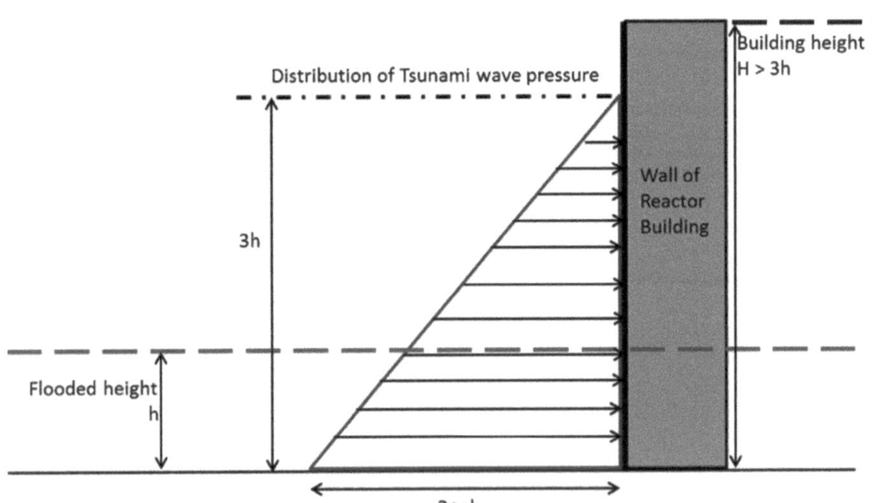

Fig. 4.168 Distribution of tsunami wave pressure (building height is higher than three times the flooded height) [110].

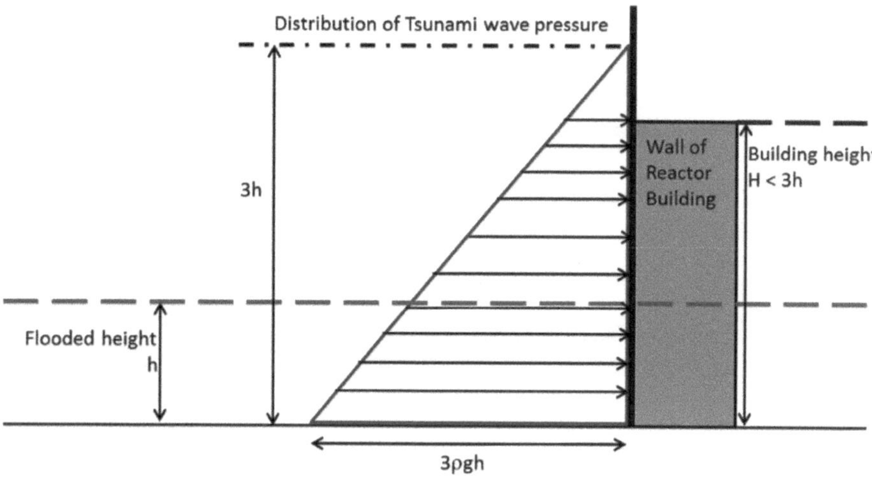

Fig. 4.169 Distribution of tsunami wave pressure (building height is lower than three times the flooded height) [110].

The building's structural integrity was evaluated using the wall member and the whole building using the abovementioned evaluation figures showing the static wave loads to the building wall. Both the bending moment and the shear force were evaluated. The building stability was evaluated using the condition equivalent to the evaluated earthquake condition. The advanced seismic isolation system was evaluated using the strains and displacement of each component against the static wave load.

The building ground level was EL 15 m from the sea level (EL 0 m) in the proposed extreme condition. The tsunami height in the proposed extreme condition was 25 m. Therefore, the flooded height of the building was 10 m. The tsunami static load was evaluated using the evaluation formula.

Table 4.31 shows the evaluation results. In these results, the wall thickness must be about a minimum of 1.3 m against the static wave load of the proposed extreme-condition tsunami.

When the reactor building is flooded, water may flow into the building. In this case, important components related to safety must be protected. Therefore, the watertight strategy was studied against floods. An example of this countermeasure is the adoption of a watertight door and hatch. However, the door and hatch cannot be completely watertight. Thus, a reservoir pit and a drain pump near the door and hatch were considered.

The floor response spectra were evaluated for cases when the advanced seismic isolation system is flooded. Fig. 4.170 shows the acceleration response spectra comparison for flooded and nonflooded conditions considering the fluid shear force using a flow equation based on the finite difference method (FDM). The black line shows the nonflooded condition, and the red one shows the flooded condition. For the horizontal direction, very little influence of the flooded versus nonflooded conditions was observed. For the vertical direction, the peak value decreased because the apparent

Table 4.31 Evaluation results for a 25-m-high tsunami [110].

	Evaluated items			Margin
SC structure	Member	Wall member (wall thickness 1250 mm)	Bending moment	1.15
			Shear force	1.31
		Building	Bending moment (EW/NS)	286/471
			Shear force (EW/NS)	16/32
	Stability	Overturning		20
		Sliding		22
		Vertical bearing capacity		11
Advanced seismic isolation system	Laminated rubber	Shear strain		3.2
		Tensile strain		402
		Compressed strain		–
	Oil damper	Displacement		1.8

building weight decreased because of buoyancy. This result shows that the reactor building and advanced seismic isolation system remain stable in case of flooding because the advanced seismic isolation system is not damaged by the tsunami load.

The response specification of the advanced isolation system does not change for short-time flooding. However, its performance might be diminished by long-term flooding. Therefore, the seismic isolation system must be protected from floods. For this case, candidate countermeasures are shown in Fig. 4.171. The canopy top is placed to protect the wall from the impact of floating materials and for flood protection, so long as the canopy is designed to withstand tsunami loads. Even in flooding conditions, the water can be drained by the reservoir pit and drain pump.

(c) Wind

The tolerance to strong winds is evaluated under the conditions discussed in Section 4.3.1. The wind load at a building height Z (m) is given by the following equation:

$$WD(Z) = qHCD(Z)GDA \tag{4.5}$$

where

WD(*Z*): wind load at Z (N).

qH: velocity pressure (N/m^2).

CD(*Z*): wind force coefficient at Z.

GD: Gast influence coefficient.

A: projected area (m^2).

Table 4.32 shows the evaluated results against strong winds. The design conditions were 70 m/s (10-min average) and 85 m/s (instantaneous). The proposed extreme

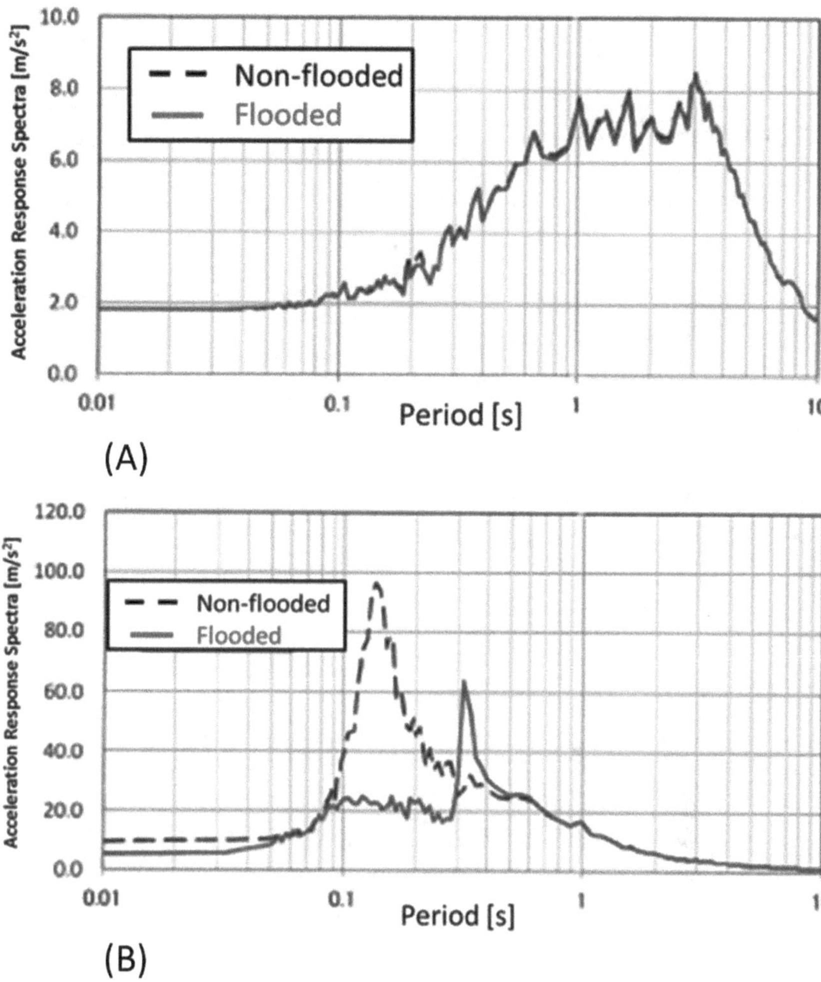

Fig. 4.170 Spectra comparison for flooded and nonflooded conditions [110]. (A) horizontal, (B) vertical.

condition was 130 m/s. The lowest margin for the entire building floor for structural integrity is shown in Table 4.32. The reactor building stability was evaluated under the condition equivalent to the evaluated earthquake condition. The evaluated results show that the wind effects on the structural integrity and stability of the reactor building are negligible. Considering the building shape, there is little influence due to the inward and downward wind effect and wind vibration.

The structural integrity of DHRS stacks is important for core safety because the final heat sink of JSFR is air. Thus, the structural integrity of DHRS stacks is evaluated for a strong wind condition. The evaluation in this case is for the DHRS stacks to be reinforced by stacking to the wall. An example of the evaluated stack wall thickness is 800 mm. In this condition, the tolerance for strong winds is about 1.2 times the margin.

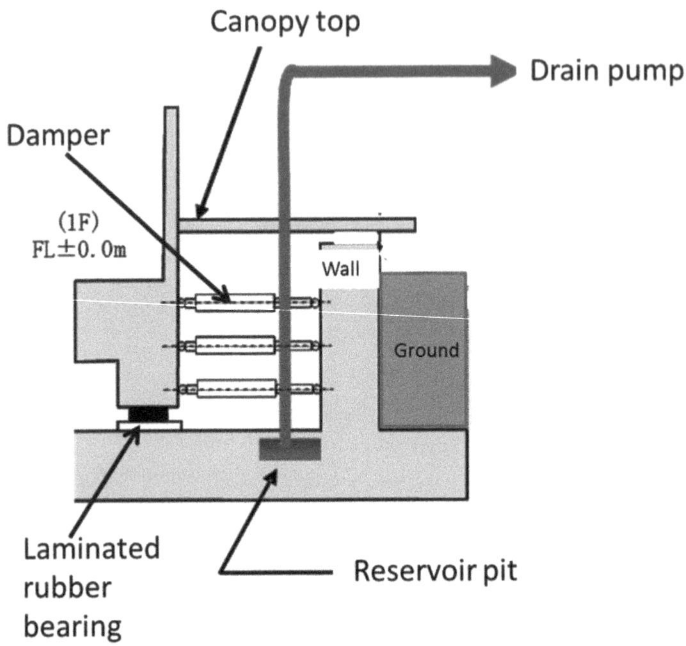

Fig. 4.171 Candidate countermeasures against floods [110].

Table 4.32 Evaluated results against strong winds [110].

Evaluated items			Margin
Structural integrity	Bending moment (4F)	DBE	366.1
		DEC	103.3
	Shear force (6F)	DBE	10.8
		DEC	3.0
Stability	Overturning	DBE	109.1
		DEC	30.5
	Sliding	DBE	261.0
		DEC	88.0
	Vertical bearing capacity	DBE	12.2
		DEC	11.7

(d) Snowpack

The snowpack height for the extreme condition was specified as 4 m according to Japanese records. In the proposed extreme condition, the snowpack must be higher than that in the design condition. However, the snowpack height can be reduced through the manual removal of snow during accident management.

As mentioned earlier, the structural integrity of the snowpack was evaluated using a 4-m snowpack load shown in Fig. 4.172. The snowpack pressure was 20 N/m^2, each 1 cm deep. Fig. 4.173 and Table 4.33 show the analytical model and the results for the truss girder at a

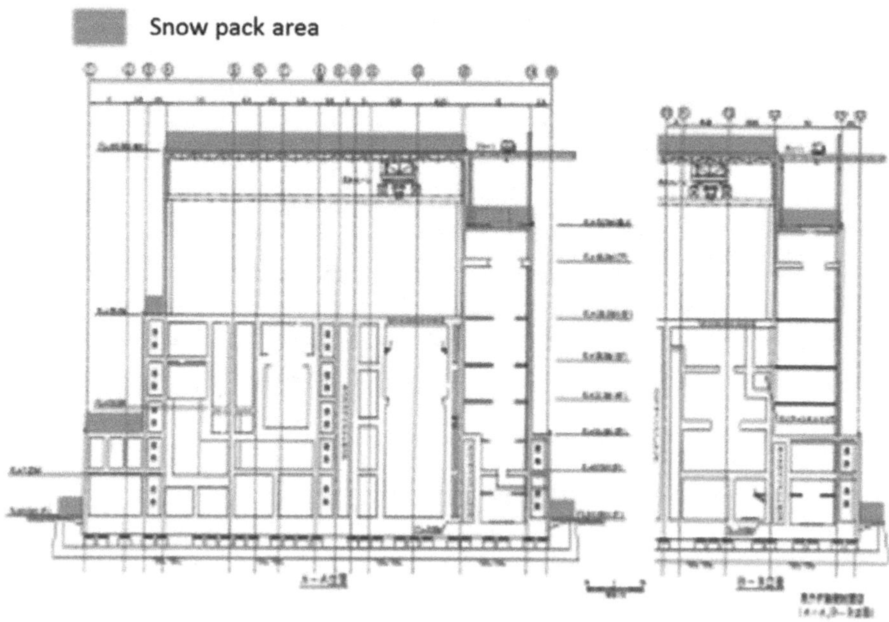

Fig. 4.172 Snowpack area on the roof of the reactor building [110].

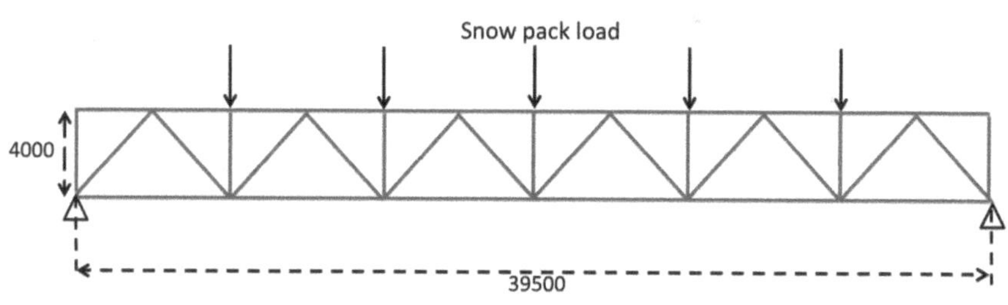

Fig. 4.173 Evaluated truss model [110].

snow height of 4 m. The evaluated results show that the structural integrity of the reactor building roof support structure is maintained by the adequate truss structure of the roof.

(e) Layout

The reactor building layout designs for aircraft attacks (and the consequential fires) and tsunamis are evaluated so it can withstand external hazards.

Fig. 4.174 shows the distributed layout concept [116]. This means that each safety facility and power equipment for the two-loop configuration is symmetrically positioned parallel to the RV. For this configuration, at least one DHRS could be expected to survive in case of an aircraft attack. The DHRS stacks are put onto different sides of the building separately. As a countermeasure to the loss of heat removal system (LOHRS) due to airplane attacks, the air inlet of the ACS is separated from those of DRACS and PRACS. The EVST cooling systems and stacks also adapt the distributed layout onto both sides

Table 4.33 Evaluated results (snowpack 4 m, truss girder height 4 m) [110].

Evaluated items	Margin	
	Bending and axial force	Shear
Upper chord member	1.14	9.76
Lower chord member	1.14	9.76
Bent	1.41	18.24
Tying material	1.43	19.68

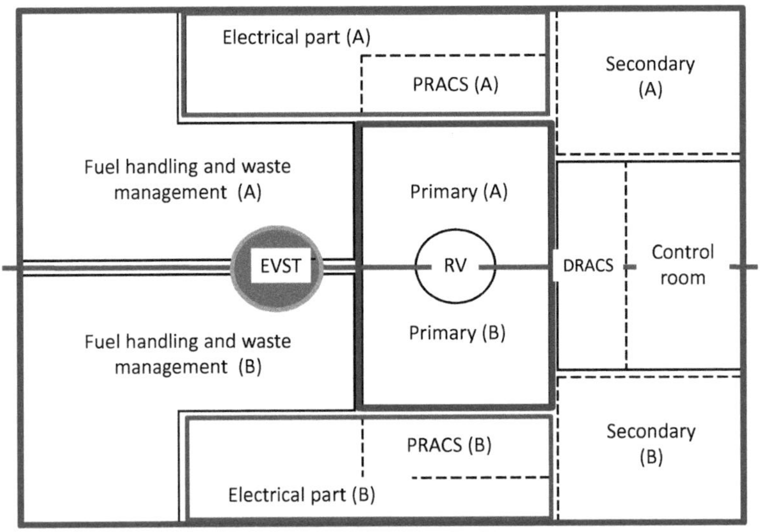

Fig. 4.174 Distributed layout concept [110].

of the building. Because of this distributed layout, plant commodities may increase. However, the impact of hazards on safety-grade components could be limited.

Fig. 4.175 shows the plot plan of JSFR. This design shows that an elevated installation is adopted for the JSFR site, and alternative power supplies are put into another isolated building. The elevation level of the reactor building depends on the site and on tsunami impacts for the protection of the alternative emergency power supply from floods. The emergency power supply can be put on the upper floor of the reactor building. In addition, the waterproof hatch can effectively mitigate the impacts of tsunamis.

These measures could avoid the local initiation of events and the simultaneous redundant failure of safety-grade systems and could decrease the risk of loss of all fixed (loss of all air heat sink) components to maintain essential power supply.

A summary of the evaluation results and countermeasures against external hazards is shown in Table 4.34. These results show that the JSFR design can withstand possible external hazards. The distributed building layout is adopted for the safety enhancement of the

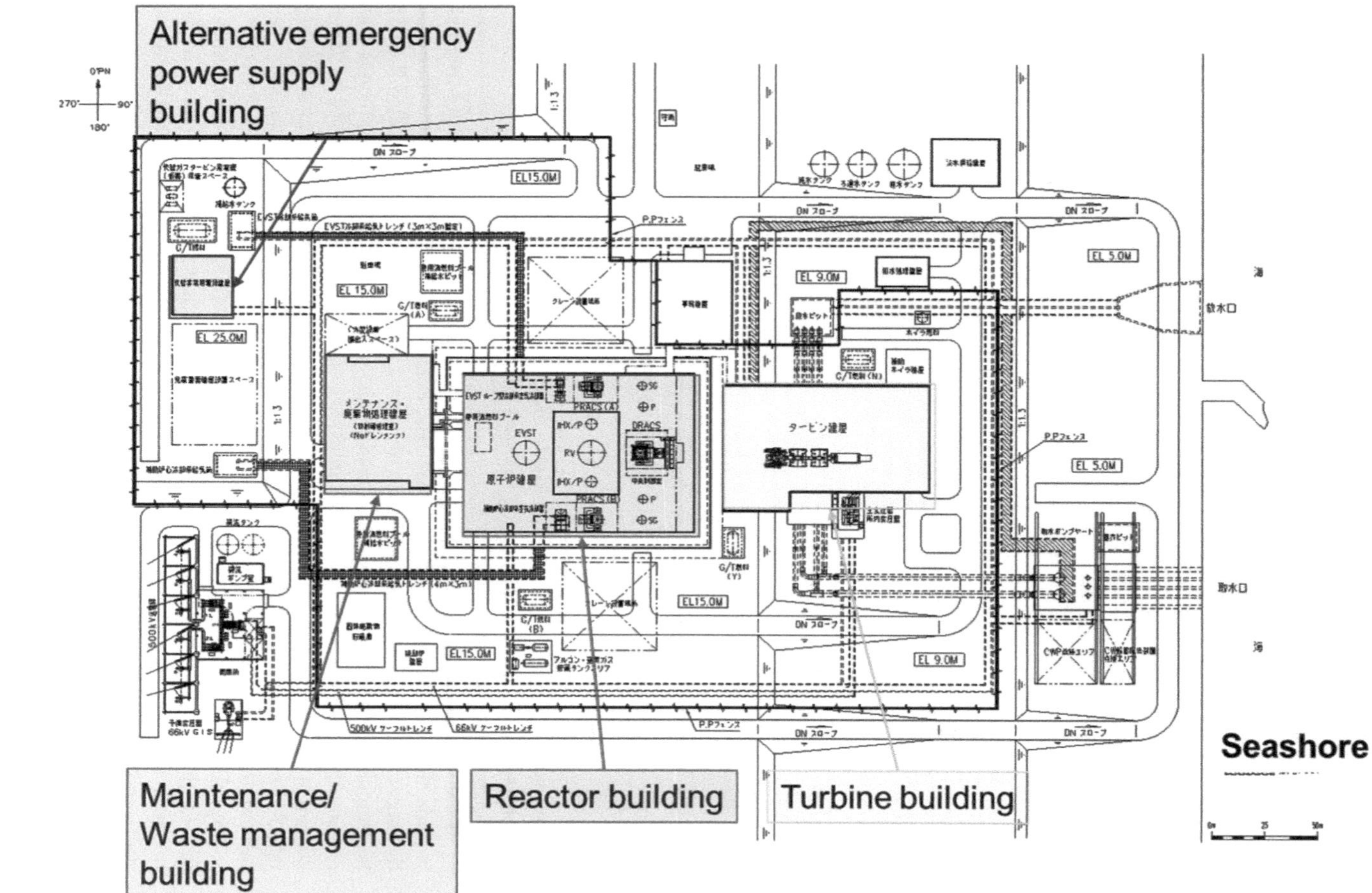

Fig. 4.175 Updated plot plan [110].

Table 4.34 Countermeasures against external hazards [110].

External hazards	Countermeasures
Seismic	1.8 × Ss margin
Tsunami	Wall thickness: 1.3 m for extremely condition Waterproof hatch Upper-level layout of emergency power supply
Strong wind	Wall thickness of DHRS stack: 0.8 m
Snowpack	No change

Table 4.35 Frequency mode without the vertical wall [112].

Mode	Previous building IC (1.5 m)	Reinforced building IC (2.0 m) Vertical wall
1st	Whole building	
	6.2 Hz	6.2 Hz
2nd (A) (B)	RB + IC (A) 17 Hz (B)20 Hz	RB 19 Hz
3rd	SC 28 Hz	SCCV+IC 26 Hz

Copyright 2016 by the American Nuclear Society, La Grange Park, Illinois.

reactor building. The layout aims to mitigate the impact of airplane attacks and tsunamis and prevent the failure of safety-grade facilities from external hazards.

(f) Reinforcement of the reactor building

Because of the changes in the seismic conditions and the countermeasures for external hazards, the outer SC walls must be thicker. The improvement of the structural stiffness of the inner concrete (IC) and the installation of vertical walls supporting the installed floor of the primary pump motor must be considered for the reactor building to reduce the vertical response acceleration. Before 2011, the thickness of the outer SC wall was 0.4 m. After 2011, it became 1.4 m as a countermeasure against external hazards. Moreover, the IC stiffness was improved. Table 4.35 shows the comparison before and after the stiffness improvement at about the natural frequency.

Figs. 4.176 and 4.177 show vibration modes for two conditions. Before the countermeasure for the IC thickness, coupled vibration between the reactor building and IC at 20 Hz occurred. However, after the countermeasure for the IC thickness and installation of vertical walls, coupled vibration between the reactor building and IC did not occur.

A 3D finite element method (3D-FEM) model of the containment vessel (CV) was built (Fig. 4.178), and a countermeasure against vertical seismic response was studied using the results of the seismic response analysis considering the effects of out-of-plane vibration. Because the seismic conditions for the primary pump motor were severe, the floor response

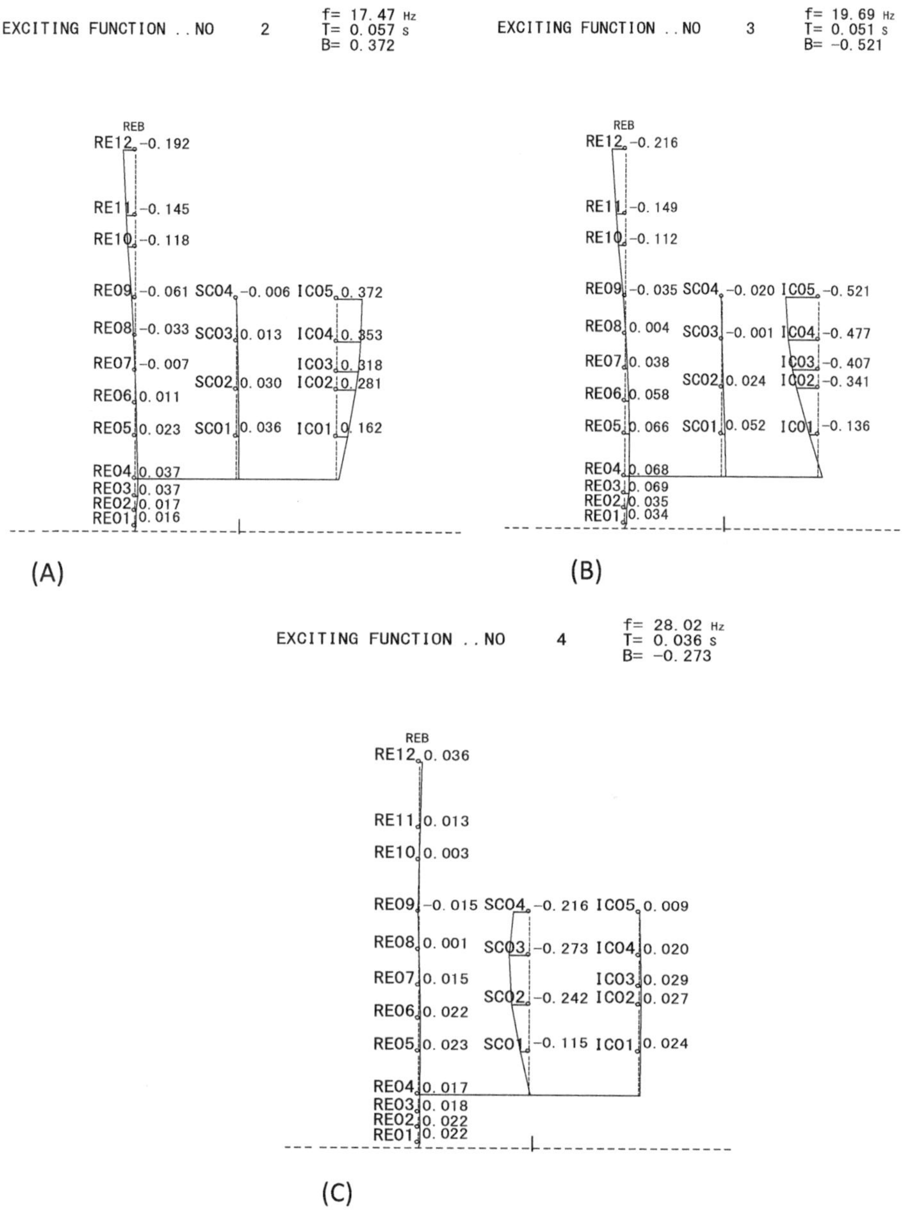

Fig. 4.176 Vibration mode for the original reactor building [112]. (A) 2nd mode (A), (b) 2nd mode (B), (C) 3rd mode.

Copyright 2016 by the American Nuclear Society, La Grange Park, Illinois.

was selected as an evaluation item. The 3D building analyses were conducted under two conditions: without and with vertical walls. These comparisons between the "without" and "with" vertical walls conditions can grasp the impact of out-of-plane vibration at the local floor.

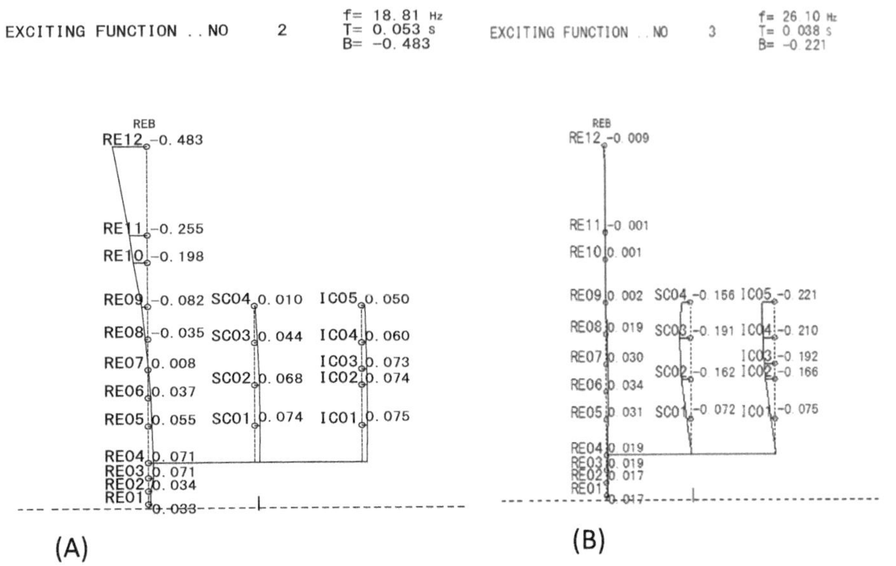

Fig. 4.177 Vibration mode for the updated reactor building [112]. (A) 2nd mode, (B) 3rd mode. Copyright 2016 by the American Nuclear Society, La Grange Park, Illinois.

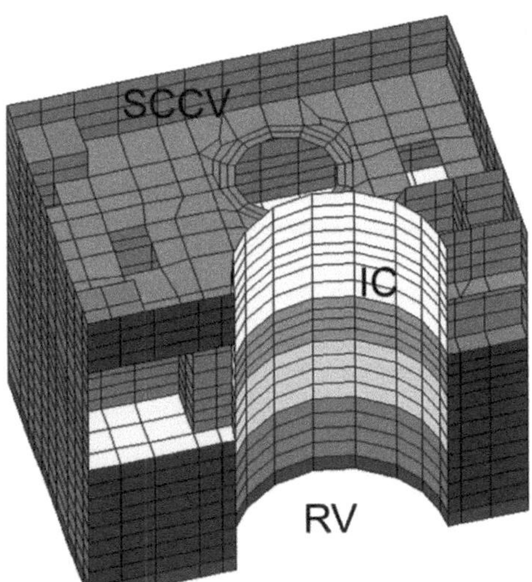

Fig. 4.178 3D-FEM model of the building [112].
Copyright 2016 by the American Nuclear Society, La Grange Park, Illinois.

Fig. 4.179 shows the vertical walls around the installed position of the primary pump motor in the 3D-FEM building model.

Fig. 4.180 shows the results of the frequency mode for the condition without vertical walls. The frequency modes 26.43 and 26.82 Hz were the natural frequency modes at

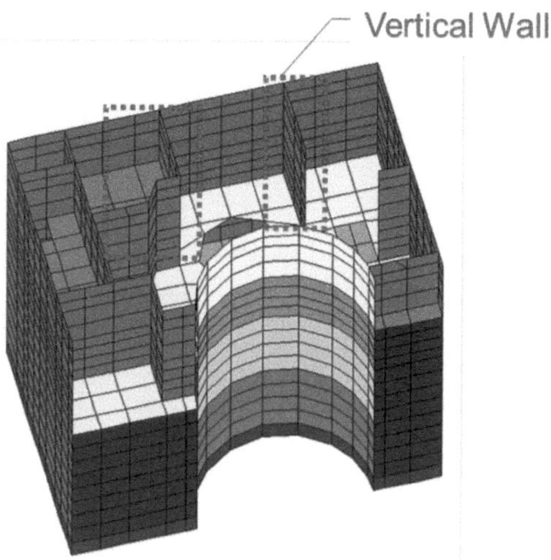

Fig. 4.179 Vertical walls around the installed position of the primary pump motor [112].
Copyright 2016 by the American Nuclear Society, La Grange Park, Illinois.

Frequency	CV mode
17.31Hz (0.58 s)	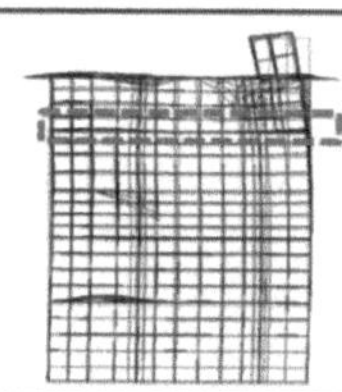
26.43 Hz (0.038 s)	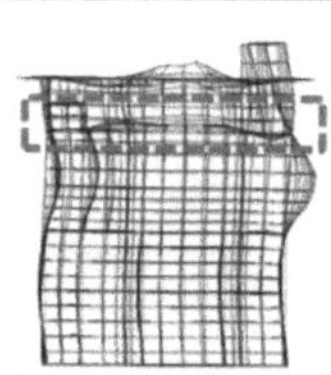
26.82 Hz (0.037 s)	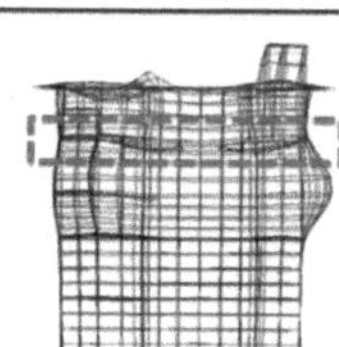

Fig. 4.180 Frequency mode without vertical walls [112].
Copyright 2016 by the American Nuclear Society, La Grange Park, Illinois.

the floor of the primary pump motor. Fig. 4.181 shows the frequency mode with vertical walls. The frequency mode 26.55 Hz was the natural frequency mode at the floor of the primary pump motor. Fig. 4.182 compares the conditions without and with vertical walls. The result of the analysis for the condition without vertical walls was over the threshold for

Fig. 4.181 Frequency mode with vertical walls [112].
Copyright 2016 by the American Nuclear Society, La Grange Park, Illinois.

Frequency	CV mode
17.31 Hz (0.058 s)	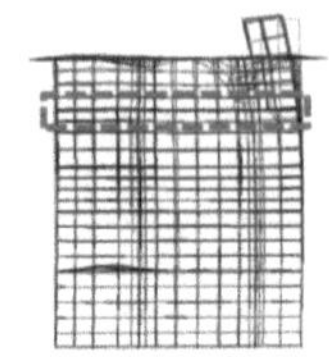
26.55 Hz (0.038 s)	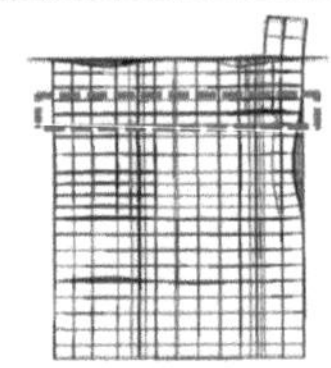

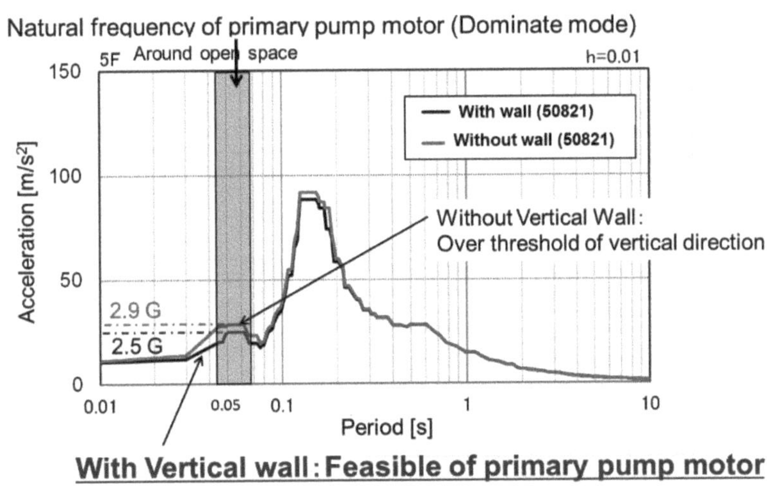

Fig. 4.182 Comparison of the conditions without and with vertical walls [112].
Copyright 2016 by the American Nuclear Society, La Grange Park, Illinois.

the vertical direction. In contrast, that for the condition with vertical walls satisfied the threshold. Therefore, the vertical walls supporting the installed floor of the primary pump motor are effective as response measures to reduce out-of-plane vibration.

The results for the mass-spring model and that for the 3D-FEM model are compared. Fig. 4.183 shows the relation between the 3D-FEM and the mass-spring models. The mass-spring model for the condition without vertical walls has a shear spring and a bending spring at the floor of the primary pump motor (Fig. 4.184). As shown in Fig. 4.185, there is little difference between the mass-spring model and the 3D-FEM model.

The mass-spring model of the condition with vertical walls is rigid at the floor of the primary pump motor (Fig. 4.186). As shown in Fig. 4.187, there is little difference between

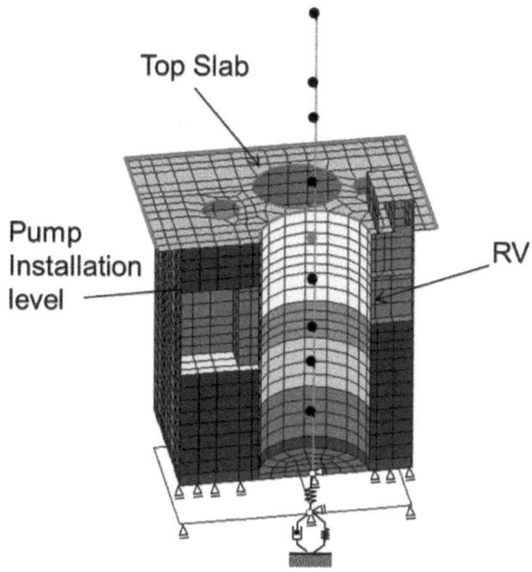

Fig. 4.183 Relation between the 3D-FEM and mass-spring models [112]. Copyright 2016 by the American Nuclear Society, La Grange Park, Illinois.

the mass-spring and 3D-FEM models (i.e., the difference is negligible). Therefore, the adequacy of the mass-spring model in the floor response spectra analysis was confirmed.

4.3.3 Reactor cooling system update

For the reactor cooling system, the component designs of the demonstration JSFR with 750 MW electric output were updated considering the SDC and SDG. For example, an alternative cooling system was installed in the reactor vessel, and the seismic design was enhanced with the updated seismic conditions. Additionally, improvements in the maintainability and reparability of the reactor vessel, piping system, and integrated IHX-pump were investigated. For the integrated IHX-pump and SG, the progress of the R&D and alternative concepts to reduce the R&D load are also described.

4.3.3.1 Reactor structure

To improve the maintainability and reparability, the following criteria were considered [117]:

- At least one side of the coolant or cover gas boundary must be available for visual or volumetric tests on welding lines without cutting boundaries.
- For parts that might have a higher risk of failure or defect, an access route for visual inspection and repair must be available.
- For parts that have the possibility of lost components, an access route for visual inspection and device recovery must be available.

Fig. 4.188 shows the inspection and checking of difficult parts. Checking of the following parts was difficult or possible only within a limited region: "the roof deck

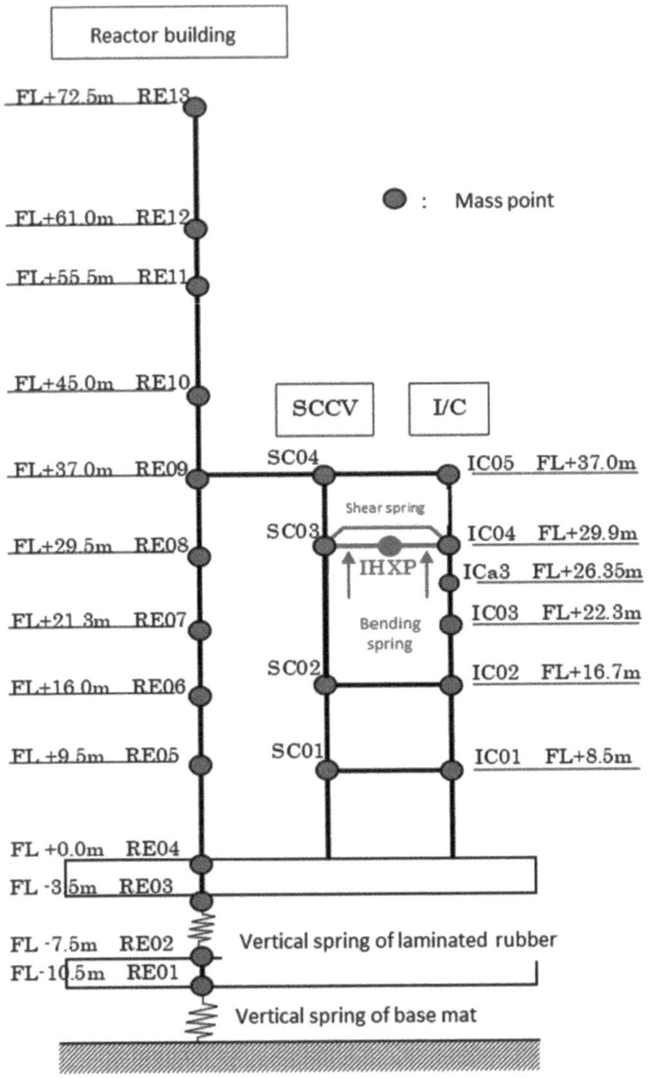

Fig. 4.184 Mass-spring model (without vertical walls) [112].
Copyright 2016 by the American Nuclear Society, La Grange Park, Illinois.

interior," "the rotating plug interior," "the control rod drive guide tube surfaces located within the UIS region," "the reactor vessel (RV) and sodium dam walls within the dam annulus," "the core support structure interior," "the upper core support plate undersurface," "the lower core support plate upper surface," "the dip plate upper surface," "the core catcher lower part," etc. The parts were mainly difficult to check because there was no access route to the parts or clearance of the access route was narrow.

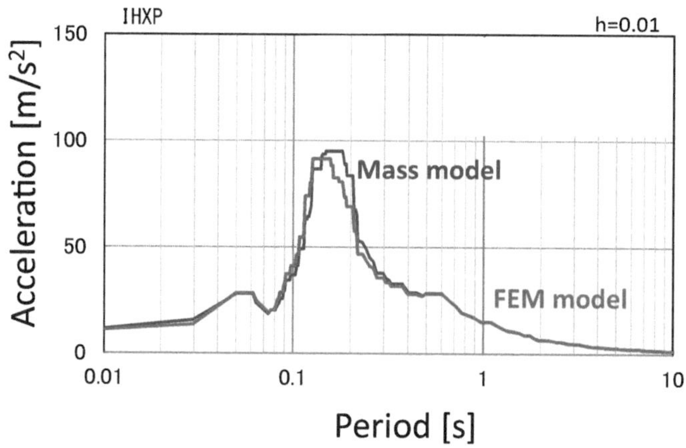

Fig. 4.185 Comparison of the mass-spring model and the FEM model (without vertical walls) [112].
Copyright 2016 by the American Nuclear Society, La Grange Park, Illinois.

Fig. 4.189 shows the points of issue and the improvement measures for the RV. Improvement measures were provided for the following parts: RV-guard vessel (GV), roof deck, rotating plug-UIS, and the core internals. These measures were mainly enlargement of the access route for inspection devices, additional access routes, and accommodation of in-vessel piping repair. The clearance between the RV and GV was enlarged from 150 mm to 200 mm considering development on inspection devices. Additional access routes were installed to allow inspection of the in-vessel structures, as shown in Fig. 4.190. To accommodate in-vessel piping repair, flanges were installed at the connection between the in-vessel and ex-vessel piping. The design of the reactor structure was modified considering all the measures. It was found that most of the parts inside the RV could be accessible and inspected using remote access devices, such as fuel handling-machine-type or periscope-type devices with ultrasonic wave sensors. Although it is possible to access and detect lost parts under the core catcher, collecting the lost parts remains a challenge. For the compact design, the UIS and rotating plug are combined and must be extracted together. The repair procedures and handling cask for the UIS and rotating plug have been investigated.

In addition to improvements in maintenance and repair, RV protection, CDA consideration, and alternative DHRSs were installed [118]. A sodium annulus was installed on the RV to protect the main vessel from the moving sodium surface during start-up and shutdown operations. With the sodium annulus, the thermal stress on the main vessel could be reduced in the event of a stagnant sodium level.

For CDA consideration, discharge of molten fuel from the CRGTs and its relocation to the core catcher were incorporated into the design. The core support structure has protection against molten fuel jets, and the distance between the core support and the core catcher is large enough to form debris. Since the structural integrity of the core catcher was of concern due to its exposure to high sodium flow rates, the outlet

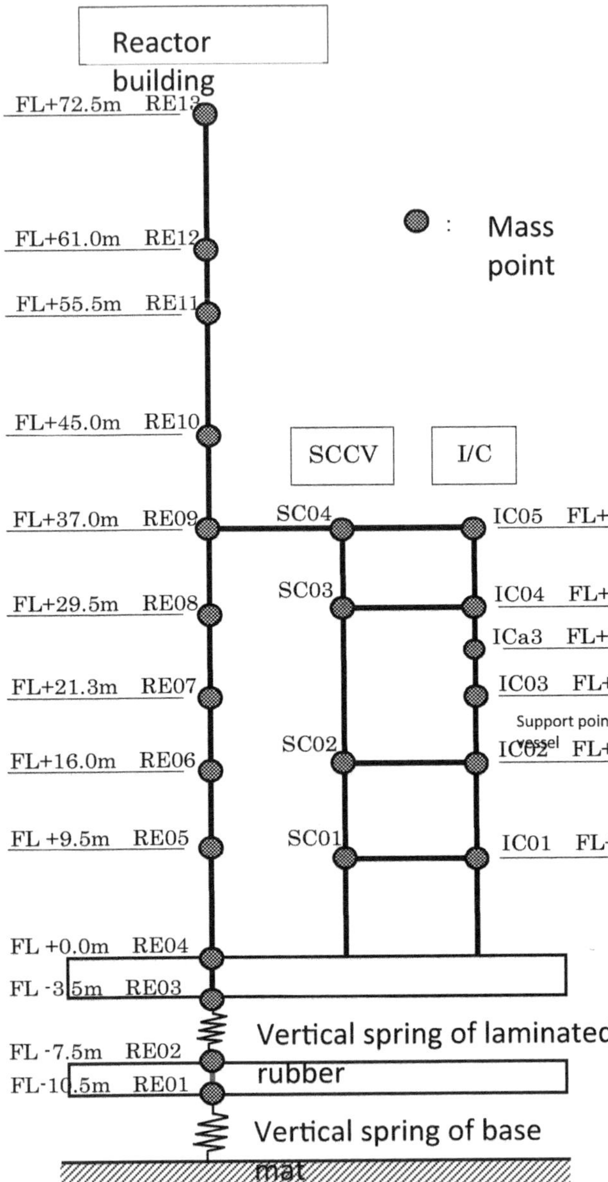

Fig. 4.186 Mass-spring model (with vertical walls) [112].
Copyright 2016 by the American Nuclear Society, La Grange Park, Illinois.

nozzle of the CL piping was modified to prevent such sodium flow from contacting the core catcher.

An auxiliary core cooling system (ACS) was recently installed as an alternative to the DHRS. Accommodating those design improvements, the inner diameter of the RV was enlarged from 9.46 m to 11.83 m, as shown in Fig. 4.191.

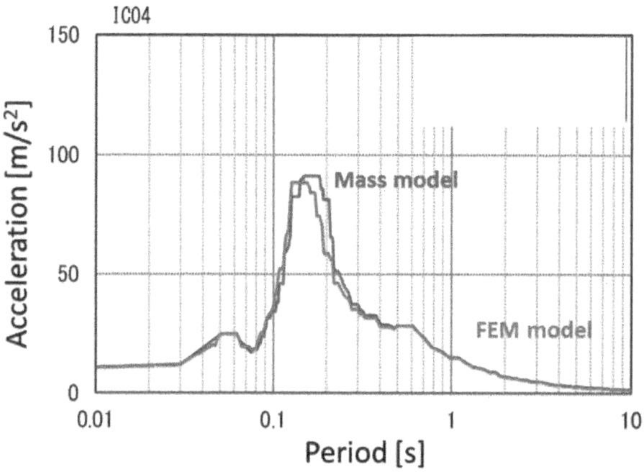

Fig. 4.187 Comparison of the mass-spring model and the FEM model (with vertical walls) [112].
Copyright 2016 by the American Nuclear Society, La Grange Park, Illinois.

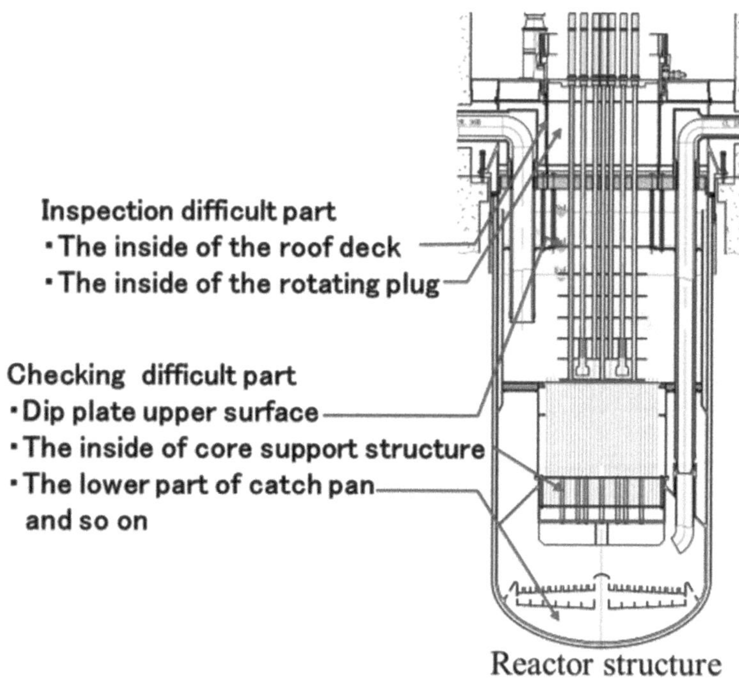

Fig. 4.188 Inspection and checking of difficult parts [117].

For seismic design, buckling was evaluated based on the revised design-basis earthquake conditions (SS). Considering the majority of domestic LWR conditions for both soft ground and hard ground, the design-basis earthquake condition was selected as 0.8 of the 2012 conditions (Fig. 4.163). For severe earthquakes beyond Ss conditions, 1.8SS was selected. The floor response spectrum at the RV support

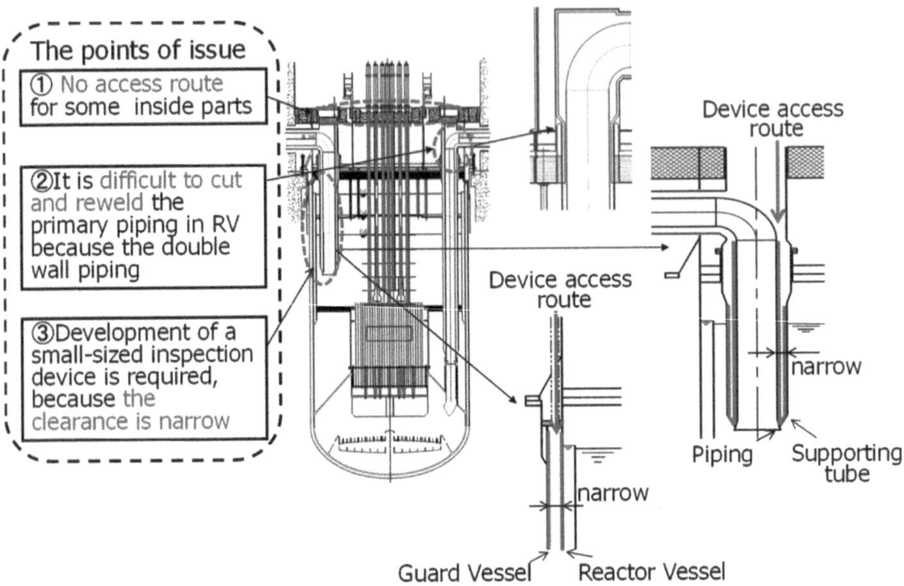

Fig. 4.189 Review of maintenance and repair improvements on reactor vessel [117].

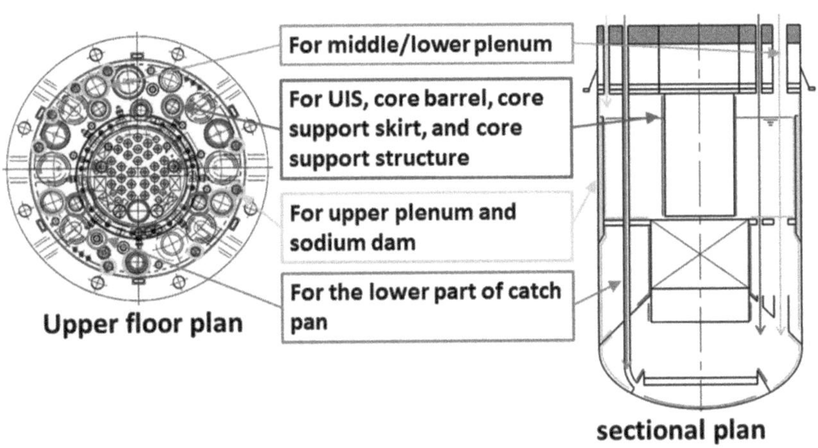

Fig. 4.190 Access to reactor structure [117].

is shown in Fig. 4.192. Based on seismic analysis, the RV could maintain its integrity with a thickness of 45 mm, even in a severe earthquake, including 1.8SS conditions. The integrity of the GV was evaluated with sodium in the annulus between the reactor and GVs. Based on seismic analysis, a GV with a 25 mm thickness could withstand the 1.8SS seismic conditions.

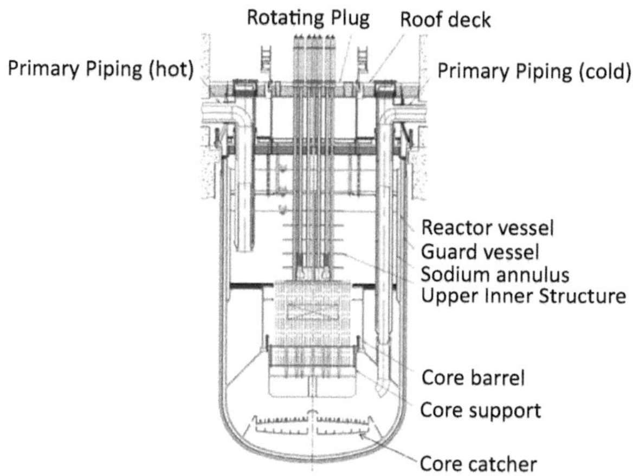

Fig. 4.191 Sketch of updated reactor structure [118].

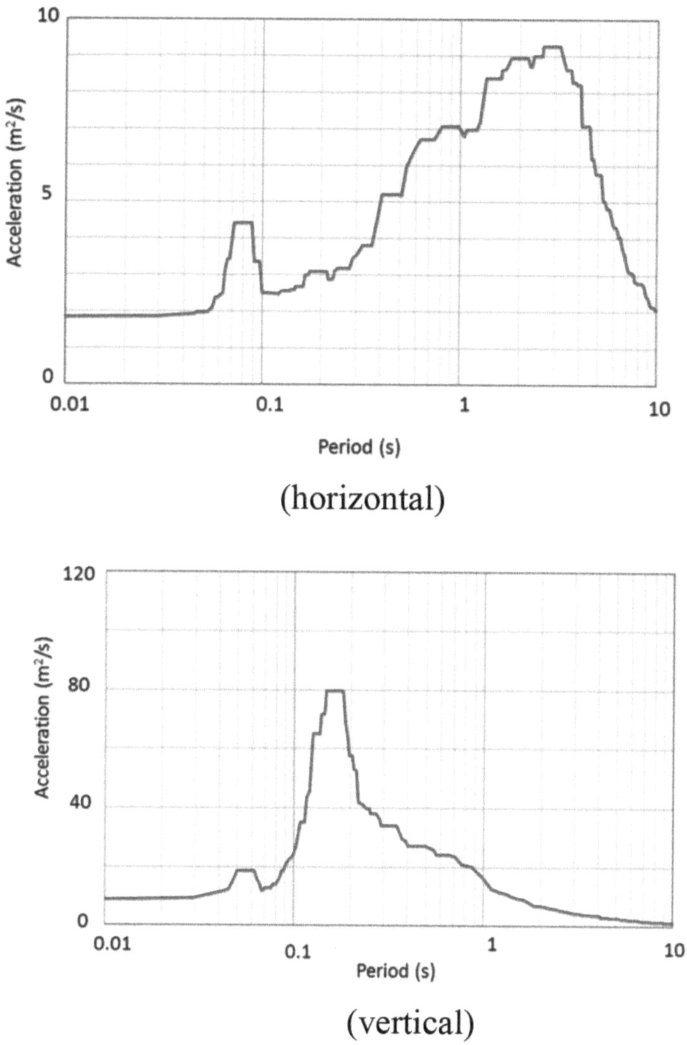

Fig. 4.192 Floor response spectrum at the reactor vessel support [118].

4.3.3.2 Primary main piping

The primary main piping was difficult to inspect, check, and repair because a double-walled piping system was applied and the clearance between the pipe and guard pipe was narrow. The double-layer structure was applied for the bellows of the horizontal guard pipe to shorten the length as much as possible. On the outside bellows, a penetration crack could be detected using a pressure resistance test; however, the inside bellows were difficult to check and repair. The primary hot-leg piping parts were difficult to check, and a sketch of the improved design is shown in Fig. 4.193 [117]. For the primary piping inspection, a survey study on the inspection devices was conducted and improvement of the inspection access was investigated. An electromagnetic acoustic transducer (EMAT) was selected as the inspection sensor. For the inspection device access, the thermal insulation was moved from inside to outside of the guard piping. The partition structure and bellows on the guard pipe design were adjusted. For repair, the work procedure and space were evaluated based on the experiences for the Mod. 9Cr—1Mo steel piping in fossil-fueled power plants. Considering the chemistry and radioactivity of primary sodium, remote control devices in a greenhouse (temporary workroom) filled with argon gas must be developed. Component arrangement must be adjusted to provide space for the repair procedure and greenhouse.

4.3.3.3 Integrated pump-IHX

(a) Design update

From the viewpoint of maintainability and reparability improvement, an integrated IHX-pump was reviewed, as shown in Fig. 4.194. Inspection of all of the parts is feasible; however, the representative parts that are difficult to access are the following: "the heat exchange tube surfaces in the IHX," "the upper tube sheet," "the upper surface of the lower

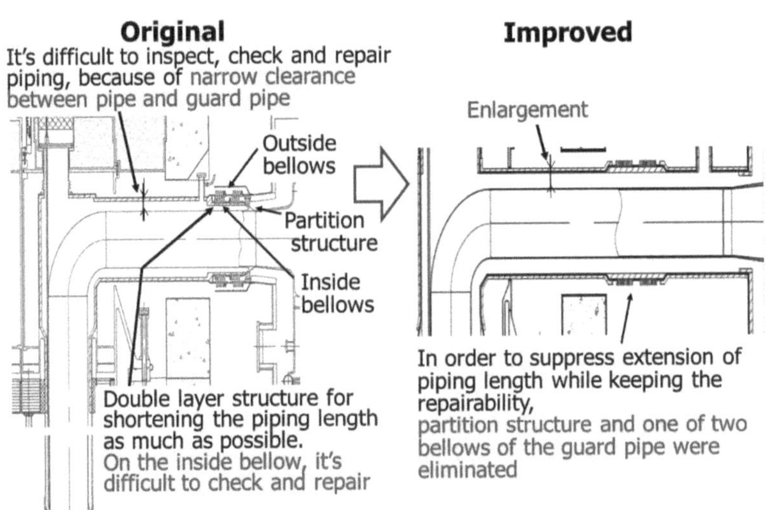

Fig. 4.193 Improvement of primary piping [117].

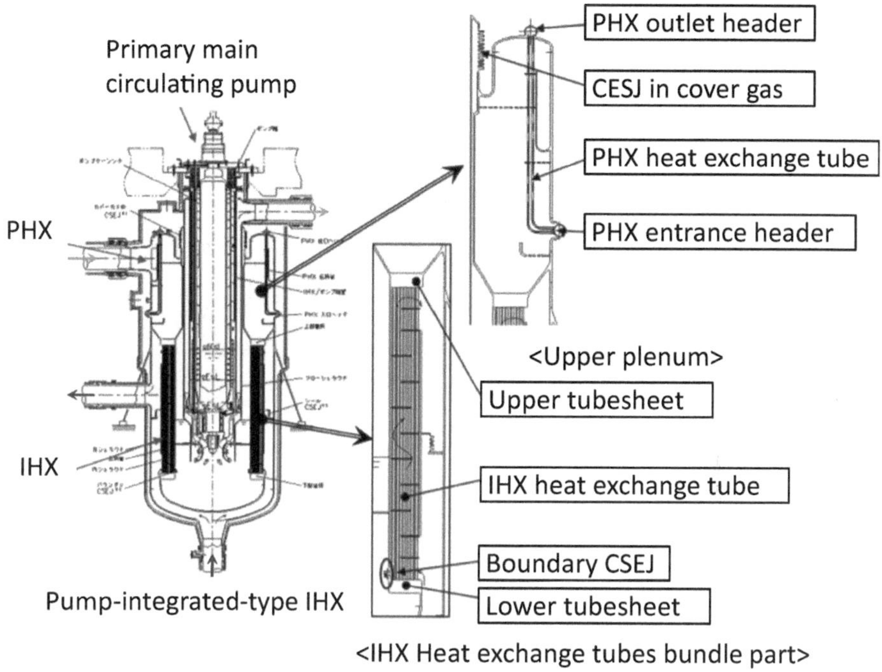

Fig. 4.194 Review of maintenance and repair on integrated IHX-pump [117].

tube sheet," "the interior of the convoluted shell expansion joint (CSEJ) in the cover gas atmosphere," "the CSEJ exterior used as boundary," "the header interior on the PHX," and "the heat exchange tubes in the PHX." For the repair work, extraction of the primary pump provides some access routes. However, a device for repairing the lower tube sheet surface must be developed. It is difficult to repair the upper tube sheet, CSEJ, PHX header, and PHX heat exchange tubes.

For the integrated IHX-pump component, inspection and repair of the PRACS heat exchange tubes located in the IHX (PHX), and repair of the IHX have been improved, as shown in Fig. 4.195. In the FaCT phase I design, the PHX tubes were built in the upper plenum of the IHX. In the improved design, the built-in-type modules were modified to plug-in-type modules. With the plug-in-type PHX, inspection and repair are possible after extraction of the PHX, and access to the IHX upper plenum was improved. For repair of the IHX itself, extraction of the entire IHX was investigated considering the development risks of the in-place repair devices. As a result, extraction and recovery of the entire IHX are feasible considering the workspace and dose ratio.

For the integrated IHX-pump, design evaluations of the improved M&R design were conducted. In the improved design, six modules of the plug-in-type PHX were located in the upper plenum of the IHX. For the plug-in-type PHX, support for the PHX was investigated considering thermal insulation and severe earthquake. Due to the newly added deck on the IHX GV, the amount of steel in the integrated IHX-pump was enlarged.

When the pump is not in operation, natural convection occurs in the cover gas around the pump shaft, which could induce asymmetric temperature distribution on the shaft periphery. This effect could cause bending of the shaft. A thermal shielding plate and convection

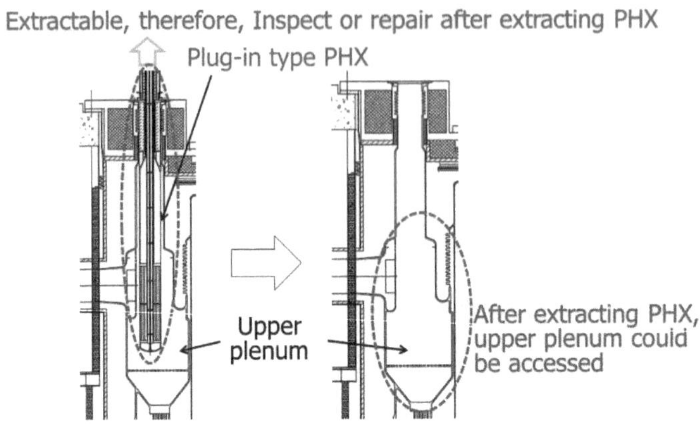

Fig. 4.195 Improved design for accessing upper plenum [117].

preventive plate were adopted as the countermeasure to minimize the effect [119]. Accounting for manufacturability, the design of these structures is shown in Fig. 4.196. The analytical model is shown in Fig. 4.197. The maximum circumferential temperature differences were found to be 2°C and 6°C in the shaft surface and pump casing, respectively, from the temperature distribution evaluation in the shaft peripheral structure (see Fig. 4.198). Displacement at the lower shaft bearing was 0.2 mm, which is sufficiently small compared to the design gap of 0.8 mm (see Fig. 4.199). Thus, one can conclude that contact between the shaft and lower shaft bearing due to the cover gas effect can be prevented.

The design requirements include enhancing the resistance of the component integrity against the latest seismic conditions. The durability of the integrated IHX-pump against

Fig. 4.196 Pump shaft peripheral structure [117].

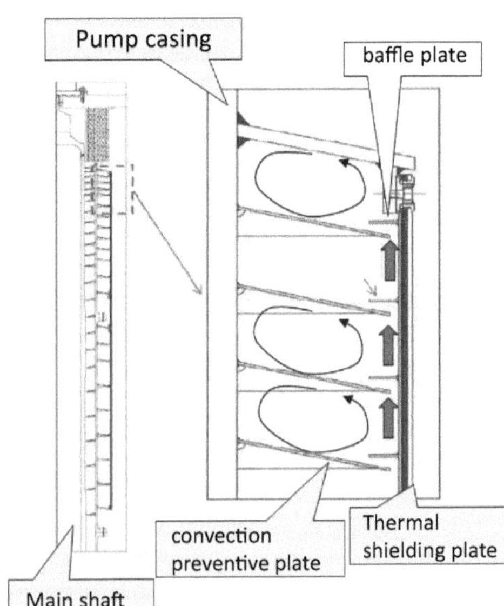

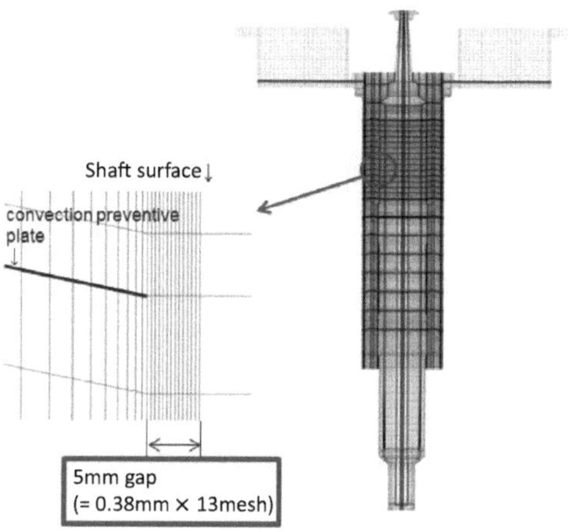

Fig. 4.197 Model for pump shaft analysis [117].

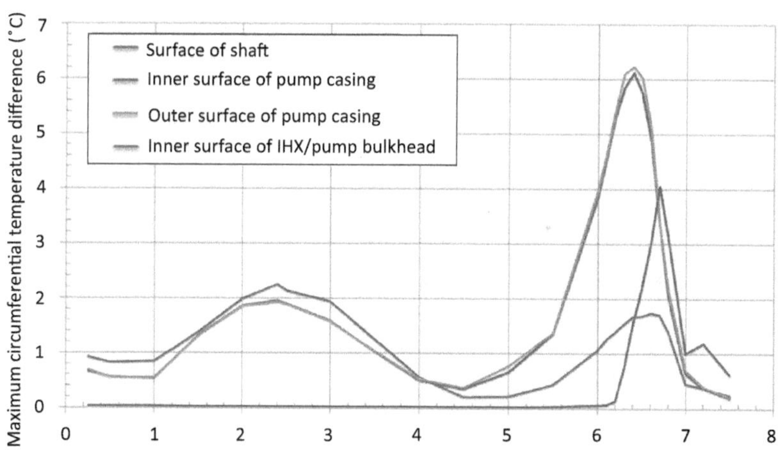

Fig. 4.198 Axial distribution of the maximum circumferential temperature difference [117].

severe seismic conditions, including the 1F accident conditions, has previously been evaluated [120]. After the 1F accident, the scope of the seismic conditions for the design study expanded to cover most of the seismic conditions at Japanese nuclear sites, as shown in Fig. 4.163. Therefore, a seismic evaluation for the most advanced integrated IHX-pump against the 2012 seismic conditions was made for boundary integrity and pump coast-down. A schematic of the integrated IHX-pump and the seismic analysis model is shown in Fig. 4.200. The analysis model was composed of a combination of shell and beam models. The top flange was a support for the pump shaft, and the skirt was a support for the IHX shell. For the boundary integrity, buckling at the Y-piece near the upper tube plate and connection between the shell and skirt were evaluated to be the high stress components. Table 4.36 shows the analyzed stresses. The evaluated buckling margins were 11.1 at the Y-piece and 11.7 at the skirt. Incorporating the required safety factor of 1.5, the buckling margins were determined to be 7.4 and 7.8 at the Y-piece and the skirt, respectively.

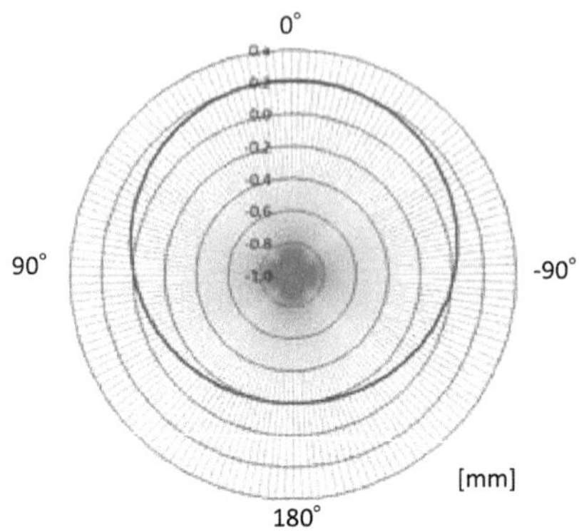

Fig. 4.199 Pump shaft displacement at the lower bearing [117].

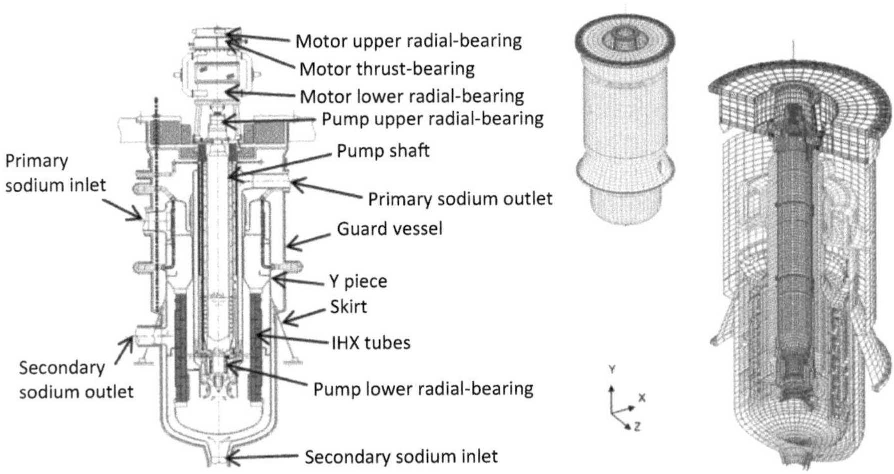

Fig. 4.200 Schematic and analysis model of integrated IHX-pump [121].

Table 4.36 Buckling evaluations on IHX boundaries [121].

Item	Y-piece	Skirt
Shear stress	6 MPa	5 MPa
Bending stress	1 MPa	1 MPa
Compressive stress	16 MPa	15 MPa
Gravity stress	5 MPa	5 MPa

In addition, the interaction between the shafts and bearings was evaluated for the pump coast-down. An upper radial bearing and a lower radial bearing exist for the pump shaft. There are also a thrust bearing, an upper radial bearing, and a lower radial bearing for the motor shaft. The evaluated seismic loads and criteria to prevent interaction between the shafts and bearings are shown in Table 4.37. The evaluated loads at the motor bearings

Table 4.37 Prevention of interaction between shaft and bearing [121].

Item	Lower pump	Upper motor	Lower motor
Evaluated load	73 kN	30 kN	81 kN
Criteria to prevent interaction	227 kN	155 kN	155 kN
Ratio of load to criteria	3.1	5.1	1.9

and the pump lower radial bearing showed sufficient margins. In contrast, the pump upper radial bearing and the motor thrust bearing were able to interact. By comparing the interacting contact surface pressure with the bearing circumferential velocity, it was determined that the pump upper radial bearing would not sear. The motor thrust bearing was improved so that it did not exceed the allowable load limit. Nevertheless, the load at the motor thrust bearing, which was intense due to the second peak of the increased vertical floor response, must be decreased by improving the buildings or evaluations rather than the component.

The heat removal function of the advanced loop-type SFR was enhanced to avoid loss of the function, even in the occurrence of internal events exceeding design-basis or severe external events. As a countermeasure against a loss of reactor level type event, double failure of the RV and GV is prevented by securing a margin for earthquake resistant performance and the reliability of the RV and GV. Therefore, the reliability of the integrated IHX-pump GV, particularly in the primary coolant leak, was evaluated. A primary sodium leak would cause high thermal stresses at the colder GV. Thermal strain with a primary sodium leak was therefore evaluated by an unsteady thermal stress analysis. Fig. 4.201 shows the evaluation section. Thermal strain at the skirt side of the connection was evaluated as a high stress integrity range part but within the allowance strain range. Fig. 4.202 shows Mises stress contour figures 30 s and 3000 s after the primary sodium leaked. The evaluated strain range was 2.04%, while the allowance strain range was 2.66% with 10 cycles. The ratio between the evaluated value and the allowance was 1.3. The GV was evaluated to be sound in earthquake conditions with the margin previously described. However, as a hypothetical case, the seismic structural soundness of the GV with the leaked

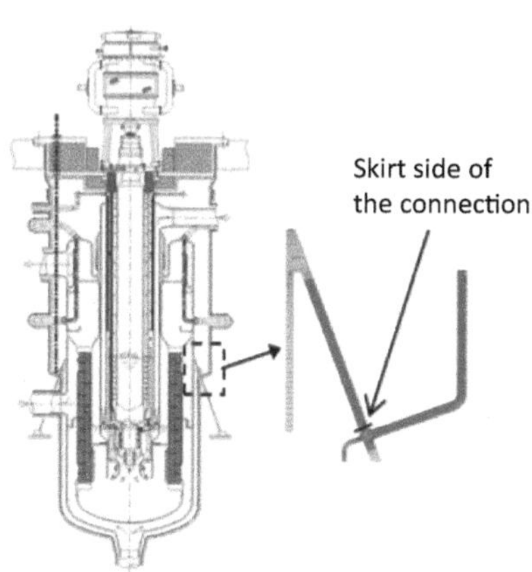

Fig. 4.201 Evaluation section for thermal stress [121].

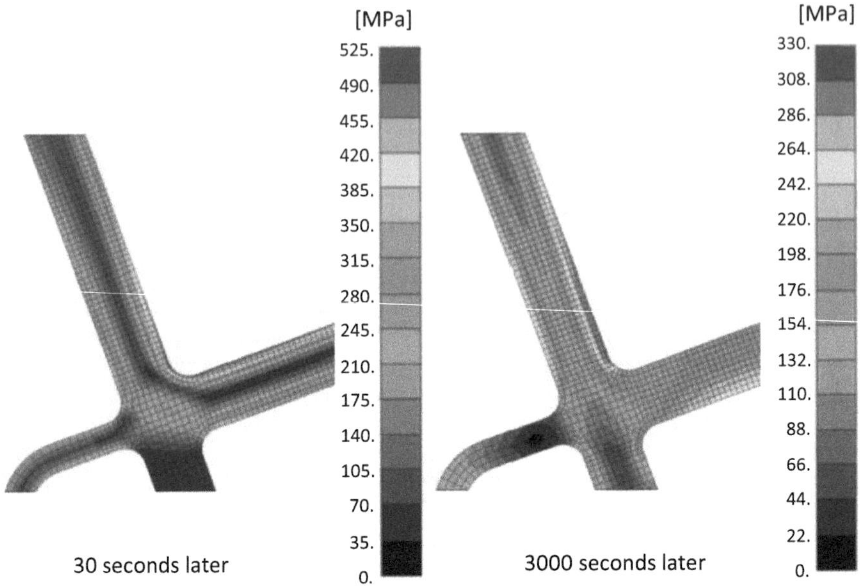

Fig. 4.202 Mises stress after sodium leak [121].

primary sodium was evaluated. Fig. 4.203 shows the analysis model and the buckling evaluating point. Buckling at the lower GV trunk was evaluated as a high stress part. Table 4.38 shows the analyzed stresses at the lower guard vessel trunk. The evaluated buckling margin was 200, while the required safety factor was 1.5. The ratio between the evaluated value and the requirement was 133.

A sodium-water reaction in the SG has been taken into account in the DECs. The CSEJs are one of the vulnerable parts in the integrated IHX-pump during the sodium-water reaction. Therefore, the CSEJs were enhanced to withstand the sodium-water reaction. The

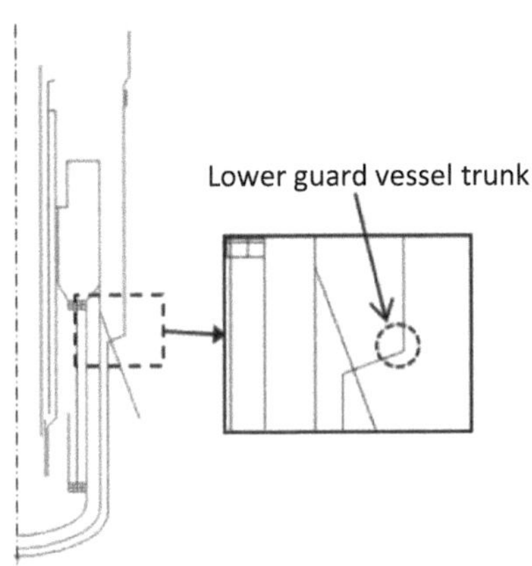

Fig. 4.203 Analysis region for buckling evaluation [121].

Table 4.38 Buckling evaluations on IHX GV with sodium leak [121].

Item	
Shear stress	0.6 MPa
Bending stress	0.1 MPa
Compressive stress	0.7 MPa
Gravity stress	0.2 MPa

CSEJ used as a primary-secondary boundary and as a secondary inlet-outlet seal is exposed to high pressure during a sodium-water reaction. These CSEJs are shown in Fig. 4.204. The boundary CSEJ is exposed to external pressure during the sodium-water reaction. The capacity of the boundary CSEJ with external pressure was evaluated by an elastic analysis. It was found that the boundary CSEJ could endure over 10 MPa of external pressure. The sodium-water reaction margin was evaluated to be 3.3, and the seismic margin was 2.4. The CSEJ has been used as a secondary inlet-outlet seal because it has a good sealing capacity. On the other hand, it has been difficult to maintain sealing in the sodium-water reaction and to maintain its performance during in-service periods due to the installation location of the CSEJ seal. Therefore, despite its lower sealing capacity than the CSEJ, the seal ring was tentatively adopted as the secondary inlet-outlet seal due to its tolerance to the sodium-water reaction, as determined by studies of the IHX heat transfer performance and structural soundness. The final adaptation will be made using several experimental data for the IHX heat transfer performance and structural soundness. Accommodating the above design

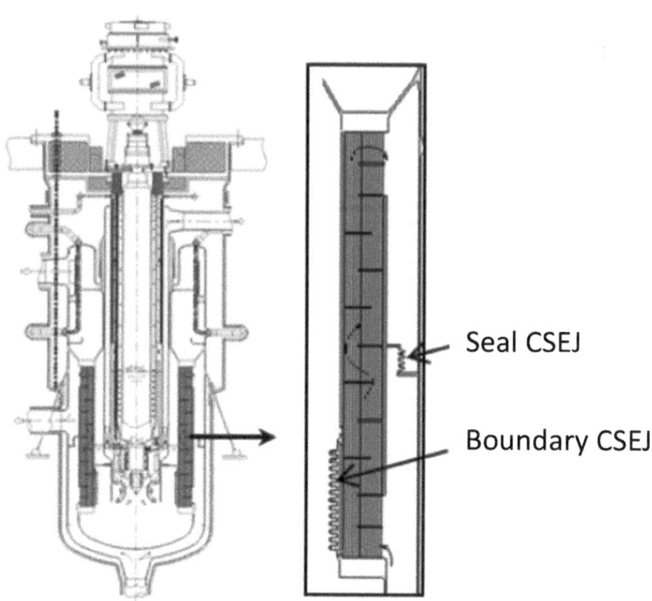

Fig. 4.204 CSEJ in integrated IHX-pump [121].

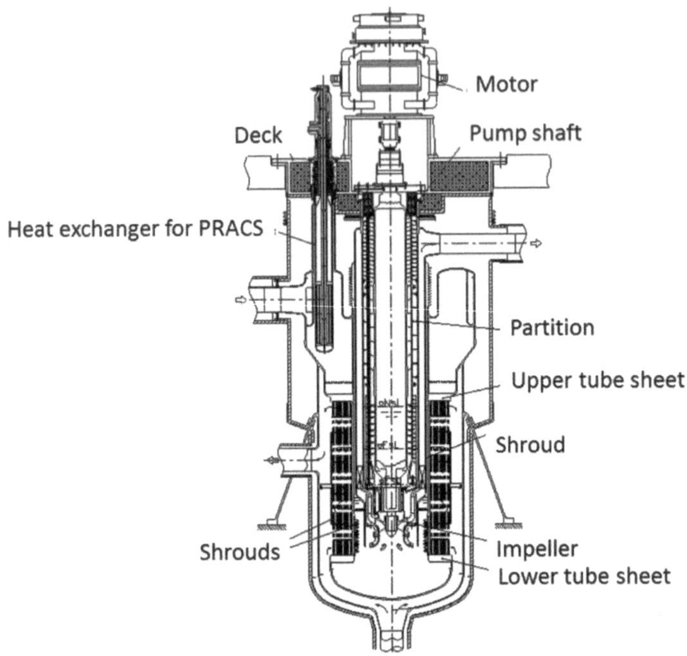

Fig. 4.205 Updated integrated IHX-pump [118].

update and requirements regarding the fabrication capability as well, the integrated IHX-pump has been modified as shown in Fig. 4.205.

(b) Evaluation of updated design

Some important evaluations have been updated, accommodating the above design update, including thermal transient, structural sympathetic vibration with pump rotation, and wear-out of the IHX tubes.

From the viewpoint of thermal transient, when an accident event of a primary pump seizure occurs, the primary coolant will flow back and the temperature drops in the accident loop due to the pressure difference between the accident loop and the sound loop. The temperature drop is intense and rapid because the advanced loop-type SFR has two primary coolant loops. Fig. 4.206 shows the temperature history around the upper tube plate of the IHX after an accident. For the temperature drop around the upper tube plate being over 200°C, high thermal stress occurs at the discontinuous part around the tube plate and makes the creep fatigue damage intense.

Moreover, the modification for the enhancement of the maintainability and reparability has made the creep fatigue damage more intense. Therefore, the weakest part of the upper tube plate against thermal transients in the integrated IHX-pump was improved to absorb the thermal transient impact:

- The base of the tube plate slinging trunk was tapered to reduce stress by dispersing the stress occurrence area.
- The tube plate slinging trunk was made from a conical to cylindrical structure to decrease the bending stress at the base of the tube plate slinging trunk.
- The discontinuous part around the tube plate was flattened by increasing or decreasing the thickness of each part to decrease the stress.

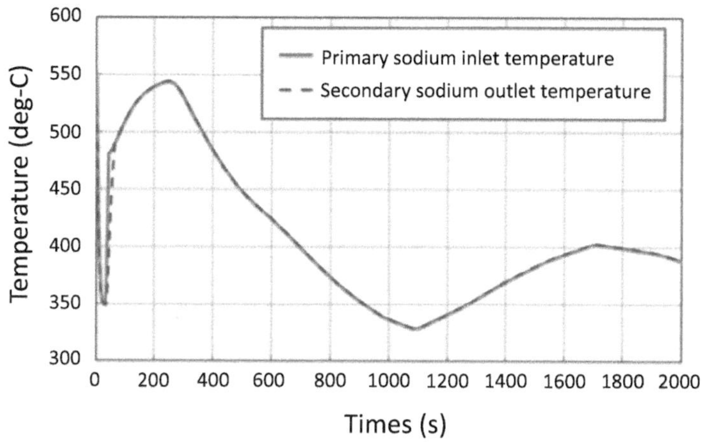

Fig. 4.206 Temperature transient around upper tube plate [121].

Based on the improved upper tube plate, the thermal stress in the accident was analyzed and the creep fatigue damage was evaluated. Fig. 4.207 shows a Mises stress contour figure when the temperature around the upper tube plate drops to 350°C in 30 s from the beginning of the accident. The creep fatigue damage in areas with the highest thermal stress, namely the base of the tube plate slinging trunk and the inner-shroud joint, was evaluated and was found to satisfy the criteria.

It is important to keep the natural frequency of the integrated IHX-pump separate from that of the pump rotation to avoid sympathetic vibrations since the pump is within the IHX. The natural frequency of the most advanced integrated IHX-pump was analyzed using an eigenvalue analysis and compared with the pump rotation. The seismic analysis model, including the plug-in type PRACS heat exchanger (PHX) and others, was used as the natural frequency analysis model, as shown in Fig. 4.200. Fig. 4.208 shows the frequency to

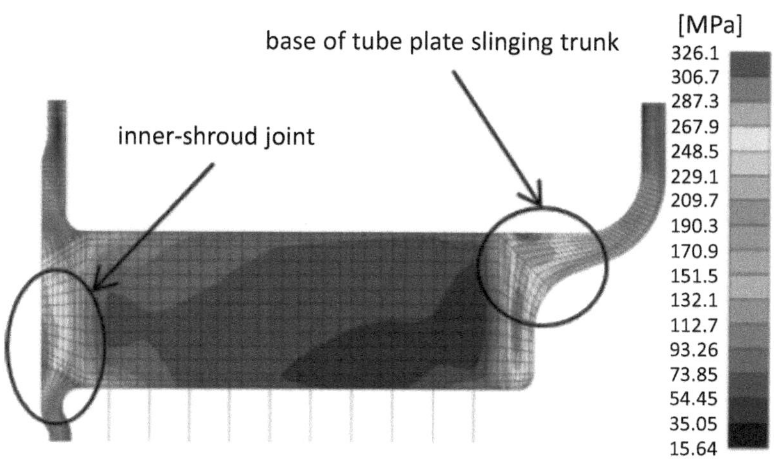

Fig. 4.207 Mises stress after 30s [121].

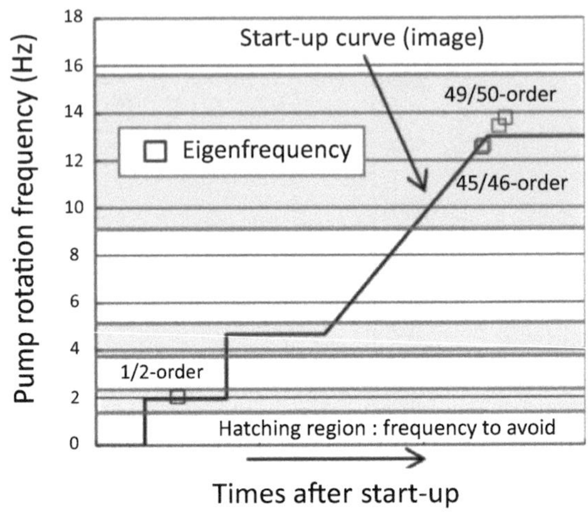

Fig. 4.208 Relationship between analyzed natural frequency and pump start-up [121].

avoid after the pump rotation start-up. The analyzed natural frequency was only within the sympathetic frequency region, corresponding to 1/2nd mode shapes; the pump shaft and casings corresponded to beam vibration mode shapes and 49/50th mode shapes; the upper and lower plenum vibration corresponded to monolithic mode. The analyzed 2nd and 49th modes are shown in Fig. 4.209. Increasing the thickness and thus the stiffness of the structures can prevent these eigenfrequencies from falling within the sympathetic frequency

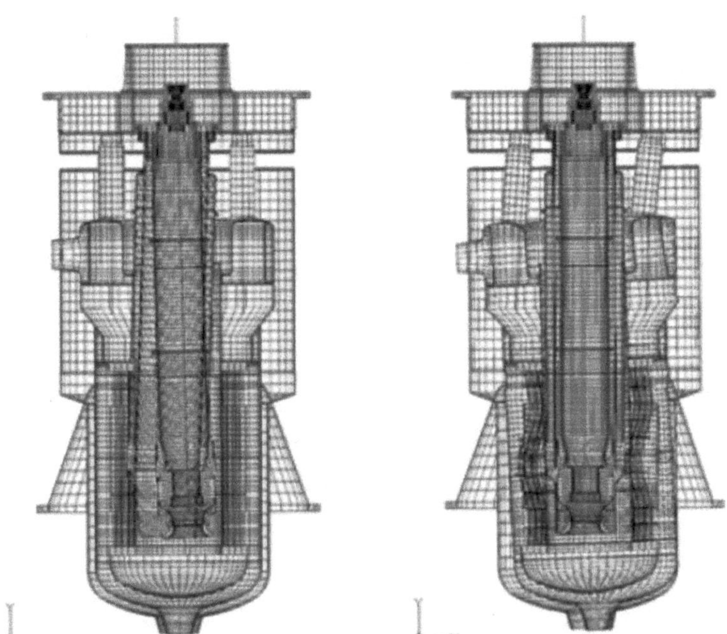

Fig. 4.209 2nd and 49th modes [121]. (A) 2nd mode (2.03 Hz), (B) 49th mode (13.44 Hz).

region. The pump shaft bending mode (45/46 order mode) was also within the sympathetic frequency region. The pump shaft bending mode can have a higher natural frequency than that analyzed by a pump rotation stability evaluation with a gyroscopic effect. The validity of the analyzed eigenfrequencies and the improvements must be confirmed by a mockup experiment in the future.

It is necessary to minimize wear-out of the IHX tubes caused by vibration responses of the baffle plates and IHX tubes. Wear-out of the IHX tubes can be evaluated by the combination of the wear property of the tube material, work rate, and operating time. The work rate, which is defined as the integration of the product of sliding length and frictional force per second, consists of the work rate of the mechanical excitation force by the pump vibration that propagates through the structure and fluid, and the work rate of the hydraulic excitation force by zigzag flow-induced vibration outside of the IHX tubes. Focusing on the hydraulic excitation force, the support pitches of the baffle plates were extended to slow the orthogonal flow to the IHX tubes and to reduce the work rate of the hydraulic excitation force. A 2D analysis with a porous model was conducted to confirm the flow velocity outside of the IHX tubes under the condition of the support pitches of the extended baffle plates. Figs. 4.210 and 4.211 show the 2D analysis range and the result, respectively. The work rate of the hydraulic excitation force was calculated using the analyzed flow velocity outside of the IHX tubes in several eccentric distances and eccentric angles of the IHX tubes as parameters, with the result that the necessary tube thickness required to accommodate wear-out was 1.14 mm. The reference tube thickness of 1.4 mm was evaluated to have a margin for the necessary thickness. In addition, a 3D analysis was conducted to comprehend the flow around the IHX tubes and check the validity of the 2D analysis

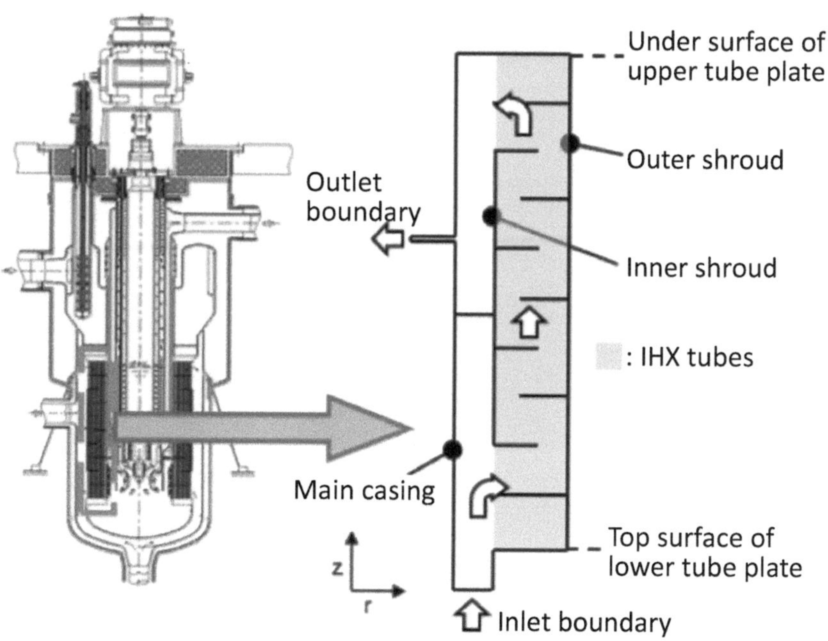

Fig. 4.210 2D flow analysis region [121].

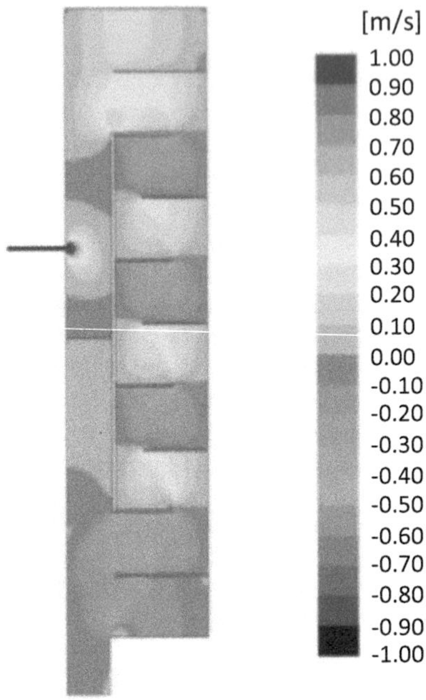

Fig. 4.211 Flow velocity results by 2D flow analysis [121].

Fig. 4.212 3D flow analysis model [121].

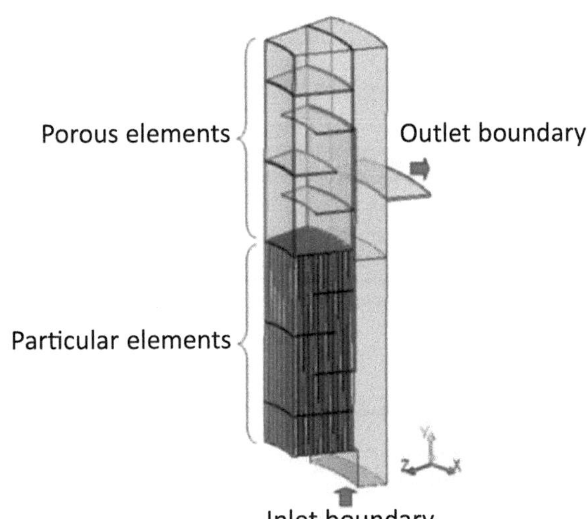

result. Figs. 4.212 and 4.213 show the 3D analysis model and the result, respectively. The flow velocity distributions around each IHX tube and more rapid flow areas were observed locally compared with the 2D analysis results. The hydraulic excitation force must be confirmed by 3D flow experiments in the future.

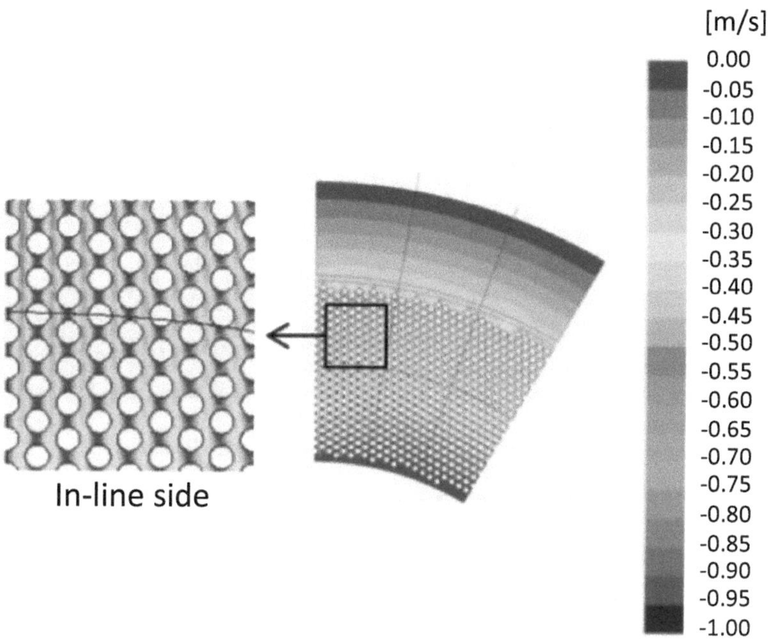

Fig. 4.213 Flow velocity results by 3D flow analysis [121].

4.3.3.4 Steam generator

(a) Design update

The commercial JSFR design adopts a double-walled tube SG for both safety and investment protection. Periodical inspections of both the inner and outer tubes are required to maintain reliable sodium-water boundaries. The JSFR double-walled tube SG can eliminate tube failure propagation. One remaining critical issue of the double-walled tube SG is the capacity of the tube fabricators. Considering the limited R&D resources and reduction of R&D risks, alternative SG concepts (coated wall tube, protective wall tube, and nickel-based alloy tube SGs) were investigated for the demonstration JSFR as a development step for the double-walled tube SG [122]. For comparison, a protective wall tube SG was studied for the demonstration JSFR [117]. The selected protective wall tube was similar to the double-walled tube for the commercial JSFR. However, the fabrication and ISI requirements were reduced due to the R&D risks. The inner and outer tubes were mechanically contacted. The inner tube was thicker than the original double-walled SG since the outer tube was not secured by ISI and acted as a protective layer against target wastage during a sodium-water reaction.

As a measure to mitigate the sodium-water reaction consequences caused by SGTR, the following systems were investigated. For tube failure detection, a water leak is detected by the following systems for the water leak signal to be sent.

- Hydrogen detection system in sodium: 2 out of 4, detects hydrogen concentration rise in sodium "high"
- Cover gas pressure gage: 2 out of 4, detects the cover gas pressure "high"

- Rupture disk: 2 out of 4 (relay circuit), detects the rupture disk "open"

For blowdown, the following water-steam blowdown systems were investigated for early termination of the sodium-water reaction.

- Outlet release valve: 2 trains and multiplexing in parallel in terms of reliability of blowdown
- Inlet release valve: 2 trains and multiplexing in parallel in terms of reliability of blowdown
- Main feed water isolation valve: 2 trains and multiplexing in series in terms of reliability of water feed stop
- Main steam isolation valve: 2 trains and multiplexing in series in terms of reliability of isolation steam
- Check valve: equipment to function dependently at the time of loss of function of the isolation valve. (The check valve is installed to the steam outlet side.)

In case of a relatively large water leak caused by the SGTR, pressure of the secondary coolant system is increased by the sodium-water reaction consequences. A relief system for the sodium-water reaction products was installed in the cover gas region of the SG in order to mitigate the pressure increase described above. A rupture disk was placed in the release system of the sodium-water reaction products, where the disk would burst due to an increase in the cover gas pressure. A pressure increase of the secondary coolant system and the cover gas system would be suppressed by bursting of the rupture disk, thus ensuring the structural integrity of the boundary between the primary and secondary coolant systems.

(b) Evaluation of property protection performance

The number of propagated failed tubes was analyzed to evaluate the property protection performance of the protective wall tube SG. In this study, functions such as wastage resistance and overheating rupture resistance were expected in the analysis. In contrast to the safety assessment, the following are expected:

- thickness of the outer tubes (considered such as corrosion allowance for 60 years and wear allowance)
- acceleration of blowdown after water leak detection (from 30 s to 20 s)

Table 4.39 shows the analysis results of the tube failure propagation. The initial water leak rate between 0.56 and 5.7 g/s resulted in tube failure propagation by wastage. Failure propagated in one tube. In other ranges, no tube failure propagated as a result of wastage or overheating rupture.

These results suggest that the sodium-water reaction consequences can be mitigated by double-walled tubes. Tube failure propagation to neighboring tubes is also suppressed significantly. Hence, it seems that extension to a large-scale leak affecting the structural integrity of the boundary between the primary and the secondary coolant systems can be prevented.

(c) Evaluation of severe sodium-water reaction

According to the SDC discussed in the GIF, measures against events more severe than the DBA are required. The analysis conditions are shown as follows:

- During a water leak accident, it is assumed that the function of water leak suppression is lost by dependent failure of the outer tubes at the time of the initial tube failure.
- The initial water leak rate is up to 1 double-ended guillotine (DEG) from the minute leak.
- The hydrogen detection system for sodium, the cover gas pressure gage, and the rupture disk are supposed to be water leak detection systems.

Table 4.39 Tube failure analysis from the viewpoint of property protection [123].

Case no.	Initial water leak rate (g/s)	Maximum water leak rate (g/s)	Detection time (s)	Termination time (s)	Propagated failure tube	
					Overheating rupture	Wastage
1	0.06	10.3	2022	2042	0	0
2	0.11	0.6	1158	1179	0	0
3	0.56	37.9	279	299	0	1
4	1.1	105	197	218	0	1
5	5.7	1100	70	90	0	1
6	11.5	55.3	70	90	0	0
7	26.7	108	70	90	0	0
8	63.3	253	70	90	0	0
9	132	500	70	90	0	0
10	244	826	70	90	0	0
11	617	1102	7.5	28	0	0
12	1157	1263	1.5	22	0	0
13	3392	3392	0.2	21	0	0

- After blowdown of water-steam, nitrogen is injected into the heat transfer tubes to prevent sodium infiltration into the water-steam system facilities. At this point, the event is deemed to have terminated.
- Tube failure propagation is evaluated assuming loss of two functions among the different water leak detection and blowdown systems.

Table 4.40 shows Case No. and results of tube failure propagation. Assuming loss of the mitigation function of the sodium-water reaction consequences, a loss of the function of the hydrogen detection system in sodium has a relatively large influence to create an overheating rupture. In cases where the initial leak rate is small, tube failure propagation continues until it is possible for the cover gas pressure gage to detect a water leak by loss of the

Table 4.40 Tube failure analysis in severe conditions [123].

Case no.	Water leak detection system			Water/steam blowdown facilities				Number of propageted failure tubes	Maximum water leak rate (kg/s)
	Hydrogen detector (2 out of 4)	Cover gas pressure gauge (2 out of 4)	Rupture disk (2 out of 4 : relay circuit)	Inlet release valve (2trains)	Outlet release valve (2trains)	Feed-water isolation valve (2trains)	Steam isolation valve (2trains)		
1	×	O	O	O	O	O	O	3	14
2	O	×	O	O	O	O	O	0	3.4
3	O	O	×	O	O	O	O	0	3.4
4	O	O	O	×	O	O	O	0	3.4
5	O	O	O	O	×	O	O	0	3.5
6	O	O	O	O	O	×	O	0	3.4
7	O	O	O	O	O	O	×	0	3.4
8	×	×	O	O	O	O	O	3	14
9	×	O	O	×	O	O	O	3	14
10	×	O	O	O	×	O	O	5	22
11	×	O	O	O	O	×	O	3	14
12	×	O	O	O	O	O	×	3	14
13	O	×	×	O	O	O	O	0	3.4
14	O	O	O	×	×	O	O	38	130
15	O	O	O	×	O	×	O	0	3.4

detection function of the hydrogen detection system in sodium. Additionally, an overheating rupture occurs.

When the blowdown function of the outlet and inlet release valve was lost, an overheating rupture caused the largest tube failure of 38 tubes. The maximum water leak rate was about 130 kg/s. In this case, it was possible to detect the water leak with the hydrogen detection system, rapture disk, etc. However, it was impossible to terminate the water leak event by water-steam blowdown because the blowdown function of the outlet and inlet release valve was lost. Since the water-steam system was isolated by actuation of isolation valves, it is presumed that the water-steam system maintained a high pressure. Accordingly, it seems that a heat transfer tube failure was easily propagated by an overheating rupture.

Combined losses of the various mitigation functions for the sodium-water reaction have been comprehensively sorted for the double-walled tube SG. Subsequently, tube failure propagation is evaluated using the tube failure propagation analysis program. From these results, it is estimated that loss of the blowdown function (inlet and outlet release valves) is the most severe supposition, causing the largest-scale tube failure.

The initial spike pressure and quasi-steady-state pressure caused by the sodium-water reaction were considered in order to evaluate the integrity of the secondary sodium boundaries including the IHX tubes. The evaluation result of the initial spike pressure in the DEC would be the same as that of the DBA because the number of initial heat transfer tube failures is one. On the other hand, the water leak scale for a quasi-steady-state pressure evaluation becomes more severe than that of the DBA since multiple losses of the mitigation function are assumed in the DEC. In this case, the number of failed tubes is 38 and the maximum water leak rate is approximately 130 kg/s. The initial water leak has been assumed to occur at the upper end of the effective heat transfer zone. Quasi-steady-state pressure is calculated using the time change of the water leak rate described above. Fig. 4.214 shows the calculation results. If a loss of the inlet and outlet release valve function is assumed (case no.14 shown in Table 4.40), the maximum water leak rate is approximately 130 kg/s.

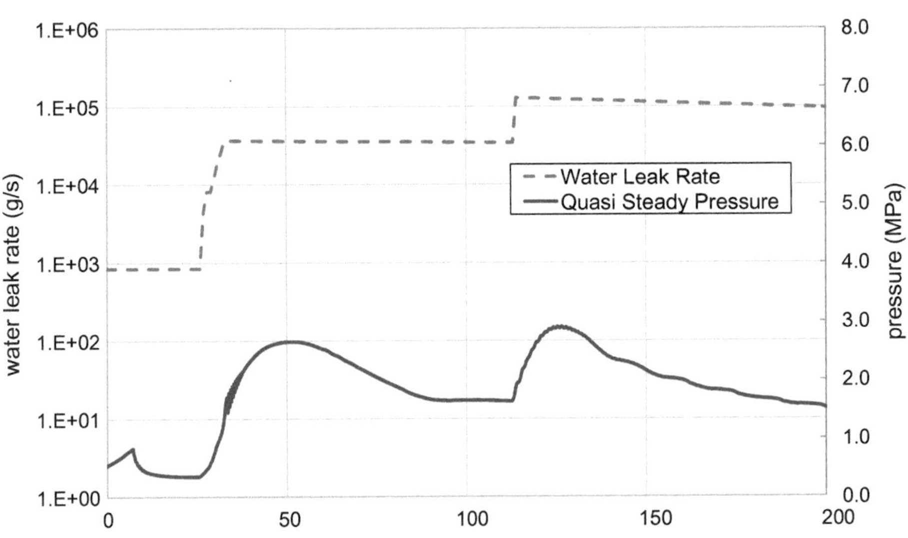

Fig. 4.214 Calculation results for quasi-steady-state pressure [123].

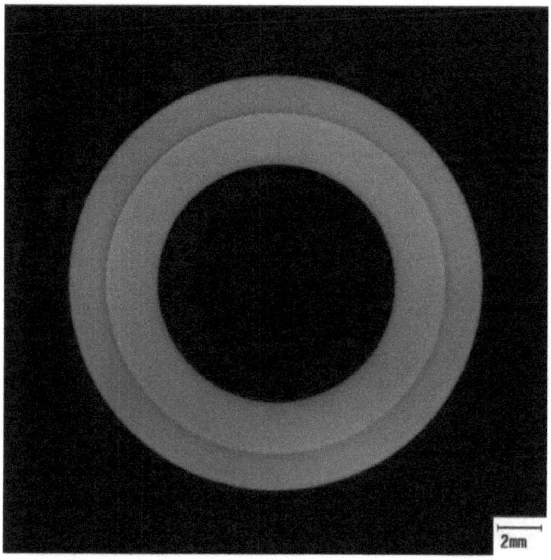

Fig. 4.215 Microscopic photograph of double-walled tube [117].

However, a pressure increase of the secondary coolant system by quasi-steady-state pressure is estimated to be about 3.0 MPa, calculated using a water leak rate based on realistic tube failure propagation behavior. The realistic tube failure propagation behavior has been assumed because it is unlikely that the heat transfer tubes will simultaneously be broken by the sodium-water reaction. It is presumed that the structural integrity of the boundary of the primary and secondary coolant systems is ensured against the above pressure increase. Moreover, CV bypass is supposed to be preventable even under DECs.

(d) Tube fabrication

Trial fabrications of the modified double-walled tube have been made in a factory under various processing conditions. The clearance between the outer and inner tubes was measured by microscopy (shown in Fig. 4.215). Buckling due to the temperature difference between the exchange tubes was a concern for the straight tube SG design. Previous studies have found that a difference of thermal contact resistance between the inner and outer tubes could cause temperature differences in the heat transfer tubes. It was also suggested that the difference of the thermal contact resistance depends on the contact conditions between the inner and outer tubes. Reduction of oxide scale at the interface, which prevents smooth contact between the inner and outer tubes, was estimated to be an effective method to decrease thermal contact resistance. In the present study, a hydrogen furnace was applied for heat treatment in the elemental tube manufacturing (normalizing and tempering) as a measure for reducing the oxide scale. By comparing tubes that did not have interfaces with reduced oxide scale, it was confirmed that reduction of the oxide scale affects the contact thermal resistance of the inner and outer tubes. Fig. 4.216 shows an overview of a prototypical manufacturing process for the double-walled tube. The target dimensions of the double-walled tube were 19 mm in diameter and 4.2 mm (inner tube: 2.4 mm, outer tube: 1.8 mm) in thickness. The target outer diameters of the elemental tubes were 20.2 mm (outer tube) and 15.6 mm (inner tube). In order to reduce the amount of oxide scale generated, the double-walled tubes were produced experimentally under the conditions as shown in Table 4.41.

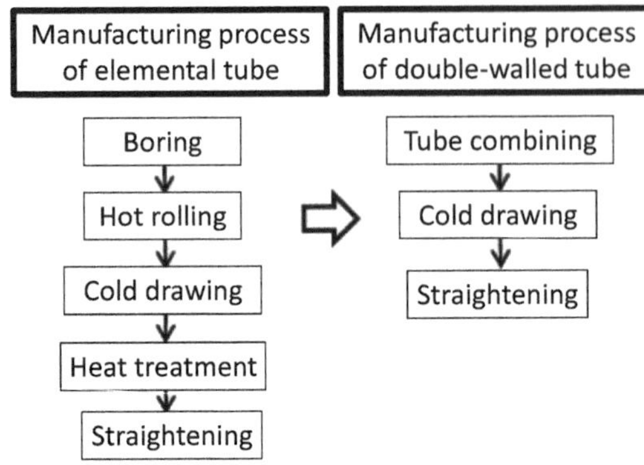

Fig. 4.216 Prototypical manufacturing process for double-walled tube [123].

Table 4.41 Test product condition of double-walled tube [123].

		Case 1	**Case 2**	**Case 3**	**Case 4**	**Not heat-treated**
Outer tube	Normalizing	Hydrogen furnace	←	←	←	Bright furnace
	Tempering	Hydrogen furnace	Bright furnace	←	←	←
	Polishing (inner face)	Not polished	←	←	←	←
Inner tube	Normalizing	Hydrogen furnace	←	←	Bright furnace	←
	Tempering	Hydrogen furnace	Bright furnace	←	←	←
	Polishing (outer face)	Not polished	Not polished +buff finished	Not polished	Polished	Not polished

To understand the heat transfer performance of the prototypical double-walled tubes, the basic data, such as contact pressure and thermal contact resistance, were acquired for the following tubes.

- Double-walled tube simulating relaxation of the contact pressure in service
- Double-walled tube without heat treatment

The double-walled tube simulating relaxation of the contact pressure in service was heat-treated at 650°C (1 h), which corresponds to the operating conditions in plants (515°C × 60 years).

Saturated steam heated the outside of the double-walled tube and a reference tube (single tube) simultaneously. Then, the cooling water flowing inside the tubes was heated. The heat transfer rates were calculated from the temperature difference of the cooling water flowing through the double-walled tube and the reference tube. The contact thermal

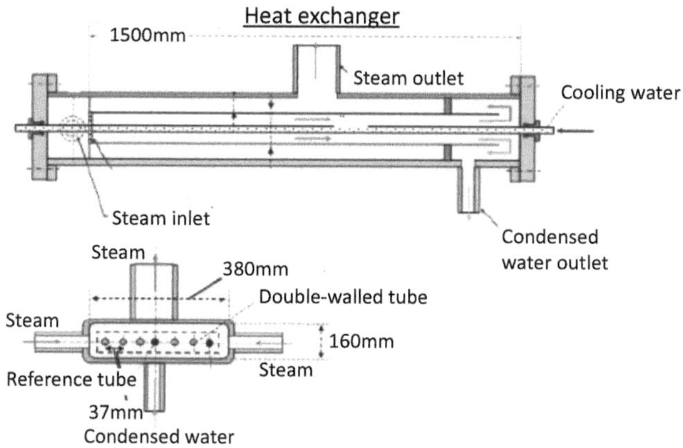

Fig. 4.217 Apparatus for contact resistance measurement [123].

resistance of the double-walled tube was calculated by comparing the heat transfer rates between the double-walled tube and the reference tube. The measurement device for the thermal contact resistance was a multitube heat exchanger structure.

Fig. 4.217 shows the concept of the testing part of the test equipment. From the temperature results of the cooling water flowing through the reference tube and the double-walled tube, the heat transfer coefficient [124] of the heat transfer tubes was calculated. The contact thermal resistance was measured under the following conditions:

- flow velocity of cooling water: 4.0 m/s
- flow velocity of steam outside tube: 26 m/s

For measurement of the contact thermal resistance after contact pressure reduction, the double-walled tubes were heat-treated (650°C, 1 h) to decrease the contact pressure similar to that after 60 years of operation.

Fig. 4.218 shows the measurement results of the contact thermal resistance. Also shown are the measurement results for the double-walled tube that was not treated to reduce the

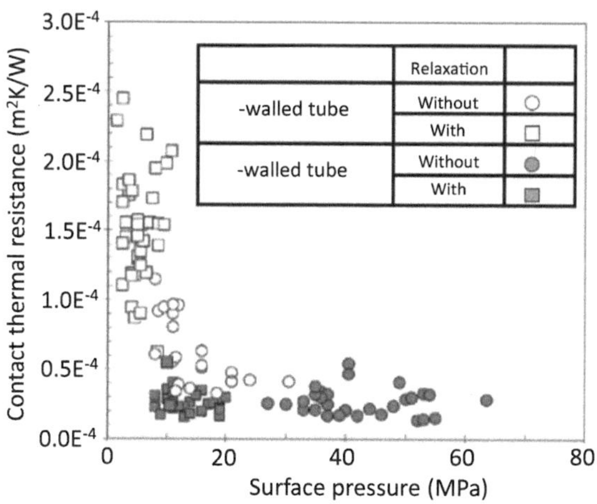

Fig. 4.218 Relationship between contact pressure and thermal resistance [123].

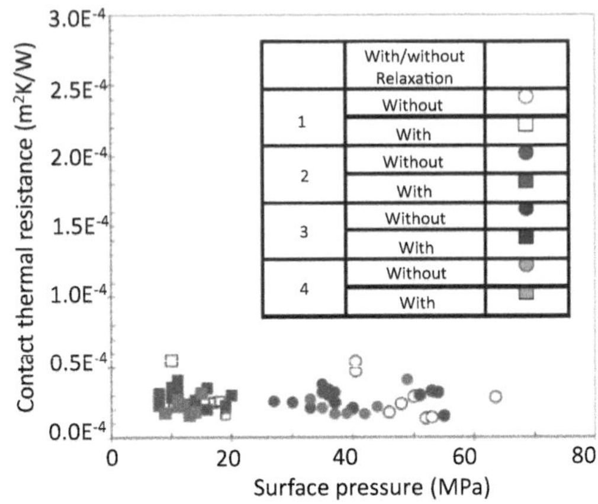

Fig. 4.219 Relationship between contact pressure and thermal resistance with oxide scale reduction [123].

oxide scale. The contact thermal resistance of the double-walled tubes that were treated to reduce the oxide scale achieved a provisional target value despite the prototypical production conditions. Fig. 4.219 shows the measurement results of the double-walled tubes that were treated to reduce the oxide scale. It was found that the prototypical production conditions had minimal influence on the contact thermal resistance. Furthermore, the heat treatment to simulate the operating conditions in the actual plant scarcely changed the contact thermal resistance.

From the above, it is confirmed that the double-walled tubes produced under all prototypical conditions achieved the provisional target value for the contact thermal resistance. The basic data measured for the double-walled tube prototype offered the following findings:

- Despite the prototypical condition and the heat treatment, the double-walled tubes with the reduced oxide scale achieved the provisional target value of the contact thermal resistance.
- Because all of the double-walled tubes achieved the provisional target value of the contact thermal resistance regardless of the conditions, Case No. 4, with reduced use of the hydrogen furnace, seems to be the best prospect for the industrial manufacturing method.

The followings are subjects for future investigations.

- A manufacturing facility, which includes a heat treatment furnace for elemental tubes, must be prepared for long double-walled tubes. The quality control method of the hydrogen furnace must be investigated in order to reduce the oxide scale evenly from the center of the inner face of the long double-walled tubes.
- Industrial manufacturing and quality control methods that guarantee the interface properties of the inner and outer tubes are needed.
- In case contact pressure control is necessary for the structural integrity of the double-walled tubes, quality control methods of contact pressure are required.

(e) Alternative SG design

As an alternative design, a helical coil tube SG has been investigated. Specifications and the sketch are shown in Table 4.42 and Fig. 4.220. The heat exchange tubes for the alternative concept were single-walled instead of double-walled.

Table 4.42 Specifications of helical coil tube SG.

Items	Values
Fluid (tube/shell)	Water-steam/sodium
Thermal capacity	375 MW
Sodium temperature (inlet/outlet)	530/345 deg-C
Sodium flow rate	5725 t/h/loop
Water-steam temperature(inlet/outlet)	245/500 deg-C
Water-steam flow rate	610.2ton/hour
Water-steam pressure	18 MPa [absolute]
Heat transfer area	2562 m2
Tube material	Mod.9Cr-1Mo
Number of tubes	330
Tube pitch (axial/radial)	50 mm/50 mm
Tube outer diameter	31.8 mm
Tube thickness	5.5 mm

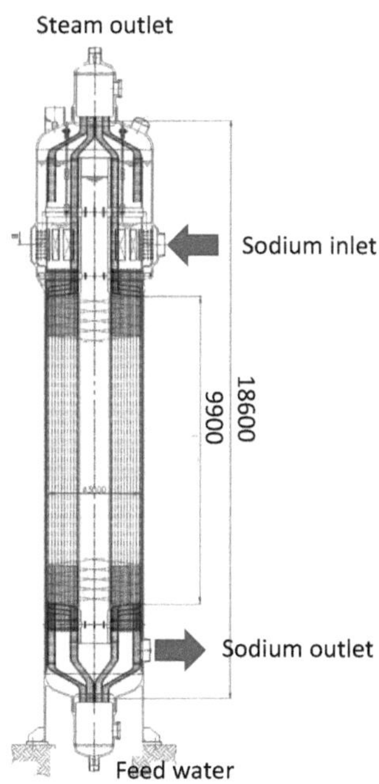

Fig. 4.220 Sketch of helical coil SG.

The detection and blowdown systems were the same as the reference system. SG leak detection in the alternative SG concept was calculated for DBA, as shown in Fig. 4.221. For DBA, an initial leak from a single tube (maximum 1 DEG) was assumed. Detection times for the hydrogen and pressure detectors are shown comparing the failure propagation times. In almost all the ranges, leaks were detected before failure propagation. For a leak rate

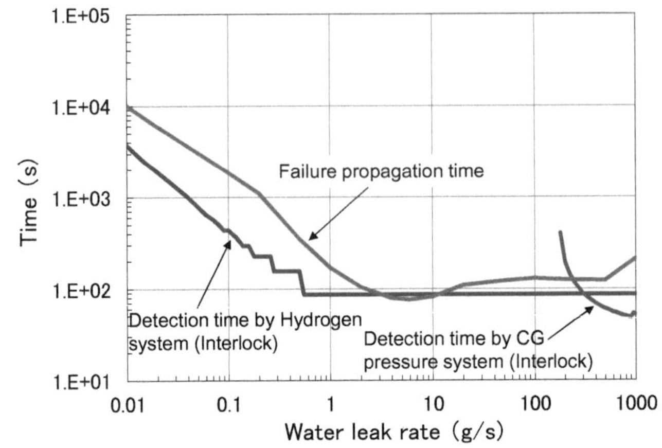

Fig. 4.221 Leak detection results for alternative SG.

Table 4.43 Analysis results for tube failure propagation in DBA (alternative).

		Propagated failure tube	
Initial water leak rate (g/s)	**Detection time (s)**	**OHR**	**Wastage**
0.05	994.5	0	0
0.1	994.5	0	0
0.5	157.0	0	0
1.0	87.0	0	0
5.0	**87.0**	**0**	**1**
10	87.0	0	0
20	87.0	0	0
50	87.0	0	0
102	87.0	0	0
261	87.0	0	0
528	6.5	0	0
1016	1.0	0	0
10,300 (1DEG)	0.1	0	0

under 300 g/s, the hydrogen detector can detect leaks before failure propagation, except for in the range from a few g/s to 10 g/s. For a leak rate more than 300 g/s, pressure monitoring can detect leaks before failure propagation. Table 4.43 shows the analysis results of the failure propagation analysis. As shown in the results, failure propagation in the range of the DBA conditions would be a result of only one additional tube failure due to wastage. The maximum leak rate, including failure propagation, is shown in Fig. 4.222. In the case of an initial leak rate of 5 g/s, the maximum leak rate was evaluated to be approximately 200 g/s, which was confirmed to be lower than 1 DEG.

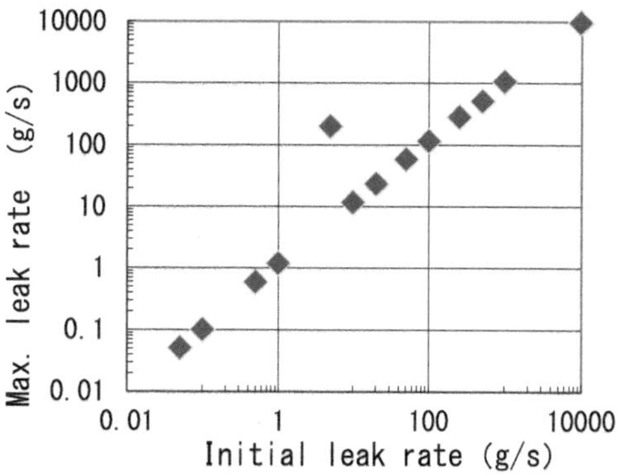

Fig. 4.222 Failure propagation analysis for alternative SG.

Table 4.44 Analysis results for tube failure propagation in DEC (alternative).

No.	Water leak detection			Water/steam blowdown				Number of propagated failure tubes	Maximum water leak rate (kg/s)
	Hydrogen detector	CG pressure gauge	Rupture disk	Inlet release valve	Outlet release valve	Feed-water isolation valve	Steam isolation valve		
1	×	○	○	○	○	○	○	3	14
2	○	×	○	○	○	○	○	0	3.4
3	○	○	×	○	○	○	○	0	3.4
12	×	○	○	○	○	○	×	3	14
13	○	×	×	○	○	○	○	0	3.4
14	○	○	○	×	×	○	○	**38**	**130**
15	○	○	○	×	○	×	○	0	3.4

For DEC evaluation, tube failure propagation was evaluated assuming loss of two functions similar to the reference concept, as shown Table 4.44. As a result, failure of the inlet-outlet release valves was assumed as the most severe condition. The maximum leak rate and quasi-steady-state pressure are evaluated as shown in Fig. 4.223. In the alternative concept, failure is mainly caused by high temperature rupture because the tube diameter and thickness are larger than the reference concept. The maximum leak rate was evaluated to be 50 kg/s.

4.3.3.5 Alternative design reducing R&D loads

After the 1F accident, JSFR development was slowed and focus was on the reduction of R&D loads. Considering those circumstances, an alternative concept to reduce R&D loads was investigated. The evaluation process shown in Fig. 4.224 was discussed. In the first step, R&D items with high R&D loads were listed. The ease of development and the status of the design standard preparation were evaluated. Technology readiness (high(H)/medium(M)/low(L)) was evaluated based on the combination of ease of development and the status of the design standard preparation. If ease of

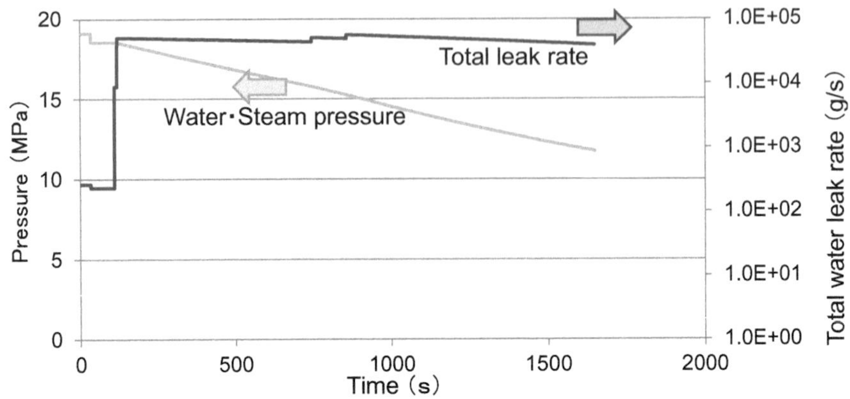

Fig. 4.223 Calculation results for quasi-steady-state pressure (alternative).

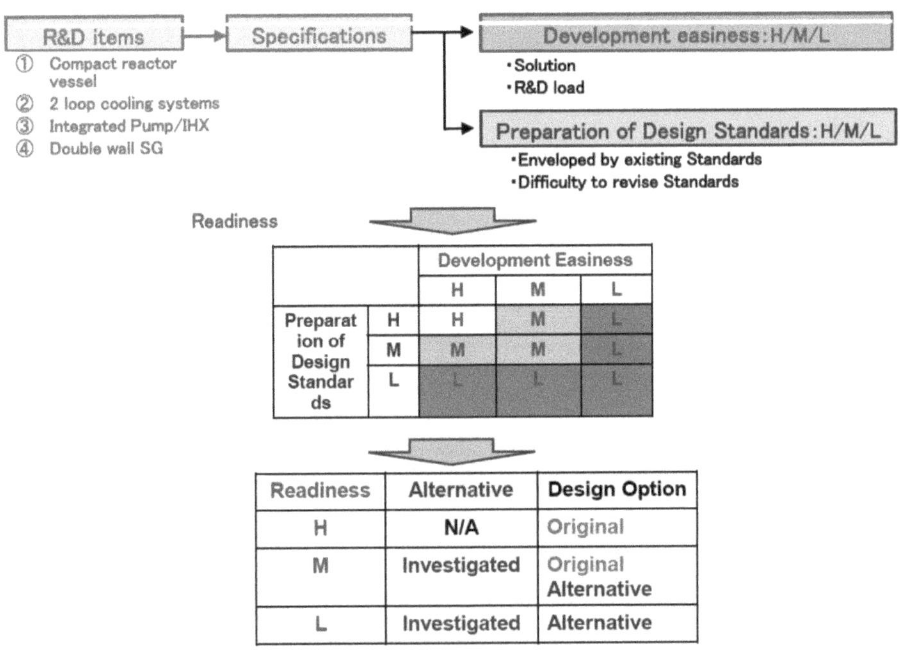

Fig. 4.224 Evaluation process for reduction of R&D load.

the evaluated technology was high, the alternative concept also adapted the original technology installed in the JSFR. However, if ease of the technology was medium or low, alternative technologies were investigated and design modification was discussed.

The compact RV, two-loop cooling system, integrated IHX-pump, and double-walled tube SG were selected as the R&D items with high load. The evaluated

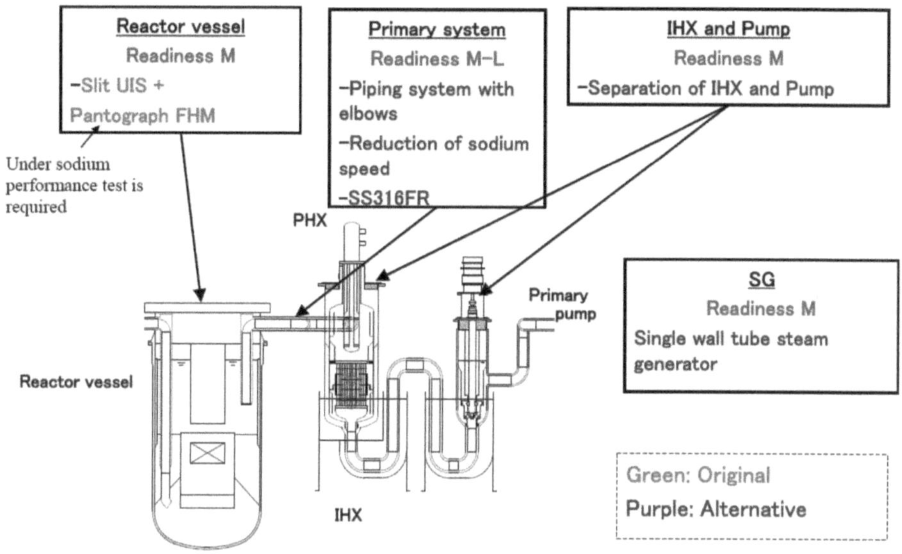

Fig. 4.225 Evaluated readiness levels of the alternative design concepts.

readiness levels for those four items are shown in Fig. 4.225. For the compact RV, the alternative concept still adapts a combination of a slit UIS and pantograph FHM, since the impact of the RV diameter with alternative technology is significant. Since the feasibility of the slit UIS was already evaluated in the FaCT R&D, the residual R&D is mainly under the sodium performance test of the pantograph FHM. For the primary piping system, an alternative piping system with elbows, reduced sodium speed, and 316FR material was adopted, since the L-shaped piping, high-speed sodium, and Mod. 9Cr—1Mo steel piping for the primary system will require significant R&D efforts and there is no preparation of standards for Mod. 9Cr—1Mo steel piping. For the integrated IHX-pump and double-walled tube SG, separation of the IHX and pump and an alternative SG were investigated, since R&D results during the FaCT project required additional R&D to confirm the feasibility of those technologies.

An innovative technology list of the alternative concept to reduce the R&D load is shown in Table 4.45. As discussed in Fig. 4.225, three items were modified from the original JSFR. For the impact of the separation of piping and separation of the IHX and pump, the reactor building arrangement was roughly evaluated as shown in Fig. 4.226. The space required for the primary system in the safety-enhanced JSFR was 45 m width and 31 m depth. On the other hand, in the alternative concept with primary piping elbows and separation of the IHX and pump, the dimensions were 45 m width and 34 m depth. The impact seemed rather limited because the primary system space was dominated by the DHRS instead of the main piping system. In the safety-enhanced JSFR, four DHRSs (DRACS, 2 PRACSs, and ACS) were arranged in that space. As for the PRACS, there were six heat exchangers for each integrated IHX-pump and the piping systems for those heat exchangers required certain spaces above the integrated IHX-pump.

Table 4.45 Innovative technologies for the alternative concept reducing R&D load.

Item	Evaluation results
High burnup fuel	OK
Safety enhancement	OK
Compact RV	OK
Two-loop cooling system	Two loops are still adopted but ▪ Reduction of sodium velocity with lager diameter piping ▪ Elbows in piping system ▪ SS316FR
Integrated IHX-pump	Separation of IHX and pump
Straight double-walled tube SG	Single-walled tube SG
Natural circulation DHRS	OK
Simple fuel handling system	OK
Steel plate-reinforced concrete CV	OK
Seismic isolation	OK

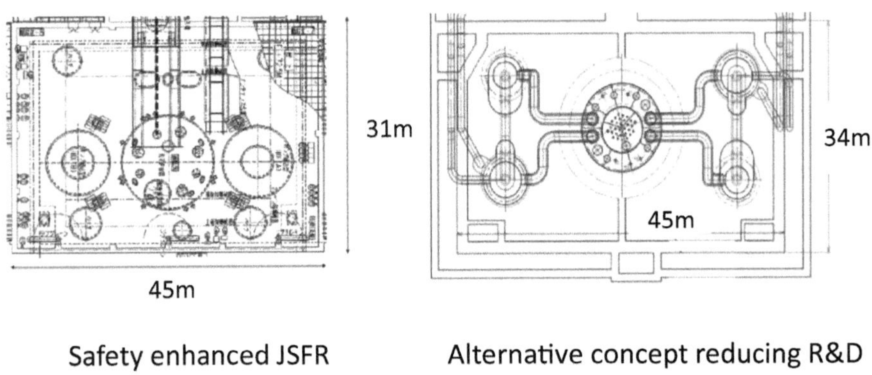

Fig. 4.226 Comparison of primary cooling system spaces.

For the alternative SG, a single helical coil tube SG was investigated, as discussed in the previous subsection. The mass of the SG is considerably reduced in the alternative concept because the alternative SG adopts a single-walled tube and the tube material shares major parts of the SG structural material.

4.4 Pool-type SFR

This chapter introduces the design activities for a pool-type SFR that addresses Japan's specific siting conditions such as earthquakes and meets safety design criteria (SDC) [125] and safety design guidelines (SDG) [126] for generation IV SFRs prepared in the framework of the generation IV International Forum (GIF). A design concept of 1500 MWt (650 MWe) class pool-type SFR was conceived by applying the design technology obtained from the design of the advanced loop-type SFR, named JSFR [10], which is equipped with the safety measures that reflect the results from the feasibility study on commercialized fast reactor cycle systems and fast reactor cycle technology development and that have improved maintainability and repairability and are based on the lessons learned from the 1F accident. In addition, the design extrapolation capability to a large-size reactor of around 1000 MWe is examined from the viewpoint of seismic-proofing reactor structures.

4.4.1 Design concept

Loop-type SFRs have been in use for many years in Japan. On the other hand, BN-800 of Russia, PFBR of India, and ASTRID of France are well-known pool-type SFRs. The pool-type reactor encloses primary pumps and IHXs located around the core barrel in a reactor vessel (RV) (Fig. 4.227). As a demonstration reactor toward future commercialization of SFRs, the pool-type design concept (JP-pool SFR) considers the possibility of increasing its output power. In addition, design parameters such as the number of cooling circuits and output power should be selected flexibly as needed because the target electric power for commercial SFRs will depend on circumstances, such as energy markets, political decisions, and regulations in the future.

Table 4.46 lists one example of the main parameters of the design concept of the JP-pool SFR. Fig. 4.228 shows a bird's-eye view of the JP-pool SFR. Two core concepts are currently under investigation: One is an axial heterogeneous core with an upper sodium plenum zone that leads to a low total sodium void reactivity effect [129], and the other is the innovative design concept called fuel subassembly with an inner duct structure (FAIDUS) that is intended to secure an early discharge of molten fuel from the core region in the event of a core disruptive accident, thereby preventing severe recriticality in the case of core damage [130]. The targets are core performances with a breeding ratio of around 1.1 and a discharged driver fuel average burnup of 150 GWd/t.

Generally, the RV diameter in pool-type reactors tends to be larger than that of loop-type reactors. Referring to the database from IAEA-TECDOC [131], Fig. 4.229 shows the differences between the two types in terms of the approximate dependency of the diameter on the thermal power of the reactor. The correlation

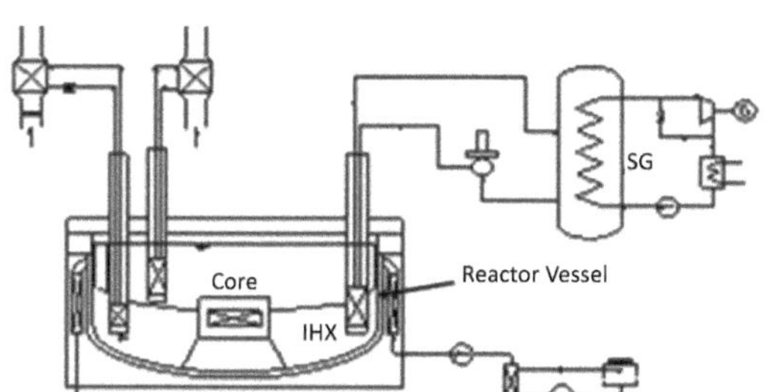

Fig. 4.227 Configuration diagram of a pool-type SFR [127]. Copyright 2018 by the American Nuclear Society, La Grange Park, Illinois.

Table 4.46 Main specifications of the JP-pool SFR [128].

Item	Value or system
Electric output	650 MWe
Core concept	MOX fuel, axial heterogeneous core or homogeneous core (FAIDUS concept)
Core inlet/outlet coolant temperature	550/400°C
Reactor design	Pool-type architecture with the applicability of a high-frequency design (HFD) for the reactor structure
Reactor shutdown systems	Two active shutdown systems with self-actuated shutdown system (SASS)
Safety design in the reactor vessel	Mitigation features of core damage
Primary pumps	Three mechanical pumps
Intermediate heat exchangers (IHXs)	Four IHXs (straight tube containing secondary coolant inside)
Secondary pumps	Four mechanical pumps
Power conversion	Four steam generators with helical coil tubes, 1 water-steam turbine
Decay heat removal systems (DHRSs)	One immersed direct reactor auxiliary cooling system (DRACS)[a] Four intermediate reactor auxiliary cooling systems (IRACSs) One penetrated-DRACS[b]
Spent fuel storage	Ex-vessel storage tank, studying an impact of in-vessel storage concept in parallel
Seismic isolation system	Horizontal seismic isolation system on the building base, studying an impact of three-dimensional seismic isolation system in parallel
Building structure	Steel concrete-reinforced structure for the reactor building and also the containment vessel

[a] It is installed in a hot plenum (HP) of the RV and removes decay heat via sodium in the HP.
[b] It is installed that it penetrates the inner vessel. In the case of decay heat removal in a design extension condition (DEC), a shutter on an inlet in the HP is opened to make a flow path to an outlet in a cold plenum (CP).

between RV diameter and thermal power is not perfectly linear because the sizes of the RV are affected by the core characteristics and layout of the primary components. From the viewpoint of economics, previous studies [132] showed that large-scale plants are advantageous in terms of economic competitiveness. Therefore, considering possible commercialization in the future, pool-type SFRs with large RV diameters should be studied. In parallel, residual risk evaluation should be conducted as required by the Japanese regulation for seismic design. IAEA safety standards [133] and GIF-SFR-SDC require providing design measures against design extension conditions. Therefore, the seismic assessment of the main components such as RV is required not only for design-basis conditions but also for design extension conditions to ensure sufficient margin against the failure of such main components of the reactor

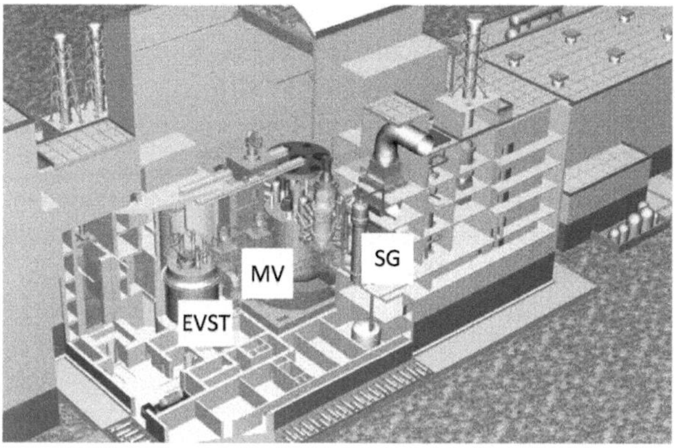

Fig. 4.228 Bird's-eye view of the JP-pool SFR in Japan [128].

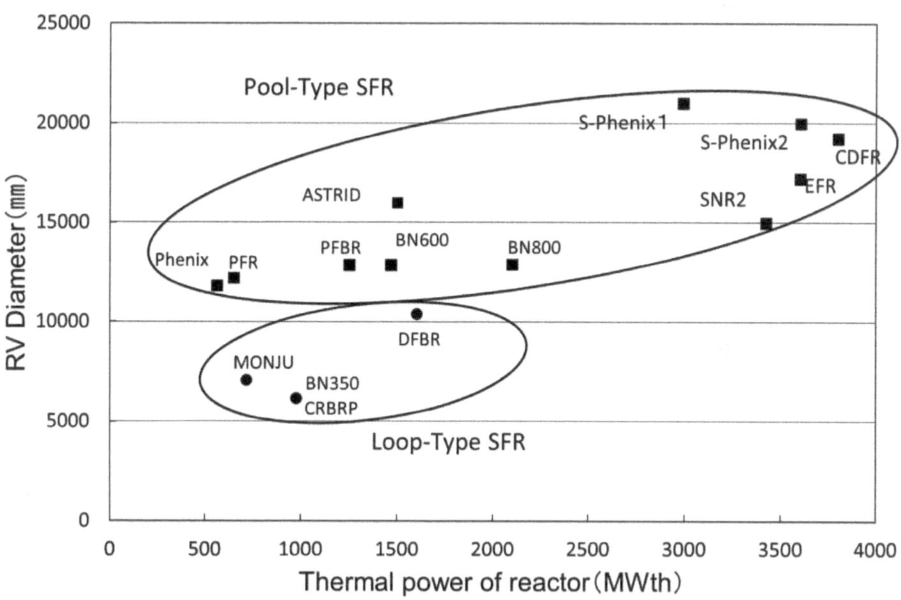

Fig. 4.229 RV diameter and thermal power of reactors [127].
Copyright 2018 by the American Nuclear Society, La Grange Park, Illinois.

plant. This point is critical, especially in Japan where the most severe earthquake conditions in the world occur.

4.4.2 Reactor structure under severe seismic conditions

For large, heavy RVs under severe seismic conditions, buckling evaluation of the cylindrical part of the vessel is significant for a feasibility study. Additionally, the vessel should be as thin as possible for SFRs because the vessel wall around the high-temperature liquid surface is subjected to severe cumulative damage due to thermal loadings. When the ratio of the diameter to the thickness of the vessel is large, it becomes difficult to achieve compatibility between the tolerance to seismic loadings with the tolerance to thermal loadings.

This chapter describes the feasibility of the large RV under severe seismic conditions as confirmed via the procedure described in the following text [127].

The seismic conditions as the design-basis earthquake are set to be severe based on the conditions in past studies on SFRs and the recent requirements for the seismic design of nuclear power plants in Japan.

With regard to the ratio of buckling limit load to seismic load, a margin for buckling strength is required as the safety factor. In this study, the safety factor is taken as 1.8 tentatively to incorporate some reliability against beyond design-basis earthquake.

From the viewpoint of compatibility of the tolerance to seismic loadings with that to thermal loadings, two RVs of different diameters were evaluated to reveal the dependency of the RV diameter on vessel thickness.

As a countermeasure to improve the feasibility, a three-dimensional (3D) seismic isolation system was developed to secure a margin against the buckling of a large RV. The concept and the performance of its system under development are introduced.

4.4.2.1 Seismic loading

As this is a preliminary study, buckling evaluations for the cylindrical part of RVs were conducted using a simplified calculation. Two types of seismic ground motions were prepared as the seismic loading conditions because the conditions were not provided by limiting to a specific site but based on comprehensive conditions to confirm feasibility. One type of seismic ground motion is the standard condition (Condition A), and the other is the more severe condition (Condition B) in Japan. The response analyses for the reactor building were conducted under these seismic conditions. As the standard option, the reactor building was equipped with the horizontal seismic isolation system for SFR [134].

As shown in Fig. 4.230, the earthquake ground motions at the free rock surface were prepared from the target spectra for conditions A and B in this study. The maximum horizontal accelerations under conditions A and B are 6.4 and 9.6 m/s^2, respectively. The maximum vertical accelerations under conditions A and B are 4.3 and 6.4 m/s^2, respectively. As this study focuses on buckling failure, the hard rock condition ($Vs = 1500$ m/s) was set so that the vertical response on the building through the ground was not reduced. In other words, the soft rock case provides easier conditions in terms of buckling failure.

The reactor building was equipped with a horizontal seismic isolation system as the standard option; this system consisted of laminated rubber bearings and oil dampers [135]. The floor responsive spectra were calculated at the floor where RV was installed. The floor should be constructed on the inner concrete above with sufficient stiffness to prevent abnormal amplifying of acceleration at the local frequency. The floor responsive spectra were analyzed for the damping coefficient 1% based on the requirement by seismic design code in Japan.

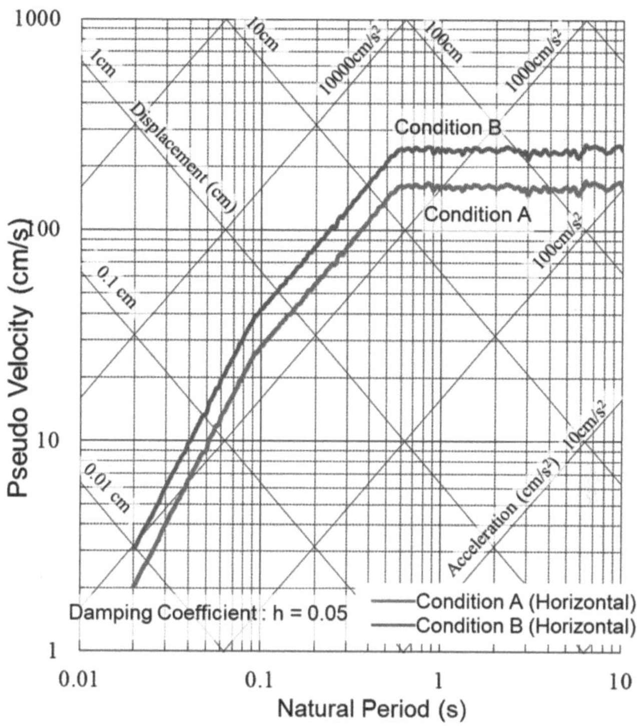

Fig. 4.230 Horizontal target spectra for seismic conditions [127].
Copyright 2018 by the American Nuclear Society, La Grange Park, Illinois.

4.4.2.2 Structural characteristics of RVs

An SFR is operated under low-pressure conditions at high temperatures and has thin-walled structures to reduce thermal stress. However, the thin walls are a drawback against seismic loads. To enhance the seismic capacity, the wall of the RV should be thickened but should be kept as thin as possible. Fig. 4.231 shows the design concept of the reactor structure with enhanced antiseismic capability. Considering the earthquake levels and regulations, we adopted a certain seismic isolation system and to reinforce a high-frequency design (HFD) for the reactor block; that is, a hard stiffness structure that has a natural frequency higher than the peak frequency of vertical acceleration on the floor response spectrum is adopted to prevent buckling distortion at the thin wall of the vessel. This design concept has an enhanced RV support structure, an enhanced conical core support structure, a thickened knuckle of the RV, and a flat plenum separator with ribs. To minimize the elastic deformation of the RV flange structure by the earthquake, the thickness of the RV flange was increased. Further, the thickened knuckle part could minimize the elastic deformation of the vessel bottom. The flat plate plenum separator, which was reinforced by attaching ribs radially on the undersurface, was used instead of a conical-shaped one. The connecting part between the core support and RV was shifted outward to increase the stiffness of the core support structure. These structures can ensure the stiffness of the reactor

Fig. 4.231 Concept of the reactor structure [136].

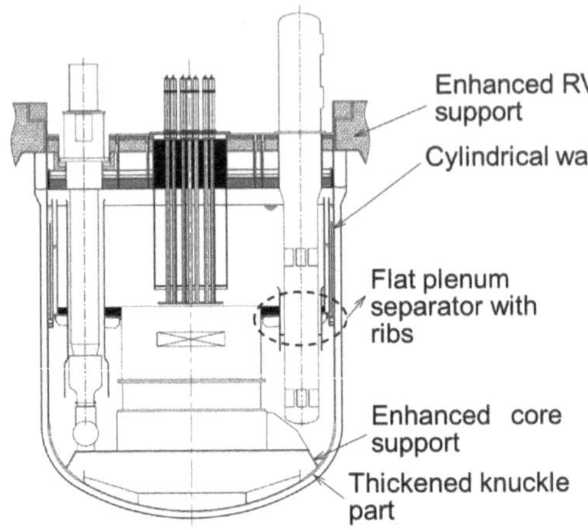

structure and reduce the vertical response. The thickness of the cylindrical part of the RV is increased as necessary.

By using a simple spring-mass model, the seismic response analysis of the reactor building was examined under the conditions of the horizontal target spectra. Vertical ground motions were assumed as two-thirds of horizontal ground motions. The reactor building had horizontal seismic isolation on the building base, and the system was equipped with laminated rubber bearings and oil dampers. To prevent buckling, we set the thickness of the RV to meet the criteria shown in Eq. (4.6) developed by Matsuura et al. based on experiments [137].

$$\left\{\frac{Q}{Q_{al}}\right\}^5 + \left\{\frac{M}{M_{al}} + \frac{F}{F_{al}}\right\}^5 < 1 \tag{4.6}$$

where

Q, $Q_{\rm al}$: transverse shear load and transverse shear buckling limit load.

M, $M_{\rm al}$: applied bending moment and limit bending buckling moment.

F, $F_{\rm al}$: applied axial force and axial buckling limit force.

Table 4.47 lists the results obtained under different output power conditions with appropriate margins. It was found that for large-scale plants (1000 and 1500 MWe)

Table 4.47 Necessary vessel thickness [136].

Spectrum	**Part**	**600 MWe (15.9 m[a])**	**1000 MWe (17.2 m[a])**	**1500MWe (19.4 m[a])**
A	Cylinder	50	55	60
	Knuckle	100	110	120
B	Cylinder	55	60	65
	Knuckle	100	120	130

[a] RV diameter.

with larger diameter RVs and more components such as primary pumps and IHXs, the thickness of the cylindrical parts should be 60 mm or more. Therefore, it is important to confirm the feasibility of enlarged RVs in terms of the structural integrity against thermal stress and manufacturability.

For a diameter of 19 m, there is concern that the thermal loading effects such as cumulative thermal fatigue and creep damage may exceed the allowable values. The thermal loadings on the RV are mainly due to repeated thermal transients during the start-up, normal shutdown, and automatic shutdown of the reactor. Because it is possible to suppress the thermal stress on the RV by extending the duration of the reactor start-up and shutdown, the feasibility of the thermal loading depends on requirements by plant owner. As an example, a trial calculation given in Fig. 4.232 shows the start-up days to suppress thermal stress to the same level for increased wall thicknesses for an RV diameter of approximately 20 m. The start-up days usually should be less than 1.5 days in conceptual studies on SFRs. From the viewpoint of economics, the acceptability for extending start-up days is only considered for the cases of thickness less than 60 mm. Accordingly, some improvements are necessary for severe earthquake conditions for an RV diameter of around 19 m.

To ensure the feasibility of a large RV, countermeasures, such as reduction in the increase in the RV diameter, reduction in seismic loads, and relaxation of thermal transients, may be required. An optimum countermeasure is to introduce a 3D seismic isolation system [138]. The system for the seismic isolation of the system in the horizontal direction is identical to the horizontal seismic isolation system. This system has a stack of disk springs and vertical oil dampers to reduce the seismic response in the vertical direction. A previous study [139] showed that the vertical seismic response of the RV is approximately 40% when a horizontal seismic isolation system with a vertical frequency of around 10 Hz is used. If this system is adopted, the required thickness of the walls of the enlarged RV will be reduced to approximately 40 mm, even for a capacity of 1500 MWe under the seismic condition of spectrum A.

The plenum separator, which is an internal structure of the RV, is also necessary to ensure seismic capability. The design concepts of the plenum separator in some countries are shown schematically in Fig. 4.233 [131]. In this study, a single plate-type

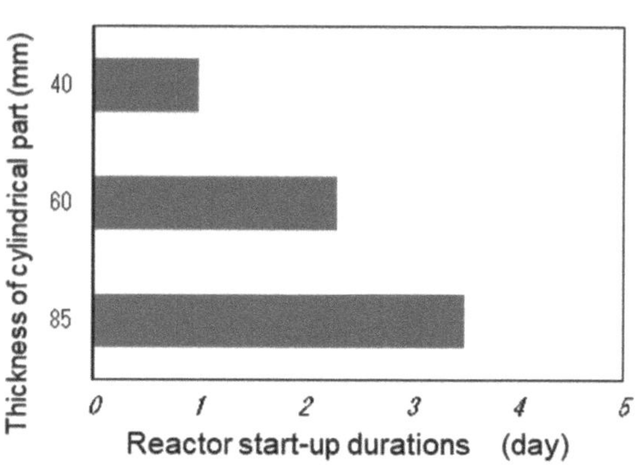

Fig. 4.232 Differences in the reactor start-up durations [127]. Copyright 2018 by the American Nuclear Society, La Grange Park, Illinois.

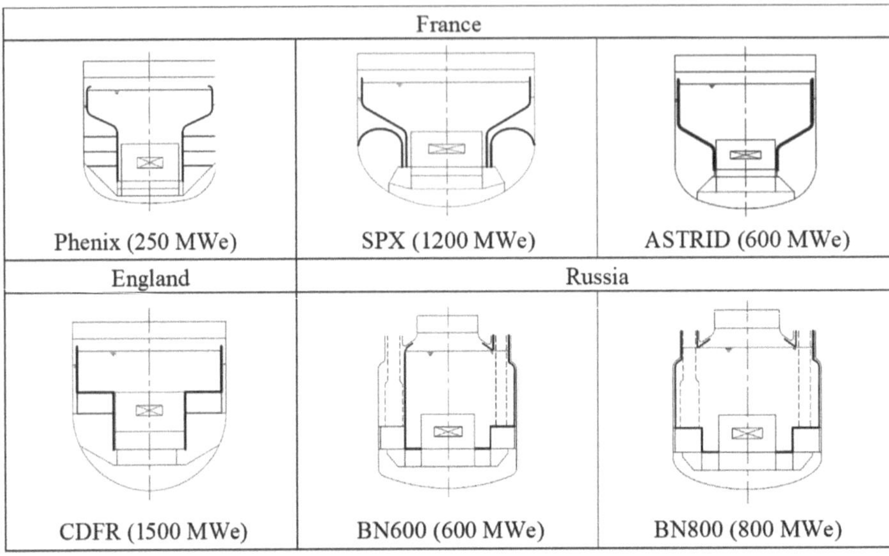

Fig. 4.233 Various design concepts of plenum separators [136].

separator, identical to the one shown in the sketch of ASTRID in the figure, was adopted because of the advantages of its simple configuration. However, this separator has the disadvantages that there exist severe temperature transients on both sides of the plate and the steel material suffers accumulative creep fatigue damage. Therefore, we adopted a flat-shaped single-plate plenum separator and attached reinforcing ribs radially on the undersurface to ensure seismic capability. In addition, layers of steel plates were installed on the hot plenum side for thermal insulation. A modal spectral analysis performed to evaluate the stiffness of the plenum separator revealed that the natural frequency was 16 Hz (0.06 s) and that the vertical seismic loads on the plenum separator were adequately suppressed (Fig. 4.234). This is because the natural frequency of the separator was higher; in other words, its natural period was sufficiently lower than the period of the peak acceleration range in the vertical floor response spectrum.

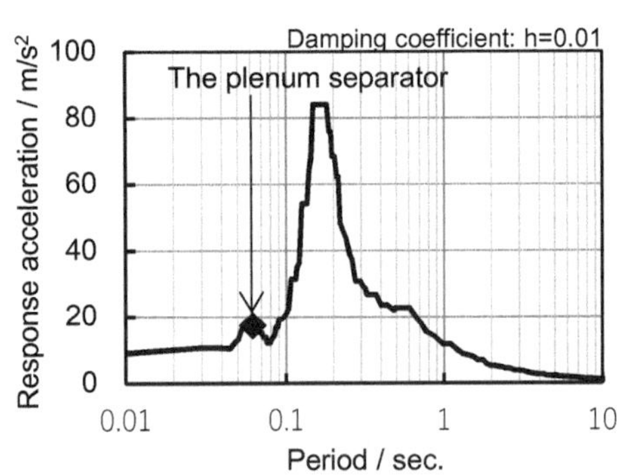

Fig. 4.234 Vertical floor response spectra at the RV support level [136].

4.4.2.3 Three-dimensional seismic isolation system

As mentioned in the previous chapter, it is necessary to scale up the RV and increase the seismic loadings for the future development of SFRs. It is not preferable that thickness of the RV should be increased considering seismic loading because the SFRs should be thin from the viewpoint of thermal loadings. If horizontal seismic isolation is applied, the vertical loading is critical in buckling evaluation. Then, as an advanced system to reduce the vertical seismic loadings, 3D seismic isolation in the base mat of the building is proposed in the following section.

A 3D seismic isolation system consists of a horizontal insulator and a vertical insulator. The horizontal insulator is formed by laminated rubber bearings, and the vertical insulator is formed by coned-disk springs, which have abundant track records in a market as a conventional technology. As shown in Fig. 4.235, the stacked coned-disk springs and vertical dampers are placed under a table that is fixed to the laminated rubber bearings. The disk springs and dampers are surrounded by four steel pillars that guide the table in the vertical movement. Hundreds of integral units are located in the base mat of the reactor building instead of the horizontal seismic isolations. The specifications of the isolation system are listed in Table 4.48.

If seismic design is based on the 3D seismic isolation system, attention should be paid to the rocking motion of the building. The rocking motion is the phenomenon in which the vertical axis of the building rotates and oscillates with respect to the horizontal axis. In this phenomenon, the following main problems arise:

- An overload exceeding the allowable pressure of the rubber bearing occurs on the side where the building rolls downward extremely.
- The tensile load is applied on the other side.

As a countermeasure to prevent the rocking motion, the method of controlling the rotational motion by using compressed air is well-known [140,141]. However, additional devices that consist of air compressors and controllers are necessary in the case of large buildings such as the NPPs. This will in turn not only raise the plant cost but

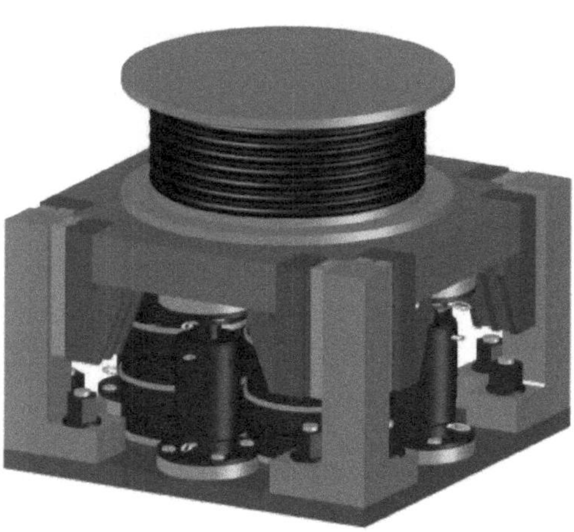

Fig. 4.235 Coned-disk springs with rubber bearing [127].
Copyright 2018 by the American Nuclear Society, La Grange Park, Illinois.

Table 4.48 Specifications of the 3D seismic isolation system [127].

Horizontal isolation	
Type Diameter Height Material Horizontal natural frequency	Rubber bearing 1.6 m 0.8 m Laminated steel plates and natural rubbers 0.29 Hz (3.4 s)
Vertical isolation	
Type Diameter Layout Height Material Vertical natural frequency Damper Type	Coned-disk spring 0.7–0.8 m Four stacks under the horizontal isolation 1.4 m Disks: (Sup10 steel for spring 6150 AISI/SAE) Pillars: steel 3 Hz Oil damper

Copyright 2018 by the American Nuclear Society, La Grange Park, Illinois.

also reduce system reliability. Accordingly, the rocking motion should be prevented without using any additional systems.

The causes of rocking motion are as follows.

(a) The rotational stiffness is low. This stiffness can be determined by relation between the vertical natural frequency and horizontal natural frequency.
(b) There is a gap between the center of gravity of the building and the center of rigidity of the isolation springs.

According to the evaluation of rotational stiffness using a simple calculation model, we see that a range of more than 3 Hz in the vertical natural frequency can provide sufficient rotational stiffness if the horizontal natural frequency is attributed to the rubber bearings [138,140]. Meanwhile, the vertical isolation system is expected to have the lowest possible frequency to reduce the loads. For the proposed 3D seismic isolation, the vertical frequency is regarded to be about 3 Hz to achieve a frequency lower than that of the rubber bearings, which have a natural frequency of about 8 Hz.

The gap described in (b) can be expressed as the eccentricity ratio defined in Eq. (4.7).

$$\text{Eccentricity Ratio} = \frac{-K_{V1} + K_{V2}}{2(K_{V1} + K_{V2})} \tag{4.7}$$

where

K_{V1} : stiffness coefficient of vertical spring at one side of building

K_{V2} : stiffness coefficient of vertical spring at another side of building to K_{V1}

In other words, Eq. (4.7) indicates that the eccentricity ratio is determined by the different stiffness coefficients of vertical springs at the two ends of building. Hundreds of isolation units contribute to the average stiffness coefficients at the two ends of large buildings such as NPPs. An example of the calculations for the eccentricity ratio is shown in Fig. 4.236. The results show that the eccentricity ratio can be suppressed by introducing a small deviation in spring stiffness and large unit number.

Furthermore, the example of calculations for the rotation angle led by seismic loadings is shown in Fig. 4.237 [142]. The figure shows that small eccentricity ratio makes it possible to suppress the rotational angle of the rocking motion. The rotational angle is quite small similar in the cases of 3D seismic isolation and horizontal seismic isolation. The results say that the rocking motion can be suppressed sufficiently by appropriate quality management for the dispersion of vertical spring performances.

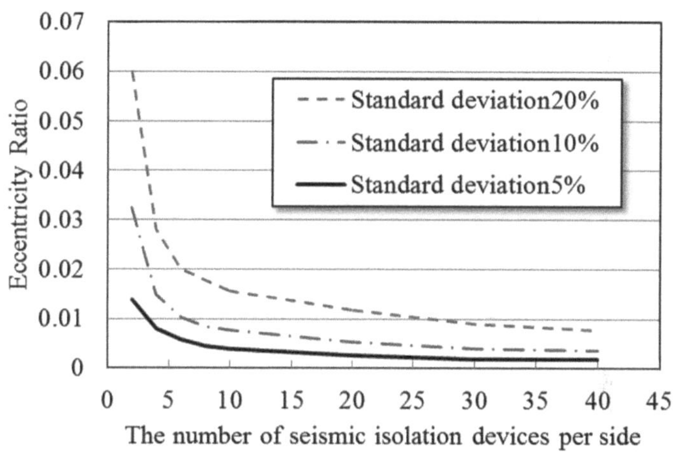

Fig. 4.236 Dependence of the eccentricity ratio on units and deviation [127]. Copyright 2018 by the American Nuclear Society, La Grange Park, Illinois.

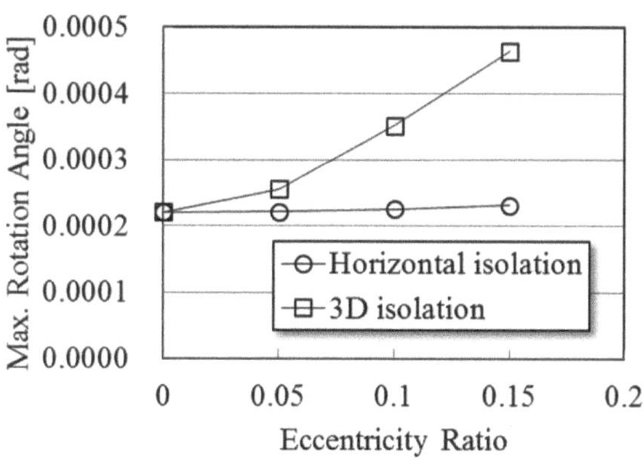

Fig. 4.237 Dependence of the maximum rotational angle on the eccentricity ratio [127]. Copyright 2018 by the American Nuclear Society, La Grange Park, Illinois.

Fig. 4.238 Oil damper for the 3D seismic isolation system [128].

The 3D seismic isolation system under development is a combination of the conventional technologies consisting of laminated rubber bearings, disk springs, and oil dampers to reduce the horizontal and vertical seismic loads. As the horizontal damping device in the system, the 3D isolation has been successfully applied with high-speed oil dampers, and a large damping force capability of around 1000 kN has been achieved [143]. Fig. 4.238 shows the oil damper. Its allowable velocity of 2.7 m/s is more than twice the ability of commercial products. By utilizing the seismic isolation system with this oil damper for horizontal damping in the SFR reactor building design, the design range will be expanded, and then, an improved earthquake resistance can be expected in the SFR design.

4.4.2.4 Structural intactness based on thermal hydraulic analysis

(a) Intactness of the RV structure

Design evaluations were conducted to ensure the structural intactness of the RV structure with regard to specific structural features of the flat plenum separator and the thermal hydraulic conditions attributed to it. Fig. 4.239 shows a computational model for the 3D fluid analysis, including the core, IHX, DRACS, and pump. The thermal insulation plates and gaps were simulated in detail because they are considered to affect the upper surface temperature. The ribs were also modeled because they affect the lower surface temperature. Approximately 3.7 million meshes were considered in the computation. Using the commercial CFD code STAR-CCM+ Ver.11, a realizable k-ε turbulent model and a second-order upwind differencing scheme for the advection term in space were applied. The design of the thermal insulation plate was optimized based on the analysis of the flow condition in the RV and the effectiveness of the thermal insulation plate under normal plant operation conditions. For example, the ribs attached at the bottom of the plenum separator have flow holes to prevent stagnant conditions around the ribs (Fig. 4.240) [136].

The boundary conditions were the heat power generated in the core, heat removal rate in the IHX, and the primary flow rate. As shown in Fig. 4.239, the core coolant flow conditions after reactor trip were set as follows:

(1) Manual trip: The flow rate changes from 100% to 10% of the primary flow after the trip.

(2) Loss off-site power: The flow rate decreases to 25% of the primary flow rate according to the flow coast-down condition and remains constant at 25% (halving time: 15 s).

In the case of manual trip from 100% primary flow, the low-temperature coolant from the core collides on the inner cylinder of the plenum separator, and then, the temperatures of

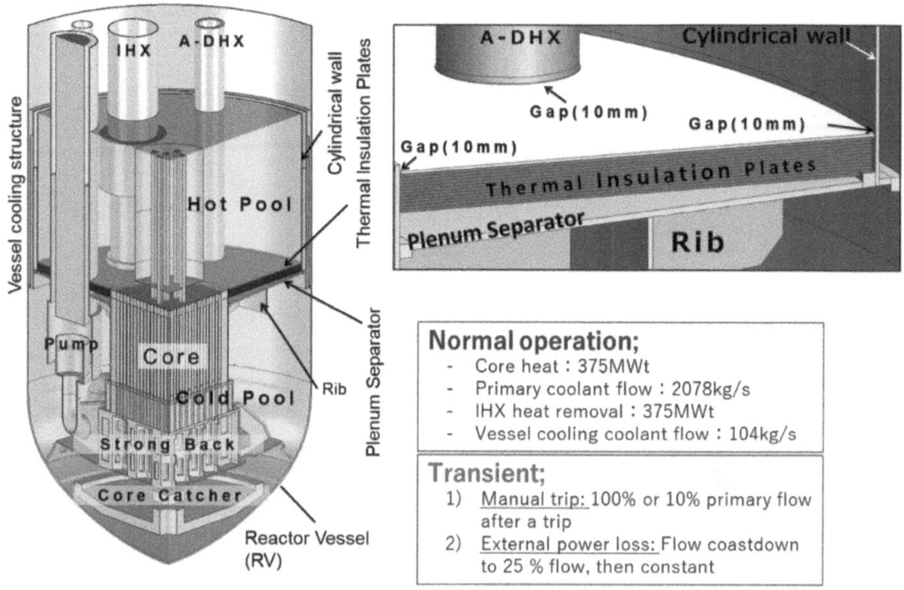

Fig. 4.239 Quadrant RV model [128].

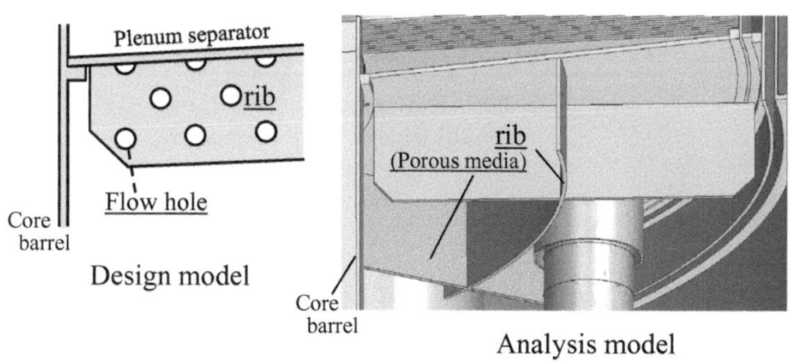

Fig. 4.240 Flow hole on the ribs [136].

the inner and outer surfaces temperatures are reversed. The axial thermal gradient is large around the temperature stratification surface. Structural intactness evaluation was performed based on the Monju standard for the thermal transient, number of occurrences, and range of stress strength as determined by the thermal transient analysis (Fig. 4.241). In the case of 10% flow after the manual trip and external power loss, structural intactness of the plenum separator can be maintained (Table 4.49).

(b) Gas entrainment analysis

In the SFR, gas entrainment from the interface between the cover gas and sodium, for instance, from the free surface of the hot pool, should be prevented. This is because gas inflow into the core may cause power fluctuations depending on the amount of gas flow. Since the RV diameter of the JP-pool SFR is larger than that of the loop-type SFR, it is difficult to cover all free surfaces by a dipped plate to prevent gas entrainment; this is unlike

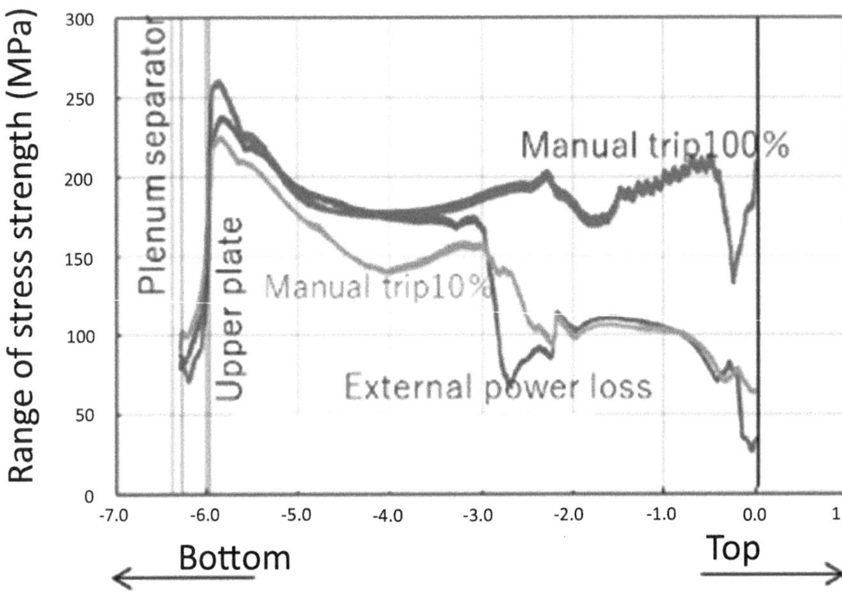

Fig. 4.241 Range of stress strength responding to the thermal transient [128].

Table 4.49 Intactness evaluation of the plenum separator [128].

	100% Flow after manual trip	**10% Flow after manual trip**	**External power loss**
A. Fatigue damage	0.122	0.030	0.017
B. Creep damage due to steady stress	0.300	0.300	0.300
C. Creep damage due to residual stress relaxation	3.97	0.02	0.03
A+(B+C)	4.388	0.355	0.343
Limit	1.00	0.614	0.615
Intactness	Failed	Assured	Assured

the JSFR, which adopted such a plate. As shown in Fig. 4.242, the ring-shaped baffle plate, which is like a partial structure of the dipped plate on the vessel wall, can be adopted as a countermeasure at present. The effectiveness of the countermeasure was confirmed by a CFD analysis and gas entrainment evaluation tool named Stream Viewer, which was developed by JAEA to evaluate the degree of gas entrainment [144]. Fig. 4.243 shows the transient of the gas core length, which is the length from the surface to the bottom edge of the concave eddy with time during normal plant operation [128]. The baffle plate could suppress the gas core length to approximately 4 m, which is greater than the distance from the surface to the IHX window, and the possibility of bubble entrainment to the IHX was not eliminated completely. Currently, design improvements to eliminate gas entrainment and

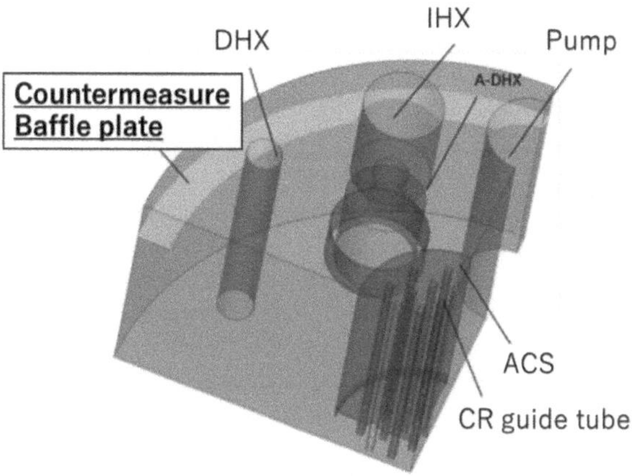

Fig. 4.242 Configuration of the baffle plate [128].

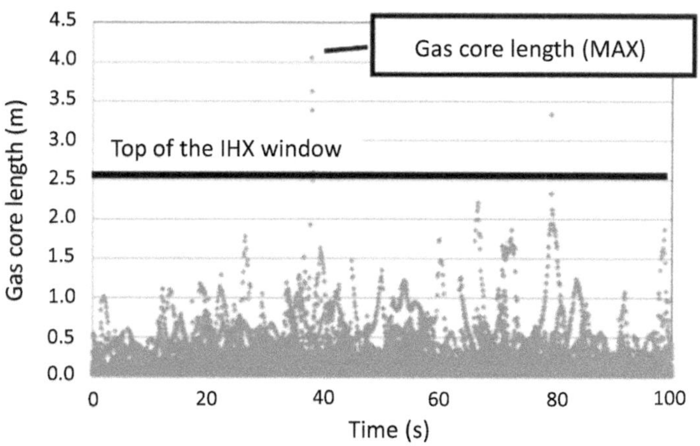

Fig. 4.243 Gas core length evaluated by Stream Viewer [128].

improve the evaluation method as well as the impact evaluation of entrained gas and evaluation of the gas volume criterion that may affect the core performance are under investigation.

4.4.3 Safety design

4.4.3.1 Measures for the safety design concept

The GIF has been developing the SDC and SDGs for SFRs. Japan has played a leading role in the development based on the design technologies accumulated through various studies on the JSFR. This chapter describes the safety design concept for the pool-type SFR based on the results of studies of the JSFR and the GIF SDC and SDGs for SFRs, considering the characteristics of pool-type SFRs. The design concept features six safety measures that are listed in Table 4.50. This chapter focuses on (i) passive

Table 4.50 Measures for the safety design concept of the pool-type SFR [136].

(i) Reactor shutdown measure	• Two active reactor shutdown systems • Passive reactor shutdown measure; the SASS using a Curie point electromagnet
(ii) Measures to achieve in-vessel retention (IVR) of damaged core	• Measures to mitigate the consequence of core damage resulting from anticipated transients without scram (ATWS) (reduction of void reactivity to approximately 6$ or less, provision of the fuel assembly with inner duct structure, transfer of re-molten fuels from the control rod guide tube, and stable cooling and retention using a core catcher)
(iii) Measures to ensure the decay heat removal	• Prevention of complete loss of decay heat removal function; the DHRS configuration considering the use of natural circulation capability, heat removal capacity, and requirements of SDC and SDG • Prevention of core uncovering and ensuring a primary coolant path for decay heat removal; practical elimination of double leakage by securing high reliability of the main vessel and the safety vessel
(iv) Formation of a containment vessel (CV) boundary	• Rectangular steel plate-reinforced concrete CV (SCCV)
(v) Measures against sodium leakage/combustion	• Double-walled structure for all sodium boundaries including secondary cooling systems • Compartmentation of the CV, and inert atmosphere • Transfer and storage of leaked sodium
(vi) Measures against sodium-water reaction	• Water-steam leak detection system, pressure relief system, means for isolation of sodium from steam-water and separation of hydrogen and other sodium-water reaction products, and quick release-reduction of hydrogen

reactor shutdown measures and (ii) measures to ensure decay heat removal because the differences between the loop-type SFRs and pool-type SFRs are remarkable in their design conditions. As for (i), the self-actuated shutdown system (SASS) is applicable for the pool-type concept in the same manner as in the JSFR. However, the cover gas pressure of the pool-type SFR is slightly lower than that of a loop-type SFR. Thus, the boiling point of sodium at the core outlet, where the SASS is installed, is lower. Therefore, the effectiveness of the SASS is very important to prevent the coolant from boiling in an abnormal event without scram. The next chapter describes the applicability evaluation of the SASS to the pool-type SFR. With regard to (ii), because the internal volume of an RV of a pool-type SFR is larger than that of a loop-type SFR in general, it is easier to install multiple heat exchangers as parts of the decay heat removal system (DHRS) inside the RV of a pool-type SFR. Section 4.4.3.3 describes the examination of the DHRS configuration considering the sufficient utilization of the natural circulation capability of the sodium coolant, heat removal capacity of each cooling system, conformance with design requirements, and recommendations of SDC and SDG such as the diversity and redundancy of components [145].

4.4.3.2 Applicability evaluation of the SASS

The SDC and SDG require that even if the function of the active reactor shutdown systems is lost, core damage should be prevented by using a passive shutdown mechanism or an inherent reactivity feedback, or a combination of these measures. The passive reactivity reduction mechanisms should

- be activated and operated by directly responding to natural phenomena (such as increased coolant temperature or reduced coolant pressure) without any active signals, activation mechanisms, or power source;
- provide sufficient and irreversible negative reactivity within an allowable time to prevent core damage; and
- cope with all types of anticipated transient without scram (ATWS) (transient overpower [TOP], loss of flow [LOF], and loss of heat sink [LOHS]).

The SASS is a passive control rod insertion mechanism using the characteristics of an electromagnet that detaches the control rods at Curie-point temperature (Fig. 4.244).

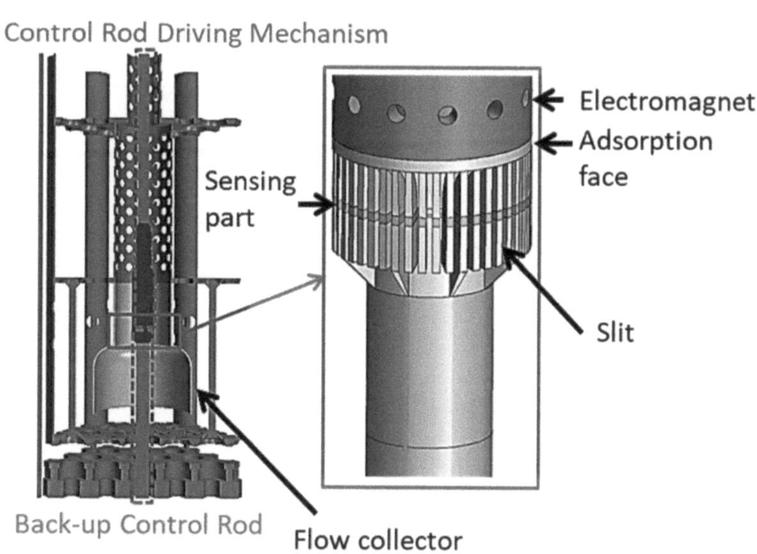

Fig. 4.244 Configuration of the SASS [136].

The cover gas pressure of the pool-type SFR is slightly lower than that of the JSFR, and therefore, the boiling point of sodium is lower in the pool-type SFR. The boiling point of sodium at the core outlet in the normal operation of the pool-type SFR is around 940°C, whereas it is around 1020°C in the JSFR. Therefore, the safety margin to coolant boiling in ATWS may be smaller than that in the JSFR [146].

Among all types of ATWS, the LOF-type ATWS will cause the most severe temperature increase. To confirm the effectiveness of the SASS, a transient analysis of a LOF-type ATWS was performed, and the margin against coolant boiling was evaluated by using a one-dimensional reactor system dynamics analysis code Super-COPD [147,148]. The detachment temperature of the SASS was set at 660°C, and the safety criterion of the maximum coolant temperature in the core was set at 940°C to prevent the coolant from boiling.

Fig. 4.245 shows the analysis results of the maximum coolant temperature in an LOF-type ATWS event under the initial condition of rated power (initial power 100%); the results for the JSFR (detachment temperature of the SASS: 650°C and safety criterion: 1020°C) are also shown for comparison. In this analysis, all primary pumps lost their power source, and flow coast-down started; further, two active shutdown systems were not activated. The nominal value of the flow-halving time was 15 s. The safety criterion was satisfied with sufficient margin even for cases with shorter flow-halving times. In the case of the JSFR, the flow-halving time was 9 s, and the coolant temperature rise was steeper than that of the pool-type SFR.

We conducted an analysis of an LOF-type ATWS event initiated under the condition of 30% partial power. In this case, the initial core outlet temperature was lower than that under the condition of 100% power, but the initial core flow rate was lower, and the response time of the SASS was longer. As shown in Fig. 4.246, the safety criterion was satisfied when the flow-halving time was 15 s. However, if the flow-halving time was shorter, the safety criterion could not be satisfied.

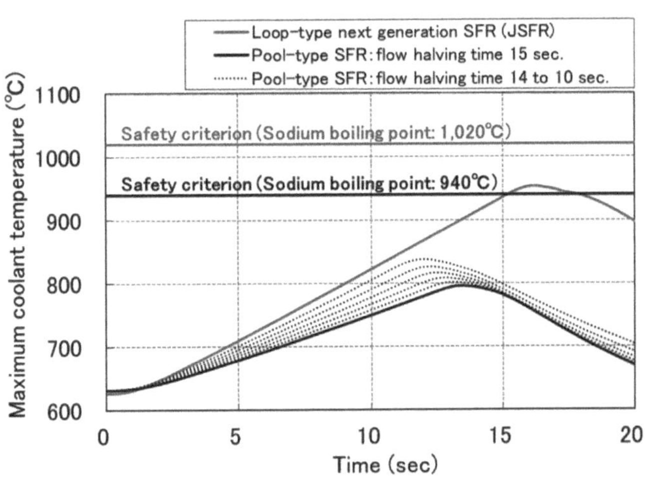

Fig. 4.245 Maximum coolant temperature in the LOF-type ATWS (initial power 100%) [136].

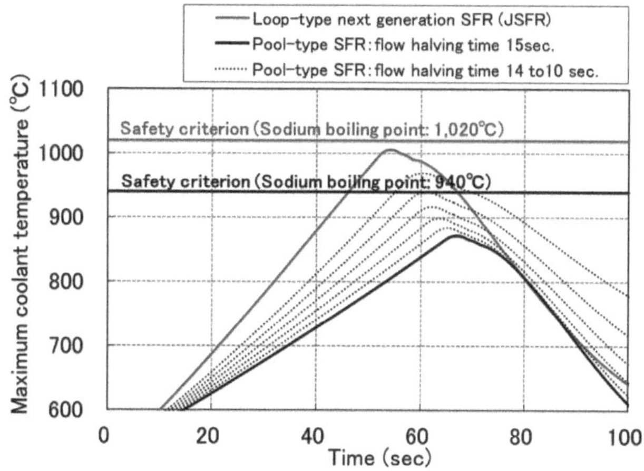

Fig. 4.246 Maximum coolant temperature in the LOF-type ATWS (initial power 30%) [136].

As discussed above, ensuring appropriate flow-halving time is essential for the SASS under an LOF-type ATWS to prevent boiling of the coolant.

4.4.3.3 Decay heat removal system

The requirements as per the SDC and SDG for DHRSs are as follows:

- To ensure that the DHRS can deal with anticipated operational occurrences (AOOs) and design-basis accidents (DBAs), they should have redundancy and/or diversity considering the loss of off-site power and single-failure criterion.
- To cope with design extension conditions (DECs), the capability of the DHRS should be extended, or alternative cooling measures should be provided.
- Natural circulation capability should be incorporated while taking advantage of the characteristics of the SFR.
- The situations of complete loss of heat removal function that could lead to core damage and failure of the reactor coolant boundary should be practically eliminated.

The types of DHRS and their conceptual drawings are shown in Fig. 4.247. Based on the evaluation of the advantages and disadvantages of DHRSs from the viewpoints of natural circulation capability, heat removal capacity, the amount of materials (cost), and applicability to the pool-type SFR (Table 4.51), the system configuration of the DHRS of the JP-pool SFR was examined.

(a) Selection of cooling systems

Aiming at the utilization of natural circulation capability, the DRACS (immersed type), DRACS (penetrated type), and the IRACS, which have favorable natural circulation heat removal capabilities, were selected as conceivable concepts. Although the primary reactor auxiliary cooling system (PRACS) and the reactor auxiliary cooling system (RACS) also have natural circulation capability, they were not selected as the combination of the IHX and PRACS forms a complicated structure with a large RV, and the RACS, which takes primary sodium out of the RV, is not suitable for a pool-type SFR.

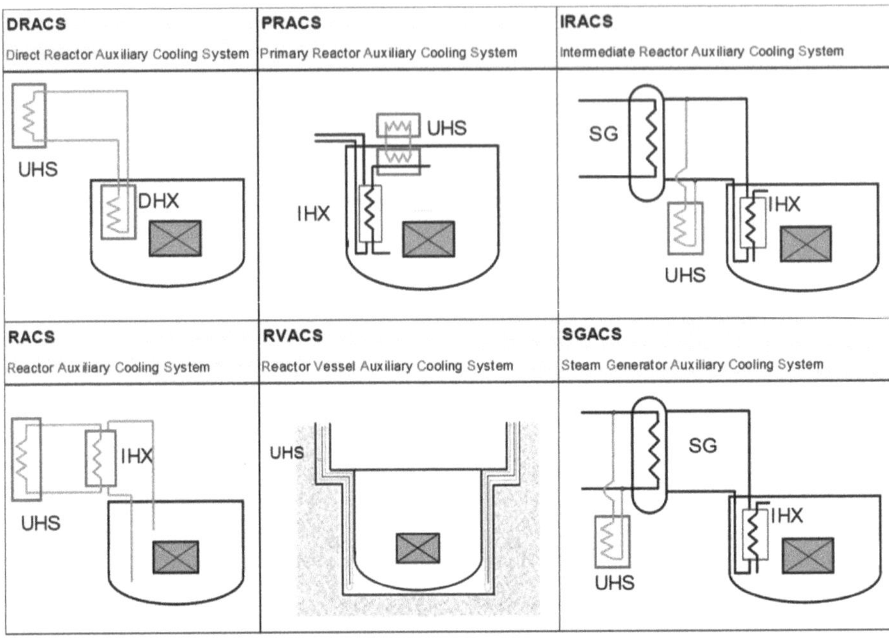

DHX: heat exchanger of DRACS, IHX: intermediate heat exchanger, PHX: heat exchanger of PRACS, SG: steam generator, UHS: ultimate heat sink, (Orange and red lines show decay heat removal systems.) DRACS is classified into two types. The first one is an immersed type whose DHX is installed in the hot pool. The second one is a penetrated type whose DHX penetrates from the hot pool to the cold pool.

Fig. 4.247 Types of decay heat removal systems [136].

(b) DHRSs to cope with AOOs and DBAs

Based on the deterministic analysis, the loss of function of two DHRSs is assumed for DHRSs to cope with AOOs and DBAs considering one DHRS failure with a single failure of another DHRS. The remaining cooling system must ensure the required heat removal capacity. Further, to reduce the probability of common cause failure, it is preferable to apply different types of cooling systems. Therefore, the DHRSs should be composed of more than three independent systems. These systems should be diversified into at least two groups.

Through the analysis, the DRACS (immersed type) and IRACS were selected for the DHRS to cope with AOOs and DBAs. Although different types of DRACSs (penetrated type) are available, the DRACS (immersed type) was selected. From the viewpoint of enhancing diversity, IRACS was selected as another type of cooling systems. The number of systems is as follows: For the DRACS (immersed type), the DHX was installed in the hot pool, and the core cooling relies significantly on the interwrapper flow (IWF) among the subassemblies in the core. The core cooling by the IRACS relies on the main flow path, i.e., the core inlet to the outlet, hot pool, IHX, and cold pool. The IWF caused by the DRACS (immersed type) may impede the main flow if it has a larger heat removal capacity. Therefore, only one DRACS (immersed type) is assigned. On the other hand, for the IRACS, four cooling systems were assigned conforming to the number of the secondary cooling systems.

Table 4.51 Evaluation of each cooling system [136].

<table>
<tr><th rowspan="2">Cooling system concept</th><th colspan="2">Advantage</th><th rowspan="2">Disadvantage</th></tr>
<tr><th>Common items</th><th>Special items</th></tr>
<tr><td>DRACS (immersed type)</td><td rowspan="5">• Enough heat removal capacity
• Natural circulation capability
• Simple air cooling system</td><td rowspan="4">• Less restriction of heat exchanger location
• Simple and high core cooling natural circulation</td><td>• Validation of interwrapper flow core cooling is necessary.</td></tr>
<tr><td>DRACS (penetrated type)</td><td>• R&D for shutter of primary sodium inlet is necessary.</td></tr>
<tr><td>PRACS</td><td>• Diameter increase of RV due to complicated structure around IHX</td></tr>
<tr><td>IRACS</td><td>• Loss of function caused by secondary heat transport system failure</td></tr>
<tr><td>RACS</td><td>• Small impact on the design of RV</td><td>• Presence of primary sodium circuit out of RV
• Slight pressurization of cover gas is necessary.</td></tr>
<tr><td>RVACS</td><td colspan="2">• Less susceptible to in-vessel events</td><td>• Difficulty in securing heat removal capacity</td></tr>
<tr><td>SGACS</td><td colspan="2">• Enough heat removal capacity when using water-steam cooling equipment</td><td>• Loss of function caused by secondary heat transport system failure</td></tr>
</table>

The IRACS has an advantage that the RV size is not affected by an increase in the number of IRACSs.

To ensure that DHRSs cope with AOOs and DBAs, the natural circulation capability is utilized in the primary, secondary, and air cooling systems.

(c) DHRS to cope with DECs

The SDC and SDG recommend providing additional cooling measures against DECs, which are independent of DHRSs, to cope with AOOs and DBAs, and to ensure the required heat removal capacity of at least one system on the assumption that the functions of DHRSs to cope with AOOs and DBAs are lost completely. Following this recommendation, one DRACS (penetrated type) was assigned as an additional cooling measure. In the DRACS (penetrated type), the DHX penetrates from the hot pool to the cold pool, and core cooling is performed by the main flow path. However, a part of the flow to the IHX is bypassed to the DRACS (penetrated type) during normal operation. Therefore, some countermeasures based on R&D, such as installing a shutter at the inlet of the DHX, are necessary to restrain the bypass flow during normal operation. As R&D for the shutter of the DRACS (penetrated type) is necessary, this DRACS is assigned as an alternative cooling measure instead of DHRSs to cope with AOOs and DBAs.

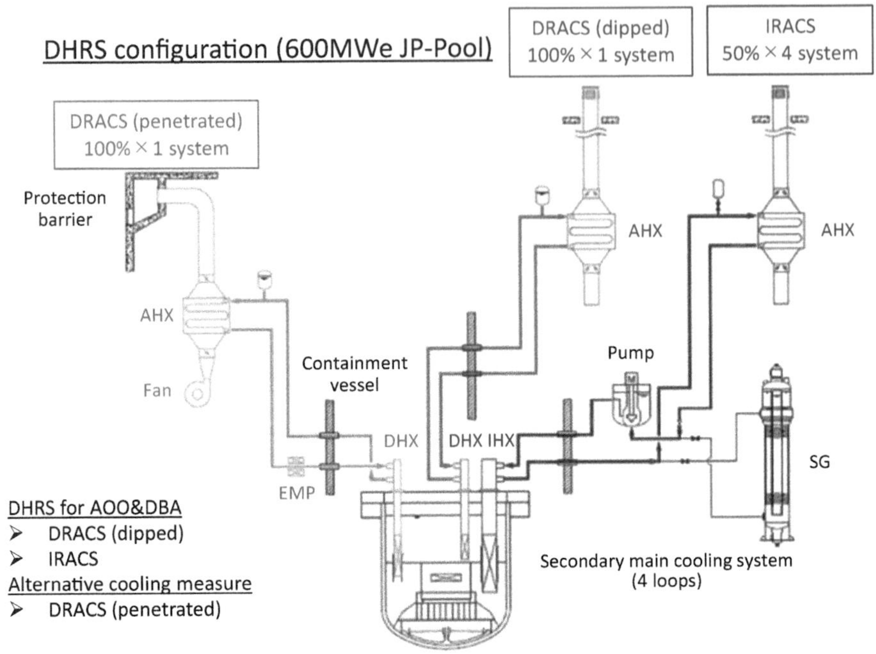

Fig. 4.248 Candidate DHRS configuration of the JP-pool SFR [136].

For the DHRS to cope with DECs, the natural circulation capability is utilized in the primary cooling system, but forced circulation is used for the secondary cooling system and air cooling system from the viewpoint of diversity.

(d) Equipment configuration of the decay heat removal system

Fig. 4.248 depicts the selected candidate of the DHRS configuration using the DRACS (immersed type) and IRACS to cope with AOOs and DBAs, and the DRACS (penetrated type) as an alternative cooling measure to cope with DECs. Both the DRACS and the IRACS have simple structures. This configuration provides sufficient heat removal capacity and natural circulation capability as well as independence and diversity to cope with severe internal and external hazards. It also meets the SDC and SDG requirements and recommendations. Physical separation is taken into account to prevent the simultaneous loss of their functions.

References

[1] K. Ito, S. Kotake, Transition and present status the 7th history of fast reactor development (II), J. Atomic Energy Soc. Jpn. 50 (2) (2008) 114–119.

[2] M. Ueta, Development of fast breeder reactor, Therm. Nucl. Power Eng. Soc. 49 (10) (1998) 194–199. Thermal and Nuclear Power. (In Japanese).

[3] Thermal and Nuclear Power Engineering Society, 5. Fast breeder reactor, in: 50 Years Progress of Thermal and Nuclear Power, 2000, pp. 302–310 (In Japanese).

[4] B. Farrar, J.C. Lefevre, et al., Fast reactor decay heat removal: approach to the safety system design in Japan and Europe, Nucl. Eng. Des. 193 (1999) 45–55.

[5] T. Nakamura, H. Kaguchi, et al., Evaluation method for structural integrity assessment in core disruptive accident of fast reactor, Nucl. Eng. Des. 227 (2004) 97–123.
[6] M. Miura, T. Inagaki, et al., Present status of Fast breeder reactor development and prospect for implementing, J. Atomic Energy Soc. Jpn. 37 (2) (1995) 112–123.
[7] Y. Sagayama, Launch of Fast REACTOR Cycle Technology Development Project in Japan, 2007. Global 2007, No. 175710, Boise, Idaho, USA.
[8] H. Niwa, K. Aoto, M. Morishita, Current Status and Perspective of Advanced Loop Type Fast Reactor in Fast Reactor Cycle Technology Development Project, 2007. Global 2007, No. 175740, Boise, Idaho, USA.
[9] Y. Sagayama, Feasibility Study on Commercialized Fast Reactor Cycle Systems (1) Current Status of the Phase-II Study, 2005. Global 2005, No. 380, Tsukuba, Japan.
[10] S. Kotake, Y. Sakamoto, M. Ando, T. Tanaka, Feasibility Study on Commercialized Fast Reactor Cycle Systems/Current Status of the FR System Design, 2005. Global 2005, No. 435, Tsukuba, Japan.
[11] S. Kotake, T. Mihara, S. Kubo, K. Aoto, M. Toda, Development of advanced loop-type fast reactor in japan (1): current status of JSFR development, in: Proc. ICAPP 2008, Anaheim, CA USA, June 8–12, 2008, 2008.
[12] M. Ichimiya, T. Mizuno, S. Kotake, A next generation sodium-cooled fast reactor concept and its R&D program, Nucl. Eng. Technol. 39 (3) (2007) 171–186.
[13] N. Kasahara, K. Nakamura, K. Inoue, Recent developments for fast reactor structural design standard (FDS), in: SMiRT18, No.F04-2, Beijing, China, 2005, 2005.
[14] T. Mizuno, H. Niwa, Advanced MOX core design study of sodium-cooled reactors in current feasibility study on commercialized fast reactor cycle systems in Japan, Nucl. Technol. 146 (2004) 155–163.
[15] T. Mizuno, et al., Advanced oxide fuel core design study for SFR in the "feasibility study" in Japan, in: Proc. GLOBAL 2005, Tsukuba, Japan, Oct. 9-13, 2005, Paper 434, 2005.
[16] M. Naganuma, et al., Minor actinide-bearing oxide fuel core design study for the JSFR, Nucl. Technol. 170 (2010) 170–180.
[17] T. Okubo, et al., Conceptual design for a large-scale japan sodium-cooled fast reactor (3) core design in JSFR, in: Proc. ICAPP 2011, Nice, France, May 2-5, 2011, 2011, pp. 479–486. Paper 11345.
[18] S. Ohki, et al., Core performance requirements and design conditions for a next-generation sodium-cooled fast reactor in Japan, Proc. ICAPP 2017, Fukui and Kyoto, Japan, April 24–28, 2017, 2017.
[19] T. Kan, et al., Core Design of the Next-Generation Sodium-Cooled Fast Reactor in Japan, idem, 2017.
[20] H. Niwa, A comprehensive approach of reactor safety research aiming at elimination of recriticality in CDA for commercialization of LMFBR, Prog. Nucl. Energy 32 (1998) 621–629.
[21] S. Maruyama, et al., TRU composition changes and their influence on FBR core characteristics in the LWR-to-FBR transition period, in: Proc. ICAPP '09, Tokyo, Japan, May 10-14, 2009, Paper 9300, 2009.
[22] K. Kawashima, et al., Fast reactor core design studies to cope with TRU fuel composition changes in the LWR-to-FBR transition period, in: Proc. ICAPP '09, Tokyo, Japan, May 10-14, 2009, Paper 9288, 2009.
[23] S. Maruyama, et al., Correlations among FBR core characteristics for various fuel compositions, J. Nucl. Sci. Technol. 49 (2012) 640–654.

[24] Advanced Nuclear System Research and Development Directorate, Japan Atomic Energy Agency, Fast reactor cycle technology development project (FaCT Project) - phase I report, in: JAEA-Evaluation 2011-003, 2011 (in Japanese).
[25] S. Maruyama, et al., Consideration of methods to determine an enrichment of commercial fast reactor fuel, in: Proc. ICAPP 2011, Nice, France, May 2-5, 2011, 2011, pp. 1635–1643. Paper 11295.
[26] K. Ohgama, et al., Analysis of fuel subassembly innerduct configurational effects on the core characteristics and power distribution of a sodium-cooled fast breeder reactor, J. Nucl. Sci. Technol. 53 (2016) 1155–1163.
[27] K. Sugino, et al., Advanced metal fuel core design study for SFR in the "feasibility study" in Japan, in: Proc. GLOBAL 2005, Tsukuba, Japan, Oct. 9–13, 2005, Paper 399, 2005.
[28] T. Ogawa, et al., A design study of the sodium cooled metal fuel core with high reactor outlet temperature aiming for the enhancement of safety, in: 2007 Fall Meeting of Atomic Energy Society of Japan, D36, 2007 (in Japanese).
[29] S. Ohki, et al., FBR core concepts in the "FaCT" Project in Japan, in: Proc. PHYSOR'08, Interlaken, Switzerland, Sep. 14-19, 2008, 2008.
[30] K. Ohgama, et al., Design study of a 750 MWe Japan sodium-cooled fast reactor with metal fuel, in: Proc. ICAPP 2017, Fukui and Kyoto, Japan, April 24–28, 2017, 2017.
[31] U.S. DOE, Nuclear energy research advisory committee and the generation IV international forum, a technology roadmap for generation IV nuclear energy systems, 2002. Tech. Rep. GIF-002-00.
[32] H. Yamano, et al., Safety design and evaluation in a large-scale Japan sodium-cooled fast reactor, Sci. Technol. Nucl. Install. 2012 (2012), 614973.
[33] T. Suzuki, K. Kamiyama, H. Yamano, S. Kubo, Y. Tobita, R. Nakai, K. Koyama, A scenario of core disruptive accident for Japan sodium-cooled fast reactor to achieve in-vessel retention, J. Nucl. Sci. Technol. 51 (4) (2014) 493–513.
[34] H. Yamano, S. Kubo, K.I. Kurisaka, Y. Shimakawa, H. Sago, Technological feasibility of two-loop cooling system in JSFR, Nucl. Technol. 170 (1) (2010) 159–169.
[35] K. Kurisaka, S. Okamura, Preliminary evaluation of JSFR achievement level to risk targets, in: Proc. of ICONE19, No. 43802, Chiba, Japan, May, 2011, 2011.
[36] K. Ichikawa, et al., Conceptual design study of JSFR (3) reactor cooling system, in: Proc. of FR09, Kyoto, Japan, 2009, 2009.
[37] Y. Chikazawa, et al., Evaluation of JSFR key technologies, Nucl. Technol. 179 (2009) 360.
[38] M. Etoh, et al., Conceptual design study of JSFR (2) reactor system, in: Proc. of FR09, Kyoto, Japan, 2009, 2009.
[39] K. Oyama, et al., Conceptual Design for Japan Sodium-Cooled Fast Reactor (2): Thermal-Hydraulic Design for Reactor Upper Sodium Plenum in JSFR, 2009. ICAPP'09 Paper 9296.
[40] N. Kimura, et al., Experimental study on flow optimization in upper plenum of reactor vessel for a compact sodium cooled fast reactor, Nucl. Technol. 152 (2) (2005) 210–222.
[41] The Nuclear Safety Commission of Japan, Regulatory Guide for Reviewing Seismic Design of Nuclear Power Reactor Facilities, 2006. NSC Decision No. 2006-D59, Appendix 1.
[42] Y. Sakamoto, et al., Development of advanced loop-type fast reactor in Japan: progress of design study based on preliminary assessment results, in: Proc. of ICONE18, No. 30322, Xi'an, China, May 2010, 2010.

[43] S. Okamura, M. Etoh, Y. Kamishima, K. Negishi, Y. Sakamoto, S. Kitamura, S. Kotake, Seismic isolation design for JSFR, in: Proceedings of International Conference on Fast Reactors and Related Fuel Cycles (FR09), Kyoto, Japan, No. IAEA-CN-176-08-28P, 2009.
[44] T. Shiraishi, H. Watakabe, H. Sago, M. Konomura, A. Yamaguchi, T. Fujii, Resistance and fluctuating pressures of a large elbow in high reynolds numbers, J. Fluids Eng. 128 (9) (2006) 1063.
[45] T. Nakamura, T. Shiraishi, Y. Ishitani, H. Watakabe, H. Sago, T. Fujii, A. Yamaguchi, M. Konomura, Flow-induced vibration of a large-diameter elbow piping based on random force measurement caused by conveying fluid (visualization test results), in: Proc. ASME Pressure Vessels and Piping Division Conference, Denver, Colorado, July 17–21, 2005, American Society of Mechanical Engineers, 2005. PVP2005-71277.
[46] I.E. Idelchik, Handbook of Hydraulic Resistance, Hemisphere, New York, 1986.
[47] T. Kawamura, T. Nakano, M. Takahashi, Reynolds number effect on turbulence downstream from elbow pipe, Trans. Jpn. Soc. Mech. Eng. B 68 (667) (2002) 645.
[48] K. Hirota, Y. Ishitani, T. Nakamura, T. Shiraishi, H. Sago, H. Yamano, S. Kotake, Flow-induced vibration of a large-diameter elbow piping in high reynolds number range; random force measurement and vibration analysis, in: Proc. 9th Int. Conf. Flow-Induced Vibrations (FIV2008), Prague, Czech Republic, June 30–July 3, 2008, No. 264, Institute of Thermomechanics, 2008.
[49] T. Handa, et al., Research and development for the integrated IHX/Pump, in: Proc. of FR09, Kyoto Japan, 2009, 2009.
[50] N. Kisohara, Study on the integrated IHX/sodium mechanical pump (evaluation of tube fretting wear), in: Mechanical Engineering Congress (2227), Japan, 2003, 2003.
[51] N. Sawa, et al., Status of Development of an Integrated IHX/Pump Component, 2005. GLOBAL 2005, Paper No. 207, Japan, October 2005.
[52] H. Hayafune, M. Konomura, et al., Development of the integrated IHX/pump component 1/4-scale vibration testing, in: ICONE14, Paper 89745, Miami, Florida, USA, July 17-20, 2006, 2006.
[53] T. Chiba, et al., Extension and maintenance of finite element nonlinear structural analysis system FINAS (I), in: JAEA, PNC-TJ9270-88-002, Japan, February 28, 1987, 1987.
[54] S. Kitamura, et al., Sloshing analysis of a reactor vessel model, in: JAEA, PNC-TN9410-87-125, Japan, September, 1987, 1987.
[55] N. Kisohara, et al., Conceptual design of study for the demonstration reactor of JSFR (5) reactor cooling system, in: Proc. of ICONE19, No. 44042, Chiba, Japan, May 2011, 2011.
[56] H. Tanabe, et al., Analysis Code for Failure Propagation of Steam Generator Tube—LEAPII, PNCTN952 82-04, Power Reactor and Nuclear Fuel Development Corp, 1982 (in Japanese).
[57] H. Hamada, H. Tanabe, A Study of the Selection of Rational Design Basis Leak, PNC TN9410 92-097, Power Reactor and Nuclear Fuel Development Corp, 1992 (in Japanese).
[58] I. Ono, A. Kurihara, Quasi-Steady Pressure Analysis in Intermediate Water Leak Rate with SWAC-13 Module in SWACS Code, JNC TN9400 2005-047, Japan Nuclear Cycle Development Institute, 2005 (in Japanese).
[59] K. Kurome, et al., Steam generator with straight double-walled tube, in: Proc. of FR09, Kyoto Japan, 2009, 2009.
[60] N. Kisohara, T. Sakai, T. Moribe, Temperature and flow distributions in sodium-heated large straight tube steam generator by numerical methods, Nucl. Technol. 164 (2008).

[61] K. Ohyama, et al., Decay heat removal system by natural circulation for JSFR, in: Proc. of FR09, Kyoto Japan, 2009, 2009.
[62] Y. Eguchi, et al., Quantitative prediction of natural circulation in an LMFR with a similarity law and a water test, Nucl. Eng. Des. 178 (3) (1997) 295–307.
[63] H. Kamide, K. Hayashi, T. Isozaki, M. Nishimura, Investigation of core thermohydraulics in fast reactors—interwrapper flow during natural circulation, Nucl. Eng. Des. 133 (2001) 77–91.
[64] F. Yamada, H. Ohira, Numerical simulation of Monju plant dynamics by super-COPD using previous startup tests data, in: Proc. ASME 2010 3rd Joint US-European Fluids Engineering Summer Meeting and 8th International conference on Nanochannels, Microchannels and Minichannels, 2010. FEDSM-ICNMM2010-30287; 2010; Montreal (Canada).
[65] V. Yakhot, S.A. Orszag, Renormalization group analysis of turbulence—I: basic theory, J. Sci. Comput. 1 (1986) 1–51.
[66] V. Yakhot, S.A. Orszag, S. Thangam, T.B. Gatski, C.G. Speziale, Development of turbulence models for shear flows by a double expansion technique, Phys. Fluids A 4 (7) (1992) 1510–1520.
[67] S. Usui, et al., Development of advanced loop-type fast reactor in Japan (4) an advanced design of the fuel handling system for the enhanced economic competitiveness, in: Proc. of ICAPP'08, Anaheim, CA, USA, June 2008, 2008.
[68] Y. Chikazawa, et al., JSFR key technology evaluation on fuel handling system, J. Nucl. Sci. Technol. 51 (4) (2014) 437.
[69] T. Mizuno, et al., Advanced oxide fuel core design study for SFR in the feasibility study in Japan, in: Proc. GLOBAL 2005; 2005 Oct; Tsukuba (Japan), 2005.
[70] S. Maruyama, S. Ohki, T. Mizuno, TRU composition changes and their influence on FBR core characteristics in the LWR-to-FBR transition period, in: Proc. ICAPP'09; 2009 May; Tokyo (Japan), 2009.
[71] S. Maruyama, et al., Study on FBR core concepts for the LWR-to-FBR transition period, in: Proc. Global 2009; 2009 September; Paris (France), 2009.
[72] K. Kawashima, et al., Fast reactor core design studies to cope with TRU fuel composition changes in the LWR-to-FBR transition period, in: Proc. ICAPP'09; 2009 May; Tokyo (Japan), 2009.
[73] A. Katoh, et al., Development of advanced fuel handling machine for JSFR, J. Nucl. Sci. Technol. 47 (7) (2010) 642.
[74] A. Katoh, et al., Development of the JSFR fuel handling system and mockup experiments of fuel handling machine in abnormal conditions, in: Proc. of ICAPP2010; 2010 June; San Diego (USA), 2010.
[75] Y. Chikazawa, et al., Heat transfer experiments on fuel subassembly transfer pot for JSFR, J. Nucl. Sci. Technol. 51 (6) (2014) 798–808.
[76] Y. Chikazawa, S. Hirata, H. Obata, Development of design evaluation tools for the JSFR fuel transfer pot, Nucl. Eng. Design. 273 (2014) 1–9.
[77] S. Toda, Y. Ieda, H. Kamide, K. Komatsuzaki, Study of heating/cooling effects on transverse temperature distribution in a FBR subassembly, in: Proc. Fall Meeting of Atomic Energy Society of Japan D37; September; Sapporo, Japan (1994), 1994 (in Japanese).
[78] Y. Chikazawa, H. Katoh Aand Obata, Development of argon gas cleaning for sodium-cooled reactor spent fuel, J. Nucl. Sci. Technol. 50 (10) (2013) 988–997.
[79] T. Narita, et al., Water corrosion resistance of ODS ferritic-martensitic steel tubes, J. Nucl. Sci. Technol. 45 (2) (2008) 99.

[80] T. Kaito, et al., High temperature oxidation behavior of ODS steels, J. Nucl. Mater. 329 (2004) 1388. 333.
[81] A. Katoh, Y. Chikazawa, H. Obata, Feasibility study on shipping cask for JSFR fresh fuel contained minor actinide, Nucl. Eng. Des. 247 (2012) 98.
[82] H. Hara, et al., Conceptual design study of JSFR (4) reactor building layout, in: Proc. of FR09, Kyoto, Japan, 2009.
[83] T. Yamamoto, et al., Design evaluation method of steel-plate reinforced concrete structure containment vessel for sodium cooled fast reactor, J. Disaster Res. 7 (5) (2012) 645.
[84] M. Ozaki, et al., Study on steel plate reinforced concrete bearing wall for nuclear power plants part 1; shear and bending loading tests of SC walls, in: Proc. of 16th SMiRT, Paper #1554, 2001.
[85] T. Fujita, et al., Development of advanced concept for shortening construction period of ABWR plant (Part 3), in: Proceedings of 13th ICONE, 2005. ICONE13-50646.
[86] A. Yamaguchi, Y. Tajima, Validation study of computer code SPHINCS for sodium fire safety evaluation of fast reactor, Nucl. Eng. Des. 219 (2003) 19–34.
[87] T. Yamamoto, et al., Conceptual design study of JSFR reactor building, in: Proceedings of ICAPP'12, June 24-28, Chicago, USA, 2012. Paper ID #12387.
[88] European Committee for Standardization (CEN), Eurocode 4- Design of composite steel and concrete structures—part 1-2: Structural fire design, in: European Prestandard, CEN, Brussels, Final Draft prENV 1994-1-2, Oct. 1993.
[89] T. Hosoya, et al., Conceptual design for japan sodium-cooled fast reactor (4): developmental study of steel plate reinforced concrete containment vessel for JSFR, in: Proc. of ICAPP'09, Paper 9418, 2009.
[90] S. Okamura, et al., Seismic isolation design for JSFR, in: Proc. of FR09," Kyoto, Japan, 2009.
[91] Japan Nuclear Safety Commission, Report of the seismic evaluation according to revision of 'Regulatory guide for reviewing seismic design of nuclear power reactor facilities', 2007. April 4, 2007. (in Japanese).
[92] Japan Atomic Energy Agency, Prototype reactor Monju, report of the seismic evaluation according to revision of 'Regulatory guide for reviewing seismic design of nuclear power reactor facilities', 2009. March 31, 2009. (in Japanese).
[93] Etoh et al, Conceptual design study of JSFR (2)—reactor system, in: FR09, Paper No. IAEA-CN-176-08-11P, Kyoto, Japan, December 7-11, 2009, 2009.
[94] Japan Electric Association, Technical guidelines on seismic base isolation system for structural safety and design of nuclear power plants, 2001. JEAG 4614-2000.
[95] J.M. Kelly, Earthquake-Resistant Design with Rubber, 3rd ed., Translated by Japan Association for Vibration Technologies, Tokyo Denki University Press, 2005.
[96] N. Nishiyama, S. Kotake, M. Uzawa, In-service inspection and repair program for commercialized sodium-cooled fast reactor, in: Proc. of FR09, Kyoto Japan, 2009, 2009.
[97] JSME, Code for Nuclear Power Generation Facilities-Rules on Fitness-for-Service for Nuclear Power Plants, 2007 (JSME S NA1).
[98] Y. Sakamoto, S. Kubo, S. Kotake, Y. Kamisima, Development of advanced loop-type fast reactor in Japan (3): easy inspection and high reliable reactor structure in JSFR, in: Proc. of ICAPP 2008, Anaheim, California, USA, 2008.
[99] A. Tagawa, T. Yamashita, Development of high resolution sensor for under sodium viewing inspection, in: 2009 Fall Meeting, Atomic Energy Society of Japan, 2009.
[100] M. Ohtaka, H. Hayashida, Study on Measurement Technique for Sodium Aerosols Based on Laser Induced Breakdown Spectroscopy, JNC technical review, No. 18, 2003, pp. 23–28.

[101] T. Sakai, S. Kotake, K. Aoto, Current status of conceptual design study toward the demonstration reactor of JSFR, in: Proc. of ICONE18, No. 30172, Xi'an, China, 2010.
[102] H. Kamide, et al., JSFR design progress related to development of safety design criteria for generation IV sodium-cooled fast reactor (1) overview, in: Proc. of ICONE-13, Chiba, Japan, May 2015, 2015.
[103] S. Kubo, et al., Safety design approach for JSFR toward the realization of GEN IV Sodium cooled fast reactor, in: Proc. of FR13, Paris, France, March 2013, 2013.
[104] R. Nakai, et al., Development of safety design criteria for the generation-IV sodium-cooled fast reactor, in: NTHAS8, Beppu, Japan, 2012.
[105] S. Okamura, et al., Seismic isolation design for JSFR, in: Proc. the Int. Conf. on Fast Reactors and Related Fuel Cycles (FR09), No. IAEA-CN-176-08-28P Kyoto, Japan, 2009.
[106] H. Hayafune, et al., Evaluation of severe external events on JSFR, in: FR13, Paris, France, 2013.
[107] S. Kondo, et al., Integrated analysis of In-vessel and Ex-vessel severe-accident sequences, in: Proc. Int. Fast Reactor Safety Meeting, 1990, Snow bird, Utah, US, 1990.
[108] A. Bayer, K. Koberlein, Risk-oriented analysis on the German prototype fast breeder reactor SNR-300, Nucl. Saf. 25 (1) (1984) 19–32. Jan.-Feb. 1984.
[109] K. Isono, et al., Studies on maintenability and reliability for Japan sodium-cooled fast reactor (JSFR), in: Proc. of ICONE22, No. 30175, Prague, Czech Republic, July 2014, 2014.
[110] T. Yamamoto, et al., Safety enhancement approach against external hazard on JSFR reactor building, Nucl. Technol. 206 (12) (2020) 1875.
[111] H. Yamano, Safety design approach for external events in japan sodium-cooled fast reactor, in: Proc. of ICAPP'12, Paper 12158, Chicago, USA, June 24-28, 2012, 2012.
[112] T. Yamamoto, et al., Study on reduction of vertical seismic response for sodium-cooled fast reactor building, in: Proc. of ICAPP2016, San Francisco, CA, April 2016, 2016.
[113] S. Kubo, Y. Shimakawa, JSFR design progress related to development of safety design criteria for generation IV sodium-cooled fast reactor (2) progress of safety design, in: Proc. of ICONE-23, Chiba, Japan, May 2015, 2015.
[114] A. Katoh, Design approach for decay heat removal systems based on the safety design criteria for Gen-IV sodium-cooled fast reactor, in: Proc. of ICAPP2014, Charlotte, USA, April 2014, 2014.
[115] The Tsunami Evaluation Subcommittee, The Nuclear Civil Engineering Committee, JSCE (Japan Society of Civil Engineers), Tsunami Assessment Method for Nuclear Power Plants in Japan, 2006. May, 2006.
[116] A. Katoh, et al., Safety improvement in building layout design to meet the safety design criteria for the Generation IV SFR, in: Proc. of ICAPP 2015, Paper 15433, Nice, France, May 03-06, 2015, 2015.
[117] Y. Enuma, et al., JSFR design progress related to development of safety design criteria for generation IV sodium-cooled fast reactor (3) progress of component design, in: Proc. of ICONE-23, Chiba, Japan, May 2015, 2015.
[118] H. Hayafune, et al., Advanced sodium-cooled fast reactor development regarding GIF safety design criteria, in: Proc. of FR17, Yekaterinburg, Russia, June 2017, 2017.
[119] N. Kisohara, et al., Conceptual design study for the demonstration reactor JSFR (5) reactor cooling system design, in: Proc. of ICONE-19, No. 44042, Chiba, Japan, May 2011, 2011.

[120] Y. Chikazawa, et al., Severe external hazard on hypothetical JSFR in 2010, Nucl. Technol. 192 (2) (2015) 111–124.
[121] K. Amano, et al., Development of the pump-integrated intermediate heat exchanger in advanced loop-type sodium-cooled fast reactor for demonstration, in: Proc. of ICONE-24, Charlotte, NC, the US, June 2016, 2016.
[122] S. Futagami, et al., A study on LMFBR steam generator design without tube failure propagation in water leak events, in: Proc. of ICAPP'09, No. 9169, Tokyo, Japan, 2009.
[123] H. Kanda, et al., A study on the straight double-wall tube steam generator–design against sodium-water reaction in DECs, in: Proc. of ICONE-24, Charlotte, NC, the US, June 2016, 2016.
[124] ASME, SECTION II Materials Part D-Properties 1992 EDITION, 1992.
[125] GIF SDC-TF, Safety Design Criteria for Generation IV Sodium-Cooled Fast Reactor System (Rev.1), 2016.
[126] GIF SDC-TF, Safety Design Guidelines on Safety Approach and Design Conditions for Generation IV Sodium-cooled Fast Reactor Systems, 2016.
[127] M. Uchita, T. Miyagawa, K. Dozaki, Y. Chikazawa, S. Kubo, H. Hayafune, T. Suzuno, T. Fukasawa, Y. Kamishima, S. Fujita, Seismic evaluation for a large-sized reactor vessel targeting SFRs in Japan, in: Proc. of ICAPP2018, 2018, pp. 380–386.
[128] A. Kato, S. Kubo, Y. Chikazawa, T. Miyagawa, M. Uchita, T. Suzuno, J. Endo, K. Kubo, H. Murakami, M. Uzawa, K. Ichikawa, R. Shimizu, Progress in conceptual design of a pool-type sodium-cooled fast reactor in Japan, in: Proc. of FR22, 2022. CN290-269.
[129] J. Rouault, P. Le Coz, J.-C. Garnier, J.-M. Hamy, H. Hayafune, T. Iitsuka, H. Mochida, Japan-France collaboration on the ASTRID program and sodium fast reactor, in: Proc. of ICAPP2015, 2015, pp. 832–837.
[130] I. Sato, Y. Tobita, K. Konishi, K. Kamiyama, J. Toyooka, R. Nakai, S. Kubo, S. Kotake, K. Koyama, Y. Vassiliev, A. Vurim, A. Zuev, A. Kolodeshnikov, Safety strategy of JSFR eliminating severe recriticality events and establishing in-vessel retention in the core disruptive accident, J. Nucl. Sci. Technol. 48 (6) (2013) 556–566.
[131] IAEA, Fast Reactor Database 2006 Update, 2006. IAEA- TECDOC-1531.
[132] A. Katoh, H. Hayafune, S. Kotake, Design features and cost reduction potential of JSFR, Nucl. Eng. Des. 280 (2014) 586–597.
[133] IAEA, Safety of Nuclear Power Plant: Design, Specific Safety Requirements, 2016. No. SSR-2/1 (Rev.1).
[134] S. Okamura, Y. Kamishima, K. Negishi, Y. Sakamoto, S. Kitamura, S. Kotake, Seismic isolation design for JSFR, J. Nucl. Sci. Technol. 48 (4) (2011) 688–692.
[135] T. Fukasawa, S. Okamura, T. Yamamoto, N. Kawasaki, T. Hirotani, E. Moriizumi, Y. Sakurai, N. Masaki, Research and development of rubber bearings for sodium-cooled fast reactor: ultimate properties of half-scale thick rubber bearings based on breaking tests, J. Press. Vessel. Technol. 140 (1) (2017). research-article.
[136] S. Kubo, Y. Chikazawa, H. Ohshima, M. Uchita, T. Miyagawa, M. Eto, T. Suzuno, I. Matoba, J. Endo, O. Watanabe, K. Higurashi, A conceptual design study of pool-type sodium-cooled fast reactor with enhanced anti-seismic capability, Mech. Eng. J. 7 (3) (2020) 19-00489_1–19-00489_16.
[137] S. Matsuura, H. Nakamura, S. Ogiso, T. Murakami, Y. Kawamoto, Formulae for evaluating buckling strength of FBR main vessels under earthquake loading, JSME Int. J. 36 (3) (1993) 485–492.
[138] T. Miyagawa, T. Watakabe, T. Yamamoto, T. Fukazawa, S. Okamura, et al., Research and development of three dimensional seismic isolation system utilized coned-disc-springs with rubber bearings, in: Proc. of ASME 2017 PVP, 2017. PVP2017- 65549.

[139] T. Fukazawa, S. Okamura, T. Somaki, T. Miyagawa, M. Uchita, T. Yamamoto, Y. Watakabe, S. Fujita, Research and development of three dimensional seismic isolation system for sodium-cooled fast reactor (Part4) proposal of optimal combination method for disc spring units and newly friction for sliding elements, in: Proceedings of ASME 2019 PVP, 2019. PVP2019- 93480.

[140] M. Morishita, K. Inoue, T. Fujita, Development of three-dimensional seismic isolation systems for fast reactor application, J. Jpn. Assoc. Earthq. Eng. 4 (3) (2004) 305–310 (Special Issue).

[141] N. Kawasaki, T. Watakabe, T. Wakai, T. Yamamoto, T. Fukasawa, S. Okamura, Development of seismic isolation systems for sodium-cooled fast reactors in Japan, in: Proc. of the ASME 2016 PVP, 2016. PVP2016-63437.

[142] N. Tomita, S. Fujita, Study on the rocking vibration induced in three-dimensional seismic isolation system, in: Proc. of the ASME 2017 PVP, 2017. PVP2017-65714.

[143] T. Fukazawa, T. Somaki, T. Miyagawa, M. Uchita, T. Yamamoto, M. Miyazaki, S. Okamura, S. Fujita, Research and development of three-dimensional isolation system for sodium-cooled fast reactor: Part 6 -Development of high speed oil damper for isolated building, in: ASME 2020 PVP, 2020. PVP2020- 21320.

[144] K. Ito, H. Oshima, Y. Nakamine, Y. Imai, Study on turbulent modeling in gas entrainment evaluation method, J. Power Energy Syst. 6 (2) (2012) 2012.

[145] S. Kubo, Y. Shimakawa, JSFR design progress related to development of safety design criteria for generation IV sodium-cooled fast reactors (2) progress of safety design, in: Proc. of ICONE23, 2015. ICONE23-1748.

[146] H. Saito, Y. Yamada, K. Oyama, S. Matsunaga, H. Yamano, S. Kubo, Safety evaluation of self actuated shutdown system for gen-IV SFR, in: Proc. of ICAPP 2017, 2017, p. 17439.

[147] S. Nakai, T. Iwasaki, Y. Shimakawa, M. Tanji, Development of Module Integrated Plant Dynamics Analysis Code-Development of Super-COPD Code- (1988), 1988. PNC Technical review No. 68-3 (in Japanese).

[148] F. Yamada, H. Ohira, K. Okusa, A. Miyakawa, Validation of plant dynamics analysis code Super-COPD by MONJU startup tests, in: Proc. of FR09, 2009. Paper No. IAEA-CN-176-06-39P.

Key technologies for future sodium-cooled fast reactors

5

Kosuke Aizawa[a], Masanori Ando[a], Yoshitaka Chikazawa[a], Norihiro Doda[a], Yasuhiro Enuma[a], Toshiki Ezure[a], Yoshitaka Fukano[a], Erina Hamase[a], Shun Hirooka[a], Tomoyuki Hiyama[a], Katsunori Ishii[a], Takeji Kaito[a], Kenji Kamiyama[a], Norihiro Kikuchi[a], Jun Kobayashi[a], Shigenobu Kubo[a], Akikazu Kurihara[a], Shuhei Maruyama[a], Kentaro Matsushita[a], Takero Mori[a], Koji Morita[a,b], Hiroaki Ohira[a], Shigeo Ohki[a], Satoshi Okajima[a], Yasushi Okano[a], Takashi Onizawa[a], Ayako Ono[a], Takayuki Ozawa[a], Hiroshi Seino[a], Takashi Takata[a,c], Shigeru Takaya[a], Masayuki Takeuchi[a], Masaaki Tanaka[a], Hiroshi Taninaka[a], Takashi Tanno[a], Kodai Toyota[a], Akihiro Uchibori[a], Tomoyuki Uwaba[a], Takashi Wakai[a], Tomoyoshi Watakabe[a], Hiroki Yada[a], Tomohiko Yamamoto[a], Hidemasa Yamano[a], Kenji Yokoyama[a], Ryuji Yoshikawa[a], Kazuo Yoshimura[a], Ryuta Hashidate[a], Yuichi Sano[a], and Akiyuki Seki[a]

[a]Sector of Fast Reactor and Advanced Reactor Research and Development, Japan Atomic Energy Agency, Ibaraki, Japan, [b]Department of Applied Quantum Physics and Nuclear Engineering, Kyushu University, Fukuoka, Japan, [c]Department of Nuclear Engineering and Management, School of Engineering, The University of Tokyo, Tokyo, Japan

Chapter outline

Sodium Fast Reactors. https://doi.org/10.1016/B978-0-12-824076-2.00005-5
Copyright © 2022 Elsevier Inc. All rights reserved.

Nomenclature

Greek symbols

$\Delta H_{298.15}$ enthalpy difference at 298.15 K (25°C)

Subscripts

(l) liquid phase
(g) gas phase
(s) solid phase

This chapter describes the latest results and future prospects of the research and development activities in ten key technologies for the future sodium-cooled fast reactors (SFRs).

1. ***Safety***: Safety is the top priority for SFR development and design. In this section, research and development on the critical safety issues specific to SFRs are discussed, including safety design criteria and safety design guidelines (SDC/SDG), self-actuated shutdown system (SASS), severe accident, sodium combustion, sodium-water reaction, and source term.
2. ***Sodium component development***: Conceptual designing of Japan Sodium-cooled Fast Reactor (JSFR) has been conducted since 2006 and several innovative sodium component designs were established. The development plan for JSFR includes large-scale sodium component experiments and system demonstrations.
3. ***Reactor core physics***: JAEA has been developing a standard reactor core physics analysis method for fast reactors for many years. This analysis method was developed on the premise that integral experimental data are reflected in the core design using a kind of data assimilation methodology. This section outlines the features of this analysis method, including not only the analytical models and computer codes but also the experimental database.
4. ***Fuel and materials***: This section introduces the results of physical properties study, irradiation behavior evaluation and performance code development of MOX fuel, and long-life core material development such as ODS steel cladding that has been implemented in Japan for practically realizing fast reactors.
5. ***Thermal-hydraulics***: Various thermal-hydraulics phenomena should be evaluated accurately to ensure the safe and reliable operation of an SFR plant, mainly by computational fluid dynamics (CFD). These include plant dynamics, natural circulation, thermal-hydraulics in fuel assemblies, and the hot plenum of a reactor vessel.
6. ***High-temperature structural materials and structural design***: Elevated temperature design methods considering time-dependent deformation and failure by creep phenomena are required for design and construction of the SFRs operated at high temperatures. This section introduces the development and codification of the Japanese unique elevated temperature design codes.
7. ***Seismic design***: This section introduces some typical research activities related to the seismic design of SFRs. Establishing the seismic design technology of structures and mechanical components is a critical challenge in realizing SFRs in Japan that has many experiences of the severe earthquake disaster.
8. ***Operation and maintenance***: This section describes the fitness-for-service code and LBB assessment guidelines for sodium-cooled fast reactors that have been newly developed by the Japan Society of Mechanical Engineers. In addition, the development of an under-sodium viewer that is a unique technology for inspections in the opaque coolant is explained.
9. ***Fuel cycle-related studies***: JAEA has promoted the development of simplified pelletizing fuel fabrication and advanced aqueous reprocessing methods to establish the fuel cycle technology for fast reactors through the Fast Reactor Cycle Technology (FaCT) development project. In this section, the summary of the developments described before is introduced as the main issues in the fuel cycle-related studies.
10. ***ARKADIA–Advanced reactor knowledge- and AI-aided design integration approach through the whole plant life cycle***: JAEA has started to develop "Advanced Reactor Knowledge- and AI-aided Design Integration Approach through the whole plant life cycle (ARKADIA)" as one of the research and development infrastructures. ARKADIA offers the best possible solutions for any challenges that could arise in the plant life cycle, including plant design and safety measures, maintenance, and decommissioning.

5.1 Safety

5.1.1 Introduction

In proceeding with design, construction, and operation of experimental reactor Joyo and prototype reactor Monju, and the conceptual design of DFBR and JSFR, knowledge acquisition and improvement of analysis tools were promoted to conduct safety design and assessments specific to sodium-cooled fast reactors (SFRs).

The core of a fast reactor is not in the most reactive configuration during normal operation because it is a critical system based on fast neutrons, which do not require neutron moderation. This means that positive reactivity might be inserted when removing coolant and structural materials such as cladding tubes from the core. In addition, the reactor cooling system can be set to low pressure without the need to pressurize because liquid metal sodium used as a coolant has a high boiling point. On the contrary, it is required to prevent and mitigate sodium leak and combustion and sodium-water reaction caused by steam generator tube leak, since sodium is chemically active. It is necessary to establish the safety design criteria (SDC) considering such characteristics of SFRs. The "Safety Evaluation Policy for liquid metal cooled fast breeder reactors" was issued in Japan by the Nuclear Safety Commission for prototype reactor Monju in 1980, and the Monju safety design and its evaluation were conducted in line with this policy. At the beginning of the 21st century, the development of SFRs entered demonstration and commercial phases in some countries, and SDC and safety design guidelines (SDGs) were established at the Generation IV International Forum in response to the need to develop safety standards as an international standard. Japan contributed to the establishment of SDC and SDGs based on the experiences and achievement of Joyo, Monju, Feasibility Study on Commercialized Fast Reactor Cycle Systems (FS), and Fast Reactor Cycle Technology (FaCT) development project.

The basic concept of safety design for SFRs is also defense in depth, and it is necessary to implement design measures for prevention and mitigation of core damage in design extension conditions, which is the fourth level of defense in depth. In particular, the measures against anticipated transient without scram (ATWS), which is considered as an initiator of core damage leading to the nuclear fission energy release, have been emphasized, and research and development have been conducted.

A passive reactor shutdown mechanism (self-actuated shutdown system—SASS), which enhances prevention capability of core damage, started to be developed for demonstration fast breeder reactor (DFBR) and was later introduced into Japan Sodium-cooled Fast Reactor (JSFR).

Many of the various phenomena occurring in the core damage situation, including fuel failure behavior, are specific to SFRs. The research was started in the United States and Europe in the 1970s, and much knowledge was acquired. Japan has proceeded to obtain experimental knowledge and develop analysis tools on various phenomena of core damage through collaboration with the United States and Europe and original research. By utilizing these knowledge and analysis tools, design measures for so-called in-vessel retention (IVR) have been established to mitigate the consequences even in core damage situations and contain degraded core materials in the reactor vessel to terminate the accident.

To mitigate the radiological consequences of severe core damage accidents, it is necessary to develop analysis tools for evaluating the release of the radioactive materials to the environment by analyzing the transport behavior of gaseous and volatile radioactive materials simultaneously with the achievement of IVR. A challenging issue for this development is quantifying the retention effect of fission products (FPs) such as iodine and cesium in sodium and the transport behavior of FPs in the gas atmosphere containing sodium aerosol.

Experimental studies and analysis tools for sodium combustion and sodium-water reaction are also under development.

This section introduces the achievements of research and development on these safety issues specific to SFRs.

5.1.2 Development of SDC/SDG

5.1.2.1 Introduction

After TEPCO's Fukushima Daiichi Nuclear Power Plant's accident (the 1F accident), activities to enhance the safety of nuclear power plants have become conspicuous. The development of Sodium-cooled Fast Reactor (SFR) is vigorously being promoted especially in the United States, Russia, China, and India, and therefore, a demand for an international standard for ensuring the safety of SFRs is growing.

In the framework of the Generation-IV International Forum (GIF), an effort to develop "safety design criteria (SDC)" for SFR systems was initiated in 2011. For this purpose, an SDC task force (SDC-TF) was formulated in GIF in July 2011, to summarize and consolidate the SDC. The SDC-TF members consist of representatives of CIAE (China), CEA (France), JAEA (Japan), KAERI, KINS (Republic of Korea), IPPE (Russia), ANL, INL, ORNL (United States), EC, and IAEA.

Fig. 5.1 shows the hierarchy of GIF safety standards, including the SDC and safety design guidelines (SDGs). The SFR SDC report was completed in 2013 and distributed to the international organizations, namely IAEA, MDEP, NEA/CNRA, and regulatory bodies of the GIF member states with active SFR development programs (China, EC, France, Japan, Korea, Russia, and the United States) [1]. In 2017, the SDC-TF finalized responses to feedback from IAEA, US NRC, China's NNSA, and France's IRSN, and the SDC report was updated based on various comments on general matters, e.g., safety approaches for Generation-IV (Gen-IV) reactor systems, the interface between safety and security (necessity of management system to take into account the potential for adverse effects on safety or security) and suggestions on specific criteria for dealing with, e.g., sodium fire, design basis accident (DBA) and design extension condition (DEC). The revised SDC also reflects the revision of IAEA SSR 2/1 [2].

In 2016, the SDC-TF completed the first SFR safety design guideline report titled "Safety Design Guidelines on Safety Approach and Design Conditions for Gen-IV Sodium-cooled Fast Reactor Systems (SDG on Safety Approach)" [3]. The guidelines

Fig. 5.1 Hierarchy of GIF safety standards, including safety design criteria and safety design guidelines [1].

are a set of recommendations on how to comply with the SFR SDC and address SFR-specific safety topics by clarifying technical issues and providing design options. The SDG on Safety Approach report was distributed to OECD/NEA's Working Group on the Safety of Advanced Reactors (WGSAR) and the IAEA for review. Leveraging the important and constructive feedback, the TF integrated the resolutions for WGSAR and IAEA comments into the report in 2018.

The SDC-TF has been developing the second safety design guidelines report, "Safety Design Guidelines on Structures, Systems and Components for Gen-IV Sodium-cooled Fast Reactor Systems (SDG on SSCs)," which provides recommendations on the design of structures, systems, and components (SSCs) important to safety and supports the practical application of the SDC and the SDG on Safety Approach to the safety-related SSC designs. The SDG on SSC specifies 14 focal points related to three fundamental systems: (1) reactor core system, (2) coolant system, and (3) containment system. These recommendations on the specific SSCs have been developed to clarify the safety requirements for the Gen-IV SFR systems. The TF intends to complete the SDG on the SSC report and issue it in 2022.

These SDC/SDG reports have been introduced at joint IAEA-GIF workshops on the safety of SFR and shared widely to GIF member countries, nonmember countries like India, regulatory bodies and associated organizations, and plant manufacturers. They were also presented at the 4th GIF symposium in 2018 [4].

This section describes the development of SDC, SDG on Safety Approach, SDG on SSCs, and SFR safety design concepts in a manner consistent with SDC/SDGs.

5.1.2.2 SDC

The GIF safety and reliability goals were established by the Policy Group in 2002 in the "Generation-IV Nuclear Energy Systems under the GIF Roadmap" [5]. The goals state that the Gen-IV systems are expected to excel in safety and reliability for Defense in Depth (DiD) Levels 1 and 2 (operational states) and to have a very low likelihood and degree of core damage for Defense-in-Depth Levels 3 and 4 (DBAs and DECs).

These goals are expected to be achieved by implementing a robust design to reduce the likelihood of abnormal occurrences and by employing design features to control the progression of an accident and mitigate its consequences. Focus is given to safety design for accident prevention and a robust safety demonstration that uses a methodology (based on "deterministic and probabilistic" with associated confidence) to address uncertainties.

The SDC follows this DiD philosophy [6] as the fundamental safety approach to assure design measures for every plant state, i.e., normal operation, anticipated operational occurrence (AOO), DBA, and DEC. The design for operational states (i.e., normal operation and AOO) and DBAs are required to be conservative with due account of uncertainties of design conditions and transient phenomena. For DECs, the design is required to prevent significant radioactive material releases to the environment based on best estimate analysis. To further ensure safety, the appropriate management of the fuel handling and storage and the radioactive waste management systems are also required.

(a) SFR-specific revisions to the general design criteria

SFR-specific revisions are included in the criteria for fuel elements and assemblies, reactor core structure and characteristics, reactor shutdown systems, primary coolant and cover gas systems, containment systems, fuel handling, and storage systems. New criteria for intermediate coolant system, decay heat-removal system (including final heat sink), sodium heating systems, sodium purification system, sodium leak detection, and sodium fire suppression system are also created.

Considerations are required for SFR fuel elements and fuel assemblies that are operated in a fast neutron spectrum under the conditions of high-power density, high burn up, and high sodium temperature. The creep and radiation effects on core structural materials are also emphasized. Since the SFR reactor core is not in the most reactive configuration under normal operating conditions and it is possible to have a positive void reactivity at the center area of the reactor core, provisions are specified for the design of the reactor core to prevent excessive reactivity insertion.

Since sodium is chemically reactive, it is necessary to manage sodium fire in contact with air or concrete to prevent its potential impact on the safety functions of systems, structures, and components. An intermediate coolant system is required for a Gen-IV SFR system to control/manage the consequences of sodium-water reaction during steam generator tube failure accidents. Provisions are also requested to address the opacity and high melting point of sodium coolant that pose challenges to in-service inspection, maintenance, and repair. Since SFRs allow the use of near-atmospheric-pressure coolant systems, application of the leak before break concept to enable continuous leakage monitoring as an inspection method for the coolant boundary is recommended and expected.

DECs also include sodium chemical reactions (e.g., combustion and sodium-concrete reactions resulting from leakage, and sodium-water reaction resulting from steam generator tube failure) since they could affect the safety of the reactor core or loss of containment function. The capability of ensuring containment integrity is required so that it can withstand thermal and mechanical loads generated from sodium combustion, sodium-concrete reaction, and potential debris-concrete interaction.

(b) Structure of the GIF SFR SDC report

Consistent with the structure of IAEA SSR-2/1, the SFR safety design criteria are categorized in several chapters. The responsibilities in the management of safety in plant design, management system for the plant design, and the safety of the plant design throughout the lifetime of the plant are addressed in the SDC report. The report also covers principal technical criteria: the fundamental safety functions (control of reactivity, heat removal, and confinement of radioactivity), radiation protection, application of DiD, interfaces of safety with security and safeguards, proven engineering practices, and provisions for waste management and decommissioning.

The SDG report addresses wide-ranging criteria relevant to the general plant design issues such as identification of plant states, design limits, postulated initiating events, internal and external hazards, common cause failures, engineering design rules, operational limits and conditions for safe operation, DBAs, DECs, principles for physical separation and independence of safety systems, design basis, safety classification and reliability of items important to safety, support service systems, calibration, testing, maintenance, repair, replacement, inspection and monitoring of items important to safety, qualification of items important to safety, aging management, design for optimal operator performance, sharing of safety systems between multiple units of a nuclear power plant, communications, emergency management, interactions between the electrical power grid and the plant, and safety analysis.

Next, the SDG report describes the remaining criteria related to the design of specific plant systems including the reactor core (fuel elements and assemblies and other core structures, reactivity control, and shutdown systems), reactor coolant system, decay heat-removal system, containment system, instrumentation and control systems, main, supplementary, and emergency control rooms, emergency power supply, fuel handling, and storage systems, as well as other supporting and auxiliary systems. A glossary covering SFR specific terminology and a number of important terms as defined in the IAEA safety standards/glossary [7] are also incorporated.

(c) Revisions to SFR SDC based on external feedback

After the approval by the GIF Policy Group in May 2013, the SDC report was circulated to international organizations (i.e., IAEA, MDEP, and OECD/NEA/CNRA) and regulatory bodies of the SFR developing states under GIF (i.e., China, EC, France, Japan, Korea, Russia, and the United States) for external feedback [8]. The comments received from the IAEA, the US NRC, China's NNSA, and the France's IRSN varied from general comments (e.g., safety approach for the Gen-IV reactor systems, differences with Gen-III systems, and the relation between safety and security) to detailed specific recommendations for individual criterion related to technical characteristics of SFRs (e.g., sodium-fire/water reactions and their consequences, accident initiators and parameters crucial to transient, DBAs, and DECs). The task force conducted a thorough analysis of the feedback received from these external reviews and summarized its resolutions in a separate TF report along with the revised version of the SDC report by the end of 2016.

5.1.2.3 SDG on safety approach

The primary focus of the "SDG on Safety Approach" is to provide SDGs for Gen-IV SFR systems as a technical supplement to the SDC report [9]. The enhancement of Gen-IV SFR safety is mainly focused on improving each level of DiD, including the 4th level, with particular attention on the robustness of safety demonstrations (practical elimination demonstrations and independence of lines of defense). Like Generation-III light-water reactors (LWRs), Gen-IV SFR safety is primarily based on the use of multiple redundant safety systems to lower the probability of accidents and to limit the consequences of AOOs and DBAs. These safety systems include independent and diverse scram systems, multiple coolant pumps and heat transport loops, decay heat-removal systems (DHRSs), and multiple barriers against the release of radioactive materials. In addition to these systems, passive/inherent features for cooling and shutdown/power reduction may also play a significant role in the safety performance of Gen-IV SFRs by improving the diversity of safety systems.

(a) Safety approach for events considered in different plants states

The events considered for the safety design are the internal events, resulting from human errors or plant component failures, and external events. For internal events, AOOs, DBAs, and DECs are defined, and preventive and mitigation measures are required to be built into the design. As for external events, design conditions are required to be established in accordance with site conditions to protect safety functions, including additional margins to the design conditions as necessary. An exhaustive approach is expected regarding the design basis against external hazards, considering the type of hazards, the combinations of loadings, and the design margins to improve the robustness of the power plant safety, and to confirm that consequences of degraded plant situations induced by extreme hazards are acceptable.

The approaches for normal operation, AOOs, and DBAs include continuous feedback on "operation/accident experience" and "maintenance/repair experience" to achieve high system reliability, enhancement of safety margins from the introduction of new technologies, and advances in inspection technology for early detection of the conditions that could lead to failures.

The approaches for DECs include providing additional measures to prevent their occurrences and mitigate their consequences along with the robust demonstration of these measures. Due consideration of the potential for common cause failures is also required. Application of passive design measures, by utilizing/enhancing favorable safety features specific to the Gen-IV SFR system, is also required for DECs with feedback from experience to improve their reliability. The selection of DBAs and DECs is required to be based on the combined use of deterministic approach based on fundamental characteristics of the reactor system, supplemented by probabilistic analysis, based on a prior operational, external event, and licensing experiences to support a risk-informed framework.

Provisions for well-balanced design measures using an appropriate combination of active and inherent/passive safety systems to enhance safety against a number of wide-ranging events are prescribed. For DBAs, the importance of characterizing the safety features of structures, systems, and components with enhanced reliability based on proven technologies and adequate redundancy and diversity is emphasized. For DECs, independence and additional diversity of the preventive design measures from those relied on during DBAs based primarily based on inherent and passive safety principles are also emphasized. Using inherent and/or passive safety features of the design is required to allow termination of accidents in the DEC category, or mitigation of their consequences, even with the postulated failure of the active safety systems.

(b) Considerations for AOOs and DBAs

As for the design for normal operation, AOOs, and DBAs, it is required that the reactor can be reliably shut down as needed, continuous effective core cooling can be maintained by the decay heat-removal systems, and the core remains covered in the case of a leak in the primary coolant boundary. Since SFRs operate at near atmospheric pressure and with a large margin to sodium boiling, coolant leakage from the primary coolant system does not lead to the type of loss of coolant accident with depressurization (i.e., concurrent losses of not only cooling function but also coolant inventory). Therefore, an emergency core cooling system for coolant injection is not required for SFRs.

Two reliable and independent shutdown systems that comply with relevant national or international codes and standards based on proven engineering practices are required to assure a safe shutdown in the event of abnormal occurrences. Reliability of the shutdown system is achieved by monitoring, testing, and maintaining the system throughout the lifetime of the plant. The shutdown systems are required to be designed with adequate shutdown margin in all operational states and DBAs.

Decay heat-removal system (DHRS) should be able to cool the reactor core immediately after reactor shutdown and for as long as needed time. The system configuration and heat-removal capacity of DHRSs should be set not to exceed design limits of AOOs and DBAs, assuming a single failure. Diversity in the system configuration and/or operation mode (forced circulation and natural circulation), as well as physical separation, should be introduced.

(c) Considerations for DECs

For DECs, additional measures to prevent core damage and maintain containment function are required in addition to the measures that are relied on for DBAs. Typically, DECs include an off-normal event followed by failure to shut down the reactor or the inability to remove heat from the core.

The failure to shutdown is paired with the three typical SFR accident sequences: loss of flow, loss of heat sink, and transient overpower. In these instances, the design is required to make use of inherent and/or passive reactivity control capabilities. Analysis of the plant response to DECs is recommended to be done using best estimate analysis, supported by probabilistic risk assessment (PRA) to ensure comprehensive coverage of postulated events and to consider their frequency and consequences. If the core damage cannot be prevented, in-vessel retention and core debris coolability are also requested to reduce the potential impact on the containment function.

The inability to remove heat from the core can result in three other DEC events: loss of coolant flow (when forced-circulation flow in the reactor core becomes disrupted), loss of primary coolant level (when core becomes uncovered), and long-term loss of heat sink (with scram). In these instances, the design is required to provide a means to prevent core damage or loss of containment function by maintaining sodium coolant level for core cooling and ensuring decay heat removal even under the conditions. Similar approaches that address the loss of heat-removal events may also apply to spent fuel storage systems.

5.1.2.4 SDG on SSCs

The SDG on SSCs makes a clear connection between the recommendations in the SDG on Safety Approach and each SSC design. In addition, the SDG on SSC provides recommendations on the requirements of the SDC report that were not elaborated in the SDG on Safety Approach. The recommendations in these guidelines include measures against ATWS for reactivity characteristics and measures for practical elimination of core uncovering and complete loss of decay heat-removal function. The recommendations, which were out of the scope of the SDG on Safety Approach, are, e.g., for fuels and materials under high-temperature and radiation conditions, measures against various hazards such as sodium fire, sodium-water reaction, and load factors on the containment system. Recommendations for addressing these issues are provided in the chapters with guidelines for reactor core, reactor coolant, and

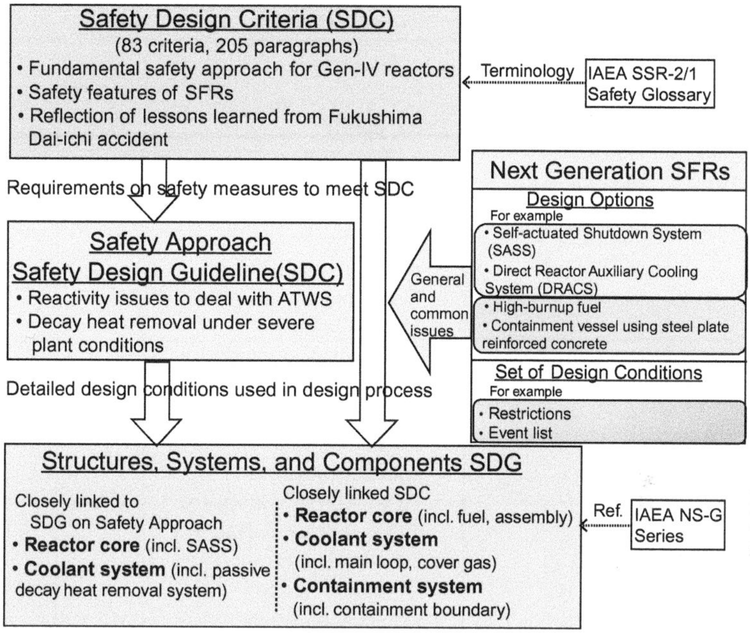

Fig. 5.2 Development process of SDG on SSC [4].

containment systems. Fig. 5.2 shows the development process of the SDG on SSC. The objective of the SDG on SSC is to provide detailed guidelines for SFR designers to support the practical application of the SDC in the design process to ensure the highest level of safety in SFR design. Although the SDG on SSC is currently focusing on issues related to the main parts of SFR, it will be extended to other issues such as fuel handling and fuel storage.

The SDG on SSC shows recommendations and guidance to comply with the SDC report and the SDG on Safety Approach with examples, which can be applied to Gen-IV SFR systems in general. The SDC-TF expects that these recommendations and examples will be appropriately considered in design according to each design characteristic. It is recommended to adopt the stated measures or equivalent alternative measures.

In the SDG on SSC, the three fundamental safety systems, core systems, coolant systems, and containment systems, together with selected 14 focal points on the systems, are described. Table 5.1 lists the SFR-specific safety features of the systems. The TF referred to design features of the Gen-IV SFR systems, and the descriptions, definitions, and formats of the IAEA SSG series [10–12] to develop recommendations on the specific SSCs.

The SDG on SSC covers the recommendations of the SDG on Safety Approach, such as active and passive reactivity reduction mechanisms, and preventive measures against significant energy release in a severe accident. It also describes design measures to maintain the reactor coolant level and coolability of the core, to utilize the natural circulation of sodium, and to ensure its reliability for decay heat removal from the core. In addition, SFR-specific measures against sodium leakage and combustion and sodium-water reaction are included as internal hazard countermeasures. Measures

Table 5.1 Fourteen focal points in the SDG on SSC [9].

Systems	Safety features	Focal points	SDC	Safety approach SDG
Reactor core systems	Integrity maintenance of core fuels	1. Fuel design to withstand high temperature, high inner pressure, and high radiation conditions	✓	
		2. Core design to keep the core coolability	✓	✓
	Reactivity control	3. Active reactor shutdown	✓	✓
		4. Reactor shutdown using inherent reactivity feedback and passive reactivity reduction	✓	✓
		5. Prevention of significant energy release during a core damage accident, in-vessel retention	✓	✓
Coolant systems	Integrity maintenance of components	6. Component design to withstand high-temperature and low-pressure conditions	✓	
	Primary coolant system	7. Cover gas and its boundary	✓	
		8. Measures to keep the reactor level	✓	✓
	Measures against chemical reactions of sodium	9. Measures against sodium leakage	✓	
		10. Measures against sodium-water reaction	✓	
	Decay heat removal	11. Application of natural circulation of sodium	✓	✓
		12. Reliability maintenance (diversity and redundancy)	✓	✓
Containment systems	Design concept and load factors	13. Formation of containment boundary and loads on it	✓	
	Containment boundary	14. Containment function of secondary coolant system	✓	

against external hazards like earthquakes are incorporated in each item individually. The design of fuel and components of the coolant systems to withstand the high-temperature conditions is also addressed.

Examples of design measures proposed by GIF SFR System Steering Committee member countries for the 14 focal points are presented in the SDC-TF. Common measures are described in the main text, while design concepts that can differ from one country to another are collected in Appendix as a part of the main text, as defined in the IAEA SSG series.

There are four chapters in the SDG on SSC.

Chapter 1 "Introduction" describes the background and objectives together with the scope and structure of the SDG on SSC.

Chapter 2 "Guidelines for reactor core systems" includes recommendations on fuel elements and fuel assemblies for the integrity maintenance of the reactor core system, active reactor shutdown system, and reactor shutdown under a DEC with inherent reactivity feedback effects and/or passive feedback or reactivity reduction mechanisms.

Chapter 3 "Guidelines for coolant systems" provides recommendations on component design and reactor cover gas and its boundary, coolant level maintenance, and measures against sodium leakage and combustion in the primary coolant system. Fundamental functions, decay heat removal under a DBA and a DEC, are presented to give recommendations on DHRS. It shows the application of natural circulation and safety considerations of tests and inspections. Measures against sodium leakage and combustion and measures against sodium-water reaction in the secondary coolant system are also described.

Chapter 4 "Guidelines for containment systems" presents containment systems and their safety functions, the general design basis of containment systems, and the design of containment systems to withstand accident conditions. Tests and inspections for the whole system are included in this chapter. This chapter also describes the confinement function of the secondary coolant system, which is one of the SFR characteristics, under an accident condition.

5.1.2.5 SFR safety design concepts in a manner consistent with SDC/SDGs

The SDC and SDGs are beginning to be introduced in the safety design of SFRs in many countries through the activities of GIF and IAEA. At the same time, these documents are being updated based on feedback from international organizations and national regulatory organizations.

This section describes recent SFR safety design concepts in the world regarding design measures against two types of DECs, namely ATWS and loss of heat-removal system. In the latest design, SFR-specific design measures, especially for prevention and mitigation of severe accidents, have been installed as consistent with the SDC and SDG. For example, the documents require or recommend that a reactor is shut down without damaging the core by using an appropriate combination of inherent reactivity feedback and passive reactivity reduction mechanisms and that even if core damage occurred, reactor coolant boundary failure is prevented (in other words, IVR is achieved) to maintain the containment function. Some recent SFRs will adopt or enhance the passive mechanisms among other design options. HSR is installed in Russian BN-1200 [13]. A self-actuated rod insertion mechanism using Curie-point magnets, as well as HSR, is introduced into ASTRID of France [14]. Similar design concepts are under review for Indian CFBR [15]. Furthermore, those countries are considering installing a core catcher in their RVs, aiming at IVR.

Regarding decay heat removal, the SDC and SDG require or recommend that the coolant level is maintained by ensuring the structural integrity of an RV and GV and that the loss of decay heat-removal function is practically eliminated by maintaining coolability through enhancement of the DHRSs or highly independent alternative cooling measures. Latest SFRs are equipped with GVs that cover whole primary coolant systems with diverse and redundant DHRSs and with natural circulation capability of the coolant to prevent core damage caused by loss of reactor level and loss of decay heat removal after reactor shutdown.

As exemplified previously, SFR-developing countries such as Russia, France, India, China, the United States, Korea, EU, and Japan are striving to enhance the safety of the next-generation SFRs in a manner consistent with the SDC and SDGs.

5.1.3 Self-actuated shutdown system (SASS)

5.1.3.1 Introduction

The reactor shutdown system of a sodium-cooled fast reactor (SFR) typically consists of two independent systems: a main reactor shutdown system (MRSS) and a backup reactor shutdown system (BRSS) as shown in Fig. 5.3. Sensors of safety protection systems are also redundant and diverse. There are several sensors equipped with thermometers or neutron instrumentation to detect abnormal events and transmit signals to the shutdown systems, activating the systems. The MRSS works with the control rod drive mechanism and uses gas pressure to accelerate rod insertion, whereas the BRSS uses gravitational force to drop the rods. This ensures improved reliability of reactor shutdown. An anticipated transient without scram (ATWS), which is an anticipated operational occurrence followed by failure of both shutdown systems, could lead to core damage in a short time, although the occurrence frequency of ATWS is low. To enhance safety, Generation IV fast reactors introduce passive safety features for the reactor shutdown [16].

There are several passive safety features, such as the installation of a mechanism that triggers negative reactivity insertion using the axial expansion of a reactor structure [17] and installation of liquid neutron absorbers that can eliminate the failure of reactivity insertion [18]. Finally, a mechanism of control rod insertion has been selected [19] because it can achieve large reactivity insertion [20,21], and rod insertion failure (i.e., rod stuck) can be eliminated by installing a stiff core barrel that can prevent excessive deformation of core components [22].

A self-actuated shutdown system (SASS) is one of the passive reactor shutdown mechanisms using rod insertion. Although several approaches have been developed for SASS [23], the use of a Curie-point electromagnet (CPEM) has been selected [19]. The electromagnet, at which control rods are detached, is installed over the core fuel assemblies. The SASS using CPEM can detect coolant temperature rise postulated in unprotected events, such as a loss-of-flow type, transient overpower type, and loss of heat sink type, and can function as an active reactor shutdown system triggered by the signals mentioned earlier.

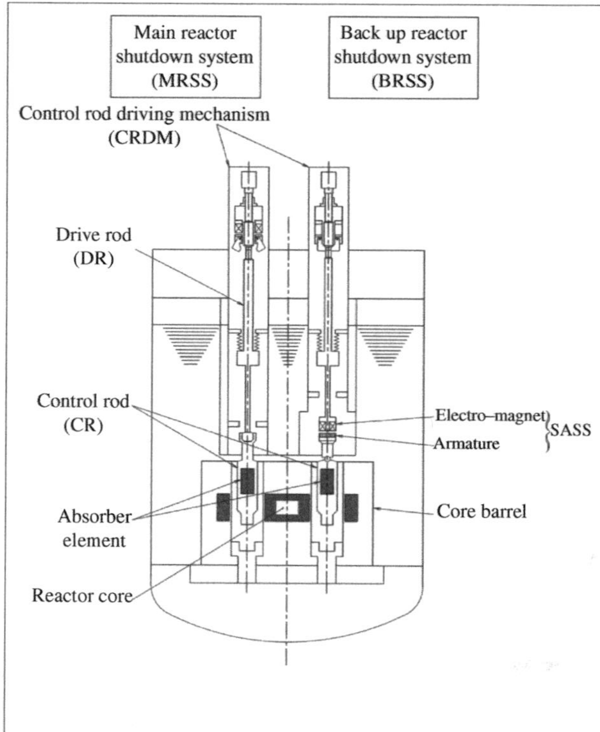

Type	MRSS		BRSS
Control rod	FCR	CCR	BCR
Quantity	6	18	6
Method of scram	Falling CR with DR		Falling CR
Acceleration method of scram	Gas		None
Required time for scram	Less than 1.2 sec		Less than 2.2 sec

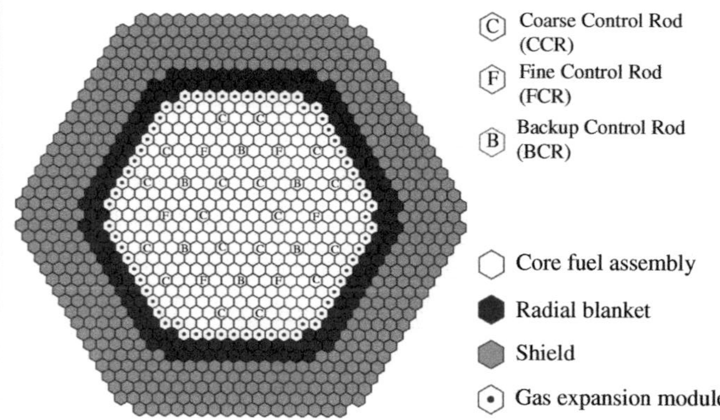

Fig. 5.3 Reactor shutdown system of a demonstration fast breeder reactor (DFBR) [9].

The SASS using CPEM is expected to hold control rods by using the magnetic force of the electromagnet (a coiled iron core) and the armature made of an iron core and a temperature-sensing alloy (Fig. 5.3). These elements will be installed in the harsh environments inside the reactor vessel (RV), such as extremely high temperatures, sodium, and neutron irradiation. Before the SASS is implemented into a fast reactor, the technical issues must be ensured: to satisfy the required functions, i.e., holding control rods and responding to the temperature rise, under the environment, and to make its design feasible. Also, the materials used in the SASS are not used as structural materials but as functional materials requiring magnetic property. These have not yet been practically used in the reactor environment in the past, so their performances and integrity over an expected lifetime must be ensured.

To achieve the design feasibility and the integrity of the SASS under the environment over its lifetime, material tests have been carried out to select the SASS materials, of which the prototypes of the iron core and the electromagnet coil were made to understand their properties [24]. Test samples made of the selected materials have been used to measure holding forces, response to temperatures, and resistance to heat, thermal shock, and radiation. The materials have also been tested under conditions similar to the actual environments (exposure to extremely high temperatures, sodium, and neutron radiation). This section describes mainly these test results, which indicate selected materials and the expected performance and integrity over the lifetime of the SASS after the definition of development programs and goals.

5.1.3.2 Development program

Since the 1970s, the passive reactor shutdown systems have been studied [23,25], which involve the SASS with CPEM [26]. They confirmed that the systems and the SASS with CPEM are feasible in principle but not enough to show their reliability and to ensure that the high technology required to achieve them is feasible for their application to an actual reactor. This has driven researchers to study the SASS to install it in a 670 MWe-scale demonstration fast breeder reactor (DFBR) in Japan [27,28].

The SASS with CPEM (Figs. 5.4 and 5.5) is required to ensure that it holds control rods during normal operation and that it naturally detects abnormal temperature rise in coolant flowing from the core assembly outlet and drops the rods within the predetermined time. To hold the rods, it must have a powerful magnetic force in the temperature range of normal operation, because the rods are suspended by attracting the armature to the electromagnet. On the contrary, it loses its magnetic force when the electric current of the electromagnet is stopped by a scram signal transmitted by the safety protection system in a scram or emergency reactor shutdown. Even if an abnormal event occurs, and the safety protection system fails to transmit the signals to stop the current, the SASS with CPEM must drop the rods by using the property of a Curie-point material. The mechanism here is as the coolant temperature rises, the magnetic resistance increases, and the Curie-point material loses its magnetic force, dropping the rods to shut down the reactor (control rod scram).

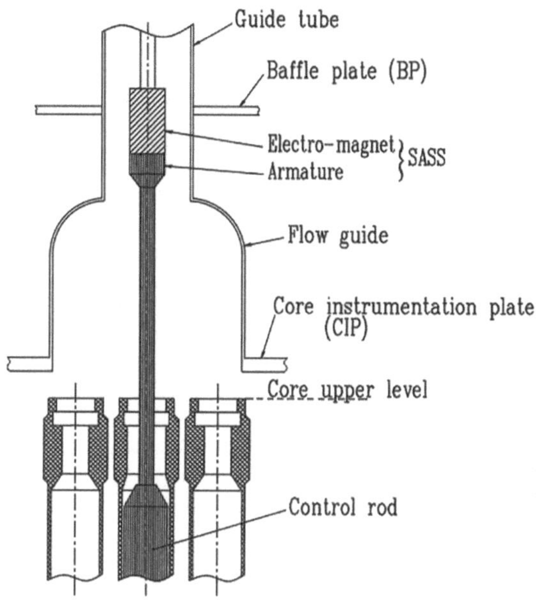

Fig. 5.4 Schematic diagram of SASS [9].

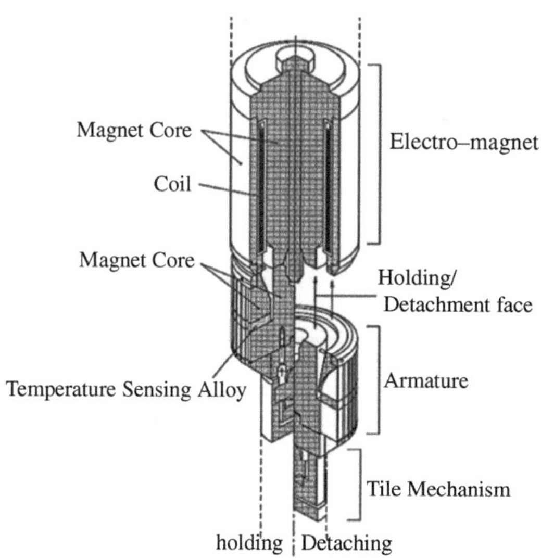

Fig. 5.5 Detailed diagram of SASS [9].

Fig. 5.6 shows nine issues in the development of the SASS and tests to be performed to verify the equipment that was developed considering these harsh environments and the performance required. From Nos. 1–4 are on the materials. The temperature-sensing alloy and the iron core are required to have stable magnetism under extremely high temperatures, to have good compatibility with sodium so that

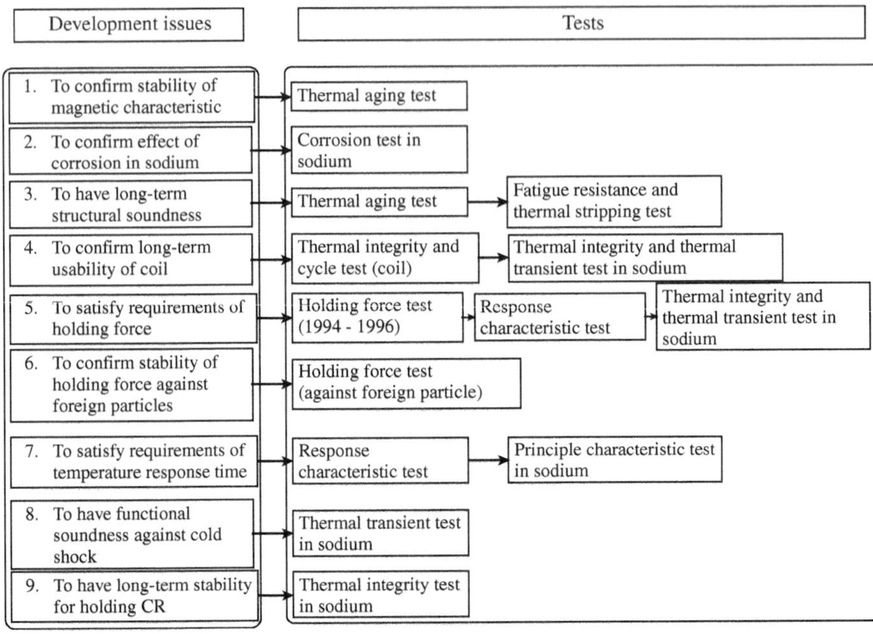

Fig. 5.6 Development issues and tests for SASS [9].

sodium corrosion will be limited, and to maintain structural integrity under high-temperature conditions with temperature fluctuations foreseen in in-vessel environments. Also, the electromagnetic coil must work over the needed duration under the temperature. To check these, various out-of-pile tests have been conducted magnetism tests, thermal aging tests, sodium corrosion tests, creep-fatigue tests, heat resistance tests, and thermal integrity and electricity cycle tests.

5.1.3.3 Development goals

The SASS will be installed to prevent significant core damage even if an ATWS occurs, so it must drop control rods before the maximum temperature of sodium flowing in the fuel assemblies reaches its boiling point (about 960°C). This rod insertion is also to limit the rise in temperature of the coolant boundary within the allowable limit (about 750°C). From the viewpoint of reliable electricity supply, the SASS also must avoid erroneous rod insertion during a rated output operation, which shuts down the reactor and results in a power outage.

In DFBR, although the average temperature of sodium at the reactor inlet is 400°C, its average temperature at the outlet is 550°C in rated power operation, and the temperature at the fuel assembly outlet is 570–600°C. The design should assume 20% higher than the rise in temperature from 400°C to 600°C, which means 640°C. The SASS, therefore, must hold the rods or avoid their erroneous insertion up to 640°C so that DFBR can have a temperature margin against the erroneous insertion and deal with the uncertainty of the fuel assembly outlet temperature caused by uncertainties of

Table 5.2 Development goals [24].

Items	Development goals
Holding force	More than 1600 N (640°C or lower) Less than 700 N (up to 680°C)
Life span	More than 24,000 h (temperature-sensing alloy) More than 36,000 h (coil element and coil for electromagnet)
Fast neutron fluence	4×10^{17} n/cm^2 or over (temperature-sensing alloy) 6×10^{18} n/cm^2 or over (coil element and coil for electromagnet)
Temperature response time (temperature time constant)	Less than 1.0 s
Detach delay time	Less than 0.2 s
Coolant transport time (100% flow rate)	Less than 1.0 s
Control rod insertion time (85% stroke insert)	Less than 2.0 s

outlet temperature distribution, coolant flow rate distribution, and coolant temperature at the outlet.

On the contrary, there is safety analysis using dynamic analysis codes conducted on temperature change under ATWS, showing that rod insertion must be done before the coolant temperature reaches 680°C to prevent sodium boiling and excessive rise in temperature of reactor coolant boundary. The holding force of the electromagnet for preventing erroneous insertion and for ensuring rod insertion must be determined. For the error prevention, the force is set at 1600 N or greater, considering uncertainties of the rod weight, seismic force (lower than the gravitational acceleration set for reactor trip), hydrodynamic force, and buoyancy based on the rod weight in the air, 800 N. For the rod insertion, the force is set at 700 N or lower, considering uncertainties of the rod weight, hydrodynamic force, and buoyancy. Other design conditions for the SASS include the time needed for coolant to flow from the assembly outlet to the temperature-sensing alloy and the time needed for the alloy to respond to temperature rise. Both are required to be 1.0 s or shorter [29].

Table 5.2 summarizes the developing goals for DFBR. Note that the coolant transport time, less than 1.0 s, and conditions for rod scram depend on the designs of the core and reactor structure. These should be determined through design studies, separately from the SASS development.

5.1.3.4 Test results

(a) Out-of-pile tests
  (i) Studied materials

The SASS consists of a temperature-sensing alloy (a Curie-point material) and materials for the iron core and the electromagnetic coil.

For the alloy, candidates must have their Curie temperatures at around 680°C since the SASS for DFBR must detach the rods before the temperature reaches 680°C. The holding force of an electromagnet is proportional to the product of the square of the density of magnetic flux flowing through the cross-sectional area of the magnetic path × the cross-sectional area of the magnetic path. Based on the cross-sectional area that can be ensured by the design of the SASS, the density of magnetic flux (Bs) is 0.5 T or greater, considering the required holding force, 1600 N, to avoid erroneous rod insertion at up to 640°C. For the holding force for the detachment, 700 N or less (Table 5.3), Bs can be 0.3 T or less at 680°C, because the holding force decreases in proportion to the square of the density.

Preliminary tests have been conducted with a commercially available vibrating sample magnetometer for 26 types of temperature-sensing alloys and selected 3 ternary alloys as candidates.

Table 5.3 Test items [24].

Temperature-sensing alloy	30Ni-32Co-Fe 23Ni-38Co-Fe 20Ni-42Co-Fe	Magnetism	Bs B–H
		Mechanical property	Hardness 0.2% offset yield strength Tensile strength Elongation
		Structure of metal	Metallographic observation Amount of austenite
		Corrosion in sodium	Weight variation, etc.
Iron core	Soft magnetic iron 2.25Cr-Mo	Magnetism	Bs B–H
		Mechanical property	Hardness 0.2% offset yield strength Tensile strength Elongation
		Structure of metal	Metallographic observation
		Corrosion in sodium	Weight variation, etc.
	Soft magnetic iron	Mechanical property	Creep rupture strength Fatigue strength
Coil cable	MI cable with sheathed core MI cable Ceramic cable	Heat resistance	Cable resistance Insulation resistance

Bs: Saturation magnetic flux density; B–H: Magnetic flux density (B)–magnetic field strength (H).

20Ni-42Co-Fe: Saturated magnetic flux density at 640°C is higher.

30Ni-32Co-Fe: Co content is lower.

23Ni-38Co-Fe: Between the aforementioned.

Two types of materials were selected for the iron core.

Electromagnetic soft iron: Saturated magnetic flux density and permeability are higher and remanent magnetism is lower.

2.25Cr-1Mo steel: Saturated magnetic flux density is relatively higher, and mechanical strength is higher.

For the electromagnetic coil, three types of materials were selected.

Unsheathed MI cable: Commonly used in cables for high temperatures

Sheathed MI cable: Expected to withstand high temperatures

Ceramic cable: Withstands high temperatures, and the number of turns can be increased

These candidate materials have never been used practically in a nuclear reactor.

(ii) Test items

Table 5.3 lists test items for solving the issues shown in Fig. 5.6. Aside from these in the list, additional tests have been performed to measure the longitudinal elastic modulus, Poisson's ratio, thermal conductivity, specific heat, and electric resistivity of the alloys and iron core materials. Table 5.4 shows the thermal conductivity and the specific heat, as an example.

Table 5.4 Measurement results of thermal characteristics [24].

	Thermal conductivity [W/(m °C)]			
	Temperature-sensing alloy (30Ni-32Co-Fe)		**Soft magnetic iron**	
Temperature	**No. 1**	**No. 2**	**No. 3**	**No. 4**
25°C (RT)	27.3	28.0	59.8	62.3
100°C	28.1	28.8	58.0	60.5
200°C	28.4	29.1	55.6	58.2
300°C	27.7	28.8	53.8	55.3
400°C	27.3	28.3	51.9	52.3
500°C	27.6	28.3	46.9	47.7
550°C	27.2	28.3	45.6	44.8
600°C	27.7	28.5	41.5	41.4
650°C	27.5	29.3	39.9	40.0
700°C	27.4	28.6	36.9	37.1
	Specific heat capacity [J/kg °C)]			
	Temperature-sensing alloy (30Ni-32Co-Fe)		**Soft magnetic iron**	
Temperature	**No. 1**	**No. 2**	**No. 3**	**No. 4**
25°C (RT)	431	440	452	452
100°C	465	469	486	484

Continued

Table 5.4 Continued

	Specific heat capacity [J/kg °C)]			
	Temperature-sensing alloy (30Ni-32Co-Fe)		Soft magnetic iron	
Temperature	**No. 1**	**No. 2**	**No. 3**	**No. 4**
200°C	498	502	531	527
300°C	536	540	575	571
400°C	565	573	619	615
500°C	603	620	669	669
550°C	628	653	703	703
600°C	670	691	741	741
650°C	712	762	787	791
700°C	607	628	866	874

DFBR was designed to have the SASS on the control rods for the backup shutdown system. The temperature-sensing alloys need to be tested longer than 24,000 h (16-month operation × 2 cycles), which is the assumed product lifetime of the control rods for the backup system, because the alloy is installed in the armature for the rods. The iron core material is used in not only the armature but also the electromagnet in the control rod activation mechanism, so ideally, it should be tested longer than 90,000 h (at least 10 years, with the aim of a 40-year reactor lifetime). However, because there is no law of acceleration on magnetism, it has been tested for 36,000 h considering limited test duration.

To check the effects of thermal aging exceeding 24,000 h, the magnetism, mechanical property, and metal composition have been measured for the temperature-sensing alloys at 12,000, and 24,000 h and for the iron core materials at 6000, 12,000, 24,000, and 36,000 h. The thermal aging test has been measured at 550°C, 600°C, and 650°C.

(iii) Temperature-sensing alloys

Figs. 5.7 to 5.9 show the results of tests conducted based on the relation of Bs before thermal aging of the three alloys [29,30]. Fig. 5.7 shows the relation between the Bs and thermal aging time, Fig. 5.8 shows the tensile strength and thermal aging time, and Fig. 5.9 shows the amount of austenite contained and thermal aging time.

The test of 20Ni-42Co-Fe shows that although the change in the amount of austenite is less significant, Bs and the tensile strength increase in 12,000 h and significantly increase in 24,000 h. This change adversely affected the magnetic coercive force, hardness, 0.2% proof stress, and elongation as well, and this showed that this alloy is unsuitable for the SASS. The other two alloys were not adversely affected by thermal aging even in 24,000 h and were found to be suitable for the SASS.

To see their compatibility with sodium, these three alloys were tested using parameters, the time and temperature in sodium, and purity of sodium. Fig. 5.10 shows the equipment used in the tests.

It has three stainless steel tanks (TP1 to TP3), each with about 160 mm in diameter and about 450 mm in height. Heaters are attached to the external surface coated with heat insulators. These were filled with sodium and connected with cold traps (sodium purifiers) through pipes to control the oxygen concentration in the sodium.

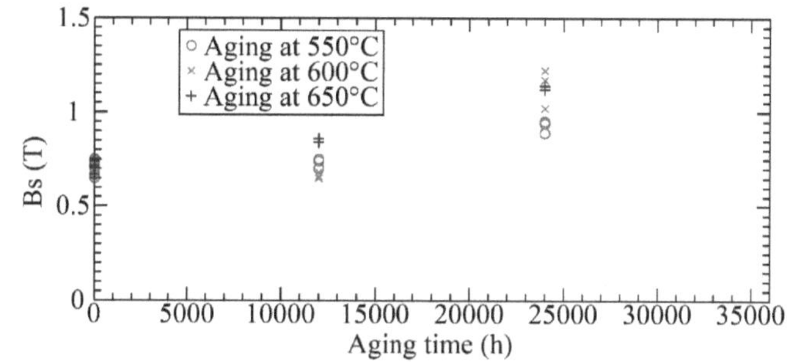

(a) 20Ni-42Co-Fe

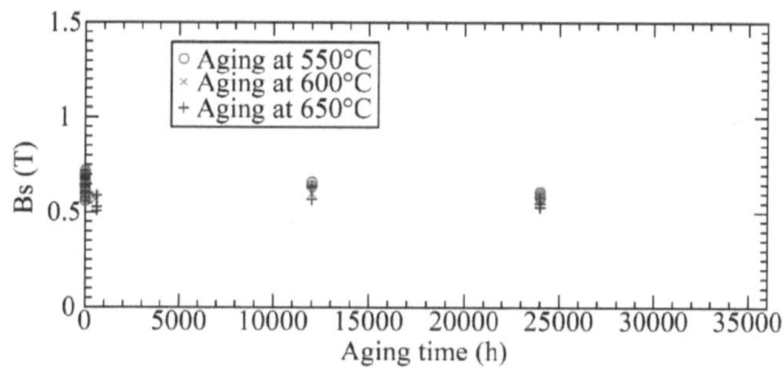

(b) 30Ni-32Co-Fe

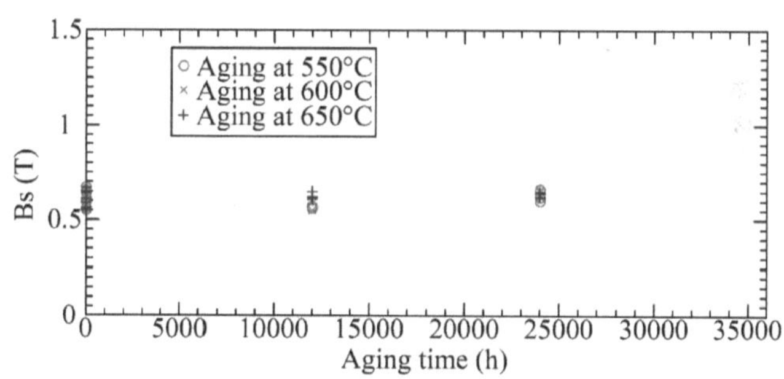

(c) 23Ni-38Co-Fe

Fig. 5.7 Results of thermal aging tests for temperature-sensing alloy (Bs). The Bs of 20Ni - 42Co-Fe increases from 12,000 h; the effect of thermal aging is observed [9].

Fig. 5.11 shows its change in weight and time in sodium, whereas Fig. 5.12 shows its corrosion depth and time in sodium for 30Ni-32Co-Fe. The reason why the increase in weight shown in Fig. 5.11 seems because impurities such as Fe, Cr, Mn, and Ni contained in the sodium intruded into the alloy, and the weight of the impurities was heavier than the weight of constituents of the alloy seeped into the sodium. The corrosion depth in Fig. 5.12 continued to increase almost linearly for 6000 h. There have been many studies of structural material corrosion in sodium [31], and

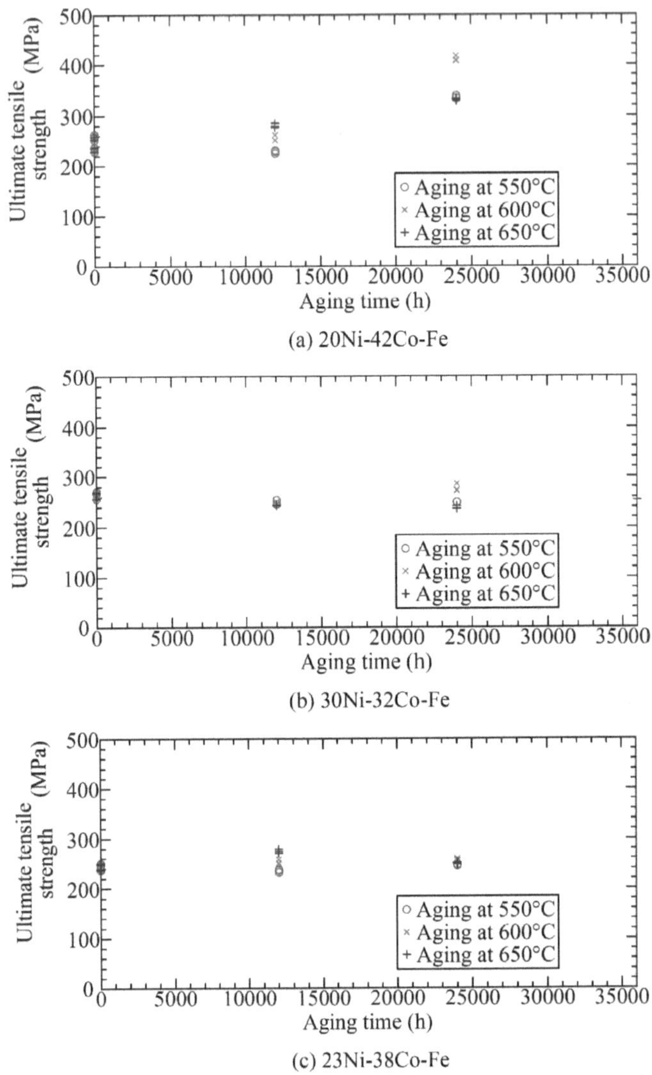

Fig. 5.8 Results of thermal aging tests for temperature-sensing alloy (tensile strength). The tensile strength of 20Ni-42Co-Fe increases from 12,000 h; the effect of thermal aging is observed [9].

these studies show that the depth of corrosion can be divided into two: the initial stage during the first several thousand hours and the steady state after that. The studies revealed that the material corrosion depth during its lifetime calculated based on the corrosion rate in the initial stage would be conservative. Based on this, it was concluded that the corrosion depth after 24,000 h (the lifetime of the SASS) will be less than 10 μm. This was calculated from the value 2.3 μm for 6000 h $\times$ 4 = 9.2 μm.

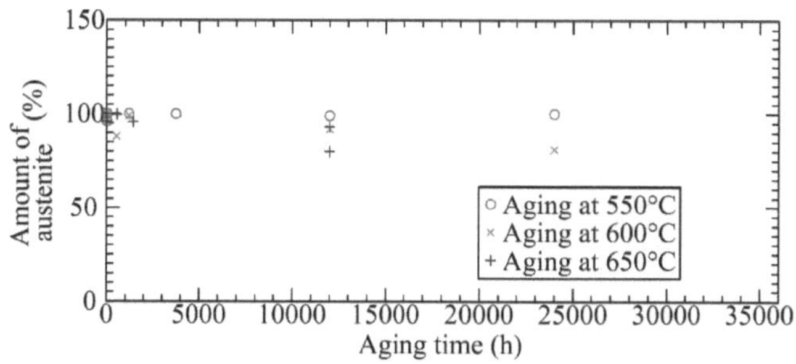

(a) 20Ni-42Co-Fe

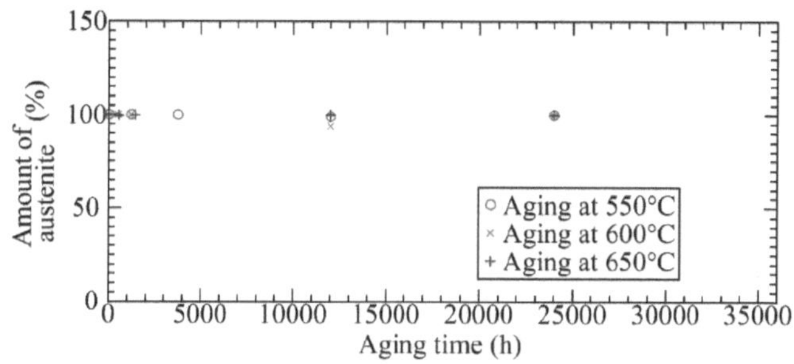

(b) 30Ni-32Co-Fe

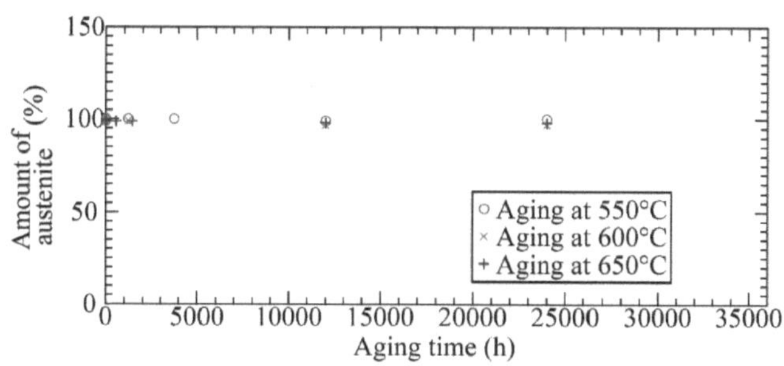

(c) 23Ni-38Co-Fe

Fig. 5.9 Results of thermal aging tests for temperature-sensing alloy (Amount of austenite). No effect of thermal aging on the amount of austenite is observed [9].

Factors that can affect the holding force (magnetic attraction) of the SASS include the changes in the area of fins of the armature and changes in the gap at the contact surface between the electromagnet and the armature (iron core).

Fig. 5.13 shows the armature with 30 fins. To improve their temperature response, the fins have a large surface area, but their heat capacity (cross-sectional area) needs to be small. On the contrary, the cross-sectional area should be large enough to maintain the folding force, which is proportional to the area. The SASS is designed to have a holding force with 20% or more margin. To limit the effects of sodium corrosion on the force to a negligible degree (about 1%), the corrosion depth must be smaller than 15 μm, considering the width of the fins, 3 mm on average. The result shows that the

Fig. 5.10 Test equipment for corrosion in sodium [9].

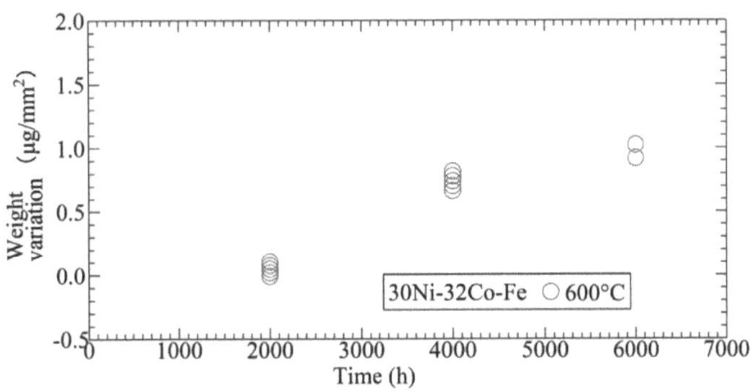

Fig. 5.11 Test results of sodium corrosion (weight variation, 3 ppm). The weight increases, and the increase tends to saturate from 4000 h [9].

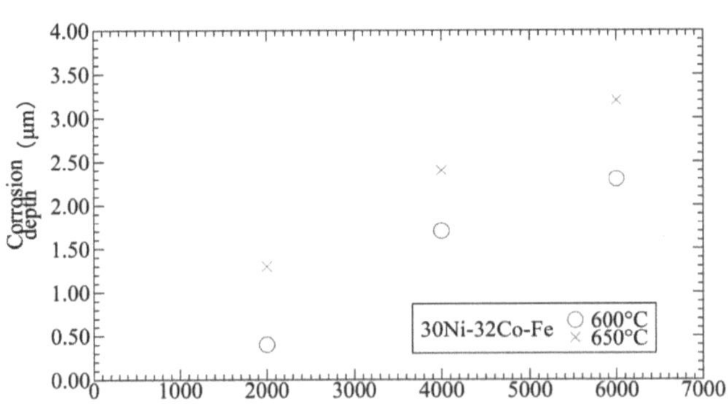

Fig. 5.12 Test results of sodium corrosion (corrosion depth, 3 ppm). The corrosion depth increases almost linearly up to 6000 h [9].

Fig. 5.13 Mock-up of armature for SASS [9].

corrosion depth is less than 10 μm; thus, the corrosion will not affect the performance. This study obtained similar results to the other two alloys; therefore, it is concluded that all the three alloys can be adopted to the reactor in terms of sodium corrosion.

As mentioned earlier, the SASS is made of functional materials, not strengthened materials. Also, the control rods suspended by the electromagnet weigh about 800 N, and the stress on the temperature-sensing alloy is small, around 0.4 MPa. There are little data on the strength of the alloys against high temperatures, so fatigue tests and creep tests were performed to obtain reference data for the design of the SASS. Based on the conjugate gradient method [32], to confirm if the strength of the alloys affects the structural integrity of the equipment, tentative design creep curves were created including tentative design fatigue curves and the results of acceleration tests at 700°C, although the number of test data is not enough, and the data reliability is not sufficient enough compared to those used as material strength criteria. As an example, Figs. 5.14 and 5.15 show the results of 30Ni-32Co-Fe.

Co contained in temperature-sensing alloys is significantly radioactivated by neutrons, so *Co* content in the alloy needs to be small to minimize its effects on the radiation shield essential in the processes after the equipment is taken out from the RV, such as sodium cleaning and storage in a water pool. Based on the results and this fact, selected was 30Ni-32Co-Fe, which contains less amount of Co than the other two candidates.

(iv) Iron core materials

Figs. 5.16 and 5.17 show the thermal aging of electromagnetic soft iron and 2.25Cr-1Mo steel (Cr—Mo). Fig. 5.16 shows Bs measured at 640°C and thermal aging time, and Fig. 5.17 shows tensile strength and thermal aging. The tests conducted for up to 36,000 h showed that the soft ion exhibits no effects of thermal aging on both Bs and tensile strength. Although the effects on its tensile strength can be seen in Cr—Mo, the tensile strength of Cr—Mo decreased by thermal aging is still larger than that of the soft ion. Additional investigation separately revealed that the thermal aging did not adversely affect the magnetic coercive force, hardness, 0.2% proof stress, and elongation. It is concluded that the two candidates can be adopted in the actual reactor in terms of thermal aging.

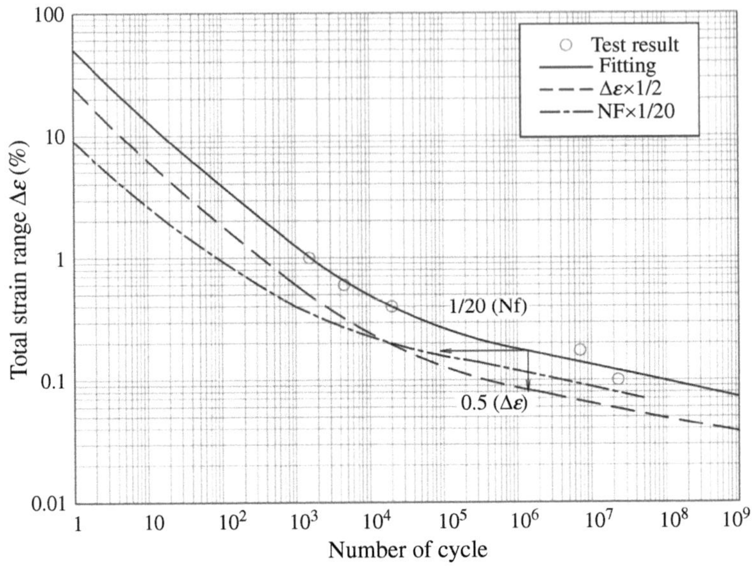

Fig. 5.14 Design fatigue curve for temperature-sensing alloy. The fatigue curve is drawn in the same way as for stainless steel with abundant data [9].

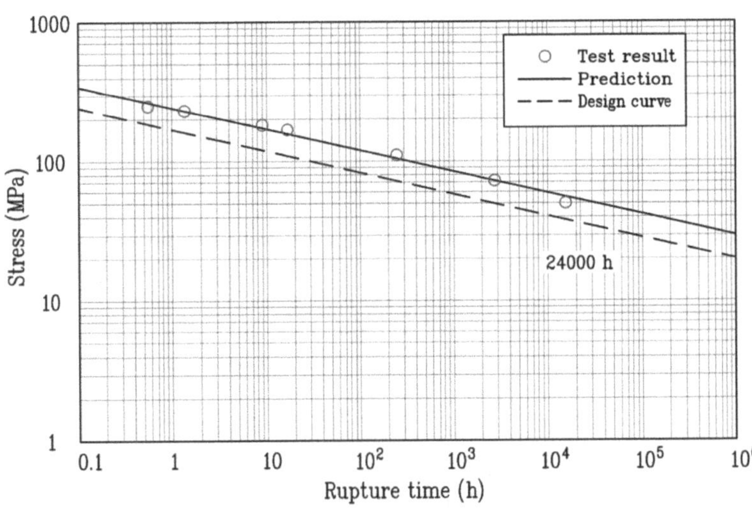

Fig. 5.15 Creep rupture curve for temperature-sensing alloy. The creep rupture curve is drawn in the same way as for stainless steel with abundant data [9].

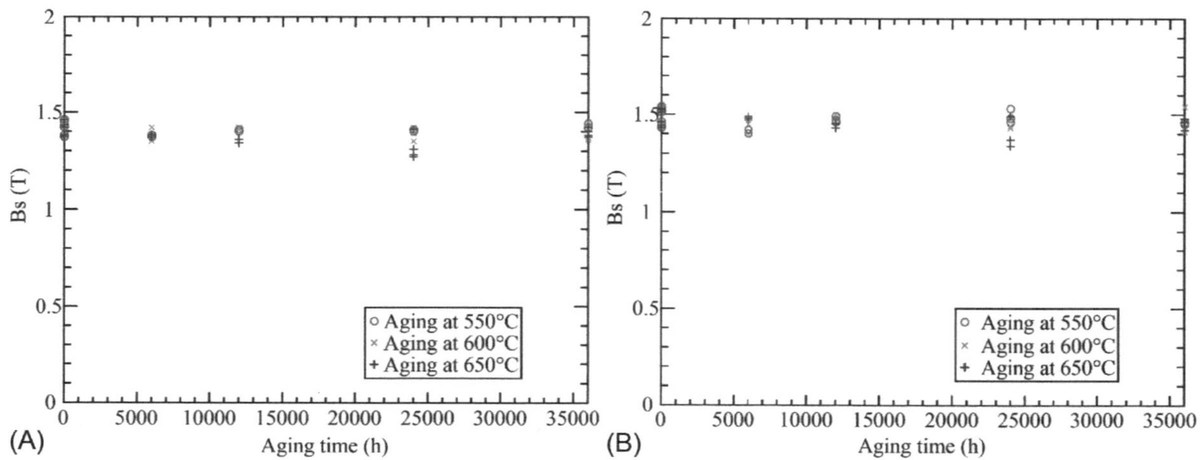

Fig. 5.16 Results of thermal aging tests for iron core (Bs). No effect of thermal aging on the Bs is observed up to 36,000 h [9]. (A) Cr-Mo, (B) soft magnetic iron.

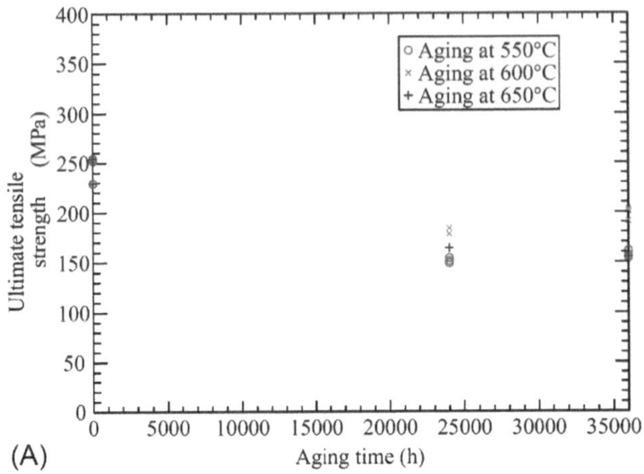

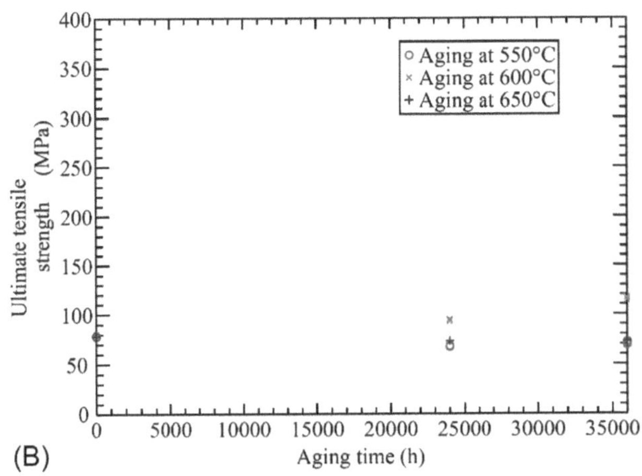

Fig. 5.17 Results of thermal aging tests for iron core (tensile strength). No effect of thermal aging on the tensile strength is observed up to 36,000 h [9]. (A) Cr-Mo, (B) soft magnetic iron.

The candidates were also tested in sodium on the time immersed, temperature, and sodium purity as parameters. Fig. 5.18 shows the relation of their corrosion depths and the time immersed in sodium. At these temperatures up to 6000 h, the depths almost linearly increased.

The SASS used at high temperatures is installed with a gap of 0.05 mm to 0.10 mm to prevent a self-bonding at the contact surfaces between iron core materials of the electromagnet and the armature, taking into account the increase in the magnetic resistance. If this gap is expanded by sodium corrosion, the magnetic resistance increases to decrease the holding force, possibly dropping the control rods. The allowable maximum gap that ensures 1600 N of holding force (see Table 5.2) is estimated to be 0.18 mm based on the test results of the relationship for gap and the folding force shown in Fig. 5.19. Therefore, the allowable limit of sodium corrosion depth is about 0.90 mm or less.

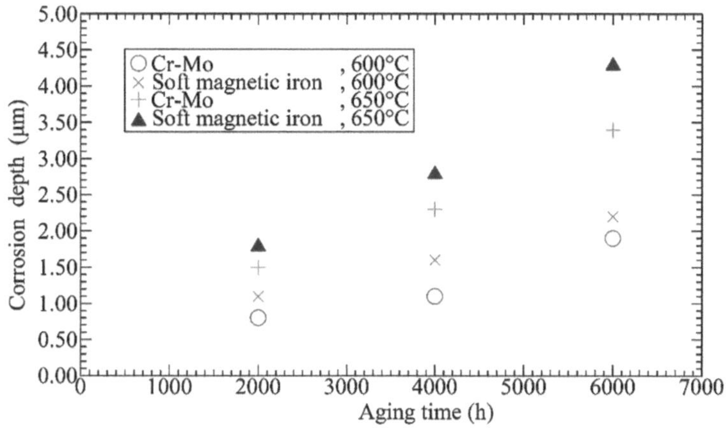

Fig. 5.18 Test results of sodium corrosion (corrosion depth, 3 ppm). The corrosion depth increases almost linearly up to 6000 h [9].

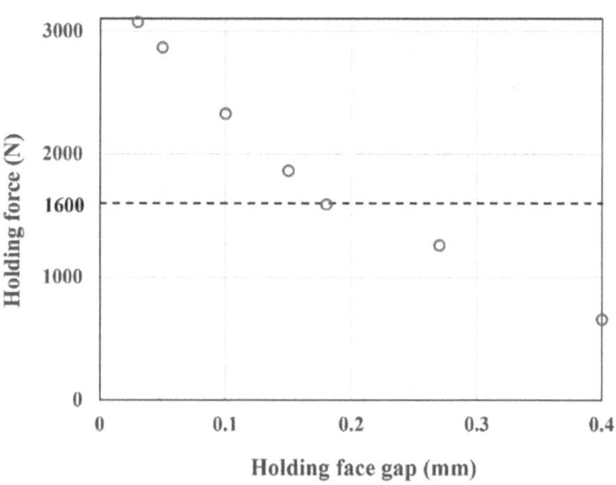

Fig. 5.19 Difference of holding force with holding face gap [9].

The ideal life of the electromagnets for the control rod driving mechanism is 40 years (around 280,000 h with an 80% operation rate), and that of the control rods (armature) is 24,000 h. Based on the corrosion depth data of the electromagnet soft iron, 2.2 μm, and Cr—Mo, 1.9 μm, at 6000 h, if the depths increase linearly as seen one in the temperature-sensing alloys, the depths in the electromagnet are 103 μm in the soft iron and 89 μm in Cr—Mo, and those in the armature side 9 μm in the iron and 8 μm in Cr—Mo. These may exceed the allowable limit, 90 μm. This requires the lifetime of the electromagnet on the rod mechanism to be 20 years, meaning that the electromagnet will be replaced during the plant lifetime, in terms of sodium corrosion. This replacement strategy gives a conclusion that both the candidate materials are applicable materials to the actual reactor. The stress on the iron core material is very small, about 0.4 MPa, similarly to the temperature-sensing alloys. On the contrary, there are little data on the strength of the electromagnetic soft iron against high

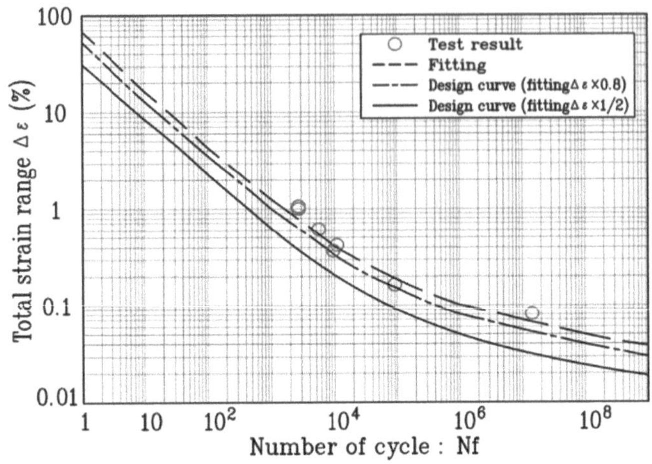

Fig. 5.20 Design fatigue curve for soft magnetic iron core. The fatigue curve is drawn in the same way as for stainless steel with abundant data [9].

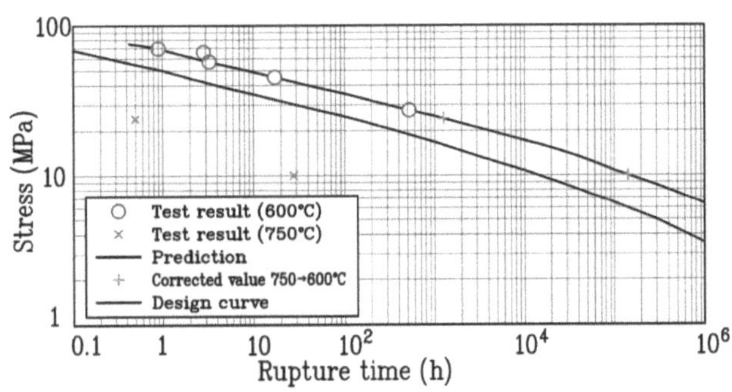

Fig. 5.21 Creep rupture curve for soft magnetic iron core. The creep rupture curve is drawn in the same way as for stainless steel with abundant data [9].

temperatures. Fatigue tests and creep tests were performed to draw tentative design fatigue curves and design creep curves to confirm if their strengths affect the structural integrity, similarly to the temperature-sensing alloys. Figs. 5.20 and 5.21 show the results. From these results, the electromagnetic soft iron that features excellent magnetism was selected for the iron core material.

(v) Feasibility analysis results of the temperature-sensing alloy (30Ni-32Co-Fe) and the iron core material (electromagnetic soft iron)

Based on the sodium corrosion test, 30Ni-32Co-Fe was selected for the temperature-sensing alloy because of its small content of Co, and the electromagnet soft iron was selected for the iron core material because of its excellent magnetism. Using these materials, a preliminary analysis of the creep-fatigue is described here. The SASS is installed over the fuel assemblies (see Figs. 5.3 and 5.4), so it may be affected by thermal striping caused by temperature fluctuations above the fuel assemblies. The coolant flowing from the backup control rod assemblies is colder than that from the neighboring fuel assemblies. However, the coolant flow rate from the backup

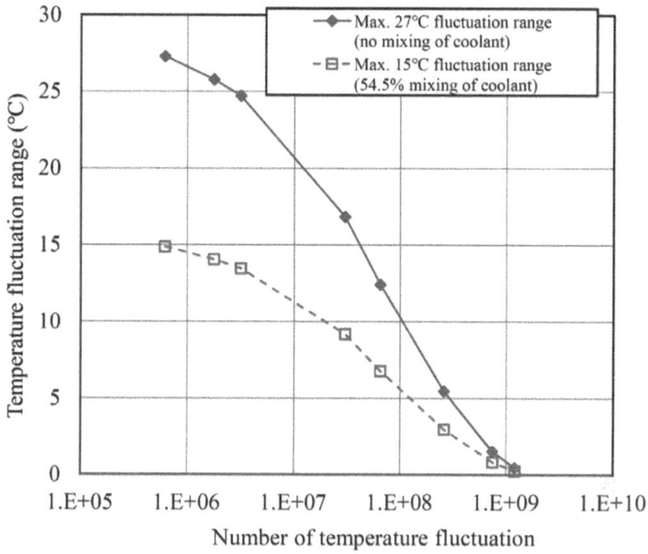

Fig. 5.22 Thermal striping condition for DFBR [9].

control rod assemblies is about 3 kg/s, which is much smaller than that from the neighboring fuel assemblies (30 kg/s). Therefore, the temperature fluctuation would not significantly cause thermal striping.

Fig. 5.22 shows the conditions of the thermal striping at the installation position of the SASS in DFBR. The vertical axes are the range of temperature fluctuation, and the horizontal axes are the cumulated number of fluctuation occurrences. The maximum range of the temperature fluctuation is 27°C.

The strain ranges of the temperature-sensing alloy and iron core material for the SASS were defined based on the fluctuation range, and the number of occurrences of strain was set as a conservative approach (Fig. 5.23). The cumulative fatigue damage Df at 600°C was calculated based on the most conservative design formula in the fatigue curves in Figs. 5.14 and 5.20 assuming the number of occurrences of strain in 24,000 h for the temperature-sensing alloy and in 280,000 h for the iron core material. The result shows that Df of the alloy is 0.03 and the iron core is 0.25 even neglecting the reduction effect of temperature fluctuation caused by the mixing effects of coolant as shown in Fig. 5.22.

Creep damage Dc of both the materials is negligibly small, below 0.01 (nearly 0) because of their small strain, 0.4 MPa or less, since these materials will be used to merely support the weight of the control rods, which is 800 N in the air. The sum of Df and Dc of each material is less than 0.3, so this is far below the allowable creep-fatigue damage, 1.0. Therefore, these results give a perspective to assure structural integrity.

(vi) Electromagnetic coil materials

Candidate materials were MI cables (one with the core sheathed and the other unsheathed) and three types of ceramic cables. To understand their thermal resistance at 600°C for 40 years, corresponding to the lifetime of the SASS, acceleration tests at 650°C, 850°C, and 950°C have been conducted with an electric furnace.

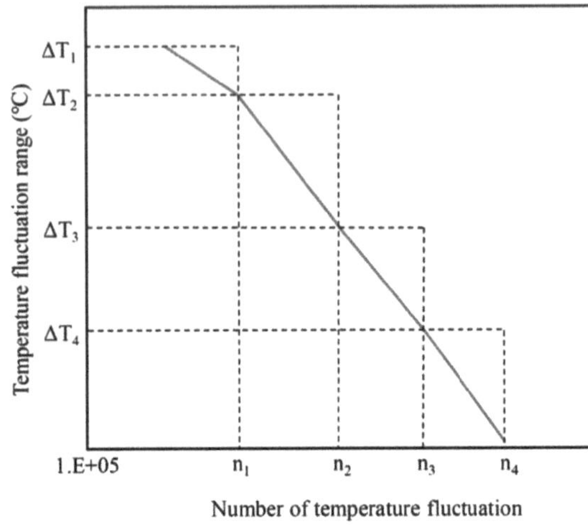

Fig. 5.23 Method for estimating number of temperature fluctuation [9].

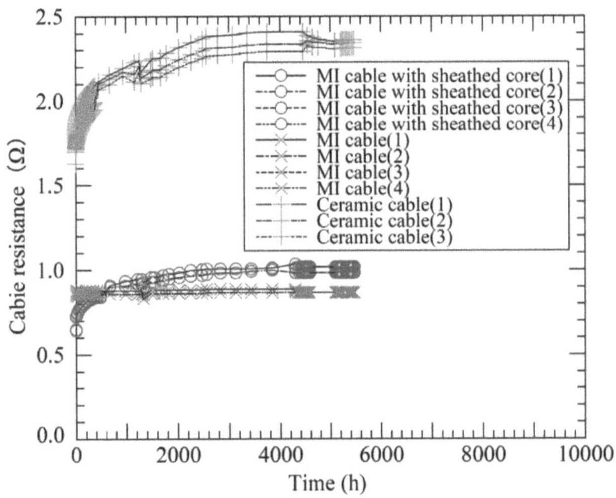

Fig. 5.24 Cable resistance (950°C) [9].

Fig. 5.24 shows the thermal resistance histories for the tested coil cables at 950°C. Results show that all the cables could maintain their integrity for over 40 years. The integrity of the coils was also confirmed against cyclic transmissions of electricity expected in service.

Therefore, the MI cable with an unsheathed core was selected for the coil cable, because of its easy availability.

(b) In-pile tests

The materials described in Section 5.1.3.4(a)(i) have been tested in the experimental fast reactor Joyo [19,33]. The detailed description of the in-pile tests appears in Section 2.2.2. The radioactive conditions in the tests correspond to the fast neutron levels for the DFBR, which are expected to be 4×10^{17} n/cm^{b} for the armature on the control rods and

6×10^{18} n/cm^b for the electromagnet on the driving mechanism. The test results have shown that the radiation effect was not significant for the magnetism (B—H magnetic hysteresis), the mechanical property (tensile strength), and the insulation resistance of the temperature-sensing alloy, the iron core material, and the electromagnetic coil.

5.1.4 Severe accident

5.1.4.1 Background and objective

The core of sodium-cooled fast reactors (SFRs) is not designed in the most reactive configuration, and this leads to the safety characteristics that insert positive reactivities by coolant boiling and the change of configuration of core materials. Therefore, from the early phase of fast reactor development, consequences of severe accidents, which are called core disruptive accidents (CDAs) in SFRs, have been assessed in the licensing procedure. In the past licensing for the experimental reactor Joyo and the prototype reactor Monju, and in the demonstration, fast breeder reactor (DFBR) program that was conducted from the late 1980s to the 1990s by the Japan Atomic Power Company (JAPC) and other Japanese electric power companies, energetic consequences caused by prompt criticality on the occurrence of an unprotected loss of flow (ULOF) accident were assessed, and the integrity of the containment was confirmed. Through these assessments, technologies to evaluate responses of structures that consist of the primary coolant boundary have been developed [34–36].

In parallel with the technology development summarized already, considering the social acceptability of SFRs toward their commercialization, a study to ensure safety in the case of core melting has begun. This study aimed for the achievement of in-vessel retention (IVR), which means that the core is designed to eliminate the power excursion driven by the insertion of positive reactivities peculiar to SFRs [37,38] and that in-vessel structures are designed to contain and cool the degraded fuel in the reactor vessel [39,40]. Therefore, instead of the conventional approach to ensuring the integrity of the primary coolant boundary against the mechanical energy by power excursions, it becomes important to show the achievement of IVR by introducing design measures to mitigate CDA consequences.

The effectiveness of design measures is evaluated through CDA analyses, which require a combination of technological developments. These are developments of computer codes designed to analyze event progression in CDA, experimental studies addressing dominant phenomena of CDA progression to obtain both phenomenological knowledge and database for validations of the computer codes, and safety assessment of the SFR into which the design measures are introduced. These technological developments are an iterative cycle in which the computer codes will be improved, and advanced, additional experimental studies will be performed and the design measures improved if safety assessments identify problems for the achievement of IVR. Event progression in CDA is generally analyzed by dividing the entire accident sequence into several phases. This is because the dominant physical phenomena, the models required by the computer codes, and experimental knowledge vary by accident progression. An accident sequence in the case of ULOF, which has been considered a representative initiator leading to CDA, is shown in Fig. 5.25 [41].

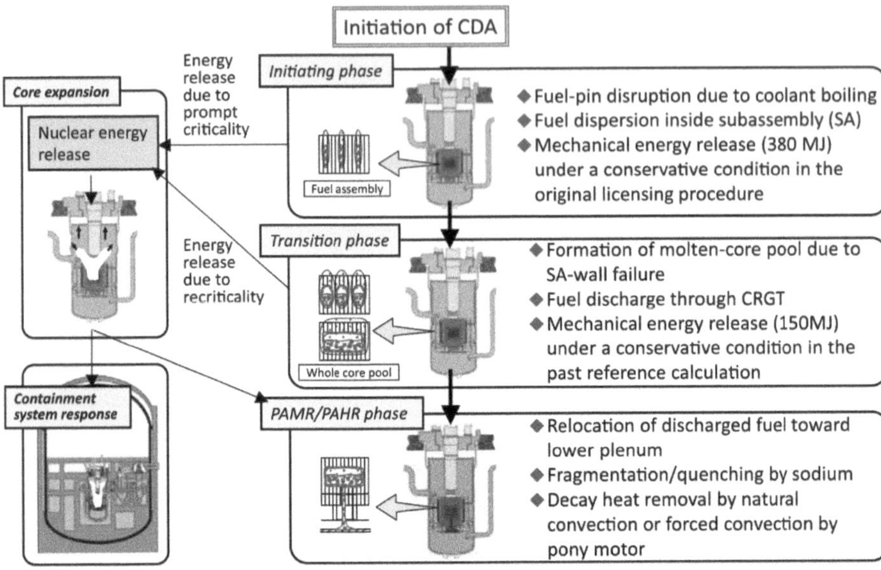

Fig. 5.25 Categorization of unprotected loss-of-flow (ULOF) sequence and outline of event progression [41].

Until the first half of the 1980s, the technology for CDA evaluations was dominated by the United States. Japan introduced computer codes through international cooperation with the United States and applied them to safety reviews of Monju for licensing. Although the development program of SFRs in the United States slowed from the mid-1980s, Japan has continued safety research on CDA even after the safety review of Monju for licensing. As a result, a series of safety evaluation technologies for all CDA sequences of SFRs with oxide fuel has been established as unique semidomestic technology. This technology has become the basis of technical cooperation with France, which has resumed SFR development in recent years.

In the present subsection, future technical issues and directions of development are summarized based on the achievements of the results of research and development to date regarding the main technologies related to CDA evaluations in Japan. Since the design measures to eliminate energetic consequences driven by power excursions will be introduced into future SFRs for commercialization, as mentioned earlier, the scope of the present subsection includes the initiating phase, the transition phase, and the material relocation and heat-removal phases.

5.1.4.2 The initiating phase

(a) Feature of the initiating phase and required evaluation technologies

The initiating phase is the early stage of the CDA, in which the core damage progresses gradually from the fuel subassemblies with the highest power-to-flow ratio as a result of coolant boiling and fuel pin melting due to the imbalance between heat generation and cooling, typically caused by the superposition of the decrease in core flow rate from the steady-state operation and the failure of reactor shutdown (ULOF). A temperature change of core materials, coolant boiling, molten cladding relocation, and molten fuel relocation during CDA all have reactivity effects. In particular, the coolant void reactivity in the center of the core, the dispersal of molten cladding from the axial center of the core, and the concentration of the fuel to the axial center of the core have positive reactivity effects in medium-to-large-size cores, including "Monju." Therefore, a method for analyzing the temperature and movement of core materials coupled with nuclear kinetics is necessary. To validate the analytical method, it is also necessary to obtain experimental knowledge on coolant boiling, molten fuel, and cladding movement.

(b) Development of evaluation technologies for the initiating phase

For the analytical method, the SAS code, which has the following features, was developed at Argonne National Laboratory in the United States [42,43]:

- The entire core is modeled as a multichannel model by ramping (grouping) subassemblies under similar power and flow conditions into the same channel, where the average fuel pin of a fuel subassembly and the associated coolant and structural materials, including wrapper tube, constitute one channel;
- Each channel is coupled hydrodynamically by upper and lower plenums and coupled nuclearly by a kinetic model.

Based on the database of out-of-pile and in-pile tests by Transient Reactor Test (TREAT) facility, in the United States, the analytical modules were developed for thermal-hydraulics, including coolant boiling, material relocation behavior after fuel melting and cladding failure including fuel-coolant interaction, molten cladding motion, fuel disruption, and fuel motion, and SAS3A with all the elemental physical models necessary for mechanistic analysis of the CDA initiating phase was completed [44].

After that, SAS3D was developed with almost the same model as SAS3A but with an increased limit of the number of channels from 10 to 33, and then, SAS4A was developed with more mechanistic modeling of material relocation after fuel pin failure [45].

The SAS4A code has been developed and improved for oxide fuels in cooperation with the Japan Atomic Energy Agency (JAEA) and European research institutes, such as the Institut de Radioprotection et de Sûreté Nucléaire (IRSN, France) and the Karlsruhe Institute of Technology (KIT, Germany), including the addition of new models such as a plenum gas release model and a jamming model of the fuel flow with solid fuel chunks, based on the findings from the in-pile CABRI test described later [46], and the fuel relocation behavior can now be properly evaluated, as shown in Fig. 5.26 [47]. In addition, in recent years, the PIRT (Phenomena Identification and Ranking Table) method has been applied to select important phenomena to be verified, and verification and validation of the SAS4A code are underway [47,48].

At the CABRI reactor in France, a number of test data with fuel melting and cladding failure were obtained by loading test fuel pins preirradiated in the Phenix reactor (France) and prototype fast reactor (PFR, UK) into a sodium loop and subjecting them to nuclear heating under various transient conditions. This is because the transient nuclear heating of fuel pins irradiated in a fast reactor spectrum is indispensable to the clarification of the transient behavior of fuel pins under CDA conditions, such as melting and relocation of fuel pins. In addition, CABRI can measure the fuel relocation behavior with a high temporal and spatial resolution by using a fast neutron hodoscope [49]. An example of the measurement is also shown in Fig. 5.26.

Among the many test results obtained in the CABRI tests related to CDA evaluations, the most important results concerning the reduction of the uncertainty range of the evaluation and the clarification of the mitigation effect against energy generation with reactivity increase are summarized as follows [50–55]:

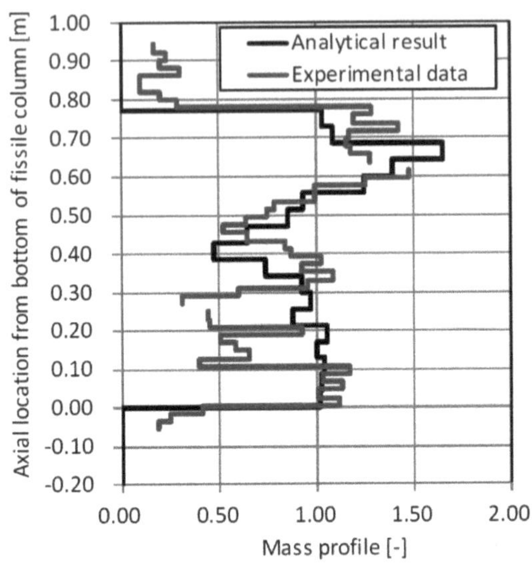

Fig. 5.26 An example of SAS4A application to a CABRI test [47]. Axial fuel mass profile (55–105 ms).

- In the early CABRI tests, emphasis was placed on clarifying transient fuel pin behavior such as the axial expansion of fuel pins during the CDA initiating phase, and the behavior of fuel pin failure and postfailure fuel pin relocation. From the results of many CABRI tests simulating power excursions due to coolant boiling, the fuel pin failure mechanism and failure conditions under various fuel and transient conditions were understood.
- Postfailure fuel dispersal is dominated by the enthalpy of the fuel pins, as shown in Fig. 5.27, and it is confirmed that this fuel dispersal effectively reduces reactivity. In other words, there is a unique mechanism to suppress the energy release in the core of the initiating phase. It was also confirmed that the FP gas released from the failed fuel pin effectively contributed to fuel dispersal.
- The relationship between the location of fuel pin failure and fuel smear density was also confirmed, as shown in Fig. 5.28. In the case of low smear density, the fuel pins failed at a position higher than the axial center of the core. This is because the deterioration of strength becomes apparent due to the temperature increase in the fuel-cladding material. This results in molten fuel relocation in the fuel pin toward the failure rip where the neutron flux is relatively low, which has a negative reactivity effect, and the increase in the void reactivity due to fuel-coolant interactions (FCIs) by released molten fuel is much lower than in the case of fuel pin failure in the center of the core.

(c) Summary

Through the experimental studies represented by TREAT and CABRI, the phenomena of fuel pin disruption and dispersal behaviors have been understood from the initiating events, and the methodology for evaluating the changes in core reactivity due to these behaviors has been established. This methodology can significantly reduce the conventional evaluation value of mechanical energy generation and can be used to evaluate the effectiveness of the measures to prevent prompt criticality, such as limiting the fuel axis length and void reactivity during the initiating phase. The evaluation results are described in detail in the reference [39]. On the contrary, the application of the methodology is limited

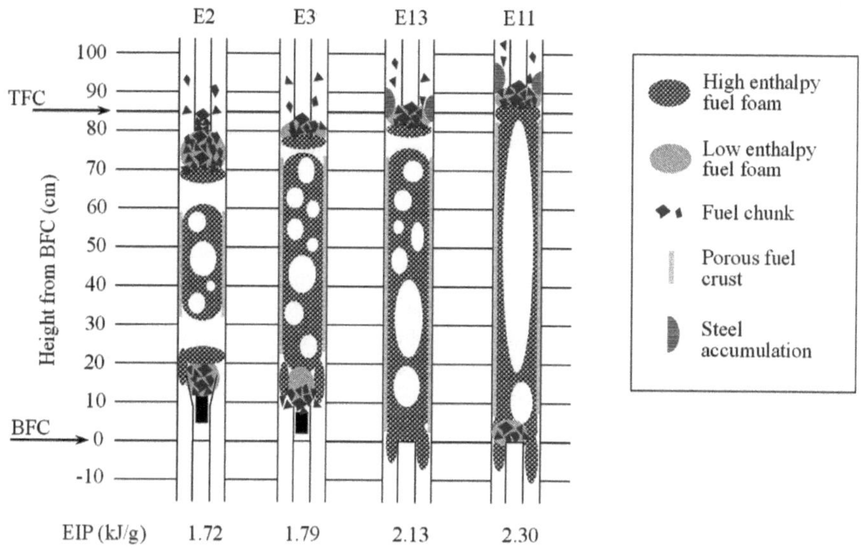

Fig. 5.27 Dependency of fuel redistribution on accumulated fuel enthalpy [55]. TFC: top of fissile column, BFC: bottom of fissile column. EIP is defined as the sum of the fuel enthalpy at a peak power node (PPN) at the transient overpower (TOP) triggering and the energy injection at PPN during TOP, assuming an intact geometry.

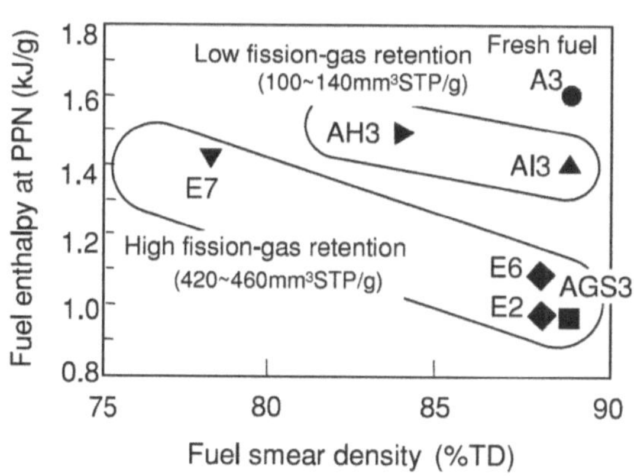

Fig. 5.28 Evaluated fuel enthalpy at observed failure time in fast or medium TOP CABRI tests with relatively low cladding temperature. Fuel enthalpy is the evaluated value at axial peak power node.

The original is in I. Sato, F. Lemoine, D. Struwe, Transient fuel behavior and failure condition in the CABRI-2 experiments, Nucl. Technol. 145 (1) (2004) 115–137, copyright © American Nuclear Society, http://www.ans.org/reprinted by permission of Taylor & Francis Ltd., http://www.tandfonline.com on behalf of American Nuclear Society, http://www.ans.org/.

to axially homogeneous cores with oxide fuel. To apply the methodology to various core designs, Japan and France must cooperate in improving the fuel pin model and for Japan and the United States to cooperate in studying the evaluation methodology for metal-fueled cores based on technology development such as the core dynamics analysis code named CANIS in the Central Research Institute of Electric Power Industry [56].

5.1.4.3 The transition phase

(a) Feature of the transition phase and required evaluation technologies

Although the core reactivity does not cause prompt criticality in the initiating phase, the core state is still active and fuel melting progresses since most of the fuel remains in the core region. In the initiating phase, the direction of molten-core material movement is restricted in the axial direction by the wrapper tube of the fuel assembly. On the contrary, in the transition phase, the core-melted region progresses radially by melting of the wrapper tube of the fuel assembly, and this enables large-scale core material motion, including in the radial direction, which leads to reactivity fluctuation.

The transition phase is characterized by core reactivity being dominated by the motion of molten-core materials because of the removal of coolant sodium from the core region. The dynamics of the motion of molten-core materials vary from one-dimensional axial motion in the fuel assembly initially to multidimensional motion when the core melt progresses to the whole core.

Therefore, evaluation technology is required that can continuously calculate core-reactivity changes coupled with core melt progression, axial and radial motion of molten-core materials, and it is also necessary to acquire experimental knowledge on dominant phenomena.

(b) Development of evaluation technologies for the transition phase

To predict the coupled neutronic and fluid-hydrodynamic behavior of SFRs during CDAs, the development of a computer code SIMMER-II, a two-dimensional, multiphase, multicomponent, and multivelocity field fluid-hydrodynamic code coupled with a space- and energy-dependent neutron dynamics model, was started at Los Alamos National Laboratory in the United States in the 1970s [57]. SIMMER-II was a remarkable achievement that enabled mechanistic evaluations of the transition phase and the core expansion phase; but at the same time, extensive worldwide code applications revealed many limitations due to the code framework as well as the need for model improvements. For this reason, to alleviate many of the limitations of the SIMMER-II models and their applications, the development of a next-generation code, SIMMER-III, was underway at JAEA from around 1990 [58,59] taking advantage of the research results and lessons learned from advanced fluid dynamics model (AFDM) development [60,61]. A conceptual overall framework and the basic geometric structure of SIMMER-III are shown in Figs. 5.29 and 5.30, respectively.

A comprehensive and systematic assessment of the SIMMER-III code was conducted in two steps: Phase 1 for fundamental assessment of individual code models and Phase 2 for integral code assessment. The Phase 1 assessment applied the code to single- and multiphase flow benchmark problems, small-scale experiments using reactor and simulant materials, and physical problems with known solutions [62–65], whereas the Phase 2 assessment involved applications of the code to integral, complex multiphase situations [66–68]. These assessment programs were conducted with the participation of experts from European research organizations: Forschungszentrum Karlsruhe (FZK) in Germany, AEA Technology in the United Kingdom, and Commissariat a l'Energie Atomique (CEA) and

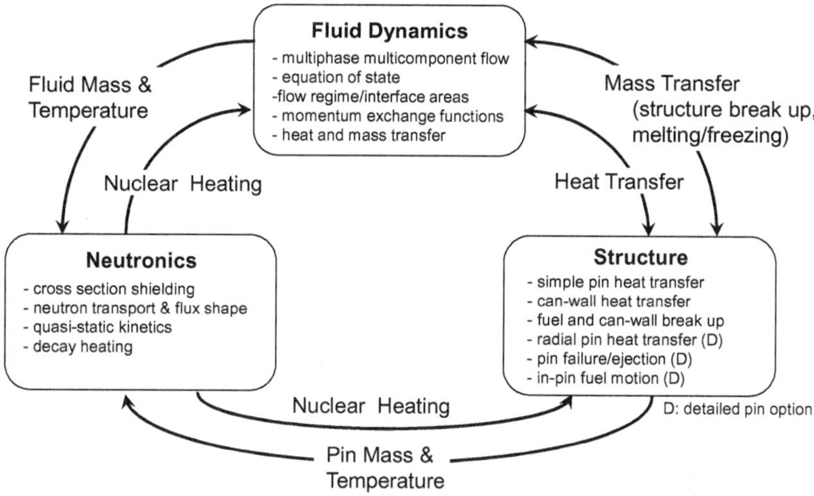

Fig. 5.29 SIMMER-III Overall Code Structure [65].

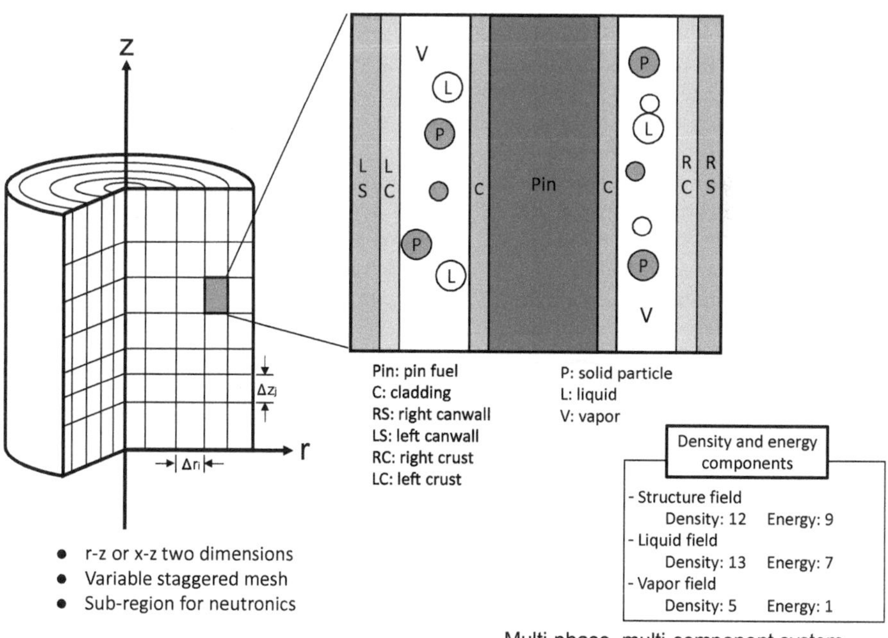

Fig. 5.30 Multiphase, multicomponent fluid dynamics model in SIMMER-III.
The base concept is presented in S. Kondo, Y. Tobita, Phase 1 code assessment of SIMMER-III, a computer program for LMFR core disruptive accident analysis, JAEA Res. (2019). 2019–009, and the number of density and energy components are updated based on the latest version.

Institute for Nuclear Safety and Protection (IPSN) in France. The lists of test problems for the Phase 1 and Phase 2 assessments of SIMMER-III are shown in Tables 5.5 and 5.6, respectively. The Phase 2 assessment program was completed in 2000 by issuing a comprehensive assessment report [69].

Following Phase 2 assessment, the SIMMER-III development has continued. Regarding discharge and freezing of molten-fuel, based on results of Phase 2 assessment, a semiempirical correlation to predict the supercooling temperature of the melt in the vicinity of the channel wall was newly derived, and the introduction of this correlation into SIMMER-III has made it possible to appropriately evaluate the penetration distances of various melts [70]. In addition, by improving the momentum exchange function, it has become possible to appropriately evaluate the penetration and solidification behavior of the melt mixed with solid particles [71]. Validations of the freezing model were conducted for more complex situations in which coolant affected the freezing behavior on the structural wall through basic experimental studies [72,73]. Furthermore, it was shown that SIMMER-III can reproduce the coolant void expansion and fuel dispersion behavior observed in the EFM1 test of the in-pile test program CABRI-FAST, which realized large-scale fuel dispersion in the axial direction [74]. For the molten-core pool behavior, model validations related to fuel-to-steel heat transfer characteristics within a molten fuel/steel mixture system were performed through CABRI-FAST TP-A2 test analysis [75]. For vaporization and condensation behavior, a model that can more generally evaluate the effects of noncondensable gas on vapor condensation was developed. Validation studies through basic experimental studies improved the evaluation accuracy and reliability of such behavior [76–78].

In parallel with the SIMMER-III development mentioned earlier, the Experimental Acquisition of Generalized Logic to Eliminate re-criticalities (EAGLE) program described later has commenced, and experimental knowledge on the discharge of molten fuel through the sodium-filled channel, which was insufficient at the time of Phase 2 assessment, has been obtained.

The EAGLE program is a series of experimental studies, which uses a test reactor named Impulse Graphite Reactor (IGR) and the out-of-pile test facility owned by the National Nuclear Center of the Republic of Kazakhstan (NNC-RK), to address large-scale molten-fuel relocation that had not been realized in the previous studies.

The EAGLE program was originally designed to validate the FAIDUS concept [38] in which a duct for molten-fuel discharge is introduced into the fuel assembly [79]. However, the behavior of the molten-core material contacting the steel duct containing sodium, melting the duct, and discharging through it is common with dominant phenomena related to discharge of molten-core materials from the core region. The major outcomes are as follows:

- The molten-core materials generated by the melting of the fuel pin bundle contain molten stainless steel derived from the cladding of the fuel pins. Mixing of stainless steel under the nuclear-heating condition increased the heat flux from the molten-core to the steel duct wall up to approximately 10 MW/m^2 as shown in Fig. 5.31. As a result, a discharge opening of the molten-core materials was formed in the duct wall in quite a short time before the cooling effect of sodium on the duct became apparent [79,80].
- Penetration of molten-core materials into the sodium-filled duct led FCIs, sodium vapor expanded inside the duct, and liquid-phase sodium was removed from the duct. The molten-core materials were discharged through the voided duct, and, therefore, the cooling effect of sodium on their discharge was limited [81,82].
- FCIs at the molten-core materials penetrating the duct also increased the pressure of the core region, and this became the driving force of their discharge. As a result, the molten-

Table 5.5 List of test problems for SIMMER-III Phase 1 assessment [62].

No	Title	Organization	Conv.	IA/MX	HTC	M/F	V/C	EOS	Str.
Category 1: Fluid convection algorithm									
1.1	Ideal gas shock tube	PNC	×						
1.2	Two-phase shock tube	FZK	×	×					
1.3	Oscillating manometer	IPSN	×						
1.4	One-dimensional sedimentation	PNC	×						
1.5	Two-dimensional sedimentation	PNC	×						
1.6	One-dimensional sodium boiling	PNC	×	×			×		
1.7	Liquid sloshing with particles	FZK	×	×					
1.8	Water hammer	PNC	×					×	
1.9	Impact of liquid slugs	FZK	×	×					
1.10	Steam expulsion by subcooled water	PNC	×						
1.11	Stability of one-dimensional bubble column	PNC	×	×					
Category 2: Interfacial area and momentum exchange models									
2.1	Zeroth-dimensional pool flow	PNC		×					
2.2	One-dimensional isothermal bubble column	CEA-G	×	×					
2.3	Two-dimensional isothermal bubble column	CEA-G	×	×					
2.4	Pressure drop in fully developed flow	CEA-G	×	×					
2.5	Momentum exchange in pipe flow	PNC	×	×					
2.6	Developing annular flow	PNC	×	×					
Category 3: Heat transfer									
3.1	Can wall heat transfer	PNC							×
3.2	Structure axial heat transfer	PNC							×
3.3	Film boiling in sodium	PNC			×		×		

Continued

Table 5.5 Continued

No	Title	Organization	Conv.	IA/MX	HTC	M/F	V/C	EOS	Str.
Category 4: Melting and freezing									
4.1	Fuel freezing: GEYSER experiments	CEA-G	×	×	×	×			×
4.2	Fuel freezing: SMPR experiments	CEA-G	×	×	×	×			×
4.3	Freezing of hot melts in tubes: THEFIS	FZK	×	×	×	×			×
Category 5: Vaporization and condensation									
5.1	Condensation of steam on droplet	PNC			×		×		
5.2	Droplet evaporation	PNC			×		×	×	
5.3	Vapor bubble collapse	PNC			×		×	×	
5.4	Rapid fuel vaporization	FZK		×	×		×	×	
5.5	Boiling in a pipe	PNC	×	×	×		×	×	
5.6	Vapor condensation on structure	CEA-G	×	×	×		×	×	×
5.7	Boiling pool with wall heat transfer	CEA-G	×	×	×		×	×	×
5.8	Two-phase blowdown: Bartak's pipe	CEA-G	×	×	×		×	×	
5.9	Two-phase blowdown: Edwards' pipe (1)	CEA-G	×	×	×		×	×	
5.10	Two-phase blowdown: Edwards' pipe (2)	PNC	×	×	×		×	×	
5.11	Thermite injection into sodium: THINA	CEA-G	×	×	×	×	×	×	

Conv.: Fluid convection algorithm; IA/MX: Interfacial area and momentum exchange function models; HTC: Heat transfer coefficient model; M/F: Melting and freezing model; V/C: Vaporization and condensation model; EOS: Equation of state model; Str.: Structure model; PNC: Power Reactor and Nuclear Fuel Development Corporation that was reorganized into Japan Nuclear Cycle Development Institute in 1998 and then into Japan Atomic Energy Agency in 2005.

Table 5.6 List of test problems for SIMMER-III Phase 2 assessment [69].

Problem #	Full title	Short title	Org.	Brief description
Category 1: Boiling pool dynamics			**CEA-G**	
1.1	Isothermal 3D bubble column	3D bubble	JNC	An isothermal bubble-column problem is recalculated by 3-D prelim. Version of SIMMER-IV to confirm problems with 2-D simulation
1.2	Two-phase flows with large liquid density	Dense flow	JNC	A simulation of two-phase flows with high liquid-to-gas density ratio, with comparison and modeling the effect of bubble shape observed
1.3	Void distribution in two-phase flow in a pipe	Void in pipe	JNC	A laminar two-phase flow in a pipe is calculated to test the effect of radial viscous diffusion terms
1.4	Analysis of vaporization and condensation in the Sebulon experiment	SEBULON	CEA-G	An internally heated boiling pool experiment is recalculated to confirm that the improved simulation is possible with recent model refinement
1.5	Analysis of two-phase flow phenomena in the Burty and Castillejos experiment	Burty	CEA-G	A couple of series of basic two-phase flow experiments is analyzed to validate momentum coupling and viscous/turbulent diffusion
1.6	Analysis of vaporization and stratification in the MMB experiment	MMB	CEA-G	A series of multiphase, multicomponent box experiments is evaluated with and without heat transfer effects
1.7	Calculations of SCARABEE APL3	APL3	CEA-G	An attempt to simulate a loss-of-flow sequence of coolant boiling, fuel-pin disruption, etc., observed in the in-pile test
1.8	Analysis of SCARABEE BF2 test	BF2/JNC	JNC	A nuclear-heated boiling fuel pool test in SCARABEE is calculated
1.9	Analysis of a boiling pool of fuel in the SCARABEE BF2 experiment	BF2/CEA	CEA-G	An independent evaluation of the same Problem 1.8
1.10	Assessment of the 1-D two-phase flow modeling of the SIMMER-III code on the LOTUS experiment	LOTUS	CEA-G	A series of steady-state two-phase flow tests is evaluated by recalculating Phase 1 problem
1.11	Rapid depressurization of a tube initially filled with water	Bartak	CEA-G	An update of the Phase 1 calculation of a highly transient experiment with steep pressure gradient

Continued

Table 5.6 Continued

Problem #	Full title	Short title	Org.	Brief description
Category 2: Fuel freezing and relocation			**JNC**	
2.1	Recalculation of a THEFIS experiment: freezing of Al_2O_3 melt in a quartz tube	THEFIS	JNC	An experiment of alumina freezing in a quart tube is recalculated with a new freezing model and cylindrical structure geometry
2.2	Analysis of THEFIS with particles	THEFIS with particles	JNC	A freezing experiment with solid particles is analyzed to study the effect of particles on freezing and plugging
2.3	Basic freezing tests in a tube and analysis	Basic freezing	JNC	Basic melt freezing experiments with simulants are analyzed with a new freezing model
2.4	Fuel freezing in steel tubes: analysis of GEYSER and BLOKKER II tests	GEYSER/ BLOKKER	JNC	Molten fuel freezing experiments in a tube are recalculated with a new freezing model
2.5	Two-phase flow freezing: interpretation of the Bullage experiment	BULLAGE	CEA-G	Two-phase simulant freezing tests are calculated to study the effects of void fraction and pressure on freezing
2.6	Analysis of the CABRI-2 E11 experiment	CABRI E11	IPSN	A postfailure extended fuel motion of a typical CABRI test is analyzed
2.7	Analysis of CAMEL C6 and C7 tests	CAMEL	JNC	An experiment of fuel relocation through a simulated control-rod channel is analyzed
2.8	Analysis of SCARABEE PV-A test	PV-A	JNC	An in-pile experiment of structure wall melt-attack and relocation is analyzed
Category 3: Fuel-coolant interactions			**FZK**	
3.1	Analysis of the behavior of thermite melt injected into sodium	THINA	JNC	An FCI experiment with molten alumina and sodium is analyzed
3.2	Analysis of high-pressure corium melt quenching test	FARO	JNC	A FARO LWR scoping experiment is calculated to study the premixing phase of steam explosion
3.3	Analysis of large-scale fuel-sodium interactions in the TERMOS T1 experiment	TERMOS	CEA-G	A large-scale FARO experiment with molten fuel and sodium is calculated

3.4	Analysis of FCI experiment in alumina/water system	KROTOS	JNC	A typical KROTOS experiment is calculated to study the propagation phase of steam explosion
3.5	Analysis of PREMIX experiment PM06	PREMIX	FZK	An experiment of alumina-jet pouring into water pool is calculated
3.6	Analysis of QUEOS experiments Q08 and Q12	QUEOS	FZK/JNC	An experiment of dropping high-temperature solid balls into water pool is calculated
Category 4: Material expansion dynamics			**CEA-C**	
4.1	Calculation of the SGI expansion phase experiment	SGI	CEA-G	Scaled experiments for postdisassembly expansion with nitrogen or steam bubble in water are calculated
4.2	Analysis of Purdue OMEGA tests	OMEGA	JNC	A test series similar to Problem 4.1 but larger in scale are calculated
4.3	Calculation of CARAVELLE 6 experiment	CARAVELLE	CEA-C	An expansion and condensation behavior of a large steam bubble in water is calculated
4.4	Analysis of VECTORS experiment	VECTORS	JNC	High-speed single- and two-phase flows through a simulated above-core structure are calculated to study energy mitigation
4.5	Analysis of developing annular flow	Annular flow	JNC	A backing analysis for Problem 4.4, by recalculating the Phase 1 problem
Category 5: Disrupted core neutronics			**FZK**	
5.1	Analysis of FCA-VIII fuel slumping experiments	FCA	JNC	Reactivity changes due to fuel motion are evaluated for the Fast Critical Assembly VIII-2 fuel slumping experiments
5.2	Analysis of space-time neutron kinetics using the improved quasistatic method in SIMMER-III	Kinetics	JNC	The neutronics time step control is investigated for a transient test problem to validate the improved quasistatic kinetics method
5.3	Analysis of FCA-VIII fuel slumping experiments with TWODANT	FCA/TWODANT	FZK	The neutron flux shape and eigenvalue calculations are compared between TWODANT and TWOTRAN for reactor and critical assembly configurations
5.4	Neutronics validation for reactor transition phase	ERANOS	CEA-G	The neutronics model is compared with the ERANOS system for typical transition phase configurations

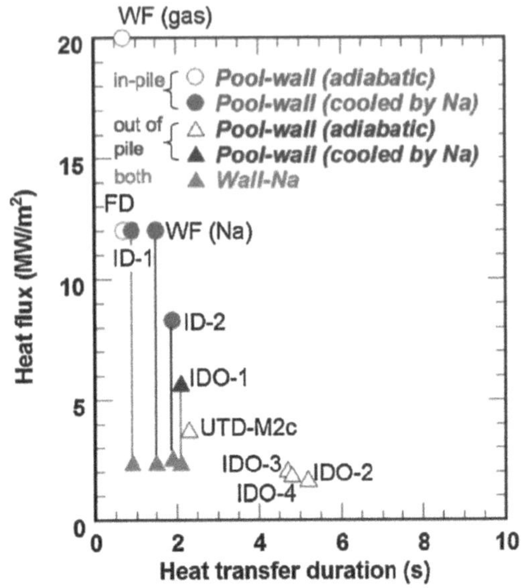

Fig. 5.31 Heat flux evaluated from EAGLE tests [80].

core materials were discharged upward against gravity in the geometry, where the lower end of the duct was closed and the upper end was opened as shown in Fig. 5.32 [83].

The duct used in the test device of the EAGLE program was connected to the region simulating the coolant plenum, where FCIs occurred. Characteristics of these FCIs are described in the following material relocation and heat-removal phases.

Based on the knowledge obtained through the EAGLE program, validations and model improvements of SIMMER-III are in progress . The evaluation accuracies for the formation of the discharge opening in the steel duct due to heat transfer from the molten-core materials and the discharge behavior of the molten-core materials through the path filled with sodium have been improved [84,85].

Since SIMMER-III is modeled in a two-dimensional (2D) system, the reactivity tends to be overestimated due to spatial simultaneity, for example, the concentration of molten fuel on the central axis. To solve this problem, the development of SIMMER-IV, which is an extension of the SIMMER-III model to a three-dimensional (3D) coordinate system, has also been promoted [86]. Analyses by 3D code require a relatively large amount of computer resources, and, therefore, the utilization of the 3D code is limited to cases where the spatial consistency significantly affects results such as the core reactivity since appropriate results are obtained by 2D code analyses in many cases. Both 2D and 3D codes are maintained while keeping the consistency of physical models [87,88].

(c) Summary

The development of computer codes SIMMER-III/SIMMER-IV and accumulated knowledge on major physical phenomena through the progress of experimental studies such as the EAGLE program have made it possible to more mechanistically and rationally evaluate the transition phase in which the motions of molten-core materials dominate the reactivity fluctuation. This set of technology can be applied to evaluations to consider the effectiveness of design measures such as FAIDUS on the elimination of prompt criticality. The application results are described in detail in Ref. [39]. For the application of these

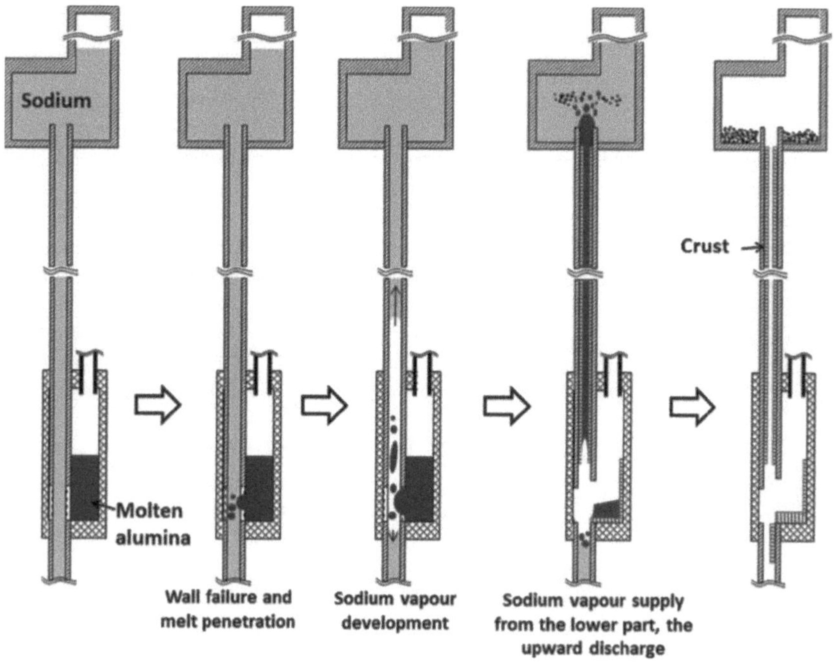

Fig. 5.32 An interpreted sequence of upward melt discharge [83].

technologies to various core designs, SIMMER-V is under development based on SIMMER-III/SIMMER-IV with research and development cooperation between Japan and France [89]. In addition, it will be beneficial to introduce the particle-based simulation method, which is the latest numerical analysis method, into the evaluation technologies for the transition phase since the particle-based simulation method can improve the accuracy of numerical analyses on multiphase, multicomponent thermo-hydrodynamics in the molten-core region [90,91].

5.1.4.4 Material relocation phase and heat-removal phase

(a) Features of the material relocation and heat-removal phases and required evaluation technologies

The material relocation and heat-removal phases are processes designed to assess long-term behavior after the termination of short-term nuclear events dominated by void reactivity and the motion of molten-core materials. The degraded fuel remaining in the core region relocates gradually, and the decay heat of the relocated fuel is removed in the material relocation and heat-removal phases. Since these two phases cannot be clearly distinguished, they will be explained together here. These phases are also important to achieve IVR, in which degraded core materials are stably retained and cooled in the reactor vessel for a long term.

During CDA event progression in the representative accident initiator of SFRs such as ULOF, the temperature rises only in the core region, and fuel melting proceeds while the level of liquid coolant sodium and the subcooling temperature is maintained. Therefore, when the molten-core materials are relocated from the core region, dynamic phenomena such as FCIs occur and these affect both the relocation behavior of molten-core materials and their state after relocation, which is the initial condition of the heat-removal phase. In addition, sodium around the core region plays an important role in removing decay heat from the relocated fuel. Decay heat from fuel relocated in the lower region of the reactor vessel is transferred directly to sodium in the coolant plenum. On the contrary, decay heat from core residual fuel is transferred to sodium both indirectly and directly. The decay heat is indirectly removed by sodium flow formed in the intact fuel assemblies, the radial blanket assemblies, and the radial shielding assemblies. Direct decay heat is removed by sodium reentry into the core region after the decrease in the core pressure. Fuel decay heat is finally transported to a heat sink such as a decay heat-removal system through the natural circulation of sodium and so on.

The major phenomena in the material relocation phase are associated with the relocation of molten-core materials, and they are common with those of the transition phase. Therefore, the evaluation technologies for the transition phase are available for the material relocation phase. However, it should be noted that long-term calculations are required because the accident progresses over a longer time than the transition phase. Since decay heat from the relocated fuel is removed by heat transfer to sodium in the reactor vessel and then transported to the heat sink by its flow, technology to evaluate these local- and plant-scale heat transports is also required. In addition, it is necessary to obtain experimental knowledge on the main phenomena that occur in these phases.

(b) Development of evaluation technologies for the material relocation and heat-removal phases

Elemental phenomena associated with the discharge of molten-core material from the core region are common with those of the transition phase, and SIMMER-III validations based on experimental data such as EAGLE are progressing. On the contrary, since the control rod guide tubes (CRGTs), which are available because the discharge path in the material relocation phase has an internal structure such as a dashpot, it is necessary to investigate the effects of this internal structure on the discharge of molten-core materials. In addition, since CRGTs are usually connected to a region in which the space height and coolant volume are limited, FCIs by the discharge of molten-core materials through CRGTs include phenomena that have not been observed in the existing FCI experiments, as mentioned later. It is also necessary to obtain the lacking data in consideration of the cooling mode with the core-material relocation. To address these issues, the EAGLE-3 program was started in 2016 [92]. Knowledge and data are acquired through the out-of-pile and in-pile tests, and model validation of SIMMER is progressing [93,94].

The molten-core materials discharged through CRGTs are relocated to the lower coolant plenum via the inlet coolant plenum. Supposed phenomena in the inlet coolant plenum may be combinations of FCIs and jet impingement of molten-core materials on the bottom structure of the inlet coolant plenum, because, as mentioned previously, the space height and coolant volume of the inlet coolant plenum are limited in general. Although experimental studies on FCIs [95,96] and the jet impingement [97,98] have been conducted, their combined phenomena under sodium-cooled conditions have not yet been studied. To study such the combined FCIs and jet impingement phenomena, basic experiments using molten-alloy and water were performed in the MELT test facility at the JAEA Oarai Research and Development Institute [99], and the fundamental behavior was observed visually. Then, to investigate the extrapolation of obtained knowledge through the basic experiments to the sodium-cooled condition, a series of experiments were conducted in the EAGLE-3 program [93].

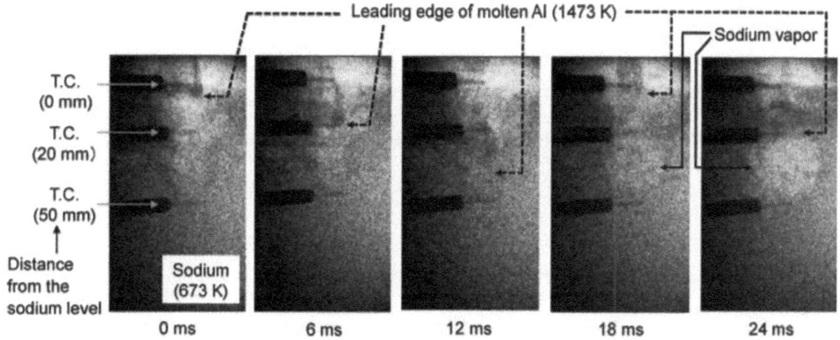

Fig. 5.33 Typical example of X-ray images of molten aluminum discharged into a sodium pool [96].

FCIs also occur when molten-core materials are discharged into the lower coolant plenum. Under sodium cooling conditions with a high subcooling temperature, a stable vapor film is not formed at the contact interface between the molten-core materials and the coolant, the the molten-core materials intruding into sodium are immediately fragmented by FCIs, and the so-called coarse-mixing conditions leading vapor explosion do not occur as shown in Fig. 5.33 [95,96]. From the viewpoint of safety evaluations, in addition to the potential of mechanical energy release against the reactor vessel boundary due to the occurrence of vapor explosions, the penetration distance of molten-core materials into the coolant is also important in evaluating the potential of jet impingement on the structures placed in the lower part of the core such as the core catcher and the reactor vessel. A semi-empirical correlation to evaluate the penetration distance of molten-core materials into the coolant has been established through experimental studies [100–103].

The solidified core materials fragmented by FCIs into particles, which are called "debris" in the field of the SFR CDA study, settle and accumulate on the bottom structures in the reactor vessel. This accumulated debris is called the "debris bed." Regarding the settlement and accumulation behavior of debris, a hybrid Eulerian-Lagrangian numerical simulation method, coupling SIMMER-IV with the discrete element method (DEM), has been developed [104]. In addition, in the debris bed, a self-leveling phenomenon was confirmed in which the debris bed shape collapsed and its thickness became uniform due to the internal sodium boiling caused by the fuel decay heat [105–107]. Two analytical methods for self-leveling phenomenon have been developed [108]: The first method is a microscopic model coupling SIMMER-III/SIMMER-IV with DEM as shown in Fig. 5.34 to enable the detailed analysis of the self-leveling behavior in a local scale of structures in the reactor vessel [108–110] and the second method is a macroscopic model that treats the debris bed as a fluid component in the SIMMER code and introduces physical models that express the contact and collision of debris to stably simulate the self-leveling behavior on a reactor scale [111]. The result of the validation study for the second model is presented in Fig. 5.35.

The relocation of core materials continues until heat removal exceeds fuel decay heat in the core region. Decay heat from core residual fuel is removed indirectly by the sodium flow around the core region and removed directly by the sodium reentry into the core region. Regarding the heat removal from the core residual fuel due to both the sodium flow around the core and the countercurrent flow by the sodium reentry from the upper core

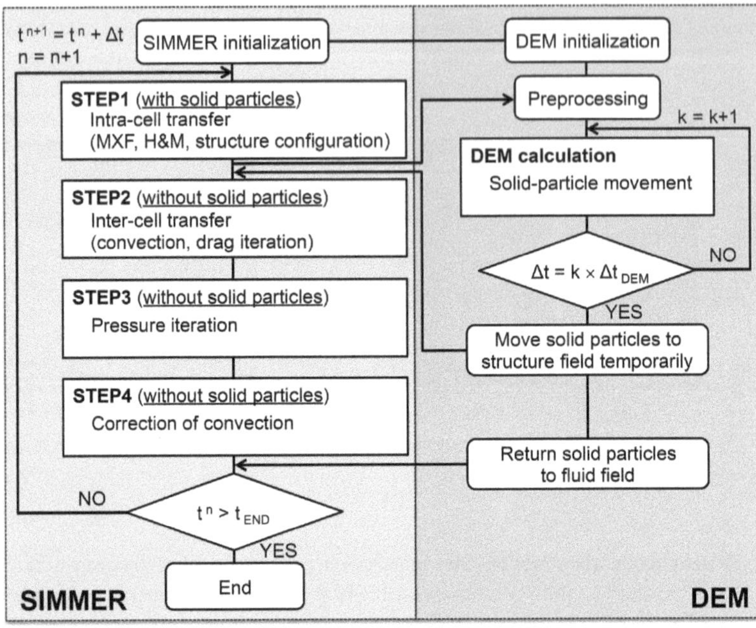

Fig. 5.34 Schematic description of multitime step algorithm in SIMMER-DEM calculation [109].

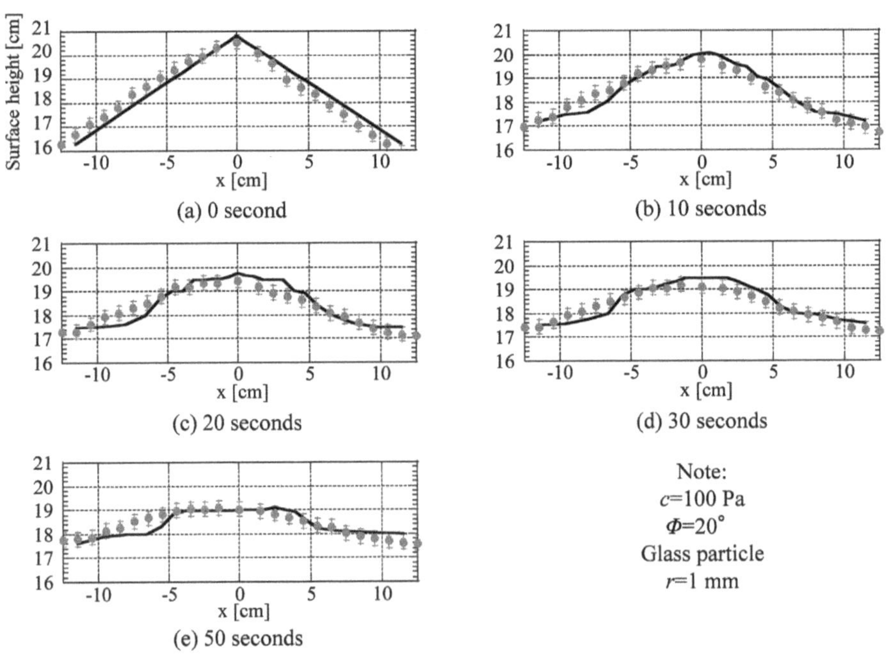

Fig. 5.35 Comparison of surface height profiles between calculation (black solid line) and experimental result (gray closed circle with error bar) from the start of gas inflow [111]. The vertical axis represents the surface height [cm] and the horizontal axis represents the distance from the center [cm].

region, a method was developed through the evaluation on Monju [41,112,113]. Heat removal by through-flow of sodium reentering from below through CRGTs and evaporating in the core region was selected as an issue for the EAGLE-3 program. Experimental data are currently being acquired and model validation of the SIMMER-III code is underway [114].

Characteristics of the heat removal from the debris-bed formed at the bottom of the reactor vessel were studied in the United States from the 1970s to the mid-1980s through out-of-pile and in-pile tests [115,116]. Although there are complex modes of heat transfer as the phenomena progress, the heat transfer from the debris bed is basically evaluated as heat conduction of the coolant sodium since FCIs lead formation of the debris bed with relatively small particle size and low fluid permeability [117,118]. Based on this knowledge, an axial one-dimensional model to evaluate heat removal from the debris bed taking boiling and dry-out of sodium in it was developed [115,116]. Furthermore, by extending this one-dimensional model, the DEBRIS-MD code was developed that enables two-dimensional calculation for heat removal from the debris bed with boiling [117]. The evaluation method for heat removal from debris beds is improved so that it can be evaluated in response to a wide range of event progressions [118,119]. To evaluate the long-term heat removal from the debris bed in a nuclear plant, as mentioned in (a) previously, it is necessary to evaluate the heat transfer from the debris bed to the heat sink through the cooling system of the plant. An analysis code was developed by combining a one-dimensional model similar to DEBRIS-MD with a plant flow-network model, and it was applied to the evaluation of debris bed coolability in a future SFR [120,121].

To cope with the evaluation of long-term behavior, SIMMER-LT, which has increased the calculation speed of long-term behavior, and MUTRAN, a computer code that can evaluate degraded core behavior over several hours to several tens of hours, have been developed [120]. In the material relocation phase, the control material behavior plays an important role in maintaining a subcritical state. Since the control material may fall in the core region and diffuse or discharge through eutectic reactions with the degraded core materials, model development of SIMMER to treat eutectic reactions and experimental studies for model validation are under progress [122,123].

(c) Summary

Since the relocation of the core materials affects core reactivity, evaluation technology that can continuously calculate core-reactivity changes coupled with relocation of the core materials is required; utilizing the SIMMER code is the most realistic and effective way to accomplish this. The EAGLE-3 program has been conducted on the major phenomena that occur during the core material relocation process, and it is necessary to validate SIMMER by the obtained data. In addition, since the evaluation of core material relocation behavior outside the core region exceeds the original design of the SIMMER code, new models that evaluate debris bed formation and self-leveling behavior are studied and are optionally available in SIMMER. Since the relocation of core materials continues until the decay heat from the core residual fuel can be sufficiently removed, it is important to evaluate the coolability of the core residual fuel. Coolant reentry into the core region is expected to have high cooling performance, and, therefore, it is necessary to develop an evaluation method through experimental studies. For the long-term removal of fuel decay heat, it is necessary to evaluate the amount of heat transferred from the relocated core materials into sodium and to evaluate the heat transfer to the heat sink. In particular, to ensure the achievement of IVR, efficient decay heat removal from the debris bed formed near the reactor vessel boundary is required, and the introduction of a design measure such as the debris tray [39] plays an important role.

5.1.4.5 Concluding remarks

Historically and common with other countries, studies on the CDAs have been regarded as one of the most important fields of SFR safety. In Japan, the development of safety evaluation technologies started with reference to the most advanced technologies of the United States and the case of CRBR safety assessments for licensing. Along with safety reviews starting from the initial licensing of Monju, research and development for the commercialization of SFRs have been undertaken, and Japan has established semidomestic safety evaluation technologies for CDAs.

In the initiating and transition phases, understanding of the dominant phenomena has progressed through the experimental studies represented by TREAT, CABRI, and EAGLE, and development of the computer codes SAS4A and SIMMER-III/SIMMER-IV has progressed. As a result, regarding the mechanical energy driven by positive reactivity insertions, the evaluated value of work energy can be reduced significantly because the conventional assumptions that were regarded as excessively conservative can be eliminated. In addition, CDA evaluation technologies have reached a level that enables assessment of the effectiveness of design measures to prevent the generation of mechanical energy.

For the achievement of IVR, technologies to evaluate CDA progression in the material relocation phase and the heat-removal phase are required. It is inherently possible for SFRs to contain and cool the degraded core materials because the core disruption and melt progresses while maintaining sodium levels and subcooling. The effectiveness of design measures is evaluated through CDA analyses using computer codes validated by phenomenological knowledge and database obtained through experimental studies. Therefore, research and development should be continued to obtain experimental knowledge and to advance the computer codes as mentioned in this subsection.

5.1.5 Sodium combustion

Since a liquid sodium has a high chemical reactivity with oxygen and water/water vapor, sodium combustion will take place when it leaks out to atmospheric conditions. During sodium combustion, a surrounding environment is heated up due to the chemical reaction. In addition, chemical reactants, such as sodium oxide (Na_2O), sodium peroxide (Na_2O_2), and sodium hydroxide (NaOH), have toxicity and will be released to the environment as aerosol when the combustion takes place. Therefore, a consideration of design against sodium combustion is necessary for both integrity of the plant itself and a circumference environment.

In this section, experimental research works, where combustion types and their theoretical and mechanical understanding have been carried out, and the development of numerical tools is described.

5.1.5.1 Chemical reaction of sodium combustion

Sodium has a chemical reaction with oxygen and water vapor as shown in the following [124]:

$$
\begin{aligned}
&2\mathrm{Na(l)}+\frac{1}{2}\mathrm{O_2(g)}=\mathrm{Na_2O(s)}, && \Delta H_{298.15}=-415\,\mathrm{kJ/mol},\\
&2\mathrm{Na(l)}+\mathrm{O_2(g)}=\mathrm{Na_2O_2(s)}, && \Delta H_{298.15}=-511\,\mathrm{kJ/mol},\\
&2\mathrm{Na(l)}+\mathrm{H_2O(g)}=\mathrm{Na_2O(s)}+\mathrm{H_2(g)}, && \Delta H_{298.15}=-173\,\mathrm{kJ/mol},\\
&\mathrm{Na(l)}+\mathrm{H_2O(g)}=\mathrm{NaOH(s)}+1/2\mathrm{H_2(g)}, && \Delta H_{298.15}=-184\,\mathrm{kJ/mol}.
\end{aligned}
\tag{5.1}
$$

where (*l*), (*g*), and (*s*) mean the liquid phase, the gas phase, and the solid phase, respectively. $\Delta H_{298.15}$ means the enthalpy difference at 298.15 K (25°C) and a negative value corresponds to a heat generation. When oxygen and water vapor exist sufficiently in an atmosphere, reactants, such as sodium oxide and hydrogen, also react as follows [124]:

$$
\begin{aligned}
&\mathrm{Na_2O(s)}+\mathrm{H_2O(g)}=2\mathrm{NaOH(s)}, && \Delta H_{298.15}=-97\ \mathrm{kJ/mol},\\
&\mathrm{Na_2O_2(s)}+\mathrm{H_2O(g)}=2\mathrm{NaOH(s)}+\frac{1}{2}\mathrm{O_2(g)}, && \Delta H_{298.15}=-49\,\mathrm{kJ/mol},\\
&\mathrm{H_2(g)}+\frac{1}{2}\mathrm{O_2(g)}=\mathrm{H_2O(g)}, && \Delta H_{298.15}=-242\,\mathrm{kJ/mol},\\
&\mathrm{Na(l)}+\frac{1}{2}\mathrm{H_2(g)}=\mathrm{NaH(s)}, && \Delta H_{298.15}=-56\,\mathrm{kJ/mol}.
\end{aligned}
\tag{5.2}
$$

5.1.5.2 Types of sodium combustion

Fig. 5.36 shows a schematic of sodium combustion and key phenomena. Since liquid sodium has a high boiling temperature (approximately 1156 K in the atmosphere), pressurization of coolant is not required. A piping system will be pressurized several times as high as an atmosphere. When liquid sodium leaks from a piping system, it will

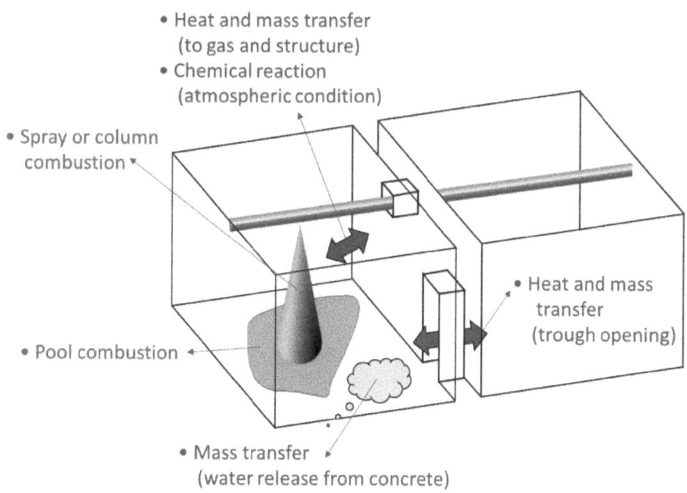

Fig. 5.36 Schematic of sodium combustion.

burn during its fall (spray or column combustion). Furthermore, it also will burn on the floor when unburnt sodium piles up on the floor (pool fire) as shown in Fig. 5.36.

(a) Spray combustion

When a leakage velocity of liquid sodium from the piping system is sufficiently high, the liquid sodium will be broken up into small particles (order of several mm in diameter). The combustion of those small particles is called "spray combustion."

In general, chemical reaction rate at the sodium combustion is much faster than the mass flux rate of oxygen and water vapor to the reaction region. Accordingly, the burning rate of the spray combustion is evaluated based on the mass transfer of oxygen and water vapor or the famous d^2-law theory considering the forced convection effect due to a droplet fall.

In the spray combustion, an interfacial area between liquid sodium and the atmosphere becomes large and all chemical reaction heat will be released to the atmosphere as well as aerosol of reactants. Consequently, a high temperature of the gas and high pressure will appear during the spray combustion. And thus, the structural integrity of the compartment will be one of the key issues.

(b) Column combustion

Since the piping system covers with multiple layers such as inner/outer packaging steel and an insulator, the velocity of leaked sodium will be moderated and thus the leaked sodium falls down like a column shape or waterfall. The sodium will burn at the surface of the column called "column combustion."

The combustion rate of the column combustion is smaller than that of the spray combustion due to a reduction of the contact area. Therefore, a detailed investigation or development of a numerical model for column combustion has not been carried out. In general, the spray combustion model is applied by coinciding with the contact area.

(c) Pool combustion

When unburnt sodium is piled upon a floor like a pool shape (see Fig. 5.36), the pool combustion takes place. The pool combustion is moderated rather than the spray combustion because it only occurs just above the pool surface. However, it will continue for a long time until the unburnt sodium runs out.

When unburnt sodium connects to a concrete structure directly, a sodium-concrete interaction (reaction) takes place resulting from heat generation and hydrogen generation as well as pool combustion. Therefore, a concrete floor is lined with a steel liner to eliminate the sodium-concrete interaction where sodium leakage may occur.

5.1.5.3 Experimental and numerical research

Fundamental experiments for each combustion model are carried out mainly in JAEA. After the sodium leakage at a secondary loop in the Japanese prototype fast breeder reactor, Monju, in 1995 (Monju incident), mock-up experiments are also carried out to investigate a detail of progression during the incident. Table 5.7 summarizes the list of the related experiments [124].

With regard to numerical research, a lumped mass model code, ASSCOPS [125] is developed based on modified SOFIRE-II [126] for the pool combustion and SPRAY-II [127] for the spray combustion. After the Monju incident, the more theoretical and mechanistic approach has been applied in JAEA such as a lumped mass zonal model SPHINCS [128], a multidimensional field model AQUA-SF [129], a

Table 5.7 Summary of sodium combustion experiments.

Category	Facility volume (m^3)	Floor or pool area (m^2)	Experimental conditions					
				Sodium leakage				
			Oxygen (vol%)	Temp. (°C)	Leakage rate (g/s)	Duration	Total amount (kg)	Leakage height (m)
Small pool	3	1.0	21	507	1.9–7.1	22–30 min	2.5–10	0.1, 1.5
Pool	21	3.8	0–5	400, 530	–	–	410–860	–
	21	0.18, 0.35	21	530	–	–	41–83	–
	2.7	0.5	21	400	–	–	162	–
	80	2.25	21	505	–	–	550	–
	3	0.07	3–8	150–400	–	–	10	–
Spray (single droplet)	0.1	–	21	300–400	4 mm in diameter (stationary)			
	0.28	–	0.21–21	500	4 mm in diameter			2.7
	2.3	–	21	450	5 mm in diameter			1.9–8
Spray	1.9	–	0–21	300–530	0.03–0.05	8–15 s	0.4	0 (Upward)
	1.9	–	0–21	530	0.04–0.05	7–10s	0.36–0.39	0 (Upward)
	21	–	0–21	500–540	0.3–0.6	18–76 s	8–38	3.0
	100	–	21	505	0.51	1800s	915	4.0
	100	–	3, 21	250	0.57, 0.59	300 s	171, 176	4.0
Column	21	–	0–5	400, 530	1.9–13	30–270 s	410–860	2.4
	21	–	21	530	1.4–2.2	30–58 s	41–83	2.4
	100		3, 21	500	0.7–2.3	300 s	210–690	1.0–4.0
Humidity effect	0.17	0.008	21	500	0.14–83	2–1200s	0.083, 0.17	0.55
Integrated (mocked up of	27	0.8	21	485, 480	17, 51	10, 20 min	20, 32	0.95
Monju incident)	100	6.6	21	480	44	91 min	241	3.59
	170	22.0	21	480	40–60	222 min	690	3.59

direct simulation of single droplet COMET [130], a moving particle method for investigating a pool spread phenomenon MPS-3D [131], and a stoichiometric analysis of sodium combustion using a Gibbs free energy minimization BISHOP [132]. Recently, an international collaborative work between JAEA and Sandia National Laboratories has been carried out to modify combustion models and make benchmark analyses as a part of the Civilian Nuclear Energy Research and Development Working Group (CNWG) [133,134].

5.1.6 Sodium-water reaction

5.1.6.1 Overview

Steam generators are installed for electrical power generation by the steam turbine in a sodium-cooled fast reactor. In steam generators, high-temperature and low-pressure sodium flows in the shell side and low-temperature and high-pressure water/steam flows inside the heat transfer tubes to exchange the heat. If a penetrating defect occurs in the heat transfer tubes due to pinholes, cracks, and so on, water/steam leaks into the sodium. The direct contact between sodium and water/steam causes the exothermic reaction and generates highly alkaline reaction products such as sodium hydroxide and sodium oxide and flammable hydrogen gas; then, a unique reaction region (reaction jet) is formed [135]. Once water leaks inside a steam generator, the heat transfer tube adjacent to the leaked heat transfer tube is worn out by erosion and corrosion (wastage). Besides, when the high-temperature reaction jet is covering the heat transfer tube, the cited tube is damaged by internal pressure due to a decrease of mechanical strength (overheating tube rupture). As the result, the affected region of heat transfer tubes expands temporally and spatially (tube failure propagation) [136]. When a large-scale water leak occurs, a spike-like pressure rises at the initial stage of the water leak (initial spike pressure) and pressure increase in the system due to the accumulation of generated hydrogen gas (quasistate pressure) come about, which affect the structural integrity of the system [137].

For these reasons, the hydrogen detectors as an equipment measure are installed in the sodium side and cover gas side of steam generators. Moreover, the progress of the event can be terminated by detecting the water leak at the initial stage and discharging the water/steam in the steam generator after the reactor shuts down. From the viewpoint of safety protection, even if the scale of water leak becomes large, the pressure inside the steam generator is released by bursting the rupture disc, and the reaction products move to the reaction product release system, the influencing area is localized. Many sodium-water reaction tests have been carried out so far, and analysis codes for pressure propagation [138] and tube failure propagation [139] have been developed based on the previously mentioned test data. These accumulated findings were reflected in the regulatory safety review of fast breeder reactor "Monju."

5.1.6.2 Research and development in recent years

In recent years, mechanistic sodium-water reaction analysis codes have been developed to improve the flexibility in the design of steam generators for the next-generation sodium-cooled fast reactors and the accountability in the regulatory safety reviews [140]. In addition, the separation effect experiments and basic tests have been carried out to elucidate the phenomena of the sodium-water reaction and to validate the analysis codes.

(a) Mechanistic sodium-water reaction analysis system

The mechanistic sodium-water reaction analysis system comprises the compressible multicomponent and multiphase flow with chemical reaction analysis method (SERAPHIM code), the heat transfer analysis method of the adjacent heat transfer tube (TACT code), and the heat transfer analysis method in the adjacent heat transfer tubes (RELAP5 code).

In the SERAPHIM code, a multifluid model is adopted as an analysis mode of multiphase flow, and the conservation formulas of mass, momentum, and energy including the formation terms of heat and substance associated with phase change and chemical reactions are solved. The modified HSMAC method is applied to the solution of the base equations. So far, we have conducted the basic verification and validation of the phenomenon of high-pressure gas ejecting into a liquid and the behavior of a chemical reaction between gas and liquid. An unstructured analysis method was developed to improve the simulation of complex structures inside the steam generator; the validity was confirmed by sodium-water reaction tests [141].

(b) Phenomena elucidation tests of sodium-water reaction

As mentioned already, in the past, many tests related to pressure wave propagation, tube failure propagation, wastage, and overheating tube rupture were carried out under the condition simulating the actual accidental phenomena for sodium-water reaction. They have contributed to the regulatory safety review of fast breeder reactor "Monju."

However, it is difficult to directly measure the sodium-water reaction field due to its inherent environment (high temperature, high velocity, and high alkalinity). Thus, the mechanism of the sodium-water reaction phenomena has not been clarified so far. The chemical reaction process of sodium-water reaction was investigated by thermal analysis and counterflow diffusion test. It was revealed that sodium hydroxide and hydrogen were generated by the initial reaction, and then sodium monoxide and hydrogen were generated by the secondary reaction between the sodium hydroxide generated by the initial reaction and the surrounding sodium within the time of the event progress of the sodium-water reaction [142].

As mentioned before, the wastage has the feature of mechanical and chemical attacks against the tube wall. Thus, it is considered that wastage is the phenomenon combined with the erosion of liquid droplets entrained by reaction jet and flow-accelerated corrosion due to high-temperature reaction products such as sodium hydroxide and sodium monoxide. The new wastage correlation equations were derived from some separation effect experiments [143].

5.1.6.3 Summary

Steam generators are a key component in the sodium-cooled fast reactor. When the water leakage from the defect heat transfer tube occurred, it is necessary to detect early and terminate the event as soon as possible. At the time of the regulatory application of construction permission for fast breeder reactor "Monju," an enormous number of tests were carried out, and evaluation models were developed on the basis of the empirical laws. In recent years, the mechanistic sodium-water reaction analysis method and phenomena elucidation experiments have been carried out to improve the rationality for many design options and accountability for the establishment of the highly versatile sodium-water reaction analysis system.

5.1.7 Source term

5.1.7.1 Overview

"Source term" is a general term for the types, quantities, properties, and release time, etc., of radionuclides released to the environment or containment vessel. To carry out a source term study on sodium-cooled fast reactors (SFRs) efficiently, JAEA has been conducting several experiments focusing on the effects of sodium while referring to knowledge of light-water reactors. The main experimental and analytical studies conducted at JAEA are introduced in the following.

5.1.7.2 Experimental studies at JAEA

(a) Volatile fission product release behavior from the liquid sodium to the gas phase

Some fission products (FPs) released from damaged fuel to sodium coolant are transported to the cover gas by sodium evaporation. Volatile FPs such as iodine, cesium, and tellurium are especially important for this evaporation release.

A gas-liquid equilibrium partition coefficient Kd (Kd is defined as "FP concentration relative to sodium in the gas phase" divided by "FP concentration relative to sodium in the liquid phase") when the sodium and the cover gas are isothermal, and nonequilibrium partition coefficient Kd' in the nonisothermal case were measured to evaluate the evaporation and release behavior of the volatile FPs from the sodium to the gas phase.

Results revealed the unlikelihood of iodine being released into the gas phase due to the retention effect because of its chemical affinity with sodium. On the contrary, since cesium is also an alkali metal, it dissolves infinitely in sodium, but it was found that it is easily released into the gas phase because its vapor pressure is higher than that of sodium [144].

(b) Nonvolatile fission product release behavior during the sodium-concrete reaction

A sodium-concrete reaction not only causes concrete penetration and hydrogen generation, but also affects the release behavior of radioactive substances. Specifically, the heat of reaction and stirring of the sodium pool with hydrogen bubbles raise the overall pool temperature and increase the evaporation rate of sodium and volatile FPs. In addition, the carryover of sodium mist and nonvolatile FPs with the bubble release increases the mass of the FPs released into the gas phase.

A retention factor RF (RF is defined as "FP concentration in sodium pool" divided by "FP concentration in aerosol released into the gas phase") of the nonvolatile FP was measured to evaluate the release behavior of the nonvolatile FP during the sodium-concrete reaction.

As a result, an empirical formula for the aerosol generation rate due to the hydrogen gas bubbling was obtained. In addition, the retention factors for strontium oxide (SrO), europium oxide (Eu_2O_3), and ruthenium (Ru), which are typical nonvolatile FPs, were obtained [145].

(c) Iodine mass transfer from xenon-iodine mixed gas bubble to liquid sodium pool

Since iodine, which is an important nuclide for exposure assessment, is a halogen element, it easily reacts with sodium, which is an alkali metal, and it can be expected to have a retention effect due to the sodium coolant. On the contrary, there is a concern that a noble gas coexisting under the accidental condition will prevent direct contact between iodine and sodium, thus reducing this retention effect. Therefore, a sodium experiment was performed using a mixture of xenon and iodine vapor.

In the experiment, by breaking a quartz glass ball, gas bubbles containing iodine vapor and xenon were generated in sodium. The mixed gas released into the cover gas was then sampled, iodine was quantified, and decontamination factor DF (DF is defined as "aerosol injection mass" divided by "aerosol collection mass") was calculated.

As a result, it was found that the iodine DF in the sodium accompanying the xenon gas bubbles was strongly influenced by the initial equivalent bubble diameter and iodine vapor concentration [146].

5.1.7.3 Analytical study at JAEA

A computer code TRACER (Transport Phenomena of Radionuclides for Accident Consequence Evaluation of Reactor) has been developed to evaluate the so-called in-vessel source term for SFRs. The TRACER code can analyze the physical and chemical behavior related to FP release and transport in a primary cooling system. Phenomena modeled in the code are FP release from fuel to coolant, bubble behavior, coolant flow, and FP transfer in cover gas.

TRACER users need to specify the fuel temperature, FP inventory in the fuel pins, coolant temperature, and coolant flow rate. As a result, TRACER can calculate time histories of FP concentrations in the cover gas, coolant, and bubbles [147].

5.1.7.4 Summary

JAEA has been conducting source term studies focusing on the effect of sodium, which is a characteristic of SFRs. However, to realize a less conservative source term evaluation, further studies, including improvement of the accuracy of the experimental data acquired, need to be done. In the future, JAEA plans to conduct experiments to acquire further data on iodine and cesium, etc., and to develop and validate the relevant analysis codes.

5.2 Sodium component development

5.2.1 Introduction

Conceptual designing and R&D for Japan Sodium-cooled Fast Reactor (JSFR) have been conducted since 2006 JFY in Japan. Several innovative sodium component designs were established in these design studies and R&D. The development plan for JSFR includes large-scale sodium component experiments and system demonstrations.

Two stages of testing are planned to mitigate the risk of innovative technology developments (e.g., the risk of large and complicated component fabrication such as straight double-walled tube steam generator [SG]) and provide component development data for conceptual design at an appropriate time. The first stage is a component development test and the second stage is a system demonstration test.

Component development tests for those two items were planned since JSFR adopted the straight double-walled tube SG and the integrated pump-intermediate heat exchanger (IHX). System demonstration tests and equipment demonstration tests were planned in addition to these tests. The sodium tests for JSFR were outlined as follows:

- Component tests
  - Straight double-walled tube SG
    - First stage: straight single-walled tube 10 MW SG (component development test)
    - Second stage: straight double-walled tube 50 MW SG (demonstration test)
  - Integrated pump-IHX
    - First stage: full-length shaft with 1/10 flow rate pump (component development test)
    - Second stage: integrated pump-IHX (demonstration tests)
- System demonstration test
  - Partially integrated system demonstration test or whole integrated system demonstration test
    - To demonstrate cooling components and system
    - To demonstrate natural circulation decay heat removal
- Equipment demonstration in sodium
  - Inspection devices
  - Fuel handling machine
  - Rotating plug
  - Control rod drive mechanism, etc.

5.2.2 Component development plan for JSFR

SG is a key innovative component for JSFR. Fig. 5.37 describes sodium experiment items. Two stages of large-scale sodium tests for SG were planned at the Advanced Technology Experiment Sodium Facility (AtheNa Facility). The first stage was a component development test. The second stage was a system and component demonstration test. The component development test was planned to verify heat transfer in a 29-m-long tube bundle and flow stability in water-steam multichannel. The system demonstration test was planned to demonstrate the cooling system for safe and stable operations.

Fig. 5.38 shows the test scale of the straight double-walled tube SG. The scale of the component development test was planned at 10 MW thermal with full-length single-walled tubes to simulate thermohydraulics (e.g., heat transfer distribution at tube wall, multichannel flow stability of two-phase flow without inlet orifice) and sodium flow field (e.g., edge effect, single tube plugged condition). The scale of the system and component demonstration test was planned at 50 MW thermal with

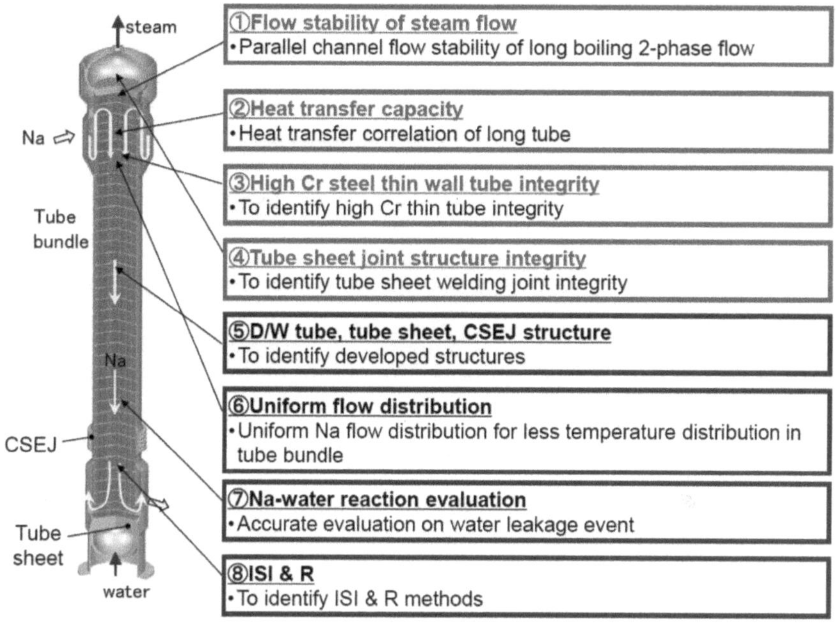

Fig. 5.37 Component test #1 (straight double-walled tube SG).

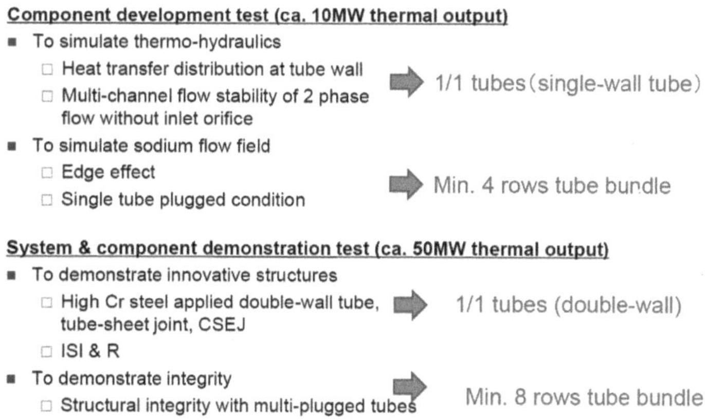

Fig. 5.38 Test scale for SG.

full-length double-walled tubes to demonstrate innovative structures (e.g., high Cr steel applied double-walled tube, tube-sheet joint, and CSEJ, ISI, and repair) and to demonstrate integrity (e.g., structural integrity with closed multiplugged tubes).

Integrated pump-IHX is another key innovative component for JSFR. Fig. 5.39 describes sodium experiment items. Two stages of large-scale sodium tests for integrated pump-IHX were also planned at the AtheNa Facility. The component development test was planned to verify the rotating stability of the long axle with a flexible

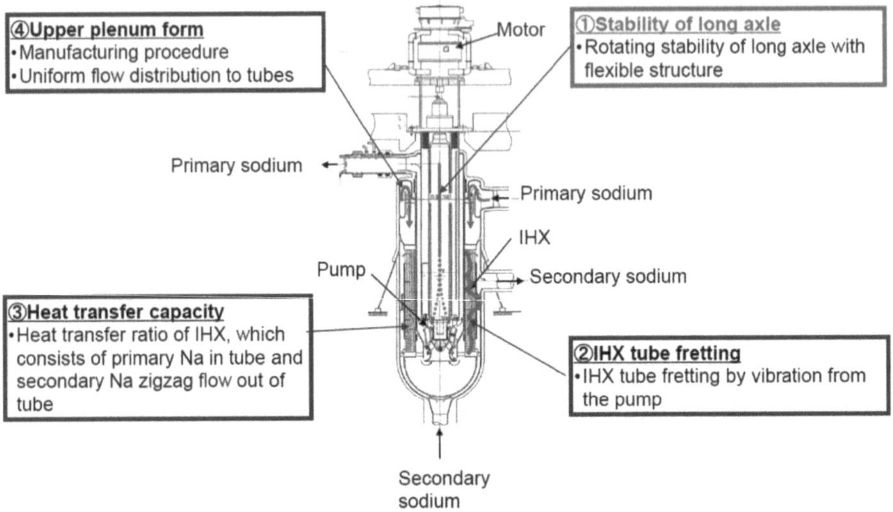

Fig. 5.39 Component test #2 (integrated pump-IHX).

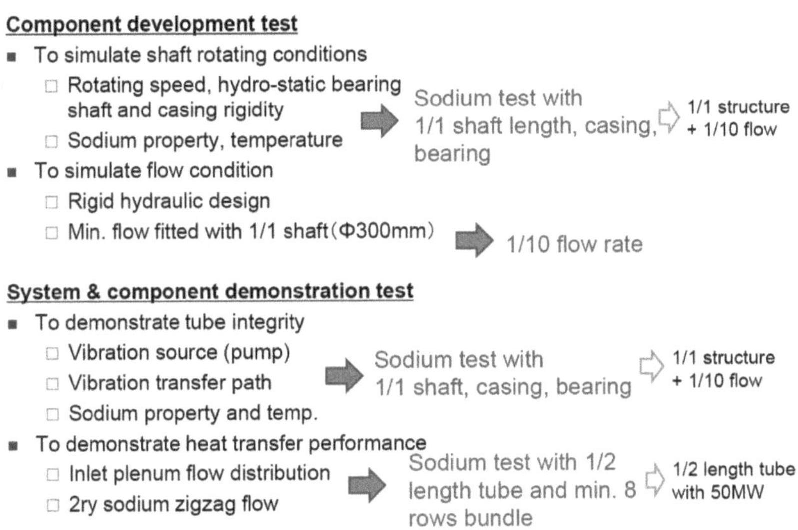

Fig. 5.40 Test scale for integrated pump-IHX.

structure. The system and component demonstration test could reveal IHX tube fretting characteristics based on the vibration from the pump rotation and also the heat transfer capacity.

Fig. 5.40 shows the test scale of the integrated pump-IHX. The scale of the component development test was planned for a full-length shaft with a 1/10 flow rate to simulate shaft rotating conditions (e.g., rotating speed, hydrostatic bearing shaft, and casing rigidity) and flow condition (e.g., rigid hydraulic design and minimum flow fitted with full-length shaft). For facility configuration, tube integrity demonstration tests (e.g., vibration source (pump), vibration transfer path) and heat transfer demonstration tests (e.g., inlet plenum flow distribution, secondary sodium zigzag flow) should be separately conducted in the system, and component demonstration tests. The scale of the tube integrity demonstration test was planned for a full-length shaft with a 1/10 flow rate. The scale of heat transfer demonstration test was planned for 50-MW thermal with a half-length tube.

5.2.3 AtheNa facility specifications

The JAEA AtheNa Facility can serve as an experimental test bed for the various cooling system and component performance tests using large sodium loops. The concept of the AtheNa Facility is described in the following.

- Flexible configuration and schedule for experimental demand
  - Shared utilities:
    - Common utilities, "mother loop" and several test rigs, "daughter loops" in each facility, concept
  - "Daughter" test loops are independent of each other
    - Several experiments at the same time
    - The construction and maintenance schedule of the test loop is on-demand
    - Low cost of construction and operation
- Flexible application for sodium test requirements
  - Another "daughter loop" can be constructed and "the mother loop" can supply sodium to them
    - Sodium inventory is 240 ton
    - The range is 570°C.

Major specifications of the facility for component development test and system and component demonstration test are shown in Figs. 5.41 and 5.42.

The facility view, inner facility view, and layout of the facility are shown in Figs. 5.43–5.45 to aid in the imaging of these tests. The AtheNa Facility has a large sodium inventory of 240 tons, thermal output of 60 MW, and a building of 55 m in height. Many other infrastructures (e.g., pressured air, nitrogen gas, argon gas, electricity supply including emergency generator, and cranes) are provided.

5.2.4 GIF collaboration

After the 1F accident, JAEA has been focusing on activities concerning the SFR safety issues and their reflections on the international safety design criteria (SDC). Activities for "the international safety design criteria" are described as follows.

- Establishment of an international SDC for SFR
  - Reevaluation of JSFR safety
  - Improvements in safety design
  - Proposal of "safety design criteria" for the Generation IV nuclear system International Forum (GIF)

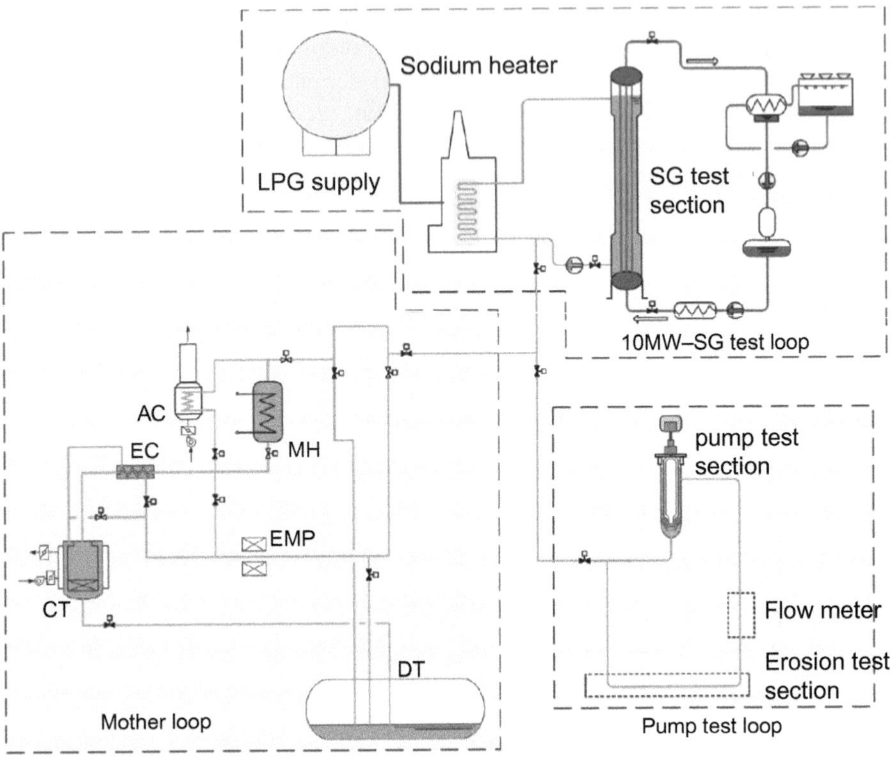

Fig. 5.41 Overview of component development tests.

- Requirements for the SDC
  - ➢ Experimental data
  - ➢ Analytical data
- Experiments for Severe Accident Measures
  - ➢ Experimental plans for a safety design against severe accidents.

Considering these situations, experiments for severe accident measures were investigated for international collaboration (Fig. 5.46). Concepts for experiments are described in the following.

- Safety design policy for severe accident measures.
  - ➢ Measures for in-vessel retention (IVR) have priority.
  - ➢ For IVR, measures against loss of heat-removal systems (LOHRS) and anticipated transient without scram (ATWS) should be reinforced.
- Candidates for measures against LOHRS.
  - ➢ Alternative cooling systems must be installed.
  - ➢ External vessel cooling and internal vessel cooling systems are important candidates.
- Phenomena to be investigated for measures.
  - ➢ Heat transfer and thermal-hydraulics in the reactor vessel and from the external vessel and steam generator.
  - ➢ Local heat-removal characteristics from the damaged core.

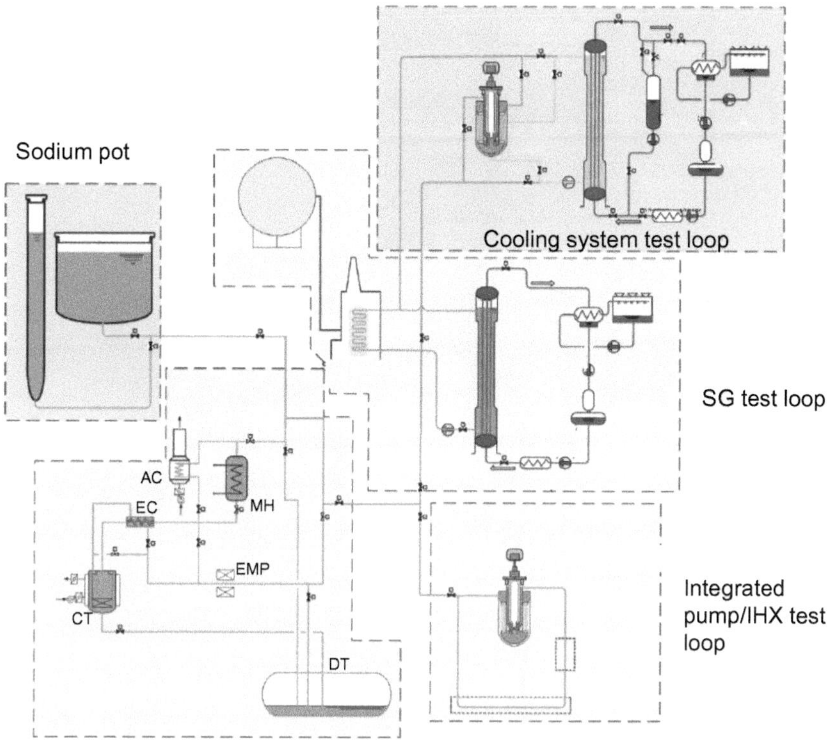

Fig. 5.42 Overview of system and component demonstration test.

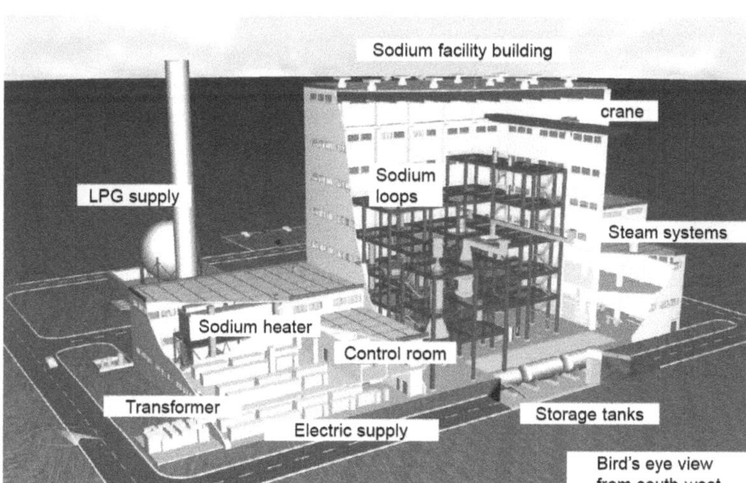

Fig. 5.43 Bird's-eye view of AtheNa Facility.

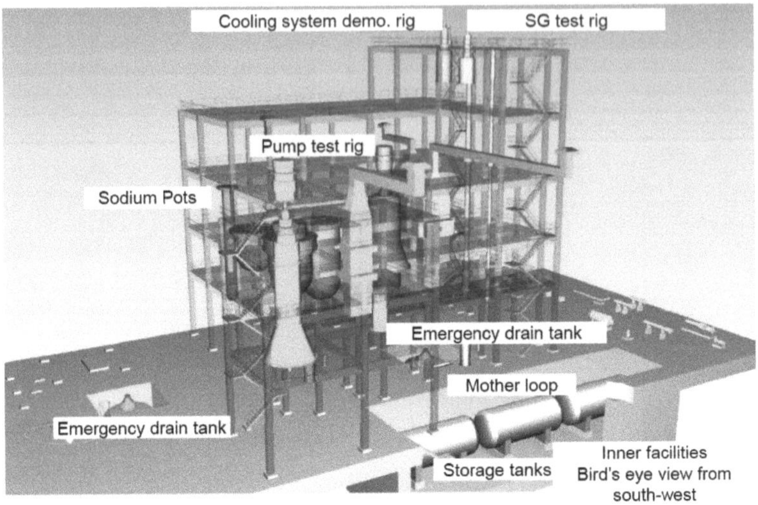

Fig. 5.44 Bird's-eye view of experimental apparatus.

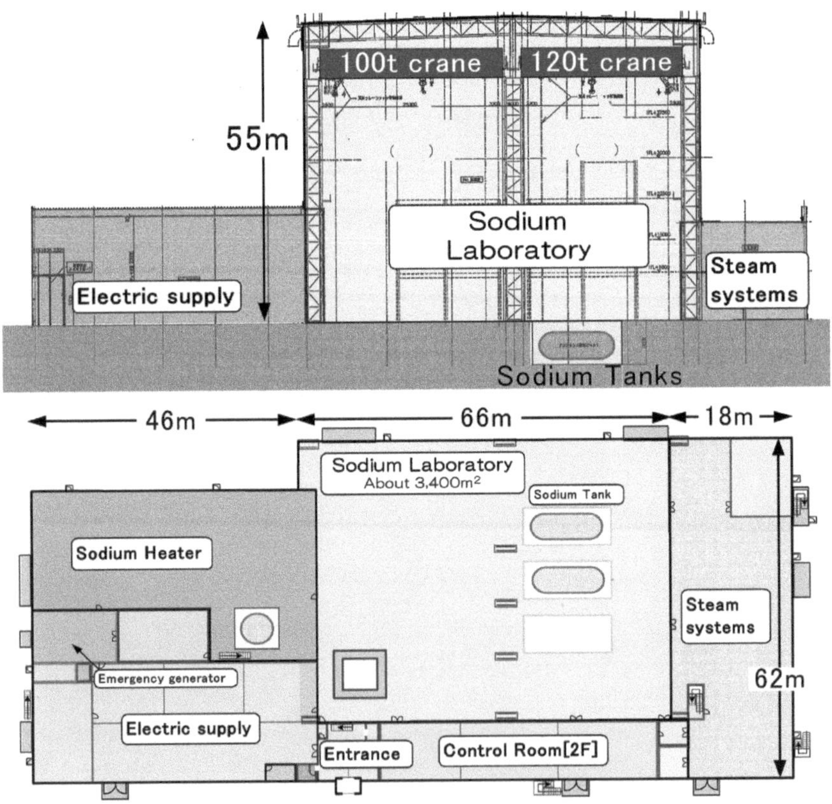

Fig. 5.45 Layout of AtheNa Facility building.

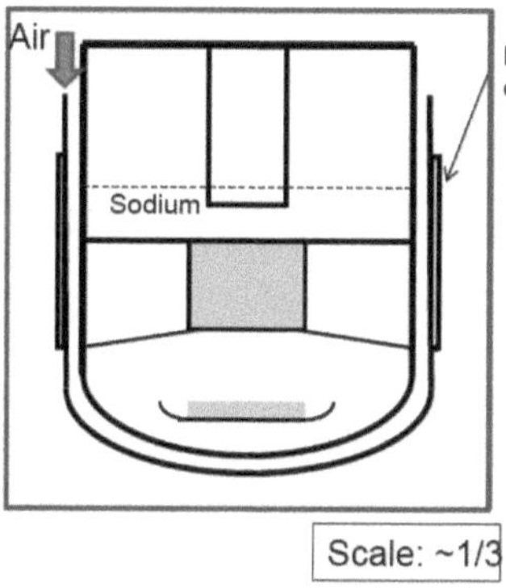

Fig. 5.46 Experiments for severe accident measures.

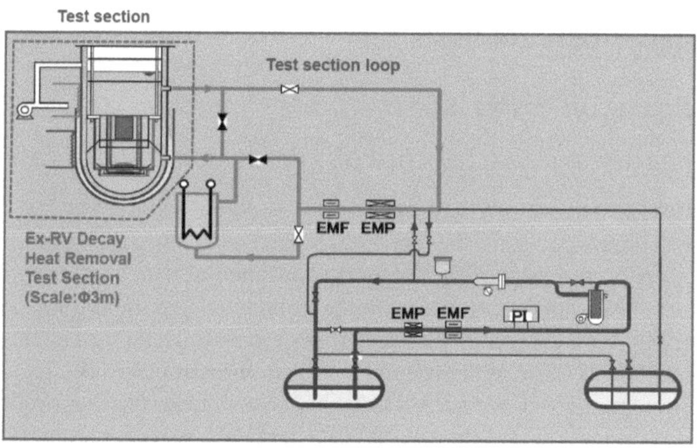

Fig. 5.47 Overview of external vessel cooling test.

After the 1F accident, international collaboration has become more important because safety design criteria/measures should be confirmed with objectivity and cross-checked. An international R&D collaboration on a decay heat-removal system under severe situations has been proposed and discussed in the GIF framework as the first experimental series in the AtheNa Facility (Fig. 5.47).

5.2.5 Current status

After the 1F accident, component development tests and system and component demonstration tests were suspended. However, construction of the AtheNa Facility was continued. A mother loop including sodium tanks was completed and 240 tons of sodium was charged. In the mother loop, the performances of the sodium purification system and monitoring system, including flowmeters, were tested. For various component tests, a sodium heater with a 60-MW capacity is under construction. The following items are being discussed for future test plans:

- Support for fast reactor development in industries (various component development tests for which industries are under development),
- International collaboration,
  - Experiments for severe accident measures as discussed in the GIF framework (multilateral coloration),
  - A bilateral international collaboration between France and the United States (common component development or demonstration tests on provided components),
- Common technology development for fast reactors (thermal-hydraulics tests of steam generator tubes and reactor vessel plug performance tests [thermal isolation, sodium vapor accumulation prevention, and cover gas seal]).

5.3 Reactor core physics

5.3.1 Calculation codes and methods

5.3.1.1 Features of reactor core physics calculation for fast reactors

The standard reactor core physics (neutronics) analysis method for fast reactors in Japan has been developed since the late 1980s based on two research aspects: the analysis and evaluation research of the integral experimental data acquired in the JUPITER program [148–150] that was jointly conducted by Japan and the United States, and the application research of the cross-section (nuclear data) adjustment theory [151] that reflects these experimental results in the neutronics design of fast reactors to improve prediction accuracy [152]. For this reason, this neutronics analysis method was originally called the "JUPITER standard analysis method." Currently, the neutronics analysis method is used not only for the JUPITER experimental analyses, but also for the analyses of the other experiments in zero-power critical facilities, and start-up tests and/or operation data in power fast reactors such as Joyo and Monju.

This Japanese standard neutronics analysis method for fast reactors is based on the so-called "two-step calculation" scheme consisting of (1) the homogenization calculation in the fuel assembly, so-called cell calculation and (2) core calculation [153]. In fast reactors, since the average energy of neutrons is high and its mean free path is long, the heterogeneity of the fuel assembly is relatively small. However, it is known that the heterogeneity effect of the wrapper tube structure of a subassembly and the fuel pin arrangement inside the wrapper tube, which is called the "double heterogeneity effect," cannot be ignored [154]. In addition, since the long mean free path forms the core-wide neutron flux distribution, the core calculation is indispensable.

To apply the cross-section adjustment theory, it is necessary to quantitatively evaluate the uncertainty induced by nuclear data, the uncertainty caused by experiments, and the uncertainty related to analytical models. For this reason, uncertainty quantification is one of the important development items. Furthermore, sensitivity analysis of nuclear data is also required, and the development of the standard neutronics analysis method requires research and development across many fields, including the preparation and processing of the nuclear data, its uncertainty, and the development of neutronics analysis codes.

Computational efficiency is an essential requirement because it is intended for application to design analysis. Therefore, the standard analysis method has a basic concept of dividing into base and correction calculations [155]. In the base calculation, a simplified analytical model with practical accuracy is used while emphasizing the calculation efficiency. In the correction calculation, detailed analysis models are adopted to calculate the coefficients for correcting the base calculation, and correction is made for all possible analytical model uncertainties. It has the advantage of shortened calculation time by alleviating the need to repeat detailed-model calculations. It is still effective as a realistic approach even now that computing power has been greatly improved compared to the beginning of development. This concept of base and correction calculations is developed for static nuclear characteristics without burnup calculation, but it can be extended to burn-up nuclear characteristics [156].

5.3.1.2 Modules and methods

This section describes the main features of the cell and core calculations in the two-step scheme. In the cell calculation, the ABBN-type cross-section set [157] is used in principle. The collision probability method is used in the cell calculation, and the Tone method [158] is adopted to evaluate the self-shielding factor. In the Tone method, the background cross section of the nuclide n for the gth energy group is given by

$$\sigma_0^n(g) = \frac{\sum_{m \neq n} \sum_j N_j^m \sigma_j^m V_j P_{ji}(g)}{\sum_j N_j^n V_j P_{ji}(g)} \tag{5.3}$$

where P_{ji} is the probability that a neutron that collides with a nuclide in region j will collide in region i, N_j^m is the atomic number density of nuclide m in region j, and σ_j^m is the microscopic total cross section of nuclide m in region j.

At the beginning of development, the resonance interference effect was considered by R parameter [159]. An extended application to fast reactors of the ultrafine group calculation was proposed [160]. We have developed the ultrafine group cell calculation code, SLAROM-UF, and the ultrafine group reactor constant set, UFLIB, to consider the resonance interference effect [161]. SLAROM-UF can calculate effective cross sections in ultrafine groups of about 100,000 below 50 keV and in fine groups (70, 175, or 900) above the energy. In the ultrafine group calculation, the slowing down equation is solved using Kier's recursive formula [162].

At the beginning of development, a two-step method was used to evaluate the double heterogeneity of the fuel assembly. Currently, a one-step method is made possible by modeling all the structures in the fuel assembly in a one-dimensional ring model

[163]. Since, with this method, double heterogeneity can be evaluated with a single cell calculation, no correction calculation using core calculation is required.

As for control rod homogenization, the reaction rate ratio preservation (RRRP) method [164] is adopted instead of the homogenization method using the neutron flux weight. In the RRRP method, the homogenized cross sections are defined to preserve the reaction rate ratio between the control rod and fuel regions in a supercell model before and after homogenization. In other words, the homogenized cross sections are calculated by the following:

$$\Sigma_{absorber}^{hom} = \frac{\int_{absorber} \Sigma^{het}(r)\phi^{het}(r)dr}{\int_{fuel} \Sigma(r)\phi^{het}(r)dr} \cdot \frac{\int_{fuel} \Sigma(r)\phi^{hom}(r)dr}{\int_{absorber} \phi^{hom}(r)dr} \tag{5.4}$$

where Σ is the macroscopic cross section, ϕ is the neutron flux, r represents the region, and the superscripts of "het" and "hom" mean before and after homogenization, respectively. In recent years, we have also improved the application of the RRRP method. By refining the boundary conditions, the control rod worth and the radial power distribution can be evaluated more accurately [165].

In the core calculation, a 70-group neutron diffusion calculation based on the finite difference method is used as the base calculation. As described before, the uncertainty due to the analytical model is reduced as much as possible by adding various corrections to the result of the base calculation. The correction factors applied as standard include those for the transport effect, spatial mesh, ultrafine group, and anisotropic diffusion coefficient.

Regarding the transport effect correction, the difference between diffusion and transport calculations is corrected using the neutron transport calculation code based on the discrete ordinate method (S_N method). The spatial mesh effect is known to be canceled out with the S_N order effect. This fact is used to simplify the correction calculation for the transport and mesh effects. Specifically, when we evaluate the cancellation between the spatial mesh effect and the S_N order effect for each reactor system, we need to determine a core-specific set of coarse calculation conditions (e.g., the angular discretization S_4 and the spatial mesh of 5 cm) that are equivalent to the finest calculation conditions (i.e., an infinite number of angular discretization and an infinitesimal spatial mesh). Using this set of coarse calculation conditions, the transport and mesh effects can be corrected simultaneously.

Regarding the ultrafine group correction, the correction factor for the energy group effect is calculated by the combined use of the cell calculation with the ultrafine groups and the core calculation with the 175 groups.

Regarding the anisotropic diffusion coefficient correction, Benoist's anisotropic diffusion coefficient [166] is adopted to consider the anisotropy of the neutron current in the multigroup neutron diffusion calculation in the base calculation. By combining the base calculation using the anisotropic diffusion coefficient and the correction calculation using the S_N transport calculation, both the anisotropy of the neutron current and the neutron transport effect can be considered. We have also proposed a new method to directly incorporate the anisotropic effect into discretized transportation theory without using a diffusion calculation [167].

The S_N transport calculation code required for the transport correction has been developed in Japan. Specifically, three-dimensional Cartesian geometry (X-Y-Z) finite difference S_N transport code, TRITAC [168], three-dimensional hexagonal geometry (Hex-Z) nodal S_N transport code, NSHEX [169], and three-dimensional hexagonal/triangular geometry (Hex-Z/Tri-Z) finite difference S_N transport code, MINIHEX/MINISTRI [170,171], have been developed.

For calculating reactivity in the core calculation, the multigroup perturbation calculation codes PERKY [172] and SNPERT/SNPERT-3D [173] are available. Using these codes, it is possible to apply the first-order and exact perturbation theory, which are used in the analysis of Doppler and sodium void reactivity. In the analysis of the sodium void reactivity, it is important to evaluate the leakage term and the nonleakage term separately based on the perturbation theory.

The exact perturbation theory used to calculate reactivity is well known in the field of reactor physics [174]. Operator A represents the transport, absorption, and scattering; operator F represents the fission in the neutron transport equation. The perturbed operators A' and B' are defined as $A' = A + \delta A$ and $F' = F + \delta F$, respectively. The effective multiplication factors before and after the perturbation are represented by k and k', and the corresponding neutron flux is represented by w_k and $\phi_{k'}$, respectively. The reactivity is expressed as follows:

$$\rho = \frac{1}{k} - \frac{1}{k'} = \frac{\left\langle \phi_k^* \left(-\delta A + \frac{1}{k'} \delta F \right) \phi_{k'} \right\rangle}{\left\langle \phi_k^* F \phi_{k'} \right\rangle} \tag{5.5}$$

where the superscript * represents the adjoint flux, and < > means the integration over space and energy.

To analyze the nuclide transmutation caused by neutron irradiation and decay, i.e., burnup calculation, the Bateman equation can be solved using the matrix exponential method. However, in this method, it is known that the calculation time increases when a nuclide with a short half-life time is included in the burnup chain. For this reason, we are using the ORIGEN2 code, which can effectively solve the Bateman equation, and developing the ORIGEN2 libraries for fast reactors that reflect the data from the Japanese evaluated nuclear data libraries ORLIB [175–177]. On the contrary, a numerical solution method for efficiently solving the burnup equation, for instance, Chebyshev rational approximation method (CRAM) [178], has recently been proposed. This numerical solution method is introduced as a burnup equation solver [179]. M represents the transition matrix of the burnup equation. The nuclide concentration vector at time t, $n(t)$, which is the solution of the burnup equation, can be expressed by the matrix exponential of Mt. Furthermore, in CRAM, the matrix exponential is approximated as follows:

$$n(t) = \exp(Mt)\,n(0) \approx \alpha_0 \prod_{i=1}^{k}\left(n(0)+ 2\,Re\left(\alpha_l(Mt-\theta_l I)^{-1} n(0)\right)\right) \quad (5.6)$$

where k is the approximation order of CRAM, α_l and θ_l are the coefficients of the lth order, and I is the identity matrix.

5.3.1.3 Cross-section sensitivity analysis

As mentioned earlier, cross-section sensitivity analysis is an important technique in the development of the standard neutronics analysis method for fast reactors from the viewpoint of applying the cross-section adjustment theory. For the cross-section adjustment method, we adopt the formulation derived by Takeda et al. based on Bayes' theorem [151]. The adjusted cross-section set and its covariance matrix are expressed using vectors and matrices, which have different dimensions depending on the number of cross sections to be adjusted n_a and the number of integral experiments n_e, in the form:

$$\boldsymbol{T}' = \boldsymbol{T}_0 + \mathbf{MG}^T\left(\mathbf{GMG}^T + \mathbf{V}_e + \mathbf{V}_m\right)^{-1}\left(\boldsymbol{R}_e - \boldsymbol{R}_c(\boldsymbol{T}_0)\right) \quad (5.7)$$

$$\mathbf{M}' = \mathbf{M} - \mathbf{MG}^T\left(\mathbf{GMG}^T + \mathbf{V}_e + \mathbf{V}_m\right)^{-1}\mathbf{GM} \quad (5.8)$$

where $\boldsymbol{T}_0$ is an $(n_a \times 1)$ vector that represents the cross-section set before adjustment, $\boldsymbol{R}_e$ is an $(n_e \times 1)$ vector that represents the measurement values of the integral experiments, and $\boldsymbol{R}_c(\boldsymbol{T}_0)$ is an $(n_e \times 1)$ vector that represents the calculation value of the integral experiments obtained by using the cross-section set $\boldsymbol{T}_0$. $\mathbf{G}$ is an $(n_e \times n_a)$ matrix that represents the sensitivity coefficients of the integral experiments with respect to the cross-section set. $\mathbf{M}$, $\mathbf{V}_e$, and $\mathbf{V}_m$ are $(n_e \times n_a)$, $(n_e \times n_e)$, and $(n_e \times n_e)$ matrices that represent the covariance matrix of the cross-section set before adjustment, the covariance matrix with respect to the experimental uncertainty, and the covariance matrix with respect to the analysis method uncertainty, respectively.

The sensitivity coefficient can be calculated directly by perturbing the cross section. However, since it is necessary to prepare a number of sensitivity coefficients for all nuclides, reactions, and energy groups, the direct calculation requires significant calculation time. In addition, as the number of integral experiments increases, the number of sensitivity calculations also increases.

In the calculation of the sensitivity coefficient for static nuclear characteristics without burnup calculation, the sensitivity analysis code SAGEP [180] based on the generalized perturbation theory [181] is used. The sensitivity coefficient is defined as

$$G = \frac{dR}{R} \Big/ \frac{d\sigma}{\sigma} \quad (5.9)$$

where R represents nuclear characteristics and σ represents cross section. Nuclear characteristics such as the reaction rate ratio and distribution are expressed in the form of the ratio of the product of cross sections Σ_i ($i = 1, 2$) and neutron flux ϕ as

$$R = \frac{\langle \Sigma_1 \phi \rangle}{\langle \Sigma_2 \phi \rangle} \tag{5.10}$$

where < > represents the integration of space and energy. In the generalized perturbation theory, the sensitivity coefficient for the nuclear characteristics expressed in the aforementioned form is written as follows:

$$G = \left(\frac{\left\langle \frac{d\Sigma_1}{d\sigma} \phi \right\rangle}{\langle \Sigma_2 \phi \rangle} - \frac{\left\langle \frac{d\Sigma_2}{d\sigma} \phi \right\rangle}{\langle \Sigma_2 \phi \rangle} - \left\langle \Gamma^* \frac{dB}{d\sigma} \phi \right\rangle \right) \sigma \tag{5.11}$$

where B is the forward operator of the neutron transport equation, and Γ^* is the generalized adjoint neutron flux. Since Γ^* is independent of nuclides, reaction types, and energy group, it only needs to be calculated once. Therefore, the sensitivity coefficient can be calculated efficiently.

Nuclear characteristics such as multiplication factor and reactivity are expressed in the form of the ratio of the product of operators H_i ($i = 1, 2$), the neutron flux ϕ, and the adjoint neutron flux ϕ^* as follows:

$$R = \frac{\langle \phi^* H_1 \phi \rangle}{\langle \phi^* H_2 \phi \rangle} \tag{5.12}$$

where < > represents the integration of space and energy. The sensitivity coefficient for the nuclear characteristics expressed in the form explained earlier is written as follows:

$$G = \left(\frac{\left\langle \phi^* \frac{dH_1}{d\sigma} \phi \right\rangle}{\langle \phi^* H_1 \phi \rangle} - \frac{\left\langle \phi^* \frac{dH_2}{d\sigma} \phi \right\rangle}{\langle \phi^* H_2 \phi \rangle} - \left\langle \Gamma^* \frac{dB}{d\sigma} \phi \right\rangle - \left\langle \Gamma \frac{dB}{d\sigma} \phi^* \right\rangle \right) \sigma \tag{5.13}$$

where Γ is the generalized neutron flux. Similarly, the generalized neutron flux needs to be calculated once.

To calculate the sensitivity coefficient for nuclear characteristics that require burnup calculation, the burnup sensitivity analysis code SAGEP-BURN [182] based on the depletion generalized perturbation theory is used. Burnup nuclear characteristics can be generally expressed as the time integral

$$R = \int_0^{t_{l+1}} dt \langle R \rangle \tag{5.14}$$

where t_{I+1} is the end of the burnup period [182,183]. The sensitivity coefficients for the burnup nuclear characteristics are written as follows:

$$G = \frac{\sigma}{R}\left(\begin{array}{c} \sum_{i=1}^{I}\left\langle \int_{t_i}^{t_{i+1}} dt \frac{\partial R}{\partial \sigma} \right\rangle + \sum_{i=1}^{I}\left\langle \int_{t_i}^{t_{i+1}} dt N^* \frac{\partial M}{\partial \sigma} N \right\rangle \\ + \sum_{i=1}^{I+1}\left\langle \Gamma_i^* \frac{\partial B}{\partial \sigma} \phi_i \right\rangle + \sum_{i=1}^{I+1}\left\langle \Gamma_i \frac{\partial B_i^*}{\partial \sigma} \phi_i^* \right\rangle - \sum_{i=1}^{I+1}\left\langle P_i \frac{\partial P_i}{\partial \sigma} \right\rangle \end{array} \right) \quad (5.15)$$

where ϕ is the neutron flux, ϕ^* is the adjoint neutron flux, N is the nuclide concentration, M is the transposition matrix of the burnup equation, B^* is the adjoint operator of the neutron transport equation, and P is the operator to evaluate the reactor power. The subscript i represents the time step.

For the sensitivity coefficient of temperature-dependent nuclear characteristics such as Doppler reactivity, a method for calculating the sensitivity coefficient for the gradient of the self-shielding factor is used [184]. Doppler reactivity is defined as follows:

$$R = \frac{1}{k_{low}} - \frac{1}{k_{high}} \quad (5.16)$$

where k_{low} and k_{high} represent the effective multiplication factors at a low and a high temperature. As a pseudocross section, a new parameter regarding the temperature T and the self-shielding factor f is defined as follows:

$$\alpha = \frac{1}{f_{low}}\left(\frac{df}{dT}\right) \quad (5.17)$$

where f_{low} is the self-shielding factor at a low temperature. The sensitivity coefficient of Doppler reactivity with respect to α can be calculated by the following:

$$G_\alpha = \frac{dR}{R} / \frac{d\alpha}{\alpha} = \left(\frac{\sigma_{high} - \sigma_{low}}{\sigma_{high}}\right) \cdot \frac{1}{R} \cdot \frac{G_{k,high}}{k_{high}} \quad (5.18)$$

where σ_{high} and σ_{low} represent the effective cross sections at a low and a high temperature, respectively. $G_{k,high}$ is the sensitivity coefficient of the effective multiplication factor at a high temperature [184].

On the contrary, the application of the cross-section adjustment method requires not only the sensitivity coefficient, but also the covariance matrix of the nuclear data. The Japanese evaluated nuclear data library JENDL-4.0 [185] provides covariance data necessary for fast reactor analysis. These covariance data are converted into the multigroup format and applied to the cross-section adjustment method and uncertainty quantification. For the nuclear data processing required for this multigroup conversion, a nuclear covariance processing code ERRORJ [186] was developed based on the ERRORR module of the NJOY code [187]. The new features of ERRORJ were then integrated into the original NJOY version [188]. As for the covariance data of the temperature-dependent nuclear characteristics, the ERRORF code is used [189].

5.3.1.4 Integrated code system

When using the analysis codes described earlier, they must be connected to each other. At the beginning of development, an interface program that sequentially converts the format of input/output files was used. However, with the increase in the number of analysis codes and calculation procedures, this interface program became extremely complicated. Similar issues have arisen in other fields of fast reactor development as well. In 2002, researchers in the relevant fields began to study the ideal form of a next-generation analysis code system [190]. From the results of this study, the concept of Engineering Modeling Language (EML), which is an analysis system as a collection of modules with physical meaning, was proposed. Furthermore, in order to realize this concept, the utilization of object-oriented technology that promotes modularization and reuse and a scripting language [191] that flexibly describes how to combine modules were proposed. In response to this proposal, the object-oriented scripting language Python was adopted, and the integration of the nuclear characteristic analysis system for fast reactors was promoted. The reasons for the adoption of Python are that it has a very simple syntax compared to other object-oriented scripting languages and that it has a relatively large number of applications to scientific computing.

In the first development step, an integration test of SAGEP-BURN [192], which was a very complicated analysis system, was conducted to evaluate the applicability of these technologies. Based on the experience of this integration test, the development of MARBLE [193], which integrates the nuclear characteristic analysis system for fast reactors, was started. MARBLE includes functionality for linking existing analysis codes and can be used as a development framework for a code linkage system [194]. In addition to reusing existing analysis codes, new solvers are also being developed by utilizing Python numerical calculation libraries, NumPy [195] and SciPy [196]. Regarding the cross-section adjustment method [197] and the burnup calculation [179], it is possible to develop a practical solver based on Python and these numerical libraries. Prototypes of neutron diffusion solver, perturbation solver, and sensitivity solver [198] including burnup sensitivity [199] have been developed in Python. Currently, these solvers run slower than the existing codes developed in FORTRAN. However, this is mainly due to the lack of development experience. By using Python and its excellent numerical libraries, the solvers can be implemented in a very short and readable program. In particular, the implementation of the numerical solution formulated in the matrix form becomes almost the same as the theoretical formula, which is very useful in improving the numerical solution. Easy maintenance ability and extensibility of the solvers are also very important advantages. The development of a numerical solver using Python suggests an ideal way for future solver development.

5.3.2 Experimental validation database

This section gives an overview of the integral experiment database for fast reactors. This integral experiment database is necessary not only for the application of the cross-section adjustment method, but also for the validation of the analysis code system [200]. Furthermore, the integral experimental database is used as a benchmark for nuclear data libraries [201]. Development of the experimental database was started from the analysis and evaluation of the experimental data measured in the JUPITER program, which is a series of critical experiments for large fast breeder reactors jointly conducted by Japan and the United States from 1978 to 1988. Comprehensive analysis and evaluation results of major nuclear characteristics measured in the JUPITER program were reported in 1996 [155]. Since then, the database has been expanded to include many other critical experimental data. It should be noted that the performance test data of the prototype fast breeder reactor Monju and the performance test and operation data of the experimental fast reactor Joyo were also included. Table 5.8

Table 5.8 Integral experimental data used for ADJ2017.

Facility	Core name	Features	Parameters
ZPPR	9, 10A–10C	600–800 MWe-class, two-region homogeneous MOX cores	keff, RR, CRW, SVR, S-DR
	13A	650 MWe-class, radially-heterogeneous MOX cores	keff, RR, CRW, SVR, S-DR
	18A, 18C, 19B	1000 MWe-class, two-region homogeneous MOX cores with enriched-uranium regions	keff, RR, CRW, SVR
ZEBRA	MZA	550-l, one-region MOX core as a clean benchmark	keff, SVR
	MZB, MZC	2300-l, two-region homogeneous MOX cores to simulate the prototype fast reactor Monju	keff, CWR, SVR
Joyo	MK-I	300-l, 50/75 MWt fast reactor with MOX and enriched-uranium fuel with blanket	keff, CWR, SVR, ZMRR, ITC, BRC
	MK-II	240-l, 100 MWt fast reactor core with MOX and enriched-uranium fuel without blanket	MA-PIE
Monju	Start-up tests (1994)	280 MWe prototype fast breeder reactor with two-region homogeneous MOX core	keff, CRW, ITC
	Restart core (2010)	Same as the start-up tests but with ~1.5% of Am-241 produced by beta decay of Pu-241	keff, CRW, ITC

Table 5.8 Continued

Facility	Core name	Features	Parameters
BFS	62-1–62-5, 66-1	3400-l, three- or four-region enriched-uranium and/or MOX-fueled cores with or without radial blankets	keff, RR, CRW, SVR
	67, 69, 66	10 kg of NpO_2-loaded cores in central MOX region with various grade Pu	keff, RR, CRW, SVR
MASURCA	ZONA-2B	380-l, core in the CIRANO experiment series, which aimed at the study of Pu burner cores	SVR, ZMRR
SEFOR	CORE-I, II	20 MWt fast reactor core fueled with MOX and cooled with sodium	W-DR
LANL	FLATTOP-Pu, -25, GODIVA, JEZEBEL, JEZEBEL-Pu240	~10-cm-diameter sphere-shaped cores with metallic fuel consisting of Pu-239, degraded Pu or U-235	keff
FCA	IX-1–7	Systematically changed neutron spectra cores with uranium and diluent (graphite and stainless steel)	TRU fission rate ratio
YAYOI	–	Fast neutron source reactor with two irradiation holes called Glory and Grazing	Np-237 capture rate ratio

keff: Criticality, RR: reaction rate, CRW: control rod worth, SVR: sodium void reactivity, S-DR: sample Doppler reactivity, W-DR: whole-core Doppler reactivity, ZMRR: zone material replacement reactivity, ITC: isothermal temperature coefficient, BRC: burnup reactivity coefficient, MA: minor actinoid, PIE: postirradiation experiment.

shows the integral experimental data used for the development of the latest adjusted cross-section set [202].

The BFS critical experiments obtained through joint research with Russia [203,204] have the second largest number of nuclear characteristics included after the JUPITER experiments. For a unique example, the BFS experiments include a critical experiment loaded with about 10 kg of Np-237 [205]. Another unique example is the evaluation result on the critical experimental data for plutonium burning fast reactor measured in the MASURCA facility in France [206].

The major JUPITER experiments [207] and Joyo MK-I performance test and operation data [208,209] included in the integral experimental database were also registered in the OECD/NEA International Handbook of Evaluated Reactor Physics Benchmark Experiments (IRPhE) [210].

In recent years, integral experimental data related to minor actinides (MA) and higher order plutonium (degraded Pu) have been added. Regarding the MA experiments, postirradiation data of the MA samples irradiated in Joyo MK-II and PFR [211,212], the criticality of Monju core with Am-241 accumulated during the long shutdown from 1995 to 2010 [213,214], and the MA fission rate measured in FCA

IX are included [215]. As for the degraded Pu experiments, degraded Pu sample reactivity of ZPPR, BFS, and ZEBRA, the fission rate ratio of Pu-241 and U-235 of ZPPR, etc., are included.

In addition to confirming the prediction accuracy of the analysis method, we are participating the IAEA neutronics benchmark of the start-up tests for CEFR, which is small-size sodium-cooled fast reactor in China [216]. In this benchmark, we applied the neutronics analysis method developed through these experimental analyses and demonstrated that it can provide excellent prediction accuracy [216].

5.3.3 Data assimilation method

As mentioned earlier, in the development of the neutronics analysis method for fast reactors in Japan, application research of the cross-section adjustment method has been conducted since the early 1990s. As another method for reflecting the integral experimental data into the neutronics design to improve prediction accuracy, there is a (conventional) bias factor method [217], in which the ratio of the experimental and calculation values is used for correction to obtain the design value. In recent years, improvement of the conventional bias factor method has been proposed. At present, both the cross-section adjustment method and the improved bias factor method are recognized as "data assimilation methods."

The concept of cross-section adjustment method [218,219] was first proposed at the 1964 IAEA conference [220]. After that, research on the cross-section adjustment method was actively conducted globally. In 1970, Dragt presented a theoretical formula in matrix form that was almost equivalent to the formula currently used [221]. In Japan, the application of the cross-section adjustment method was promoted based on the theoretical formula derived by Takeda et al. based on Bayes' theorem in 1989 [151]. Based on the cross-section adjustment method, an adjusted cross-section set that reflects the integral experimental data can be created. The use of the adjusted cross-section set for nuclear design calculation improves the prediction design accuracy. Although the adjusted cross-section set is called "tou-gou-ro-josu" in Japanese, or a unified cross-section set in English, in the sense that the integral experimental data and the differential nuclear data are unified, it is often called an "adjusted cross-section set" in English.

Table 5.9 shows the adjusted cross-section sets developed in Japan. As shown in the table, the first version is ADJ91, which was developed in 1991 [153]. The next version ADJ98 [222] only changed the basic nuclear data library from ADJ91, but it was used as the standard cross-section set for nuclear design in the feasibility study on commercialized fast reactor cycle systems (FS) started in 1999. The further next version ADJ2000 [184], which is an extension of the integral experiment data, was developed in 2000, and its revised version, ADJ2000R [223], was developed in 2002. Both were continuously used in the FS. After that, following the release of JENDL-4.0 [185] in 2011, ADJ2010 [224,225] was developed by reevaluating the integral experimental data with JENDL-4.0 and adding new integral experimental data. ADJ2010 has been used as the standard cross-section set for nuclear design in the Fast Reactor Cycle Technology (FaCT) development project and the succeeding fast reactor projects.

Table 5.9 Features of adjusted cross-section sets developed in Japan.

	ADJ91/98	ADJ2000	ADJ2010	ADJ2017
Basic library	JENDL-2/3.2 (1989/1994)	JENDL-3.2 (1994)	JENDL-4.0 (2010)	JENDL-4.0 (2010)
Number of adjusted parameters	σ_∞: 32/37 χ: 2 β: 6	σ_∞: 41 χ: 2 β: 6 SSF: 1	σ_∞: 155 χ: 2 β: 11 SSF: 1 PFP: 4	σ_∞: 179 χ: 19 β: 20 SSF: 1 PFP: 4
Number of energy groups	18 (70)[1]	18 (70)[1]	70	70
Covariance of nuclear data	Rough estimation	JENDL-3.2[2]	JENDL-4.0	JENDL-4.0
Number of integral experimental data	82/185 (ZPPR only)	237 (+ FCA, Joyo, BFS, MASURCA and LANL)	488 (+ ZEBRA, SEFOR, Monju (1994), Joyo (MA), and BFS (Np))	620 (+ ZPPR (D-Pu worth, F41/F25), BFS (D-Pu worth), Monju (2010), ZEBRA (D-Pu fission), FCA (MA fission), Joyo (MA (D-Pu ratio)), PFR (MA sample), YAYOI (Np capture))

1: Adjusted in 18 groups and expanded to 70 groups; 2: the covariance data were evaluated after the release of JENDL-3.2; σ_∞: infinitely dilute cross section, χ: fission spectrum, β: delayed neutron fraction, SSF: self-shielding factor, PFP: pseudofission product, D-Pu: degraded plutonium.

In the development of ADJ2010, significant improvement was made to the method of evaluating uncertainty and correlation coefficient, i.e., covariance data, of the integral experiments. Specifically, the "error factor correlation method" was newly proposed as a new methodology for the quantification of the correlation coefficient between the integral experimental data [226]. This methodology has been adopted in an OECD/NEA report [227] as a standard method for evaluating the correlation coefficient between the integral experimental data when applying the cross-section adjustment method. The latest adjustment cross-section set is ADJ2017 [202]. In the development of ADJ2017, the aforementioned integral experimental data regarding MA and degraded Pu were added to improve the design accuracy of the core loaded with MA and/or degraded Pu.

As a recent international research trend, research on the cross-section adjustment method has been active since the mid-2000s. One reason is that the importance of the cross-section adjustment method in nuclear design was pointed out in the activity of Sub Group (SG)26 (2005–2009) [228] by the OECD/NEA Working Party on International Nuclear Data *Co*-operation (WPEC). Since then, SG33 (2009–2013) [227], SG39 (2013–2018) [229], and SG46 (2018-) [229] have been approved by the OECD/NEA WPEC, and the activities related to the cross-section adjustment method are ongoing.

In Japan, the generalized bias factor method [230] was proposed in 2006, and the extended bias factor method [231] was proposed in 2007. Similar to the cross-section adjustment method, these new bias factor methods can improve the accuracy of nuclear design prediction using integral experimental data. In 2012, these methods led to the proposal of the extended cross-section adjustment method that unifies the extended bias factor method and the conventional cross-section adjustment method [232]. In 2016, the conventional and extended adjustment methods were derived based on the same concept as the extended bias factor method, and it was shown that the extended bias factor method and the cross-section adjustment method can be discussed in the same theoretical framework [233]. In 2019, it was shown that these methods can be formulated in a unified manner [234], including another methodology called the dimension-reduced cross-section adjustment method [235]. At present, the cross-section adjustment method is regarded as a kind of data assimilation method. In addition, the extended bias factor method and the cross-section adjustment method are basically discussed in the framework of Bayes' theorem and the minimum variance method, but now, it is shown that they can also be applied in the framework of the random sampling method [236,237].

5.3.4 Future developments

For future development, it will be necessary to upgrade the analysis methods and the simulation V&V (verification and validation) methodology by utilizing the latest research results and achievements. For example, the current two-step calculation method, which consists of the cell calculation and the core calculation, is a realistic option from the viewpoint of design analysis. However, considering the remarkable development of current computer technology, a one-step calculation that directly performs core calculation without homogenization in fuel subassemblies may be feasible in the near future even for design analysis. For instance, a new deterministic method [238] has been proposed that can accurately simulate three-dimensional geometry and solve the neutron transport equations. On the contrary, the role of numerical simulation in nuclear application has been increasing in general, and discussions on methodology to assess its credibility are in progress. In 2015, a standard for guidelines for the assessment of credibility of nuclear simulations was proposed by the Atomic Energy Society of Japan [239]. As mentioned before, we have collected and evaluated a significant number of integral experimental data for fast reactors, which can be used to assess the validity of the calculation methods. However, when designing new types of reactors, it is not always possible to cover the entire applicable range of the calculation methods with experiments. Therefore, an applicable range must be extrapolated by simulations. At present, although it involves statistical uncertainty in solving the neutron transport equation, the Monte Carlo method has been used practically for the verification of calculation results, and data assimilation technology that can improve the simulation accuracy using experimental data has been significantly developed in recent years. For this reason, it is expected that the design prediction values will be verified and validated with a new approach such as by substituting experiments with numerical simulations. Developing these technologies into more reliable and practical technologies will be an important research subject in the future.

5.4 Fuel and materials

5.4.1 Introduction

For practically realizing fast reactors, improvement of economic performance is required in addition to ensuring safety. Two important technological tasks are involved in the improvement of the economic performance: the development of high-burn-up fuel to attain the average burn-up of 150 GWd/t and a high-temperature plant system to attain the coolant outlet temperature of 550 °C [240]. This average burn-up is equivalent to 250 dpa with the maximum neutron dose, and the cladding temperature reaches approximately 700 °C. Oxide dispersion-strengthened (ODS) ferritic steels are being developed as the most promising cladding material for use in such severe environments. In addition, MOX fuel with large-diameter annular pellets is being explored to improve the internal conversion ratio, increase the fuel melting margin, and reduce the fuel-cladding mechanical interaction (FCMI).

Fast reactors can use long-lived and highly harmful minor actinides (MA) contained in high-level radioactive wastes as a fuel and can transmute them to short half-life radionuclides or stable nuclides. Thus, MA-bearing MOX fuel is developed to reduce the volume and decrease the toxicity of radioactive wastes. In addition, the fast reactors cannot only breed plutonium (Pu) but can also burn it depending on the presence or absence of the blanket fuel and Pu content of the core fuel. Therefore, Pu can be stabilized by burning, and the degraded Pu recovered from Pu-thermal fuel can also be used. To burn Pu effectively, MOX fuel with highly increased Pu content (for example, Pu content of 40 wt%) is being developed.

5.4.2 MOX fuel development

Fuel properties play crucial roles in understanding the mechanism of irradiation behaviors and predicting them using simulation techniques. Investigating fuel properties targeting extremely high-temperature region exceeding 2000°C, which, MOX fuels experience in fast reactors, is challenging given the difficulties in Pu handling. Experimental data are emphasized for validating the reliability; the limited number of facilities in Japan and other countries is leveraged for performing the experiments.

Fuel melting has to be avoided as a basic requirement of fuel design. The melting temperature and tools to simulate the temperature profile in fuel pellet, where one-dimensional heat transfer in the radial direction is often considered for cylindrical symmetry, are necessary to demonstrate that the fuel is below the melting temperature during irradiation. The thermal conductivity is the most critical property for determining the temperature profiles. Thus, numerous researchers have studied these properties.

Thermal conduction in MOX usually takes place through phonons and electrons. As MOX temperature increases, conduction by phonons decreases, whereas the conduction by electrons increases [241]. The increase is attributed to the increased concentration of electronic defects, which are quantitatively analyzed by oxygen potential study [242]. The switching of the types of dominant thermal conduction mechanism with increasing temperature results in minimum thermal conductivity in the region of 1300–1800°C. Other than the temperature, the Pu and MA contents and oxygen-to-metal (O/M) ratio are critical parameters that decrease the thermal conductivity, with the consideration of these parameters as impurities incorporated into UO_2 [243–245]. In addition, pores decrease thermal conduction because the gas phase in pores has an extremely low thermal conductivity compared with MOX. Experimental approach for investigating thermal conductivity usually involves the evaluation of thermal diffusivity through laser flash technique, specific heat, and density, followed by the multiplication of the three quantities. Kato et al. proposed a semiempirical model based on experimental data to calculate the thermal conductivity with the parameters of density, Pu and MA contents, and O/M ratio [246]. Fig. 5.48 shows the temperature dependence of thermal conductivity on different MA (Am and Np) contents and O/M ratios calculated using the model. MA fraction of 5% was observed, and the addition of Am, rather than Np, largely decreased the thermal conductivity. The effect of O/M ratio was significant because of the expected dynamic redistribution of O/M ratio along the radial direction of MOX pellets during irradiation [247].

Melting temperature is used as a milestone of fuel design in simulating temperature profile. Because actinide oxides have an extremely high melting temperature, the development of measurement techniques for melting temperature is an important part of the study. Empirically, MOX with high Pu content/O/M ratio is known to be chemically active, which increases experimental errors. Thermal arrest technique was developed for accurately observing thermal arrest in heating MOX specimens and measuring temperature. A tungsten or rhenium capsule is used to encapsulate

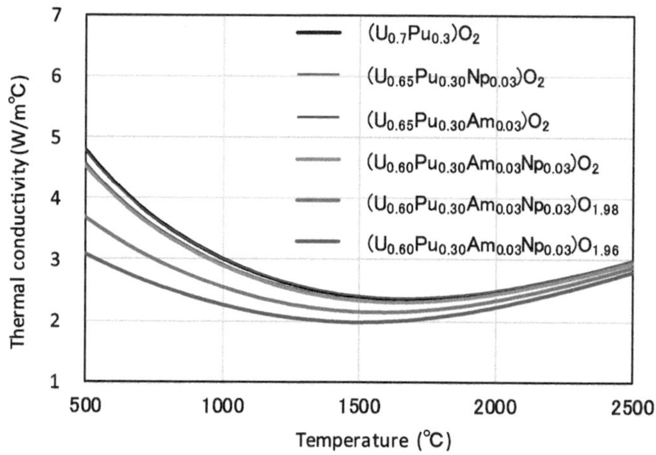

Fig. 5.48 Thermal conductivity with influences of MA addition and change in O/M ratio.

MOX under reduced atmosphere to reduce interactions between the capsule and MOX. Kato et al. measured the melting temperature of MOX with various parameters and reported that the increase in Pu/Am content in MOX decreased the melting temperature [248,249], whereas a decrease in O/M ratio increased the melting temperature. A laser heating method developed at the Institute for Transuranium Elements (ITU) was employed by Bruycker et al.; this method can retain the chemical stability of MOX under container-less conditions through pulse laser heating. They reported that the melting temperature of PuO_2 is 2744°C [250], which is notably higher than the data in the literatures [248,249]. Then, Böhler et al. reassessed the phase diagram of the binary system UO_2-PuO_2 and the influence of O/M ratio [251]. Further, Calabrese et al. proposed the TRANSURANUS model to calculate the melting temperature of MOX by reviewing the existing literatures [252].

Fig. 5.49 shows the relation between oxygen potential and thermoequilibrium O/M ratio. An increase in Pu and MA contents increased the oxygen potential. The effect of Pu and MA addition to increase the oxygen potential presented the following order from largest to smallest: Am (2.5 times larger than Pu), Pu, and Np (0.6 times larger than Pu) [242]. The O/M ratio changes due to the occurrence of oxygen-related defect reactions. Certain defect reactions and the equilibrium constants with Arrhenius-type energies were evaluated from the experimentally obtained relations between oxygen potential and O/M ratio; thus, a model describing the relations was proposed, as shown in Fig. 5.49 [242]. This model is useful in adjusting the O/M ratio of MOX for studying properties such as thermal conductivity and evaluating oxygen partial pressure in pores regarding vapor pressure analysis.

Although the density of as-fabricated fuel pellet is homogeneous, a steep temperature gradient drastically changes the local density and properties. The vaporization and condensation rates of actinide species such as U, UO, UO_2, UO_3, Pu, PuO, and PuO_2 are different between the cold and hot areas in a pore that lies along the temperature gradient in radial direction. Such differences result in the transportation of

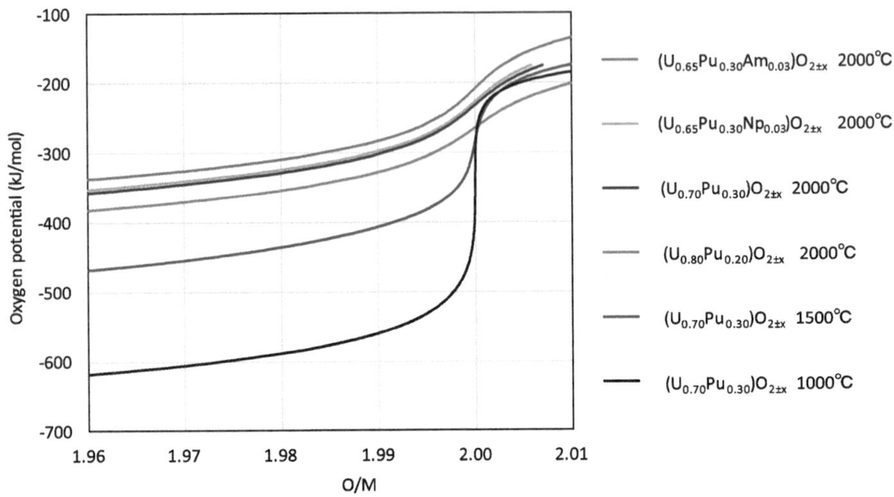

Fig. 5.49 Oxygen potential at 1000°C, 1500°C, and 2000°C with influences of MA addition and change in temperature.

actinide species and changing local density [253]. Thus, the vapor pressure is required to evaluate changes in local density and can be calculated using the Rand-Marking model based on the Clausius-Clapeyron relation and the related Gibbs energies of reaction, e.g., UO_2 to UO_3 [254–256]. As shown in Fig. 5.50, UO_2 and UO_3 are the dominant vapor species. In the calculation of vapor pressure, oxygen partial pressure was used, which can be evaluated from the equilibrium relation of the oxygen potential. Thus, an increase in Pu and MA contents increased the oxygen partial pressure and vapor pressure of UO_2 and UO_3, consequently accelerating the fuel restructuring.

Modeling the properties with crucial parameters is necessary for fuel technology. Such mechanistic models are useful not only for simulating the irradiation behaviors of MOX pellets but also for predicting the properties with uninvestigated parameters. For further improvement of fuel technology, the high-temperature region exceeding 2000°C, as a unique field of MOX fuels for fast reactors, should be explored.

5.4.3 MOX fuel performance

5.4.3.1 Irradiation behavior

Fuel restructuring, as shown in Fig. 5.51, occurs in MOX fuels irradiated with SFR (SFR-MOX). This phenomenon is caused by the higher linear heat rate and steeper temperature gradient due to smaller pin diameter than those of LWR fuel pins. Fuel structure of the irradiated MOX consists of as-fabricated region present in the circumferential region, equiaxed grain region, where grains isotopically grow, and columnar

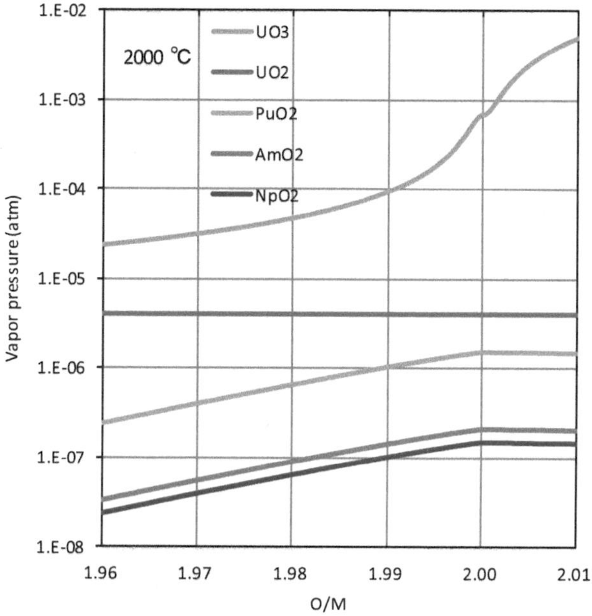

Fig. 5.50 O/M ratio dependence of vapor pressure of $(U_{0.64}Pu_{0.30}Am_{0.03}Np_{0.03})O_{2\pm x}$.

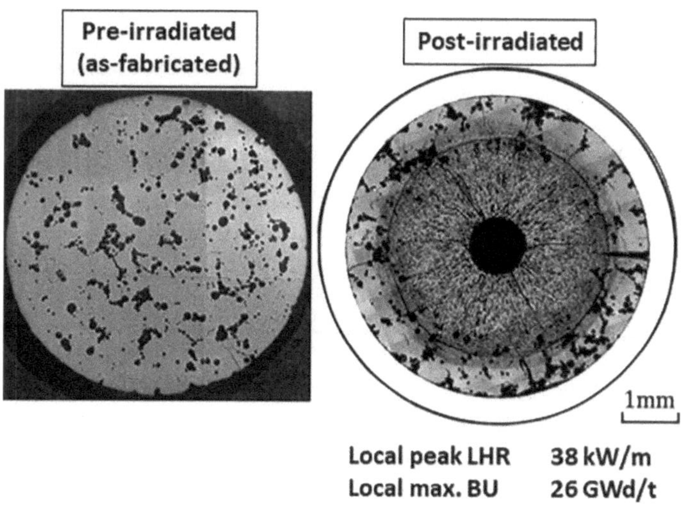

Fig. 5.51 MOX fuel restructuring.

grain region, where as-fabricated pores radially migrate toward the pellet center. The fuel thermal conductivity improves due to densification and central-hole formation caused by fuel restructuring. This improvement allows the mitigation of fuel temperature rise during irradiation. Furthermore, the annular fuels that have an as-fabricated central-hole and high pellet density are expected to be used for implementing higher linear heat rate in the futuristic SFR technology.

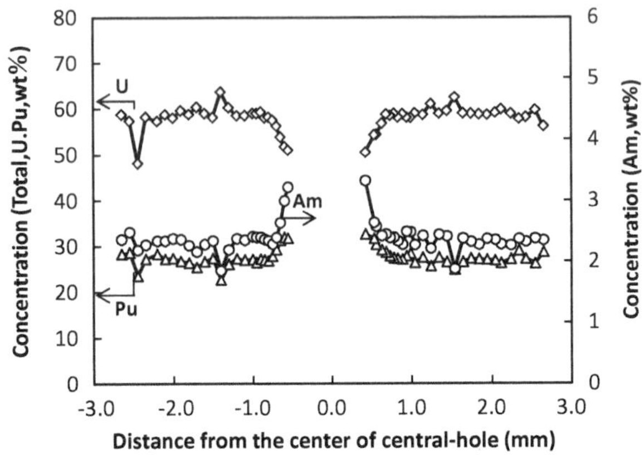

Fig. 5.52 Radial distribution of U, Pu, and Am concentration in postirradiated SFR-MOX.

Pore migration is caused by an evaporation-condensation phenomenon occurring in a pore around the temperature gradient. Plutonium (Pu) and americium (Am) concentration at the pellet center increases due to discrepancies in vapor pressures of each species, which are distributed through the pore migration process in material. Fig. 5.52 shows the radial distribution of concentration of U, Pu, and Am observed in postirradiation examinations (PIEs) for MOX fuels irradiated in the experimental fast reactor, Joyo. The radial redistribution around the temperature gradient is expected to be caused by thermal diffusion-driven vapor-phase and solid-state transportation processes. In particular, vapor-phase transport is dominant at the beginning of irradiation when pores are actively migrating. Fuel restructuring and redistribution behavior are remarkable in MOX with high oxygen-to-metal (O/M) ratio.

As the rates of diffusion and release of fission product (FP) gas, such as krypton and xenon, are accelerated in SFR-MOX fuels having comparably higher burnup and higher fuel temperature than LWR fuels, FP gas release rate of SFR-MOX fuels and the amount of FP gas released to gas the plenum during irradiation are higher than LWR fuels. Fig. 5.53 shows the FP gas release rate of Joyo driver MOX fuels and uninstrumented fuel irradiation subassembly (UNIS) experimental MOX fuels related to the pin-averaged burnup. The FP gas release rate was observed to be ~50% up to 70 GWd/t. However, in high-burnup MOX fuel pins, the FP gas release rate was more than 80% above 100 GWd/t.

Fuel-to-cladding gap width changes due to the deformation of fuel pellet and cladding during irradiation. Fig. 5.54 shows the residual gap width that was normalized by as-fabricated gap size of irradiated SFR-MOX organized by the maximum linear heat rate. Residual gap width decreased with increasing local burnup due to the dominant effect of pellet swelling but enlarged above the local burnup of ~80 GWd/t. This enlargement was caused by pellet shrinkage due to the separation of cesium molybdate (Cs_2MoO_4) over the local burnup of ~80 GWd/t. The JOG formation in fuel-to-cladding gap is expected to mitigate the fuel-cladding mechanical interaction (FCMI)

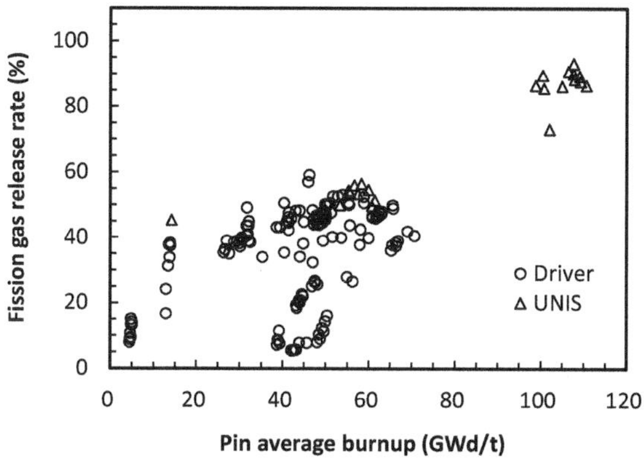

Fig. 5.53 FP gas release rate of SFR-MOX.

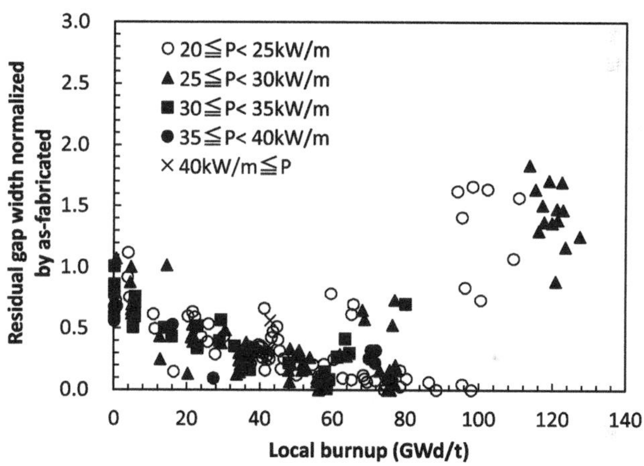

Fig. 5.54 Residual gap width of irradiated SFR-MOX.

and improve heat transfer through a gap. The mechanism of JOG formation should be investigated to ensure the pin integrity up to high burnup.

The stainless steel, which has superior strength at high temperatures, is used as cladding material in SFR because of the high temperatures of cladding and fuels used in SFR. As fast neutron flux in SFR is about one order larger than that in LWR, the radiation damage-resistive ability is required for the SFR core material. Particularly, the material swelling due to fast neutron irradiation has to be restricted. An advanced austenitic stainless steel, PNC316, was developed to acquire superior strength at high temperature and swelling resistance by doping a small amount of phosphorus, niobium, titanium, etc., and 20% cold working. Fig. 5.55 shows the dependence of fast neutron fluence on PNC316's swelling rate. The swelling rate in fuel pin irradiation is greater than that in material irradiation due to the effects of temperature changes and pin internal pressure increase during irradiation.

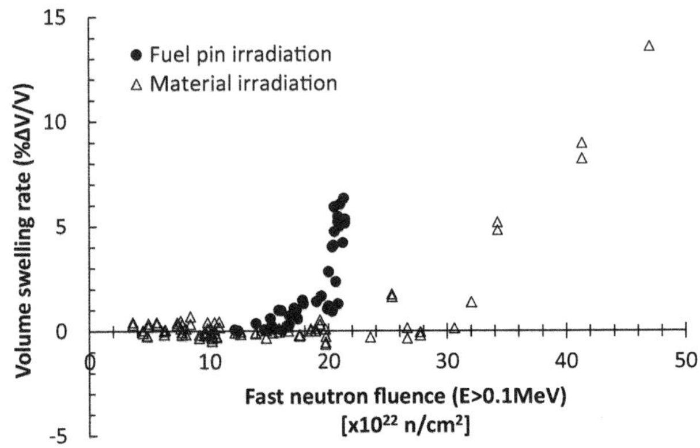

Fig. 5.55 Dependence of fast neutron fluence on PNC316 swelling rate.

5.4.3.2 *Fuel performance code*

As an analytical tool for evaluating the irradiation behaviors of a MOX pellet-type fuel pin, the CEDAR computer code was developed by introducing mechanistic models of irradiation behaviors into the one-dimensional (1D) axisymmetric description of a fuel pin [257]. The code performs thermal and mechanical analyses of the fuel pin irradiation behavior, such as pellet restructuring, fission gas release, fuel pellet gas swelling, and FCMI. CEDAR was coupled with a specialized code for analyzing FP cesium behaviors in a MOX fuel element. The coupled code system allowed the analysis of radial and axial Cs migrations, the generation of Cs chemical compounds, and fuel swelling due to Cs-fuel reactions in association with the thermal and mechanical behaviors of the fuel element. The coupled code analysis was applied to analyze the irradiation performance of SFR-MOX attaining high burnup. The analysis facilitated discussion of the axial distribution of Cs, FCMI resulting from the Cs-fuel reactions, and Cs-Mo-O compound formed in the fuel-to-cladding gap, as shown by the comparison of calculated results with experimental data obtained from PIEs [258]. CEDAR is currently used as a part of the integrated code system consisting of irradiation behavior analysis of fuel pins, a fuel pin bundle, and coolant thermal-hydraulics in a fuel subassembly [259]. In the system, each code provides data required by other codes, and the analyzed results are shared among them. The system allows for the analyses of thermal, chemical, and mechanical behaviors in a fuel subassembly. The system can provide transverse cross-sectional images of the fuel subassembly and three-dimensional (3D) images of a fuel pin, as shown in Figs. 5.56 and 5.57, to visualize the calculation results.

Fig. 5.58 shows the outline of CEPTAR code [260] that has been developed for application in annular fuel pin design up to high burnup by improving CEDAR. This code calculates the radial temperature distribution in fuel and cladding by solving 1D heat conduction equation and stress-strain in fuel and cladding using the generalized plane strain analysis method. Based on mass conservation law, radial pore migration and radial redistribution of Pu, Am, and oxygen during irradiation, which are specific behaviors in SFR-MOX, can be calculated. The radial migration of pore and materials

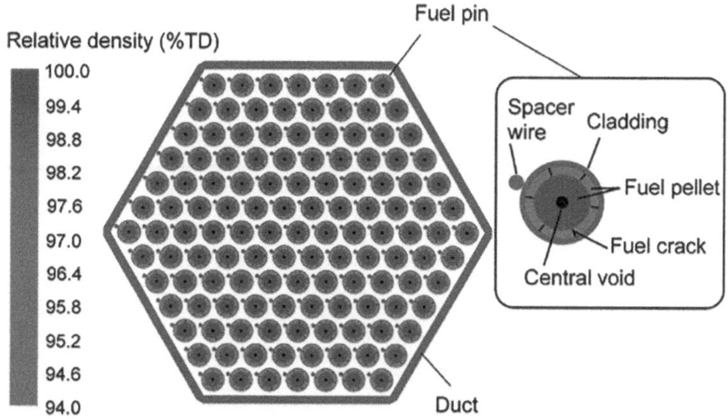

Fig. 5.56 Transverse cross-sectional image of fuel subassembly for fuel pellet density distributions.

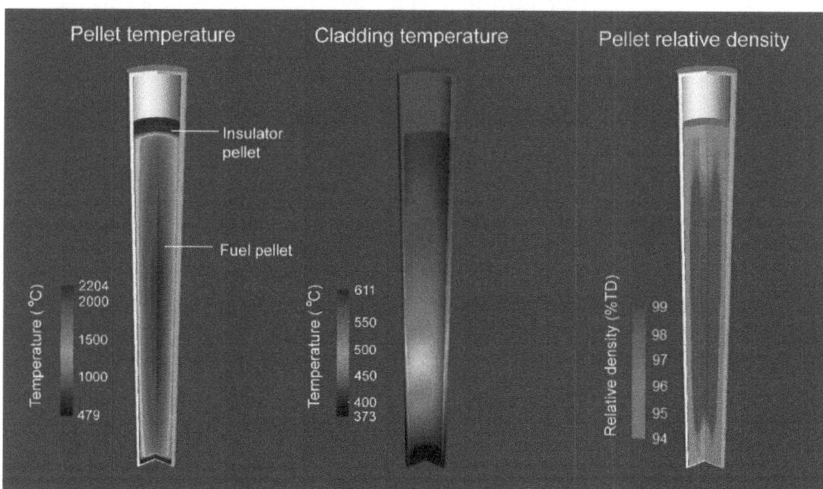

Fig. 5.57 3-D images of a fuel pin model (pellet and cladding temperatures and pellet theoretical density.

is calculated using the vapor pressure applied in Rand-Markin's model based on fuel composition, O/M ratio, and temperature. The heat transfer in a fuel-to-cladding gap, called the gap conductance, is calculated from the sum of conductions due to heat diffusion in mixture gas, face-to-face radiation, and solid-solid contact. The thickness of the JOG layer, which forms in fuel-to-cladding gap of high-burnup SFR-MOX, is calculated from fuel's outer temperature, O/M ratio, and the amount of Cs in the outermost mesh of fuel pellets. The effect of JOG on gap conductance can be considered in the thermal analysis for fuel temperature calculation by considering the JOG thermal conductivity with conduction due to heat diffusion in the mixture gas. Furthermore, a

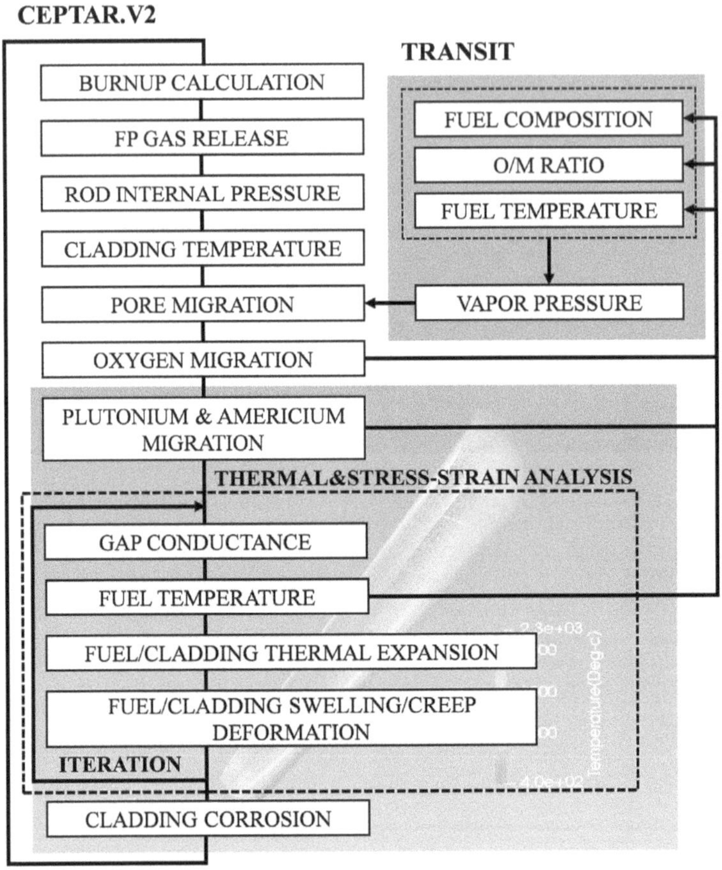

Fig. 5.58 Outline of CEPTAR code.

fuel pellet is shrunk based on the amount of Cs released from its outer surface. This allows to consider the effect on FCMI mitigation with central-hole shrinkage due to pellet's internal stress in stress-strain analysis. Thus, the CEPTAR has been developed aiming at implementing high-performance SFR-MOX fuel design through a rational analysis for MOX irradiation behavior.

A multidimensional fuel performance analysis code for SFR-MOX was developed under the collaboration of JAEA and INL, which were established by the Civil Nuclear Energy Working Group between Japan and the United States. JAEA has leveraged its experiences with MOX properties and models to extend the fuel performance capabilities using BISON code [261], an engineering-scale fuel performance application developed at INL. MOX fuel properties and models for thermal conductivity [246], oxygen potential [262], and vapor pressure [263] were installed in BISON [263a] to represent the thermal and restructuring behavior of SFR-MOX.

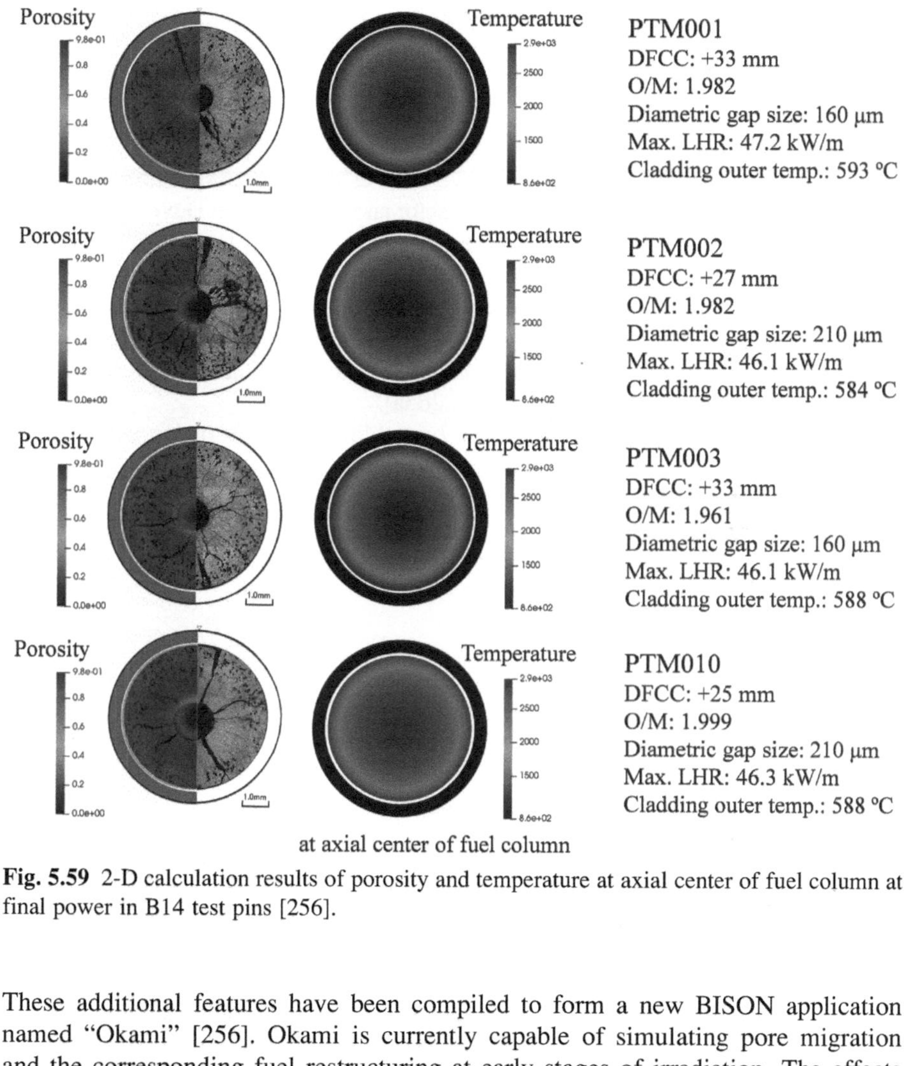

Fig. 5.59 2-D calculation results of porosity and temperature at axial center of fuel column at final power in B14 test pins [256].

These additional features have been compiled to form a new BISON application named "Okami" [256]. Okami is currently capable of simulating pore migration and the corresponding fuel restructuring at early stages of irradiation. The effects of diametric gap size and O/M ratio, observed in PIE for MOX fuels irradiated in B14 test in Joyo, and the influence of pellet eccentricity were compared with the two-dimensional (2D) Okami calculations (Fig. 5.59). Okami successfully accounts for the effects of the gap width and O/M ratio on central void formation by considering the O/M influence on vapor pressure in calculating pore migration velocity. In addition, the models for the redistribution of Pu, MA, and oxygen, which is another specific behavior of SFR-MOX, should be applied to fuel performance analysis at the early stage of irradiation. In the next phase, the inclusion of a fission gas release model [264], mechanical models for pellet deformation and burnup-dependent properties should be developed to apply fuel performance analyses in long-term irradiation.

5.4.4 Core material development

5.4.4.1 Modified-type 316 stainless steel

Type 316 stainless steels have been used as core material (fuel cladding tubes, wrapper tubes, etc.) of fast reactors (FRs) in the world owing to their good high-temperature strength and corrosion resistance; they have been established as structural materials based on industrial achievements. Power Reactor and Nuclear Fuel Development Corporation (PNC, current: Japan Atomic Energy Agency (JAEA)) developed PNC316 as a modified-type 316 stainless steel for the core material. PNC316 was designed in consideration of void swelling resistance and high-temperature strength. Several techniques, including 20% cold work inducing dislocations and micronizing carbides due to addition of phosphorus, boron, titanium (Ti) and niobium (Nb), were applied to PNC316 to enhance its advantages [265,265a].

PNC316 has been applied to Joyo and Monju driver fuels, and several postirradiation examinations (PIEs) have been performed. However, PNC316 has a limit in dimensional changes due to the specific swelling of austenitic matrix, as shown in Fig. 5.60 [266,267].

5.4.4.2 Ferritic steel

Fast demonstration reactors should apply an advanced nuclear fuel that can achieve a high burnup to reduce the cost in the fuel cycle (fabrication, fuel exchange, reprocessing, etc.). The target is 5×10^{27} n/m^{b} (E > 0.1 MeV) corresponding to 250 displacements per atom (dpa) [240]. It was necessary to develop another material superior to PNC316 in terms of swelling resistance.

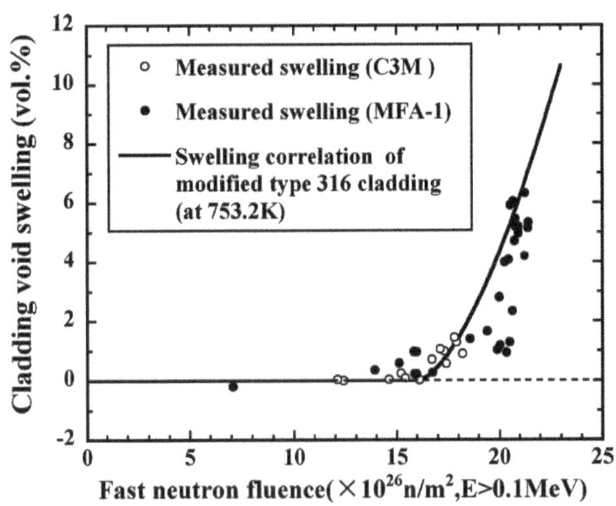

Fig. 5.60 Void swelling of modified-type 316 cladding as a function of fast neutron fluence [267].

Ferritic steels are promising core materials because their swelling resistance is better than that of austenitic steels in general. Dimensional deformation due to the swelling and irradiation creep, which cause extension, bending, and bulging, is a limiting factor in the life duration of the wrapper tube composing the fuel subassembly. Therefore, ferritic-martensitic steels (FMSs), which are superior in dimensional stability against high-dose irradiation, have been developed [268,269].

Nevertheless, FMSs are inferior compared with austenitic steels in terms of high-temperature strength. Thus, the enhancement of high-temperature strength is one of the major issues in the development of FMSs. PNC developed PNC-FMS, and JAEA performed its evaluation. In the development of PNC-FMS, the chromium content was set in the range of 10–12 wt% to avoid decarburization and aging embrittlement. The contents of other additional elements were considered to strengthen the PNC-FMS. Molybdenum (Mo) is a solution-strengthening element, which it stabilizes $M_{23}C_6$-type carbide. However, excess Mo leads to an increase δ ferrite fraction and/or Laves phase precipitation. For the prevention of these events, a part of Mo was replaced with tungsten (W), and 1.5-wt% equivalent Mo content (Mo + 0.5 W) was selected to balance the high-temperature strength and toughness. In addition, minor elements, including 0.2-wt% vanadium (V) and 0.05-wt% Nb to precipitate fine carbide and 0.05-wt% nitrogen (N) to strengthen fine nitrides, were added. Finally, Fe-0.12C-11Cr-0.5Mo-2 W-0.4Ni-0.2 V-0.05Nb-0.05 N was selected as the nominal chemical composition of PNC-FMS [270,271]. Cold work, which introduces dense dislocations to enhance swelling resistance, was not applied to PNC-FMS as opposed to PNC316 because the recovery of the introduced dislocations and recrystallization occurred easily in ferritic steels. Heat treatments for the wrapper tube of PNC-FMS for normalization were performed at a relatively low temperature, i.e., at 1050°C for 10 min and at 700°C for 1 h for guaranteed tensile strength considering certain situations, such as handling of the subassembly and earthquake.

5.4.4.3 Oxide dispersion-strengthened steel

The target temperature at the hot spot of fuel cladding tube of the fast demonstration reactor was 700°C. ODS steels were developed as advanced materials that have superior swelling resistance and high-temperature strength exceeding that of legacy FMSs. The important feature of ODS steels is their strengthening by a number of very fine and stable oxide particles. The development of ODS steels started in the United States [272,273], and the extreme advantage in swelling resistance was confirmed in the PIEs of FFTF/MOTA irradiation as shown in Fig. 5.61 [274]. In Japan, PNC started the development of ODS steels in the late 1980s, and JAEA has been developing them in the original way in Japan for three decades. ODS steels with high Cr concentration, such as 14Cr-ODS steels, have been developed. However, JAEA regards the 9Cr tempered martensitic ODS steel as the main material for cladding tubes of FRs in Japan, and its base chemical composition is Fe-9Cr-2 W-0.2Ti-0.35Y_2O_3 [275–278]. The 9Cr concept aims to avoid the risk of aging and/or irradiation-induced embrittlement. In addition, 11Cr-ODS steel has been developed as a material that has importance in corrosion resistance [279].

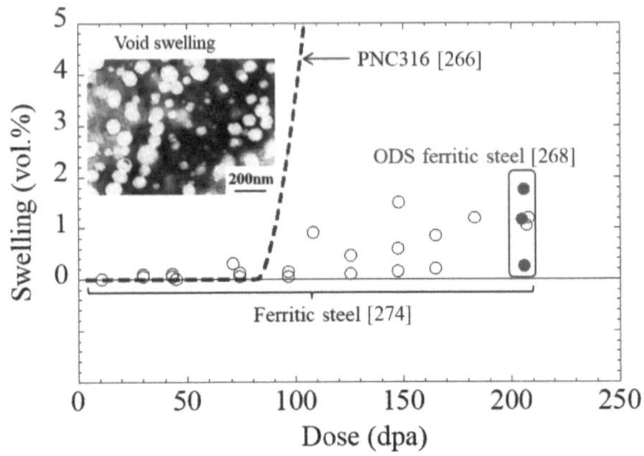

Fig. 5.61 Void welling of austenitic (PNC316) and ferritic (ODS and not ODS) steels as a function of neutron dose.

Certain technologies are necessary to disperse very fine nanoscale oxide particles, which cannot dissolve into the matrix, to realize the high-temperature strength of ODS steels. The dispersion should be uniform, and the density must be sufficiently high for strengthening. Industrial materials include inevitable heterogeneity and inclusions. The application of ODS steels to cladding tubes requiring particularly high homogeneity necessitated the upgrading of the powder metallurgical process. The raw material powders with specified amount were prealloyed in the melting process, and a mechanical alloying (MA) process was adopted to add Y_2O_3 using an attritor, which is a high-energy agitator. The MA powder was hot-extruded, and the added Y_2O_3 was reprecipitated. The nanoscale structure control technology was established as aforesaid, as shown in Fig. 5.62. In addition, several improvements in the powder handling process and equipment were made based on the failure caused by heterogeneity in previous irradiation tests [280,281]. These improvements successfully reduced inclusions to a minimum. Therefore, ODS steels have the same ductility as the ordinary cast ferrite steel [282] while having a high hardness, as shown in Fig. 5.63. The stable rolling of cladding tubes with lengths of 1 m or more has become possible owing to these improvements in manufacturing processes and the optimization of rolling tools and intermediate heat treatment.

The strength of ODS steel cladding tubes is superior to that of conventional steel, and Japanese ones are currently at the highest level in the world. Regarding creep strength, the rupture life is 500 times longer than that of PNC-FMS under the condition with hoop stress of 100 MPa at 700°C, as shown in Fig. 5.64. Furthermore, recent research has revealed that ODS steel has three times the tensile strength of conventional steel at an ultrahigh temperature of 1200°C. Thus, the ODS steel cladding tubes have a great advantage in improving the economic efficiency by increasing the burnup and safety in severe accidents [283].

In addition, pressurized resistance welding technology as a joining technology for ODS steel cladding tube and end plugs has already established the basic technology [284,285]. JAEA is proceeding with the development of mass production technology for cladding tubes, and it is ahead of the rest of the world in the development of ODS

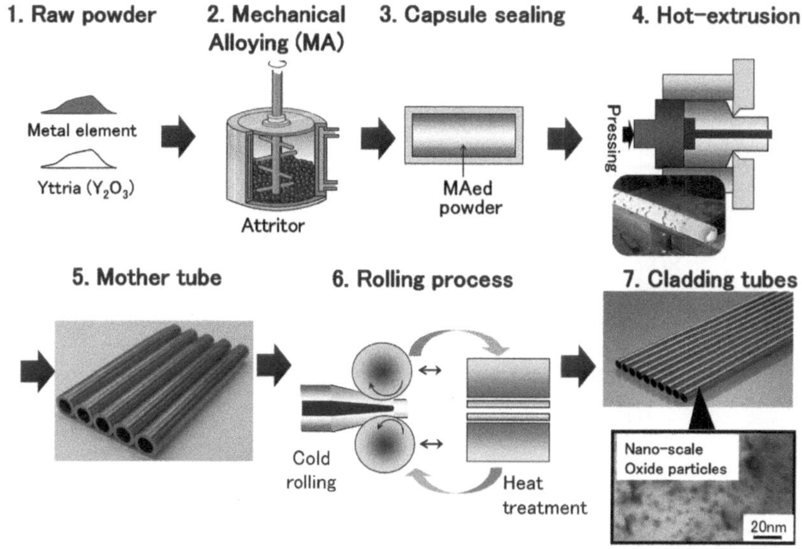

Fig. 5.62 Manufacturing process of ODS steel fuel cladding tubes.

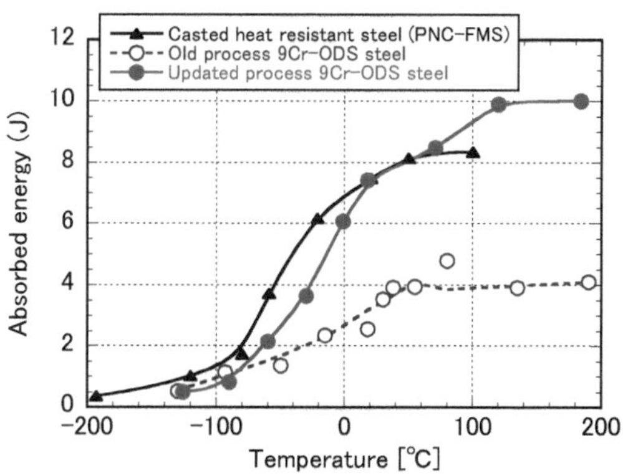

Fig. 5.63 Charpy impact properties of 9Cr-ODS steels manufactured by old and updated processes.

steel cladding tubes. JAEA is planning to evaluate the effects of irradiation, establish material strength standards, and demonstrate performance.

5.5 Thermal-hydraulics

The foremost concern in the thermal-hydraulics design of SFRs is to ensure advantages in plant safety, reliability in plant operation, and economic efficiency by optimizing the heat transport system. To construct the SFR plant fitting to these features, thermal-hydraulics issues arising in plant design must be predicted accurately for static and dynamic behavior in various plant conditions, including normal operation, anticipated operational occurrence condition (AOO), design basis accident (DBA) and

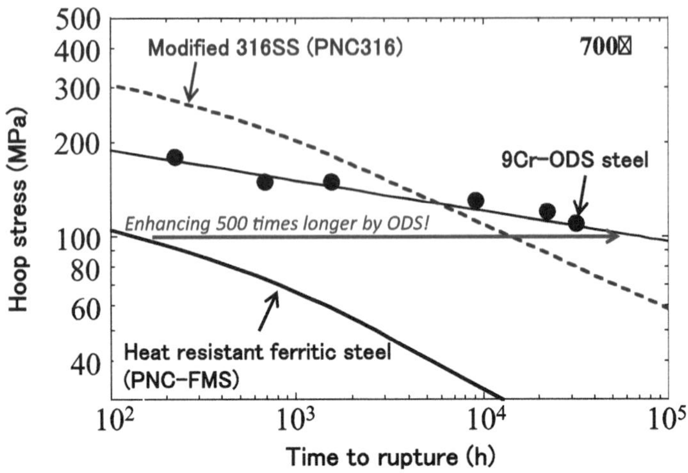

Fig. 5.64 Comparison of creep strength of cladding tubes of 9Cr-ODS steel and other materials.

beyond-design basis accident (BDBA), and severe accident (SA). In this section, R&D activities mainly performed in JAEA to solve the issues of thermal-hydraulics in SFR plant dynamics within the area of the DBA are outlined.

In Section 5.5.1, thermal-hydraulic issues and evaluation methodologies for plant dynamics are described. For the base approach in plant design and safety assessment, activities regarding plant dynamics analysis are explained. Recently, multilevel and multiphysics approaches are applied to plant dynamics analysis with the integrated approach ARKADIA written in Section 5.10. Thus, recent activities developing the coupling methodology from a multilevel approach using the one-dimensional (1D) plant analysis code and the multidimensional computational fluid dynamics (CFD) codes are introduced. The coupling methodology from a multiphysics approach regarding core design using the analysis codes for nucleonics, thermal-hydraulics in the core, and core deformation is also introduced.

Section 5.5.2 describes the thermal-hydraulics issues and evaluation methodologies for fuel subassembly. The flow in the fuel subassembly in SFR is characterized by the wire spacer wound around the fuel pin. Specifications of thermal-hydraulics in the fuel subassembly are introduced, and the R&D activities regarding experimental investigations and numerical analysis are outlined. Moreover, R&Ds for local fault issues in the fuel subassembly are mentioned as an issue in safety assessment.

Section 5.5.3 describes the thermal-hydraulics issues and their evaluation methodologies for the upper plenum of the reactor vessel where fluid mixing at low and high temperatures is concerned. As an issue affecting the structural integrity in plant design because of the specific features in thermal properties of the sodium as a coolant, thermal stratification and thermal striping phenomena caused by fluid mixing at high and low temperatures are outlined. The R&D activities for these issues, interacting the fluid and structural fields mutually, are introduced. Moreover, the R&Ds regarding gas entrainment issue on the free surface caused by the fluctuation of the free surface of the sodium in the upper plenum are introduced.

In Section 5.5.4, as an important phenomenon in plant design, the thermal transient behavior from the normal operation to natural circulation conditions for decay heat removal accompanied by the thermal stratification in the upper plenum is explained. The R&D activities regarding the experimental investigations and the development of the numerical analysis method to analyze the mutual thermal interaction of the thermal-hydraulics phenomena in the core and plenum are explained. As an advanced method for simulating the thermal-hydraulics in the reactor vessel during decay heat removal, the status of the multidimensional CFD model development, which can connect the core and the upper plenum regions seamlessly to predict the mutual interaction accurately, is introduced. For applying plant dynamics analysis in natural circulation conditions for decay heat removal, core hot spot evaluation in natural circulation is introduced.

5.5.1 Plant dynamics thermal-hydraulics

5.5.1.1 Plant dynamics analysis

(a) Thermal-hydraulics in plant dynamics

In the plant design phase and existing plant operation, a plant dynamics analysis code is used to evaluate and confirm the thermal-hydraulics behavior under various plant conditions. Preserving the heat balance between the core power and cooling capacity of the heat sink during steady and transient conditions is essential in plant thermal design. Transient behavior must be smoothed by controlling the core power and flow rate in the heat transport systems transporting the heat generated at the core to the heat sink because the temperature change rate in each component must be lower than that of the limitation condition in the structural design. The core design is an important process for plant safety. The core dynamics are evaluated using the plant dynamics simulation code by considering the core reactivity feedback due to the control rod operation and the deformation depending on the temperature transients of fuel, sodium as a coolant, and core structure. The complex transient response characteristics of SFR comprising many components and controlling system are shown due to the mutual interaction of factors with different time coefficients, e.g., reactivity change in the core nucleonics by control rod insertion and the thermal-hydraulics transient in components and piping after the pump trip. During safety assessment, the plant design must meet safety and radiological criteria. By using the plant dynamics analysis code, the sufficient margin to prevent the core degradation using the appropriate measures is evaluated on the indices required in each evaluation scenario, e.g., the maximum temperature of the fuel cladding, the fuel material, or the reactor vessel.

(b) Brief history of the development of plant dynamics analysis code

The plant dynamics analysis code named COPD [286] was early developed based on mock-up test results of Monju and validated by the test results of the experimental fast reactor Joyo and other related experiments. The COPD was used for evaluating the safety assessment for the licensing approval of Monju. For the core dynamics analysis, a core transient analysis code named HARHO-IN [287], was developed and used for evaluating the safety assessment for licensing approvals of Monju and Joyo. Super-COPD [288,289] was developed by unifying these two codes to evaluate the plant dynamics for the events in the normal operation, the AOO, the DBA, and the BDBA. In the super-COPD, the analytical models for thermal and hydraulic calculations were highly modularized to reinforce the versatility of the code for multipurpose, and the function of the plant control system was equipped as a calculation module in the code. Consequently, it became possible to construct arbitrary flow-network models for entire SFR plants comprising the main components and control system and analyze the entire behavior of the plant.

(c) Model of super-COPD

A plant dynamics analysis code named super-COPD is based on the flow-network models for the components and piping in the plant. It can calculate the thermal-hydraulics for the design and safety assessments of SFR according to the transient scenario considering the logic sequence of the operating control system [290]. To construct a flow-network model for the target plant, the super-COPD is equipped with thermal and hydraulic calculation models for the core, the primary heat transport system (PHTS), the intermediate heat exchanger (IHX), the secondary heat transport system (SHTS), the auxiliary cooling system, the air cooler, the steam generator, and water-steam system. Furthermore, the upper plenum thermal-hydraulics model, point reactor neutron kinetics model, calculation models for active equipment (the pump and the valves), and plant control system models for the plant protection system and the interlocking system are also installed.

(d) Representative validations and application study of super-COPD

The super-COPD was validated using the test results of Monju, Joyo, and other related experiments [289].

Yamada et al. [291,292] performed a numerical simulation of plant dynamics in Monju using system startup test data for the validation of the super-COPD. Plant trip tests from 40% rated power and natural circulation tests in PHTS and SHTS using heat released from primary pumps instead of the core decay heat were evaluated. And also, the natural circulation test in Joyo Mk-II [293] was employed for validation. Fig. 5.65 shows a typical flow-network model for hydraulic analysis of the super-COPD for Monju [292]. Fig. 5.66 shows the comparison of the temperature transients at the inlet and outlet of the reactor vessel (RV) between numerical results using the super-COPD and measured data as a typical result. From numerical analysis and comparison with the measured data of Monju, the plant dynamics model of the super-COPD could simulate the entire plant dynamics with good accuracy.

Hiyama et al. [294] performed a numerical analysis using the super-COPD, and the results were compared with the experimental results from the natural circulation test in Joyo Mk-II [293]. Focusing on the thermal-hydraulics behavior in the PHTS, a flow-network model from the reactor core to the IHX was constructed. From the numerical analysis, modeling the mixing behavior in the upper plenum of the RV considerably influenced the natural circulation behavior calculated using the super-COPD. Fig. 5.67 shows comparisons of the center fuel subassembly outlet coolant temperature and primary flow rate in the short term as typical results of the super-COPD. The upper plenum thermal-hydraulics model in the super-COPD can consider the effect of thermal stratification in the upper plenum. The predicted transient results of the coolant temperature at the core outlet and the primary flow rate showed good agreement with the test results. Nabeshima et al. [295] also performed a numerical analysis to validate the calculation models for natural circulation in the super-COPD by numerically analyzing the natural circulation test in the Joyo Mk-II. Most Joyo components were modeled in the flow-network model and the core model comprising all subassemblies was used in the analysis as advancement from Hiyama's work [294]. From the simulation, the transient behavior predicted by the super-COPD showed good agreement with the experimental data.

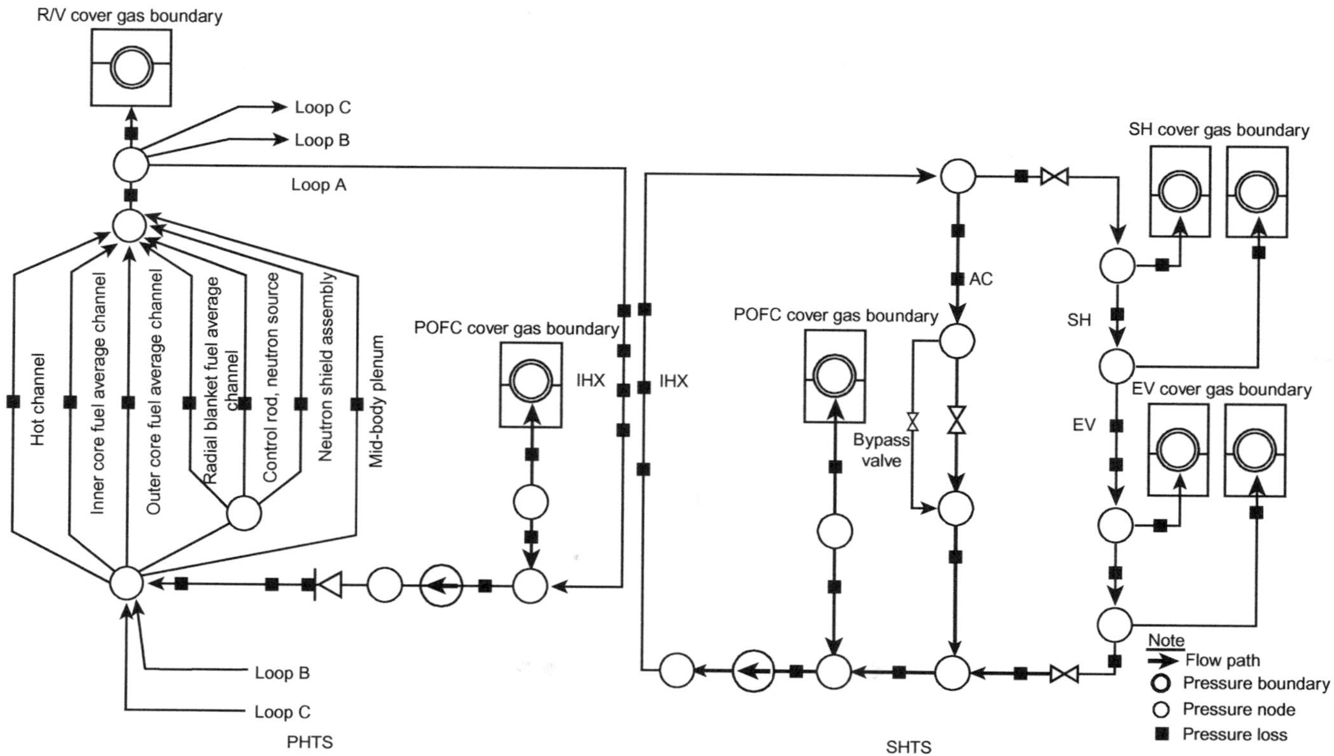

Fig. 5.65 Flow-network model for hydraulic calculation of super-COPD for Monju [292].

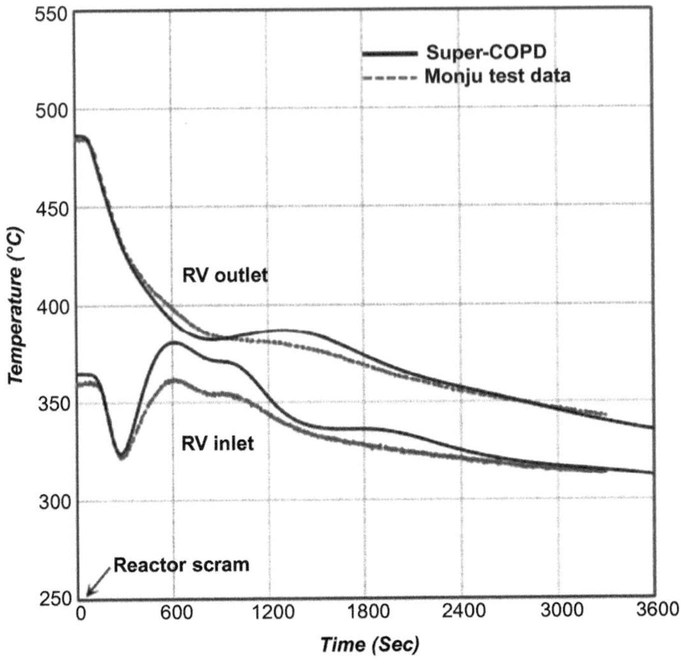

Fig. 5.66 Typical results of temperature at reactor vessel (RV) inlet and outlet in Monju test analysis by super-COPD [292].

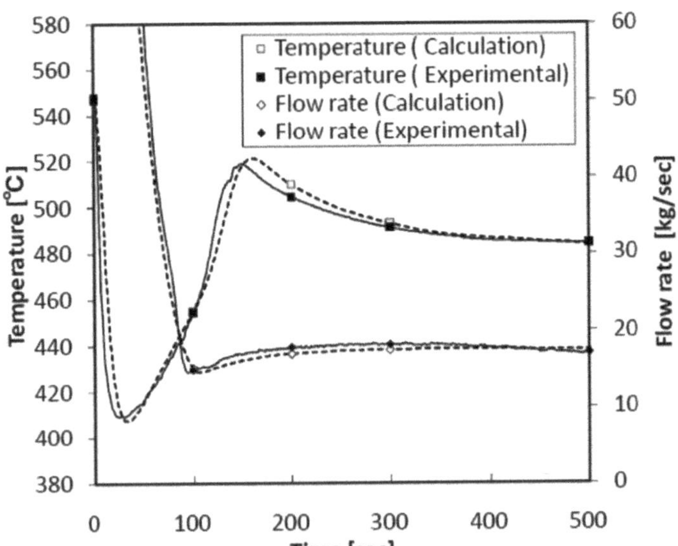

Fig. 5.67 Typical results of center fuel subassembly outlet temperature and primary flow rate in Joyo test analysis by super-COPD [294].

Doda et al. [296] performed a numerical analysis of the shutdown heat-removal test (SHRT-17) [297] in the experimental breeder reactor-II (EBR-II). The numerical results by the super-COPD could demonstrate the entire thermal-hydraulic behavior in the plant during natural circulation decay heat-removal operations.

For the application study, Yamada et al. [292] developed an analytical model for natural circulation decay heat removal in the super-COPD. The core cooling ability of natural circulation has been evaluated for a station blackout (SBO) event induced by an earthquake and subsequent tsunami. Consequently, it was concluded that the natural circulation of the sodium coolant could remove the decay heat from the core under such an SBO condition.

5.5.1.2 Multilevel simulation by 1D—3D coupling analysis

Conventionally, the thermal responses in the entire plant and local multidimensional behavior were evaluated independently using the 1D plant dynamics analysis code and the multidimensional analysis code, respectively. Therefore, the mutual influences among the components in the plant and the local behavior in each component were considered only using the conservative boundary conditions between them. Then, the multilevel simulation (MLS) system enabling the consistent evaluation from the entire behavior in the plant to local phenomena was developed [298].

Yoshimura et al. [299] applied a coupling (1D-CFD coupling) method to the unprotected loss of flow test of SHRT-45R in EBR-II to validate the coupling method. A 1D code (super-COPD) is combined with a CFD code (FLUENT) for local thermal-hydraulics analysis in SFRs using the sequential two-way coupling method [300]. Fig. 5.68A and B shows the plant layout of the EBR-II and CFD model for the upper plenum in the reactor vessel and Z-shaped piping (Z-pipe) from the outlet of the reactor vessel to the inlet of the IHX. During the coupling analysis, 1D code is calculated using the boundary conditions calculated by the CFD code, and the CFD code is calculated using the boundary conditions calculated by the 1D code, sequentially, in a one-time step. By repeating this procedure, the time step proceeds. Fig. 5.69A and B, respectively, shows the instantaneous temperature distribution of the CFD result at 100 s and the transient profile comparison of the temperature in the upper plenum between the measured data and 1D-CFD coupling analysis results. The 1D-CFD coupling analysis results followed the trend of the measured data. In the first 100 s, hot sodium flowed into the upper plenum from the middle of the core, where the driver fuels are mostly located, and thermal stratification appeared. Then, the hot sodium accumulated in the upper area of the upper plenum was almost stagnant and cooled by heat transfer to the surrounding structure. At 200 s, thermal stratification in the upper plenum disappeared due to heat transfer to the surrounding structure.

5.5.1.3 Neutronics-related issues

In SFRs, thermal deformation of the core fuel subassembly causes reactivity feedback due to a temperature rise during the AOO or accidents. The core deformation reactivity is a combined phenomenon of neutronics, thermal-hydraulics, and structural mechanics. Because a large uncertainty is considered in evaluating the reactivity, negative feedback effects are usually neglected in the conventional safety evaluation to

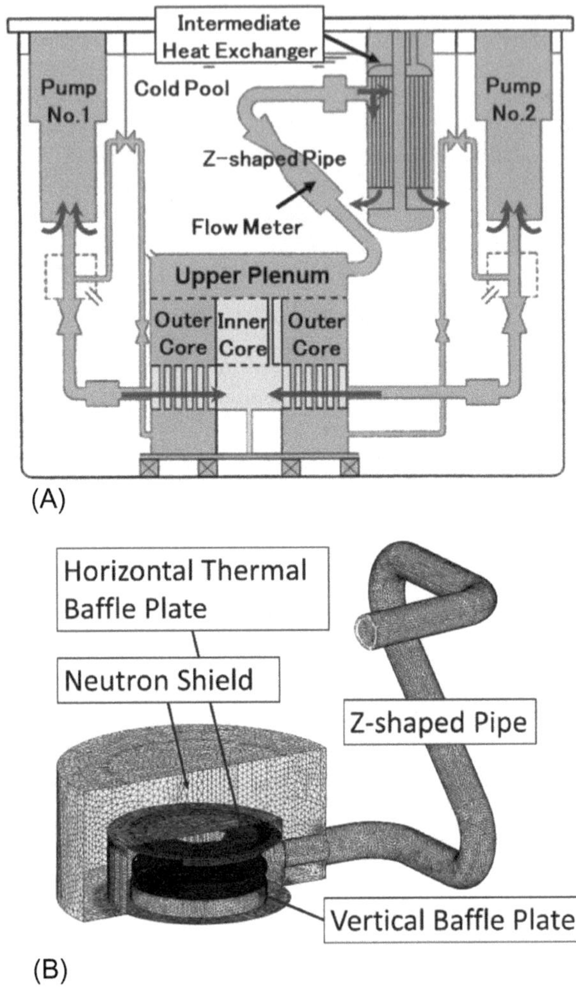

Fig. 5.68 Plant overview of EBR-II and model of upper plenum and Z-pipe for CFD calculation. [299].
(A) Overview of EBR-II plant. (B) 3D model for CFD calculation.

maintain conservativeness. To consider the core deformation reactivity as an inherent safety feature and eliminate excessive conservativeness in core design safety evaluation, an evaluation method for core deformation reactivity feedback has been developed [301,302]. This method combines neutronics calculation, core thermal-hydraulics analysis, and structural mechanics analysis of the fuel subassembly. The core thermal-hydraulics model in a plant dynamics analysis code estimates the temperature distribution of the fuel subassembly, and the finite element method estimates the deformation of fuel subassembly. The obtained temperature and deformation information is used to analyze the core deformation reactivity based on a diffusion perturbation theory. By applying this coupling method to an unprotected loss of flow condition test of SHRT-45R in the EBR-II, the reactivity due to core deformation had a negative feedback effect during the transient. The negative feedback was caused by

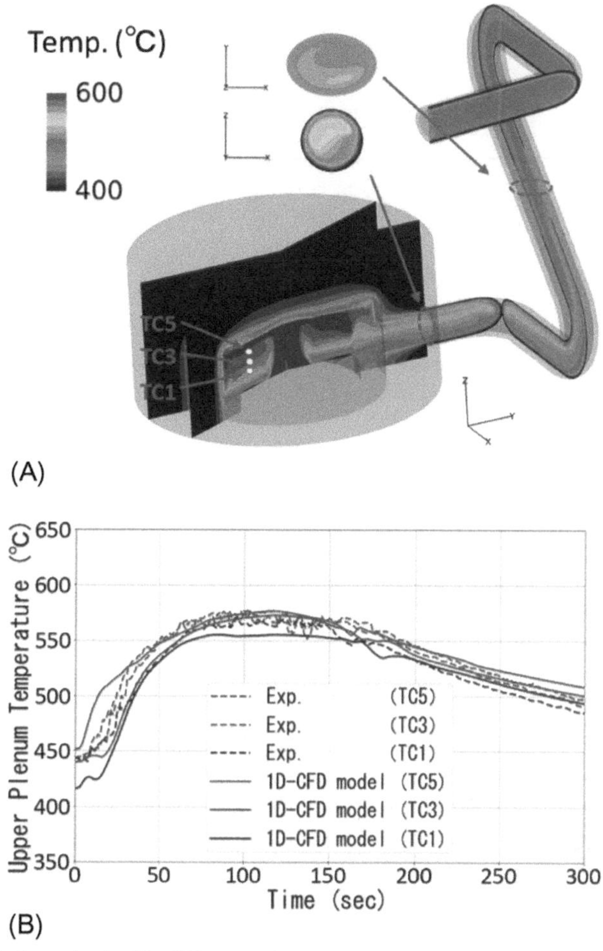

Fig. 5.69 Typical results in 1D-CFD coupling analysis [299]. (A) Instantaneous temperature distribution of CFD result at $t = 100$ s, (B) Thermocouple (TC) trees in the upper plenum.

the outward displacement of the reflector subassemblies surrounding the fuel region in the center of the core and fuel subassembly deformation. Fig. 5.70 shows the instantaneous distribution of the temperature and deformation at 100 s in the EBR-II core.

5.5.2 Fuel subassembly thermal-hydraulics (FS-TH)

5.5.2.1 Specifications of FS-TH

Improving the economic performance of the reactor requires installing a high-performance core with a high burnup ratio and power density. To confirm the design feasibility of such a high-performance core, thermal-hydraulics phenomena in fuel subassemblies must be clarified under normal, transient, or deformed geometry

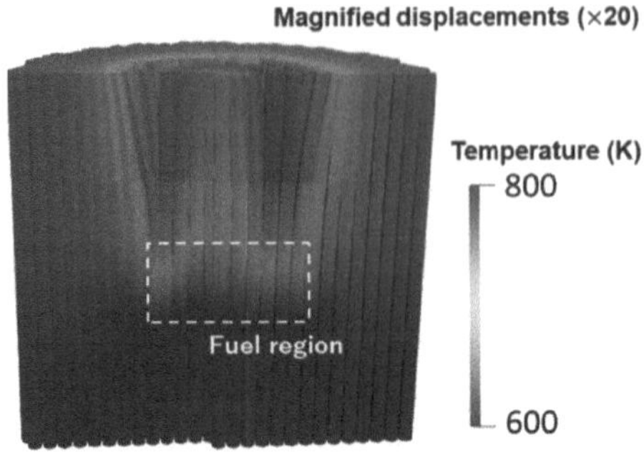

Fig. 5.70 Typical results of instantaneous distribution of whole-core temperature and deformation at 100 s.

operating conditions from the viewpoint of fuel pin structural integrity assessment. Typical SFR fuel subassembly comprises a fuel pin bundle with a wire spacer helically wound around the pin and a hexagonal wrapper tube in which the fuel pin bundle is inserted. Fig. 5.71 shows a typical fuel subassembly for SFR [303]. Sodium flows in the fuel subassembly through the subchannels among the fuel pins, wrapper tube wall, and structures. The helically wound wires enhance coolant mixing by generating a transverse flow between the subchannels and maintaining the flow area of the coolant. The flow regime in a fuel subassembly varies from laminar at low flow rate conditions under the decay heat-removal operation in natural circulation to turbulent at high flow rate conditions under the rated power operation. The thermal-hydraulics issue in the fuel subassembly concerns pressure drop, temperature, and velocity distributions. Evaluating the pressure drop in the fuel subassembly is critical for appropriate flow rate distribution through each fuel subassembly to ensure adequate core cooling. A set of subchannel friction factor models based on the empirical correlations applicable to bare and wire-wrapped hexagonal pin bundles was developed and calibrated [304–307]. Furthermore, the power distribution in the axial and radial locations, the flow rate distribution for each subchannel, and the flow mixing effect of wire spacers among adjacent subchannels determine the temperature distribution in the fuel subassembly. During natural circulation, the flattened temperature distribution, which differs from the convex-shaped distribution in rated operations, appears in the fuel subassembly due to the flow redistribution by buoyancy. Specifically, the heat removal from the wrapper tube wall by the interwrapper flow could significantly influence the temperature and velocity distributions in the fuel subassembly.

In developing an advanced SFR, a fuel subassembly with an inner duct structure named FAIDUS is adopted as a design measure for realizing early molten fuel discharge to mitigate a core-disruptive accident [308]. Fig. 5.72 shows a conceptual view of FAIDUS [308]. Because the fuel pins were arranged asymmetrically by installing the inner duct in FAIDUS, the temperature distribution under various operating conditions must be estimated to avoid the occurrence of a hot spot around the inner duct and confirm the feasibility of FAIDUS. Kikuchi et al. [308,309] performed a numerical analysis of the thermal-hydraulics in FAIDUS using an in-house subchannel

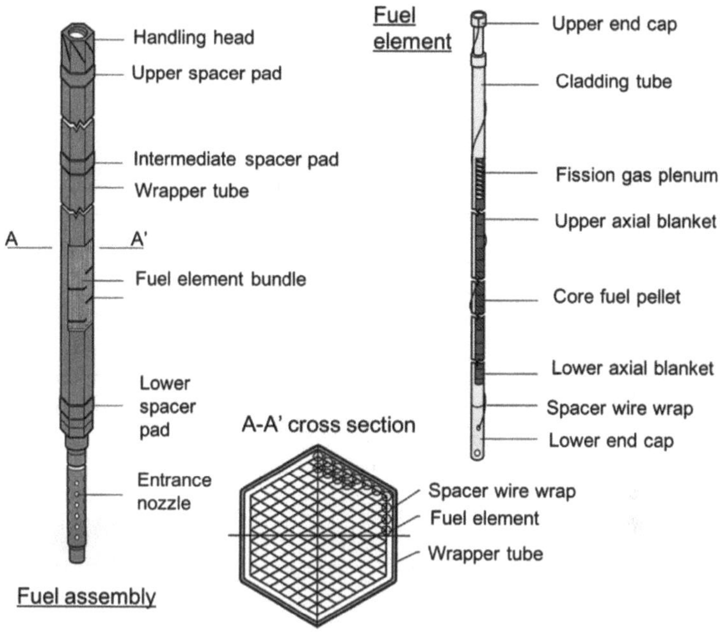

Fig. 5.71 Illustration of typical fuel subassembly [303].

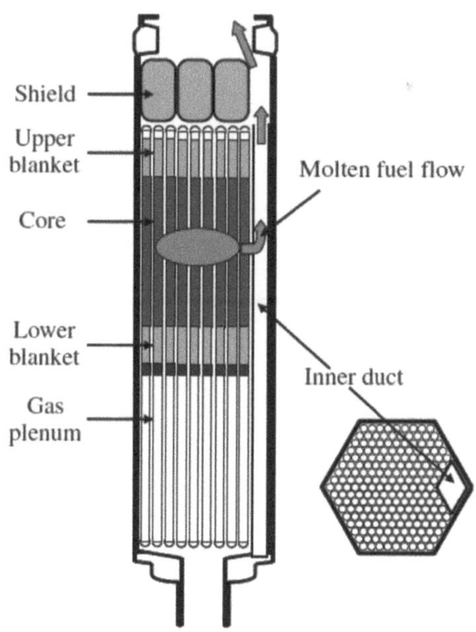

Fig. 5.72 Illustration of a fuel subassembly concept with inner duct (FAIDUS) [308].

analysis code named ASFRE [310] and an in-house CFD code with a finite element method code named SPIRAL [311].

5.5.2.2 Experiments for FS-TH

A fundamental water experiment using a wire-wrapped 3-pin bundle model was conducted to obtain detailed velocity distributions in subchannels [312–314] to investigate the flow field in a pin bundle under the intact and deformed pin configurations and validate data in developing numerical analysis methods. Fig. 5.73 shows a diagram of the test section comprising three wire-wrapped pins [314]. Each pin in the test section comprised stainless steel pipes in lower and upper sections and the transparent pipe between them was made by fluorine resin to match the refractive index to that of the water as working fluid for the particle image velocimetry (PIV) method. Using index matching techniques, the velocity distribution in the interior subchannel surrounded by wire-wrapped pins was measured using the PIV method. Typical flow velocity conditions in the pin bundle were at 0.36 m/s ($Re = 2700$) and 1.6 m/s ($Re = 13{,}500$). In addition to the 3-pin bundle test, the water experiment using a wire-wrapped 7-pin bundle model [315] was conducted to investigate the velocity field in the subchannel under the condition of Re = 6000. The velocity distribution in an interior subchannel was measured using the PIV method. In the experiment, the specific flow pattern and crossflow along wire spacers around the pin in vertical and horizontal velocity distributions, respectively, were observed.

Using the plant dynamics test loop (PLANDTL) facility, the 37-pin bundle sodium experiment (PLANDTL-37) simulating the fuel subassembly of the SFR was

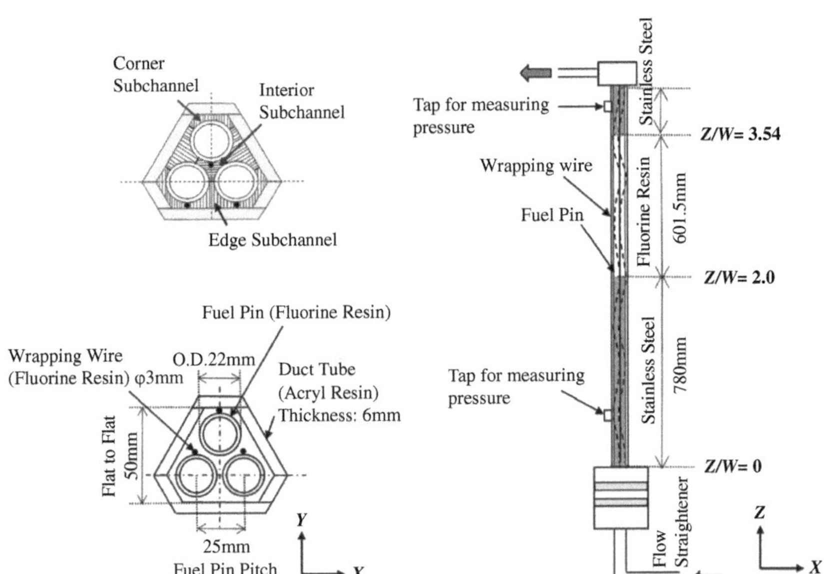

Fig. 5.73 Schematic diagram of a wire-wrapped three-pin bundle test section [314].

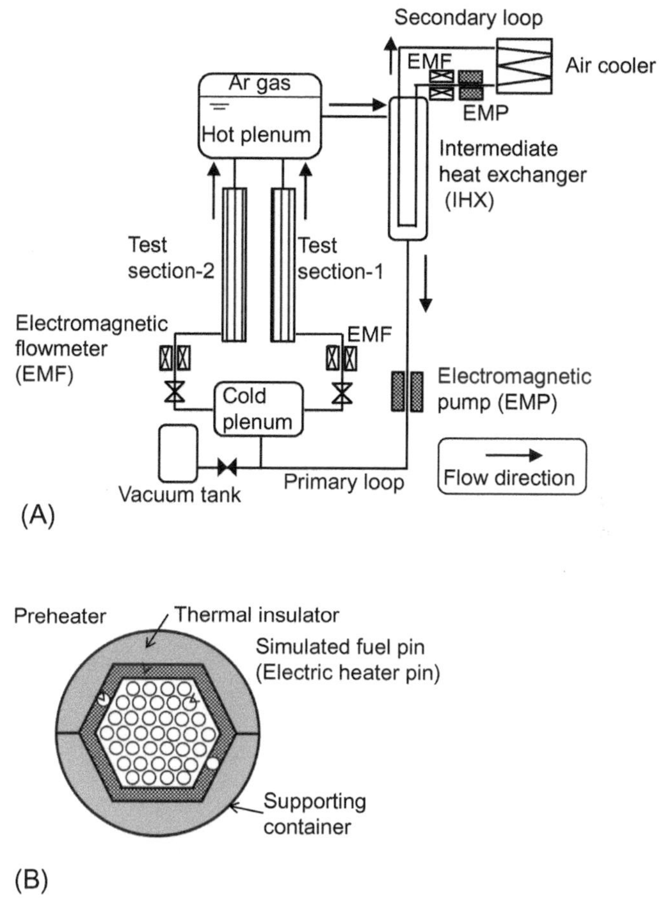

Fig. 5.74 Flow diagram of PLANDTL-37 and horizontal cross section of 37-pin bundle. (A) Flow diagram of experimental setup, (B) Cross section of the 37-pin bundle.

conducted [308]. Fig. 5.74A and B shows the test section flow diagram of the PLANDTL-37 and the horizontal cross section of the 37-pin bundle, respectively. The sodium experiment investigated sodium boiling behavior in the fuel subassembly through the static boiling in decay heat condition, transient boiling in plant dynamics, parallel channel boiling, and computer code validation tests [316,317]. Thermal-hydraulics data in the liquid single-phase flow conditions obtained as the initial condition before the sodium boiling test for several test cases were used for developing numerical simulation programs [308].

Modifying the test section of the PLANDTL facility, sodium experiment (PLANDTL-DHX) with a simulated core model comprising seven subassemblies and a direct heat exchanger (DHX) were conducted in the sodium test loop PLANDTL [318,319] to simulate the complex phenomena in the core. Fig. 5.75A and B shows the

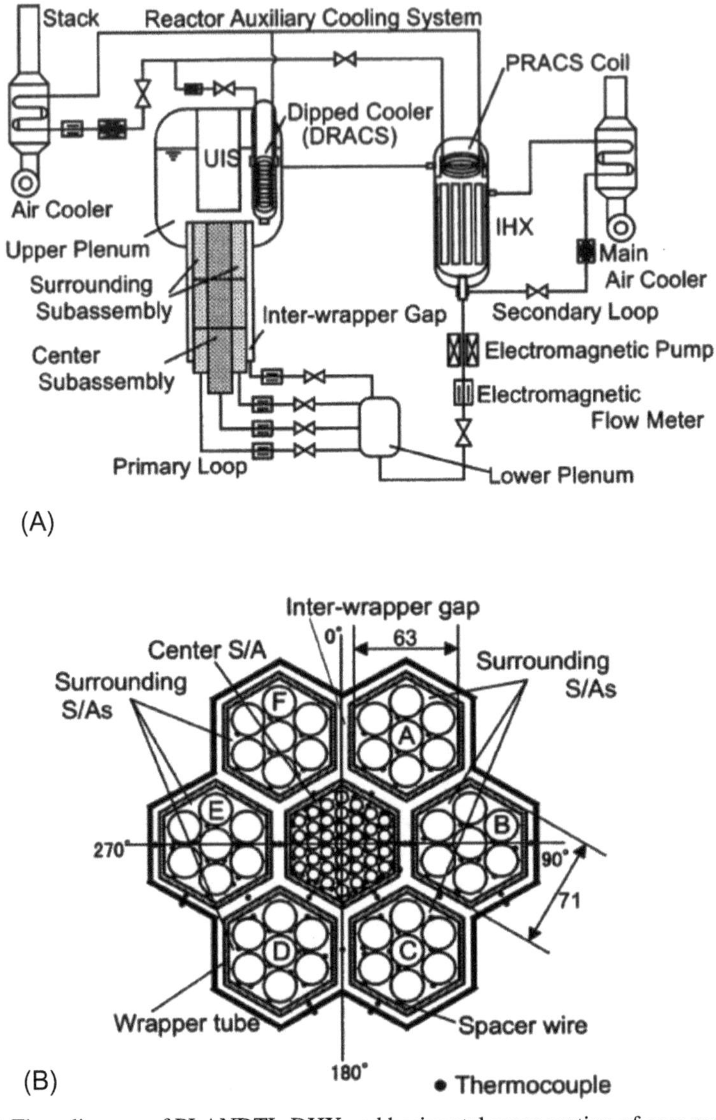

Fig. 5.75 Flow diagram of PLANDTL-DHX and horizontal cross section of core model [319]. (A) Flow diagram of PLANDTL-DHX. (B) Cross section of the simulated core.

test section of the flow diagram of PLANDTL-DHX and the horizontal cross section of the core model, respectively [319]. The center subassembly comprises a 37-pin bundle, and the surrounding subassemblies have a 7-pin bundle each. Electrical heaters heated each simulated fuel pin. The heater pin in the center subassembly simulated an SFR core fuel pin in full scale. Transient behavior in the subassemblies was studied under conditions with and without inter-subassembly heat transfer.

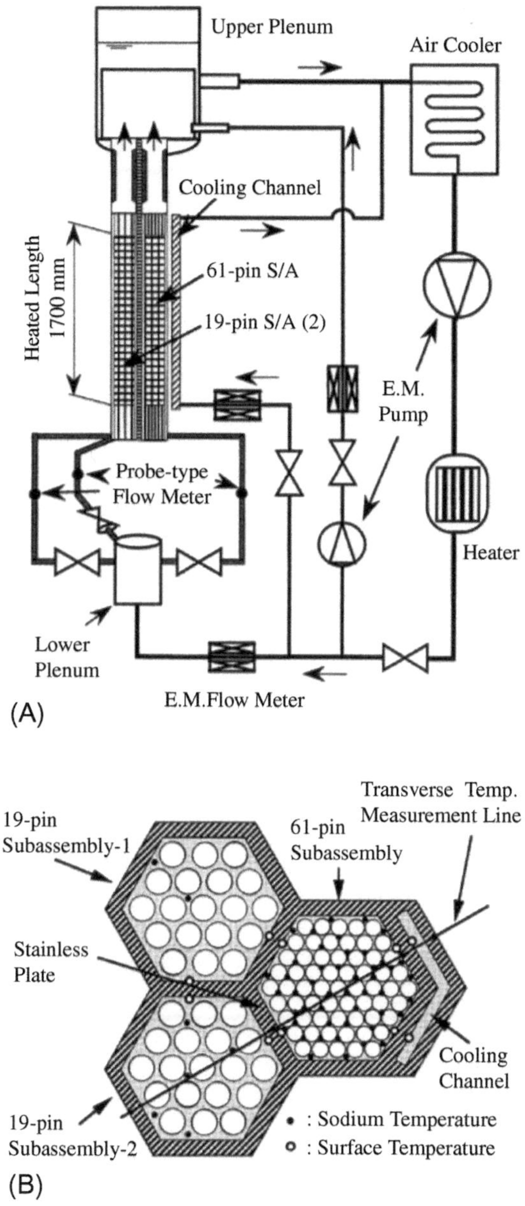

Fig. 5.76 Flow diagram of core component test loop (CCTL-CFR) and its test section [320]. (A) Flow diagram of CCTL. (B) Cross section of the simulated core.

Temperature profiles of the subassemblies accompanied by inter-subassembly heat transfer were measured in core component test loop (CCTL-CFR) sodium experiments [320]. Fig. 5.76A and B shows flow diagram of the CCTL-CFR and its test section [320]. The test section comprised a 61-pin bundle and two 19-pin bundles. A solid steel slab bounds each bundle to the others to measure the heat flux due to the inter-subassembly heat transfer using thermocouples embedded on the slab surfaces through the sodium in the small width gaps between the bundles in the real reactor. A cooling channel was also equipped in the 61-pin bundle behind the wrapper wall opposite the 19-pin bundle side to provide various thermal boundary conditions simulating inter-subassembly heat transfer. Sodium test data from PLANDTL-DHX and CCTL-CFR facilities have been used for validation of the numerical analysis codes for FS-TH [321].

5.5.2.3 *Numerical simulation programs for FS-TH*

ASFRE is the subchannel code developed in JAEA to analyze FS-TH with reasonable computational load for efficient parametric analyses in design studies [310,322]. The mesh arrangements on the horizontal cross section comprise three cells: interior, edge, and corner subchannels. Fig. 5.77 shows the mesh arrangement of the 37-pin bundle fuel subassembly for ASFRE [310]. The interior subchannels are triangular cells surrounded by the surfaces of three fuel pins. The edge subchannels are rectangular cells surrounded by the surfaces of two fuel pins and a wrapper tube. The corner subchannels are quadrilateral cells surrounded by the surface of one fuel pin and a wrapper tube near the corner. The governing equations of mass, momentum, and energy were discretized in the control volumes. The mass and energy equations and the pressure gradient terms in the momentum equation are solved implicitly, and other terms in the equations are solved explicitly. The incomplete lower-upper biconjugate gradient (ILUBCG) method [323] is used to solve the pressure equation. To predict the effects of the crossflow around the fuel pin and the swirl flow along the inner surface of the wrapper tube caused by the wire spacer on the fuel pin, the distributed resistance model (DRM) [324] is used for evaluating the pressure drop. As a simulation option of the pressure drop model in ASFRE, Cheng and Todreas's model [304], a commonly used model, can be selected instead of the DRM. Todreas and Turi's model

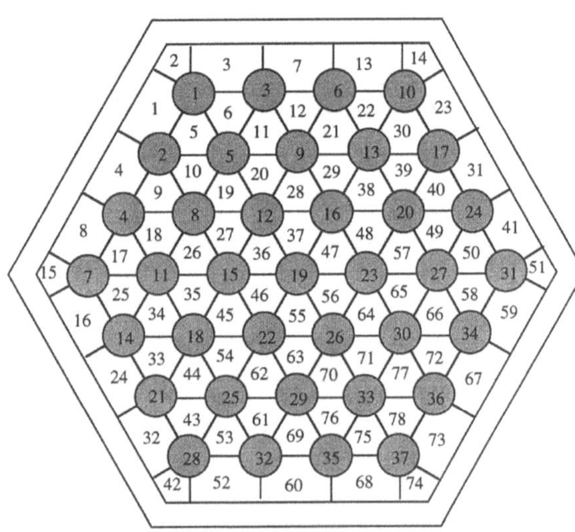

Fig. 5.77 Cross-sectional view of subchannel arrangements in 37-pin bundle for ASFRE [310].

(TTM) [325] is used for evaluating the thermal diffusion caused by the flow between interacted subchannels. The heat transfer between the fuel pin and fluid in the subchannel is evaluated using the fast flux test facility (FFTF) fuel or modified Schad correlations [326]. The numerical results of ASFRE in the typical fuel subassembly were validated by comparing the water and sodium experiment results [308]. The pressure drops over various Reynolds numbers were calculated using ASFRE with the DRM and correlated with the results in the simulated typical fuel subassembly on water experiments [308]. The coolant temperature distributions under high and low flow rate conditions were calculated using ASFRE with the DRM and the TTM and correlated with the results in the simulated typical fuel subassembly on the sodium experiments [308].

SPIRAL is a finite element method CFD code developed in JAEA to analyze thermal-hydraulics in the fuel subassembly in detail [311,327]. For numerical stabilization, a streamlined upwind method [328] was applied to the convection terms in the Navier-Stokes equation, the heat conduction equation, and the transport equations of turbulence models. Because of the wire spacer, the velocity distribution in the fuel subassembly differs, depending on local positions. Even though the flow in the center region of the subchannel has a high Reynolds number turbulent condition, the flow behind the wire spacer has a low Reynolds number turbulent condition. To suppress overestimating turbulent kinetic energy in the low Reynolds number region around the surface of the wire spacer and fuel pin, the hybrid k-ε/k_θ-ε_θ turbulence model is used. In the hybrid model, the damping functions [329] are applied to outer layer elements from the low Reynolds number model, and the wall functions are applied to wall-adjoining elements from the high Reynolds number model. For wall-adjoining elements, wall boundary conditions for k, ε, k_θ, and ε_θ were modified considering low *Re* number effects [311]. SPIRAL with the hybrid model was validated by comparing the pressure drop and temperature field in the wire-wrapped fuel subassemblies on several tests in water and sodium [311,327,330–332]. Fig. 5.78 shows a partial view of mesh arrangement in a typical fuel subassembly for SPIRAL [311]. Fig. 5.79 shows the temperature field in a typical fuel subassembly by SPIRAL [311].

5.5.2.4 Local blockage issue

A local blockage is an event that should be evaluated for the safety of SFRs because of the narrow coolant flow area, the high power density, and the high fuel burnup in the fuel subassembly of the SFR compared with those of light-water reactors. In the SFR fuel subassembly, the local blockage as a conceivable mode of local fault issues was investigated as a presumed condition for the safety evaluation. For safety analysis, the maximum cladding and coolant temperatures must be evaluated in the fuel subassembly. Studies concerning the formative process of the porous blockage were reported [333–335]. From existing studies, a porous-type blockage comprising foreign particles around 1 mm in diameter, smaller than the coolant flow area in a subchannel, would be the most possible form of local blockage in a wire spacer-type fuel subassembly. Moreover, continuous porous blockages over the subchannels could not be formed, but those at each subchannel in a staggered arrangement on the cross section

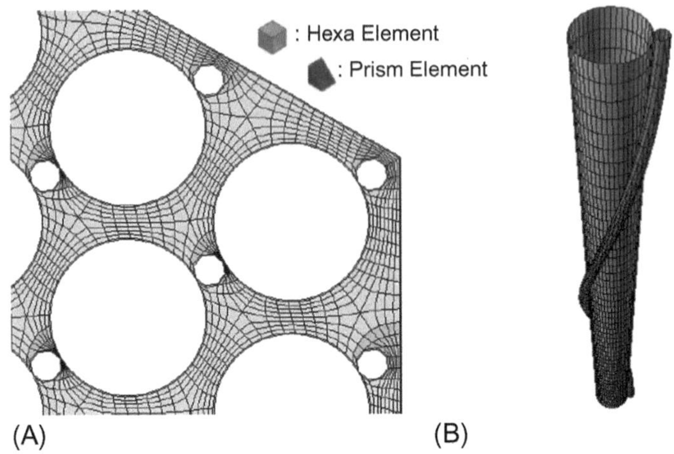

Fig. 5.78 Partial view of mesh arrangement in typical fuel subassembly for SPIRAL [311]. (A) Horizontal cross-section, (B) fuel pin surface with wire-spacer.

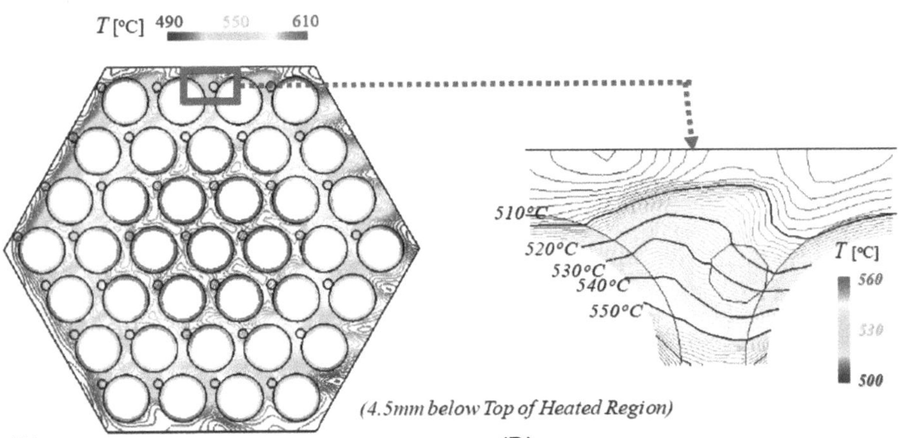

Fig. 5.79 Temperature field in typical fuel subassembly by SPIRAL [311]. (A) Horizontal cross-section, (B) local edge subchannel area.

could be possible. The shape of each blockage was column configuration with a certain height within 1/6 length of the vertical winding pitch of the wire spacer. The SCARLET [336], MOL/7C [337], and two subchannels' experiments [338] showed the temperatures in the fuel subassembly containing a porous blockage. However, information on the temperature distribution and the heat transfer mechanism in the porous blockage was limited. Kikuchi et al. [339] summarized out-of-pile experiments to evaluate the mechanical and thermal effects of fission gas ejection on adjacent pins and a wrapper tube to measure local temperature rises due to fuel pin contact and local flow blockages.

JAEA conducted three scale experiments on the porous blockage: the particle-level blockage test in a packed spheres configuration [340,341], subchannel-level blockage test with four-subchannel configured test section [342], and the full-pin bundle-level blockage with a 37-pin bundle fuel subassembly configured test section [343]. The experiments investigated the heat transfer mechanism and predicted the temperature distribution inside the blockage [344]. Multidimensional analyses with fine mesh subdividing a subchannel were applied to the four-subchannel configured test [344] and the 37-pin bundle fuel subassembly configured test [343] to clarify the temperature and velocity distributions and predict the position and degree of the maximum temperature in the fuel subassembly using the CFD code as the mezzo-scale view. Subchannel analysis with triangular mesh representing each subchannel was conducted using the in-house code ASFRE [345–347] as the macroscale view.

When a random pin failure occurs, the fission product (FP) gas in the gas plenum of the fuel pin could be released into the sodium. This is a safety issue because the released gas might degrade the heat transfer around the fuel pin in the fuel subassembly [339,348,349]. As the local blockage is formed in a fuel subassembly, a high-temperature region appears around the blockage. Thus, the released FP gas from breached cladding could cause an additional temperature increase in the surrounding pin surfaces. It is critical to estimate the temperature rise under general geometrical and operational conditions during gas release around local blockages.

5.5.3 Thermal-hydraulics in reactor vessel

5.5.3.1 Thermal stratification

Thermal stratification occurs in the RV and piping during the transient after the reactor is shut down. Thermal stratification in the upper plenum of the RV is the most typical problem. After the reactor is shut down, the core power decreases quickly by inserting the control rod, and the flow rate through the core decreases slowly with a time delay due to the pump flow coast down of the primary pump. In the upper plenum of the RV, hot sodium remains in the upper part of the plenum due to the density difference, whereas cold sodium flows into and accumulates at the bottom of the plenum. Because the temperature fluctuation around the thermal stratification interface might transfer to the wall surface of the RV and the cyclic thermal stress might be induced on the wall, it is necessary to evaluate the thermal stratification accurately.

Ieda et al. [350] summarized water and sodium experiments with scaled models to study the in-vessel thermal stratification in a loop-type SFR and investigated results on the thermal stratification threshold, the rising rate of the stratification interface, and the flow regime above the interface. Experimental studies on the thermal stratification in the hot plenum in the RV of a Japanese demonstration SFR investigated the magnitudes of the temperature gradient and interface ascending speed of thermal stratification [351–353]. Sodium experiments have been performed using PLANDTL-DHX to study the rising behavior and the temperature gradient of the interface of the thermal stratification [354,355]. In the development of the JSFR investigated as part of the Fast Reactor Cycle Technology (FaCT) development project, thermal stratification in the RV with an upper inner structure (UIS) comprising perforated horizontal plates and control rod guide tubes, with a radial slit for fuel handling was investigated using a 1/10 scale water test model simulating the upper plenum of the RV [356,357]. Fig. 5.80 shows the test section of the 1/10 water experiment [358]. In the JSFR, the jet goes through the UIS due to the discharged flow from the fuel subassemblies. The experiment investigated the influence of jet flow on thermal stratification behavior.

Numerical investigation on thermal stratification using a general-purpose multi-dimensional code named AQUA with the QUICK/FRAM scheme [359] was performed using the experimental results of the scaled test apparatuses simulating the upper plenum of the SFR [355,360]. Investigations establishing the numerical evaluation method for thermal stratification phenomena using the commercial CFD code were performed using fundamental thermal stratification sodium tests [358,361]. Fig. 5.81 shows the sodium test using the cylindrical vessel simulating the upper plenum of the SFR [358]. Fig. 5.82 compares the experimental and calculated results of the numerical analyses using standard k-ε (SKE), RNG k-ε (RNG), and the Reynolds stress models on the temperature gradient at the stratification interface. No remarkable difference was observed in the numerical simulation results using the turbulence models.

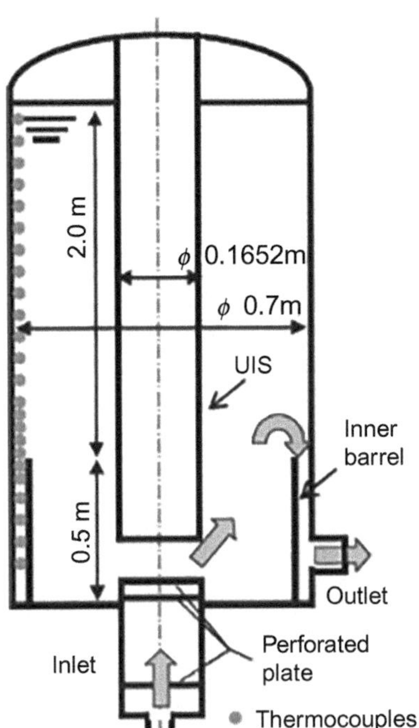

Fig. 5.80 Overview of test vessel for sodium test [358].

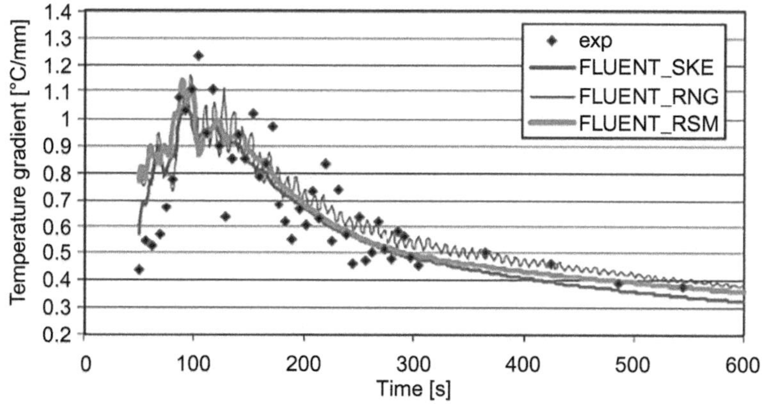

Fig. 5.81 Influence of turbulence model on thermal stratification behaviors [358].

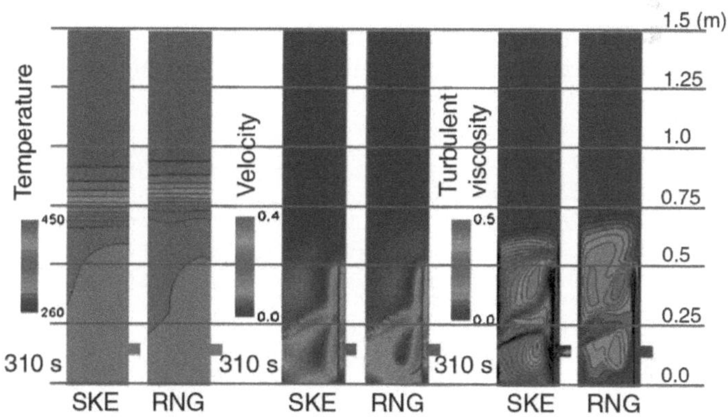

Fig. 5.82 Instantaneous distributions of velocity, temperature, and turbulent viscosity with different turbulence models of standard k-ε model (SKE) and RNG k-ε model (RNG) at 310s [358].

5.5.3.2 Thermal striping

Temperature fluctuation is induced in the region where hot and cold fluids are mixed, and it might cause thermal fatigue in the structure. This phenomenon is called thermal striping. When such high-cycle thermal stress acts in the structure, crack initiation and eventually crack penetration might occur under certain conditions. As a unique characteristic of SFR, liquid sodium metal is used as a coolant. Since the thermal conductivity of sodium is large, temperature fluctuations in the fluid caused by the mixing can be transmitted easily to the surface of the structure with little decline in its amplitude. Thus, thermal striping is the most critical phenomenon to be evaluated for the structure design in SFR.

From the early 1980s, thermal striping had been studied by Wood [362], Brunings [363], Betts [364], and Moriya [365]. Tenchine et al. [366,367] investigated the mixing phenomenon of coaxial sodium and air jets. JAEA has conducted a series of water experiments of triple-planar parallel jets mixing, named WAJECO to investigate the thermal mixing process among the jets and the temperature fluctuation attenuation process from fluid to structure [368]. Also, a series of sodium experiments, named PLAJEST [369], with the test section corresponding to that of WAJECO, had been conducted. Fig. 5.83 shows the test section of WAJECO-IV and PLAJEST. In the test section, a stainless-steel plate with thermocouples was set along the flow to compare the temperature fluctuation transfer characteristics from fluid to structure with that in the water experiment of WAJECO-IV. From the experiments, the temperature fluctuation transfer characteristics from fluid to structure could be evaluated using a heat transfer coefficient obtained from a transfer function between temperature fluctuations in the fluid and structure. Numerical simulations for the benchmark [370–373] using the WAJECO-IV and PLAJEST results were reported.

A T-junction pipe (T-pipe) comprising principal and branch pipes is used to mix the different temperature fluids. The T-pipe is also where thermal striping might be concerned. Water experiments, named WATLON, had been performed to reveal thermal striping and obtain validation results for developing the numerical evaluation method [374–376]. Fig. 5.84 shows the test section of the WATLON [375]. From the WATLON flow visualization test [374], the flow pattern of branch pipe can be

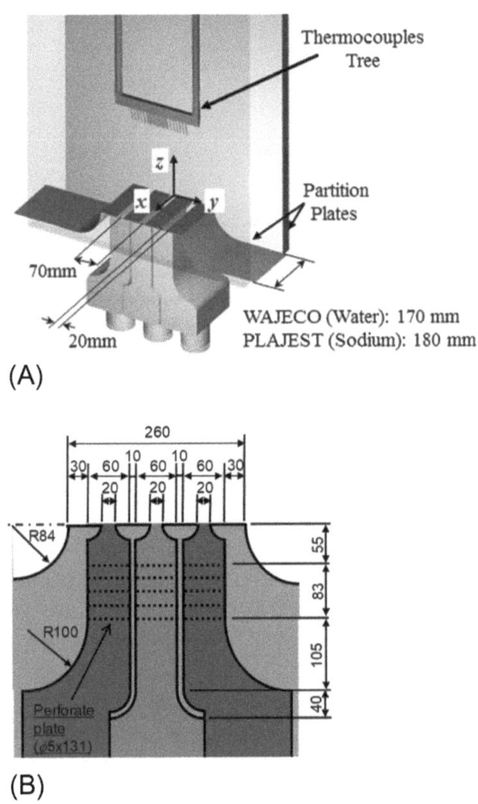

Fig. 5.83 Overview of test section of WAJECO and PLAJEST [370]. (A) Test section, (B) configuration of converged nozzles.

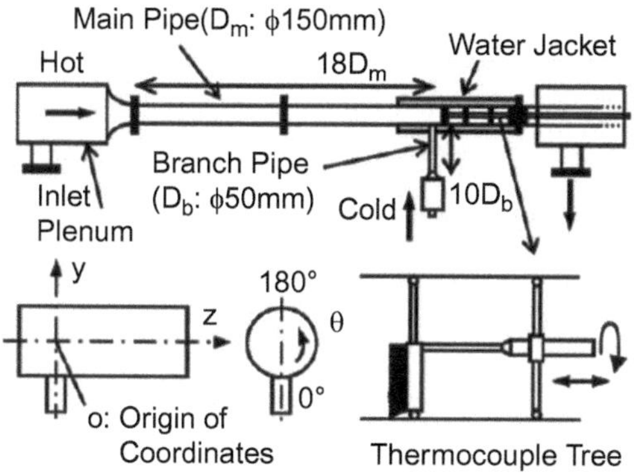

Fig. 5.84 Schematic view of test section of WATLON [375].

classified into (A) impinging jet, (B) deflecting jet, and (C) wall jet as shown in Fig. 5.85 by the bulk momentum ratio of the main pipe flow to the branch pipe jet. Using the WATLON results, a numerical evaluation method using the large eddy simulation (LES) approach for the thermal striping has been developed [377,378]. When the elbow pipe is set on the upstream side of the T-pipe, the flow fluctuation due to the secondary flow induced through the elbow might affect the velocity profiles in the main pipe before the branch pipe and the mixing behavior in the T-pipe. Kimura et al. [379] conducted a water experiment using the WATLON test section and installed an elbow pipe upstream of the T-pipe. Yuki et al. [380] experimented with the T-pipe installed an elbow on the upstream side, simulating a part of the secondary cooling system of the French prototype fast reactor, named Phenix, to investigate the secondary flow effect on turbulence mixing characteristics in the T-pipe. IAEA launched a benchmark exercise project [381] in the secondary loop of Phenix where an initial crack in a T-pipe was detected during inspections.

In the JSFR, thermal striping is a critical issue because hot sodium from the fuel subassembly can mix with cold sodium from the control rod channel at the core outlet [382]. The temperature fluctuation caused by mixing fluids at different temperatures from the control rod channels and core fuel subassemblies might touch the core instrument plane (CIP) at the bottom of the UIS and cause high-cycle thermal fatigue there [382]. A 1/3-scale five-jet water test, named FIWAT, was conducted to investigate thermal striping around the CIP [383]. In the FIWAT, the test section simulated a control rod channel, four adjacent fuel subassemblies, and a part of the CIP. The central jet discharging the fluid at low temperatures modeled the control rod channel. The four jets (hot jets) around the central jet modeled the core fuel subassemblies discharging the fluids at high temperatures. Experimental parameters were the inlet temperature difference between the cold and hot jets, the flow rate ratio of the cold jet to the hot jet, and the absolute velocity magnitude of jets. The flow rate ratio of the center jet to the neighbor jet simulates that of JSFR. The experiment indicated the fluid temperature fluctuation characteristics, including a spike-like fluctuation around the structure, the attenuation behavior characteristics of the temperature fluctuation, and heat transfer characteristics on the structure. The amplitude of such spike-like temperature

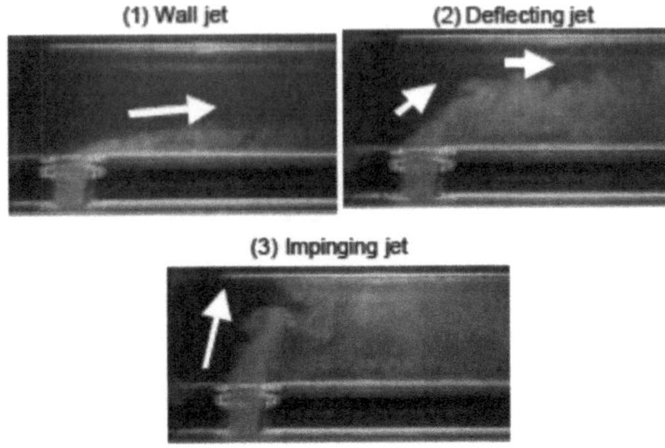

(A)

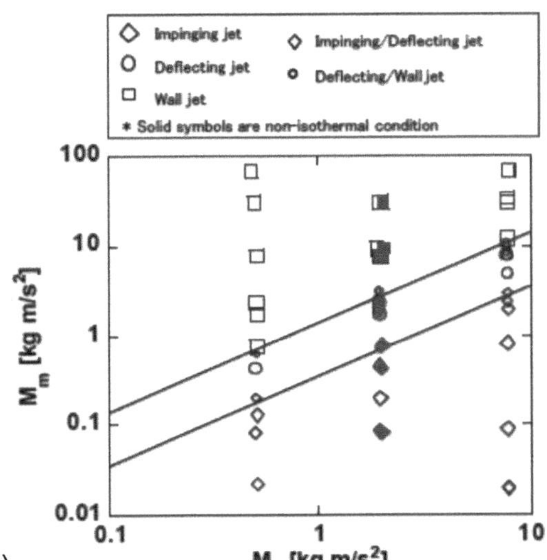

(B)

Fig. 5.85 Flow patterns of branch pipe flow in main pipe in T-junction pipe. (A) Pictures in flow visualization test. (B) Flow pattern map.

fluctuations in the fluid was mitigated on the surface of the structure. Another water experiment using the JSFR-scaled upper-plenum model as shown in Fig. 5.86 was conducted to obtain temperature fluctuation characteristics near the core outlet and confirm the effect of countermeasures to mitigate temperature fluctuations [384–387]. To mitigate temperature fluctuations in the structure near the control rod outlet, countermeasures such as reducing the outflow rate of low-temperature fluid and increasing the distance between the core and the structure were devised. Fig. 5.87 compares the temperature transient at the primary and backup control rod channel outlet. The differences in temperature fluctuations after applying the countermeasure structure were recognized. Thus, this measure could mitigate the temperature fluctuations.

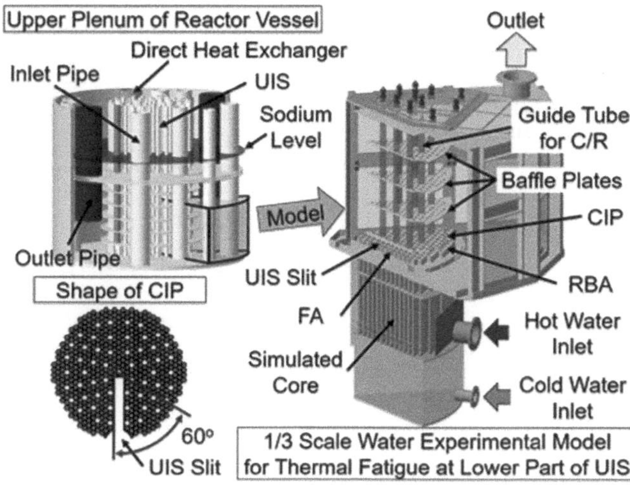

Fig. 5.86 Test section of 1/3-scaled partial test section [384].

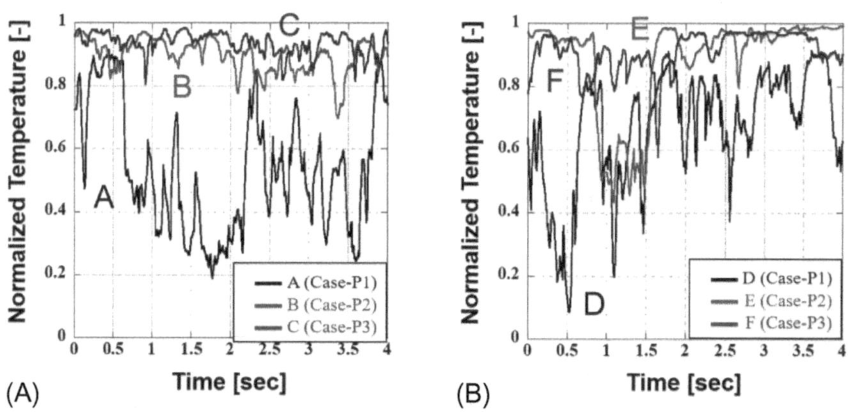

Fig. 5.87 Time trend of temperature fluctuation at primary control rod (PCR) and backup control rod (BCR) outlet [384]. (A) PCR, (B) BCR.

5.5.3.3 Gas entrainment and cavitation

In SFRs, argon gas covers a large free-surface coolant in the RV. At the surface, the cover gas entrainment due to the free-surface flow is a flow optimization issue in the reactor design because entrainment gas bubbles might disturb the reactivity when passing through the core. Gas entrainment in a vessel is categorized by free-surface vortex, submergence of the free surface, and a breaking wave [388]. From these three regimes, gas entrainment due to free-surface vortex as shown in Fig. 5.88 [389] is focused on considering the challenges of design countermeasures [390]. This type

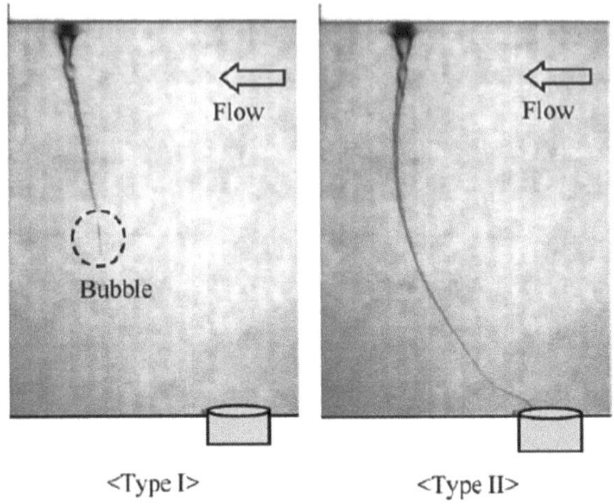

Fig. 5.88 Gas entrainment due to free surface vortex (Type I shows bubble detachment from gas core tip and Type II shows the arrival of gas core at suction nozzle).

of gas entrainment is caused by the surface deformation due to the pressure decrease in the vortex center. Therefore, an evaluation method for free-surface vortex-type gas entrainment has been developed [391]. In this method, the velocity distribution of the concerned area is obtained using numerical analysis with a single-phase and rough mesh size. The pressure decrease at the vortex center is predicted by the Burgers vortex model assuming the velocity distribution near the center. The consistent evaluation results to experimental results in the large-scale water experiment using this model were observed [391]. For predicting the pressure decrease, a modified evaluation model regarding the influences of turbulent viscosity and surface tension force was presented [392,393].

A primary system of a loop-type SFR is designed to have relatively low pressurization conditions [394]. If the compact reactor design is oriented, the flow velocity at the intakes of the hot-leg pipe increases, causing a strong vortex-type cavitation [395,396]. This vortex cavitation is a significant issue to be addressed from viewpoint of the structural integrity of JSFR because the coolant velocity in the hot leg piping in JSFR reaches 9.2 m/s [394]. Fig. 5.89 shows the vortex cavitation at the intake of the suction pipe in the scaled water model test simulating a pool-type SFR. To develop an evaluation method for the vortex-type cavitation, applying the vortex model-based evaluation approach, developed for gas entrainment [391], is also examined [397].

5.5.3.4 Flow-induced vibration (FIV)

In the engineering plant, structural integrity might be damaged due to the vibration induced by the flow fluctuation or vortex shedding in the stream around the structure due to mechanical resonances at specific frequencies, called flow-induced vibrations (FIV), and could occur at structures placed in the stream, including suction pipes, heat exchanger tubes, piping supports, penetration components in the plenum, blades in

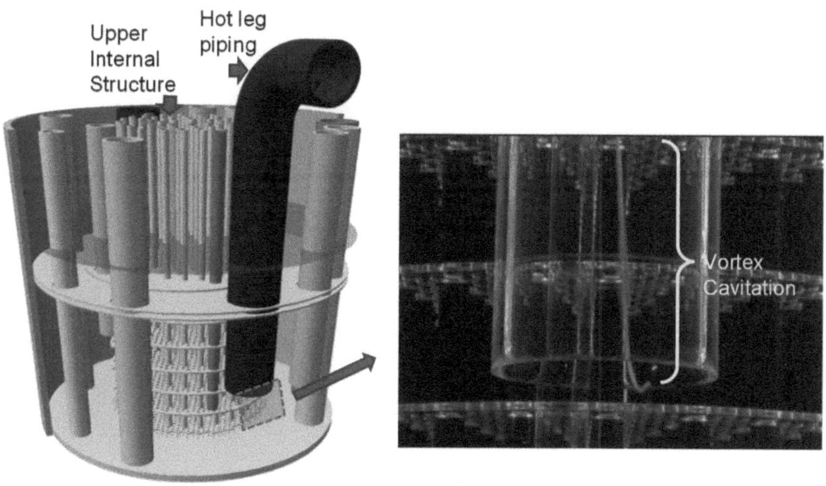

Fig. 5.89 Vortex cavitation at suction pipe intake of scaled model of pool-type SFR. (A) Scaled model of upper plenum, (B) vortex cavitation at hot leg intake.

rotating machinery, and thermocouple wells and instrumentation inserted in the flow. FIV is a critical issue in plant design because it can cause catastrophic failure to the structure.

JAEA had promoted the JSFR conceptual design study in the framework of the FaCT project [398]. In the hot-leg piping between the reactor upper plenum to the IHX, a two-loop cooling system was employed to reduce the plant construction cost, and a large-diameter piping system in which the diameter of the pipe is approximately 1.3 m is adopted to maintain the necessary flow rate for core cooling. The coolant velocity through the hot-leg piping is remarkably high at about 4×10^7 Reynolds number. Moreover, an elbow with a short curvature radius (short elbow) in which the curvature radius was equivalent to the piping diameter was used in the hot-leg and cold leg piping systems to configure the compact plant design. Such a high-velocity and thin piping system in the JSFR could cause FIV [398,399]. Fig. 5.90 shows the configuration of the primary piping system in JSFR [399]. Fig. 5.91 shows the images of flow separation in the short elbow observed in a 1/3-scale flow visualization test [400]. In the short elbow, flow separation occurs inner pipe wall at a short curvature radius, and velocity fluctuation causes the FIV downstream of the elbow outlet. To investigate the generation mechanism of fluctuating flow in the short elbow including the influence of the flow conditions upstream of the elbow, water experiments to measure the velocity and pressure fluctuations in the short elbow using 1/10-scale [401], 1/8-scale [402,403], and 1/3-scale [404] test sections of the hot-leg piping and 1/7-scale test section [405] of the cold leg piping system had been conducted. Analytical studies using the unsteady Reynolds averaged Navier-Stokes equation approach [400] and the large eddy simulation approach [406,407] investigating the FIV characteristics regarding fluid force induced by turbulence flow and FIV response in the piping with the short elbow, FIV evaluation methodology for JSFR

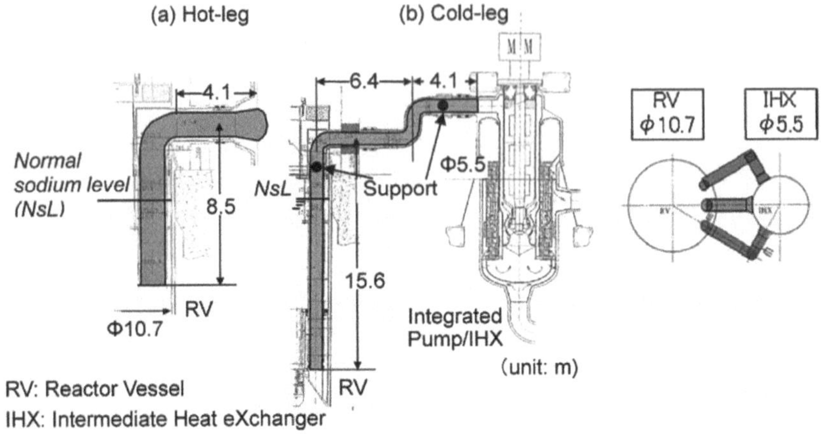

Fig. 5.90 Illustration of JSFR primary piping system [399].

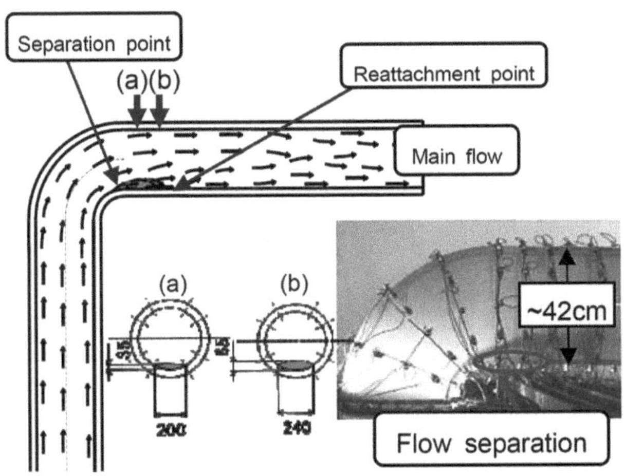

Fig. 5.91 Images of flow separation observed in 1/3-scale flow visualization test [400].

piping had been developed [408,409]. Then, a proposal for the JAEA internal guideline for large-diameter piping FIV evaluation was documented [410,411].

The important thermal-hydraulics problem regarding the FIV that occurred in the Monju is mentioned here though it was not in the RV. In December 1995, a thermocouple well installed on the main pipe broke, and liquid sodium leaked out of the secondary heat transport system of Monju during a 40% power operation performance test [412]. Fig. 5.92 shows the breakage of the thermocouple and the sodium leak path [412]. The thermocouple well suffered from FIV, leading to high-cycle fatigue failure. This vibration was an in-line oscillation associated with symmetric vortex shedding [413]. Fig. 5.93 shows analytical results of the symmetric vortex generated at the back

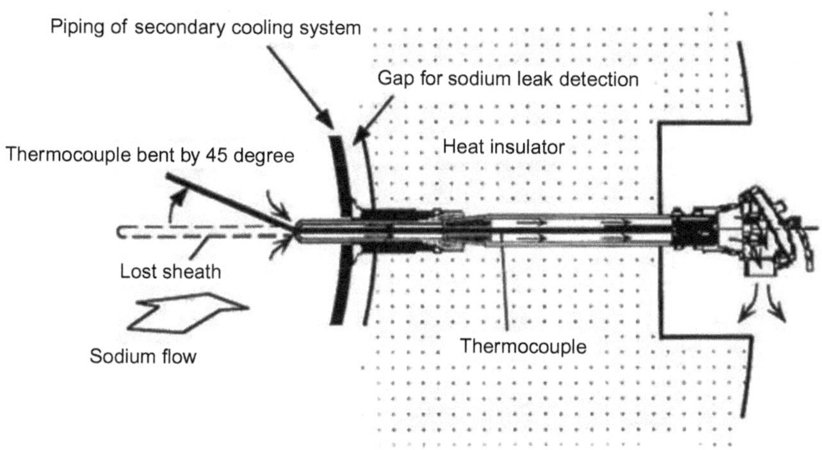

Fig. 5.92 Breakage of thermocouple and sodium leak path [412].

of the thermocouple well [413]. A design guide to preventing the recurrence of thermometer well failure by FIV was proposed [414], and a supplemental document for background information [415] was documented to specify the evaluation methods and design criteria on structural integrity against FIV for thermometer wells inserted into the pipes. Further experiments for determining the design criteria for preventing thermometer well failures by FIV were conducted [416,417] and validating the improved thermometer well design [418].

5.5.4 Thermal-hydraulics in natural circulation decay heat removal

5.5.4.1 Core-plenum thermal interaction

The decay heat removal (DHR) by natural circulation (NC) driven by buoyancy force depending on the large temperature difference between the core as heat source located at a lower position and the air cooler as final heat sink located at a higher position can be a realistic approach in SFRs. A DHR system in NC is significant passive safety measures against accident conditions, e.g., station blackout, and improves SFR reliability by enhancing the heat-removal capability. As one of the most effective DHR systems, the dipped-type DHX (D-DHX) using a heat exchanger installed in the upper plenum has been considered [419,420]. The sodium cooled by the heat exchanger in the D-DHX flows down to the top of the core and penetrates to the fuel subassemblies and the narrow gaps between the fuel subassemblies due to negative buoyancy. During the D-DHX operation, the penetration flow into the fuel subassembly, the interwrapper flow (IWF), and the radial heat transfer between the neighboring fuel subassemblies are observed [421–423]. These specific thermal-hydraulics phenomena are

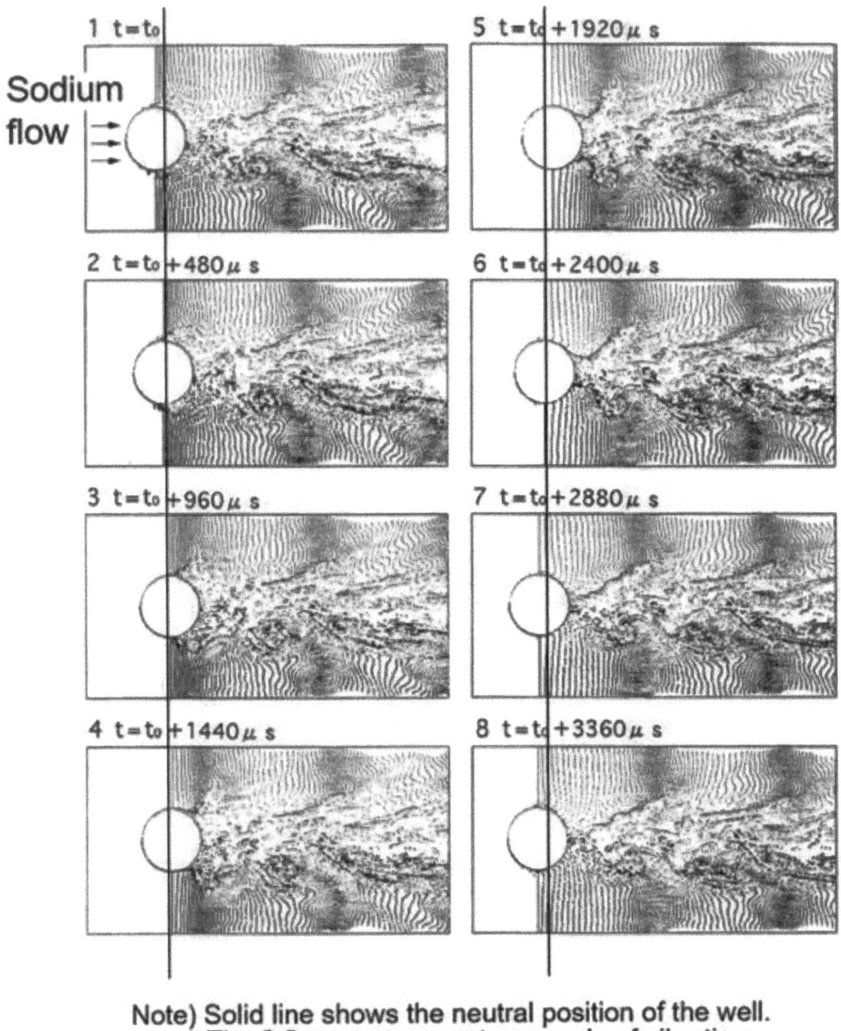

Fig. 5.93 Symmetrical vortex generated at the backside of Tc well [412].

caused by the thermal interactions between the core and upper plenum (core-plenum thermal interaction) and can decrease the maximum temperature in the core. The IWF [318,319,424,425] can remove the heat from the core through the wrapper tubes and the radial heat transfer through the sodium between wrapper tubes occurs due to temperature difference between the fuel subassemblies. To predict these complicated thermal-hydraulics phenomena and evaluate the core-cooling performance during the D-DHX operation in NC conditions, the establishment of an appropriate thermal-hydraulic analysis model in the reactor vessel is strongly demanded.

5.5.4.2 Experimental investigations in NC-DHR

Sodium experiments using PLANt Dynamics Test Loop with Direct Heat Exchanger (PLANDTL-DHX) were implemented to investigate the thermal-hydraulics characteristics during NC-DHR operation using direct reactor auxiliary cooling system (DRACS) and primary reactor auxiliary cooling system (PRACS) as shown in Fig. 5.75 [318,319]. The subassemblies are connected to the upper plenum installing a UIS and the D-DHX for DRACS. The primary loop has an IHX, an electromagnetic pump, flow control valves, and a lower plenum. The secondary loop of the IHX has an air cooler and a pump. As part of the DHR system, the PRACS is also installed on the primary side of the inlet plenum of the IHX. PLANDTL-DHX had a core model comprising seven subassemblies to investigate the thermal-hydraulics in the fuel subassembly and the IWF through the gap between subassemblies. The center subassembly had 37 electrically heated pins, and the six outer subassemblies each had seven heater pins. The heater pins of the center subassembly simulate the core fuel pins of a full-scale reactor, including the spacer wires. From the experiments in PLANDTL-DHX, the steady-state experiments to find the basic characteristics of the inter-subassembly heat transfer and the transient experiments to simulate transitions of the SFR plant from forced to natural circulation operations had been performed.

A project to develop a methodology for evaluation of the thermal-hydraulics behavior under transient conditions from forced to NC operations had been performed [419,426]. In the project, a water experiment on the NC-DHR system was conducted using a 1/10-scale water model for the primary circuit of a loop-type SFR [427]. Fig. 5.94 shows the flow diagram of the 1/10-scale water experiment. From the experiment satisfying the similarity lows regarding the Richardson number, nondimensional heating rate, and local Euler number, the effectiveness of NC-DHR in JSFR was shown for four representative DBE events. A sodium experiment (PLANDTL-1) simulated a part of the primary system and a DHR system with about a 1/7-scale model of the JSFR had been conducted in the project [428,429]. Fig. 5.95 shows the flow diagram of the PLANDTL-1 experiment [428]. In the PLANDTL-1 experiment, the straight-tube (L-shape)-type PRACS was newly equipped in the IHX instead of the helical-tube-type PRACS equipped in the former PLANDTL-DHX, although the simulated reactor vessel remained the main test section. The experimental result simulating the scram transient showed a smooth start-up. The system robustness by the NC-DHR using PRACS in PLANDTL-1 was evaluated [428–430]. The safety analysis method using super-COPD was validated using the results of the 1/10-scale water test simulating the primary system of the SFR [427].

The water experiment PHenomena clarification Experimental Apparatus for Severe AccideNT (PHEASANT) was launched to simulate the RV under severe accident conditions to investigate the cooling path of the coolant fluid between the D-DHX and the core catcher through the degraded core using flow visualization and PIV measurements [431]. A core catcher below the core to hold and cool the debris of the molten core during a severe accident is designed in an SFR. Fig. 5.96A and B shows the test section and typical flow visualization test results,

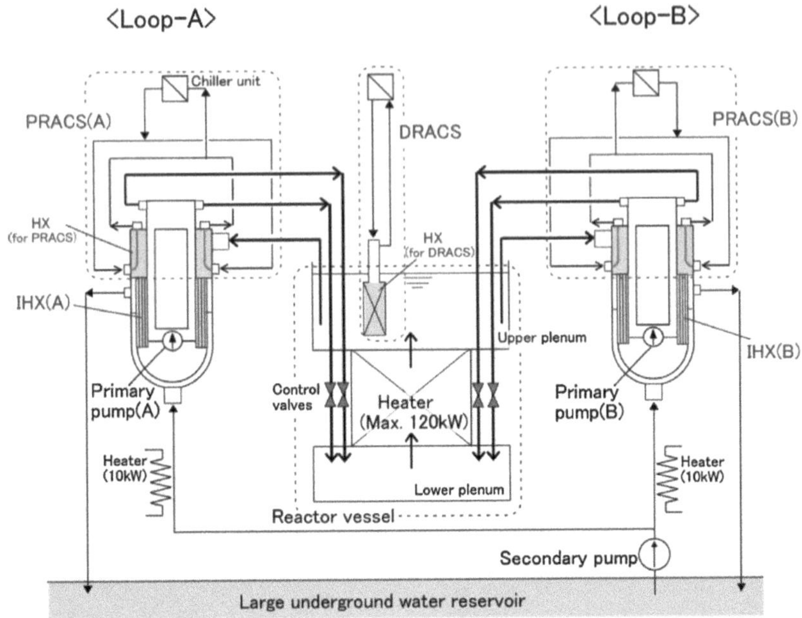

Fig. 5.94 Flow diagram of 1/10-scale water experiment [427].

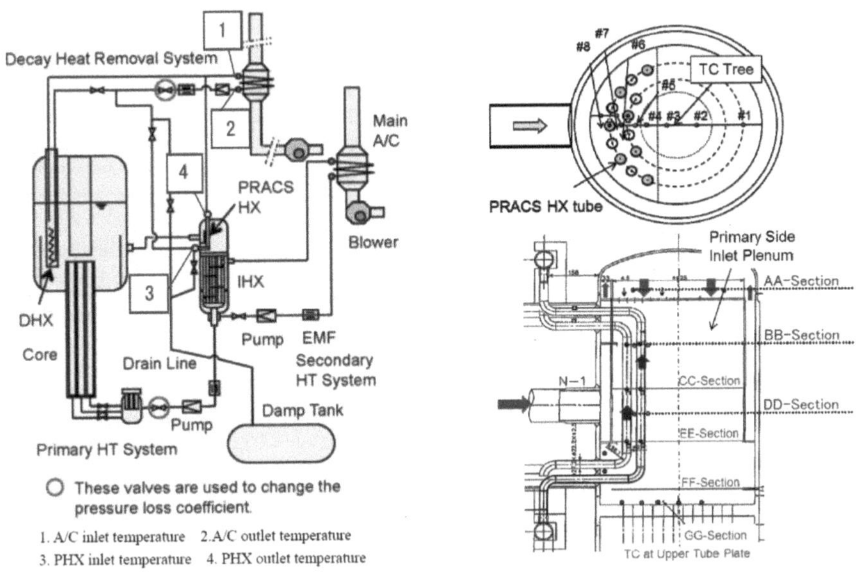

(A) (B)

Fig. 5.95 Flow diagram of PLANDTL-1 experiment and configuration of PRACS [430]. (A) Flow diagram of PLANDTL-1, (B) PRACS in PLANDTL-1.

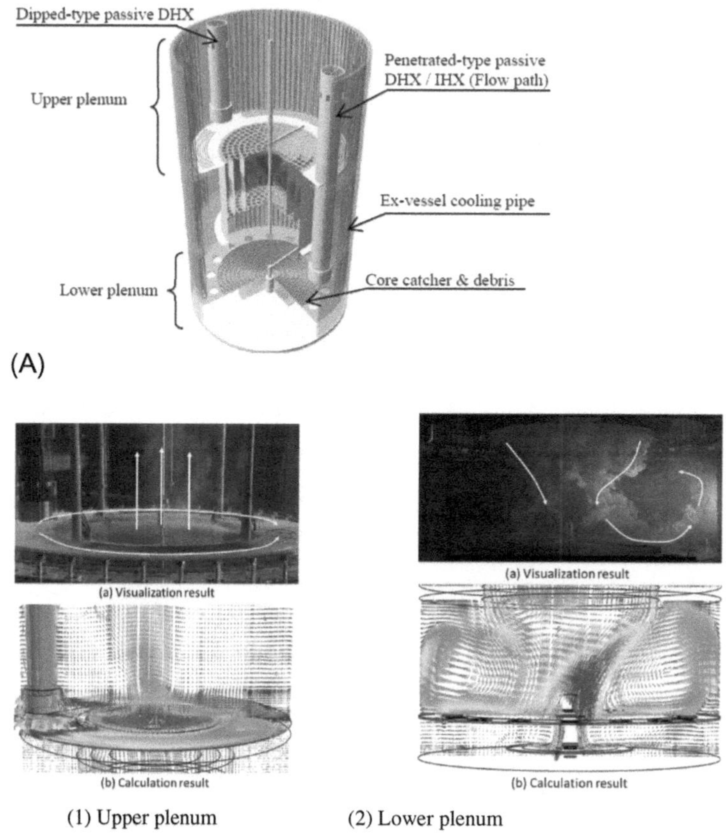

Fig. 5.96 Test section and typical flow visualization test results [431]. (A) Test section of PHEASANT, (B) Comparisons of flow behavior in upper and lower plena.

respectively [432]. The heat sources are distributed in three parts in the RV: the bottom of the hot plenum, core catcher, and core [432]. PHEASNT conducts experiments using various DHR systems: the dipped-type DHX installed in the upper plenum, the penetrated-type DHX connecting the hot and cold plena, the DHX installed in the lower plenum, and ex-vessel cooling system. The flow paths and thermal-hydraulics in the reactor vessel have been investigated, e.g., up and downward flows through the core, thermal stratifications in hot and cold plenum, and interactions of multiple DHR systems during the plant operating conditions using different DHR system(s).

Another NC-DHR sodium experiment (PLANDTL-2) has been conducted in JAEA [433]. Fig. 5.97 shows the PLANDTL-2 test section. The establishment of DHR systems, which are considered to be reliable in severe accident situations, is recognized as one of the key issues in enhancing the safety of SFRs. The test section includes a simulated core and an upper plenum area with a D-DHX. The simulated core comprises 30 heated channels with electric heaters and 25 nonheated channels simulating the fuel

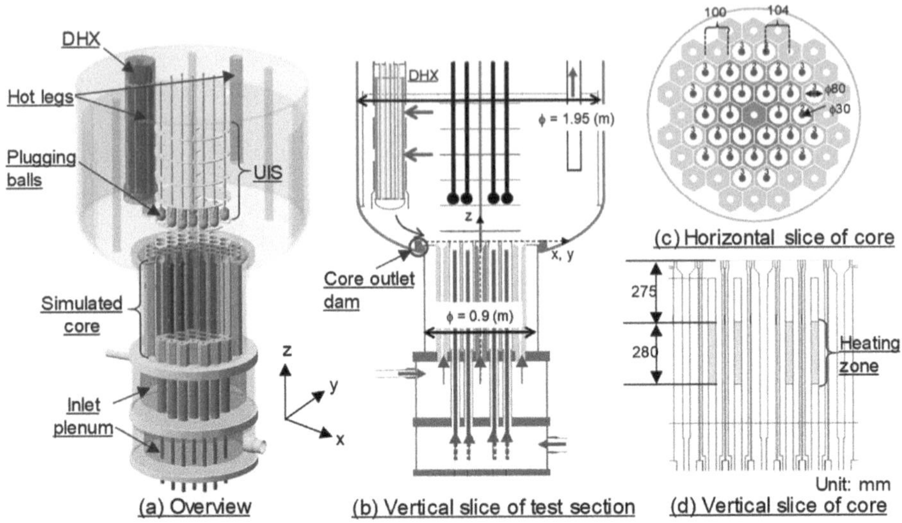

Fig. 5.97 Overview of test section of PLANDTL-2 [433].

subassembly and the reflector subassemblies, respectively. This test section has replaced the former test section of PLANDTL-DHX and PLANDTL-1 in early 2017. The core-plenum thermal interactions during NC-DHR using the D-DHX, e.g., IWF and penetration flow from the upper plenum toward the core region, are investigated.

Fig. 5.98 shows representative temperature distributions at core outlet and center height of heating zone in PLANDTL-2 [433]. In the experiment, cold coolant flowing out from the DHX goes into the core from the outside region and temperature in the peripheral region is low. The cold coolant goes through the gaps between the wrapper tubes (inter-wrapper flow) and removes the heat of the core. Thus, temperature distributions in the core have been quantitatively investigated under a situation where cold coolant from DHX penetrated the core.

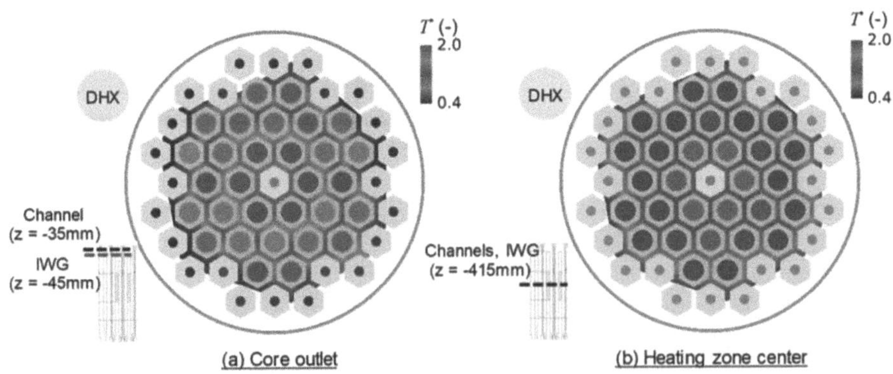

Fig. 5.98 Representative temperature distributions at core outlet and center height of heating zone in PLANDTL-2 [433].

5.5.4.3 Core hot spot evaluation in NC-DHR

A statistical method has been developed for evaluating hot spot temperature in the core to ensure the safety of SFRs [434–438]. A 1D safety analysis using super-COPD evaluated the core hot spot temperature considering the temperature-flattening effect in the core, a 3D fluid flow analysis evaluated the thermal-hydraulics for local convections and thermal stratifications in the primary system and DHRs, and a statistical safety method evaluated the hot spot temperature in the core. Factors, such as geometry, fluid properties, thermal-hydraulics characteristics of reactor components, simulation models, and thermal-hydraulics correlations influence NC behavior and hot spot temperatures corresponding to the highest temperature in the core. Further, some factors influence each other via the buoyancy force. Thus, a statistical method considering variations of influential factors is a powerful tool to evaluate the hot spot temperature with a reliability coefficient. Fig. 5.99 shows the cumulative density function (CDF) of the maximum fuel-cladding temperature at the secondary peak in the loss-of-offsite-power case of the JSFR. In this example, uncertainties for 62 inputs, selected based on phenomena identification and ranking table (PIRT) evaluation, e.g., decay heat, gap conductance between fuel pellet and cladding, and pressure-loss coefficient of fuel pin bundles, were considered [438]. These input uncertainties can be categorized into the known uncertainty group described as probability density functions, and the unknown uncertainty group described only as variability ranges. Monte Carlo sampling for the known and unknown inputs was independently conducted to provide 100 known and 10 unknown input sets. A CDF was estimated using the plant dynamics analysis results using the 100 known input sets with one of the 10 unknown input sets. In total, 1000 cases of plant dynamics analyses were performed, and 10 CDFs were obtained. The core hot spot temperature can be evaluated as the temperature at 95%

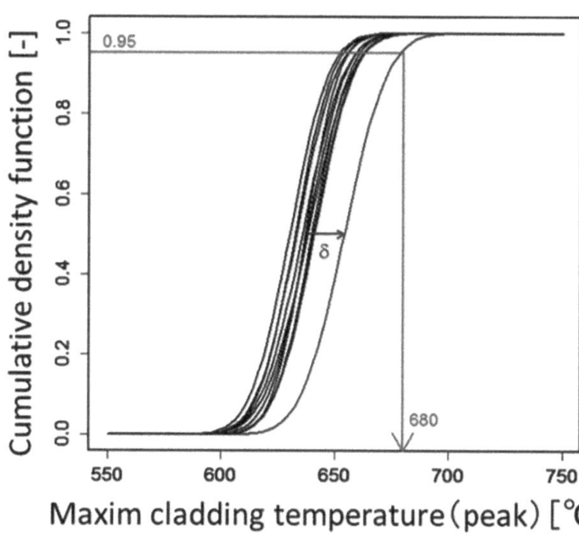

Fig. 5.99 Statistical evaluation of core hot spot temperature during natural circulation operation [438].

cumulative probability of the right-end separated curve (blue colored), obtained by shifting the rightmost CDF curve to the right by modeling errors δ.

5.5.4.4 Multidimensional CFD model

During the operation of the DRACS with the D-DHX, complicated thermal-hydraulics phenomena, core-plenum thermal interactions, were observed as mentioned in Section 5.5.4.1. A multidimensional thermal-hydraulics analysis model in the RV has been developed to evaluate the core-cooling performance under such complicated phenomena. Efforts have been made to develop an analysis model in the RV using the multidimensional thermal-hydraulics code AQUA [318,439,440]. In the model of AQUA, the upper plenum and the simulated core were modeled with a rectangular mesh arrangement in the Cartesian coordinate system and the subassembly was modeled using the concept of subchannel analysis. Numerical simulations for the core comprising seven subassemblies installed in the PLANDTL-DHX facility were performed using the AQUA model in a 180° symmetric sector to investigate the cooling capability by the IWF [318]. Furthermore, numerical simulation for the test rig of the IWF (TRIF) experiment [439,440] was performed. The TRIF experiment modeled the 30° sector of the reactor core and upper plenum to clarify the flow field of IWF in the whole core. For flow visualization using the PIV technique, the wrapper tube of the subassembly was made of transparent resin and the glass heater panels coated by tin-oxide were set at the wrapper tube wall to simulate heating by core fuel subassemblies. These two calculation results reproduced the overall features of the temperature distribution in the core; however, discrepancies were seen in interwrapper gaps compared with experimental results. Because the actual geometry of interwrapper gaps was converted to be rectangular and the symmetric flow field was assumed in the AQUA model, challenges occurred in accurately predicting the IWF, and a full sector model with actual geometry was required. As another analysis model in the RV, the whole-core thermal-hydraulics analysis model using the ACT code was developed [441]. In the ACT model, the different mesh systems of the upper plenum model using AQUA and a subchannel model using ASFRE [311], and the interwrapper gaps model developed based on the ASFRE were coupled. The different mesh interfaces between the upper plenum and the simulated core caused challenges in strictly conserving the mass, momentum, and energy in the interpolation and averaging processes. Thus, a seamless model between the upper plenum and the simulated core was required.

Based on the modeling requirements from previous studies, the analysis model in the RV using the RV-CFD model is constructed. For the design study, the RV-CFD model necessitates simulating with reasonable calculation costs while maintaining accuracy. Then, the subchannel analysis method using the CFD code for fuel subassemblies (subchannel CFD model) to the RV-CFD model was applied to develop an appropriate CFD model of the RV for various operating conditions, including NC-DHR [442]. Fig. 5.100 shows the subchannel CFD model. In the model, the porous media approach was used to consider the fuel subassembly's local geometry, and effective heat conductivity coefficients in the diffusion term of the energy equation were set to fit the actual radial thermal diffusion between subchannels. Fig. 5.101 shows the mesh arrangement of the PLANDTL-1 test section for CFD

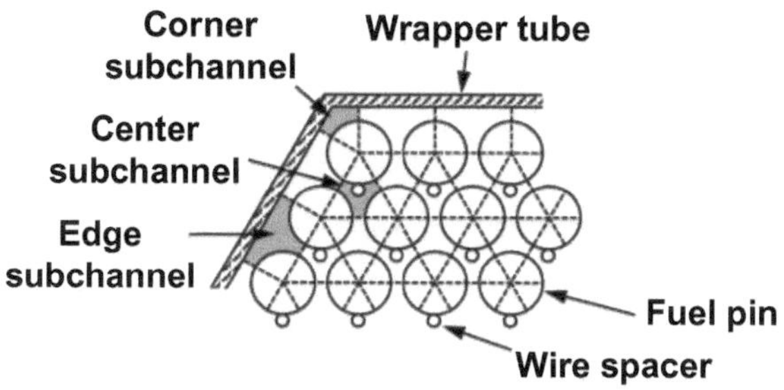

Fig. 5.100 Mesh arrangement for Subchannel CFD model [442].

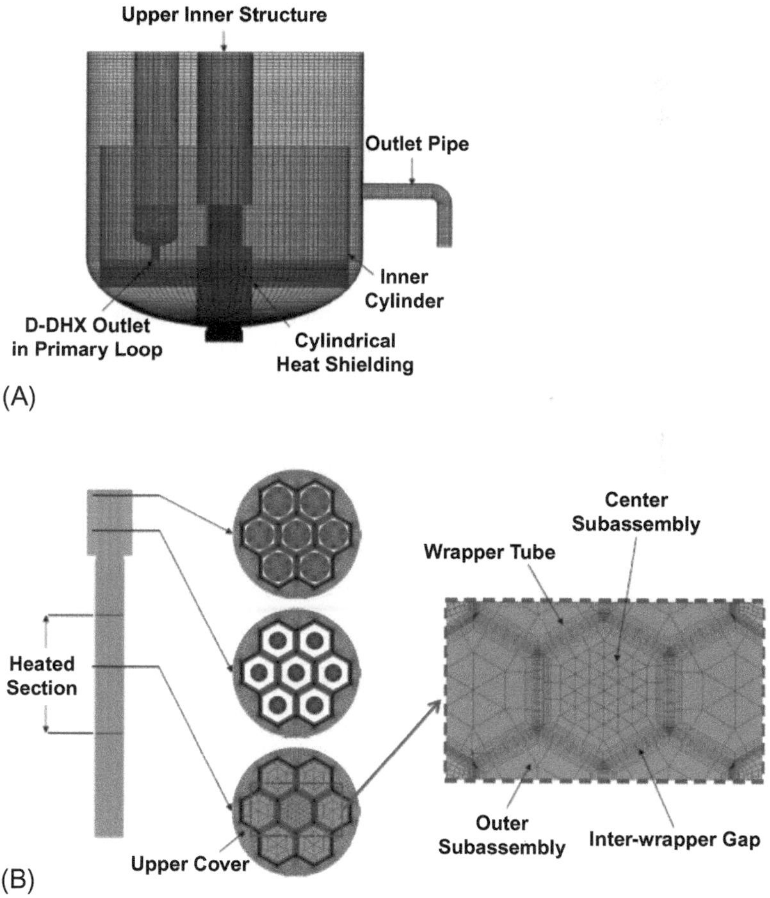

Fig. 5.101 Mesh arrangement of test section in PLANDTL-1 [442]. (A) Upper plenum, (B) Simulated core.

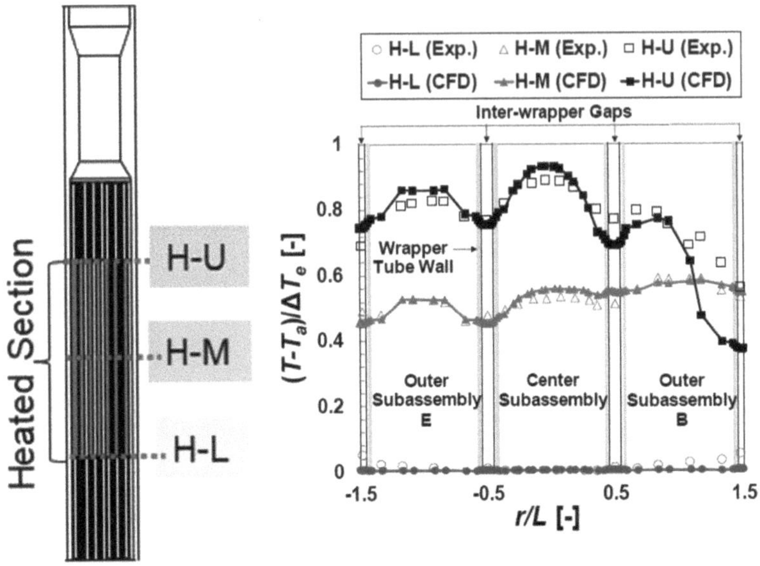

Fig. 5.102 Comparisons of transverse temperature distributions between numerical and experimental results [442].

calculations. Fig. 5.102 shows the transverse temperature distribution comparison in the heated region. Numerical simulation results were compared to experimental data obtained in PLANDTL-1, and the calculated sodium temperature in the core correlated well with the experimental results. The RV-CFD model with the subchannel CFD model applied core thermal-hydraulics analysis during the DRACS with the D-DHX operation under NC conditions. The RV-CFD model of the PLANDTL-2 was developed to understand the thermal-hydraulics characteristics and establish an appropriate model of numerical analysis to reproduce the thermal-hydraulics behavior in NC-DHR [443].

5.6 High-temperature structural materials and structural design

5.6.1 Japanese elevated temperature design code

5.6.1.1 Historical background

Elevated Temperature Structural Design Guide for Class 1 Component of Prototype Fast Breeder Reactor (ETSDG) was constructed to design the Japanese prototype sodium fast reactor (SFR) "Monju" [444,445]. In those days, although codes, standards, and guides had been prepared by the Japanese regulatory authority, nearly all of these documents were only for light-water reactors (LWRs). Power Reactor and Nuclear Fuel Development Corporation developed a structural design guide for Class 1 components of prototype SFR for operation at elevated temperatures, and this guide provided the original design rules developed in Japan with reference to ASME Code Case N-47 [446]. Since the "Monju" was designed as a loop-type SFR, the original elevated temperature code focused on and was applicable to SFRs. Therefore, the original design rules for "Monju" were developed considering operations in the temperature range where creep deformation occurs under low-pressure conditions.

An outstanding feature of the original design rules for "Monju" is their consideration of elastic follow-up phenomena, which allowed the evaluation of the inelastic behavior of structural components subjected to elevated temperatures that bring creep phenomena by utilizing the results of conventional elastic analysis. Specifically, the rules applied to evaluate the strain limit and creep-fatigue damage limit incorporated elastic follow-up concepts [447]. Fig. 5.103 shows the results of mock-up testing to verify the elastic follow-up deformation of the piping system. The concept of elastic follow-up was originally incorporated into ETSDG and then its concept was followed by the RCC-MR [448] code in French and ASME Boiler and Pressure Vessel Code (ASME BPVC) Section III Division 1 Subsection NH [449], which is the predecessor of Section III Division 5 [450]. ETSDG also adopted an original concept of creep damage estimation during stress relaxation at elevated temperature considering low

Fig. 5.103 Piping test for validation of elastic follow-up behavior at elevated temperature.

mechanical loads during the operation of SFRs. Therefore, although the main structures are similar, the detailed rules and evaluation procedures described in the elevated temperature design code of Japan and those in ASME BPVC Section III Division 5 differ from each other, especially in terms of the rules for creep-fatigue damage limits.

In parallel with the development of the original design rules for "Monju," a material strength standard was also developed on the basis of the statistical analysis of domestic material test data. Standard values determined by properties, such as tensile strength, creep strength, and fatigue strength, were developed for use with the design rules for four materials, namely SUS304 (SA-240304), SUS316 (SA-240316), SUS321, and 2¼ Cr—Mo (SA-387 22 cl.1) steels. The design fatigue curves of each material were provided by three tables and figures to consider the loading speed of the estimated event. In addition, the original Japanese rules accounting for the effect of irradiation and a sodium environment on material strength were implemented to apply the design of "Monju."

5.6.1.2 Development of the JSME code

(a) Code for the design and construction of SFR

The first version of the JSME Code, Rules on Design and Construction for Nuclear Power Plants, Section II, SFR (JSME D&C SFR Code) was published in 2005 [451] based on the ETSDG developed for "Monju." The JSME D&C SFR Code was nearly identical to the ETSDG, except for a number of editorial modifications. Compared with other elevated temperature design codes [448,450], the JSME D&C SFR Code was unique because it provided detailed commentaries on its rules, especially those related to elevated temperature design and material strength standards. These commentaries explained the concepts and background of each rule in the code to help users understand and apply them correctly. These commentaries included equations for material behaviors (e.g., tensile, creep, and fatigue curves), as well as detailed descriptions of procedures for setting allowable stress values.

After the launch of "Monju" studies to realize a Japanese demonstration reactor followed [452,453]. Based on these studies, the rules and allowable stress values in the JSME D&C SFR Code have been updated since its first publication. One of the outstanding updating was registration of two candidate materials for the next-generation reactors, 316FR steel and Mod.9Cr-1Mo steel, in the 2012 edition [454]. This registration was based on the results of research and development performed to adopt 316FR steel for reactor vessels and primary/secondary circuits and Mod.9Cr-1Mo steel for steam generators in the demonstration SFR [455–460].

316FR steel was originally developed for SFRs in Japan [461–463]. The chemical composition of 316FR steel was modified from normal 316 stainless steel to enhance its creep rupture strength; specifically, nitrogen (0.06 mass%-0.12 mass %) was added to the base steel instead of carbon (maximum 0.020 mass%) and the phosphorus content was adjusted (0.020 mass%-0.045 mass%). Because the coordination of phosphorus is a unique concept of 316FR steel, the creep rupture strength of this material is superior to similar low-carbon nitrogen-added 316 stainless steels. In fact, the newly replaced intermediate heat exchanger for the Japanese experimental SFR "Joyo" adopted 316FR steel as its main structural material. Therefore, 316FR steel has a track record of being adopted and used in elevated temperature operation of SFR [464].

Mod.9Cr-1Mo steel is basically equivalent to ASTM/ASME Grade 91 steel, which was developed as a candidate material for steam generators of SFRs at the Oak Ridge National Laboratory in the United States; this material demonstrated superior elevated temperature tensile and creep strength compared with conventional 2¼Cr-1Mo steel [465]. ASTM/ASME Grade 91 steel has been widely used in many thermal power plants although this material was originally developed for SFRs.

Similar to those for conventional materials, material strength standards for 316FR and Mod.9Cr-1Mo steels were developed on the basis of the statistical analyses of domestic material test data. In fact, the applicability of the rules in the JSME D&C SFR Code to these materials was validated to register these two materials [455–457]. In particular, the appropriateness of creep-fatigue damage evaluation procedure was demonstrated using the material and structural test data of these materials [458,459,466–471]. Several tests were performed using a large-scale sodium test loop called the Thermal Transient Test Facility for Structures (TTS) at the Oarai Engineering Institute at the Japan Atomic Energy Agency (JAEA). A schematic of the TTS is shown in Fig. 5.104. The TTS includes two independent sodium circuits. One is a hot sodium circuit with an electric heater and the other is a cold circuit with an air cooler. Since electromagnetic pumps installed in each circuit could mix 600°C and 250°C sodium at various proportions by precisely controlling each pump, sodium with complex temperature waveform could be supplied to the test model structure with a constant flow rate, and test model structure could be subjected to programmed cyclic thermal transient. An interesting result observed in the test with a welded vessel model is shown in Fig. 5.105 [467]. In this model, the inner shell is constructed from welded plates of SUS304 and 316FR steel. The crack situation of this inner shell is also shown in Fig. 5.105. Half of the inner shell in the vertical direction is made of SUS304 (left side of Fig. 5.105), while the other half is made of 316FR steel (right side of Fig. 5.105). Many cracks were

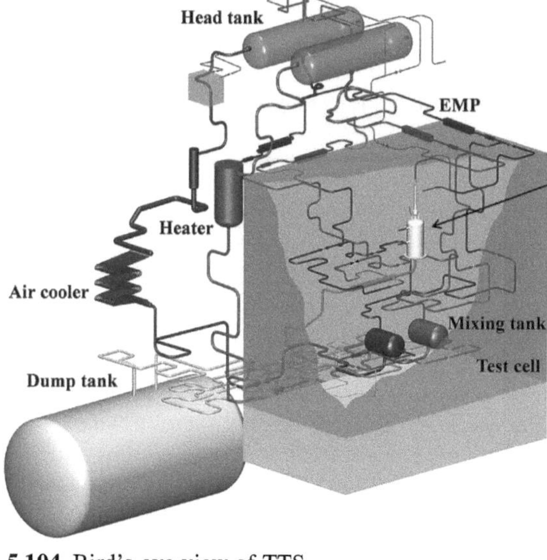

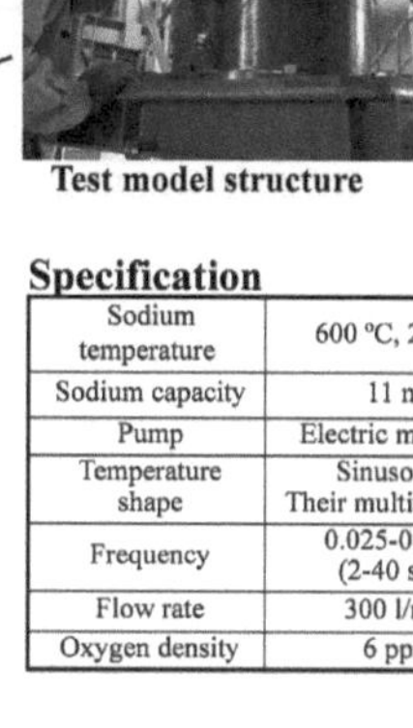

Specification

Sodium temperature	600 °C, 250 °C
Sodium capacity	11 m³
Pump	Electric magnetic
Temperature shape	Sinusoidal, Their multiplication
Frequency	0.025-0.5 Hz (2-40 sec)
Flow rate	300 l/min
Oxygen density	6 ppm

Fig. 5.104 Bird's-eye view of TTS.

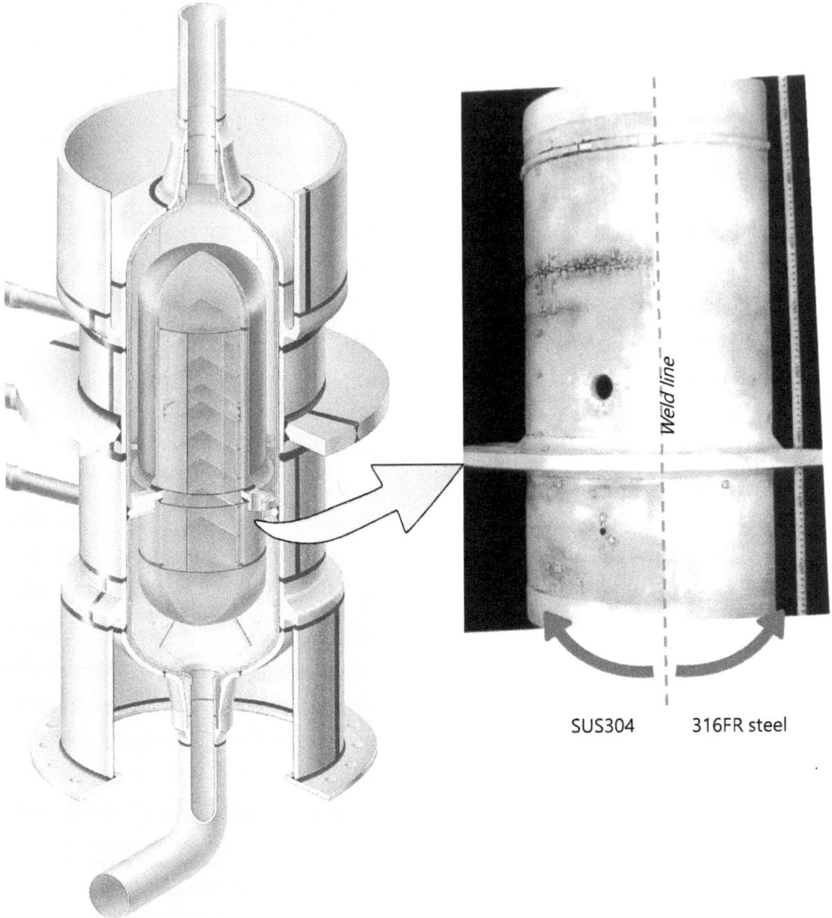

Fig. 5.105 Crack situation of welded vessel model tested by TTS.

observed at the SUS304 side, but no cracks were found at 316FR steel side, most likely because the creep rupture strength of 316FR steel is superior to that of SUS304.

The rules and allowable stress values in the JSME D&C SFR Code have been updated based on the results of various material and structural tests with evaluation of finite element analysis. For example, the effects of a sodium environment were confirmed via the testing of the creep, fatigue, creep-fatigue, and specimen immersion in a sodium environment [472]. A new methodology to predict the long-term creep strength of a structural material was adopted when Mod.9Cr-1Mo steel was registered in the 2012 version of the code [457]. Timewise design factors for creep rupture strength to define S_R, design creep-rupture stress intensity, were optimized by adopting the region-splitting method. The timewise design factors of 10 and 5 for short- and long-term regions were determined for Mod.9Cr-1Mo steel based on the statistical analysis of the collected creep rupture data. These timewise design factors give equivalent reliability to these in conventional design factors for conventional materials. This analysis represents the first application of the region-splitting method to define the allowable stress values of creep strength values; since then, the method has also been applied to ASME BPVC Section III Division 5 to estimate the creep rupture strength of ASTM/ASME Grade 91 steel [450].

In parallel with the JSME D&C SFR Code's registration of Mod.9Cr-1Mo steel, ASME BPVC Section III Code Case N-812 adopted the same criteria in the diagram of creep-fatigue damage envelope to limit the creep-fatigue damage of ASTM/ASME Grade 91 steel [473]. This code case provided an alternate creep-fatigue damage envelope for ASTM/ASME Grade 91 steel. Moreover, the fatigue design curves for ASTM/ASME Grade 91 steel in the ASME BPVC were improved by using the best-fit fatigue curve equation specified in the JSME D&C SFR Code [474]. This example is a notable case demonstrating the introduction of the research and development results for the JSME D&C SFR Code to the ASME BPVC.

Some updates to the JSME D&C SFR Code are based on the ASME BPVC. The original buckling charts for ferritic steels of 2¼Cr-1Mo and Mod.9Cr-1Mo steel were developed adopting the concept and procedure in the ASME BPVC Section II Part D Mandatory Appendix 3 using the material properties in the JSME D&C SFR Code [475].

(b) Code for welding

The first version of the JSME Code on Welding for SFRs was published in 2010 as Code Case of Welding for SFRs [476]. It was developed on the basis of the rules developed for "Monju." The original rules developed for "Monju" basically referred to the rules for LWRs at the time and modified some rules considering the features of SFRs, such as thin vessels and pipes, subjected to low pressure in a sodium environment. The JSME code on the welding for SFR also included detailed commentaries on its rules. This code described differences between the rules for SFRs and LWRs with the corresponding reasons.

In accordance with the registration of 316FR and Mod.9Cr-1Mo steel in the JSME D&C SFR Code, the Welding Code for SFR was updated and published as an independent Code on Welding for SFRs [477]. In the updated code, several rules were added to support new materials, especially Mod.9Cr-1Mo steel. One important rule added described postweld heat treatments for Mod.9Cr-1Mo steel sorted into martensite steel.

5.6.2 Structural materials

5.6.2.1 Features of the material properties in the JSME code

The material strength standard in the JSME D&C SFR Code provides allowable stress values, as shown in Table 5.10, as well as the physical properties of thermal expansion, elastic modulus, and Poisson's ratio. The effects of environmental factors, such as irradiation and sodium, on the allowable stress values were described in Appendices of the rule. In addition, material characteristic equations (e.g., equations for creep rupture, creep strain, and the best-fitting curve for a low-cycle fatigue life) necessary to determine the allowable stress values were provided in the commentary.

5.6.2.2 Path forward the structural material development

(a) Time-dependent allowable stress values to ensure a 60-year design life

Time-dependent allowable stress values were determined for up to 300,000 h for all materials in the current JSME D&C SFR Code. Design studies for a next-generation

Table 5.10 Allowable stresses and their definitions.

Allowable stresses	Definitions	
S_y: Design yield stress	S_Y (Specified minimum yield strength at RT) $\times$ R_Y (ratio of the average temperature-dependent trend curve value of yield strength to the RT yield strength)	
S_u: Design tensile strength	1% Failure probability of tensile strength	
S_m: Design stress intensity (time independent)	Min.	(a) S_T (Specified minimum tensile strength at RT) $\times$ 1/3 (b) $S_u \times 1/3$ (c) $S_Y \times 2/3$ (d) $S_y \times 2/3$(Austenite stainless steel: $S_y \times 0.9$)
S^*: Allowable tensile stress	Min.	(a) $S_T \times 1/3.5$ (b) $S_u \times 1/3.5$ (c) $S_Y \times 2/3$ (d) $S_y \times 2/3$ (Austenite stainless steel: $S_y \times 0.9$)
S_O: Maximum allowable stress intensity	Min.	(a) $S_T \times 1/3.5$ (b) $S_u \times 1/3.5$ (c) $S_Y \times 2/3$ (d) $S_y \times 2/3$(Austenite stainless steel: $S_y \times 0.9$) (e) 100% of the average stress to produce a creep rate of 0.01%/1000 h (f) 80% of the minimum stress to generate rupture at the end of 100,000 h (g) 67% of the average stress to generate rupture at the end of 100,000 h
S_t: Design stress intensity (time dependent)	Min.	(a) $2/3 \times S_R$ (b) 80% of the minimum stress to generate the onset of tertiary creep (c) Minimum stress to generate 1% total strain
S_R: Design creep-rupture stress intensity	The minimum stress to generate rupture (not to exceed S_u)	
S_r: Design relaxation strength	Stress calculated with $1.5S_m$ as initial stress by the creep strain equation using strain hardening law	
ε_t: Allowable strain range	Best-fit fatigue curve for strain rate 10^{-3}–10^{-8} mm/mm/s with lesser one of design factor 2 on strain range and 20 on number of cycles to failure	

RT: room temperature.

SFR have specified a 60-year plant design life, which corresponds to 500,000 h, to enhance economic competitiveness and reduce radioactive waste. Long-term creep behaviors have been studied to determine time-dependent allowable stress values for 316FR and Mod.9Cr-1Mo steels to ensure this design life. Figs. 5.106 and 5.107 illustrate the creep rupture test data and curves that have been developed for a 500,000-h design for 316FR and Mod.9Cr-1Mo steels, respectively [478]. Times of obtained rupture data at 550°C are up to 166,160 and 175,800 h for 316FR steel and Mod.9Cr-1Mo steel, respectively. The region-splitting

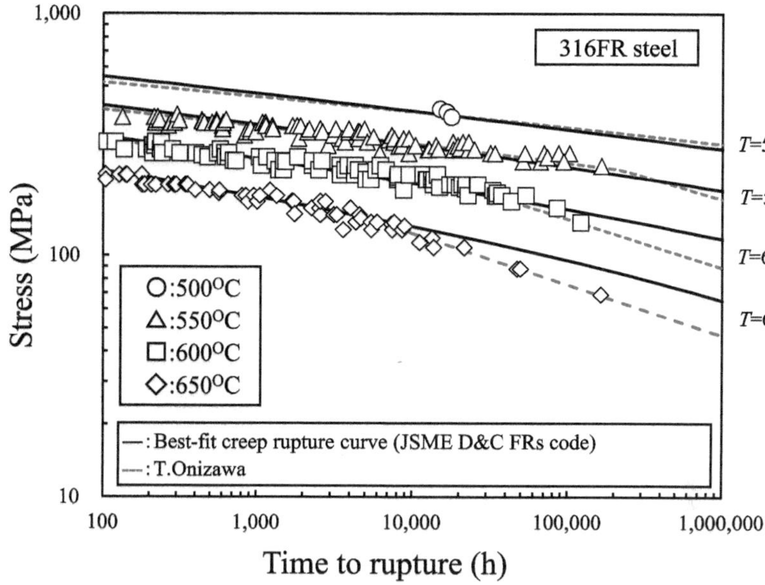

Fig. 5.106 Relationship between creep data and creep rupture equations of 316FR steel [478].

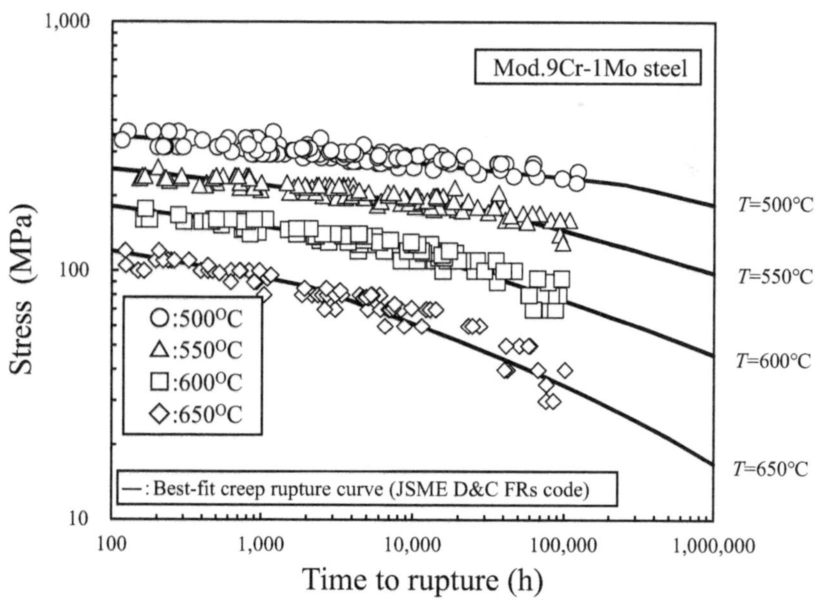

Fig. 5.107 Relationship between creep data and creep rupture equation of Mod.9Cr-1Mo steel [478].

method will also be applied to 316FR steel to improve the accuracy of predictions in long-term regions. The creep strain equations will also be revised considering the adoption of the region splitting method in the JSME D&C SFR Code. Design factors to develop the S_R values will also be optimized in the same manner as those adopted for Mod.9Cr-1Mo steel. These curves, which are applicable for up to 500,000 h, will be introduced to the forthcoming edition of the JSME D&C SFR Code. Feasibility studies of the extension of the time-dependent allowable stress values for the conventional materials SUS304 and SUS316 are also ongoing.

(b) High-cycle fatigue

The design fatigue curves in the current JSME D&C SFR Code can be applied for only up to 10^6 cycles. The improvement of the design fatigue curves on number of failure cycles is required to achieve high-cycle loading designs for a 60-year plant life under mechanical and flow-induced vibrations, thermal striping, and departure from nucleate boiling vibration loading in steam generators. High-cycle fatigue tests with advanced fatigue machines have been performed on 316FR and Mod.9Cr-1Mo steels prior to other materials to improve the design fatigue curves [479]. These tests have been conducted using laser- and ultrasonic-controlled fatigue test machines instead of conventional direct strain measure-controlled fatigue tests, and the quality assessment of the data obtained from these tests is ongoing. In addition, the effect of mean stress on the fatigue life has been considered to enhance the design fatigue curves. Fig. 5.108 shows the fatigue data of SUS304, which were obtained to validate the adequacy of the test data collected by using a laser-controlled test machine under high-frequency conditions.

(c) Product forms

The material strength standards described in the JSME D&C SFR Code provide the allowable stress values depending on the product form. Table 5.11 shows the relationship between the standardized product form and materials in the 2020 edition of the code [480]. Since the supported product form depends on the component specification, the tube forms

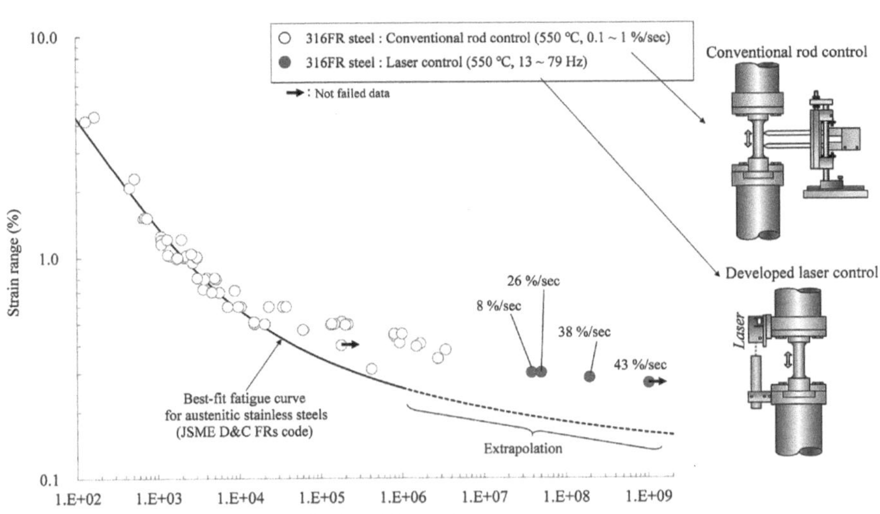

Fig. 5.108 Relationship between rod-controlled test data and laser-controlled test data.

Table 5.11 Codified product forms of materials.

Material	Product form			
	Plate	Forging	Pipe	Tube
SUS304	○	○	○	○
SUS316	○	○	○	○
316FR steel	○	○	△	△
SUS321	–	○	–	○
$2^1/_4$Cr-1Mo steel	○	○	–	○
Mod.9Cr-1Mo steel	○	○	△	○

○: Codified; △: Under preparation; –: Not supported.

of SUS321 and 2¼Cr-1Mo steel were registered as heat exchanger tubes for the steam generators of "Monju." According to the specifications of the next-generation SFR, the pipe form of 316FR and Mod.9Cr-1Mo steels and the tube form of 316FR steels, which may potentially be used in direct heat exchangers of decay heat-removal system, will be registered after the verification of the applicability of the standard values.

(d) Weld joint reduction factor

The current design rules do not provide detailed evaluation procedures for the primary stress limit of weld joints. When the "Monju" was designed and constructed, all the weld joints were located far from the evaluation section where the expected stress is not negligible. Although they increased the construction cost, large forgings were applied to many components to achieve this concept. For example, a large-diameter (>7000 mm) forging ring, measuring 50 mm in thickness, was supplied for the gas-liquid transition area of the reactor vessel [481]. Ordinarily, weld joints have higher tensile and short-term creep strength than base metals; however, this relation may not be guaranteed for long-term creep strength. In fact, the creep strength of the weld joints of Mod.9Cr-1Mo steel is inferior to that of the base metal because of the occurrence of Type IV cracking, which has been considered in the design rules of thermal power plants [482]. The concept of weld joint reduction factor (WJRF) has been developed for 316FR and Mod.9Cr-1Mo steels well ahead of conventional materials to prevent creep failure during the long-term operation of SFR plants. In addition to creep testing, metallurgical and theoretical studies, such as thermodynamic equilibrium calculation, have been employed to predict long-term creep strengths and establish the Japanese original WJRF for both steels.

(e) Welding specification

While 316FR steel was originally developed to enhance creep strength in Japan, a welding method must be developed to maintain this advantage in the evaluation of weld joints. Several welding procedures and pre –/posttreatments have been surveyed via parametric studies in weld joint manufacturing, and the results of these works will be assessed and incorporated into the Welding Code for Fast Reactors.

No special issues regarding the production of welded joints for Mod.9Cr-1Mo steel are expected because sufficient experience with this type of steel has been collected from thermal power plants. Several rules have been developed for adequate repair welding. When any defect is detected during nondestructive inspection after welding, the location of the root remachining and rewelding area should be restricted because Type IV cracking occurs in the heat-affected zone generated by welding.

5.6.3 Structural design

5.6.3.1 Features of the design rules in the JSME D&C SFR code

The basic aim of the design rules in the JSME D&C SFR Code is to prevent failure due to critical failure modes. Four short-term modes prevented by the rules in the JSME D&C Code for LWRs are also prevented by the rules in the JSME D&C SFR Code. These critical failure modes are:

- Ductile rupture from short-term loading
- Excessive plastic deformation
- Fatigue failure
- Elastoplastic buckling due to short-term loadings

In addition to the aforementioned modes, four other modes detailed later are prevented by the rules in the JSME D&C SFR Code that account for creep behavior during long-term operation under elevated temperature conditions.

- Creep rupture from long-term loading
- Excessive creep deformation
- Creep-fatigue failure
- Creep buckling due to long-term loadings

Fig. 5.109 compares the main structures described in the design rules in the JSME D&C SFR Code with those for LWRs [483].

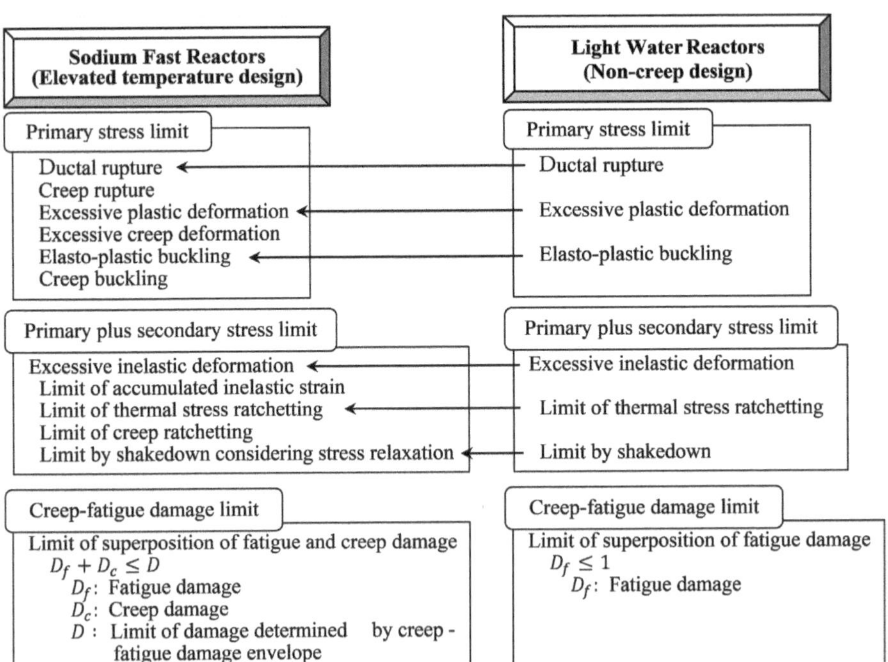

Fig. 5.109 Structure of design rules in JSME D&C FRs Code.

5.6.3.2 Path forward the design rule development

(a) Precise evaluation of creep-fatigue damage

The concept of elastic follow-up is used in the JSME D&C SFR Code to predict inelastic behavior from elastic analysis results, e.g., stress relaxation behavior due to creep in creep-fatigue damage evaluation. When the primary stress is low, the current rules define the elastic follow-up coefficient "q" as equal to 3, which is supposed to provide a conservative prediction of inelastic behavior. Therefore, $q = 3$ could estimate larger creep damage during stress relaxation than that expected for most components subject to specific evaluation conditions, i.e., the elastic follow-up coefficient of a specific component subject to a specific evaluation condition may be further optimized. Fig. 5.110 shows an outline of the optimization of the elastic follow-up coefficient. Fatigue and creep damage are evaluated based on the strain range and time history of stress relaxation, respectively. The strain range and time history are calculated using the elastic follow-up coefficient with the results of elastic analysis. Thus, more precise values of fatigue and creep damage can be evaluated by using the optimized value of q.

Robust procedures based on inelastic finite element analysis have been developed to estimate practical values (i.e., <3) of the elastic follow-up coefficient [484]. Based on the developed technique, the rules governing the concrete calculation procedures, including the material property definitions for inelastic analysis, will be established as a Code Case of the JSME D&C SFR Code.

(b) Creep-fatigue damage limit of weld joints

The development of design evaluation procedures for weld joints is an important challenge for elevated temperature codes. In elevated temperature plant systems, such as

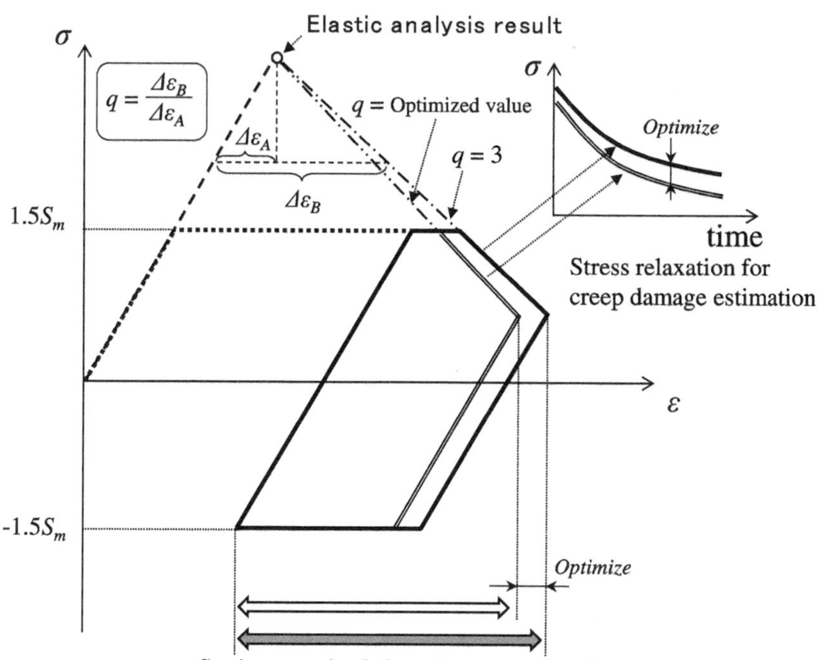

Fig. 5.110 Outline of optimization of elastic follow-up coefficient.

thermal power plants, various weldments are used for pressurized components. In general, load-controlled stress exerts an important influence on the structural design for weldments in conventional pressurized systems. Therefore, the tensile and creep strength of each welded joint should not be lower than the standard values for the base metal to prevent early failure during weldment. Conversely, cyclic deformation-controlled stress, such as cyclic thermal loading, exerts a dominant effect on the SFR components. The stress/strain distribution due to metallurgical discontinuities is believed to cause local stress/strain concentration in the weld joint area. Differences in mechanical properties of tensile and creep strength between base metal, weld metal, and heat-affected zone (HAZ) cause stress/strain concentration in the weld jointed area, and it brings inferior creep-fatigue strength to that of base metal and/or weld metal. Fig. 5.111 shows the distribution of the Vickers hardness on a weld joint of Mod.9Cr-1Mo steel; here, metallurgical discontinuities may clearly be observed. The Vickers hardness of the weld metal is higher than that of the base metal. Moreover, the hardness of the HAZ of the base metal gradually changes, and a softer area with inferior hardness relative to the base metal may be identified.

Evaluation methods of the creep-fatigue life of weld joints of SUS304, 316FR, and Mod.9Cr-1Mo steels have been developed by considering the stress/strain redistribution due to metallurgical discontinuities [485–487]. Based on these studies, investigations to establish adequate rules for the JSME D&C SFR Code are underway. Previous studies proposed creep-fatigue evaluation methods considering the effect of metallurgical discontinuities in weld joints by giving penalty factors at several steps of the evaluation procedure. However, these evaluation procedures are complicated and require the identification of the location of the weld joint prior to evaluation. The concept of assigning penalty factors for each fatigue and creep damage only has been developed to simplify the evaluation procedure and detect feedback easily during the design phase. Fig. 5.112 shows the concept of the evaluation method developed for weld joints. Rules, as well as the WJRF, for 316FR and Mod.9Cr-1Mo steels will be developed prior to those for other materials. In the

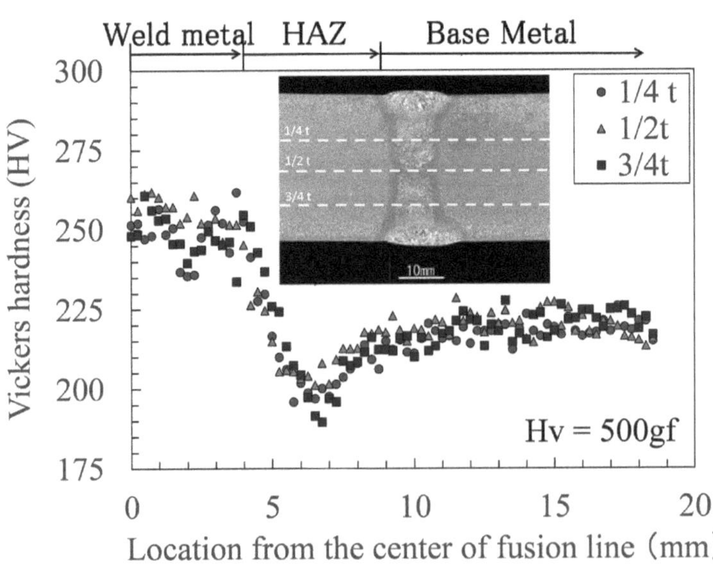

Fig. 5.111 A sample of weld joint picture of Mod.9Cr-1Mo steel with Vickers hardness test.

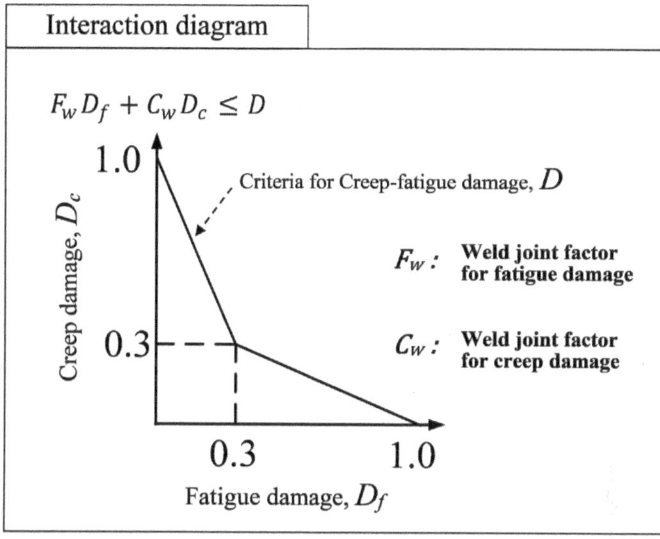

Fig. 5.112 Concept of evaluation method of weld joint to be developed.

development program, the results of material and structural tests and inelastic analyses of weld joints under cyclic loading collected to verify the adequacy of the rules to be established.

(c) Buckling evaluation

The buckling strength evaluation prescribed in the current JSME D&C SFR Code is applicable to cylindrical vessels subjected to axial and bending loads. Rules using this evaluation mainly focus on the plastic buckling of thick cylindrical vessels, and the applicable configuration is limited by the ratio of the outer diameter (D_o) to the thickness (t), i.e., D_o/t, using the material properties of elastic modulus (E) and design yield stress (S_y). According to the applicable limits, the current rules for buckling are applicable to vessels made of Mod.9Cr-1Mo and austenitic steels (e.g., SUS304 and 316FR steel) with D_o/t of up to approximately 250 and 600, respectively. However, the diameter of cylindrical vessels of the next-generation SFR will be increased to enable greater generation capacity, and the vessels are expected to become relatively thinner. Moreover, given the increase in design seismic loads in Japan, adopting seismic base isolation systems into the next-generation SFR is necessary. When horizontal seismic base isolation is adopted, cylindrical vessels are subject to cyclic vertical seismic loads with long-period horizontal seismic loads although the current rules do not explicitly consider shear loads.

A demonstration test and a research program on the buckling of thin-walled cylindrical vessels under seismic loads were conducted by the Central Research Institute of Electric Power Industry of Japan in the 1990s [488]. This research proposed buckling strength evaluation equations that considered elastoplastic axial compression, bending and shear buckling, and their interaction. Fig. 5.113 shows the relationship between the buckling test results of austenitic steels and the criteria developed without design factors. All of the data located the outside of the envelope line and it indicates that the criterion is a conservative estimation. Sophisticated buckling rules have been developed as a Code Case of the JSME D&C SFR Code based on the results of this study. Because basic studies mainly focus on austenitic steel vessels under seismic loads, the applicability to the results of Mod.9Cr-1Mo

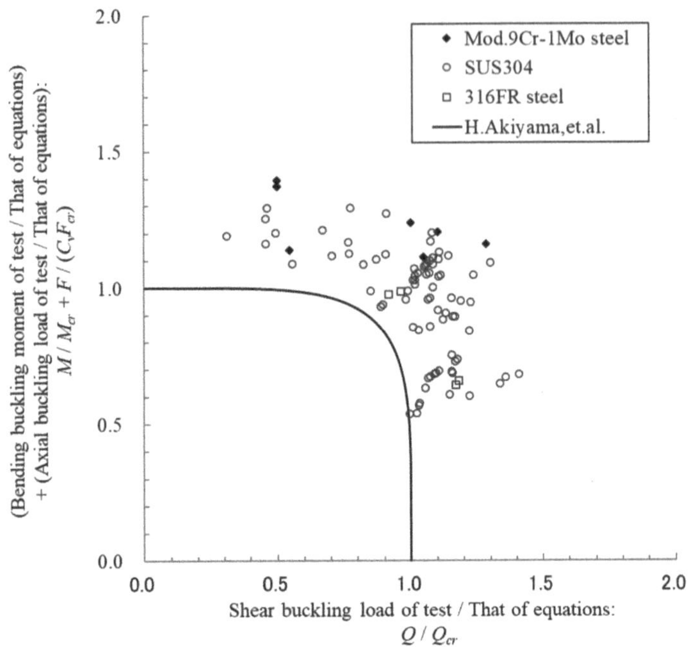

Fig. 5.113 Relationship between buckling test and criteria of developed equation [489].

steel, which has high yield stress and a low tangent modulus after yielding, will be verified and modified as necessary. The developed rules will consider several concepts ranging from buckling to elastoplastic axial compression, bending, and shear buckling (and their interactions) in both high- and low-yield stress materials, as well as larger cyclic vertical loads compared with horizontal ones under the horizontal seismic base isolation system [489].

Elastoplastic buckling analyses have been conducted to confirm the applicability and modification needs of the buckling evaluation equations to Mod.9Cr-1Mo steel and the reduction effect of the buckling strength under a cyclic axial load. In addition, a series of buckling tests and finite element analyses corresponding to the test conditions of Mod.9Cr-1Mo and austenitic steels were carried out to examine the buckling mode, strength data, and to verify the finite element analyses. Fig. 5.114 shows the relationship between the results of the buckling test and estimated deformation in the finite element analysis. The deformation profile observed in the test clearly agrees well with the analysis results [490].

(d) Inelastic analysis

Users familiar with finite element analysis may also be familiar with inelastic analysis. Many constitutive equations and models to simulate material stress-strain relationship have been investigated, and some of these have been implemented in the commercial program of finite element analysis. Direct utilization of the results of inelastic analysis is useful in evaluations seeking to optimize component design. However, incorporating practical and useful evaluation procedures to the design rules by using inelastic analysis is difficult because many undefined conditions (e.g., detailed characteristics of the purchased material and the actual order of loading by transients in the plant life) may exist in the design phase. Constitutive equations developed for the practical modeling of the inelastic behavior of materials require detailed information on these conditions for accurate simulation.

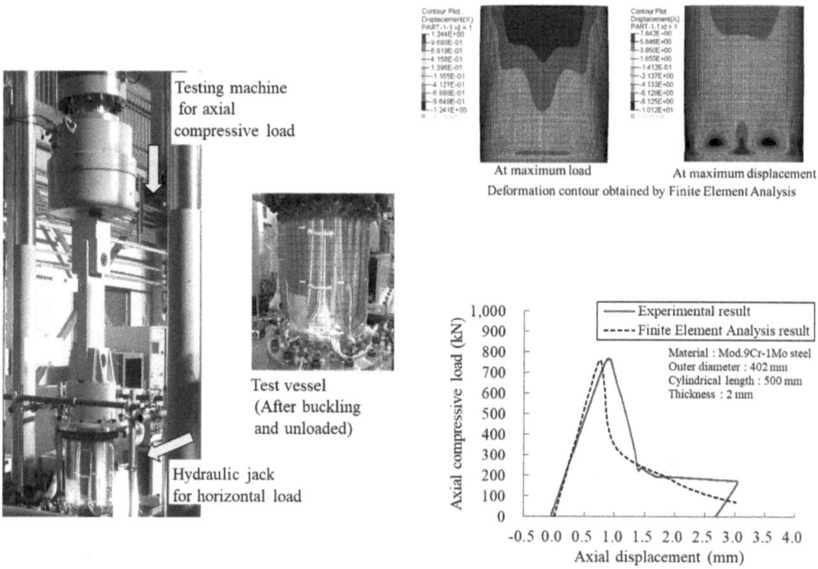

Fig. 5.114 Relationship between buckling test and analysis deformation [490].

Therefore, a consideration on how to consider conservativeness for the uncertainty on analysis conditions is required to establish the design rules using the inelastic analysis.

A guideline for inelastic design analysis has been proposed to resolve this issue [491]. This guideline was developed for the design of components made of austenitic steels in SFRs. Some unique techniques in this guideline may be applied to enable the use of the results of inelastic analysis. For example, in the creep-fatigue damage evaluation, only the strain range is obtained through inelastic analysis. In general, the stress and strain ranges are used for creep and fatigue damage evaluation, respectively; however, defining constitutive equations that could provide conservativeness for stress and strain at the same time is impossible. Therefore, in this guideline, only the strain range is estimated using inelastic analysis via a constitutive equation that provides conservativeness for strain range. Fatigue damage is estimated from this strain range and the design fatigue curves. Creep damage is estimated by the time fraction rule considering stress relaxation, the initial stress of which is determined from the cyclic stress-strain curve entered at conservative strain range calculated by inelastic analysis. This concept of initial stress determination is similar to the current JSME D&C SFR Code, which is based on elastic analyses. The JSME D&C SFR Code provides evaluation procedures for creep damage derived from peak stress relaxation in which the initial stress is determined from the cyclic stress-strain curve and the conservative strain range is predicted considering the elastic follow-up concept. Therefore, the guideline harmonizes the JSME D&C SFR Code and useful for practical use.

This guideline and the inelastic analysis techniques applied to estimate suitable elastic follow-up factors (Section 5.6.3.2(a)) are the basis behind the development of a new original guideline for inelastic design analysis. All of the rules in the guideline have been reevaluated, and its practical utility for component design will be confirmed in design studies. Developments in usability and applicability are important, and additional consideration for the incorporation of Mod.9Cr-1Mo steel to the guideline is required.

5.6.4 Structural integrity evaluation in beyond-design base events

5.6.4.1 Material testing

After the 1F accident, the structural integrity evaluation of components in beyond-design base events became an important issue for nuclear plant systems. During beyond-design base events, temperatures and pressures may exceed the range supported by the design and construction codes, and the functions required may differ from those specified in the design. Material properties corresponding to beyond-design base event conditions are required in the evaluations of whether the components maintain their performance under such conditions. Thus, material tests under temperature conditions beyond the upper limit of the design temperature have been carried out [492]. Fig. 5.115 shows the creep test results of austenitic steels collected at temperatures of up to 1000°C. Besides these data, many other types of material data, such as tensile and creep data over 1000°C, weld joint data, and oxidation effects, are required for component evaluation in beyond-design base events. In addition to those on austenitic steels, tests on Mod.9Cr-1Mo steel are underway. The environmental effect of sodium on the material properties in beyond-design base events may also be investigated for SFRs.

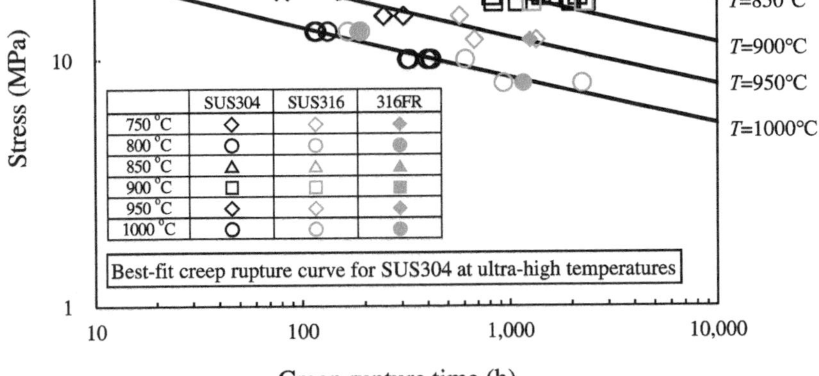

Fig. 5.115 Creep test result at temperatures of up to 1000°C in austenitic steels [492]. Figure courtesy of IASMiRT.

5.6.4.2 Structural testing

A useful tool to evaluate the structural integrity of components in beyond-design base events is finite element analysis. It could simulate the deformation behavior and stress generated in various components in the case of beyond-design base events. However, no detailed procedure for plant safety assessment has yet been authorized. The conservativeness required for plant safety assessment may differ from that applied for plant design and construction. The most likely simulation obtained by inelastic analysis without conservativeness may be applied to the predictions of event transition, although this strategy depends on the purpose of the simulation. Implicit/explicit conservativeness is required for criterion whether demanded functions of components are maintained in beyond-design base events.

Few studies evaluating the structural integrity of components in the case of beyond-design base events at ultrahigh temperatures have been reported for plant safety assessment. Therefore, the development of structural integrity evaluations for plant safety assessment in case of beyond-design base events is a challenging issue. Investigations on this topic are essential to improve the design and construction of nuclear plants, including SFRs and LWRs, especially after the 1F accident. Collecting various material and structural data, for example, from a pressurized burst test with ultrahigh-temperature heating, is necessary to verify the simulation techniques and the failure criteria. Through these experiences, a variety of simulation procedures and the technique of material properties setting should be developed.

Investigation of the following issues may be important to advance these developments: equations, including the relevant statistical information, of material properties, criteria considering the triaxiality of stress/strain, and the treatment of the contact problem of each component under ultrahigh temperatures.

5.7 Seismic design

This section introduces R&D activities related to the seismic design of SFRs. Establishing the seismic design of structures and mechanical components is a critical challenge in realizing SFRs in Japan that has many experiences of severe earthquake disasters. As there are typical examples of seismic design-related SFRs, this subsection introduces piping seismic evaluation by inelastic response analysis and seismic response analysis code of core. Developing a three-dimensional isolation system is another important R&D activity and is introduced in Section 4.4 Pool-type SFR.

5.7.1 Piping seismic evaluation by inelastic response analysis

In comparison with existing pressurized water reactors, SFRs operate under elevated temperatures. To mitigate thermal stress, the mechanical components of SFRs are designed with relatively thin-walled structures that are flexible. Meanwhile, thick-walled design is preferable in the viewpoint of seismic design. Therefore, balancing elevated temperature design and seismic design is important when developing SFRs' structural designs.

Large earthquakes that exceeded the design level have struck numerous nuclear power plants in Japan in the last decade, including the Tohoku earthquake off the Pacific coast in 2011 and the Niigata-ken Chuetsu in 2007. To ensure the seismic integrity of nuclear power plants, the important issues that have been focused on are confirming the ultimate strength under a high-level earthquake beyond the design condition and clarifying the design safety margin up to the ultimate strength.

A series of dynamic failure tests were conducted to clarify the ultimate strength of piping systems, which are important mechanical components retaining coolant material. The dynamic failure test for SFRs, as shown in Fig. 5.116, was performed based on the knowledge that stress is easily concentrated in elbows in piping [493]. High-level excitation significantly exceeding the design level was needed to cause the elbow to fail in the test. Finally, cracks in the axial direction were observed on both sides of the crown part of the elbow (Fig. 5.117). Therefore, it was confirmed from the test that the failure mode of the elbow was low-cycle fatigue with crack initiation, propagation, and through-wall penetration. No structural instability such as plastic collapse was observed, even for the thin wall piping for SFRs. Furthermore, finite element analysis (FEA) was used to confirm that inelastic analysis could reproduce the failure behavior of the elbow (Fig. 5.118).

So far, various dynamic tests, including the test mentioned earlier, have been performed on piping systems and piping components up to failure, and the results have shown that piping is significantly robust to dynamic loading, which is attributed to the increasing hysteretic energy due to plasticity at a local part of the piping [494,495]. Furthermore, it has been widely recognized that the dominant mode of failure is not

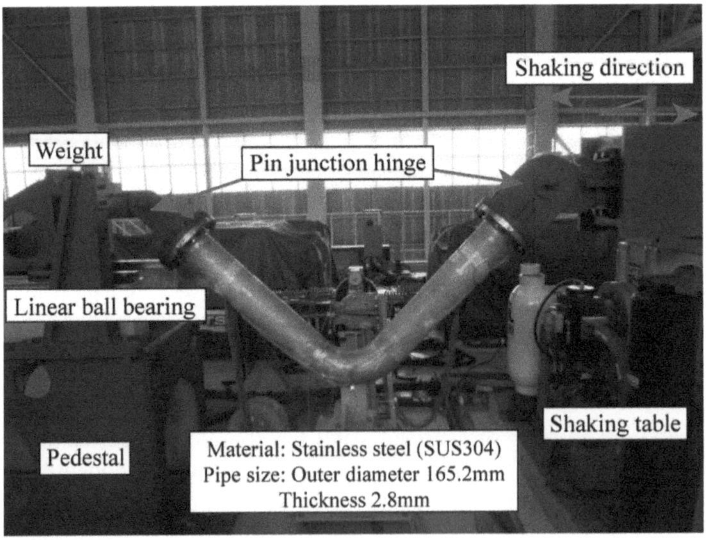

Fig. 5.116 Test setup for dynamic failure test of thin wall elbow.

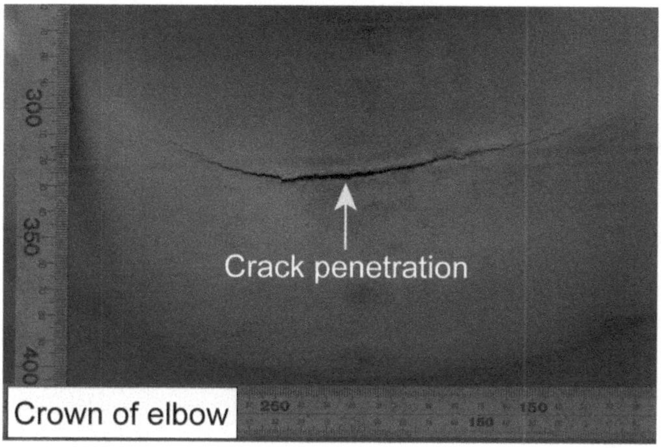

Fig. 5.117 Test observation at the crown part of elbow.

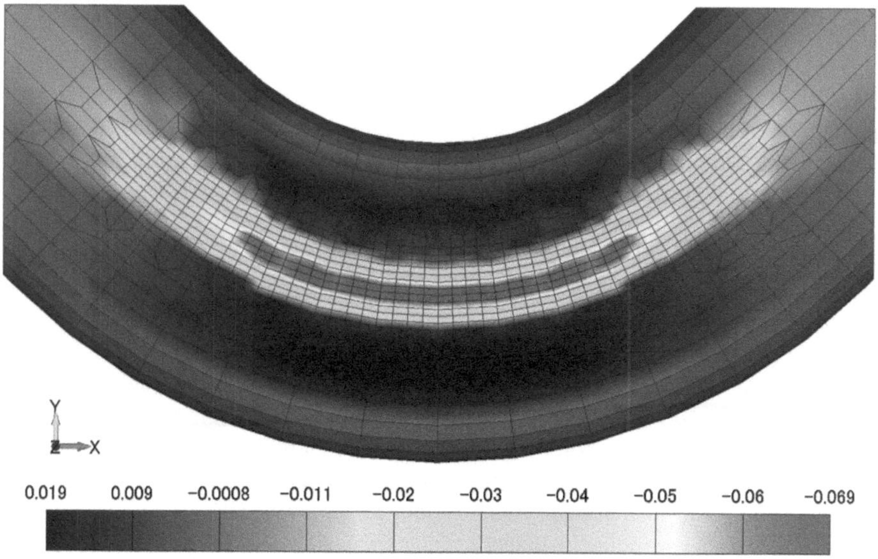

Fig. 5.118 Analysis result using FEA at the crown part of thin wall elbow.

plastic collapse but low-cycle fatigue. However, the current design rules on the seismic design of piping systems assume the linear response without response reduction due to plasticity and provide a stress-based evaluation method with linear response analysis using beam models and stress indices. Conventional response analysis due to the linear assumption with beam models leads to a larger strain than true strain as shown in Fig. 5.119 [496]. This is attributed not to considering the effect of the

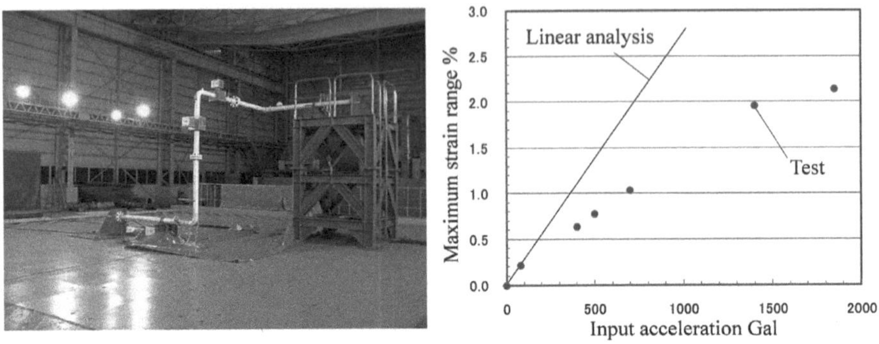

Fig. 5.119 Comparison with piping system test and its linear analysis in strain range [496].

increasing hysteretic damping due to plasticity of piping on linear analysis. Furthermore, stress (strain) of the left side is conveniently calculated by the multiplication of the moment in a linear analysis result and conservative stress indices as shown in the following equation.

$$S_p = \frac{K_2 C_2 M_{is}}{Z_i} \tag{5.19}$$

where S_p is the peak stress, K_2 and C_2 are stress indices given in the design codes, M_{is} is the moment of pipe due to mechanical load and inertia load, and Z_i is the section modulus of pipe. Although current seismic design methodologies of piping can evaluate conveniently and conservatively, the evaluation results include a large margin.

Based on such recognition, a Code Case in the framework of the Nuclear Codes and Standards of Japan Society of Mechanical Engineers (JSME) was developed and published as JSME S NC1, NC-CC-008 in June 2019 [497,498]. This code case provides rules and criteria for fatigue evaluation of piping systems using detailed inelastic FEA as an alternative design rule to the conventional rule, which leads to a more rational seismic design evaluation by directly considering the response reduction caused by plasticity energy absorption. Mainly, fatigue evaluation procedure to prevent fatigue failure was developed based on past studies knowledge that failure mode under seismic loading is low-cycle fatigue. Fatigue damage is evaluated based on Miner's rule from the relationship of the equivalent strain at the local part of piping which is directly calculated on inelastic analysis using FEA and fatigue design curve. Using FEA, a series of benchmark analyses were conducted to develop a concrete inelastic analysis methodology of piping due to inelastic analysis [499]. Based on the results and findings from the benchmark analyses, concrete modeling methods of piping, such as element type of piping, mesh size and material property definition, and damping ratio definition were also provided as guidelines.

The current code case assumes the seismic supports to be linear, and the allowable limit is the yield stress. It is recognized that a certain amount of plasticity of seismic supports does not lead to failure and loss of the supporting capability immediately. The seismic analysis method for piping systems and accompanying design rules must be developed so that the effect of plasticity on seismic supports is taken into account. The discussion for an update of this code case is in continuation in JSME activity.

5.7.2 Seismic response of core and its analysis code

A sodium fast reactor (SFR) core is a group system of core elements in a hexagonal arrangement, as shown in Fig. 5.120 [500]. A core element consists of a cylindrical entrance nozzle (E/N) part and a main part with a hexagonal section. The upper part of the core elements is free to avoid the stress by restriction of thermal expansion and swelling of core elements, and not to interfere with fuel exchanging by remote operation. The E/N of the lower end of a core element is inserted into the cylindrical receptacle tube. The receptacle tube seat (conical surface at the upper end of the receptacle tube) and the spherical seat (a hemisphere at the lower end of core element main part) contact, and the core element self-stands. The external forces that act on a core element are summarized in Fig. 5.120.

The core elements collide with each other in the horizontal direction and jump up from the support plate in the vertical direction under seismic loading. As the design earthquake condition in Japan is very severe, it is important from the viewpoint of seismic safety of the core to confirm how a large displacement and impact force of each core element affect control rod insertability and its structural integrity. Thus, considering horizontal and vertical forces (impact forces and fluid forces) acting on the core elements during an earthquake, a core seismic analysis method has been developed to evaluate the three-dimensional (3D) vibration behavior of the core assemblies considering fluid-structure interaction and vertical displacements (uplifting) [501].

Fig. 5.121 shows an example of the analysis model of the hexagonal bundle model [502]. As the boundary condition of a core element, the receptacle tube seat support receives the spherical sheet in the vertical direction, and the fitting portion of the receptacle tube receives the upper and lower parts of the E/N in the horizontal direction. In addition, the adjacent core elements collide on each pad. The pads, the spherical seat, and the E/N, on which collisions occur, are modeled by the impact stiffness, the impact damping, and the gap, respectively. Moreover, while the pads and the E/N contact with their surrounding structures, friction forces in the vertical direction occur. Each impact parameter is derived from a drop test using partial test models. An outline of the evaluation model newly incorporated to deformation of a core element is given in Fig. 5.122 [503].

Vibration tests were conducted to understand the characteristics of three-dimensional vibration behavior, including vertical jumping, and to verify the developed analysis code. The tests were carried out step-by-step by gradually increasing the number of elements from a single element to a few (32–37) and a large number of elements (313). A test result using 313 core elements is introduced as the example in Fig. 5.123 [504].

Fig. 5.124 shows an example of test results and its simulation results by REVIAN-3D [504]. Through a series of tests, the vertical jumping behavior has strong nonlinearity, and its time history behavior changes greatly depending on the difference

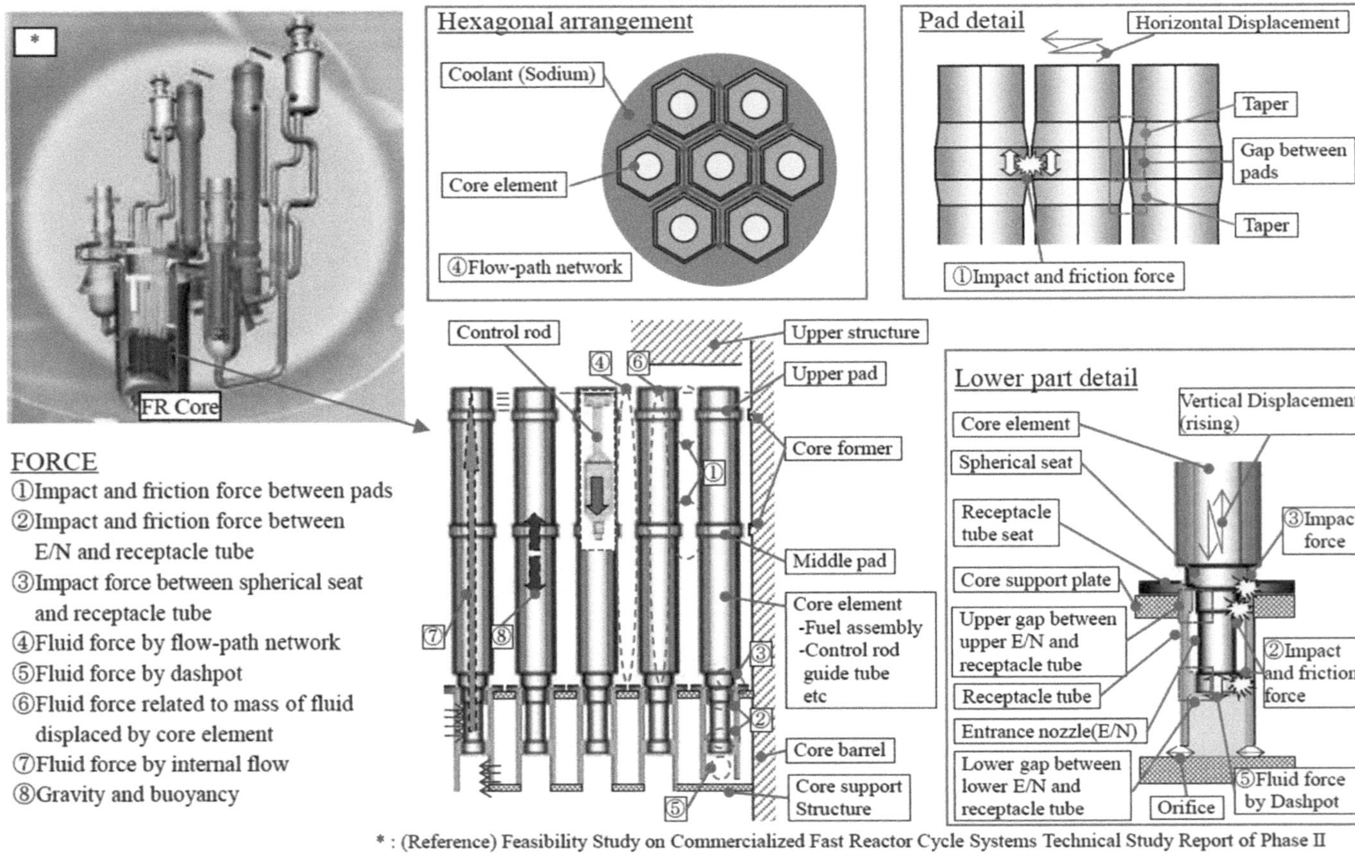

Fig. 5.120 SFR core and core element [500].

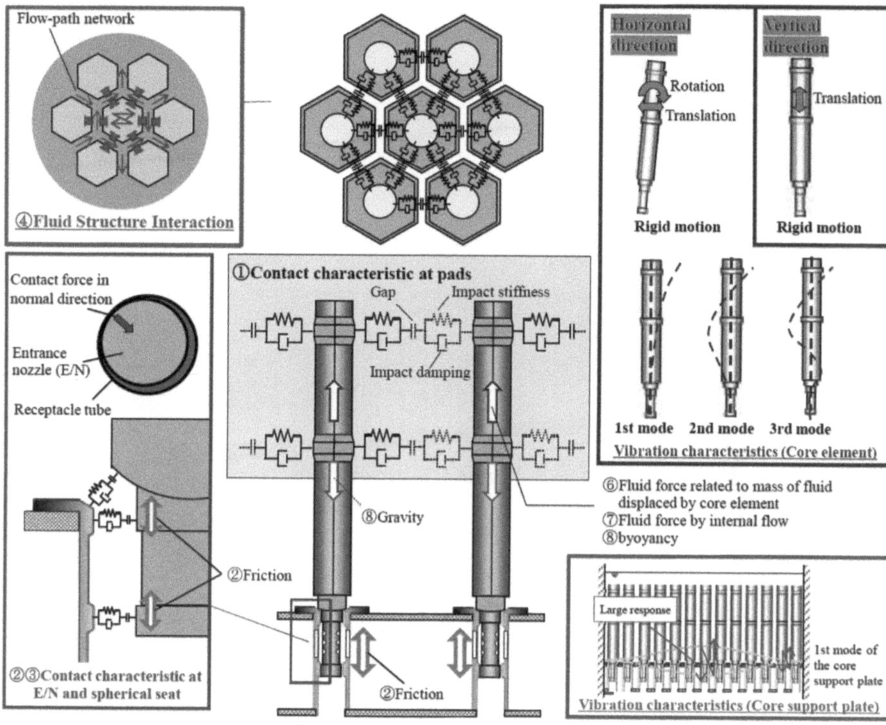

Fig. 5.121 Analysis model of SFR core [502].

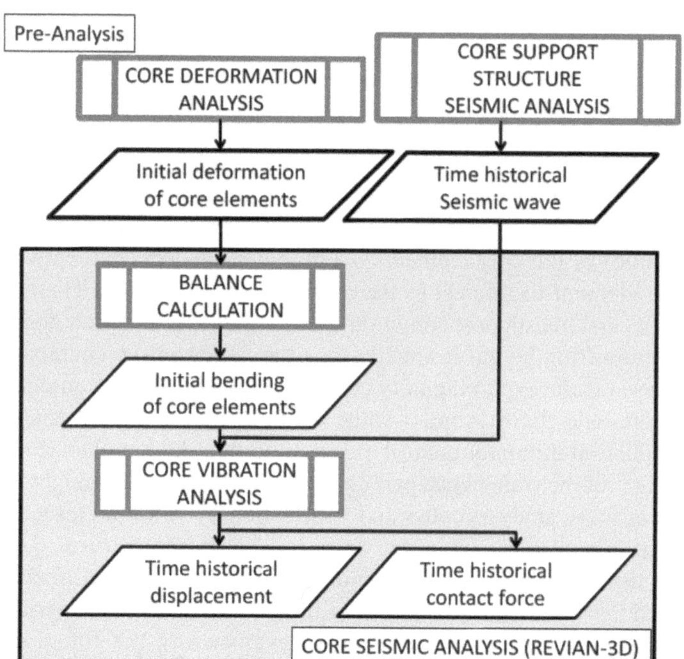

Fig. 5.122 Analysis flow for core seismic analysis considering core deformation [503].

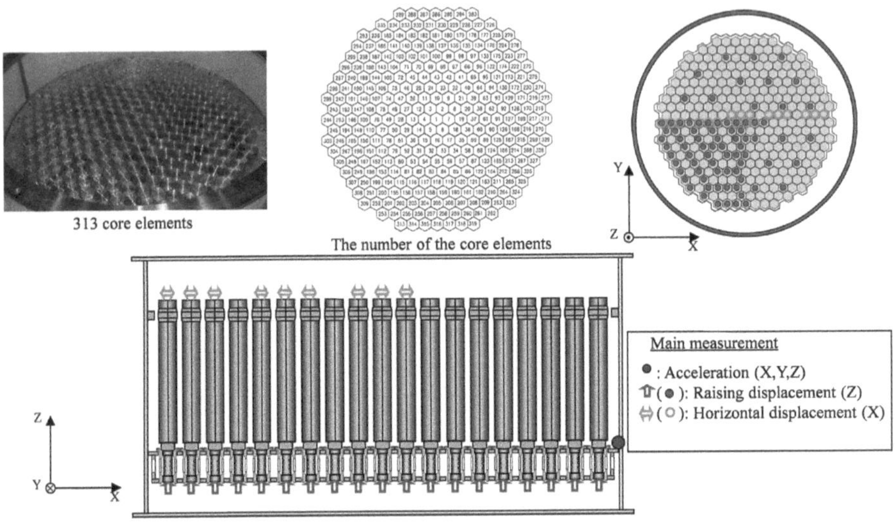

Fig. 5.123 Shaking table test of 313 core elements [504].

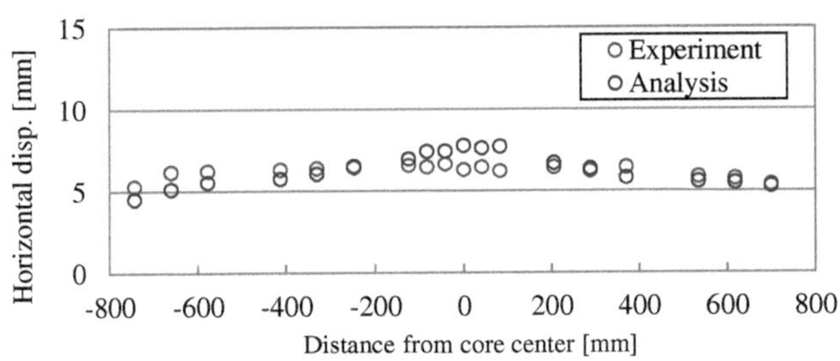

Fig. 5.124 Example of experimental and analytical results [504].

in the initial value. However, these results confirmed that the maximum value could be evaluated with a certain degree of accuracy. Further, the horizontal displacement (range of motion) of the core elements is limited by the total gap between the pads from the core element of interest to the outermost fixed portion. Therefore, even if the uplifting occurs, the support condition in the horizontal direction does not change as long as the uplifting height is smaller than the height of the contact pad portion. Moreover, these results experimentally confirmed that the support condition changed for a short time, and the maximum value of horizontal displacement, which is an important evaluation index for control rod insertability, did not change significantly, even if the height of the contact pad part uplifts up and exceeds the height. Through the development of these analysis codes and verification by vibration tests, the phenomenon of three-dimensional vibration behavior, including vertical jumping, was grasped, and the validity of the evaluation by analysis was confirmed. From this, it was obtained that the control rod insertability and the structural integrity of the core during an earthquake could be appropriately evaluated in the future development of SFRs.

5.8 Operation and maintenance

5.8.1 Development of a fitness-for-service (FFS) code for SFRs

In-service inspection (ISI) requirements provided by an FFS code are important to maintain the structural integrity of structures, systems, and components in nuclear power plants for safety operation. The ISI requirements are also necessary for the design phase so that new plants can be designed considering the inspectability for efficient maintenance and smooth operation.

In Japan, an FFS code for the light-water reactors published by the Japan Society of Mechanical Engineers (JSME) already exists, but it is not directly applicable to SFRs because the characteristics of SFRs are not considered in the code. ISI requirements should be determined by taking account of relevant characteristics of an individual reactor type to ensure the effectiveness and practicality of ISI. The key characteristics of SFRs to be considered in an FFS code include the following:

- Using sodium, which is opaque and chemically active, as a coolant material;
- Limited accessibility to components containing sodium;
- Excellent compatibility of structural materials with purity-controlled sodium;
- Low-pressure coolant systems that preclude the possibility of a loss of coolant due to flashing;
- Safety measures by design in the event of coolant leakage, including guard vessels for the limitation of the total amount of leakage and liners for the prevention of sodium-concrete reactions.

In the world, Section XI, Division 3 of the American Society of Mechanical Engineers (ASME) Boiler and Pressure Vessel Code (BPVC) providing ISI rules for liquid-metal cooled reactors existed [505]. It was developed as a part of the Clinch River Breeder Reactor Plant Project. However, no major revisions were made since the first issue in 1980 because of the cancellation of the project. Therefore, several parts of the code, including the acceptance standards for the examination of Class 1 and 2 components, were retained during preparation.

Thus, there was no FFS code practically applicable to SFRs; a new process for determining ISI requirements, which enables to reflect the characteristics of SFRs, was proposed based on the System Based Code (SBC) concept [506,507]. The proposal was discussed in the joint task group for SBC between ASME and JSME, which ended up with the publication of Code Case N-875 providing alternative ISI requirements for liquid-metal reactor passive components for the ASME BPVC Section XI, Division 3 in 2017 [508,509]. The SBC concept comprises three parts: (1) design and maintenance to meet reliability target throughout the service life; (2) margin exchange among various technical areas of concern, such as design, inspection, fabrication, and maintenance; and (3) expansion of technical options through the timely adoption of newly developed technologies not being present in current codes and standards [510,511]. Such a flexible and consistent concept contributed to deriving the ISI requirements appropriate for SFRs. Fig. 5.125 shows the logic flow diagram to determine the applicability of alternative ISI requirements in the code case based on the SBC concept. The flow consists of Stage I and II evaluations that have complementing individual objects. Stage I is a structural reliability evaluation that considers the

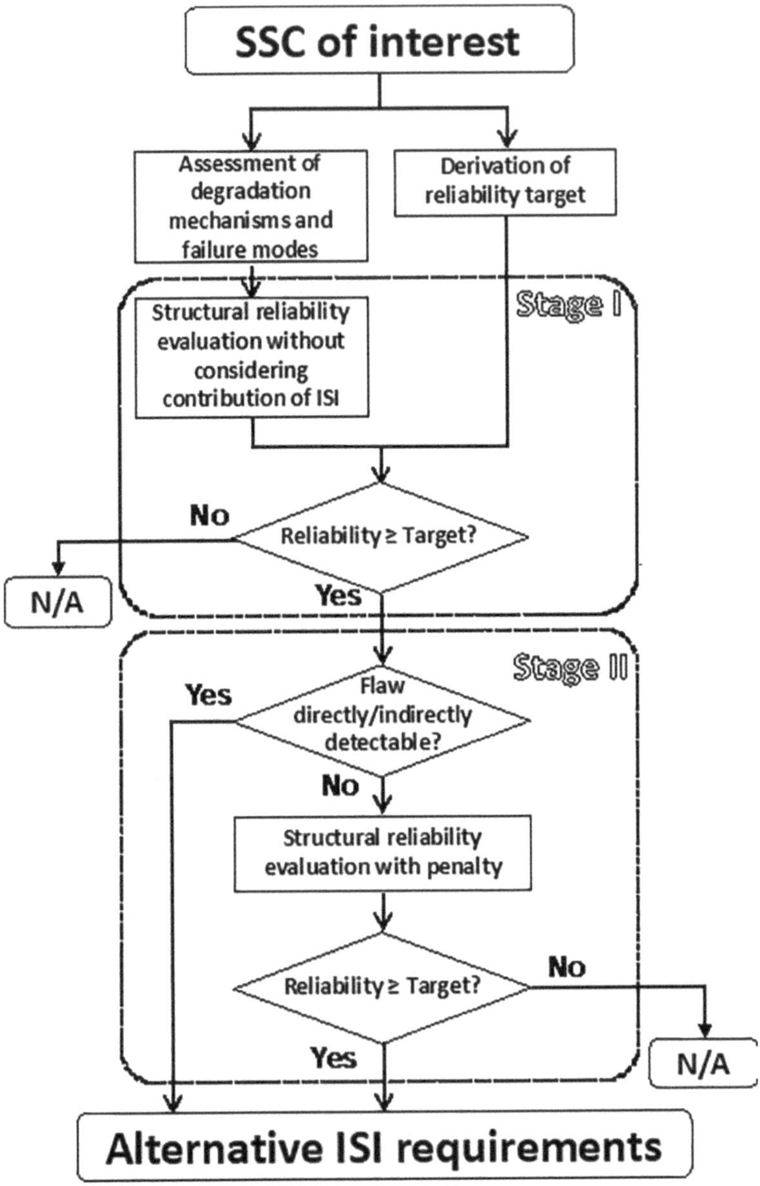

Fig. 5.125 Logic flow diagram for determining applicability of alternative ISI requirements based on SBC concept.

component-level structural integrity and probability of failure of the component under design basis conditions. Stage II is the evaluation of detectability of postulated flaws. Flaws may be directly or indirectly detected. If a postulated flaw is not reliably detectable, the detection may be exempted. However, additional margins in structural integrity are required to justify the exemption. If results at both stages are acceptable, the alternative ISI requirements in the code case, which are continuously monitoring for liquid-metal-retaining components and none for core internals, are permitted to be used. The logic itself is applicable to any other types of reactors. The key parts of Code Case N-875, including the logic, are incorporated into ASME BPVC Section XI, Division 2, which is a new technology-neutral ISI standard [512].

Based on the established process of determining ISI requirements, JSME also has developed an FFS code for SFR that will be published in 2021. The first edition of the code includes reactor coolant boundary (Class 1) components and their support structures. Continuous monitoring is required foEr Class 1 components excluding certain parts such as welded attachments, whereas visual examinations are required to determine the general mechanical and structural conditions for their support structures. Continuous monitoring is an effective and efficient ISI method for Class 1 components because of the following: As mentioned already, sodium is highly compatible with structural materials if the purity is controlled properly. In addition, corrosion from the outside is suppressed in the inert gas environment where Class 1 components are installed. Other potential degradation mechanisms, such as fatigue-creep interaction damage, are also prevented by design. Deviations from the design intentions will be checked by monitoring with process instrumentations. Furthermore, low-pressure systems and safety measures by the aforementioned design can provide sufficient time to shut down the reactor and remove the decay heat after detecting leakage. Besides, for components containing sodium, no leakage occurs during normal operation because not only flanged connection but also welded one is employed, which is suitable for continuous monitoring. On the contrary, considering the limited accessibility to components containing the sodium coolant and severe conditions in terms of inspection such as high-temperature environment, traditional volumetric and surface examinations are not as easily performed as in light-water reactors. Therefore, continuous monitoring is required for Class 1 components. One of the features of the code is to introduce two types of continuous monitoring of sodium leakage, namely CM-1 and CM-2. CM-1 is applied to components that could considerably affect the plant safety if the structural integrity is damaged, whereas CM-2 is applied to other components containing sodium. CM-1 requires determining the sensibility of leak detectors based on leak-before-break (LBB) assessment to detect leakage early enough to perform necessary actions, such as reactor shutdown. LBB assessment guidelines are also developed concurrently in JSME. The details of the guidelines are given in the next subsection. The code will be expanded to include ISI requirements for core internals and other class components in the future.

5.8.2 Development of LBB assessment procedure

The LBB concept can be described as follows: despite an initial crack, which exists or is postulated to exist in a component containing fluid, that grows and then penetrates the structure, a catastrophic break can be avoided through leak detection and the subsequent appropriate countermeasures. Fig. 5.126 shows a schematic of the LBB concept. If LBB is demonstrated, a plant can be safely shut down after the leak detection. Hence, an FFS code for the SFRs newly developed by JSME requires for determining

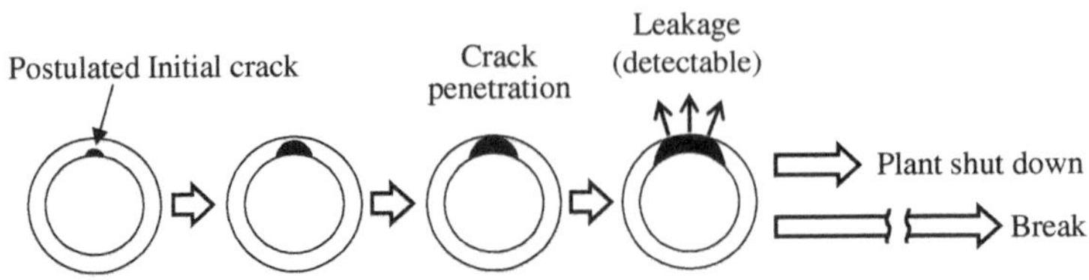

Fig. 5.126 Schematic image of LBB concept.

the sensibility of leak detectors applied to important components so that LBB can be demonstrated.

In general, LBB is likely to be demonstrated for components in SFRs because SFRs have the following characteristics: excellent compatibility of structural materials with sodium, low-pressure coolant systems, and absence of sodium outside of the components in normal conditions. However, no guidelines were provided for the quantitative LBB assessment procedure for SFRs in Japan. Thus, LBB assessment guidelines for SFRs have been developed along with the FFS code for SFRs in JSME.

The key features of the guidelines are as follows:

- In addition to austenitic stainless steels, relatively low-ductility and small-work hardening coefficient materials, such as Mod.9Cr-1Mo steel that is a candidate material for future SFRs, are applicable.
- The contribution of thermal expansion stress to unstable fracture was considered because thermal expansion stress is dominant in SFRs.
- The necessity of creep crack growth on crack growth calculation in LBB assessment was considered because SFRs are operated in creep temperature regions.

Fig. 5.127 shows the LBB assessment flow of the guidelines. The LBB assessment comprises three main evaluations, which are "critical crack length evaluation," "detectable crack length evaluation," and "penetrated crack length evaluation." An overview of LBB assessment procedure is provided here. The details of each evaluation procedure are described elsewhere [513–515].

"Critical crack length evaluation" is conducted to determine the critical crack length, C_{cr}, above which unstable fracture occurs under the design loading conditions. The guidelines provide three types of methods to evaluate critical crack length: formula-based limit load method, finite element analysis-based limit load method, and two-parameter method (R6 method). An appropriate evaluation method should be selected by considering material characteristics, fracture modes, and geometry of the component of interest. Limit load methods are applied to austenitic steel components. The two-parameter method that can evaluate crack stability is available and is applicable to low-ductility and small-work hardening coefficient materials such as Mod.9Cr-1Mo steel. One of the features of LBB guidelines is that thermal expansion stress is not considered when it is classified as a secondary stress through an elastic follow-up evaluation. Thermal expansion is dominant stress in SFRs; therefore, rational treatment is essential. The elastic follow-up evaluation is the same as that in JSME design and construction code for SFRs [516].

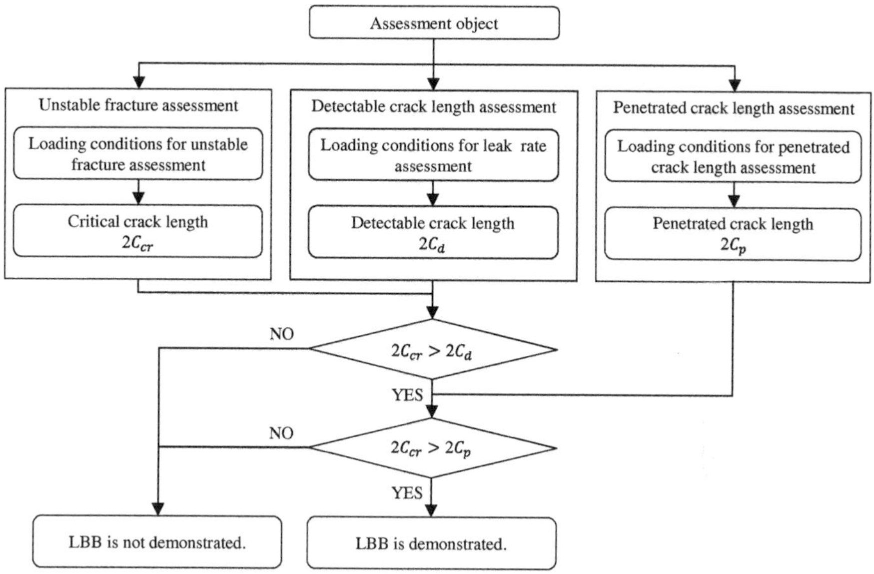

Fig. 5.127 LBB assessment flowchart.

"Detectable crack length evaluation" is performed to determine detectable crack length, C_d, above which sodium leakage can be detected using installed sodium leak detectors. C_d is obtained from crack-opening displacement and detectable equivalent hydraulic diameter. The crack-opening displacement assessment method is based on that proposed by Electric Power Research Institute (EPRI) [517]. The EPRI method was expanded to evaluate thin-walled large-diameter cylindrical structures like the SFR components operated under low pressure and include relatively large yield stress and small-work hardening coefficient materials [518,519]. A detectable equivalent hydraulic diameter evaluation method for sodium leakage proposed by Miura et al. [520] was adopted in the guidelines.

"Penetrated crack length evaluation" is performed to determine C_p, which is the length of a crack penetrated by transient loads hypothetically repeated at the crack initiation side. The length can be calculated through crack growth analysis or using a simplified method. The simplified method provides master curves, depending on the geometry and material of the component. C_p can be obtained from the ratio of membrane and bending stresses using the curves developed based on parametric crack growth analysis [521–523]. Penetrated crack length is evaluated by considering only fatigue crack growth mechanism because penetrated crack length becomes longer without considering creep effects [521].

As the result of the aforementioned three evaluations, if C_{cr} is longer than both C_d and C_p, LBB is demonstrated. The LBB evaluation methods were prepared for a cylindrical structure covering the anticipated dimensions of piping and vessels for SFRs. Thus, the developed LBB assessment procedure is applicable to both piping and vessels.

5.8.3 *Under-sodium viewer development*

The inspection in the opaque liquid-metal coolant is one of the major challenges in ISI for SFRs. Various under-sodium viewers (USVs) that use ultrasonic technology have been developed in several research institutes and countries to facilitate operations and maintenance activities.

The United Kingdom Atomic Energy Authority (UKAEA) developed a USV for prototype fast reactor (PFR) built in Dounreay, Scotland [524]. The developed USV had an 11-m-long and 25-cm-diameter tube with 12 echo-sounding piezoelements at the bottom end. The fuel subassembly outlet was imaged using the USV, and each fuel subassembly could be identified.

In addition, UKAEA developed a USV for SNR-300 built in Kalkar, Germany [525]. Unfortunately, SNR-300 was not operated. Fig. 5.128 shows the schematic of the USV for SNR-300. The nine piezoelements were equipped with focusing steel membranes. The central piezoelement transmits and receives ultrasonic waves. The eight piezoelements located around the central piezoelement receive ultrasonic waves. The frequency of the nine piezoelements is 4 MHz. The focusing lengths of the USV are 40 and 130 mm. The lateral resolution was <1 mm.

Toshiba Corporation developed a USV for demonstration fast breeder reactor (DFBR) [526]. The details of DFBR are shown in Section 4.2. The structure of the matrix-arrayed transducer was adopted. The number of piezoelements is 1296

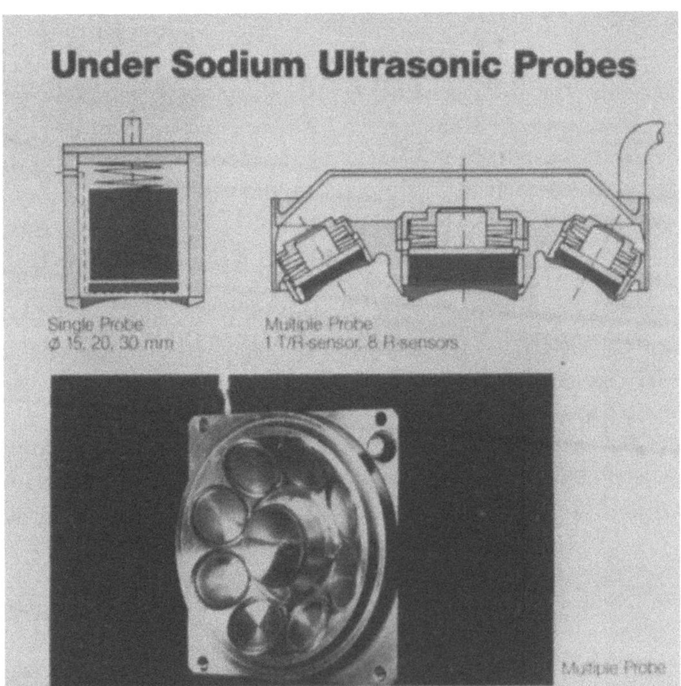

Fig. 5.128 Schematic of USV for SNR-300 [525].

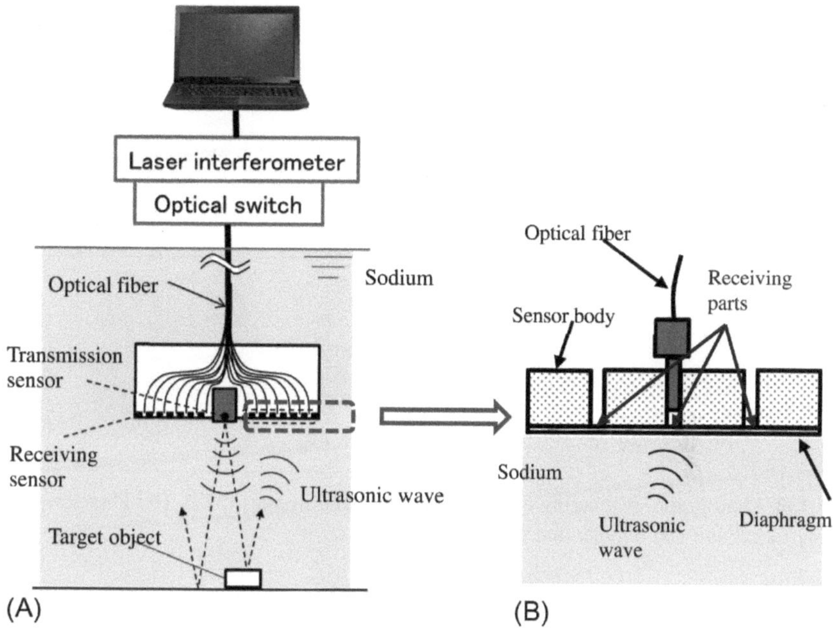

Fig. 5.129 Configuration of USV for the next-generation SFRs [527]. (A) USV system and (B) receiving sensor.

(36 × 36). Each piezoelement cannot only transmit ultrasonic waves into sodium but also receive ultrasonic waves reflected from a target under sodium. The frequency of the piezoelements is 5 MHz. The under-sodium tests confirmed that the matrix-arrayed transducer exhibits good sensitivity and signal-to-noise ratio.

JAEA and Mitsubishi Heavy Industries, Ltd., developed a USV for the next-generation SFRs [527,528]. The target resolution is equivalent to the requirement of the VTM-3 that is a visual examination conducted to detect discontinuities and imperfection, such as a deformation of parts, looseness of bolt, and loss of parts. Fig. 5.129 shows the configuration of the USV. The optical receiving system was adopted to receive ultrasonic waves. The optical receiving system uses laser beams to detect vibration displacement of the diaphragm at each of the receiving parts. Therefore, the receiving part serves instead of the piezoelement. Fig. 5.130 shows the photograph of imaging target and regenerated images in water and sodium. One of the lines in the target is 3 mm wide, and the narrowest line is 2 mm wide. As shown in the figure, the USV can identify 2-mm-wide lines under sodium.

5.9 Fuel cycle-related studies

JAEA has promoted the development of a simplified pelletizing fuel fabrication and advanced aqueous reprocessing methods to establish the fuel cycle technology for fast reactors through the Fast Reactor Cycle Technology (FaCT) development project. In

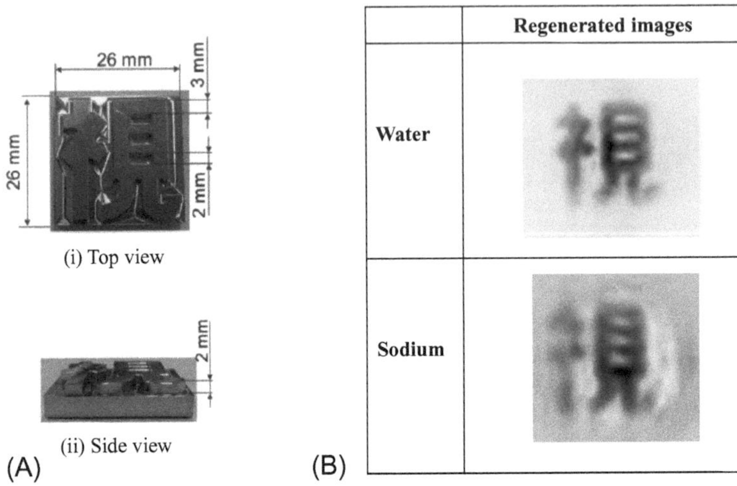

Fig. 5.130 Photograph of imaging target and regenerated images [527]. (A) Photograph of imaging target and (B) regenerated images.

this section, the summary of the developments described earlier is introduced as main issues in the fuel cycle-related studies.

5.9.1 *Fuel fabrication technology*

JAEA has been conducting fundamental research on the physical chemistry of nuclear materials and developing MOX fuel fabrication technology since 1966; it provided MOX fuel for the advanced thermal reactor "Fugen," the experimental fast reactor "Joyo," and the prototype FBR "Monju."

Fig. 5.131 indicates the process flow diagram of MOX fuel fabrication in the present process of JAEA and in the simplified pelletizing process [529]. The simplified pelletizing process is a relatively new fuel fabrication technology. The number of steps related to powder handling is reduced by employing three innovative technologies, i.e., (1) Pu concentration adjustment by mixing of Pu nitrate and uranyl nitrate solution, (2) improvement of MOX powder flowability by tumbling granulator without an organic binder, and (3) die-wall lubrication pressing technology, to improve the economic efficiency. In addition, the simplified fuel pelletizing process is appropriate for the fabrication of minor actinide (MA)-bearing MOX fuel because organic matter, which can be degraded by high heat generation in MA, is not added in the process.

By employing a tumbling granulator with a batch size of 600 g, MOX granules appropriate for die filling with flowability indices of 70 or higher could be produced [530]. Performance of a scaled-up tumbling granulation system with a batch size of 5 kg was also confirmed with simulated powder (WO_3) [531].

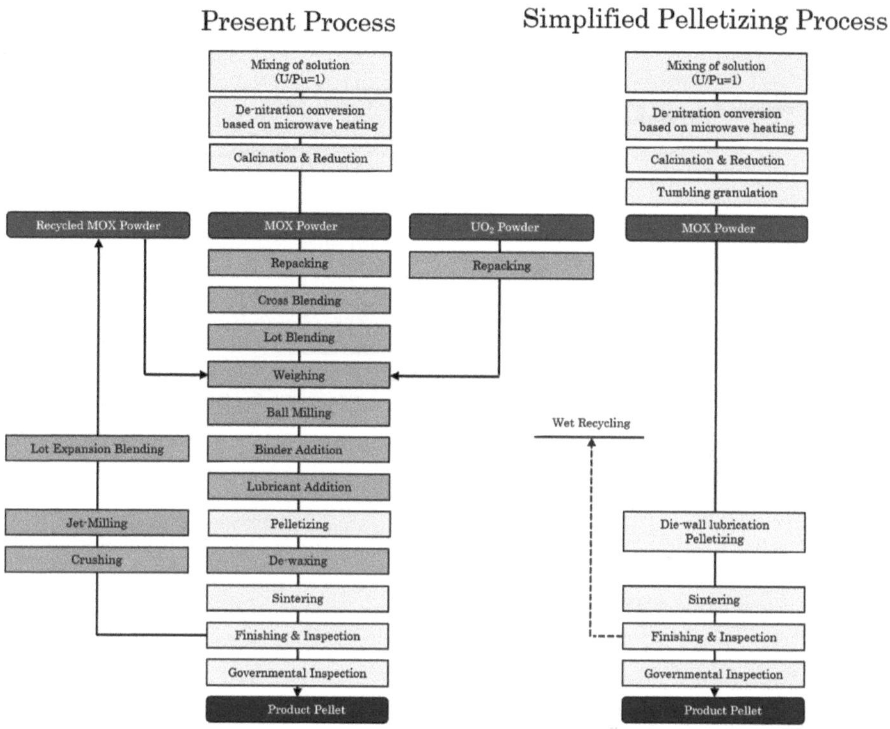

Fig. 5.131 Process flow diagram of MOX fuel fabrication in the current process of JAEA and in simplified pelletizing process.

Regarding the die-wall lubrication pressing technology, the pressing rate of 7.5 cycle/min was achieved, which can be applied to an engineering-scale process. The MOX granules, produced by the tumbling granulator, were pressed into annular preen pellets with a good appearance. Sintered pellets of more than 95% T.D. were obtained by heating the green pellets at 1750°C for 4 h [532].

To investigate the feasibility to produce MA-bearing MOX fuel by the simplified pelletizing process, R&Ds on countermeasures against high radio-activity and high heat generation on the MA-bearing MOX fuel were conducted [533]. Now, we are considering the feasibility of applying advanced technologies, such as additive-free dry granulation technology and microwave sintering, to simplify the pelletizing process.

5.9.2 Reprocessing technology

Fig. 5.132 shows the process of New Extraction System for TRU Recovery (NEXT). The NEXT is an advanced reprocessing system based on aqueous reprocessing with several innovative technologies [534].

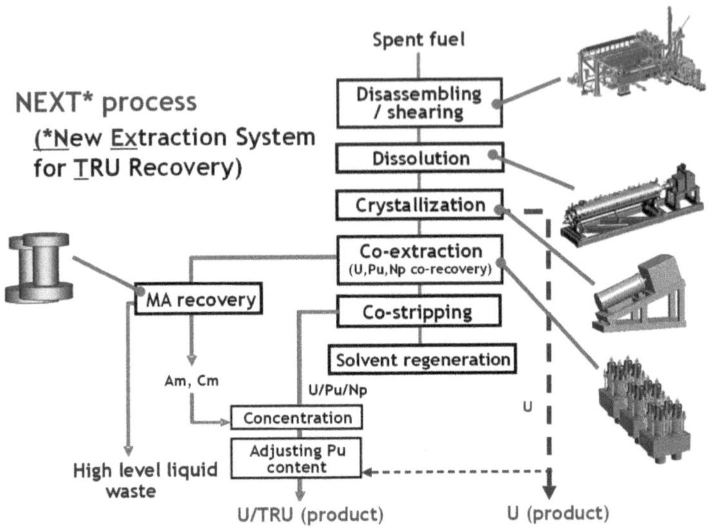

Fig. 5.132 Structure of NEXT process.

In the head-end process, a wrapper tube was first removed from the fuel assembly. Then, the fuel pin bundle was sheared into approximately 1 cm in length. The mechanical method for cutting a wrapper tube with a grindstone was evaluated, and its technical feasibility was confirmed in the FaCT project [535]. In the shearing of fuel pin bundles, the fuel pulverization by short-stroke was increased to obtain high U concentration dissolution efficiently. This goal was achieved by applying a horizontal shearing, which was already used in the reprocessing for spent LWR fuel.

The fuel component in shearing pieces was dissolved into nitric acid, and dissolver liquor with high U concentration was obtained in this process. The rotary drum-type continuous dissolver was used to ensure the criticality control and dissolution performance. Various assessments, including hot tests using spent Joyo fuel and engineering-scale *U* tests, were conducted, and no problems were observed in dealing with high U concentration dissolution [536].

The excess amount of U in the dissolver liquor was recovered via crystallization, which has a simple principle. The U was recovered as uranyl nitrate hexahydrate crystal by controlling the temperature of the solution without any reagents. Approximately 70% of U must be recovered from the dissolver liquor to adjust the ratio of residual U to Pu in the solution, which is controlled as the equivalent of the composition of the core fuel. A rotary kiln-type crystallizer was selected as the most suitable type from the viewpoints of continuous processing, criticality control, and crystal handling, and its operational stability was tested and validated [537]. The behaviors of U, Pu, and fission product (FP) in the crystallization were investigated by hot tests, but the improvement of decontamination factor (DF) for several FP elements accompanying as solid impurities must be discussed.

In the co-extraction process, Pu, U, and neptunium (Np) were recovered and separated from the FP with a single-cycle flow sheet based on the PUREX process, which is a well-established solvent extraction method in LWR reprocessing. We discussed the application of a centrifugal contactor system for the solvent extraction process because it has various advantages, including compactness, small holdup volume, and high separation performances. Good recovery yields of U, Pu, and Np and DF for FP were obtained by several hot tests using laboratory-scale centrifugal contactors. Besides, the system and durability performances of centrifugal contactors were confirmed through the engineering-scale tests [538].

MA elements, such as americium and curium, were recovered from the raffinate of the solvent extraction using extraction chromatography technique. This method is superior to solvent extraction in terms of the amount of aqueous waste and treatment cost. Although the process and safety data have been collected via several assessments, including MA recovery tests from genuine raffinate (Fig. 5.133) [539], the improvement of MA recovery yields and DF for FP elements and the establishment of safe and robust systems must be discussed.

Salt-free technology eliminates reagents containing metal salts in the reprocessing process and has the potential to reduce significantly the total amount of waste. In return, it will result in the high efficiency of waste liquid concentration and classification of various types of waste liquid into two types: the high-level liquid waste that can be vitrified and the very-low-level liquid waste that can be discharged into the ocean. The applicability of hydrazine carbonate and oxalate instead of sodium hydroxide and carbonate, respectively, to the solvent washing reagents and their decomposition condition with electrolytic oxidation was confirmed [540]. However, further experimental data should be collected.

We are attempting to evaluate the system performance of each process and equipment in the NEXT for practical use.

Fig. 5.133 MA recovery test from genuine raffinate using extraction chromatography.

5.10 ARKADIA–advanced reactor knowledge- and AI-aided design integration approach through the whole plant life cycle

5.10.1 Overview

5.10.1.1 Purpose

Japan Atomic Energy Agency (JAEA) has started to develop "Advanced Reactor Knowledge- and AI-aided Design Integration Approach through the whole plant life cycle (ARKADIA)" as one of the research and development infrastructures required in the Fast Reactor Strategic Roadmap. ARKADIA offers the best possible solutions for any challenges that could arise in the plant life cycle, including plant design and safety measures, maintenance, and decommissioning. ARKADIA consists of state-of-the-art numerical simulation technologies, artificial intelligence (AI) and related technologies, and a knowledge base that includes development experiences from the experimental reactor, Joyo, and the prototype reactor, Monju, and related research and development. ARKADIA will be a key technology to evaluate various advanced reactor concepts including sodium-cooled fast reactors (SFRs). Considering the aforementioned advantageous features, ARKADIA is a powerful tool for educating and training young scientists and engineers.

ARKADIA will be developed in two phases. ARKADIA-Design for the design optimization of a reactor core, reactor structures, and a maintenance plan and ARKADIA-Safety for evaluation of severe accident (SA) events and providing feedback to a safety design are developed independently in the first phase. Sections 5.10.2 and 5.10.3 describe the scope of application, analytical methods, and example applications of ARKADIA-Design and ARKADIA-Safety, respectively. In the following phase, they will then be integrated into a single system that will apply to a wide range of reactor concepts, coolants, configurations, and output levels.

5.10.1.2 Required function

ARKADIA integrates accumulated knowledge including past experiences, state-of-the-art numerical simulation technologies, and AI. This integration enables automatic optimization of a variety of conditions, including safety, risk response, economics, and environmental compatibility. The flowchart in Fig. 5.134 depicts a four-step evaluation for automatic optimization: (1) set objective, (2) gather and organize relevant information and set evaluation condition, (3) carry out the evaluation, and (4) assess achievement of the objective. ARKADIA's autonomous evolution is aided by the knowledge gained during this evaluation. The knowledge gained from application to a real plant is saved in the knowledge base. The expanded knowledge base is applied to new designs and safety improvements. Details of each step are described in the following.

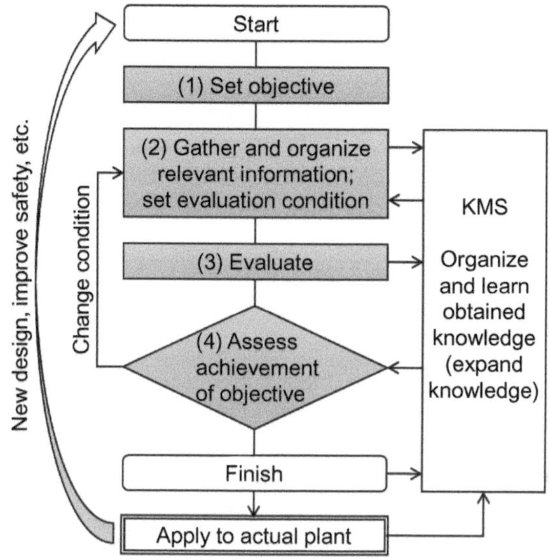

Fig. 5.134 Evaluation flowchart of ARKADIA.

In step (1), ARKADIA identifies elements related to the objective given by users. The element e depends on the parameter p as follows:

$$e_i = \text{func}\left(p_{i,j} | j \in n_i\right), \tag{5.20}$$

where the subscript i denotes the number of the components and j denotes the number of the parameters. The subscript j is a member of the set n_i. Examples of elements are design condition (e_1), safety (e_2), economics (e_3), and social acceptability (e_4). The design condition (e_1) includes parameters such as reactor output ($p_{1,1}$), temperature ($p_{1,2}$), and pressure ($p_{1,3}$). ARKADIA then constructs an objective function, which is written by either of the following two equations:

$$\text{object} = \text{func}(e_i | i \in N), \tag{5.21}$$

$$\text{object} = \text{func}\left(p_{i,j} | i \in N, j \in n_i\right), \tag{5.22}$$

where the subscript i is a member of the set N. There is a case where one component depends on another. The purpose of the evaluation shown in Fig. 5.134 is to minimize this objective function.

In step (2), ARKADIA extracts required data from the knowledge base and determines the evaluation condition. The required data are identified through the hierarchy of predetermined database tags in the knowledge base. AI allows for the automatic identification of the necessary data and evaluation conditions. Input data for numerical analysis in the following step are generated automatically based on the selected condition.

Numerical analysis is performed to evaluate the element functions in step (3). Numerical analysis is not required in some cases such as maintenance optimization.

ARKADIA evaluates the objective function given by Eq. (5.21) or (5.22) in step (4) to assess whether it satisfies a constraint condition and reaches a minimum value. According to this assessment, the process returns to step (2), and the evaluation condition is changed by a system controlling optimization. To satisfy the objective, ARKADIA provides multiple solutions in which the objective function becomes minimum or close to a minimum value.

5.10.1.3 System structure

To realize the automatic optimization described in the previous section, ARKADIA consists of a platform and three interconnected systems: Virtual plant Life System (VLS), Knowledge Management System (KMS), and Enhanced and AI-aided optimization System (EAS). The platform receives the user's requirements via a graphical user interface and controls the three systems to achieve purposes. The structure of ARKADIA is depicted in Fig. 5.135. Each system's roles are described later.

(a) EAS

EAS identifies elements and parameters from given inputs at the start of the evaluation and then creates an objective function in step (1). In step (2), relevant information is extracted from the knowledge base using a database tag hierarchy. EAS also selects appropriate evaluation conditions and generates input data for numerical analysis that is performed by VLS. In step (4), EAS assesses the achievement of the objective by evaluating the value of the objective function based on the analytical result. According to the assessment, EAS changes the evaluation conditions in order to find the better solution. AI technology is used in EAS for highly efficient optimal solution retrieval. The aforementioned functions of EAS are summarized later.

- Create an objective function in step (1),
- Extract related information from the knowledge base in step (2),
- Select evaluation condition and create input data in step (2),
- Calculate objective function and assess achievement of objective in step (4),
- Change evaluation conditions during iterations to find the optimum solution.

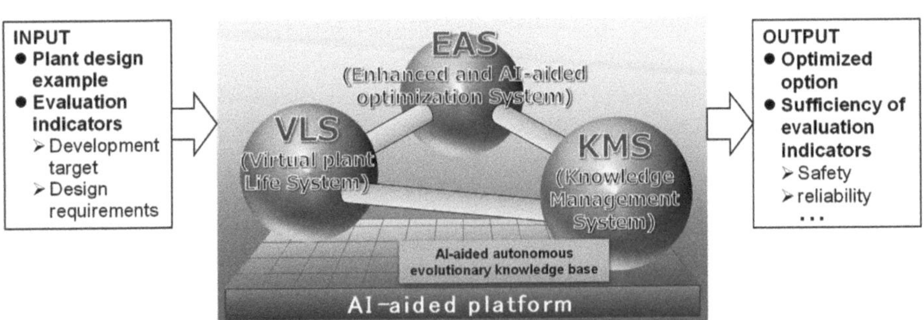

Fig. 5.135 Three systems that comprise ARKADIA.

(b) VLS

VLS evaluates all possible events including maintenance-related phenomena during a plant life cycle. VLS applications include both design basis accidents and beyond-design basis accidents. This application range encompasses all levels of defense in depth, except off-site emergency response.

VLS includes a thermal-hydraulics analysis code as a base module to evaluate plant status quantities such as pressure, temperature, and velocity. Single-phase liquid sodium flow is dominant in normal operation or abnormal transient conditions in SFRs. During SAs, on the contrary, compressible multiphase flow with phase change, transport of fission products, and chemical reaction appear. The SA phenomenon must be taken into account in the VLS's base module. Furthermore, coupling of multidimensional analysis and plant dynamics analysis is required to predict a plant's overall status. This coupling should be implemented by combining pressure equations for accurate evaluation.

A probabilistic risk assessment (PRA), which evaluates accident scenarios by quantifying branching probabilities, is also a core function of VLS. Moreover, a dynamic PRA, which considers time dependence of event progress by thermal-hydraulics analysis, will be realized by VLS for more realistic predictions. The basic functions of VLS are summarized as follows:

- Evaluate plant status quantities from thermal-hydraulics analysis, which couples multidimensional and plant dynamics analysis and
- Perform dynamic PRA.

Physical properties of materials and coolants, safety evaluation methods such as dynamic PRA, and state-of-the-art numerical analysis methods will all be integrated into VLS's base module. To meet required accuracy, VLS will have a connecting function with other individual codes or evaluation methods. From this function, other organizations or manufacturers can develop VLS for various types of reactors.

(c) KMS

The application range of KMS is a whole plant life cycle from design to decommissioning. KMS provides knowledge obtained from existing and future research and development. KMS consists of information infrastructure, a database group, and a knowledge group. Details of the structure are described later.

5.10.2 ARKADIA-Design

ARKADIA-Design supports mainly the conceptual design of core, plant structure, and maintenance plan in normal operating conditions and design basis events. Fig. 5.136 depicts a conceptual view of the functional elements in ARKADIA-Design. ARKADIA-Design consists of three major systems: the EAS for evaluation and optimization support, the VLS for virtual plant life cycle analyses, and the KMS for providing the necessary information to EAS and VLS.

In the EAS, objective functions, constraint functions, limitations in design requirements, plant operation conditions, and design specifications are initially determined to solve the optimization problem. The EAS also generates the input data for the analysis code that will be used in VLS. The KMS is used to correct the necessary information for the inputs. The VLS is required to perform numerical analysis and give the results to the EAS. Based on the analytical results, the EAS modifies specific plant parameters to maximize performance according to the objective functions.

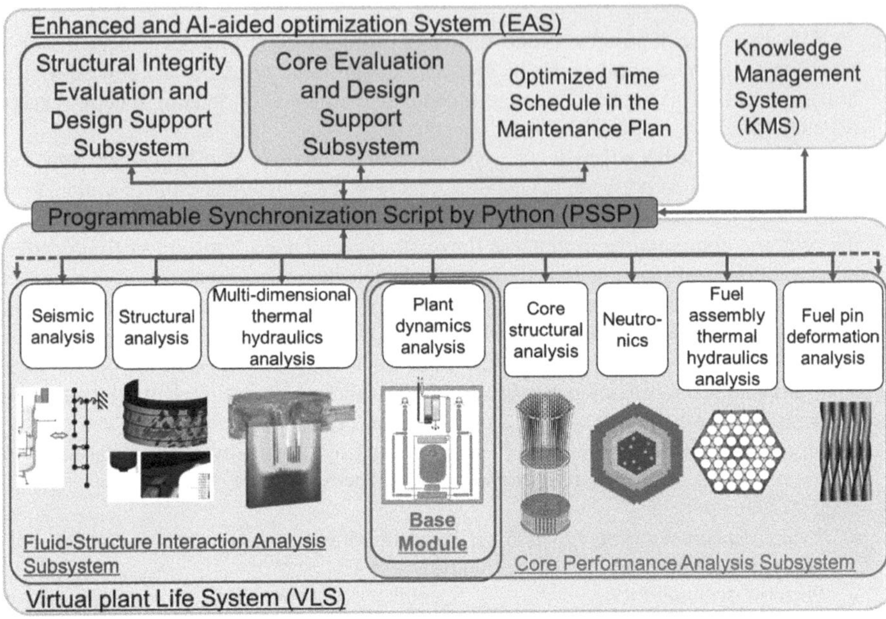

Fig. 5.136 Conceptual view of module structure in ARKADIA-Design.

In the VLS, the multilevel (ML) simulation technique [541,542] is used to build a virtual plant model. Usually, a simplified model in the one-dimensional (1D) plant dynamics analysis code is used for the simulation of whole plant behavior and a detailed model in three-dimensional (3D) analysis code is used for the simulation of local phenomena in the component. An analysis coupling the 1D and 3D codes was performed in the ML approach, with a virtual plant model constructed by combining the models at different degrees of resolution, from a simplified model for the 1D analysis to a detailed model for the 3D analysis. The coupled analysis is carried out via the PSSP interface (Programmable Synchronization Script by Python). The PSSP is in charge of the execution of analysis codes, the interpretation of data to advance each calculation from one to the next, and the synchronization of data transferred between analysis codes [541]. The PSSP is also in charge of the communication between the modules in EAS, VLS, and KMS. For the core and plant structure designs, two representative optimization schemes connecting the EAS and the VLS with the PSSP are developed in the first step of development. The two schemes will then be unified, and an optimization scheme for whole plant design will be developed. The final goal of ARKADIA-Design is to establish the optimization scheme, which can output the optimized design of the plant from the design requirements using AI technologies.

5.10.2.1 Core design

In conventional core design procedures, experts need to previously determine the boundary condition for the analysis on neutronics, thermal-hydraulics, and fuel designs, and then perform a series of analyses to evaluate the core performance. The analyses should be repeated by trial and error to obtain a specification with high performance that also meets the constraint conditions in the core design. In some cases, the conventional procedure provides excessive conservativeness because the result of each analysis must satisfy the corresponding predetermined boundary condition for the next analysis. In addition, after determining the tentative core specifications, the plant dynamics analysis was finally performed to evaluate the core's integrity. If core damage occurs, the design parameters must be changed based on expert experience and analyzed again using the neutronics calculation. As a result, achieving the optimized design took a long time.

As shown in Fig. 5.137, therefore, ARKADIA-Design is to be applied to the optimization process of the core design to improve the efficiency of the optimization process for the core design. The EAS's Core Evaluation and Design Support Subsystem is being developed to optimize the neutronics design, fuel design, thermal-hydraulics design, and core integrity evaluation with the use of an AI-aided algorithm. This subsystem is intended to automate the optimization process for core specifications using the expert experiences provided by the KMS database. The Core Performance Analysis Subsystem in the VLS is being developed to predict core performance through coupled analysis using codes for neutronics, thermal-hydraulics, and core deformation due to thermal effect. This VLS subsystem is intended to predict reactivity insertion

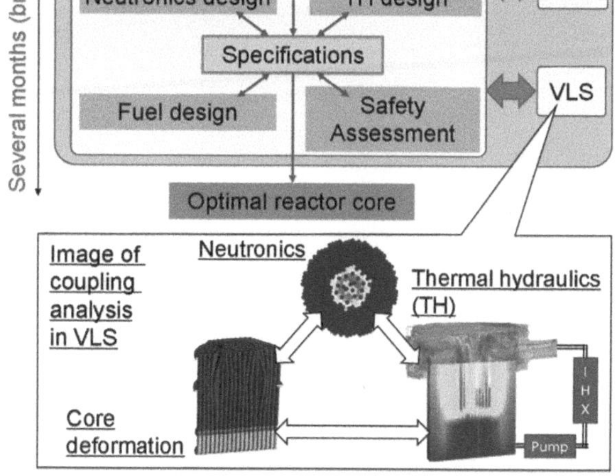

Fig. 5.137 Conceptual view of core design optimization.

caused by control rod withdrawal and negative reactivity feedback caused by core deformation during transient events. The latter effect is not considered in the conventional safety evaluation for the core design. In addition, another subsystem in the VLS for fuel assembly design is being developed by coupling the analysis codes for the fuel pin deformation, the thermal-hydraulics, and the fuel pin irradiation behavior.

5.10.2.2 Plant structure design

In the conventional design procedure for plant structure, evaluations for the thermal transition in the plant system, the heat transfer in components, and the structural integrity during plant life are performed individually. Because the conservative assumption is used to determine the boundary condition for each evaluation, the total conservativeness of this evaluation procedure becomes excessive after unifying the margins from each evaluation. This excessive conservative evaluation does not necessarily improve plant safety, but it may raise the plant's construction costs.

Fig. 5.138 shows the conceptual view of the schemes for the plant structure design optimization. For example, in a reactor vessel design, ARKADIA-Design may provide optimized design candidates under the conditions of component layout (equipment and piping in a vessel), size, material, and plant operation (operation temperature, lifetime, and number of assumed transitions) via an iterative process between the EAS and the VLS.

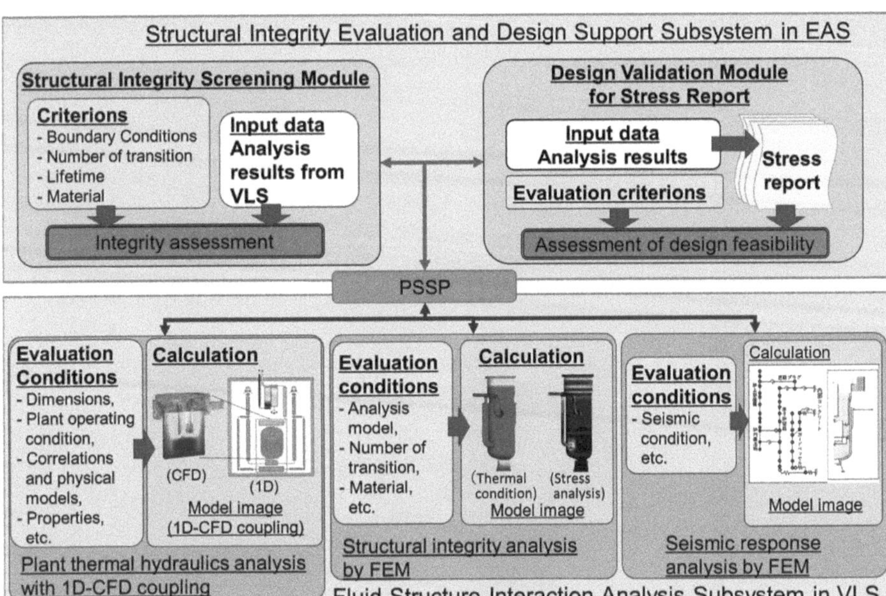

Fig. 5.138 Conceptual view of plant structure design optimization.

In the EAS, the Structural Integrity Evaluation Subsystem evaluates the adequacy and adaptability of the designed components to the given design criterion. The subsystem consists of two modules. One is the Structural Integrity Screening Module to evaluate the applicability of the candidate specification of the components to the design criterion. The other is the Design Validation Module for Stress Report, which evaluates applicability to all of the design criteria defined in the elevated temperature design code [543], and generates stress reports automatically based on the results. In the validation process, the former module comes before the latter. The numerical analysis of the virtual plant is built in the VLS using numerical analysis codes based on the ML approach. In the Fluid-Structure Interaction Analysis Subsystem, for example, 1D plant dynamics analysis and 3D computational fluid dynamics analysis are linked. To optimize the plant structure design, a combination of thermal-hydraulics analysis and structural integrity analysis is to be performed.

5.10.2.3 Plant design considering maintenance

Maintenance has an impact on the economic efficiency of an advanced nuclear reactor as well as its safety. The maintenance plan for the prototype fast breeder reactor Monju was developed by referring to that for light-water reactors (LWRs) in Japan, which resulted in a large burden because the effectiveness of the plan was not sufficiently investigated in advance. It is critical to consider the characteristics of individual advanced nuclear reactors when developing maintenance plans. In the conventional design process of a plant, much attention is not paid to maintainability in comparison with other issues such as plant safety and construction costs, but the efficiency of maintenance will be significantly improved if plant maintainability is included in the main issues in the design process. Therefore, a maintenance optimization scheme has been proposed [509] and included in ARKADIA-Design.

An overview of the proposed maintenance optimization scheme is shown in Fig. 5.139. The scheme's basic concept is the same as that of ASME Code Case N-875, which provides alternative in-service inspection requirements for liquid-metal reactor passive components based on the System Based Code Concept [544], but the proposed scheme's scope is not limited to passive components in liquid-metal reactors. Reliability targets for components are specified in terms of plant safety and others, such as stable operation, and maintenance plans are then developed to meet the reliability targets while accounting for potential degradation mechanisms. Furthermore, maintenance costs will be estimated, and necessary changes to the design will be made to improve the efficiency and effectiveness of maintenance. The analysis conditions required for the reliability assessment will be evaluated by the VLS.

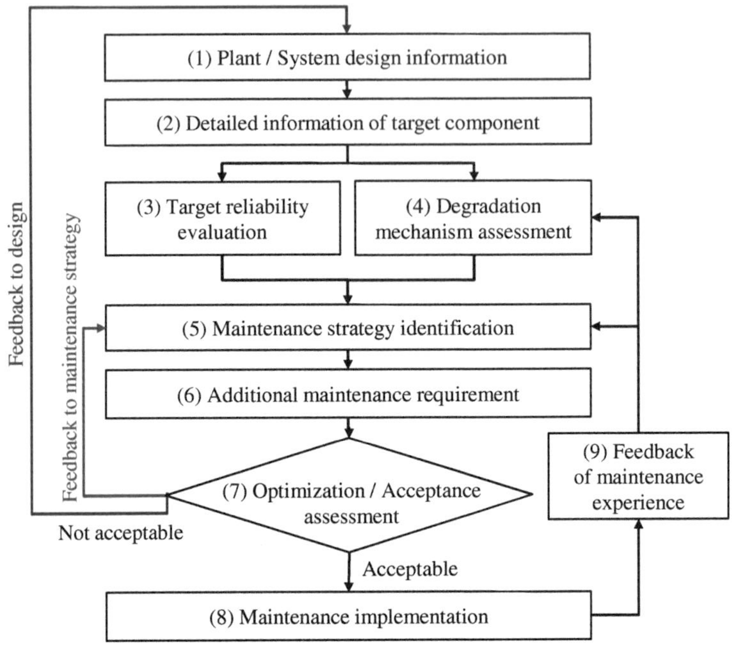

Fig. 5.139 Overview of a proposed maintenance optimization scheme.

5.10.3 ARKADIA-Safety

5.10.3.1 Scope of application

ARKADIA-Safety will realize automatic optimization of SA management and its feedback to a plant design. This goal is accomplished through the use of AI and a knowledge base built from data obtained from future research and development as well as previous SFR development projects. ARKADIA-Safety can perform entire in- and ex-vessel numerical analysis of SAs, statistical safety evaluation, and dynamic PRA. At this time, the primary focus of ARKADIA-Safety is SFRs. The sections that follow describe the integrated SA analysis code, Severe-accident PhEnomenological computational Code for TRansient Assessment (SPECTRA), as a VLS base technology.

5.10.3.2 Severe accident analysis method

There are several computational codes to evaluate plant safety of LWRs, such as MAAP [545], MELCOR [546], and IMPACT/SAMPSON [547]. These codes compute consistently in- and ex-vessel phenomena during SAs. Computational codes for SFRs have also been developed, including SAS-4A [548] for an initiating phase of core disruption, SIMMER [58] for a transition phase, and CONTAIN-LMR [549] for ex-vessel phenomena. The applicability of these codes has been well demonstrated through numerous validation analyses. Users should pay attention to temporal and spatial connections between the codes to realize the entire evaluation from in- to ex-vessel phenomena. Transferring boundary conditions between codes provides excessive conservativeness in some cases, making many parametric analyses difficult to perform. Using a single code, such as the aforementioned safety evaluation codes for LWRs, is the best way to complete the entire in- and ex-vessel evaluation. Some research projects were started to develop integrated codes for SFRs. ASTEC-Na [550] is being developed in a JASMIN project in Europe. This code is an extension of ASTEC, which is a safety analysis code for LWRs. MELCOR developed by Sandia National Laboratories is also being enhanced for sodium chemistry appearing in SFRs [133,134,551–554].

JAEA also started to develop the computational code, SPECTRA [555,556], to compute in- and ex-vessel phenomena integrally and to evaluate various SA scenarios in SFRs. Fig. 5.140 shows elemental phenomena to be considered in SPECTRA. These phenomena were identified from an importance analysis.

SPECTRA consists of in- and ex-vessel modules as illustrated in Fig. 5.141. As a base component, each module includes a thermal-hydraulics model. At each time step, the in- and ex-vessel modules are coupled by exchanging their boundary parameters. The amount of sodium that leaks from a failed pipe, for example, is calculated using the area of the failure hole and the pressure difference between the inside and outside of the pipe. As described later, several models for elemental physical phenomena are integrated into the thermal-hydraulics model. Each module will be expanded by improving the current models and integrating other models.

(a) In-vessel module

In-vessel coolant flow during a SA includes cover gas and vaporized sodium in some cases. The molten core should also be considered in the evaluation of the in-vessel coolant flow. The in-vessel module computes behaviors of liquid sodium, gas, and molten core.

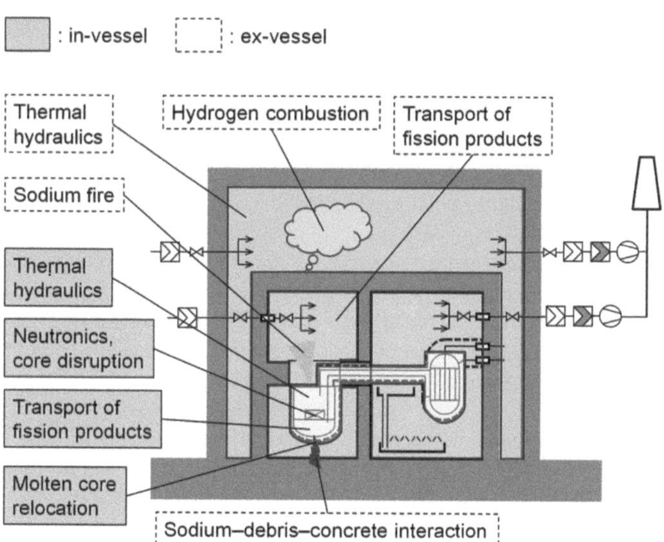

Fig. 5.140 Phenomena to be evaluated by SPECTRA.

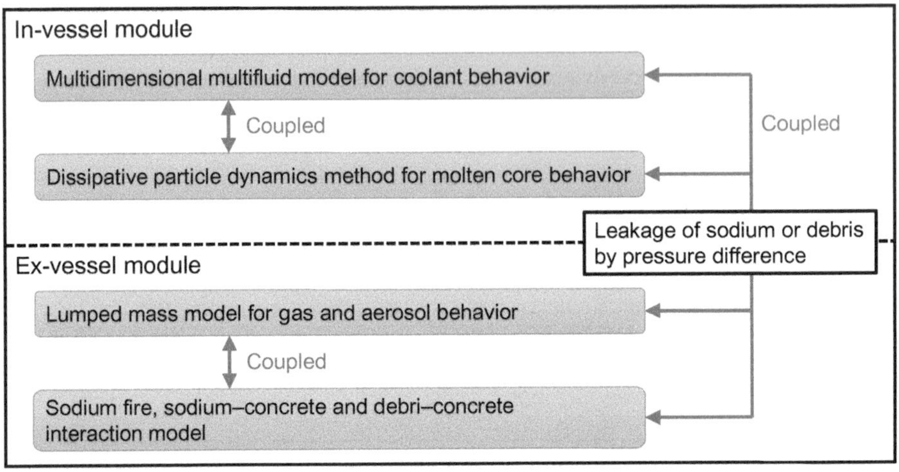

Fig. 5.141 Current structure of modules and models in SPECTRA.

For the analysis of two-phase coolant flow, a multidimensional multifluid model is used. By taking compressibility and volume conservation of gas and liquid phases into account, a method for solving mass, momentum, and energy conservation equations was developed. To enable stable computation with a large time step width, a fully implicit scheme is applied to this model. The Dissipative Particle Dynamics method, a Lagrange-based particle method, is used to compute the relocation of a molten core. This method was chosen due to its low computational load when compared to other particle methods. Using empirical parameters, the method can compute the behavior of a high-viscosity fluid. The molten-core analysis model is coupled with the multifluid model mentioned earlier by considering porosity and permeability. Momentum and heat are exchanged between the two models.

(b) Ex-vessel module

The ex-vessel thermal-hydraulics model evaluates the state of the atmosphere such as pressure, temperature, and concentration in each cell and velocity between the cells by using a lumped parameter model. This model incorporates mass, momentum, and energy conservation equations for a multicomponent atmosphere that includes aerosols. The momentum conservation equation takes into account the area, length, and pressure loss of a piping system. As with the in-vessel thermal-hydraulics model, these equations were discretized using the fully implicit method for fast computation.

The ex-vessel module can consider any number of a wall and compute convection and radiation heat transfer between the atmosphere and the wall. Users can select one- or two-dimensional heat conduction equation for each wall.

Sodium fire is one of the key phenomena during SAs of SFRs. The sodium fire model in SPECTRA consists of a spray fire model and a pool fire model [555]. The spray fire model is based on the multidimensional computational fluid dynamics code, AQUA-SF [129,557–560]. The SPECTRA model has some changes for a lumped mass model, such as the elimination of spray-droplet spreading and the effect of gas-phase velocity on spray-droplet motion. The pool fire model, on the contrary, is based on the model in the lumped parameter code, SPHINCS [561,562]. The pool and floor structure are divided into two-dimensional ring-shaped regions in radial and vertical directions. This ring model can be used to simulate pool spreading during sodium leakage and sedimentation on the floor.

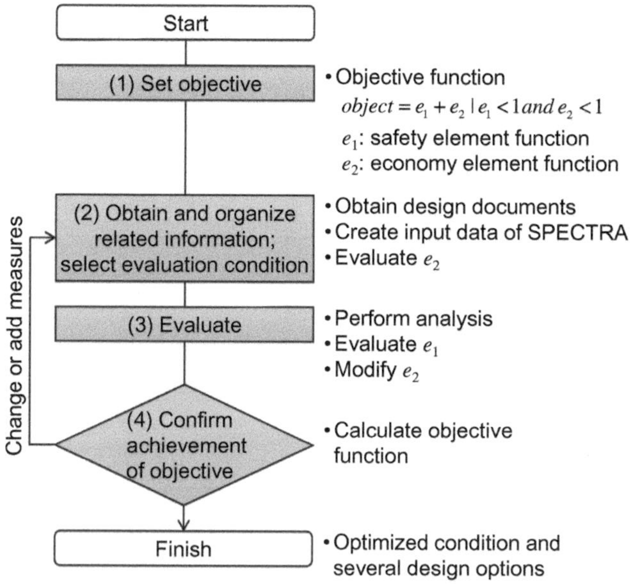

Fig. 5.142 Example optimization problem of ARKADIA-Safety.

During a SA with the discharge of molten core and coolant in the reactor pit, a concrete floor is ablated by a chemical reaction with a sodium pool and debris. The sodium-concrete, and debris-concrete interaction models were constructed based on the models in CONTAIN-LMR.

5.10.3.3 Example problems for application

Applicability of ARKADIA-Safety will be confirmed through application to an example containment vessel (CV) optimization problem considering SA phenomena. Fig. 5.142 depicts the optimization flowchart for this problem. In this problem, we use SPECTRA to evaluate a sodium fire and fission product leakage during a SA. The CV size and sodium fire mitigation measures such as CV vent, nitrogen injection, compartmentalization, and laying liner are considered as sensitive parameters. The purpose of this example problem is to minimize an objective function given by the summation of a safety element function and an economic element function.

5.10.4 Knowledge base

KMS systematically organizes and provides the results of research and development of SFR carried out at JAEA Oarai Research and Development Institute, etc., and the results of technological development obtained from the design, construction, test run, and maintenance of "Monju" [563]. KMS will thus support the succession of technologies for the next reactor design, various feasibility assessments, facility operations,

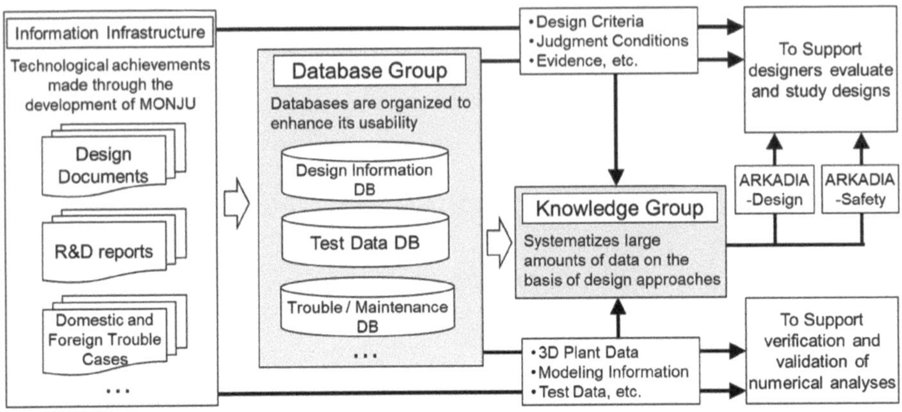

Fig. 5.143 Conceptual structure of KMS.

and other activities. Furthermore, KMS aims to support design process optimization by structuring the correlation of technical information based on the design and various evaluation processes. A knowledge base that can appropriately provide aggregated results is required to develop KMS that enables such supports. Fig. 5.143 shows the structure of KMS and its core knowledge base.

The knowledge base consists of three groups: the information infrastructure, the database group, and the knowledge group. The information infrastructure collects and organizes technical documents and reports related to SFR. SFR research and development has progressed from the experimental reactor "Joyo" to the prototype reactor "Monju," encompassing a wide range of fields including nuclear data, material development, equipment development, sodium handling, and operator training. In particular, valuable experiences and lessons, such as technical issues, accidents, and troubles, have been accumulated for the prototype reactor "Monju," whose design and development validity were confirmed through trial operation up to 40% output. Documents and reports relating to the technological development outcomes of these SFR can be accessed through the information infrastructure.

The database group aggregates technical information such as design information, various test data, and trouble/maintenance cases. This technical information is organized to improve usability by summarizing the individual results in the information infrastructure. In the future, the group will add search indexes such as "development purpose," "background," "state of the solution," and "current goal" for each technical issue peculiar to SFR and be able to search for explicit/tacit knowledge according to the development stage.

The knowledge group is then prepared in order to make the knowledge aggregated in these information infrastructures and database groups available from ARKADIA-Design and ARKADIA-Safety. This knowledge group is a collection of data that systemizes a large amount of technical data from the other two groups. Eventually, we hope to use these systematic data to create a system capable of rational judgment and evaluation without the need for human intervention.

5.10.5 Summary

The development will be done in two phases; ARKADIA-Design and ARKADIA-Safety, which deal with design and safety, respectively, will be developed in the first phase. ARKADIA will be completed in the second phase as a system that fully integrates numerical simulations, a knowledge base of fast reactor development experience, and plant life cycle optimization functions. Throughout the development process, up-to-date systems will be gradually made available. This integration will enable to make more rational optimization of plant design by automatically improving various conditions, such as safety, economic efficiency, and environmental acceptability.

References

[1] GIF, Safety Design Criteria for Generation IV Sodium-cooled Fast Reactor System, Rev.1, 2017. GIF SDC-TF/2017/02.

[2] IAEA, Safety of Nuclear Power Plants: Design, 2016. SSR-2/1 Rev.1.

[3] GIF, Safety Design Guidelines on Safety Approach and Design Conditions for Generation IV Sodium-cooled Fast Reactor Systems (Rev.1), 2019. SDC-TF/2019/01.

[4] S. Kubo, R. Nakai, T. Sofu, Development of safety design guideline on structures, systems and components for generation IV sodium-cooled fast reactor systems, in: Proceedings of the 4th Generation-IV International Forum (GIF) Symposium, Paris, France, 2018, pp. 485–494. Oct. 16–17.

[5] USDOE & GIF, A Technology Roadmap for Generation-IV Nuclear Energy Systems, 2002. GIF-002-00.

[6] IAEA, Defence in Depth in Nuclear Safety, INSAG-10, A Report by the International Nuclear Safety Advisory Group, IAEA, Vienna, 1996.

[7] IAEA, IAEA Safety Glossary, Terminology Used in Nuclear Safety and Radiation Protection, 2018 ed., 2019.

[8] T. Sofu, Y. Okano, The safety design criteria development and summary of its update for the generation-IV SFR systems, in: Proceedings of the International Conference on Fast Reactors and Related Fuel Cycles (FR17), Yekaterinburg, Russia, 2017. June 26–29, IAEA-CN-245-46.

[9] R. Nakai, The safety design guideline development for generation-IV SFR system, in: Proceedings of the International Conference on Fast Reactors and Related Fuel Cycles (FR17), Yekaterinburg, Russia, 2017. June 26–29, IAEA-CN-245-15.

[10] IAEA, Design of the Reactor Core for Nuclear Power Plants, 2019. SSG-52.

[11] IAEA, Design of the Reactor Coolant System and Associated Systems in Nuclear Power Plants, 2020. SSG-56.

[12] IAEA, Design of Reactor Containment and Associated Systems for Nuclear Power Plants, 2019. SSG-53.

[13] E. Marova, S. Shepelev, Y. Ashurko, V. Dekusar, V. Korobeynikov, V. Korobeynikov, B. Tikhomirov, Results of BN-1200 Design Assessment for the Compliance with the Requirements of Generation IV and INPRO, 2017. FR17, IAEA-CN-245-399.

[14] F. Varaine, G. Rodriguez, J.M. Hamy, S. Kubo, H. Mochida, U. Yukinori, J.P. Helle, A. Remy, T. Chauveau, J. Mazel, M. Libessart, R.P. Benard, M. Fukuie, D. Settimo, V. Gautier, Y. Lhor, M. Lefrancois, ASTRID Project General Overview and Status Progress, ICAPP 2018 Paper ID#23960, 2018.

[15] P. Puthiyavinayagam, K. Devan, S.R. Aithal, S. Athmalingam, R. Vijayashree, S. Raghupathy, K. Velusamy, N. Theivarajan, S. Usha, A.K. Bhaduri, Advanced Design Features of MOX Fuelled Future Indian SFRs, 2017. FR17, IAEA-CN-245-300.
[16] H. Yamano, S. Kubo, Y. Shimakawa, K. Fujita, T. Suzuki, K. Kurisaka, Safety design and evaluation in a large-scale Japan sodium-cooled fast reactor, in: Science and Technology of Nuclear Installations, vol. 2012, 2012. ID614973.
[17] K. Satoh, S. Kotake, T. Hoshi, S. Aoi, H. Takahashi, K. Nakai, H. Endo, S. Kasai, A study of enhanced safety systems for LMFBRs, in: International Conference on Design and Safety of Advanced Nuclear Power Plants, 1992. ANP-92.
[18] M. Kambe, M. Uotani, Design and development of fast breeder reactor passive reactivity control systems: LEM and LIM, Nucl. Technol. 122 (1998) 179–195.
[19] S. Nakanishi, T. Hosoya, S. Kubo, S. Kotake, M. Takamatsu, T. Aoyama, I. Ikarimoto, J. Kato, Y. Shimakawa, K. Harada, Development of passive shutdown system for SFR, Nucl. Technol. 170 (2010).
[20] S. Kotake, N. Uto, K. Aoto, S. Kubo, Safety design features for Japan sodium-cooled fast reactor (JSFR)—passive safety and CDA mitigation, Int. J. Nucl. Power (2010) 386–393. atw55, June.
[21] S. Kotake, H. Yamano, Y. Sagayama, Safety design approach for a large-scale Japan sodium-cooled fast reactor (JSFR), Trans. Fusion Sci. Technol. 61 (2012) 137–143.
[22] S. Kotake, N. Uto, K. Aoto, S. Kubo, Safety design features for JSFR, in: Annual Meeting on Nuclear Technology, 2009.
[23] T.R. Lancet, R.E. Specht, M. Cooper, R. Adkins, S.H. Bailey, L.A. Grants, A. Millunzi, Inherently safe reactor design, in: Proceedings of the International Conference Organized by the British Nuclear Energy Society in London, 1977. November.
[24] J. Kato, I. Ikarimoto, S. Matsunaga, S. Kotak, S. Kubo, Development of self-actuated shutdown system for sodium-cooled fast reactor (Study on component materials suitable for fast reactors), in: Transactions of the JSME (in Japanese), vol. 87, 2021. No. 19-00420.
[25] S.E. Sowa, P.W. Barthold, T.D. Eggen, R.P. Huebotter, J. Josephson, A.P. Pizzica, B.R. Turski, H.J. Van, LMFBR self-actuated shutdown systems, in: International Meeting on Fast Reactor Safety and Related Physics, 1976. October.
[26] B.R. Tupper, A.W. Brummond, K.R. Paschall, S.S. Borys, Development of a self-actuated shutdown system, in: ASME 81-JPGC-NE-8, 1981. October.
[27] M. Miura, T. Inagaki, M. Hamada, K. Muramatsu, Present status of DFBR design in Japan, in: Proceedings of the International Conference on Design and Safety of Advanced Nuclear Power Plants, 1992. ANP-92.
[28] T. Nakamura, T. Inagaki, Y. Shibata, K. Tarutani, K. Okada, The development of demonstration fast breeder reactor (DFBR), in: Proceedings of a Technical Committee Meeting Held in Kalpakkam, India, 1995. IAEA-TECDOC-907.
[29] T. Hoshi, K. Harada, H. Endo, I. Ikarimoto, Selection study of self-actuated shutdown system for a large scale FBR, in: Proceedings of 3rd International Conference on Nuclear Engineering, 1995. ICONE-3.
[30] M. Morihata, H. Aono, M. Ariyoshi, I. Ikarimoto, Development of self-actuated shutdown system for FBR in Japan, in: Proceedings of 5th International Conference on Nuclear Engineering, 1997. ICONE-5.
[31] E. Yoshida, T. Furukawa, Corrosion issues in sodium-cooled fast reactor (SFR) systems, in: Nuclear Corrosion Science and Engineering, 2012, pp. 773–806. No. 22.

[32] S.S. Manson, Fatigue: a complex subject-some simple approximations, Exp. Mechan. 5 (7) (1965) 193–226.
[33] M. Takamatsu, T. Sekine, T. Aoyama, M. Uchida, S. Kotake, Demonstration of control rod holding stability of the self actuated shutdown system in joyo for enhancement of fast reactor inherent safety, J. Nucl. Sci. Technol. 44 (3) (2007) 511–517.
[34] M. Saito, S. Yoshie, M. Ishikawa, Development of evaluation methods on shock structural integrity of reactor vessel—experimental and analytical study on shock structural response of scaled models of reactor vessel, PNC Tech. Rev. 53 (1985) 78–84 (in Japanese).
[35] Y. Onoda, K. Matusba, Y. Tobita, T. Suzuki, Preliminary analysis of the post-disassembly expansion phase and structural response under unprotected loss of flow accident in prototype sodium cooled fast reactor, Mech. Eng. J. 4 (3) (2017) 1–14.
[36] T. Nakamura, H. Kaguchi, I. Ikarimoto, Y. Kamishima, K. Koyama, S. Kobo, S. Kotake, Evaluation method for structural integrity assessment in core disruptive accident of fast reactor, Nucl. Eng. Des. 227 (2004) 97–123.
[37] H. Niwa, A comprehensive approach of reactor safety research aiming at elimination of recriticality in CDA for commercialization of LMFBR, Progr. Nucl. Energy 32 (3/4) (1998) 621–629.
[38] H. Endo, S. Kubo, S. Kotake, I. Sato, S. Kondo, H. Niwa, K. Koyama, Elimination of recriticality potential for the self-consistent nuclear energy system, Progr. Nucl. Energy, 40(3/4), pp. 577-586.
[39] T. Suzuki, K. Kamiyama, H. Yamano, S. Kubo, Y. Tobita, R. Nakai, K. Koyama, A scenario of core disruptive accident for Japan sodium-cooled fast reactor to achieve in-vessel retention, J. Nucl. Sci. Technol. 51 (4) (2014) 493–513.
[40] S. Kobo, K. Kamiyama, T. Suzuki, K. Morita, The status of safety research in the field of sodium-cooled fast reactors in Japan, 2017. Proc. International Conference on Fast Reactors and Related Fuel Cycles: Next Generation Nuclear Systems for Sustainable Development (FR17), Yekaterinburg, Russian Federation, 26–29 June IAEA-CN-245-28.
[41] T. Suzuki, Y. Tobita, K. Kawada, H. Tagami, J. Sogabe, K. Matsuba, K. Ito, H. Ohshima, A preliminary evaluation of unprotected loss-of-flow accident for a prototype fast-breeder reactor, J. Nucl. Eng. Technol. 47 (3) (2015) 240–252.
[42] J.C. Carter, G.J. Fischer, T.J. Heames, D.R. MacFarlane, N.A. McNeal, W.T. Sha, C.K. Sanathanan, C.K. Youngdahl, SAS1A, A Computer Code for the Analysis of Fast-Reactor Power and Flow Transients, 1970. ANL-7607, October, 1970.
[43] F.E. Dunn, G.J. Fischer, T.J. Heames, P.A. Pirrica, N.A. McNeal, W.R. Bohl, S.M. Prastein, The SAS2A LMFBR Accident-Analysis Computer Code, 1974. ANL-8138, October, 1974.
[44] M.G. Stevenson, W.R. Bohl, F.E. Dunn, T.J. Heames, C. Hoppner, L.L. Smith, Current status and experimental basis of the SAS LMFBR accident analysis code system, in: Proc. Fast Reactor Safety Meeting, CONF-740401-P3, Beverly Hills, California, U.S. A, 1974, pp. 1303–1321. April 2-4, 1974.
[45] A.M. Tentner, D.P. Weber, G. Birgersson, G.L. Bordner, L.L. Briggs, J.E. Cahalan, F.E. Dunn, K. Kalimullah, J. Miles, F.G. Prohammer, The SAS4A LMFBR whole core accident analysis code, in: Proc. Int. Top. Mtg. on Fast Reactor Safety, Knoxville, Tennessee, U.S.A, 1985. April 21-25, 1985.
[46] H. Niwa, Improvement of Material Motion Model in SAS4A FBR Safety Analysis Code - Improvement of the Hydrodynamics and Jamming Model of the Fuel Flow with Solid Fuel Chunks, 1993. PNC TN9410 93-23, November, 1993 (in Japanese).

[47] S. Isida, K. Kawada, Y. Fukano, Validation study of SAS4A code for the unprotected loss-of-flow accident in an SFR, Mech. Eng. J. 7 (3) (2020) 1–17.

[48] K. Kawada, I. Sato, Y. Tobita, W. Pfrang, L. Buffe, E. Dufour, Development of PIRT (phenomena identification and ranking table) for SAS-SFR (SAS4A) validation technical publication, in: Proc. 22th Int. Conf. on Nuclear Engineering (ICONE-22), July 7-11, 2014, Prague, Czech Republic, Paper No. ICONE22-30679, 2014.

[49] K. Baumung, A.H. Lumpkin, Fuel motion measurement with the CABRI hodoscope, in: Proc. Int. Conf. on Science and Technology of Fast Reactor Safety, Guernsey, Channel Islands, UK, May 12-16, 1986, 1986, pp. 141–146.

[50] N. Nonaka, F. Kasahara, H. Niwa, I. Sato, In-pile experiment analyses relevant to initiating-phase energetics, in: Proc. Int. Conf. on Science and Technology of Fast Reactor Safety, Guernsey, Channel Islands, UK, May 12–16, 1986, 1986, pp. 413–418.

[51] N. Nonaka, I. Sato, Improvement of evaluation method for initiating-phase energetics based on CABRI-1 in-pile experiments, Nucl. Technol. 98 (1992) 54–69.

[52] Y. Onoda, I. Sato, Post-failure fuel relocation behavior in the CABRI-FAST EFM1 test, JNC Tech. Rev. (7) (2000) 71–81. JNC TN1340 2000-002, June, 2000 (in Japanese).

[53] I. Sato, F. Lemoine, D. Struwe, Transient fuel behavior and failure condition in the CABRI-2 experiments, Nucl. Technol. 145 (1) (2004) 115–137.

[54] I. Sato, Y. Tobita, T. Suzuki, K. Kawada, Y. Fukano, S. Fujita, K. Kamiyama, N. Nonaka, M. Ishikawa, S. Usami, Analysis of ULOF accident in monju reflecting the knowledge from cabri in-pile experiment and others, JAEA Res. (2007). 2007-055, May, 2007 (in Japanese).

[55] Y. Onoda, Y. Fukano, I. Sato, C. Marquie, B. Duc, Three-pin cluster CABRI tests simulating the unprotected loss-of-flow accident in sodium-cooled fast reactors, J. Nucl. Sci. Technol. 48 (2) (2011) 188–204.

[56] N. Ueda, Development of Analytical Code for Core Disruptive Accident of Metallic Fuel Core and Evaluation of Upper Limitation of Coolant Void Reactivity, 2001. CRIEPI T01002, June, 2001 (in Japanese).

[57] W.R. Bohl, L.B. Luck, SIMMER-II: A Computer Program for LMFBR Disrupted Core Analysis, Los Alamos National Laboratory Report, 1990. LA-11415-MS.

[58] S. Kondo, Y. Tobita, K. Morita, N. Shirakawa, SIMMER-III: an advanced computer program for LMFBR severe accident analysis, in: Proc. Int. Conf. on Design and Safety of Advanced Nuclear Power Plants (ANP'92), Tokyo, Japan, October 25-29, 1992, 1992, pp. 40.5.1–40.5.11.

[59] S. Kondo, Y. Tobita, K. Morita, N. Shirakawa, Development of SIMMER-III: a coupled neutronics-thermohydraulics computer code for FBR safety analysis, PNC Tech. Rev. 89 (1994) 19–29. PNC TN1340 94-001 (in Japanese).

[60] W.R. Bohl, D. Wilhelm, J. Berthier, F.R. Parker, The AFDM program: scope and significance, in: Proc. the 1990 Int. Fast Reactor Safety Mtg., Snowbird, Utah, U.S.A, vol. II, 1990, pp. 155–164. August 12-16, 1990.

[61] W.R. Bohl, D. Wilhelm, F.R. Parker, J. Berthier, L. Goutagny, H. Ninokata, AFDM: an advanced fluid-dynamics model, in: Los Alamos National Laboratory Report, 1990. LA-11692-MS.

[62] S. Kondo, Y. Tobita, Phase 1 code assessment of SIMMER-III, a computer program for LMFR core disruptive accident analysis, JAEA Res. (2019). 2019-009.

[63] K. Morita, D.J. Brear, Y. Tobita, K. Kamiyama, S. Kondo, Current status and achievement of SIMMER-III development for fast reactor safety analysis, PNC Tech. Rev. 101 (1997) 21–28. PNC TN1340 97-001 (in Japanese).

[64] S. Kondo, Y. Tobita, K. Morita, D.J. Brear, K. Kamiyama, H. Yamano, S. Fujita, W. Maschek, E.A. Fischer, E. Kiefhaber, G. Buckel, E. Hesserschwerdt, P. Coste, S. Pigny, Current status and validation of the SIMMER-III LMFR safety analysis code, in: Proc. 7th Int. Conf. on Nuclear Engineering (ICONE-7), Tokyo, Japan, 1999. April 19-23, ICONE-7249.

[65] S. Kondo, et al., Status and achievement of assessment program for SIMMER-III, a multiphase, multicomponent code for LMFR safety analysis, in: Proc. 8th Int. Top. Mtg. on Nuclear Reactor Thermal-Hydraulics (NURETH-8), Kyoto, Japan, September 1997, 1997.

[66] K. Morita, S. Kondo, Y. Tobita, D.J. Brear, SIMMER-III applications to fuel–coolant interactions, Nucl. Eng. Des. 189 (1999) 337–357.

[67] Y. Tobita, S. Kondo, H. Yamano, S. Fujita, K. Kamiyama, Computer code system for the R&D of nuclear fuel cycle with fast reactor, I. SIMMER-III: fast reactor safety analysis code, JNC Tech. Rev. 11 (2001) 19–29. JNC TN1340 2001-006, (in Japanese).

[68] Y. Tobita, S. Kondo, H. Yamano, K. Morita, W. Maschek, P. Coste, T. Cadiou, The development of SIMMER-III, an advanced computer program for LMFR safety analysis, and its application to sodium experiments, Nucl. Technol. 153 (2006) 245–255.

[69] S. Kondo, H. Yamano, Y. Tobita, S. Fujita, K. Kamiyama, W. Maschek, P. Coste, S. Pigny, J. Louvet, Phase 2 Code Assessment of SIMMER-Ill, A Computer Program for LMFR Core Disruptive Accident Analysis, 2000. JNC TN9400 2000-105.

[70] K. Kamiyama, D.J. Brear, Y. Tobita, S. Kondo, Establishment of freezing model for reactor safety analysis, J. Nucl. Sci. Technol. 43 (10) (2006) 1206–1217.

[71] M. Aoyagi, K. Kamiyama, Y. Tobita, Improvement of a physical model for blockage formation of solid-liquid mixture flow with freezing for core safety evaluation of SFRs, J. Nucl. Sci. Technol. 55 (5) (2018) 530–538.

[72] M.M. Rahmana, Y. Ege, K. Morita, K. Nakagawa, K. Fukuda, W. Maschek, Simulation of molten metal freezing behavior on to a structure, Nucl. Eng. Des. 238 (2008) 2706–2717.

[73] M.K. Hossain, Y. Himuro, K. Morita, K. Nakagawa, T. Matsumoto, K. Fukuda, W. Maschek, Simulation of molten metal penetration and freezing behavior in a seven-pin bundle experiment, J. Nucl. Sci. Technol. 46 (8) (2009) 799–808.

[74] H. Yamano, Y. Tobita, Experimental analyses by SIMMER-III on fuel-pin disruption and low-energy disrupted core motion, J. Power Energ. Syst. 4 (1) (2010).

[75] H. Yamano, Y. Onoda, Y. Tobita, I. Sato, Transient heat transfer characteristics between molten fuel and steel with steel boiling in the CABRI-TPA2 test, Nucl. Technol. 165 (2) (2009) 145–165.

[76] K. Morita, T. Matsumoto, R. Akasaka, K. Fukuda, T. Suzuki, Y. Tobita, H. Yamano, S. Kondo, Development of multicomponent vaporization/condensation model for a reactor safety analysis code SIMMER-III: Theoretical modeling and basic verification, Nucl. Eng. Des. 220 (2003) 224–239.

[77] T. Suzuki, Y. Tobita, H. Yamano, S. Kondo, K. Morita, T. Matsumoto, R. Akasaka, K. Fukuda, Development of multicomponent vaporization/condensation model for a reactor safety analysis code SIMMER-III: Extended verification using multi-bubble condensation experiment, Nucl. Eng. Des. 220 (2003) 240–254.

[78] K. Morita, T. Matsumoto, K. Fukuda, Y. Tobita, H. Yamano, I. Sato, Experimental verification of the fast reactor safety analysis code SIMMER-III for transient bubble behavior with condensation, Nucl. Eng. Des. 238 (2008) 49–56.

[79] K. Konishi, J. Toyooka, K. Kamiyama, I. Sato, S. Kubo, S. Kotake, K. Koyama, A.D. Vurim, V.A. Gaidaichuk, A.V. Pakhnits, Y.S. Vassiliev, The result of a wall failure in-pile experiment under the EAGLE project, Nucl. Eng. Des. 237 (2007) 2165–2174.

[80] I. Sato, Y. Tobita, K. Konishi, K. Kamiyama, J. Toyooka, R. Nakai, S. Kubo, S. Kotake, K. Koyama, Y. Vassiliev, A. Vurim, V. Zuev, A. Kolodeshnikov, Safety strategy of JSFR eliminating severe recriticality events and establishing in-vessel retention in the core disruptive accident, J. Nucl. Sci. Technol. 48 (4) (2011) 556–566.

[81] K. Konishi, S. Kubo, I. Sato, K. Koyama, J. Toyooka, K. Kamiyama, S. Kotake, A.D. Vurim, V.A. Gaidaichuk, A.V. Pakhnits, Y.S. Vassiliev, The EAGLE project to eliminate the recriticality issue of fast reactors—progress and results of in-pile tests, in: Proc. Fifth Korea-Japan Symposium on Nuclear Thermal Hydraulics and Safety, Jeju, Korea, November 26-29, 2006, NTHAS5-F001, 2006.

[82] K. Kamiyama, M. Saito, K. Matsuba, M. Isozaki, I. Sato, K. Konishi, V.A. Zuyev, A.A. Kolodeshnikov, Y.S. Vassiliev, Experimental study on fuel-discharge behaviour through in-core coolant channels, J. Nucl. Sci. Technol. 50 (6) (2013) 629–644.

[83] K. Kamiyama, K. Konishi, I. Sato, J. Toyooka, K. Matsuba, V.A. Zuyev, A.V. Pakhnits, V.A. Vityukc, A.D. Vurim, V.A. Gaidaichuk, A.A. Kolodeshnikov, Y.S. Vassiliev, Experimental studies on the upward fuel discharge for elimination of severe recriticality during coredisruptive accidents in sodium-cooled fast reactors, J. Nucl. Sci. Technol. 51 (9) (2014) 1114–1124.

[84] J. Toyooka, K. Kamiyama, Y. Tobita, T. Suzuki, Improvements to the SIMMER code model for steel wall failure based on EAGLE-1 test results, in: Tenth Korea-Japan Symposium on Nuclear Thermal Hydraulics and Safety; Kyoto, Japan, November 27-30, 2016, N10P1116, 2016.

[85] K. Kamiyama, Y. Tobita, T. Suzuki, K. Matsuba, Development of an evaluation methodology for fuel discharge in core disruptive accidents of sodium-cooled fast reactors, J. Energ. Power Eng. 8 (2014) 785–793.

[86] S. Kondo, H. Yamano, Y. Tobita, S. Fujita, K. Morita, M. Mizuno, S. Hosono, T. Kondo, SIMMER-IV: A Three-Dimensional Computer Program for LMFR Core Disruptive Accident Analysis—Version1.B Model Summary and Program Description, 2000. JNC TN9400 2001-003.

[87] H. Yamano, S. Fujita, Y. Tobita, S. Kondo, K. Morita, M. Sugaya, M. Mizuno, S. Hosono, T. Kondo, SIMMER-IV: A Three-Dimensional Computer Program for LMFR Core Disruptive Accident Analysis—Version 2.A Model Summary and Program Description, 2003. JNC TN9400 2003-070.

[88] H. Yamano, S. Fujita, Y. Tobita, K. Kamiyama, S. Kondo, K. Morita, E.A. Fischer, D.J. Brear, N. Shirakawa, X. Cao, M. Sugaya, M. Mizuno, S. Hosono, T. Kondo, W. Maschek, E. Kiefhaber, G. Buckel, A. Rineiski, M. Flad, T. Suzuki, P. Coste, S. Pigny, J. Louvet, T. Cadiou, SIMMER-III: A Computer Program for LMFR Core Disruptive Accident Analysis—Version 3.A Model Summary and Program Description, 2003. JNC TN9400 2003-071.

[89] S. Kobo, F. Payot, H. Yamano, F. Bertrand, A. Bachrata, L. Saas, C. Journeau, S. Gosse, A. Quaini, A. Shibata, T. Iitsuka, O. Czarny, France-Japan collaboration on the SFR severe accident studies: outcomes and future work program, in: Proc. International Conference on Fast Reactors and Related Fuel Cycles: Sustainable Clean Energy for the Future (FR22), Vienna, Austria, April 25–28, 2022. (to be published), 2022.

[90] K. Morita, S. Zhang, S. Koshizuka, Y. Tobita, H. Yamano, N. Shirakawa, F. Inoue, H. Yugo, M. Naitoh, H. Okada, Y. Yamamoto, M. Himi, E. Hirano, S. Shimizu,

M. Oue, Detailed analyses of key phenomena in core disruptive accidents of sodium-cooled fast reactors by the COMPASS code, Nucl. Eng. Des. 241 (2011) 4672–4681.
[91] T. Zhang, K. Funakoshi, X. Liu, W. Liu, K. Morita, K. Kamiyama, Numerical simulation of heat transfer behavior in EAGLE ID1 in-pile test using finite volume particle method, Ann. Nucl. Energ. 150 (2021), 107856.
[92] K. Matsuba, K. Kamiyama, J. Toyooka, V.A. Zuev, D.A. Ganovichev, A.A. Kolodeshnikov, A recent experimental program to evidence in-vessel retention by controlled material relocation during core disruptive accidents of sodium-cooled fast reactors, in: Proc. The Tenth Korea-Japan Symposium on Nuclear Thermal Hydraulics and Safety (NTHAS10), Kyoto, Japan, 2016. November 27-30, 2016, N10P1100.
[93] K. Matsuba, K. Kamiyama, D.A. Ganovichev, V.V. Baklanov, Results of an out-of-pile experiment for fragmentation of a simulated molten core material discharged into a shallow sodium pool, in: Proc. The Eleventh Korea-Japan Symposium on Nuclear Thermal Hydraulics and Safety, Busan, Korea, November 18-21, 2018, N11P0111, 2018.
[94] S. Kato, K. Matsuba, K. Kamiyama, D.A. Ganovichev, V.V. Baklanov, Study on the discharge behavior of molten-core through the control rod guide tube in the core disruptive accident of SFR, in: Proc. ICAPP 2019: International Congress on Advances in Nuclear Power Plants, Juan-les-pins, France, 2019. May 12-15, 2019.
[95] S. Kondo, K. Konishi, M. Isozaki, S. Imahori, A. Furutani, D.J. Brear, Experimental study on simulated molten jet-coolant interactions, Nucl. Eng. Des. 155 (1995) 73–84.
[96] K. Matsuba, M. Isozaki, K. Kamiyama, Y. Tobita, Distance for fragmentation of a simulated molten-core material discharged into a sodium pool, J. Nucl. Sci. Technol. 53 (5) (2016) 707–712.
[97] M. Saito, K. Sato, A. Furutani, M. Isozaki, S. Imahori, Y. Hattori, Melting attack of solid plates by a high temperature liquid jet—effect of crust formation, Nucl. Eng. Des. 121 (1990) 11–23.
[98] K. Sato, A. Furutani, M. Saito, M. Isozaki, K. Suganuma, S. Imahori, Melting attack of solid plates by a high temperature liquid jet [II]—erosion behavior by a molten metal jet, Nucl. Eng. Des. 132 (1991) 171–186.
[99] Y. Imaizumi, K. Kamiyama, K. Matsuba, M. Isozaki, T. Suzuki, Y. Emura, Fundamental experiments of jet impingement and fragmentation simulating the fuel relocation in the core disruptive accident of sodium-cooled fast reactors, in: Proceedings of ICAPP 2017, April 24-28, 2017, Fukui and Kyoto, Japan, 2017, p. 17144.
[100] K. Matsuba, M. Isozaki, K. Kamiyama, Y. Tobita, T. Suzuki, Fundamental experiment on the distance for fragmentation of molten core material during core disruptive accidents in sodium-cooled fast reactors, Nucl. Saf. Simul. 4 (4) (2013).
[101] K. Matsuba, K. Kamiyama, J. Toyooka, Y. Tobita, V.A. Zuyev, A.A. Kolodeshnikov, Y. S. Vassiliev, Experimental discussion on fragmentation mechanism of molten oxide discharged into a sodium pool, Mech. Eng. J. 3 (3) (2016). Paper No.15-00595.
[102] K. Matsuba, K. Kamiyama, J. Toyooka, V.A. Zuyev, A.A. Kolodeshnikov, An experimental study on the fragmentation and accompanying cooling behaviors of a simulated molten oxide fuel penetrating into a sodium pool, in: 17th Int. Top. Mtg. on Nuclear Reactor Thermal Hydraulics (NURETH-17), Xi'an, China, 2017. September, 2017.
[103] K. Matsuba, M. Isozaki, K. Kamiyama, T. Suzuki, Y. Tobita, An empirical correlation to predict the distance for fragmentation of simulated molten-core materials discharged into a sodium pool, in: NUTHOS-11: 11th Int. Top. Mtg. on Nuclear Reactor Thermal Hydraulics, Operation and Safety, Gyeongju, Korea, 2016. October 9-13, 2016, N11A0089.

[104] M.A.R. Sheikh, X. Liu, T. Matsumoto, K. Morita, L. Guo, T. Suzuki, K. Kamiyama, Numerical simulation of the solid particle sedimentation and bed formation behaviors using a hybrid method, Energies 13 (2020) 5018.
[105] J.D. Gabor, E.S. Sowa, L. Baker Jr., J.C. Cassulo, Studies and experiments on heat removal for fuel debris in sodium, in: Proc. of Fast Reactor Safety Meeting, Beverly Hills, CA, U.S.A, 1974, pp. 823–844. April, 1974.
[106] V.K. Dhir, I. Catton, Prediction of dryout heat fluxes in bed of volumetrically heated particles, in: Proc. of Int. Mtg. on Fast Reactor Safety and Related Physics, Chicago, Illinois, 1976, pp. 2026–2035. October, 1976.
[107] E.S. Sowa, J.D. Gabor, L. Baker Jr., J.R. Pavlik, J.C. Cassulo, W. Holloway, Studies of the formation and cooling of particulate fuel debris beds in sodium, in: Proc. of Int. Mtg. on Fast Reactor Safety and Related Physics, Chicago, Illinois, 1976, pp. 2036–2044. October 5-8, 1976.
[108] Y. Tobita, K. Kamiyama, H. Tagami, K. Matsuba, T. Suzuki, M. Isozaki, H. Yamano, K. Morita, L. Guo, B. Zhang, Development of the evaluation methodology for the material relocation behavior in the core disruptive accident of sodium-cooled fast reactors, J. Nucl. Sci. Technol. 53 (5) (2016) 698–706.
[109] L. Guo, K. Morita, H. Tagami, Y. Tobita, Numerical simulation of a self-leveling experiment using a hybrid method, Mech. Eng. J. 1 (4) (2014).
[110] L. Guo, K. Morita, Y. Tobita, Numerical simulations of gas–liquid–particle three-phase flows using a hybrid method, J. Nucl. Sci. Technol. 53 (2) (2016) 271–280.
[111] H. Tagami, S. Cheng, Y. Tobita, K. Morita, Model for particle behavior in debris bed, Nucl. Eng. Des. 328 (2018) 95–106.
[112] J. Sogabe, Y. Suzuki, Y.T. Wada, In-vessel retention of unprotected accident in a fast reactor; Assessment of material-relocation and heat-removal behavior in ULOF, in: 10th Japan-Korea Symposium on Nuclear Thermal Hydraulics and Safety (NTHAS-10), Kyoto, Japan, 2016. November, 2016, N10P1114.
[113] J. Sogabe, Y. Suzuki, Y.T. Wada, Study on in-vessel retention (IVR) of unprotected accident for fast reactor (2) assessment of PAMR/PAHR phase in ULOF, Trans. JSME 83 (848) (2017) (in Japanese).
[114] Y. Imaizumi, M. Aoyagi, K. Kamiyama, K. Matsuba, D.A. Canovichev, V.V. Baklanov, Development of evaluation method for in-place cooling of residual core materials in core disruptive accidents of SFRS, in: Proc. ICONE27: 27th International Conference on Nuclear Engineering, Ibaraki, Japan, 2019. May 19-24, 2019, ICONE27-1768.
[115] R.J. Lipinski, A model boiling and dryout in particulate beds, Sand (1982). 82-0765, NUREG/CR-2646.
[116] R.J. Lipinski, A coolability model for postaccident nuclear reactor debris, Nucl. Technol. 65 (1) (1984).
[117] H. Nakamura, R.J. Lipinski, A method to calculate boiling and dryout in debris bed using a heat conduction code, in: Proc. Int. Top. Mtg. on Fast Reactor Safety, Knoxville, Tennessee, U.S.A, 1985, pp. 767–773. April 21-25, 1985.
[118] E. Matsuo, K. Sasa, Y. Abe, A. Shibata, Study on high-speed calculation of debris bed coolability for sodium-cooled fast reactors, Mech. Eng. J. 7 (2) (2020) 19–00457.
[119] E. Matsuo, K. Sasa, H. Saito, Y. Abe, Study on application of artificial neural network to debris bed coolability calculations for sodium-cooled fast reactors, Mech. Eng. J. 7 (3) (2020). 19-00541.
[120] R. Nakai, T. Suzuki, H. Yamano, H. Seino, H. Ishikawa, K. Kamiyama, K. Koyama, K. Morita, "Development of severe accident evaluation technology (level 2 PSA) for

sodium-cooled fast reactors (1) overview of evaluation methodology development," Proc. 2009 International Congress on Advances in Nuclear Power Plants, Tokyo, Japan, May 10-14, 2009, Paper 9117.

[121] K. Koyama, Y. Yamada, S. Hayakawa, M. Watanabe, O. Watanabe, Development of severe accident evaluation technology (level 2 PSA) for sodium-cooled fast reactors (4) identification of dominant factors in core material relocation and heat removal phases, in: Proc. 2009 International Congress on Advances in Nuclear Power Plants, Tokyo, Japan, May 10-14, 2009, Paper 9126, 2009.

[122] H. Yamano, T. Takai, T. Furukawa, S. Kikuchi, Y. Emura, K. Kamiyama, H. Fukuyama, H. Higashi, T. Nishi, H. Ohta, K. Morita, K. Nakamura, Study on eutectic melting behavior of control rod materials in core disruptive accidents of sodium-cooled fast reactors: (1) project overview and progress until 2019, in: Proc. 28th International Conference on Nuclear Engineering, Virtual Conference, 2021. Online, August 4-6, 2021, ICONE28-63301.

[123] X. Liu, K. Morita, H. Yamano, Modeling of eutectic reaction between molten stainless steel and B_4C for severe accident simulations, in: Proc. 12th International Topical Meeting on Nuclear Reactor Thermal-Hydraulics, Operation and Safety (NUTHOS-12), Qingdao, China, 2018, p. 572. October 14-18, 2018.

[124] S. Miyahara, S. Ohno, H. Ishikawa, N. Dota, Sodium leak and fire control in fast breeder reactors, J. Combust. Soc. Jpn. 45 (133) (2003) 141–151 (in Japanese).

[125] S. Ohno, T. Matsuki, Development and validation of sodium fire analysis code ASSCOPS, JNC Tech. Rev. 11 (2001) 93–104 (in Japanese).

[126] P. Beiriger, J. Hopenfeld, M. Silberberg, R.P. Johnson, L. Baurmash, R.L. Koontz, SOFIRE II user report, AEC Res. Dev. Rep. (1973). AI-AEC-13055.

[127] P.R. Shire, SPRAY Code User's Reprot, Hanford Engineering Development Laboratory, HEDL-TME, 1979, pp. 76–94.

[128] A. Yamaguchi, Y. Tajima, Validation study of computer code SPHINCS for sodium fire safety evaluation of fast reactor, Nucl. Eng. Des. 219 (2003) 19–34.

[129] T. Takata, A. Yamaguchi, I. Maekawa, Numerical investigation of multi-dimensional characteristics in sodium combustion, Nucl. Eng. Des. 220 (2003) 37–50.

[130] Y. Okano, A. Yamaguchi, Direct numerical simulation of liquid sodium droplet combustion in forced convection air flow, in: Proc. of 7th International Conference on Nuclear Engineering (ICONE7), ICONE-7239, Tokyo, Japan, 1999. April 19-23.

[131] M. Iida, Three dimensional particle method code for sodium leakage analysis in liquid metal reactor, in: Proc. of 7th International Conference on Nuclear Engineering (ICONE7), ICONE-7202, Tokyo, Japan, 1999. April 19-23.

[132] Y. Okano, Development of Bi-Phase Sodium-Oxygen-Hydrogen Chemical Equilibrium Calculation Program (BISHOP) using Gibbs Free Energy Minimization Method, JNC Technical Report, JNC TN9400 99-071, 1999 (in Japanese).

[133] D.L.Y. Louie, A. Uchibori, Sodium Fire Collaborative Study Progress-CNWG Fiscal Year 2019, Sandia Report, SAND2019-15043, 2019.

[134] D.L.Y. Louie, M. Aoyagi, Sodium Fire Collaborative Study Progress-CNWG Fiscal Year 2020, Sandia Report, SAND2021-0136, 2021.

[135] M. Hori, Sodium-water reactions in steam generators of liquid metal fast breeder reactors, At. Energy Rev. 18 (3) (1980).

[136] H. Tanabe, et al., Results of Failure Propagation Tests in the Steam Generator Safety Test Facility (SWAT-3) Repot No. 5, PNC Report, PNC-TN9410-86-104, 1986 (in Japanese).

[137] H. Hiroi, et al., Test Results of Run-5 in Steam Generator Safety Test Facility (SWAT-3) Repot No. 10; Large Leak Sodium-Water Reaction Test, PNC Report, PNC-TN941-79-04, 1978 (in Japanese).
[138] O. Miyake, et al., Propagated Pressure Analysis on Large-scale Sodium-water Reaction. (Analysis of SWAT-3 Run-3, -5, and -7 Tests by SWACS Code)—Large Leak Sodium-Water Reaction Analysis (Report No. 11), PNC Report, PNC-TN941-81-15, 1981 (in Japanese).
[139] H. Tanabe, Failure Propagation Analysis of LMFBR Steam Generator Tube (Analysis of SWAT-3 Runs 14 and 15 by LEAP-II Code), PNC Report, PNC-TN941-82-100, 1982 (in Japanese).
[140] T. Takata, A. Yamaguchi, A. Uchibori, H. Ohshima, Computational methodology of sodium-water reaction phenomenon in steam generator of sodium-cooled fast reactor, J. Nucl. Sci. Technol. 46 (6) (2009) 613–623.
[141] A. Uchibori, A. Watanabe, T. Takata, H. Ohshima, Development of unstructured mesh-based numerical method for sodium-water reaction phenomenon, Nucl. Technol. 25 (2019) 119–127.
[142] S. Kikuchi, H. Seino, A. Kurihara, H. Ohshima, Kinetic study of sodium-water surface reaction by differential thermal analysis, J. Power Energ. Syst. 7 (2) (2013).
[143] A. Kurihara, R. Umeda, K. Shimoyama, S. Kikuchi, Evaluation of target-wastage in consideration of sodium-water reaction environment formed on the periphery of an adjacent tube in steam generator of sodium-cooled fast reactor, Trans. JSME 84 (859) (2018) (in Japanese).
[144] K. Haga, Y. Nishizawa, T. Watanabe, S. Miyahara, Y. Himeno, Equilibrium and non-equilibrium partition coefficients of volatile fission products between liquid sodium and the gas phase, Nucl. Technol. 97 (1992) 177–185.
[145] S. Miyahara, K. Haga, Y. Himeno, Sodium aerosol release rate and nonvolatile fission product retention factor during a sodium-concrete reaction, Nucl. Technol. 97 (1992) 212–226.
[146] S. Miyahara, N. Sagawa, K. Shimoyama, Iodine mass transfer from xenon-iodine mixed gas bubble to liquid sodium pool, (I) experiment, J. Nucl. Sci. Technol. 33 (2) (1996) 128–133.
[147] D. Toyohara, S. Ohno, T. Matsuki, H. Hamada, S. Miyahara, In-Vessel Source Term Analysis Code TRACER Version 2.3 User's Manual, JNC Report, JNC TN9520 2004-004, 2005 (in Japanese).
[148] M.J. Lineberry, S.G. Carpenter, C.L. Beck, H.F. McFarlane, P.J. Collins, D.N. Olsen, G. L. Grasseschi, R.E. Kaiser, D.W. Maddison, Experimental studies of large conventional LMFBR cores at ZPPR, in: Proc. of Fast Reactor Physics 1979, Aix-en-Province, IAEA-SM-244/86, 1979, pp. 187–204.
[149] T. Inoue, K. Shirakata, K. Kinjo, T. Ikegami, M. Yamamoto, Current status of JUPITER program ('JUPITER Keikaku no Genkyo' in Japanese), J. At. Energ. Soc. Jpn. 23 (5) (1981) 310–317 (in Japanese).
[150] T. Sanda, F. Nakashima, T. Yamamoto, K. Shirakata, Outline of JUPITER-III program, J. At. Energ. Soc. Jpn. 31 (12) (1989) 1324–1331 (in Japanese).
[151] T. Takeda, A. Yoshimura, T. Kamei, K. Shirakata, Prediction uncertainty evaluation methods of core performance parameters in large liquid-metal fast breeder reactors, Nucl. Sci. Eng. 103 (1989) 157–165.
[152] M. Ishikawa, H. Hayashi, T. Kamei, T. Sanda, T. Kawakita, Improvement of nuclear design method for large LMFBR cores by means of cross-section adjustment, Power

Reactor and Nuclear Fuel Development Corporation, 1991, pp. 92–96. PNC Technical Review No. 77 [in Japanese].

[153] M. Ishikawa, T. Hoshi, T. Sanda, T. Kamei, T. Kawakita, Improvement of nuclear design method for large LMFBR cores using the cross-section adjustment, in: Proc. Joint Int. Conf. on Mathematical Methods and Super Computing Nuclear Applications, M&C +SNA'93, 1993. 19-13 April 1993. Karlsruhe, Germany.

[154] T. Kamei, T. Yoshida, Fuel pin and subassembly heterogeneity effect on neutronics properties of fast power reactor, J. Nucl. Sci. Technol. 18 (1) (1981) 74–76.

[155] M. Ishikawa, Consistency evaluation of JUPITER experiment and analysis for large FBR cores, in: Proc. Int. Conf. on the Physics of Reactors, PHYSOR'96, vol. 2, 1996. Session E, E-36, September 16-20, 1996, Mito, Ibaraki, Japan.

[156] K. Takino, K. Sugino, K. Yokoyama, T. Jin, S. Ohki, Investigation of the core neutronics analysis conditions for evaluation of burn-up nuclear characteristics of next-generation fast reactors, in: Proc. of ICAPP 2018, Charlotte, North Carolina, April 8-11, 2018, 2018.

[157] L.P. Abagyan, N.O. Bazazyants, I.I. Bondarenko, M.N. Nikolaev, "Group Constants for Nuclear Reactor Calculations," Authorized Translation from the Russian, The Original Russian Text, Published for the Power Physics Institute by Atomizdat in Moscow in 1964, Consultants Bureau, New York, 1964.

[158] T. Tone, A numerical study of heterogeneity effects in fast reactor critical assembly, J. Nucl. Sci. Technol. 12 (8) (1975) 467–481.

[159] H. Takano, Y. Ishiguro, Y. Matsui, TIMS-1: A Processing Code for Production of Group Constants of Heavy Resonant Nuclei, Japan Atomic Energy Research Institute, JAERI 1267, 1980, p. 1267.

[160] S. Okajima, H. Oigawa, T. Mukaiyama, Resonance interaction effect between hot sample and cold core in analysis of Doppler effect measurement, J. Nucl. Sci. Technol. 31 (10) (1994) 1097–1104.

[161] T. Hazama, G. Chiba, K. Sugino, Development of a fine and ultra-fine group cell calculation code SLAROM-UF for fast reactor analyses, J. Nucl. Sci. Technol. 43 (8) (2006) 908.

[162] P.H. Kier, A.A. Robba, RABBLE, A Program for Computation of Resonance Absorption in Multiregion Reactor Cells, Argonne National Laboratory, 1967. ANL-7326.

[163] K. Sugino, T. Iwai, Advances in Methods of Commercial FBR Core Characteristics Analyses—Investigations of a Treatment of the Double-heterogeneity and a Method to Calculate Homogenized Control Rod Cross Sections, Power Reactor and Nuclear Fuel Development Corporation, 1998. PNC TN9410 98-067 [in Japanese].

[164] T. Kitada, S. Kosaka, T. Takeda, New control rod homogenization method for fast reactors, J. Nucl. Sci. Technol. 31 (7) (1994) 647–653.

[165] K. Takino, K. Sugino, S. Ohki, An investigation on the control rod homogenization method for next-generation fast reactor cores, Ann. Nucl. Energy 162 (2021), 108454.

[166] P. Benoist, Theory of the diffusion coefficient of neutrons in a lattice containing cavities, in: Report CEA-R 2278 of the Center d'Etudes Nucleares de Saclay, 1964. 21 January 1964.

[167] S. Maruyama, S. Ohki, Development of discrete transport theory with anisotropic neutron streaming effect, in: Proc. of 2018 Annual Meeting of the Atomic Energy Society of Japan, 2018, p. 1F11 (in Japanese).

[168] T. Yamamoto, Three-dimensional transport correction in fast reactor core analysis, J. Nucl. Sci. Technol. 23 (10) (1986) 849–858.

[169] K. Sugino, T. Takeda, An improvement of the transverse leakage treatment for the nodal S_N transport calculation method in hexagonal-Z geometry, J. Nucl. Sci. Technol. 33 (8) (1996) 620–627.
[170] K. Sugino, Development of Neutron Transport Calculation Codes for 3-D Hexagonal Geometry—Development of NSHEX, MINIHEX and MINISTRI Codes, Japan Atomic Energy Agency, JAEA-Data/Code 2011-018, 2012, (in Japanese).
[171] K. Sugino, K. Takino, Development of Neutron Transport Calculation Codes for 3-D Hexagonal Geometry (2) Improvement and Enhancement of the MINISTRI Code, Japan Atomic Energy Agency, 2019. JAEA-Data/Code 2019-011 [in Japanese].
[172] S. Iijima, H. Yoshida, H. Sakuragi, Calculation Program for Fast Reactor Design, 2 (Multi-dimensional Perturbation Theory Code Based on Diffusion Approximation: PERKY), Japan Atomic Energy Research Institute, 1977. JAERI-M 6993 [in Japanese].
[173] K. Numata, W. Sato, K. Yokoyama, K. Sugino, A Transplantation of the Core Parameter Analytical System for Fast Reactors to the Workstation Environment (II), Japan Nuclear Cycle Development Institute, 2000. JNC TN9440 2000-009.
[174] G.I. Bell, S. Glasstone, Nuclear Reactor Theory, Robert E. Krieger Publishing Company, Malabar, Florida, 1970.
[175] K. Suyama, J. Katakura, Y. Ohkawachi, M. Ishikawa, Libraries Based on JENDL-3.2 for ORIGEN2 Code: ORLIBJ32, Japan Atomic Energy Research Institute, 1999. JAERI-Data/Code 99-003 [in Japanese].
[176] J. Katakura, M. Kataoka, K. Suyama, T. Jin, S. Ohki, A Set of ORIGEN2 Cross Section Libraries Based on JENDL-3.3 Library: ORLIBJ33, Japan Atomic Energy Research Institute, 2004. JAERI-Data/Code 2004-015 [in Japanese].
[177] K. Okumura, K. Sugino, K. Kojima, T. Jin, T. Okamoto, J. Katakura, A Set of ORIGEN2 Cross Section Libraries Based on JENDL-4.0: ORLIBJ40, Japan Atomic Energy Agency, 2012. JAEA-Data/Code 2012-032 [in Japanese].
[178] M. Pusa, Higher-order Chebyshev rational approximation method and application to burnup equations, Nucl. Sci. Eng. 182 (2016) 297–318.
[179] K. Yokoyama, S. Lahaye, Benchmarks of depletion and decay heat calculation between MENDEL and MARBLE, in: Proc. of Joint Int. Conf. on Supercomputing in Nuclear Applications + Monte Carlo 2020, Japan, 2020, pp. 109–116. No. 3294033.
[180] A. Hara, T. Takeda, Y. Kikuchi, SAGEP: Two-Dimensional Sensitivity Analysis Code Based on Generalized Perturbation Theory, Japan Atomic Energy Research Institute, 1984. JAERI-M 84-027 [in Japanese].
[181] L.N. Usachev, 1963, "Perturbation theory for the breeding ratio and for other number ratios pertaining to various reactor processes," J. Nucl. Energ. A/B, Vol. 18, pp.571-583 translated by D. L. Allan from Atomnaya Energiya 15, 472 (1963).
[182] T. Takeda, T. Umano, Burnup sensitivity analysis in a fast breeder reactor—part I: sensitivity calculation method with generalized perturbation theory, Nucl. Sci. Eng. 91 (1985) 1–10.
[183] M.L. Williams, Development of depletion perturbation theory for coupled neutron/nuclides fields, Nucl. Sci. Eng. 70 (1979) 20–36.
[184] M. Ishikawa, K. Sugino, W. Sato, K. Numata, Development of a Unified Cross-section set ADJ2000 based on adjustment technique for fast reactor analysis, J. Nucl. Sci. Technol. (Suppl. 2) (2002) 1073–1076.
[185] K. Shibata, O. Iwamoto, T. Nakagawa, N. Iwamoto, A. Ichihara, S. Kunieda, S. Chiba, K. Furutaka, N. Otuka, T. Ohsawa, T. Murata, H. Matsunobu, A. Zukeran, S. Kamada, J. Katakura, JENDL-4.0: a new library for nuclear science and engineering, J. Nucl. Sci. Technol. 48 (1) (2011) 1–30.

[186] G. Chiba, M. Ishikawa, Revision and application of the covariance data processing code, ERRORJ, in: Proc. of Int. Conf. on Nuclear Data for Science and Technology, 2004. ND2004, 26 September – 1 October 2004, Santa Fe, New Mexico.
[187] R.E. MacFarlane, A.C. Kahler, Methods for processing ENDF/B-VII with NJOY, Nucl. Data Sheet 111 (2010) 2739–2890.
[188] A.C. Kahler, NJOY User Group Meeting, Nuclear Energy Agency Issy-les-Moulineau, France, 2007. 26-28, November, 2007. *https://www.oecd-nea.org/dbdata/meetings/nov2007/njoy/jefdoc-1225.pdf*. Accessed2021-07-28.
[189] N. Otuka, A. Zukeran, H. Takano, G. Chiba, M. Ishikawa, Covariance analyses of self-shielding factor and its temperature gradient for U-238 neutron capture reaction, J. Nucl. Sci. Technol. 45 (3) (2008) 195–210.
[190] K. Yokoyama, H. Hosogai, N. Uto, N. Kasahara, F. Nagura, M. Ohira, M. Kato, M. Ishikawa, Development of the Next Generation Code System as an Engineering Modeling Language (I), Japan Nuclear Cycle Development Institute, 2002. JNC TN9420 2002-004 [in Japanese].
[191] J.K. Ousterhout, Scripting: higher-level programming for the 21st century, IEEE Comput. Mag. 31 (3) (1998) 23–30.
[192] K. Yokoyama, M. Ishikawa, M. Tatsumi, H. Hyoudou, Restructuring of burnup sensitivity analysis code system by using an object-oriented design approach, in: Mathematics and Computation, Supercomputing, Reactor Physics and Nuclear and Biological Applications (M&C2005), Avignon, France, American Nuclear Society, 2005. September 12-15, 2005, on CD-ROM.
[193] K. Yokoyama, Y. Hirai, M. Tatsumi, H. Hyoudou, G. Chiba, T. Hazama, Y. Nagaya, M. Ishikawa, MARBLE: a next generation neutronics analysis code system for fast reactors, in: Proc. of Int. Conf. on the Physics of Reactors (PHYSOR2008), Interlaken, Switzerland, 2008. September 14-19, 2008.
[194] K. Yokoyama, T. Hazama, K. Numata, T. Jin, Development of comprehensive and versatile framework for reactor analysis, MARBLE, Ann. Nucl. Energ. 66 (2014) 51–60.
[195] C.R. Harris, K.J. Millman, S.J. van der Walt, R. Gommers, P. Virtanen, D. Cournapeau, E. Wieser, J. Taylor, S. Berg, N.J. Smith, R. Kern, M. Picus, S. Hoyer, M.H. van Kerkwijk, M. Brett, A. Haldane, J. Fernández del Río, M. Wiebe, P. Peterson, K. Pe Gérard-Marchant, T. Sheppard, W. Reddy, H. Weckesser, C.G. Abbasi, T.E. Oliphant, Array programming with NumPy, Nature 585 (2020) 357–362.
[196] P. Virtanen, R. Gommers, T.E. Oliphant, M. Haberland, T. Reddy, D. Cournapeau, E. Burovski, P. Peterson, W. Weckesser, J. Bright, S.J. van der Walt, M. Brett, J. Wilson, K. Jarrod Millman, N. Mayorov, A.R.J. Nelson, E. Jones, R. Kern, E. Larson, C. Carey, İ. Polat, Y. Feng, E.W. Moore, J. Van der Plas, D. Laxalde, J. Perktold, R. Cimrman, I. Henriksen, E.A. Quintero, C.R. Harris, A.M. Archibald, A.H. Ribeiro, F. Pedregosa, P. van Mulbregt, SciPy 1.0 Contributors, Scipy 1.0: Fundamental algorithms for scientific computing in python, Nat. Methods 17 (2020) 261–272.
[197] K. Yokoyama, K. Numata, T. Hazama, M. Ishikawa, Development of neutronics design accuracy evaluation solver for next generation reactor physics analysis code system MARBLE, in: 16th Pacific Basin Nuclear Conference (16PBNC), Aomori, Japan, 2008. October 13-18, 2008, Paper ID P16P1168.
[198] K. Yokoyama, T. Jin, Y. Hirai, T. Hazama, Development of the Versatile Reactor Analysis Code System, MARBLE2, Japan Atomic Energy Agency, 2015. JAEA-Data/Code 2015-009.

[199] K. Yokoyama, Development and verification of three-dimensional Hex-Z burnup sensitivity solver based on generalized perturbation theory, in: Proc. Int. Conf. on the Physics of Reactor, PHYSOR, 2014. 28th September–3rd October, 2014, Kyoto, Japan, Paper ID 1106037.
[200] K. Ohgama, K. Ikeda, M. Ishikawa, T. Kan, S. Maruyama, K. Yokoyama, K. Sugino, Y. Nagaya, S. Ohki, Model verification and validation procedure for a neutronics design methodology of next generation fast reactors, in: Proc. of ICAPP 2017, April, 24-28, 2017, Fukui and Kyoto, Japan, 2017.
[201] G. Chiba, K. Okumura, K. Sugino, Y. Nagaya, K. Yokoyama, T. Kugo, M. Ishikawa, S. Okajima, JENDL-4.0 benchmarking for fission reactor applications, J. Nucl. Sci. Technol. 48 (2) (2011) 172–187.
[202] K. Yokoyama, K. Sugino, M. Ishikawa, S. Maruyama, Y. Nagaya, K. Numata, T. Jin, Development of the Unified cross-section set ADJ2017, Jpn. At. Energ. Agency (2018). JAEA-Research 2018-011 [in Japanese].
[203] A. Shono, K. Sugino, T. Hazama, M. Ishikawa, Experimental analysis results on BN-600 mock-up core characteristics at the BFS-2 critical facility, in: Proc. of Int. Conf. on the Physics of Reactors, PHYSOR2002, Seoul, Korea, October 7-10, 2002, 2002.
[204] A. Shono, T. Hazama, M. Ishikawa, G. Manturov, Reduction of cross-section-induced errors of the BN-600 hybrid core nuclear parameters by using BFS-62 critical experiment data, in: Proc. Int. Conf. on Physics of Reactors, Chicago, Illinois, 2004. April 25-29, 2004, on CD-ROM.
[205] M. Ishikawa, T. Hazama, Validation of Np-237 cross sections by analysis of BFS critical experiments with massively neptunium-loaded cores, in: Proc. Int. Conf. on Nuclear Data for Science and Technology, 2007, https://doi.org/10.1051/nddata:07210. 2007, ND2007.
[206] S. Ohki, T. Iwai, Analysis of the CIRANO experiment for plutonium burning fast reactor using JNC's calculation scheme, in: Proc. Int. Conf. on Nuclear Data for Science and Technology, 2002, pp. 888–891. ND2001, October 7-12, 2001, Tsukuba, Japan, J. Nucl. Sci. Technol., Supplement 2.
[207] M. Ishikawa, T. Ikegami, T. Sanda, ZPPR benchmarks for large LMFBR core physics from JUPITER cooperative program between United States and Japan, Nucl. Sci. Eng. 178 (2014) 335–349.
[208] K. Yokoyama, M. Ishikawa, Reevaluation of experimental data and analysis on experimental fast reactor Joyo MK-I performance tests, Nucl. Sci. Eng. 157 (2007) 249–263.
[209] K. Yokoyama, M. Ishikawa, Evaluation of burnup reactivity coefficients measured in experimental fast reactor Joyo MK-I duty power operation cycles, Nucl. Sci. Eng. 178 (2014) 350–362.
[210] OECD Nuclear Energy Agency, International Handbook of Evaluated Reactor Physics Benchmark Experiments, 2019. NEA/NSC/DOC(2006)1 (July 2019).
[211] K. Sugino, T. Takeda, Development of a fast reactor for minor actinides (MAs) transmutation (3)—evaluation of measurement data with MA transmutation, in: Proc. Int. Conf. of Global 2015, 2015. September 20-24, 2015, Paris, France, Paper 5094.
[212] K. Sugino, K. Numata, M. Ishikawa, T. Takeda, Cross-section-induced uncertainty evaluation of MA sample irradiation test calculations with consideration of dosimeter data, Ann. Nucl. Energy 130 (2019) 118–123.
[213] T. Hazama, A. Kitano, Y. Kishimoto, Criticality evaluation for the monju restart core, Nucl. Technol. 179 (2011) 250–265.
[214] K. Yokoyama, S. Maruyama, K. Numata, M. Ishikawa, Development of a fast reactor for minor actinides transmutation—improvement of prediction accuracy for MA-related

integral parameters based on cross-section adjustment technique, in: Proc. Int. Conf. on the Physics of Reactor (PHYSOR 2016), Sun Valley, ID, 2016, pp. 1906–1915. May 1-5, 2016.

[215] M. Fukushima, A. Oizumi, H. Iwamoto, Y. Kitamura, Establishment of Benchmark Problems for TRU Fission Rate Ratios of FCA-IX Assemblies, Japan Atomic Energy Agency, 2014. JAEA-Data/Code 2014-030 [in Japanese].

[216] C. Batra, V. Kriventsev, A. Gomez, T.K Kim, M. Jarrett, U. Davis, E. Fridman, Y. Zheng, X.Du, D. Lee, T. Quoc Tran, J. Choe, B. Batki, I. Pataki, M. Toth, J. Bodi, K.Mikityuk, H. Taninaka, M. Szogradi, P. Darilek, V. Giusti, S. Di Pasquale, A. Petruzzi, P. Sciora, 2022, "Neutronics benchmark of CEFR start-up tests: an IAEA coordinated research project," Proc. Int. Conf. on Fast Reactors and Related Fuel Cycles FR22, Beijing, China, April 25-28, 2022. [to be published].

[217] T. Kamei, T. Yoshida, Error due to nuclear data uncertainties in the prediction of large liquid-metal fast breeder reactor core performance parameters, Nucl. Sci. Eng. 84 (1983) 83–97.

[218] G. Cecchini, U. Farinelli, A. Gandini, M. Salvatores, Analysis of integral data for few-group parameter evaluation of fast reactor, in: Proc. of 3rd Int. Conf. of Peaceful Uses Atomic Energy, 1964. 31 August–9 September, 1964, Geneva, Switzerland.

[219] M. Humi, Y.Y. Wagshal, Multi-group constants from integral data, in: Proc. of 3rd Int. Conf. of Peaceful Uses of Atomic Energy, Geneva, Switzerland, 1964. 31 August–9 September, 1964.

[220] C. de Saint-Jean, E. Dupont, M. Ishikawa, G. Palmiotti, M. Salvatores, Assessment of Existing Nuclear Data Adjustment Methodologies, OECD Nuclear Energy Agency, 2011. NEA/WPEC-33, NEA/NSC/WPEC/DOC(2010)429.

[221] J.B. Dragt, Statistical consideration on techniques for adjustment of differential cross sections with measured integral parameters, in: STEK, The Fast-Thermal Coupled Facility of R.C.N. at Petten, A Collection of Reports Describing the Status of the Facility, the Program, and the Experiments, as of February 1970, RCN-122, Reactor Centrum Nederland, 1970, pp. 85–105.

[222] K. Yokoyama, K. Numata, M. Ishikawa, Development of the Adjusted Nuclear Cross-section Library Based on JENDL-3.2 for Large FBR, Japan Nuclear Cycle Development Institute, 1999. JNC TN9400 99-042.

[223] T. Hazama, G. Chiba, K. Numata, W. Sato, Development of the Unified Cross-section set ADJ2000R for Fast Reactor Analysis, Japan Nuclear Cycle Development Institute, 2002. JNC TN9400 2002-06 [in Japanese].

[224] K. Sugino, M. Ishikawa, K. Yokoyama, Y. Nagaya, G. Chiba, T. Hazama, T. Kugo, K. Numata, T. Iwai, T. Jin, Development of a unified cross-section set ADJ2010 based on adjustment technique for fast reactor core design, J. Korean Phys. Soc. 59 (2) (2011) 1357–1360.

[225] K. Yokoyama, M. Ishikawa, Use and impact of covariance data in the Japanese latest adjusted library ADJ2010 based on JENDL-4.0, Nucl. Data Sheets 123 (2015) 97–103.

[226] K. Sugino, M. Ishikawa, K. Numata, T. Iwai, T. Jin, Y. Nagaya, T. Hazama, G. Chiba, K. Yokoyama, T. Kugo, Development of a Standard Data Base for FBR Core Design (XIV)—Analysis of Extensive FBR Core Characteristics Based on JENDL-4.0, Japan Atomic Energy Agency, JAEA-Research, 2012. 2012-013 [in Japanese].

[227] M. Salvatores, G. Palmiotti, G. Aliberti, R. McKnight, P. Archier, C. De Saint Jean, E. Dupont, M. Herman, M. Ishikawa, K. Sugino, T. Ivanova, E. Ivanov, S.-J. Kim, I. Kodeli, A. Trukov, G. Manturov, S. Pelloni, C. Perfetti, B.T. Rearden, A. Plompen, D. Rochman, W. Wang, H. Wu, W.-S. Yang, Methods and Issues for the Combined

Use of Integral Experiments and Covariance Data, OECD Nuclear Energy Agency, 2013, p. 445. NEA/WPEC-33, NEA/NSC/WPEC/DOC (2013).

[228] M. Salvatores, G. Aliberti, M. Dunn, A. Hogenbirk, A. Ignatyuk, M. Ishikawa, I. Kodeli, A.J. Koning, R. McKnight, R.W. Mills, P. Oblozinsky, G. Palmiotti, A. Plompen, G. RImpault, Y. Rugama, P. Talou, W.S. Yang, Uncertainty and Target Accuracy Assesment for Innovative Systems Using Recent Covariance Data Evaluation, OECD/NEA, 2008. Nuclear Science NEA/WPEC-26, ISBN 978-92-64-99053-1.

[229] M. Salvatores, G. Palmiotti, Method and approaches to provide feedback from nuclear and covariance data adjustment for improvement of nuclear data files: major findings of the NEA WPEC Subgroup 39, in: Proc. of Int. Conf. on Nuclear Data for Science and Technology, ND2019, EPJ Web of Conferences, vol. 239, 2020. Article Number 13001.

[230] T. Sano, T. Takeda, Generalized bias factor method for accurate prediction of neutronics characteristics, J. Nucl. Sci. Technol. 43 (12) (2006) 1465–1470.

[231] T. Kugo, T. Mori, T. Takeda, Theoretical study on new bias factor methods to effectively use critical experiments for improvement of prediction accuracy of neutronic characteristics, J. Nucl. Sci. Technol. 44 (12) (2007) 1509–1517.

[232] K. Yokoyama, M. Ishikawa, T. Kugo, Extended cross-section adjustment method to improve the prediction accuracy of core parameters, J. Nucl. Sci. Technol. 49 (12) (2012) 1165–1174.

[233] K. Yokoyama, A. Yamamoto, Cross-section adjustment methods based on minimum variance unbiased estimation, J. Nucl. Sci. Technol. 53 (10) (2016) 1622–1638.

[234] K. Yokoyama, T. Kitada, Generalized formulation of extended cross-section adjustment method based on minimum variance unbiased linear estimation, J. Nucl. Sci. Technol. 56 (1) (2019) 87–104.

[235] K. Yokoyama, A. Yamamoto, T. Kitada, Dimension-reduced cross-section adjustment method based on minimum variance unbiased estimation, J. Nucl. Sci. Technol. 55 (3) (2018) 319–334.

[236] T. Watanabe, T. Endo, A. Yamamoto, Y. Kodama, Y. Ohoka, T. Ushio, Cross section adjustment method based on random sampling technique, J. Nucl. Sci. Technol. 51 (5) (2014) 590–599.

[237] T. Endo, A. Yamamoto, T. Watanabe, Bias factor method using random sampling technique, J. Nucl. Sci. Technol. 53 (10) (2016) 1494–1501.

[238] A. Yamamoto, A. Giho, Y. Kato, T. Endo, GNESIS: a three-dimensional heterogeneous transport solver based on the Legendre polynomial expansion of angular flux method, Nucl. Sci. Eng. 186 (2017) 1–22.

[239] Atomic Energy Society of Japan, Guideline for Credibility Assessment of Nuclear Simulations: 2015, Atomic Energy Society of Japan, 2015. AESJ-SC-A008:2005.

[240] M. Ichimiya, T. Mizuno, S. Kotake, A next generation sodium-cooled fast reactor concept and its R&D program, Nucl. Eng. Technol. 39 (3) (2007) 171–186.

[241] J.J. Carbajo, G.L. Yoder, S.G. Popov, V.K. Ivanov, A review of the thermophysical properties of MOX and UO2 fuels, J. Nucl. Mater. 299 (2001) 181–198.

[242] M. Kato, M. Watanabe, T. Matsumoto, S. Hirooka, M. Akashi, Oxygen potentials, oxygen diffusion coefficients and defect equilibria of nonstoichiometric $(U,Pu)O_{2\pm x}$, J. Nucl. Mater. 487 (2017) 424–432.

[243] K. Morimoto, M. Kato, M. Ogasawara, M. Kashimura, T. Abe, Thermal conductivities of $(U,Pu,Am)O_2$ solid solutions, J. Alloys Compd. 452 (2008) 54–60.

[244] K. Morimoto, M. Kato, M. Ogasawara, M. Kashimura, Thermal conductivities of hypostoichiometric $(U, Pu, Am)O_{2-x}$ oxide, J. Nucl. Mater. 374 (2008) 378–385.

[245] K. Morimoto, M. Kato, M. Ogasawara, M. Kashimura, Thermal conductivity of (U,Pu, Np)O_2 solid solutions, J. Nucl. Mater. 389 (2009) 179–185.
[246] M. Kato, K. Maeda, T. Ozawa, M. Kashimura, Y. Kihara, Physical properties and irradiation behavior analysis of Np- and Am-bearing MOX fuels, J. Nucl. Sci. Technol. 48 (2011) 646–653.
[247] K. Lassmann, The oxired model for redistribution of oxygen in nonstoichiometric uranium-plutonium oxides, J. Nucl. Mater. 150 (1987) 10–16.
[248] M. Kato, K. Morimoto, H. Sugata, K. Konashi, M. Kashimura, T. Abe, Solidus and liquidus temperatures in the UO_2-PuO_2 system, J. Nucl. Mater. 373 (2008) 237–245.
[249] M. Kato, K. Morimoto, H. Sugata, K. Konashi, M. Kashimura, T. Abe, Solidus and liquidus of plutonium and uranium mixed oxide, J. Alloys Compd. 452 (2008) 48–53.
[250] F. De Bruycker, K. Boboridis, R.J.M. Konings, M. Rini, R. Eloirdi, C. Gueneau, N. Dupin, D. Manara, On the melting behaviour of uranium/plutonium mixed dioxides with high-Pu content: a laser heating study, J. Nucl. Mater. 419 (2011) 186–193.
[251] R. Böhler, M.J. Welland, D. Prieur, P. Cakir, T. Vitova, T. Pruessmann, I. Pidchenko, C. Hennig, C. Gueneau, R.J.M. Konings, D. Manara, Recent advances in the study of the UO_2-PuO_2 phase diagram at high temperatures, J. Nucl. Mater. 448 (2014) 330–339.
[252] R. Calabrese, D. Manara, A. Schubert, J. van de Laar, P. Van Uffelen, Melting temperature of MOX fuel for FBR applications: TRANSURANUS modelling and experimental findings, Nucl. Eng. Des. 283 (2015) 148–154.
[253] D.R. De Halas, G.R. Horn, Evolution of uranium dioxide structure during irradiation of fuel rods, J. Nucl. Mater. 8 (1963) 207–220.
[254] M.H. Rand, T.L. Marking, Some Thermodynamic aspects of (U,Pu)O_2 solid solutions and their use as nuclear fuels in thermodynamics of nuclear materials, in: Symposium Proceedings, IAEA, Vienna, 1967, pp. 637–650.
[255] Y. Ikusawa, K. Maeda, M. Kato, M. Uno, Oxide-metal ratio dependence of central void formation of mixed oxide fuel irradiated in fast reactors, Nucl. Technol. 199 (2017) 83–95.
[256] T. Ozawa, S. Hirooka, M. Kato, S. Novascone, P. Medvedev, Development of fuel performance analysis code, BISON for MOX, named Okami: analyses of pore migration behavior to affect the MA-bearing MOX fuel restructuring, J. Nucl. Mater. 553 (2021), 153038.
[257] T. Uwaba, T. Mizuno, J. Nemoto, I. Ishitani, M. Ito, Development of a mixed oxide fuel pin performance analysis code "CEDAR": models and analyses of fuel pin irradiation behavior, J. Nucl. Eng. Des. 280 (2014) 27–36.
[258] T. Uwaba, J. Nemoto, I. Ishitani, M. Ito, Coupled computer code study on irradiation performance of a fast reactor mixed oxide fuel element with an emphasis on the fission product cesium behavior, J. Nucl. Eng. Des. 331 (2018) 186–193.
[259] T. Uwaba, J. Nemoto, M. Ito, I. Ishitani, N. Doda, M. Tanaka, S. Ohtsuka, "Development of an Integrated Computer Code System for Analyzing Irradiation Behaviors of a Fast Reactor Fuel", https://doi.org/10.1080/00295450.2020.1810977.
[260] T. Ozawa, T. Abe, Development and verifications of fast reactor fuel design code CEPTAR, Nucl. Technol. 156 (2006) 39–55.
[261] R.L. Williamson, J.D. Hales, S.R. Novascone, M.R. Tonks, D.R. Gaston, C.J. Per-Mann, D. Andrs, R.C. Martineau, Multidimensional multiphysics simulation of nuclear fuel behavior, J. Nucl. Mater. 423 (2012) 149–163.

[262] S. Hirooka, T. Matsumoto, M. Kato, T. Sunaoshi, H. Uno, T. Yamada, Oxygen potential measurement of (U,Pu,Am)$O_{2\pm x}$ and (U,Pu,Am,Np)$O_{2\pm x}$, J. Nucl. Mater. 542 (2020), 152424.

[263] D.R. Olander, Fundamental Aspects of Nuclear Reactor Fuel Elements, TID-26711-P1, Energy Research and Development Administration, US, 1976, pp. 155–160.

[263a] S. Novascone, P. Medvedev, J.W. Peterson, Y. Zhang, J. Hales, Modeling porosity migration in LWR and fast reactor MOX fuel using the finite element method, J. Nucl. Mater. 508 (2018) 226–236.

[264] F. Verdolin, S. Novascone, D. Pizzocri, G. Pastore, T. Barani, L. Luzzi, Modelling fission gas behavior in fast reactor (U,Pu)O_2 fuel with BISON, J. Nucl. Mater. 547 (2021), 152728.

[265] I. Shibahara, N. Akasaka, S. Onose, H. Okada, S. Ukai, Swellling of advanced austenitic stainless steels developed for the environment of heavy neutron exposure, J. Nucl. Mater. 212–215 (1994) 487–491.

[265a] T. Inoue, I. Yamagata, T. Asaga, Material for nuclear energy systems – towards high reliability (2); Modified SUS316 stainless steel for fast breeder reactors, J. At. Energ. Soc. Jpn. 53 (9) (2011) 638–642 (in Japanese).

[266] T. Uwaba, S. Ukai, The Formulation of the PNC316 Swelling Design Equation, JNC TN9400 2003-007, 2003. (in Japanese).

[267] T. Uwaba, M. Ito, K. Maeda, Diametral strain of fast reactor MOX fuel pins with austenitic stainless steel cladding irradiated to high burnup, J. Nucl. Mater. 416 (2011) 350–357.

[268] M.B. Toloczko, F.A. Garner, Irradiation creep and void swelling of two LMR heats of HT9 at ~400°C and 165 dpa, J. Nucl. Mater. 233-237 (1996) 289–292.

[269] P. Yvon, F. Carré, Structural materials challenges for advanced reactor systems, J. Nucl. Mater. (2009) 217–222.

[270] S. Shikakura, S. Nomura, S. Ukai, I. Seshimo, Y. Kano, Y. Kuwajima, T. Ito, K. Tutaki, T. Fujita, Development of high-strength ferritic/martensitic steel for FBR core materials, J. At. Energ. Soc. Jpn. 33 (12) (1991) 1157–1170 (in Japanese).

[271] A. Uehira, S. Ukai, T. Mizuno, T. Asaga, E. Yoshida, Tensile properties of 11Cr-0.5Mo-2W, V, Nb stainless steel in LMFBR environment, J. Nucl. Sci. Technol. 37 (9) (2000) 780–786.

[272] M.L. Hamilton, D.S. Gelles, R.J. Lobsinger, G.D. Johnson, W.F. Brown, et al., "Fabrication Technological Development of the Oxide Dispersion Strengthened Alloy MA957 for Fast Reactor Application", PNNL-13168, 2000.

[273] R.L. Klueh, P.J. Maziasz, I.S. Kim, L. Heatherly, D.T. Hoelzer, N. Hashimoto, et al., Tensile and creep properties of an oxide dispersion-strengthened ferritic steel, J. Nucl. Mater. 307–311 (2002) 773–777.

[274] D.S. Gelles, Microstructural examination of commercial ferritic alloys at 200 dpa, J. Nucl. Mater. 233–237 (1996) 293–298.

[275] S. Ukai, T. Kaito, S. Ohtsuka, M. Fujiwara, T. Kobayashi, Development of 9Cr-ODS ferritic-martensitic steels for fuel cladding, Mater. Jpn. 45 (1) (2006) 48–50 (in Japanease).

[276] S. Ohtsuka, S. Ukai, M. Fujiwara, T. Kaito, T. Narita, Improvement of creep strength of 9CrODS martensitic steel by controlling excess oxygen and titanium concentrations, Mater. Trans. 46 (3) (2005) 487–492.

[277] M. Inoue, T. Kaito, S. Ohtsuka, Research and development of oxide dispersion strengthened ferritic steels for sodium cooled fast breeder reactor fuels, in: Material Issues for Generation IV Systems, 2008, pp. 311–325.

[278] S. Ohtsuka, T. Kaito, S. Kim, M. Inoue, T. Asayama, M. Ohnuma, J. Suzuki, Improvement of creep strength of 9CrODS martensitic steel by controlling excess oxygen and titanium concentrations, Mater. Trans. 50 (7) (2009) 1778–1784.
[279] T. Tanno, M. Takeuchi, S. Ohtsuka, T. Kaito, Corrosion behavior of ODS steels with several chromium contents in hot nitric acid solutions, J. Nucl. Mater. 494 (2017) 219–226.
[280] T. Kaito, Y. Yano, S. Ohtsuka, M. Inoue, K. Tanaka, et al., ODS cladding fuel pins irradiation tests using the BOR-60 reactor, J. Nucl. Sci. Technol. 50 (4) (2013) 387–399.
[281] S. Ohtsuka, T. Kaito, Y. Yano, S. Yamashita, R. Ogawa, et al., Investigation of the cause of peculiar irradiation behavior of 9Cr-ODS steel in BOR-60 irradiation tests, J. Nucl. Sci. Technol. 50 (5) (2013) 470–480.
[282] T. Tanno, S. Ohtsuka, Y. Yano, T. Kaito, K. Tanaka, Effects of manufacturing process on impact properties and microstructures of ODS steels, J. Nucl. Mater. 455 (2014) 480–485.
[283] Y. Yano, T. Tanno, H. Oka, S. Ohtsuka, T. Inoue, et al., Ultra-high temperature tensile properties of ODS steel claddings under severe accident conditions, J. Nucl. Mater. 487 (2017) 229–237.
[284] M. Seki, S. Kono, Development of pressurized resistance welding technology for ODS ferritic steel cladding tube, Trans. At. Energ. Soc. Jpn. 3 (1) (2004) 95–105 (in Japanese).
[285] M. Seki, K. Hirako, S. Kono, Y. Kihara, T. Kaito, S. Ukai, Pressurized resistance welding technology development in 9Cr-ODS martensitic steels, J. Nucl. Mater. 329–333 (2004) 1534–1538.
[286] M. Tanji, Y. Abe, H. Endo, A. Izumi, Development and Verification of Thermo-Hydraulic Simulation Code for System Transient in "Monju" (COPD code), Mitsubishi Heavy Ind. Tech. Rev. 23 (6) (1986) 17–25 (in Japanese).
[287] R. Nakai, M. Itoh, K. Terata, Y. Kani, K. Maeda, H. Endo, S. Kondo, K. Aizawa, Y. Ohmori, Computer Codes for Safety Analysis of LMFBR, PNC TN241 81-28, 1981 (in Japanese).
[288] S. Nakai, T. Iwasaki, Y. Shimakawa, M. Tanji, Development of Module Integrated Plant Dynamics Analysis Code -Development of Super-COPD Code-, PNC Technical Review, No. 68, PNC TN1340 88-004, 1988, pp. 52–62 (in Japanese).
[289] A. Ohtaki, H. Ohira, Development of Computer Code in PNC 3. Development of Super-COPD Code for Plant Dynamics, PNC Technical Review, No.76, PNC TN1340 90-004, 1990, pp. 27–36 (in Japanese).
[290] T. Iwasaki, S. Nakai, Y. Shimakawa, M. Tanji, Super-COPD Module Integrated Plant Dynamics Analysis Code Code Users Manual, 1989. PNC TN9520 89-001, (in Japanese).
[291] F. Yamada, H. Ohira, Numerical simulation of MONJU plant dynamics by super-COPD using previous startup tests data, in: Proc. of 3rd Joint US-European Fluids Engineering Summer Meeting and 8th International Conference on Nanochannels, Microchannels and Minichannels, Montreal, Canada, 2010. FEDSM-ICNMM2010-30287.
[292] F. Yamada, Y. Fukano, H. Nishi, M. Konomura, Development of natural circulation analytical model in super-COPD code and evaluation of core cooling capability in Monju during a Station Blackout, Nucl. Technol. 188 (2014) 292–321.
[293] Y. Maeda, T. Aoyama, T. Odo, S. Nakai, S. Suzuki, Distinguished achievements of a quarter-century operation and a promising project named MK-III IN JOYO, Nucl. Technol. 150 (2005) 16–36.
[294] T. Hiyama, N. Doda, H. Ohshima, T. Iwasaki, Numerical analysis of JOYO MK-II natural circulation test with fast reactor plant dynamics code super-COPD, Trans. JSME 78 (787) (2012) 468–470 (in Japanese).

[295] K. Nabeshima, N. Doda, H. Ohshima, T. Mori, H. Ohira, T. Iwasaki, Analysis of natural circulation tests in the experimental fast reactor Joyo, in: Proc. NURETH-16, 2015, pp. 1041–1049.
[296] N. Doda, H. Ohira, H. Kamide, Benchmark analysis of EBR-II shutdown heat removal test-17 using of plant dynamics analysis code and subchannel analysis code, in: Proc. ICAPP2016, 2016, pp. 1618–1625.
[297] IAEA, Benchmark Analysis of EBR II Shutdown Heat Removal, IAEA-TECDOC-1819, 2017.
[298] H. Ohshima, T. Asayama, T. Furukawa, M. Tanaka, T. Takata, Y. Enuma, ARKADIA—for the innovation of advanced nuclear reactor design, in: Proc. ICONE28, 2021. ICONE2021-64525.
[299] K. Yoshimura, N. Doda, T. Fujisaki, S. Murakami, M. Tanaka, Development of multi-level simulation system for sodium-cooled fast reactor—application of coupled 1D-CFD simulation to ULOF test of EBR-II, in: Proc. MECJ-20, 2021, p. S08112 (in Japanese).
[300] R. Hu, J.W. Thomas, T.H. Fanning, Strategy for Multi-Scale Single-Phase Flow Coupling, 2013. ANL/NE-13/4.
[301] N. Doda, T. Uwaba, T. Nemoto, K. Yokoyama, M. Tanaka, Development of neutronics, thermal-hydraulics, and structure mechanics coupled analysis method on integrated numerical analysis for design optimization support in fast reactor, in: Proc. Computational Engineering Conference of JSCES, 26, 2021, p. A-12-03 (in Japanese).
[302] N. Doda, T. Uwaba, K. Yokoyama, T. Nemoto, M. Tanaka, Development of evaluation method for core deformation reactivity feedback in sodium-cooled fast reactor by coupled analysis approach, in: Proc. NURETH-19, 2021, Log nr.: 35413.
[303] H. Ohshima, S. Kubo, Sodium-Cooled Fast Reactor, Handbook of Generation IV Nuclear Reactors, Woodhead Publishing Series in Energy, Elsevier, 2016, pp. 97–118. Chapter 5.
[304] S.K. Cheng, N.E. Todreas, Hydrodynamic models and correlations for bare and wire-wrapped hexagonal rod bundles—bundle friction factors, subchannel friction factors and mixing parameters, Nucl. Eng. Des. 92 (1986) 227–251.
[305] S.K. Chen, N.E. Todreas, N.T. Nguyen, Evaluation of existing correlations for the prediction of pressure drop in wire-wrapped hexagonal array pin bundles, Nucl. Eng. Des. 267 (2014) 109–131.
[306] S.K. Chen, Y.M. Chen, N.E. Todreas, The upgraded correlation for pressure drop in hexagonal wire-wrapped rod bundles, Nucl. Eng. Des. 335 (2018) 356–373.
[307] S.K. Chen, Y.M. Chena, N.E. Todreas, Corrigendum to "the upgraded cheng and todreas correlation for pressure drop in hexagonal wire-wrapped rod bundles" (Nucl. Eng. Des. 335 (2018) 356-373), Nucl. Eng. Des. 340 (2018) 414.
[308] N. Kikuchi, Y. Imai, R. Yoshikawa, N. Doda, M. Tanaka, H. Ohshima, Subchannel analysis of thermal-hydraulics in a fuel assembly with inner duct structure of a sodium-cooled fast reactor, J. Nucl. Eng. Radiat. Sci. 5 (2) (2019) 021001-1–021001-12.
[309] N. Kikuchi, Y. Imai, R. Yoshikawa, N. Doda, M. Tanaka, Investigation of applicability of subchannel analysis code asfre on thermal hydraulics analysis in fuel assembly with inner duct structure in sodium cooled fast reactor, in: Proc. ICONE28, 2021. ICONE28-65662.
[310] M. Otaka, H. Ohshima, H. Ninokata, H. Narita, Validation of Single-Phase Subchannel Analysis Code ASFRE-III, 1996. PNC TN9410 96-212 (in Japanese).
[311] H. Ohshima, Y. Imai, Numerical simulation method of thermal-hydraulics in wire-wrapped fuel pin bundle of sodium-cooled fast reactor, in: Proc. FR17, IAEA-CN245-453, 2017.

[312] M. Nishimura, H. Kamide, H. Ohshima, J. Kobayashi, H. Sato, Investigation on velocity distribution in an inner subchannel of wire wrapped fuel pin bundle of sodium-cooled fast reactor, in: Proc. ICONE19, 2011. ICONE19-43857.
[313] M. Nishimura, H. Sato, H. Kamide, H. Ohshima, K. Nagasawa, Y. Imai, Investigation on velocity distribution around the wrapping wire in an inner subchannel of fuel pin bundle, 2012. Proc. ICONE20-POWER2012. ICONE20-POWER2012-54404.
[314] K. Aizawa, T. Hiyama, M. Nishimura, A. Kurihara, K. Ishida, Velocity distribution in the subchannels of a pin bundle with wrapping wire (evaluation of reynolds number dependence in a three-pin bundle), Mech. Eng. J. 8 (4) (2021). Paper No.20-00547.
[315] H. Sato, J. Kobayashi, H. Miyakoshi, H. Kamide, Study on velocity field in a wire wrapped fuel pin bundle of sodium cooled reactor—detailed velocity distribution in a subchannel, in: Proc. NURETH-13, 2009. N13P1406.
[316] M.E. Kabir, H. Hayafune, Study of Thermohydraulic Behavior within the Fuel Bundle under a Loss of Flow Condition, 1992. PNC TN9410 92-018.
[317] H. Hayafune, Summary of Sodium Boiling Experiment and Computer Code Development and Future Plan, 1995. PNC TN9410 95-238 (in Japanese).
[318] H. Kamide, K. Hayashi, T. Isozaki, M. Nishimura, Investigation of core thermohydraulics in fast reactors—interwrapper flow during natural circulation, Nucl. Technol. 133 (2001) 77–91.
[319] M. Nishimura, H. Kamide, K. Hayashi, K. Momoi, Transient experiments on fast reactor core thermal-hydraulics and its numerical analysis—inter-subassembly heat transfer and inter-wrapper flow under natural circulation conditions, Nucl. Eng. Des. 200 (2000) 157–175.
[320] H. Kamide, K. Hayashi, S. Toda, An experimental study of inter-subassembly heat transfer during natural circulation decay heat removal in fast breeder reactors, Nucl. Eng. Des. 183 (1998) 97–106.
[321] M. Nishimura, H. Ohshima, H. Kamide, K. Yamaguchi, A. Yamaguchi, The Status of Studies on Fast Reactor Core Thermal Hydraulics at PNC, IAEA-TECDOC-1157, 2000, pp. 23–40.
[322] H. Ninokata, A subchannel analysis code ASFRE-III and its application to a local blockage experiment in a wire-wrapped pin bundle, in: Proc. of 14th Meeting of the Liquid Metal Boiling Working Group, 1991, pp. 287–301.
[323] J.A. Meiherink, H.A. Van Der Vorst, Guidelines for the usage of incomplete decompositions in solving sets of linear equations as they occur in practical problems, J. Comput. Phys. 44 (1981) 134–155.
[324] H. Ninokata, A. Efthimiadis, N.E. Todreas, Distributed resistance modeling of wire-wrapped rod bundles, Nucl. Eng. Des. 104 (1) (1987) 93–102.
[325] N.E. Todreas, J.A. Turi, Interchannel mixing in wire wrapped liquid metal fast reactor fuel assemblies, Nucl. Technol. 13 (1) (1972) 36–52.
[326] A.E. Waltar, D.R. Todd, P.V. Tsvetkov, Fast Spectrum Reactors, Springer, US, 2012, pp. 257–259.
[327] H. Ohshima, Y. Imai, Development of thermal hydraulic analysis program for wire-wrapped fuel pin bundle of fast reactor, in: Proc. MECJ-10, 2010, pp. 153–154 (in Japanese).
[328] A.N. Brooks, T.J.R. Hughes, Streamline Upwind/Petrov-Galerkin formulations for convection dominated flows with particular emphasis on the incompressible Navier-Stokes equations, Comput. Methods Appl. Mech. Eng. 32 (1982) 199–259.

[329] K. Abe, T. Kondoh, Y. Nagano, A new turbulence model for predicting fluid flow and heat transfer in separating and reattaching flows—I. Flow field calculations, Int. J. Heat Mass Transfer 37 (1) (1994) 139–151.

[330] H. Ohshima, Y. Imai, Validation study of thermal-hydraulic analysis program "SPIRAL" for fuel pin bundle of sodium-cooled fast reactor, in: Proc. NURETH-11, 2005. Paper: 423.

[331] H. Ohshima, Y. Imai, Numerical simulation of flow field in wire-wrapped fuel pin bundle of sodium cooled fast reactor using "SPIRAL", in: Proc. NTHAS5, 2006. NTHAS5-F011.

[332] R. Yoshikawa, Y. Imai, N. Kikuchi, M. Tanaka, A. Gerschenfeld, Validation study of finite element thermal-hydraulics analysis code spiral to a large-scale wire-wrapped fuel assembly at low flow rate condition, in: Proc. SNA+MC2020, 2020, pp. 73–80.

[333] K. Koyama, K. Satoh, F. Bando, Y. Tsukui, S. Kuroki, H. Endo, F. Kasahara, Y. Hashiguti, K. Kumasaka, Study on local blockage in FBR fuel assembly, in: Proc. FR91, vol. III, 1991, pp. 3.5-1–3.5-10.

[334] G.L. Fiorini, J. Rion, R. Vidil, The ABACUS program: experimental study of phenomenology involving subassembly blockage; computer codes, in: Proc. of the Liquid Metal Fast Breeder Reactor Safety Topical Meetings, vol. IV, 1982, pp. 239–249.

[335] C.V. Miden, G.F. Shultheiss, Upon Local Blockage Formations in LMFBR Fuel Rod Bundles with Wire-Wrapped Spacers, GKSS 82/E/50, Forschungszentrum Geesthacht GmbH, 1982.

[336] J. Olive, P. Jolas, Internal blockage in a fissile super-phenix type subassembly: the scarlet experiments and their interpretation by the Cafca-NA3 Code, Nucl. Energ. 29 (4) (1990) 287–293.

[337] G. Karsten, W. Kramer, K. Schleisiek, G.F. Schultheiss, Investigations of local disturbances in LMFBR fuel assemblies, in: Proc. ANS Topical Meeting on Reactor Safety Aspects of Fuel Behavior, vol. 1, 1981, pp. 1-253–1-264.

[338] K. Itoh, K. Satoh, H. Endo, Y. Hashiguchi, 1992, Basic study on local blockage in FBR fuel subassembly, Proc. NURETH-5, pp.1529-1536.

[339] Y. Kikuchi, H. Mizuta, H. Yoshikawa, M. Hori, K. Tanaka, Evaluation of Local Accidents in LMFBR, PNC-TN941, 1976, pp. 76–128.

[340] M. Tanaka, J. Kobayashi, H. Kamide, K. Saito, H. Monji, J. Sakakibara, G. Matsui, Visualization of flow field in a cubic packing of spheres, in: Proc. NURETH-9, 1999.

[341] G. Matsui, K. Mishiro, H. Monji, M. Tanaka, H. Kamide, Flow characteristics in channel with local blockage packed with spheres, in: Proc. ExHFT-5, 2001, pp. 1675–1682.

[342] M. Tanaka, J. Kobayashi, T. Isozaki, M. Nishimura, H. Kamide, An experimental investigation on heat transfer in a subchannel with a porous blockage—the influence of flow on temperature distribution inside the porous blockage, in: Proc. AJTE99, 1999. AJTE99-6434.

[343] H. Kamide, H. Miyakoshi, N. Kawamata, H. Ohshima, Study on mixing due to transversal flow in a fuel subassembly of fast reactor (sodium experiment using a 37-pin subassembly model), in: Proc. ICONE12, 2004. ICONE12-49199.

[344] M. Tanaka, H. Kamide, H. Miyakoshi, J. Kobayashi, Study on thermal-hydraulics in a porous blockage in a fuel subassembly of LMFBR, in: Proc. ICONE11, 2003. ICONE11-36362.

[345] H. Ninokata, Summary of LMFBR Subassembly Blockage Studies at PNC, 1990. PNC-TN9420-90-004.

[346] H. Ohshima, Y. Ieda, Thermohydraulic aspects of porous blockages in a fuel subassembly of fast breeder reactors, in: Proc. of International Topical Meeting on Sodium Cooled Fast Reactor Safety, vol. 2, 1994, pp. 2-280–2-289.

[347] N. Nishimura, Y. Fukano, Local flow blockage analysis with checkerboard configuration in a wire wrapped fuel subassembly using the ASFRE code, in: Proc. NUTHOS-10, 2014. NUTHOS10-1294.

[348] K. Haga, Y. Kikuchi, Experimental investigation of coolability degradation by fission gas release into flowing sodium in a fuel pin bundle, Nucl. Technol. 70 (2) (1985) 220–234.

[349] K. Haga, K. Yamaguchi, F. Namekawa, Investigation of effect of gas release on coolability of locally blocked LMFBR fuel subassemblies, J. Nucl. Sci. Technol. 22 (8) (1985) 619–636.

[350] Y. Ieda, I. Maekawa, T. Muramatsu, S. Nakanishi, Experimental and analytical studies of the thermal stratification phenomenon in the outlet plenum of fast breeder reactors, Nucl. Eng. Des. 120 (1990) 403–414.

[351] S. Moriya, N. Tanaka, N. Katano, A. Wada, Effect of Reynolds number and Richardson number on thermal stratification in hot plenum, Nucl. Eng. Des. 99 (1987) 441–451.

[352] N. Tanaka, S. Moriya, S. Ushijima, T. Koga, Y. Eguchi, Prediction method for thermal stratification in a reactor vessel, Nucl. Eng. Des. 120 (1990) 395–402.

[353] T. Koga, K. Yamamoto, M. Takakuwa, H. Kajiwara, O. Watanabe, K. Akamatsu, Experimental study on thermal stratification in a reactor hot plenum of a Japanese demonstration LMFBR, in: Proc. of NURETH-8, 1997, pp. 1704–1711.

[354] N. Kimura, M. Nishimura, K. Hayashi, H. Kamide, Experimental Study on Thermal Stratification Phenomena in Reactor Vessel in Fast Reactors—Evaluation of Rising of Stratification Interface using Multi-Dimensional Thermal-Hydraulic Code, JNC TN9400 99-075, 1999 (in Japanese).

[355] N. Kimura, M. Nishimura, K. Hayashi, H. Kamide, An investigation of thermal stratification phenomena in fast reactors—sodium experiments and analysis, in: Proc. of ICONE-7, 1999. Paper No. ICONE-7107.

[356] N. Kimura, H. Miyakoshi, H. Kamide, Experimental study on thermal stratification in a reactor vessel of innovative sodium-cooled fast reactor—mitigation approach of temperature gradient across stratification interface, J. Nucl. Sci. Technol. 47 (9) (2010) 829–838.

[357] N. Kimura, T. Onojima, H. Kamide, Experimental study on thermal stratification in a reactor vessel of innovative sodium cooled fast reactor—characteristics of stratification interface under natural circulation operation, in: Proc. of NUTHOS-9, 2012. Paper No. N9P0177.

[358] S. Ohno, H. Ohki, A. Sugahara, H. Ohshima, Validation of a computational simulation method for evaluating thermal stratification in the reactor vessel upper plenum of fast reactors, J. Nucl. Sci. Technol. 48 (2) (2011) 205–214.

[359] I. Maekawa, Numerical diffusion in single-phase multi-dimensional thermal-hydraulic analysis, Nucl. Eng. Des. 120 (1990) 323–339.

[360] H. Tjahijono, M. Toshiharu, Investigation of In-vessel Thermal Stratification Characteristics for Large Scale Liquid Sodium Cooled Fast Breeder Reactors with General Purpose Multi-Dimensional Code AQUA, JNC-TN9400-2001-023, 2001.

[361] S. Ohno, A. Sugahara, H. Ohki, H. Ohshima, Investigation of temperature gradient characteristics for thermal stratification interface in upper plenum of fast reactors, in: Proc. of NTHAS7, 2010. Paper No. N7P0043.

[362] D.S. Wood, Proposal for design against thermal striping, Nucl. Energ. 19 (6) (1980) 433–437.
[363] J.E. Brunings, LMFBR thermal-striping evaluation, Interim report, Research Project 1704-11, Prepared by Rockwell International Energy Systems Group, 1982. EPRI-NP-2672.
[364] C. Betts, C. Boorman, N. Sheriff, Thermal striping in liquid metal cooled fast breeder reactors, in: Proc. of NURETH-2, 1983, pp. 1292–1301.
[365] S. Moriya, S. Ushijima, I. Ohshima, Prediction of thermal striping in reactors, in: Proc. of FR91, vol. 1, 1991, pp. 10.6.1–10.6.10.
[366] D. Tenchine, J.P. Moro, Experimental and numerical study of coaxial jets, Proc. NURETH 8 (3) (1997) 1381–1387.
[367] D. Tenchine, S. Vandroux, V. Barthel, O. Cioni, Experimental and numerical studies on mixing jets for sodium cooled fast reactor, Nucl. Eng. Des. 263 (2013) (2013) 263–272.
[368] N. Kimura, H. Miyakoshi, H. Kamide, Experimental investigation on transfer characteristics of temperature fluctuation from liquid sodium to wall in parallel triple-jet, Int. J. Heat Mass Transfer 50 (2007) (2007) 2024–2036.
[369] N. Kimura, H. Kamide, P. Emonot, K. Nagasawa, Study on thermal striping phenomena in triple-parallel jet—transfer characteristics of temperature fluctuation in sodium and water based on conjugated numerical simulation, in: Proc. of NUTHOS-7, 2008.
[370] J. Kobayashi, M. Tanaka, S. Ohno, H. Ohshima, H. Kamide, Proposal of benchmark problem of thermal striping phenomena in planar triple parallel jets tests for fundamental code validation in sodium-cooled fast reactor development, in: Proc. of NURETH-16, 2015, pp. 6664–6677.
[371] M. Tanaka, K. Nagasawa, Benchmark analysis of thermal striping phenomena in planar triple parallel jets tests for fundamental validation of fluid-structure thermal interaction code for sodium-cooled fast reactor, in: Proc. of NURETH-16, 2015, pp. 6650–6663.
[372] Y.Q. Yu, E. Merzari, J.W. Thomas, A. Obabko, S.M. Aithal, Steady and unsteady calculations on thermal striping phenomena in triple-parallel jet, Nucl. Eng. Des. 312 (2017) (2017) 429–437.
[373] P.-E. Angeli, Verification and validation of LES of a triple parallel jet flow in the context of a thermal striping investigation, Nucl. Eng. Des. 353 (2019), 110210.
[374] M. Igarashi, M. Tanaka, H. Kamide, Study on fluid temperature fluctuation and transfer to wall in a mixing tee, in: Proc. of ICONE-11, 2003. ICONE11-36299.
[375] H. Kamide, M. Igarashi, S. Kawashima, N. Kimura, K. Hayashi, Study on mixing behavior in a tee piping and numerical analyses for evaluation of thermal striping, Nucl. Eng. Des. 239 (2009) (2009) 58–67.
[376] N. Kimura, A. Ono, H. Miyakoshi, H. Kamide, Experimental study on high cycle thermal fatigue in T-junction, in: Proc. of NURETH-13, 2009. Paper No. N13P1169.
[377] P. Coste, P. Quemere, P. Roubin, P. Emonot, M. Tanaka, H. Kamide, Large eddy simulation of highly fluctuational temperature and velocity fields observed in a mixing-tee experiment, Nucl. Technol. 164 (2017) 76–88.
[378] M. Tanaka, Y. Miyake, Numerical simulation of thermal striping phenomena in a T-junction piping system for fundamental validation and uncertainty quantification by GCI estimation, Mech. Eng. J. 2 (5) (2015). Paper No.15-00134.
[379] N. Kimura, H. Ogawa, H. Kamide, Experimental study on fluid mixing phenomena in T-pipe junction with upstream elbow, Nucl. Eng. Des. 240 (2010) 3055–3066.

[380] K. Yuki, S. Yoshimasa, S.M. Hosseini, H. Hashizume, S. Toda, M. Tanaka, T. Muramatsu, Influence of secondary flow generated in a 90-deg. Bend on the thermal-hydraulic characteristics in a mixing tee, Nucl. Sci. Eng. 158 (2008) 194–202.
[381] IAEA, Validation of Fast Reactor Thermomechanical and Thermohydraulic Codes Final Report of a Co-Ordinated Research Project 1996–1999, IAEA-TECDOC-1318, 2012.
[382] K. Ohyama, O. Watanabe, H. Yamano, Conceptual design for japan sodium-cooled fast reactor (2): Thermal-hydraulic design for reactor upper sodium plenum in JSFR, in: Proc. of ICAPP '09, 2009. Paper 9296.
[383] K. Aizawa, J. Kobayashi, T. Onojima, M. Tanaka, S. Ohno, H. Kamide, K. Nagasawa, Investigation on thermal striping phenomena in five jets modelled water test (FIWAT) simulating sodium-cooled fast reactor, in: Proc. of NUTHOS-10, 2014. NUTHOS10-1163.
[384] J. Kobayashi, K. Aizawa, T. Ezure, A. Kurihara, M. Tanaka, Water experiments on thermal striping phenomena at the core outlet of an advanced sodium-cooled fast reactor (1) Proposal of countermeasures to mitigate temperature fluctuations around control rods, Maintenology 20 (3) (2021) 89–96 (in Japanese).
[385] J. Kobayashi, N. Kimura, A. Tobita, H. Kamide, O. Watanabe, K. Ohyama, Study on thermal striping at UIS of advanced loop type fast reactor (water experiment using a 1/3 scale 60 degree sector model), in: Proc. of ICONE-17, 2009. Paper No. ICONE17-75575.
[386] J. Kobayashi, T. Ezure, H. Kamide, K. Ohyama, O. Watanabe, Water experiments on thermal striping in reactor vessel of japan sodium-cooled fast reactor—countermeasures for significant temperature fluctuation generation, in: Proc of ICONE-23, 2015. Paper No. ICONE23-1577.
[387] J. Kobayashi, K. Aizawa, T. Ezure, A. Kurihara, M. Tanaka, Water experiments on thermal striping phenomena at the core outlet of an advanced sodium-cooled fast reactor (2) Proposal of countermeasures to mitigate temperature fluctuations around radial blanket fuel assemblies, Maintenology 20 (3) (2021) 97–101 (in Japanese).
[388] Y. Eguchi, K. Yamamoto, T. Fukuda, N. Tanaka, S. Moriya, K. Tanimoto, K. Obura, T. Suzuki, I. Maekawa, Gas Entrainment in the IHX vessel of top-entry loop-type LMFBR, Nucl. Eng. Des. 146 (1994) (1994) 373–381.
[389] T. Ezure, N. Kimura, M. Miyakoshi, H. Kamide, Experimental investigation on bubble characteristics entrained by surface vortex, Nucl. Eng. Des. 241 (2011) 4575–4584.
[390] H. Ohshima, N. Tanaka, Y. Eguchi, M. Nishimura, T. Kunugi, A. Uchibori, K. Ito, T. Sakai, Standard for prevention of gas entrainment phenomena in fast reactors, 1; Validations of CFD methods for reproducibilities of gas entrainment phenomena, Trans. AESJ 11 (4) (2012) 316–328 (in Japanese).
[391] H. Ohshima, Y. Eguchi, T. Kunugi, H. Kamide, T. Sakai, K. Ito, Standard for prevention of gas entrainment phenomena in fast reactors, 2; Proposal of gas entrainment evaluation method, Trans. AESJ 11 (4) (2012) 329–339 (in Japanese).
[392] K. Ito, T. Sakai, Y. Eguchi, H. Ohshima, A. Uchibori, Y. Xu, Improvement of gas entrainment prediction method—introduction of surface tension effect, J. Nucl. Sci. Technol. 47 (9) (2010) 771–778.
[393] K. Ito, H. Ohshima, Y. Nakamine, Y. Imai, Study on turbulent modeling in gas entrainment evaluation method, J. Power Energ. Syst. 6 (2) (2012) 151–164.
[394] R. Mitsumoto, T. Hazama, K. Takahashi, S. Kondo, Prototype fast breeder reactor Monju; Its history and achievements (Translated document), JAEA Technol. (2020). 2019-020.

[395] N. Kimura, K. Hayashi, H. Kamide, M. Ito, T. Sekine, Experimental study on flow optimization in upper plenum of reactor vessel for a compact sodium-cooled fast reactor, Nucl. Technol. 152 (2005) 210–222.
[396] H. Hagiwara, T. Ezure, K. Ito, H. Kamide, Study on vortex cavitation in scaled upper plenum model of Japan sodium-cooled fast reactor—(2nd report) investigation of effective cavitation suppressor, in: Proc of NTHAS-9, 2014. N9P0072.
[397] T. Ezure, K. Ito, H. Kamide, T. Kunugi, Study on behavior of vortex cavitation around suction pipes in sodium-cooled fast reactor geometry, Therm. Sci. Eng. 24 (3) (2016) 31–38.
[398] T. Sakai, H. Yamano, M. Tanaka, A. Ono, H. Ohshima, T. Kaneko, K. Hirota, H. Sago, Y. Xu, Y. Iwamoto, S. Ebara, T. Murakami, T. Nishihara, A. Yamaguchi, and T. Nakamura, Development of proposed guideline of flow-induced vibration evaluation for hot-leg piping in a sodium-cooled fast reactor, Proc. of NURETH-15, Pisa, Italy, May 12-17, 2013, NURETH15-042.
[399] H. Yamano, M. Tanaka, N. Kimura, H. Ohshima, H. Kamide, O. Watanabe, Development of flow-induced vibration evaluation methodology for large-diameter piping with elbow in Japan sodium-cooled fast reactor, Nucl. Eng. Des. 241 (2011) 4464–4475.
[400] K. Aizawa, S. Nakanishi, H. Yamano, S. Kotake, S. Hayakawa, O. Watanabe, K. Fujimata, Study on flow-induced-vibration evaluation of large-diameter pipings in a sodium-cooled fast reactor, 1; Sensitivity analysis of turbulent flow models for unsteady short-elbow pipe flow, in: Proc. of NTHAS-6, Okinawa, Japan, 2008. November 24-27, N6P1053.
[401] Y. Iwamoto, H. Yamano, Unsteady wall pressure characteristics of a 90 degree elbow in high Reynolds numbers, J. Fluid Sci. Technol. 9 (3) (2014).
[402] A. Ono, N. Kimura, H. Kamide, A. Tobita, Influence of elbow curvature on flow structure at elbow outlet under high Reynolds number condition, Nucl. Eng. Des. 241 (2011) 4409–4419.
[403] A. Ono, M. Tanaka, J. Kobayashi, H. Kamide, Influence of inlet velocity condition on unsteady flow characteristics in piping with a short elbow under a high-Reynolds-number condition, Mech. Eng. J. 4 (1) (2017) 1–10.
[404] H. Yamano, H. Sago, K. Hirota, S. Hayakawa, Y. Xu, M. Tanaka, T. Sakai, Effect of Swirl inflow on flow pattern and pressure fluctuation onto a single-elbow pipe in japan sodium-cooled fast reactor, J. Fluid Sci. Technol. 7 (3) (2012) 329–344.
[405] S. Ebara, H. Takamura, H. Hashizume, H. Yamano, Characteristics of flow field and pressure fluctuation in complex turbulent flow in the third elbow of a triple elbow piping with small curvature radius in three-dimensional layout, Int. J. Hydrogen Energy 41 (2016) 7139–7145.
[406] Y. Eguchi, T. Murakami, M. Tanaka, H. Yamano, A finite element LES for high-Re flow in a short-elbow pipe with undisturbed inlet velocity, Nucl. Eng. Des. 241 (2011) 4368–4378.
[407] M. Tanaka, H. Ohshima, Numerical investigation on large scale eddy structure in unsteady pipe elbow flow at high Reynolds number conditions with large eddy simulation approach, J. Power Energ. Syst. 6 (2) (2012) 210–228.
[408] T. Shiraishi, H. Watakabe, H. Sago, H. Yamano, Pressure fluctuation characteristics of the short-radius elbow pipe for FBR in the postcritical Reynolds regime, J. Fluid Sci. Technol. 4 (2) (2009) 430–441.
[409] H. Sago, T. Shiraishi, H. Watakabe, Y. Xu, K. Aizawa, H. Yamano, Study on flow induced vibration evaluation for a large scale JSFR piping: (3) pressure fluctuation

characteristics in 1/3 scale hot-leg piping experiments under deflected inflow conditions due to UIS structures, Proceedings of ICONE19, May 16-19, 2011, Chiba, Japan, ICONE19-43712.

[410] JAEA, Proposed guideline of flow-induced vibration evaluation for hot-leg piping of primary cooling system in sodium-cooled fast reactor, JAEA Res. (2011). 2011-021 (in Japanese).

[411] JAEA, Proposed guideline of flow-induced vibration evaluation for hot-leg piping of primary cooling system in sodium-cooled fast reactor-supplemental interpretation, JAEA Res. (2011). 2011-044 (in Japanese).

[412] JAEA edited by F. Nakashima, et al, Prototype Fast Breeder Reactor Monju; Its History and Achievements (Translated Document From JAEA-Technology 2019-007), JAEA-Technology-2019-020, 2019. Chapter 11.

[413] A. Yamaguchi, Numerical simulation and water experiments of flow induced vibration of a thermocouple well, in: Proc. of NURETH-8, vol. 2, 1997, pp. 1130–1131.

[414] Special Working Group on Thermometer Design Guide, Proposed Flow-Induced Vibration Design Guide for Thermometer Wells, PNC TN9410 97-042, 1997 (in Japanese).

[415] K. Iwata, M. Morishita, A. Yamaguchi, Y. Wada, Flow-Induced Vibration Design Guide for Thermometer Wells; Background Document, JNC TN9400 2000-088, 2000.

[416] T. Sakai, K. Iwata, M. Morishita, S. Kitamura, Vortex-induced vibration of a cantilever circular cylinder in super-critical Reynolds number flow and its suppression by structure damping, JSME Int. J. B 44 (4) (2001) 712–720.

[417] K. Iwata, M. Morishita, T. Sakai, A. Yamaguchi, K. Ogura, Evaluation of turbulence-induced vibration of a circular cylinder in supercritical Reynolds number flow, JSME Int. J. B 44 (4) (2001) 721–728.

[418] T. Sakai, M. Morishita, K. Iwata, S. Kitamura, Experimental study on the avoidance and suppression criteria for the vortex-induced vibration of a cantilever cylinder, J. Press. Vessel Technol. 124 (2) (2002) 187–195.

[419] O. Watanabe, K. Oyama, J. Endo, N. Doda, A. Ono, H. Kamide, T. Murakami, Y. Eguchi, Development of an evaluation methodology for the natural circulation decay heat removal system in a sodium cooled fast reactor, J. Nucl. Sci. Technol. 52 (9) (2015) 1102–1121.

[420] H. Kamide, Y. Ieda, K. Hayashi, K. Satoh, Core-plenum thermohydraulics under decay heat removal by natural circulation—sodium experiments for seven-subassembly model, in: Proc. of an International Topical Meeting on Sodium-Cooled Fast Reactor Safety, vol. 4, 1994, pp. 6-64–6-73.

[421] H. Kamide, Investigation on Thermal Hydraulics Phenomena in a Sodium-Cooled Fast Reactor Core during Natural Circulation Decay Heat Removals, Doctoral Thesis, Tokyo Institute of Technology, 2010.

[422] H. Kamide, J. Kobayashi, K. Hayashi, Sodium experiments of buoyancy-driven penetration flow into low-power subassemblies in a sodium-cooled fast reactor during natural circulation decay heat removal, Nucl. Technol. 175 (2011) 628–640.

[423] J. Kobayashi, A. Tokuhiro, H. Kamide, H. Ohshima, Y. Ieda, An experimental investigation on penetration of buoyancy-driven cold liquid into a vertical channel with hot, forced-flow, Nucl. Eng. Des. 177 (1997) 91–104.

[424] K. Momoi, K. Hayashi, H. Kamide, M. Nishimura, N. Kokaki, Experimental Study on Inter-Wrapper Flow Phenomena during Natural Circulation Decay Heat Removal in Fast Reactors—Effects on Natural Circulation Flow Rate and Core Cooling, 1997. PNC Report TN9410 97-047 (in Japanese).

[425] H. Kamide, N. Kimura, K. Hayashi, M. Nishimura, 1998, Experimental study on inter-wrapper flow during natural circulation in fast reactors—influence on natural circulation flow and core temperatures, Proc. ICONE-6, ICONE-6197.
[426] K. Oyama, J. Endo, N. Doda, A. Ono, T. Murakami, O. Watanabe, Development of natural circulation analysis methods for a sodium cooled fast reactor, J. Nucl. Sci. Technol. 53 (3) (2016) 353–370.
[427] T. Murakami, Y. Eguchi, K. Oyama, O. Watanabe, Reduced-scale water test of natural circulation for decay heat removal in loop-type sodium-cooled fast reactor, Nucl. Eng. Des. 288 (2015) 220–231.
[428] A. Ono, H. Kamide, J. Kobayashi, N. Doda, O. Watanabe, An experimental study on natural circulation decay heat removal system for a loop type fast reactor, J. Nucl. Sci. Technol. 53 (9) (2016) 1385–1396.
[429] H. Kamide, J. Kobayashi, A. Ono, N. Kimura, O. Watanabe, Sodium experiment on fully natural circulation systems for decay heat removal in japan sodium-cooled fast reactor, in: Proc. NURETH-14, 2011. NURETH14-179.
[430] A. Ono, J. Kobayashi, H. Kamide, O. Watanabe, Sodium experiment on fully natural circulation systems for decay heat removal in Japan sodium-cooled fast reactor, in: Proc. of NTHAS8, N8P1084, 2012.
[431] A. Ono, A. Kurihara, M. Tanaka, H. Ohshima, H. Kamide, Y. Miyake, M. Ito, S. Nakane, Study on reactor vessel coolability of sodium-cooled fast reactor under severe accident condition—water experiments using a scale model, Proc. ICAPP 2017 (2017) 17230.
[432] H. Kamide, H. Ohshima, T. Sakai, M. Tanaka, Progress of thermal hydraulic evaluation methods and experimental studies on a sodium-cooled fast reactor and its safety in Japan, Nucl. Eng. Des. 312 (2017) 30–41.
[433] T. Ezure, T. Onojima, Y. Kameyama, M. Tanaka, Study on decay heat removal in sodium-cooled fast reactors - in-vessel cooling characteristics by direct heat exchanger, in: Proc. National Symposium on Power and Energy Systems, 2019, p. B112 (in Japanese).
[434] N. Doda, H. Ohshima, H. Kamide, Effects of wire spacer contact and pellet-cladding eccentricity on fuel cladding temperature under natural circulation decay heat removal conditions in sodium-cooled fast reactor, in: Proc. NUTHOS-8, N8P0266, 2010.
[435] N. Doda, H. Ohshima, H. Kamide, O. Watanabe, Y. Ohkubo, Development of core hot spot evaluation method for natural circulation decay heat removal in sodium cooled fast reactor, in: Proc. NURETH-14, NURETH14-170, 2011.
[436] N. Doda, H. Ohshima, H. Kamide, O. Watanabe, Development of PIRT for fast reactor under natural circulation decay heat removal operations, in: Proc. ICONE-19, 2011. ICONE19-44124.
[437] N. Doda, H. Kamide, H. Ohshima, O. Watanabe, Global sensitivity analysis for core hot spot evaluation under natural circulation decay heat removal in sodium-cooled fast reactor, in: Proc. NUTHOS-9, N9P0227, 2012.
[438] N. Doda, H. Ohshima, H. Kamide, O. Watanabe, Development of core hot spot evaluation method of a loop type fast reactor equipped with natural circulation decay heat removal system, in: Proc. NTHAS10, N10P1050, 2016.
[439] H. Kamide, K. Nagasawa, N. Kimura, H. Miyakoshi, Evaluation method for core thermohydraulics during natural circulation in fast reactors (numerical predictions of inter-wrapper flow), JSME Int. J. Ser. B 45 (3) (2002) 577–585.
[440] H. Kamide, K. Nagasawa, N. Kimura, H. Miyakoshi, Evaluation method for core thermohydraulics during natural circulation in fast reactors-numerical predictions of inter-wrapper flow, in: Proc. ICONE-9, 2001.

[441] H. Ohshima, M. Ohtaka, Development of computer program for whole core thermal-hydraulic analysis of fast reactors, in: Proc. ICONE10, 2000. ICONE10-22034.
[442] E. Hamase, Y. Miyake, Y. Imai, N. Doda, A. Ono, M. Tanaka, Core thermal-hydraulic analysis during dipped-type direct heat exchanger operation in natural circulation conditions, in: Proc. ICONE28, 2021. ICONE28-63380.
[443] A. Ono, M. Tanaka, Y. Miyake, E. Hamase, T. Ezure, Preliminary analysis of sodium experimental apparatus PLANDTL-2 for development of evaluation method for thermal-hydraulics in reactor vessel of sodium fast reactor under decay heat removal system operation condition, Mech. Eng. J. 7 (3) (2020). Paper No.19-00546.
[444] K. Iida, Y. Asada, K. Okabayashi, T. Nagata, Construction codes developed for prototype FBR Monju, J. Nucl. Eng. Des. 98 (1987) 283–288.
[445] K. Iida, Y. Asada, K. Okabayashi, T. Nagata, Simplified analysis and design for elevated temperature components of Monju, J. Nucl. Eng. Des. 98 (1987) 305–317.
[446] American Society of Mechanical Engineers (ASME), ASME Boiler & Pressure Vessel Code, Code Case N-47 (1592), Class 1 Components in Elevated Temperature Service, Section III, Division 1, ASME, New York, 1977.
[447] N. Kasahara, T. Nagata, K. Iwata, H. Negishi, Advanced creep-fatigue evaluation rule for fast breeder reactor components: generalization of elastic follow-up model, Nucl. Eng. Des. 155 (3) (1995) 499–518.
[448] AFCEN, Design and constriction rules for mechanical components of FBR nuclear Islands, RCC-MR, Paris, 1985.
[449] American Society of Mechanical Engineers (ASME), ASME Boiler & Pressure Vessel Code, Section III, Division 1, Subsection NH, ASME, New York, 2012.
[450] American Society of Mechanical Engineers (ASME), ASME Boiler & Pressure Vessel Code, Section III, Division 5, ASME, New York, 2021.
[451] Japan Society of Mechanical Engineers (JSME), Code for Nuclear Power Generation Facilities, -Rules on Design and Construction for Nuclear Power Plants- Section II Fast Reactor Codes, JSME S NC2-2005, JSME, Tokyo, 2005 (In Japanese).
[452] N. Kawasaki, K. Takakura, T. Ohtani, M. Hayashi, Y. Yamada, Recent Design Improvements of Elevated Temperature Structural Design Guide for DFBR in Japan, SMiRT15, Div. F, F04/4, 1990, p. IV. 161-168.
[453] K. Iwata, A. Imazu, Y. Wada, T. Nagata, Recent achievements at PNC in the development of high temperature structural design method for FBR components, Nucl. Eng. Des. 138 (1992) 269–282.
[454] Japan Society of Mechanical Engineers (JSME), Code for Nuclear Power Generation Facilities, -Rules on Design and Construction for Nuclear Power Plants- Section II Fast Reactors Code, JSME S NC2-2012, JSME, Tokyo, 2012 (In Japanese).
[455] T. Asayama, K. Dozaki, T. Otani, T. Inatomi, M. Ando, Development of 2012 Edition of JSME Code for Design and Construction of Fast Reactors—(1) Overview, Proceedings of ASME Pressure Vessels & Piping Conference 2013, PVP-2013 98061, 2013.
[456] T. Onizawa, Y. Nagae, S. Takaya, T. Asayama, Development of 2012 Edition of JSME Code for Design and Construction of Fast Reactors—(2) Development of the Material Strength Standard of 316FR Stainless Steel, Proceedings of ASME Pressure Vessels & Piping Conference 2013, PVP-2013 97608, 2013.
[457] T. Onizawa, Y. Nagae, S. Takaya, T. Asayama, Development of 2012 Edition of JSME Code for Design and Construction of Fast Reactors—(3) Development of the Material Strength Standard of Modified 9Cr-1Mo Steel, Proceedings of ASME Pressure Vessels & Piping Conference 2013, PVP-2013 97611, 2013.

[458] S. Takaya, Y. Nagae, T. Asayama, Development of 2012 Edition of JSME Code for Design and Construction of Fast Reactors—(4) Creep-Fatigue Evaluation Method for Modified 9Cr-1Mo Steel, Proceedings of ASME Pressure Vessels & Piping Conference 2013, PVP-2013 97609, 2013.
[459] Y. Nagae, S. Takaya, T. Asayama, Development of 2012 Edition of JSME Code for Design and Construction of Fast Reactors—(5) Creep-Fatigue Evaluation Method for 316FR Stainless Steel, Proceedings of ASME Pressure Vessels & Piping Conference 2013, PVP-2013 97806, 2013.
[460] M. Ando, S. Watanabe, K. Kikuchi, T. Otani, K. Satoh, K. Tsukimori, T. Asayama, Development of 2012 Edition of JSME Code for Design and Construction of Fast Reactors (6) Design margin assessment for the new materials to the rules, Proceedings of the ASME Pressure Vessels & Piping Conference 2013, PVP2013-97803, 2013.
[461] Y. Asada, M. Ueta, T. Kanaoka, M. Sukekawa, T. Nishida, Current status of the development of advanced 316-steel for FBR structures, in: Stress Classification, Robust Methods, and Elevated Temperature Design, Proceedings of ASME Pressure Vessel & Piping Conference 1992, 1992, pp. 61–65. PVP-Vol.230.
[462] T. Nakazawa, H. Kimura, K. Kimura, H. Kaguchi, Advanced type stainless steel 316FR for fast breeder reactor structures, J. Mater. Process. Technol. 143–144 (2003) 905–909.
[463] M. Ueta, T. Nishida, H. Koto, M. Sukekawa, K. Taguchi, Creep-fatigue properties of advanced 316-steel for FBR structure, in: Proceedings of ASME Pressure Vessel & Piping Conference 1995, 1995, pp. 423–428. PVP-Vol.313-2.
[464] K. Isozaki, T. Ashida, K. Sumino, S. Nakai, Upgrade of cooling system heat removal capacity of the experimental fast reactor Joyo, Nucl. Technol. 150 (2005) 56–66.
[465] M. Fujimitsu, History of power plants and progress in heat resistant steels, ISIJ Int. 41 (2001) 612–625.
[466] T. Otani, H. Wada, H. Koto, T. Fujioka, The evaluation method of creep-fatigue damage of 316FR stainless steel weld joint, in: ASME Pressure Vessel & Piping Conference 1998, 1998, pp. 257–268. PVP-Vol.365.
[467] N. Kasahara, M. Kikuchi, Application of elastic follow-up model to evaluate cyclic thermal transient strength of welded vessels, in: Proceedings of ASME Pressure and Vessel Piping Conference, vol. 2, 1995, pp. 481–492. PVP-Vol.313-2.
[468] T. Otani, et al., The evaluation method of creep-fatigue damage of 316FR stainless steel weld joint, in: Proceedings of ASME Pressure Vessels & Piping Conference 1998, 1998, pp. 257–268. PVP-Vol.365.
[469] M. Ando, S. Hasebe, S. Kobayashi, N. Kasahara, A. Toyoshi, T. Ohmae, Y. Enuma, Thermal transient test and strength evaluation of a tubesheet structure made of Mod.9Cr–1Mo steel. Part I: test model design and experimental results, Nucl. Eng. Des. 275 (2014) 408–421.
[470] M. Ando, S. Hasebe, S. Kobayashi, N. Kasahara, A. Toyoshi, T. Ohmae, Y. Enuma, Thermal transient test and strength evaluation of a tubesheet structure made of Mod.9Cr–1Mo steel. Part II: creep-fatigue strength evaluation, Nucl. Eng. Des. 275 (2014) 422–432.
[471] M. Ando, S. Hasebe, S. Kobayashi, N. Kasahara, A. Toyoshi, T. Ohmae, Y. Enuma, Thermal transient test and strength evaluation of a thick cylinder model made of Mod.9Cr-1Mo steel, Nucl. Eng. Des. 255 (2013) 296–309.
[472] Woodhead Publishing Limited, Nuclear Corrosion Science and Engineering, Part VI, Chapter 21, Corrosion issues in Sodium-cooled Fast Reactor (SFR) System, 2012, pp. 773–806.
[473] American Society of Mechanical Engineers (ASME), ASME Boiler and Pressure Vessel Code, Code Case N-812, Alternate Creep-Fatigue Damage Envelope for 9Cr-1Mo-V Steel, ASME, New York, 2011.

[474] M. Ando, K. Toyota, R. Hashidate, T. Onizawa, Evaluation of the Japanese fatigue test data in Gr.91 for elevated temperature design, in: Proceedings of ASME Pressure Vessels & Piping Conference 2021, 2021. PVP2021-60773.
[475] M. Ando, S. Okajima, K. Imo, Development of the external pressure chart of 2¼Cr-1Mo and Mod.9Cr-1Mo steel for elevated temperature design, in: Proceedings of ASME Pressure Vessels & Piping Conference 2019, 2019. PVP2019-93777.
[476] Japan Society of Mechanical Engineers (JSME), Code for Nuclear Power Generation Facilities, -Rules on Welding for Nuclear Power Plants- Fast Reactors Code, JSME S NB1-CC-002, JSME, Tokyo, 2010 (In Japanese).
[477] Japan Society of Mechanical Engineers (JSME), Code for Nuclear Power Generation Facilities, -Rules on Welding for Fast Reactor Nuclear Power Plants-, JSME S NB2-2017, JSME, Tokyo, 2017 (In Japanese).
[478] T. Onizawa, R. Hashidate, Development of creep property equations of 316FR stainless steel and Mod.9Cr-1Mo steel for sodium-cooled fast reactor to achieve 60-year design life, Mech. Eng. J. 6 (2019). Paper No. 18-00477.
[479] K. Toyota, T. Onizawa, T. Wakai, R. Hashidate, S. Kato, Material data acquisition activities to develop the material strength standard for sodium-cooled fast reactors, in: Proceedings of International Conference on Fast Reactors and Related Fuel Cycles (FR22), IAEA-CN-291/106, 2022.
[480] Japan Society of Mechanical Engineers (JSME), Code for Nuclear Power Generation Facilities, -Rules on Design and Construction for Nuclear Power Plants- Section II Fast Reactors Code, JSME S NC2-2020, JSME, Tokyo, 2020 (In Japanese).
[481] JSW Technical Review, History and Future Prospects of Technologies and Products of the Muroran Plant—Focusing on Energy and Preservation of Environment, No. 19, 2008, pp. 17–35.
[482] K. Kimura, M. Yaguchi, Re-evaluation of stress rupture factors for Grade 91 weldments based on the extended database with the data collected in JAPAN, in: Proceedings of the ASME Pressure Vessels & Piping Conference 2019, PVP2019-93331, 2019.
[483] Japan Society of Mechanical Engineers (JSME), Code for Nuclear Power Generation Facilities, -Rules on Design and Construction for Nuclear Power Plants- Section I Light Water Reactors Code, JSME S NC1-2020, JSME, Tokyo, 2020 (In Japanese).
[484] S. Watanabe, K. Kubo, S. Okajima, T. Wakai, Study of the calculation method for the elastic follow-up coefficient by inelastic analysis, in: Proceedings of the JSME Material and Mechanics Division Conference, OS614, 2017, pp. 581–585 (In Japanese).
[485] Y. Wada, T. Asayama, S. Hasebe, N. Kasahara, Study on mechanical behavior of weldments of type 304SS under creep-fatigue loading, Proceedings of ASME Pressure Vessels & Piping Conference 1995, PVP-Vol.313-2, pp.487-492.
[486] T. Asayama, Creep-fatigue evaluation of stainless steel welded joints in FBR class 1 components, Nucl. Eng. Des. 198 (2000) 25–40.
[487] M. Ando, S. Takaya, Creep-fatigue evaluation method for weld joint of Mod.9Cr-1Mo steel Part II: Plate bending test and proposal of a simplified evaluation method, Nucl. Eng. Des. 310 (2016) 217–230.
[488] H. Akiyama, H. Ohtsubo, M. Yamada, H. Nakamura, S. Matsuura, Y. Hagiwara, Y. Ooka (Eds.), Outline of the seismic buckling design guideline of FBR main vessels, Transactions of the 13th Conference on Structural Mechanics in Reactor Technology (SMiRT13), Div.E, E013, 1995, pp. 445–456.
[489] T. Okafuji, K. Miura, H. Sago, H. Murakami, M. Ando, M. Miyazaki, Development of the buckling evaluation method for large scale vessel by the testing of Gr.91 vessel subjected to horizontal and cyclic vertical loading, in: Proceedings of ASME Pressure Vessels & Piping Conference 2021, 2021. PVP2021-60672.

[490] T. Okafuji, K. Miura, H. Sago, H. Murakami, K. Kubo, K. Sato, Takashi Wakai and Kenta Shimomura, Study on the seismic buckling evaluation method for Mod. 9Cr-1Mo steel cylindrical vessel, in: Proceedings of the Materials and Mechanics Conference 2017, OS0616, 2017, pp. 591–595.
[491] Y. Tanaka, H. Shibamoto, K. Inoue, N. Kasahara, Development of the guideline on inelastic analysis for design, in: Proceedings of ASME Pressure Vessels & Piping Conference 2004, 2004. PVP-2004-2259.
[492] T. Onizawa, T. Wakai, Development of extremely high temperature material property equations and physical property values on austenitic stainless steel, in: Transactions of the 25th Conference on Structural Mechanics in Reactor Technology (SMiRT25), Division XI, 2019.
[493] T. Watakabe, et al., Ultimate strength of a thin wall elbow for sodium cooled fast reactors under seismic loads, ASME J. Press. Vessel Technol. 138 (2016). Paper No. 021801.
[494] K. Suzuki, et al., Seismic providing test of ultimate piping strength, in: Proceedings ASME PVP2003, Seismic Engineering, vol. 2, 2003. PVP2003-2954.
[495] I. Nakamura, et al., Failure behavior of piping systems with wall thinning under seismic loading, ASME J. Press. Vessel Technol. 126 (2004) 85–90.
[496] A. Otani, et al., Application and consideration of code case and analysis guideline, in: JSME Mechanical Engineering Congress, Paper No. F08105, 2019 (in Japanese).
[497] M. Morishita, et al., A JSME code case on piping seismic design based on inelastic response analysis and strain-based fatigue criteria, ASME J. Press. Vessel Technol. 142 (2020). Paper No. 021203.
[498] JSME, An Alternative Rules on Seismic Design of Seismic S Class Piping by Elastic - Plastic Response Analysis, Japan Society of Mechanical Engineers, Tokyo, Japan, 2019. Standard No. JSME NC-CC-008 (in Japanese).
[499] I. Nakamura, et al., Benchmark analyses on the elastic-plastic behavior of carbon steel pipes under seismic load, ASME J. Press. Vessel Technol. 142 (2020). Paper No. 021901.
[500] A. Iwasaki, et al., Core seismic experiment and analysis of full scale single model for fast reactor, in: Proceedings ASME PVP2017, PVP2017-65329, 2017.
[501] A. Iwasaki, et al., Development of seismic analysis method of fast reactor core, Trans. JSME 82 (839) (2016), https://doi.org/10.1299/transjsme.16-00093 (in Japanese).
[502] A. Iwasaki, et al., Core seismic experiment and analysis of hexagonal bundle model for fast reactor, in: Proceedings ASME PVP2017, PVP2017-65354, 2017.
[503] S. Matsubara, et al., Fast reactor core seismic analysis for verification of assessment model considering deformation of core elements, in: Proceedings ASME PVP2019, PVP2019-93778, 2019.
[504] A. Iwasaki, et al., Core seismic experiment and analysis of a large number of element models for fast reactor, in: Proceedings ASME PVP2017, PVP2017-65357, 2017.
[505] The American Society of Mechanical Engineers, 2001, "Boiler and Pressure Vessel Code, Rules for Inservice Inspection of Nuclear Power Plant Components, Section XI, Division 3: Rules for Inspection and Testing of Components of Liquid-Metal Cooled Plants", 2001 ed., ASME BPVC. XI-2001.
[506] S. Takaya, T. Asayama, Y. Kamishima, H. Machida, D. Watanabe, S. Nakai, M. Morishita, Application of the System Based Code concept to the determination of in-service inspection requirements, ASME J. Nucl. Rad. Sci. 1 (2015). NERS-14-1020.
[507] S. Takaya, Y. Kamishima, H. Machida, D. Watanabe, T. Asayama, Determination of in-service inspection requirements for fast reactor components using system based code concept, Nucl. Eng. Des. 305 (2016) 270–276.

[508] The American Society of Mechanical Engineers, Boiler and Pressure Vessel Code, Code Cases, Nuclear Components, N-875 Alternative Inservice Inspection Requirements for Liquid-Metal Reactor Passive Components, 2017. ASME BPVC.CC.NC-2017.

[509] S. Takaya, T. Asayama, H. Yada, A.T. Roberts, F.J. Schaaf Jr., Development of in-service inspection rules for liquid-metal cooled reactors using the System Based Code concept, J. Press. Vessel Technol. 142 (2020). PVT-19-1071.

[510] Y. Asada, M. Tashimo, M. Ueta, System based code—principal concept, in: Proc. 10th ICONE, ICONE10-22730, 2002.

[511] Y. Asada, M. Tashimo, M. Ueta, System based code—basic structure, in: Proc. 10th ICONE, ICONE10-22731, 2002.

[512] The American Society of Mechanical Engineers, Boiler and Pressure Vessel Code, Rules for Inservice Inspection of Nuclear Power Plant Components, Section XI, Division 2: Requirements for Reliability and Integrity Management Programs for Nuclear Power Plants, 2021 ed., 2021. ASME BPVC. XI.2-2021.

[513] H. Yada, T. Wakai, T. Miyagawa, H. Machida, Development of leak before break assessment guidelines for sodium cooled fast reactors in Japan, in: Proc. PVP2021, PVP2021-61942, 2021.

[514] H. Yada, et al., Proposal on LBB evaluation conditions for sodium cooled fast reactor pipes and effects of pipe parameters, Trans. JSME 84 (2018). No.859 (in Japanese).

[515] T. Wakai, et al., Demonstration of leak-before-break in Japan Sodium cooled Fast Reactor (JSFR) pipes, Nucl. Eng. Des. 269 (2014) 88–96.

[516] The Japan Society of Mechanical Engineers, Codes for Nuclear Power Generation Facilities: Rules on Design and Construction for Nuclear Power Plants, Section II Fast Reactor Standards, 2018 ed., 2018. JSME S NC2-2018 (in Japanese).

[517] Electric Power Research Institute, Crack-Opening Area Calculations for Circumferential Through-Wall Pipe Cracks, 1988. EPRI NP-5959-SR.

[518] H. Machida, M. Arakawa, T. Wakai, Effect of local plastic component on crack opening displacement and on j-integral of a circumferential penetrated crack, in: Proc. 27th ICONE, ICONE27-1143, 2019.

[519] T. Wakai, H. Machida, M. Arakawa, S. Yanagihara, R. Suzuki, M. Matsubara, Development of a crack opening displacement assessment procedure considering change of compliance at a crack part in thin wall pipes made of modified 9Cr-1Mo steel, in: Proc. 26th ICONE, ICONE26-82619, 2018.

[520] N. Miura, Y. Takahashi, Y. Nakayama, T. Shimakawa, Evaluation Method of Sodium Leak for FBR Components, 2000. CRIEPI REPORT, ET99001.

[521] T. Wakai, H. Machida, S. Yoshida, T. Tokiyoshi, K. Kikuchi, Y. Xu, K. Tsukimori, Development of LBB assessment method for japan sodium cooled fast reactor (JSFR) pipes (5) crack growth assessment method for pipes made of Mod.9Cr-1Mo steel, in: Proc. PVP2011, PVP2011-57513, 2011.

[522] T. Wakai, H. Machida, K. Satoh, A study on evaluation method of penetrate crack length for LBB assessment of fast reactor pipes, in: Proc. M&M2015, 2015 (in Japanese).

[523] T. Wakai, H. Machida, M. Arakawa, Improvement of penetrate crack length evaluation method for LBB assessment of sodium-cooled fast reactor components, in: Proc. MECJ-18, 2018 (in Japanese).

[524] L.M. Barrett, J.A. McKnight, J.R. Forthergill, Ultrasonic viewing in fast reactors, Phys. Technol. 15 (1984) 308–314.

[525] J. Wessels, Inservice inspection of the reactor block of sodium-cooled fast breeder reactors, Nucl. Eng. Des. 130 (1991) 33–42.

[526] H. Karasawa, M. Izumi, T. Suzuki, S. Nagai, M. Tamura, S. Fujimori, Development of under-sodium three-dimensional visual inspection technique using matrix-arrayed ultrasonic transducer, J. Nucl. Sci. Technol. 37 (9) (2000) 769–779.
[527] K. Aizawa, Y. Chikazawa, K. Ara, M. Yui, K. Jinno, T. Hiramatsu, Development of under sodium viewer for the next generation sodium-cooled fast reactor, in: Proc. ICAPP2019, 2019, p. 103.
[528] K. Aizawa, Y. Chikazawa, K. Ara, M. Yui, Y. Uemoto, M. Kurokawa, T. Hiramatsu, Development of under sodium viewer for next generation sodium-cooled fast reactors, in: Proc. FR17, IAEA-CN-245-267, 2017.
[529] K. Asakura, K. Takeuchi, T. Makino, Y. Kato, Feasibility study on a simplified MOX pellet fabrication process, the short process, for fast breeder reactor fuel, Nucl. Technol. 167 (2009) 348–361.
[530] Y. Kato, N. Takahashi, M. Morita, K. Yoshimoto, H. Furuya, Oxide fuel fabrication technology development of the fact project (2) small scale hot tests on agitating granulation method using water for MOX pellet production, in: Proceedings of the 10th international conference, GLOBAL 2011, 2011. Paper No. 440091.
[531] K. Ishii, T. Segawa, M. Suzuki, Development of granulation system for simplified MOX pellet fabrication process, in: Proceedings of 2019 International Congress on Advances in Nuclear Power Plants (ICAPP 2019), May, Juan-Les-Pins, France, 2019.
[532] K. Sudo, T. Takano, K. Takeuchi, Y. Kihara, M. Kato, Development of pressing machine with a die wall lubrication system for the simplified MOX pellet fabrication method in the FaCT project, in: Proceedings of the 10th International Conference, GLOBAL 2011, 2011. Paper No. 387512.
[533] T. Namekawa, Y. Yamada, A. Kitamura, T. Hosogane, K. Kawaguchi, Handling Technology of Low Decontaminated TRU Fuel for the Simplified Pelletizing Method Fuel Fabrication System, 2009. FR-09, Dec.2009, Kyoto, Japan.
[534] W. Tadahiro, T. Koizumi, T. Koyama, FaCT phase-I evaluation on the advanced aqueous reprocessing process (1) summary of the advanced reprocessing technology development, in: Proceedings of the 10th International Conference, GLOBAL 2011, 2011. Paper No. 392040.
[535] M. Takeuchi, T. Kitagaki, H. Higuchi, M. Fukushima, T. Washiya, T. Kobayashi, FaCT phase-I evaluation on the advanced aqueous reprocessing process (2) Development of mechanical disassembly and short stroke shearing systems for FBR fuel reprocessing, in: Proceedings of the 10th International Conference, GLOBAL 2011, 2011. Paper No. 392040.
[536] H. Ikeuchi, K. Katsurai, Y. Sano, T. Washiya, FaCT phase-I evaluation on the advanced aqueous reprocessing process (3) highly effective dissolution technology for FBR MOX fuels, in: Proceedings of the 10th International Conference, GLOBAL 2011, 2011. Paper No. 392227.
[537] A. Shibata, K. Yano, Y. Sanbonmatsu, M. Nakahara, M. Takeuchi, T. Washiya, M. Nagata, T. Chikazawa, FaCT phase I evaluation on the advanced aqueous reprocessing process (5) Research and development of uranium crystallization system, in: Proceedings of the 10th International Conference, GLOBAL 2011, 2011. Paper No. 391745.
[538] Y. Koma, H. Ogino, A. Sakamoto, H. Nakabayashi, A. Shibata, M. Nakahara, T. Washiya, FaCT phase-I evaluation on the advanced aqueous reprocessing process (4), solvent extraction simplified for FBR fuel reprocessing, in: Proceedings of the 10th International Conference, GLOBAL 2011, 2011. Paper No. 391547.
[539] S. Watanabe, T. Senzaki, A. Shibata, K. Nomura, Y. Koma, Y. Nakajima, MA recovery experiments from genuine HLLW by extraction chromatography, in: Proceedings of the 10th International Conference, GLOBAL 2011, 2011. Paper No. 387433.

[540] Y. Sano, N. Kaji, A. Shibata, M. Takeuchi, T. Washiya, Salt-free technique for solvent washing process in NEXT process, in: Proceedings of the 10th International Conference, GLOBAL 2011, 2011. Paper No. 387558.
[541] N. Doda, E. Hamase, N. Kikuchi, M. Tanaka, Development of platform for design optimization in fast reactor based on coupling analysis codes, in: Proceedings of the Computational Engineering Conference of JSCES, Vol. 24, E-10-03 [CD-ROM], 2019 (in Japanese).
[542] N. Doda, K. Yokoyama, M. Tanaka, T. Takata, H. Ohshima, Development of multi-level, multi-scenario simulation systems for sodium cooled fast reactor (15) Development of multi-level simulation system, in: Proceedings of the Atomic Energy Society of Japan 2020 Fall Meeting, 1G09, 2020. September Online [in Japanese].
[543] Japan Society of Mechanical Engineers, Codes for Nuclear Power Generation Facilities, Rules on Design and Construction for Nuclear Power Plants. Section II Fast Reactor Standards, JSME S NC2-2020, 2020 (in Japanese).
[544] H. Yada, S. Takaya, H. Enuma, Proposal of inspection rationalization method and application for sodium cooled fast reactor, in: Proceedings of the ICONE2020, 2020, https://doi.org/10.1115/ICONE2020-16735. August, Online, ICONE2020-16735.
[545] Fauske & Associates, MAAP4 Modular Accident Analysis Program for LWR Power Plants User's Manual. Project RP3131-02, 1994–2005.
[546] L.L. Humphries, B.A. Beeny, F. Gelbard, D.L.Y. Louie, J. Phillips, MELCOR Computer Code Manuals Vol. 1: Primer and Users' Guide Version 2.2.9541 2017, Sandia Report No. SAND2017-0455, 2017.
[547] H. Ujita, N. Satoh, M. Naitoh, M. Hidaka, N. Shirakawa, M. Yamagishi, Development of severe accident analysis code SAMPSON in IMPACT project, J. Nucl. Sci. Technol. 36 (11) (1999) 1076–1088.
[548] D.P. Weber, G. Birgersson, G.L. Bordner, L.L. Briggs, J.E. Cahalan, F.E. Dunn, Kalimullah, K.J. Miles, F.G. Prohammer, A.M. Tentner, J.A. Verklan, The SAS4A LMFBR whole core accident analysis code, in: Proc. ANS/ENS Fast Reactor Safety Meeting, Knoxville, TN, 1985, pp. 989–998. April 21-24.
[549] K.K. Murata, D.E. Carroll, K.E. Washington, F. Gelbard, G.D. Valdez, D.C. Williams, K. D. Bergeron, User's Manual for CONTAIN 1.1: A Computer Code for Severe Nuclear Reactor Accident Containment Analysis, 1989. Sandia Report No. SAND87-2309.
[550] N. Girault, L. Cloarec, L. Herranz, G. Bandini, S. Perez-Martin, L. Ammirabile, Ongoing activities in the European JASMIN project for the development and validation of ASTEC-Na safety simulation code, in: Proceedings of the ICAPP 2015, Nice, France, 2015, May.
[551] D.L.Y. Louie, L.L. Humphries, Development of a MELCOR Sodium Chemistry (NAC) Package—FY17 Progress, 2018. Sandia Report No. SAND2018-1252.
[552] M. Aoyagi, A. Uchibori, T. Takata, D.L.Y. Louie, A.J. Clark, Sodium fire analysis using a sodium chemistry package in MELCOR, in: Proceedings of the 2020 International Conference on Nuclear Engineering (ICONE 2020), 2020, August. Online.
[553] M. Aoyagi, D.L.Y. Louie, Study of sodium pool fire model improvement in MELCOR for SFR, Trans. Am. Nucl. Soc. 124 (1) (2021) 824–827.
[554] M. Aoyagi, D.L.Y. Louie, A. Uchibori, T. Takata, D. Luxat, Study of recent sodium pool fire model improvements for MELCOR code, in: Proceedings of the 28th International Conference on Nuclear Engineering (ICONE 28), 2021. Online.
[555] M. Aoyagi, A. Uchibori, T. Takata, H. Ohshima, Sodium fire models for in- and ex-vessel safety analysis code SPECTRA, Trans. Am. Nucl. Soc. 122 (1) (2020) 862–865. June.

[556] A. Uchibori, M. Aoyagi, T. Takata, H. Ohshima, Development of ex-vessel phenomena analysis model for multi-scenario simulation system, SPECTRA, in: Proceedings of the 2020 International Conference on Nuclear Engineering (ICONE 2020), 2020. Online.

[557] M. Sonehara, M. Aoyagi, A. Uchibori, T. Takata, H. Ohshima, A.J. Clark, M.R. Denman, Multi-dimensional numerical investigation of sodium spray combustion; benchmark analysis of SNL T3 experiment, in: Proceedings of the 11th Korea-Japan Symposium on Nuclear Thermal Hydraulics and Safety (NTHAS-11), Busan, Korea, 2018.

[558] M. Sonehara, M. Aoyagi, A. Uchibori, T. Takata, H. Ohshima, A.J. Clark, M.R. Denman, Multi-dimensional numerical benchmark analysis of SNL T3 sodium spray combustion experiment with AQUA-SF code, in: Proceedings of the 27th International Conference on Nuclear Engineering (ICONE 27), Tsukuba, Japan, 2019.

[559] M. Sonehara, M. Aoyagi, A. Uchibori, T. Takata, H. Ohshima, A.J. Clark, D.L.Y. Louie, Numerical validation of AQUA-SF in SNL T3 sodium spray fire experiment, in: Proceedings of the 2020 International Conference on Nuclear Engineering (ICONE 2020), 2020. Online.

[560] M. Aoyagi, T. Takata, M. Uno, Numerical modeling of radiation heat transfer from combusting droplets for a sodium fire analysis, Nucl. Eng. Des. 380 (2021), 111258.

[561] A. Yamaguchi, T. Takata, Y. Okano, Numerical methodology to evaluate fast reactor sodium combustion, Nucl. Technol. 136 (3) (2001) 315–330.

[562] A.J. Clark, M.R. Denman, T. Takata, H. Ohshima, SNL/JAEA Collaboration on Sodium Fire Benchmarking, 2017. Sandia Report No. SAND2017-12409.

[563] F. Nakashima, S. Kondo, S. Usami, O. Yamazaki, Y. Kaneko, T. Hazama, R. Mitsumoto, K. Takahashi, S. Kondo, Prototype fast breeder reactor monju; its history and achievements, JAEA Technol. (2020) 2019–2020.

Index

Note: Page numbers followed by *f* indicate figures and *t* indicate tables.

K

L

FSC
www.fsc.org
MIX
Papier | Fördert gute Waldnutzung
FSC® C083411

Kolibri 360 GmbH
Ferdinand-Jühlke-Straße 7
99095 Erfurt, Deutschland
produktsicherheit@kolibri360.de